MRI Principles of the Head, Skull Base and Spine

A Clinical Approach

Springer-Verlag France S.A.R.L

J. C. Tamraz, C. Outin, M. Forjaz Secca, B. Soussi

MRI Principles of the Head, Skull Base and Spine

A Clinical Approach

Foreword by G. Salamon

Springer

Jean C. Tamraz
Professor and Chairman
Department of Neuroradiology
CHU Hôtel-Dieu de France
Université Saint-Joseph
Beirut, Lebanon

Mario Forjaz Secca
Auxiliary Professor of Biophysics
Departamento de Fisica
Faculdade de Ciências e Tecnologia
Universidade Nova de Lisboa
2825-114 Monte de Caparteo, Portugal

Claire Outin
Chief Neuroimaging Department
Trad Hospital & Medical Center
53, rue du Mexique
Beirut, Lebanon

Bassam Soussi
Professor and Director
NMR Research Lab & Bioenergetics Grp
Wallenberg Laboratory
Sahlgrenska University Hospital
Gothenburg University
SE-413 45 Göteborg, Sweden

ISBN 978-2-8178-0756-0 ISBN 978-2-8178-0754-6 (eBook)
DOI 10.1007/978-2-8178-0754-6

Originally published by Springer-Verlag France, Paris, in 2003
Softcover reprint of the hardcover 1st edition 2003

SPIN : 10898468

To Claire,
Caroline and Eve

Contents

Foreword

This new textbook is divided into three main parts. The first one is devoted to the brain. The second one is devoted to cranio-facial pathology. The last one concerns the spine and spinal cord. Every chapter is illustrated in a very rich and elegant manner. Every image is very cleverly discussed. This textbook will certainly be very attractive not only for Neuroradiologists but also for Neurologists, Neurosurgeons, Orthopedists, Ophthalmologists, ENT specialists and, in general, all specialists interested in the precise diagnosis offered by MR imaging. This new textbook is especially attractive because it has three key qualities. It is extremely clear and easy to read, and specific topics are easily found for consultation. Furthermore, this clarity is enhanced by the superb iconography, which is the trademark of Springer-Verlag. This book is also attractive because it is not a textbook based only on imaging. Every case is very well documented clinically and the discussion concerns clinical symptoms and signs related to the diagnosis and related case material in similar cases. The third key strength of the text is the anatomical knowledge that is imparted to the reader by the author Jean Tamraz, and co-author Youssef Comair, in this magnificent textbook "Atlas of regional anatomy of the brain using MRI". Everybody from the first year fellow to the most experienced attending Neuroradiologist will take great pleasure reading and using in practice these two new textbooks, which are complementary each other. I am sure that many young Neuroradiologists and trainees in other related specialties will benefit greatly from using this textbook during their training. Neuroimaging is based on excellent iconography, just as the best and the most recent staining methods are mandatory for our colleagues working in Neuropathology. Neuroimaging is also more attractive and effective when based on strong correlations with clinical Neurology and Neuroanatomy.

Jean Tamraz has received excellent training and experience at Salpetriere Hospital in Paris in Neurology before starting his education in Radiology. He spent 15 years in Neuroimaging in the best MRI centers devoted to craniofacial, brain and spinal cord pathology. He is now in Lebanon as the Head of a beautiful Department of Neuroimaging, which is a leading place in Europe and the Middle East, after having been in France as an attending in the National Hospital des Quinze-Vingts. Dr. Tamraz is also now the Secretary-General for Neuroradiology in the World Federation of Neurology. His regular participation in international meetings like the regular ones organized in Cleveland on Epilepsy, and the high quality of his own Department of Neuroimaging, are for me the best elements for a successful reception of this textbook, which I hope will find an international audience and a big success.

I wish also, in writing these lines, to take this opportunity to express to Jean Tamraz all my admiration and my friendship.

Georges Salamon
Northwestern University
Chicago, September 2000

Preface

This comprehensive text-atlas is intended to serve as an introduction to the field of neuroimaging. The investigative methodologies stressing on the major indications of MRI and its current place in the diagnosis and the management of the diseases of the central nervous system, are aimed by this monography. It does not pretend to replace the excellent and exhaustive textbooks that are presently available in this field. Rather, it is intended to serve as a companion to all neuroscientists and imaging specialists involved in the neurological diagnosis and therapy.

Following an overview of the fundamental principles of Magnetic Resonance Imaging and Spectroscopy, an exhaustive atlas illustrating the major individual pathologies of the nervous system is proposed. The work is divided into three parts each including the specific methodological approach followed by chapters providing a concise, yet pertinent, review of the salient clinical and neuropathological findings of the disease process with neuroimaging correlates, using high-resolution iconographical plates. This didactical case presentation format is used for most diseases of brain and skull, base of the skull and face, and spine and cord, that are encountered in everyday clinical practice. For this purpose, the inclusion of 320 selected pathologies represented in over 750 high resolution MR images was necessary. The relationship between fundamental clinical data and topographical and imaging findings and their common and synergic contribution to the final diagnosis are emphasized. The great majority of the reported cases have been verified pathologically.

Given its scope, we hope that this book will emphasize the multidisciplinary approach that is necessitated for optimal management of patients and hence will serve as a useful reference and drain interest from the radiologists involved in neuroimaging interpretation, neuroradiologists seeking an up-to-date review, neurologists, neurosurgeons, neuropathologists and other specialists working or in training in the fields of diagnostic and therapeutic clinical neuroscience.

The combined clinical-radiologic and pathological approaches of this teaching text-atlas is the result of the experience acquired since 1981 in Paris, initially in clinical neurology at La Salpetriere Hospital, Clinique des Maladies du Systeme Nerveux, and subsequently, in neuroradiology in the Department of Neuroradiology of the Centre Hospitalier National d'Ophtalmologie des Quinze-Vingts. With this atlas, I wish to pay tribute to my mentors in neuroradiology and particularly Professors Emmanuel Cabanis, Georges Salamon and Luc Picard, who provided me with the skills of neuroradiology and with whom I have had the privilege and pleasure of working. I have been fortunate to train and practice under Professors Dominique Laplane, René Marteau, Marie-Germaine Bousser and Olivier Lyon-Caen, in clinical neurology. The experience acquired during the clinical-imaging confrontations added undoubtedly essential insight into image interpretation and disease diagnosis, minimizing patient risk and improving cost-effectiveness.

I am deeply indebted for their outstanding contributions, to Professors Mario Forjaz Secca and Bassam Soussi for a most comprehensive approach to the basic aspects of Magnetic Resonance physics and technique, these intentionally made concise and up-to-date, in order to encourage and help the readers to grasp the fundamentals of such complex topics. This work would be hard to achieve without the more specific collaboration of Doctor Claire Outin-Tamraz, providing part of the substance of this book.

I would like to express also my gratitude to the medical staff of Hotel-Dieu de France University Hospital and more especially to our collegues from the Department of Neurosciences at Saint-Joseph's University, for their confidence in providing us their patients. They are the motive of our continuous improvement.

The authors wish to express their deep gratitude to General Electric Medical Systems-Europe and in particular to M. Guillermo Zannoli for his unflagging educational and research support in the fields of MR and neuroimaging. We also gratefully acknowledge our publisher, Springer-Verlag, especially Mrs Stephanie Van Duin and her co-workers for their help and unfailing patience during the preparation and publication of this text-atlas.

Jean Tamraz

Basic notions of MRI in neurosciences

M. Forjaz Secca

Introduction

Nuclear Magnetic Resonance (NMR) was originally a field of Physics, which overflowed into Chemistry, Biochemistry and Medicine. Several Nobel prizes have been won in this field. In the prehistory of NMR Isidor Rabi won the prize in Physics in 1944, because of his work on a resonance method for the registration of the magnetic properties of atomic nuclei. Then, in 1952 Felix Bloch and Edward Purcell won the prize in Physics for the discovery of nuclear magnetic resonance in solids. Finally, in 1991 Richard Ernst won the prize in Chemistry for his contributions to the development of the methodology of high resolution nuclear magnetic resonance spectroscopy. There is only one Nobel missing for NMR and it is in Medicine. But that does not say anything about the importance of Magnetic Resonance Imaging (MRI), only that it was developed by collaborative work and that it is impossible to single out one or two names from the hundreds of brilliant scientists and engineers responsible for the present state of MRI. The ideas are fascinating and the technology is fantastic, but it was built step by step.

One of the questions a physicist working in MRI is constantly asked by medical doctors is: "what is Nuclear Magnetic Resonance Imaging and how does it work?" The physicist normally answers rhetorically: "how much Physics do you know? Or, how far are you prepared to go?" Like everything else in life MRI can be explained at different levels, advancing more and more as the physical knowledge progresses; a bit like peeling an onion.

Starting from the outer layer we could explain how Magnetic Resonance Imaging works in a few lines, as follows.

The body is made up of atoms, a large proportion of which is Hydrogen. The nuclei of Hydrogen, which have only one proton, because they are charged and spin about themselves, behave like little magnets. These little magnets, when placed in a magnetic field align with it and rotate around the axis of the field in a movement called precession, similar to spinning tops on a table. This precession, or turning, movement is faster the higher the magnetic field. If electromagnetic radiation, like radio waves, at exactly the same frequency of the precessing nuclei is emitted near them, they can absorb this radiation, which is said to be at resonance, and they flip, becoming aligned in the opposite direction of the field. When the radiation is switched off, the nuclei get rid of the energy they absorbed by emitting back the radiation. Each tissue of the body, because of its different chemical composition and physical state, re-emits radiation at a different rate, known as the tissue relaxation time. This radiation is picked up by an antenna, transforming it into electrical current, which is then used to construct the image we want. Because nuclei are used in a magnetic field and absorb radiation at resonance, the method is called Nuclear Magnetic Resonance Imaging. However, because of the bad connotations of the word "nuclear", this has been dropped from the name and the method is usually know as Magnetic Resonance Imaging.

It is not the intention of this book to go too deeply into the Physics of MRI; so, to go to next layer, we will keep things simple and show only the essential and easily understood equations.

Basic principles of MRI

The nucleus and the spin

All matter is composed of molecules, which in turn are composed of atoms. These atoms are constituted of a positively charged nucleus, made up of protons and neutrons, surrounded by negatively charged electrons. In the case of hydrogen the nucleus has only one proton and is surrounded by only one electron.

The nucleus rotates upon itself, it has spin. And because it is charged it produces a small magnetic field, behaving like a tiny magnet. This produces what is called in Physics a nuclear magnetic moment.

The magnetic field

The magnetic field is a disturbance in space produced either by a permanent magnet or by the passage of current through a loop of wire. It is a field of attractive or repulsive forces generated by moving or spinning electric charges and can be described as a set of imaginary lines that indicate the direction a compass needle would point at a particular position in space. Magnet fields strengths are measured in either Tesla or Gauss, where 1 T equals 10,000 Gauss. In the case of MRI most magnets are made up of loops of superconducting wire, which, because of their virtual zero electrical resistance at low temperature, can withstand very high currents in very thin cross-sections, without dissipation of energy, thus allowing very high fields in compact volumes. The main MRI field is commonly designated by B_0.

The central equation in MRI

When a nuclear spin is placed in a magnetic field it tends to align itself with the main field. But, as in the case of a spinning top, it doesn't align completely, forming an angle with the field and rotating about the axis of the main field. This particular rotation is called precession and the angular frequency of precession is called the Larmor frequency and designated ω{|IN|}0.

If there is only one equation, one should remember from Magnetic Resonance it has to be the Larmor equation:

$$\omega_0 = 2 \cdot f_0 = \gamma B_0 \quad \text{or} \quad f_0 = \omega_0/2 \cdot = \gamma B_0/2\cdot$$

This equation basically states that, for a particular nucleus, the higher the magnetic field, the higher the frequency of the precessing nucleus. And it is this dependence of the frequency on the magnetic field that makes MRI possible. γ is a constant called the gyromagnetic ratio that has a particular value for each nucleus. For example, for Hydrogen, $\gamma = 2.68 \times 10^8$ rad/s/Tesla, and its precessing frequencies for the more common magnetic fields are given in the following table.

Field	Frequency
0.5 T	21.3 MHz
1.0 T	42.6 MHz
1.5 T	63.9 MHz
2.0 T	85.2 MHz
3.0 T	127.8 MHz

These frequencies are in the range of VHF radio waves, from short wave to FM; hence, they are referred to as radio frequencies. Another consequence of this equation is that for a 1 ppm (part per million) increase in the magnetic field in a 1.5 T magnet the frequency will change from 42,600,000 Hz to 42,600,042.6 Hz. It is changes like this that allow an image to be acquired as we will see later on.

It is possible to observe Nuclear Magnetic Resonance with many different isotopes, as is the case in the human body, in decreasing order of abundance, with Hydrogen, ^{1}H; Fluor, ^{19}Fl; Sodium, ^{23}Na, and Phosphor, ^{31}P. However, because of its natural predominant abundance, Hydrogen is the nucleus of choice for MRI. Each isotope has a different γ, therefore, for the same field their frequencies will be different. For example, for a field of 1.5 T the Larmor frequencies of the previous nuclei are shown in the following table.

Nucleus	Frequency
^{1}H	63.9 MHz
^{19}Fl	60.1 MHz
^{23}Na	16.9 MHz
^{31}P	25.9 MHz

The alignment of spins

In practice, the nuclear spins do not all align with the magnetic field because of the thermal energy associated with temperature. In the simplest case of Hydrogen, with only two spin quantum states, the laws of physics state that there are only two possibilities for the spins: either aligned, in the direction of the field, or anti-aligned, in the opposite direction. Because the anti-aligned spins have an energy slightly superior to the energy of the aligned spins, and this energy difference is of the order of the thermal energy for temperatures around body temperature, the spins are almost evenly distributed among the two energy levels. However, there is a slight difference, and it is this difference that allows us to obtain an image from the body. For example, for a field of 1.5 T the spin excess, the difference between the aligned and the anti-aligned spins, at a temperature of 37° C (98° F), is only of the order of 5 ppm. And it is this fraction that contributes to the signal. Fortunately, because of the very large number of nuclei, there are still enough spins to obtain a signal. As an example, there are around 7×10^{19} Hydrogen nuclei in a volume of $1 \times 1 \times 1$ mm^3 of water, and a fraction of 5 ppm still yields a spin excess of 3×10^{14}.

The net magnetization

The net magnetization is then the sum of the magnetic moments of all the spins. In equilibrium it will be pointing in the direction of the magnetic field. It is this net magnetization that will produce the MR signal and to simplify the approach to MRI we will concentrate on the behavior of this magnetization as a single varying vector rather than the behavior of the individual spins. This is normally referred to as the classical approach as opposed to the quantum mechanical approach where we look at the individual spins.

The relaxation times T1 and T2

Now, if the spins are left on their own nothing in particular happens, but if electromagnetic energy, of the same magnitude as the difference in energy between the spin levels, is emitted near the spins in the form of radiation, they will absorb that energy, disturbing their equilibrium. Once that electromagnetic radiation is switched off the magnetization will return to equilibrium by re-emitting the energy absorbed. The return to equilibrium is not instantaneous and it can be decomposed in two parts, one along the direction of the main field and the other in the plane perpendicular to the main field. In equilibrium the magnetization has only a longitudinal component in the direction of the field and no transverse components in the perpendicular plane, but after the disturbance the longitudinal component is smaller in general and the transverse component appears. As the disturbance is switched off these components return separately to their equilibrium states, one returning back to the maximum net magnetization value and the other returning to zero. They both vary exponentially: the longitudinal component increases exponentially with a characteristic time known as T1 and the transverse component decreases exponentially with a characteristic time known as T2. The longitudinal time T1 is known as the "spin-lattice" relaxation time because of its origin in the interactions of the spins with their surroundings. The transverse time T2 is known as the "spin-spin" relaxation time because of its origin in the interactions between the spins, which cause a dephasing between them. The important point in imaging is that different tissues, because of their different chemical constitutions and different physical states, will have different relaxation times. In other words, T1 and T2 contain tissue information.

*T2**

In practice, the transverse component is affected by external field inhomogeneities which cause additional dephasing and suppression of the signal and destroy tissue information. This means that, due to the faster transverse relaxation, T2 should be replaced by a smaller relaxation time designated T2*. Fortunately, T2 can be extracted by recourse to a method known as spin echo.

Faraday's law

Now that we have covered the basic physical principles of Magnetic Resonance, we have to understand the practical side of it (the engineering), that is, how to obtain a signal, and for that we have to understand the law of Faraday. It states that if one moves a magnetic field inside a coil winding in the direction of the main axis of the coil, an electric current is produced at the terminals of the winding. In equilibrium no signal can be detected because the net magnetization is constant, but by placing a coil perpendicular to the main field it is possible to monitor the disturbance as a varying transverse magnetization vector moving inside the coil. This will produce an electrical current that will be taken as the MR signal.

Spatial encoding

To construct an image it is essential to be able to tell what part of the body the signal is coming from. This is achieved by spatially encoding the signal. To understand spatial encoding we will start with a simple image. Let us suppose that we have a choir where each singer sings only one note and the singers for each octave are standing on the same line, with the lower octaves at the back and the higher octaves at the front; also, within each line, the singer on the left sings the lower note and the one next to him on the right sings the next semitone higher. If someone falls, and sings as he does, we can know exactly where it happened, even without looking, because we have encoded the frequencies. If the intensity of the singing is related to the speed at which the singer stands up again, just by listening to the sounds we have a way of telling what is happening to the choir.

In imaging this encoding is achieved in practice by making use of magnetic field gradients and taking into account the Larmor equation. The gradients will produce a variation of the magnetic field along a particular direction of space, which will cause a variation of frequencies in that direction. Thus, each point in space will produce a signal with a characteristic frequency which will enable the computer to separate the different points in space and reconstruct the image. For practical reasons the gradients in the different directions are not all switched on at the same time, but they are pulsed as the measuring sequence proceeds. The gradients applied in the three main axis are called the slice selection (SS) gradient, the frequency encoding (FE) gradient and the phase encoding (PE) gradient.

To make the image acquisition technically possible the signal is acquired line by line until a full slice is obtained, going on then for the next slice.

Proton density

As the image is reconstructed, it is divided in small equal volumes where all the spins inside each of them contribute to only one signal value. This means that within each of these volume elements (voxels) the signal assigned to it is an average of the signals of all the individual spins inside it. Each of these voxels will have a different quantity of spins inside it because of the different tissues it encompasses. The consequence of this is that different voxels will have initially different signals as a result of the particular concentration of spins, or protons of Hydrogen, hence the designation proton density, which reflects the number of protons contributing to the signal that exist within a voxel.

Basic techniques and sequences of MRI

RF pulses

In practice, the emission of RF frequency radiation is achieved in the form of a temporary variation of a magnetic field perpendicular to the main field. This field rotates at the resonant frequency and causes the net magnetization to rotate about it, thus disturbing it from its equilibrium state. This burst of radio frequency energy is called an RF pulse. The longer the RF pulse is switched on, the more the magnetization will rotate about it. In this way, by adjusting the pulse time width or intensity, it is possible to achieve a rotation of 90°, 180° or any other angle desired. In MRI a group of RF pulses used to produce a specific type of signal is called a pulse sequence.

FID

The Free Induction Decay (FID) is the simplest signal obtained from a Magnetic Resonance sequence. By applying a 90° pulse the magnetization is rotated from the axis of the main field to a plane perpendicular to it. Because it continues to rotate around the main field in the plane perpendicular to B_0, it induces a current in the detecting coil and as the magnetization relaxes back to equilibrium, the signal decreases exponentially. This decay however is dominated by T2*, and not T2, dependant on the field inhomogeneities, which makes the FID very short in general.

Spin echo

There is a way of recovering the T2 information masked by T2* and that is done by using a technique called spin echo. In this sequence a 180° pulse is applied some time (TE) after the original 90° pulse and some of the signal, originally lost after the FID, is recovered as a kind of echo that comes and goes. The best way to understand this process is to look at the system from the individual spin perspective. Right after the 90° pulse the spins, which are originally in phase, begin to dephase, some turn faster and others turn slower due to the spatial variations in magnetic field, causing the net magnetization to disappear quickly. However, when the 180° pulse is applied some time later the spins are reversed, that is the faster ones which moved further away from the origin now have a longer way to go to reach back the origin, while the slower ones which didn't move so far have a shorter way to reach back the origin. Because this movement of coming into phase is a mirror image of the dephasing movement, they all reach the starting point at the same time and then restart to dephase. It is this temporary coming into phase that is called the spin echo. The main characteristic of the spin echo is that it retains the physical information about T2 and eliminates the influence of T2*, therefore an image based on the signal height of the echo is said to be T2 weighted. If, however, TE is very short, T2 relaxation has no time to occur, so the signal height will be proportional to proton density. This means that a spin echo with a short TE will be DP weighted and with a longer TE will be T2 weighted. The spin echo sequence implies that for a DP weighted sequence, the higher the DP the brighter the signal, and for a T2 weighted sequence, the longer the T2 the brighter the signal (Fig.1).

Repetition time

The repetition time (TR) of a sequence is the time between repetitions of the basic sequence of the imaging sequence. The sequence can be repeated for instance if it is necessary to improve the signal to noise.

The repetition time is also the time taken to obtain each phase line measurement within a single slice; therefore, the time required to produce an image is determined by the product of TR by the number of phase encoding steps. The TR can affect image contrast, making it one of the image parameters.

T1 imaging

To obtain a T1 weighted image the spin echo sequence is repeated, but with a short TR. This means that the

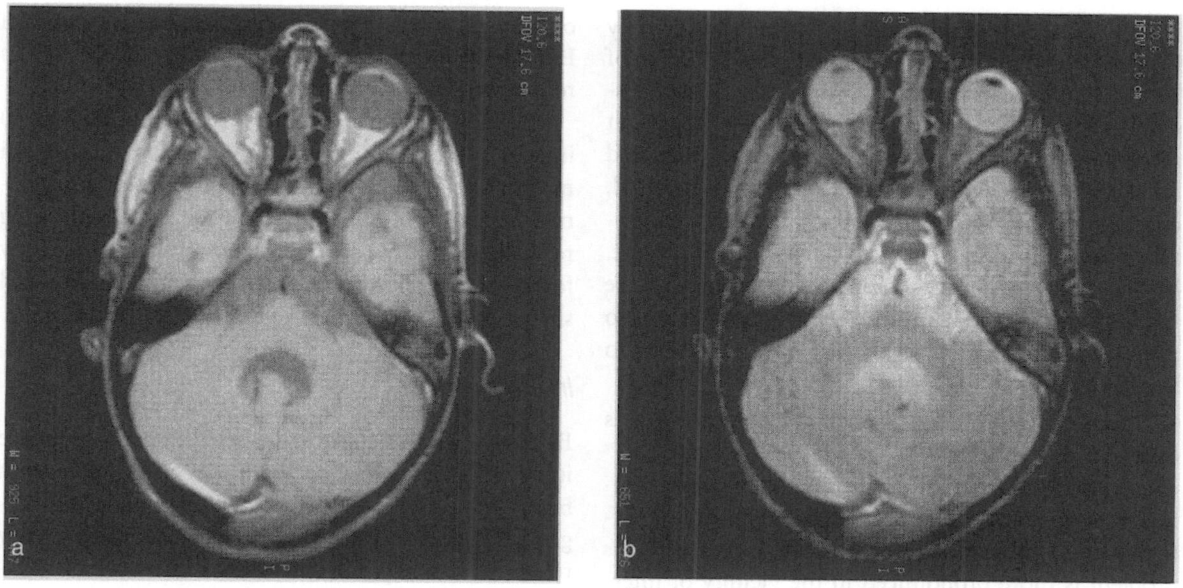

Fig. 1. Two axial images acquired in the same plane: **a)** DP weighted image with a TE = 20 msec; **b)** T2 weighted image with a TE = 90 msec

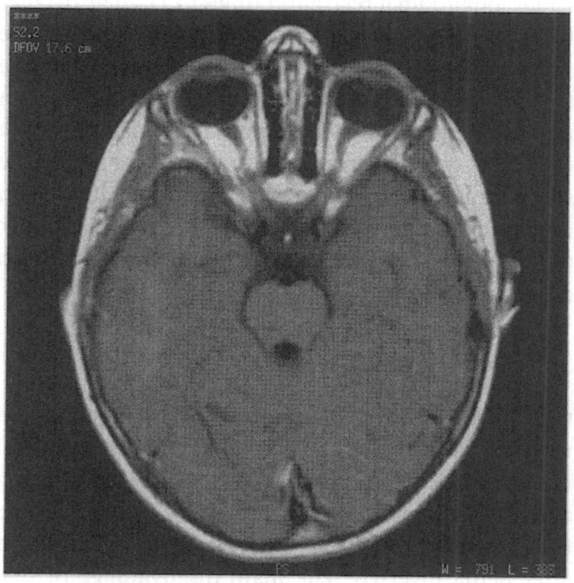

Fig. 2. A spin echo T1 weighted axial image with a TR = 400 msec

net magnetizations of the tissues with short T1's will have time to recover to equilibrium, but the magnetizations of the tissues with long T1's will have no time to recover. In this way tissues with shorter T1 will show brighter signals than the tissues with longer T1's. By varying the value of TR the contrast between the tissues with different values of T1 can be adjusted (Fig. 2).

Gradient echo

The spin echo sequence requires that the net magnetization recovers to its equilibrium position along the direction of the main field before repeating the sequence, and for tissues with a long T1 this can greatly increase the acquisition time. The incomplete recovery of the magnetization implies that there will be a signal loss. The gradient echo method allows a much faster acquisition by a combination of two techniques: a rotation of the magnetization less than 90° and a faster way of producing an echo. The purpose of rotating the magnetization less than 90° can be seen in the diagram shown in Fig. 3.

Let us consider several rotations of the magnetization, known as the flip angle, as presented in Fig. 3. For a flip angle of 90° the magnetization only has components in the XY direction and no component in the Z direction, taking its full time to recover (approximately five times T1). For a flip angle of 45° the components in the XY and Z directions are equal, that is, about 71% of the net magnetization. For a flip angle of 30°, the component in the XY direction is 50%, but the Z component is 87% of the net magnetization. For a flip angle of 15°, the component in the XY direction is 25%, but the Z component is 97% of the net magnetization. As can be seen, the use of a small flip angle can produce a magnetization in the XY direction which is sufficiently large to yield a detectable signal but reduce very little the longitudinal magnetization, implying that it will recover to equilibrium more rapidly and allowing a much shorter TR.

The faster way of producing an echo is achieved by the use of a special gradient rather than a sequence of 90°-180° pulses. In this case a strong dephasing gradient, which will produce a dephasing faster than that caused by the external field inhomogeneities, is applied for a short time. At some point the gradient is reversed, causing the spins to rephase and go through a temporary echo a time TE after the original flip angle pulse. In the gradient echo it is the reversing polarity of the readout gradient that produces the echo, having no need for a 180° pulse. This allows the minimum TE to be reduced.

One of the problems with the gradient echo, as opposed to the spin echo, is that it does have T2* effects from the external field inhomogeneities, manifesting themselves as a loss of signal and geometric distortion which becomes worse as TE increases. This problem is more pronounced in the vicinity of interfaces with different magnetic susceptibilities, like air-tissue.

The increased sensitivity of gradient echo pulse sequences to susceptibility effects makes them the methods of choice for perfusion imaging and brain functional imaging.

Echo planar imaging

Echo planar imaging (EPI) is a fast imaging technique that acquires an entire image within a single TR period. To fully understand this technique it is important to understand the strange concept of k-space as opposed to the image space.

Let us start with an example: if we look at a cylinder from the top we see a circle, but if we look at it from the side we see a rectangle. One perspective shows us the roundedness of the cylinder while the other shows us its squareness. The object we are looking at is the same but we are highlighting its different properties by looking at it from different angles. And that is exactly what the image space and the k-space do: they look at the MRI from different perspectives, but describe the same phenomena. Fortunately, there is a mathematical tool that allows us to convert from one perspective to the other and that is called the Fourier transform. The image itself is in the image space and the k-space is equivalent to the space defined by the frequency and phase encoding directions.

While conventional sequences acquire one line of k-space for each phase encoding step, which occurs every TR seconds, EPI acquires all lines of k-space in a single TR period. This is achieved by cycling the phase and frequency encoding directions so as to cover the k-space of the image. There are several methods to obtain echo planar images, which include conventional EPI, spiral EPI and square-spiral EPI, whose names refer to the way k-space is covered.

Because of the possibility of obtaining 15 to 30 images per second, depending on the acquisition matrix, one of the important applications of echo planar imaging is in obtaining ultra-fast images allowing real time acquisitions. There is, however, a price to pay for the extra speed and it is the image quality and sharpness.

Inversion recovery

By applying a 180° pulse at the beginning of a sequence it is possible to invert the alignment of the spins from being aligned with the magnetic field to being anti-aligned. If the spins are then left to themselves, they return back to equilibrium which is the aligned position. As they do that the magnetization goes from negative, through zero, to positive, that is, the magnetization recovers from inversion, hence the name Inversion Recovery. Because different tissues have different longitudinal relaxation times (T1), their magnetizations will go through zero at different times. If one starts a 90°-180° spin echo sequence at exactly the time the magnetization of a specific tissue is going through zero, then that tissue will produce no signal. This time interval between the inversion pulse and the rest of the sequence is called the inversion time or TI. The main reason for using inversion recovery sequences is either to increase T1 contrast or to eliminate the signal from a particular tissue.

STIR (short TI inversion recovery)

One of the specific inversion recovery sequences is used to eliminate the signal from fat and is called STIR (short TI inversion recovery). This is achieved by using a short TI of around 150 to 180 msec, which is the time the protons from fat take to reach zero magnetization after being inverted. The main disadvantage of this sequence is the low signal to noise ratio because the magnetization of all the other tissues is also close to zero.

FLAIR

The other common inversion recovery sequence is used to achieve heavy T2 weighting without signal from the CSF. This sequence is called FLAIR (fluid attenuated inversion recovery).

The signal from the CSF can be attenuated using a TI around 2000 msec, which produces a heavy T2 weighting of the images without virtually any signal

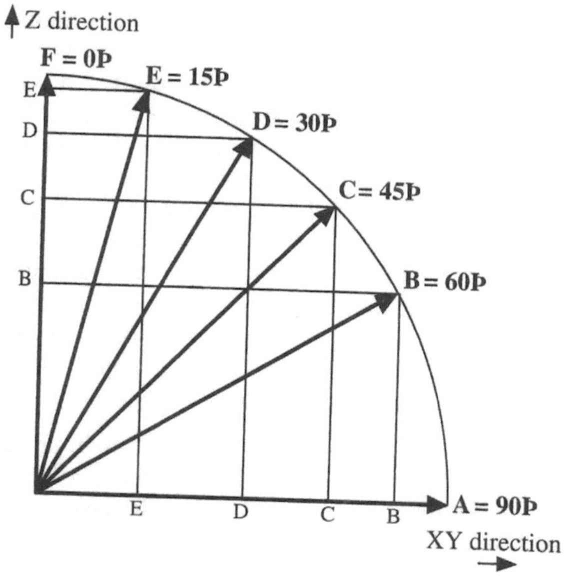

Fig. 3. Variation of the XY and Z components of magnetization with flip angle

from the CSF. The main disadvantage of this sequence is the necessity of very long TRs to allow the CSF to relax completely.

MR angiography

MR angiography (MRA) is referred to as the ensemble of techniques that allows MR to image the flowing fluids in the body. In the past angiography was only possible with the injection of contrast in the blood vessels, but it would not distinguish between flowing and stationary blood. The methods of time-of-flight and phase contrast MRA, however, are sensitive to the flow of blood.

Time-of-Flight angiography

The Time-of-Flight (TOF) method makes use of the movement of blood through the imaging plane. For instance, for the spin echo acquisition of a slice through which a blood vessel passes, a 90° pulse affects the whole slice. However, when the 180° pulse is

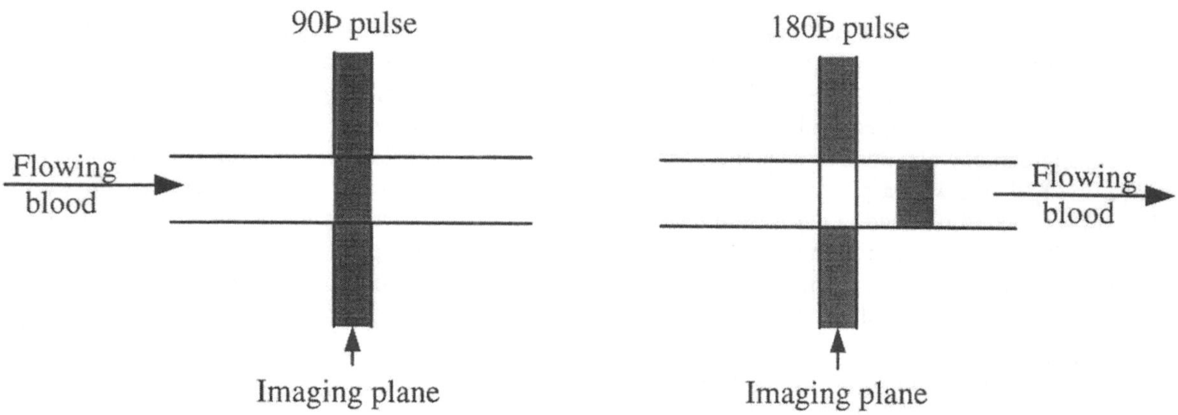

Fig. 4. Movement of blood magnetization away from the imaging plane

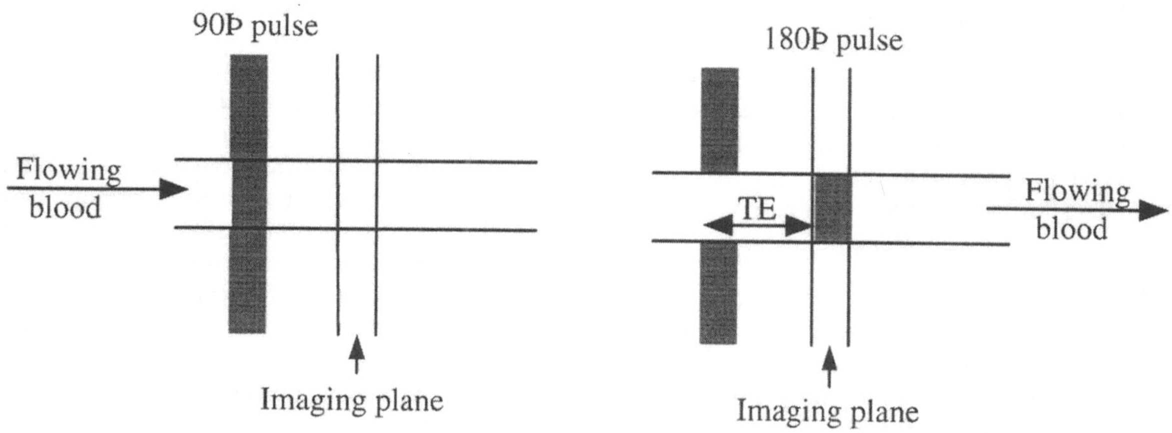

Fig. 5. Movement of blood magnetization into the imaging plane

applied, the blood that has experienced the 90° pulse is already out of the slice so it does not contribute to the signal leaving a signal void, as can be seen in Fig. 4.

This can be used for angiography by applying a 90° pulse outside the imaging plane, so that only the blood that flows into the plane within a time TE of the 90° pulse is prepared for the 180° pulse and will produce an echo. The rest of the slice will produce no signal (Fig. 5).

This method only works for flow into the plane and will not take into account flow in the plane.

Phase contrast angiography

The phase contrast (PC) method works on a different principle, making use of the dephasing produced on the spins by a non-linear bipolar magnetic field gradient, that is, one which has two lobes, one positive and one negative.

If the positive lobe comes first, the bipolar gradient is said to be positive and if the negative lobe comes first the gradient is negative. The positive lobe of the gradient will dephase the spins in one direction and the negative lobe will dephase in the opposite direction. If the spins are stationary the total dephasing will be zero, that is, the stationary spins will not be an affected. But if the spins have a velocity component in the direction of the gradient, the dephasing of the different lobes of the gradient will not be compensated.

In PC angiography two imaging sequences are performed, the first one with a positive bipolar gradient pulse and the second one with a negative bipolar gradient pulse. Then the raw data from the two is subtracted. The signals of the stationary spins cancel and the moving spins have a net signal, producing an image of the flowing spins.

To obtain the optimum signal, the spins of the fastest flowing blood should acquire 90° of phase after each bipolar gradient pulse, or 180° in total. All the other spins with slower velocities will acquire smaller phase shifts. Only those spins with a component in the direction of the bipolar gradient will produce a signal.

With PC angiography it is possible to obtain quantitative measurements of velocities both for vascular flow and for CSF flow, by means of adequate software, that will convert phase measurements into velocity values. It is also possible, by measuring vessel areas, to obtain fluid flow rates (Fig. 6).

Magnetization transfer

The Hydrogen nuclei in the body exist not only in water and fat but also in other macromolecules like proteins. However, these protons do not contribute to the MR signal because they have a very short T2 relaxation time, since they are tightly bound. They are in fact excited at the same time as the water protons, but their signal decays in less than a millisecond. In MR a system that has a short T2 responds to a very large range of frequencies, and a system with a large T2 responds to a narrow range of frequencies. This means that the protons bound to the macromolecules can respond to an RF pulse shifted, for instance, 1500 Hz, from the resonant frequency of the water protons, without affecting these. However, the protons of water bound to these macromolecules will interact with them and will become partly saturated In this way the signal from highly proteinated tissues, like brain, liver and muscle, will become suppressed (Fig. 7).

Magnetization transfer is used normally to improve the suppression of the signal from brain and muscle when performing MRA TOF, but it can also be used to obtain information on the protein contents of some tissues.

Contrast agents (Gadolinium)

Although MRI is a very powerful imaging technique not all pathologies are clearly contrasted using only proton density or relaxation times weighting. For example, some meningiomas and small metastatic lesions do not show on normal imaging. Furthermore, considering that some of these intracranial lesions have an abnormal vascular bed or a breakdown of the blood-brain barrier, a magnetic contrast agent that distributes throughout the extracellular space become an obvious choice to improve image contrast.

Some purists believe that the fact that MRI is a noninvasive method is one of its strengths and should be kept that way, but the clinical efficacy of the paramagnetic contrast is more than proven to amply justify its use.

All the common contrast agents used in MRI are Gadolinium chelates, that are not directly imaged but produce an effect, which is imaged. Gadolinium is the element of choice because of its high number of seven unpaired electrons. Each unpaired electron has a magnetic moment 657 times bigger than that of a proton, so seven unpaired electrons can induce relaxation a million times better than an isolated proton. This implies that both T1 and T2 are reduced, although the enhancement caused by the shortening of T1 is stronger than the signal loss caused by the shortening of T2; and that is why with Gadolinium contrast the images obtained are normally T1 weighted. The actual amount of T1 shortening is dependent on the concentration of Gadolinium injected, and the signal enhancement depends also on TE and TR.

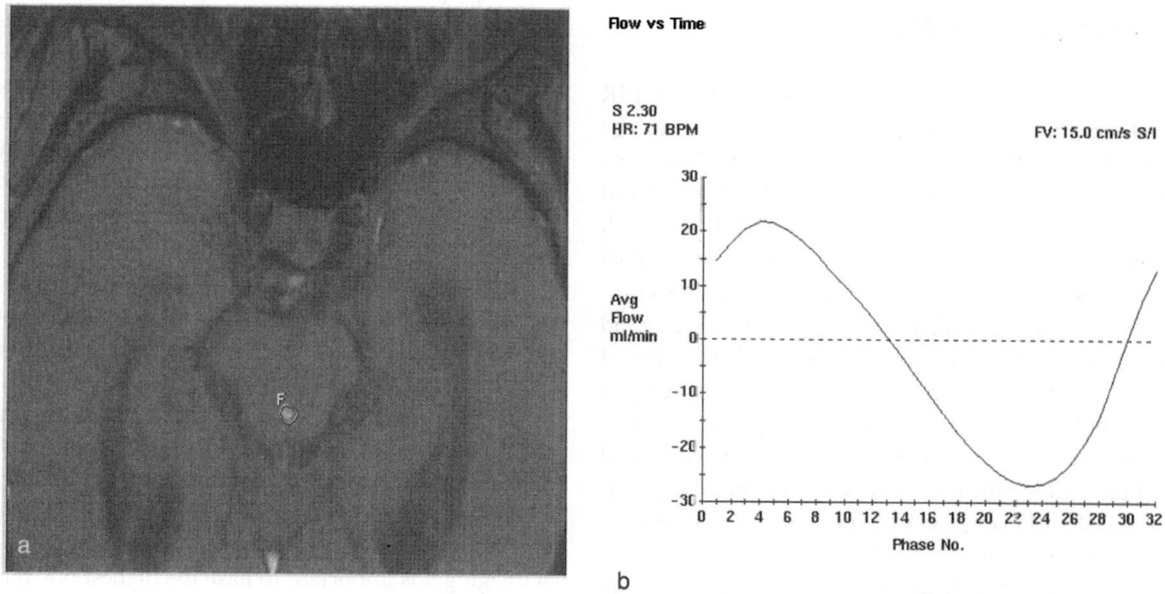

Summary Report

S 2.30
HR: 71 BPM
SlThick: 5.00 mm

FV: 15.0 cm/s S/I

Velocity /Cycle: cm/s	
Avg	-0.30
peak	11.31
Flow Rate/Cycle: ml/min	
Avg	-2.40
Pos	6.50
Neg	-8.90
Flow Volume/Cycle: ml	
Pos	0.09
Neg	-0.13

c

Fig. 6. **a**) Axial oblique plane perpendicular to the Aqueduct showing the flow area measured; **b**) graph of the average CSF flow through the cardiac cycle, and **c**) summary table of the flow parameters calculated

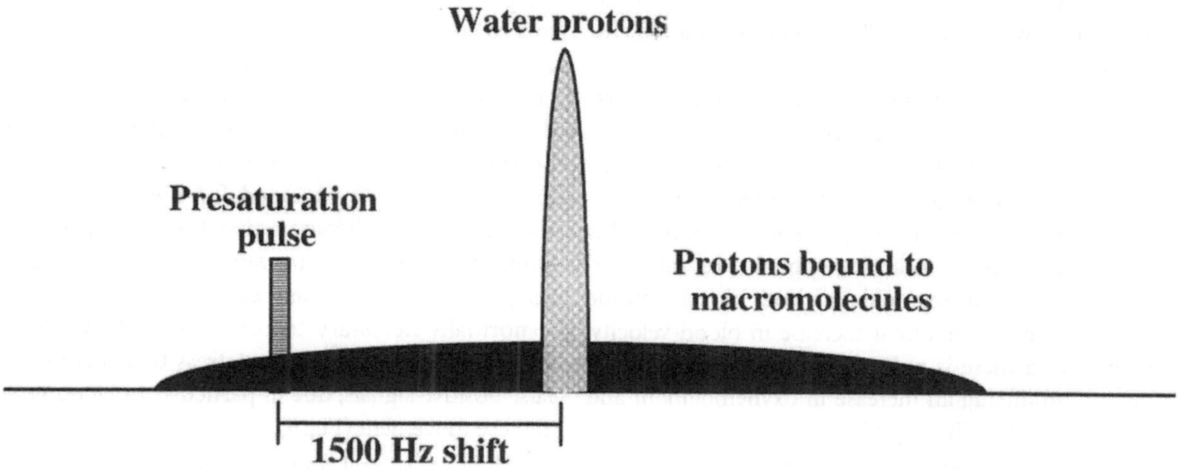

Fig. 7. Diagram for the magnetization transfer process

Contrast enhanced MRA

One of the recent uses of MR contrast agents is in MR angiography. The injection of the contrast into the blood reduces the T1 relaxation time in the blood vessels relative to surrounding tissues, therefore a rapid volume imaging sequence with a short TR value will produce a large signal for blood and a very small signal for the long T1 tissues surrounding the blood vessels.

This technique enables the acquisition of very good vessel images without recourse to the flow properties of blood. It works in the same way as digital angiography but is not selective. One of the advantages is the possibility of imaging vessels with awkward geometries and turbulent flow, which are difficult to obtain using standard TOF or PC angiography. Because of the high quality of contrast enhanced MRA (CE-MRA) images it is becoming the modality of choice in MR angiography (Fig. 8).

Functional imaging: principles and techniques

The term "functional imaging" in MRI is a very general term that covers any technique that gives functional information rather than just anatomical information. That is, any technique that acquires time dependent imaging data should be called functional imaging. Flow, perfusion, diffusion, tagging and brain activation belong to this category. However, when functional magnetic resonance imaging (fMRI) is mentioned it is normally referred to brain activation. In this section we are concentrating on this latter technique.

Brain activation can be studied either by direct methods, those that measure directly the electrical activity of neurons, like EEG (electrical effect) and MEG (magnetic effect) or by indirect methods, those that measure the hemodynamic response to the neuronal electrical activity, like ^{15}O PET (blood flow) and fMRI (BOLD effect).

The indirect method used by fMRI can be understood by following the chain of physiological events that describes it. When a set of neurons fire, there is a local increase in glucose consumption which in turn produces an increase in oxygen consumption. This induces an increase in both regional cerebral blood flow (rCBF) and regional cerebral blood volume (rCBV), with a consequent increase in blood velocity. In the blood there is a decrease in oxygen extraction fraction producing an increase in oxyhemoglobin and a decrease in deoxyhemoglobin.

In this sequence of events the most common approach used in fMRI is the Blood Oxygen Level Dependent (BOLD) contrast. The decrease in deoxyhemoglobin, because of its high paramagnetism, produces a decrease in local microscopic field gradients, which in turn produces an increase in T2*. This corresponds to an increase in signal, which is measured by the MR equipment. The ideal sequence to use is a rapid sequence with T2* sensitivity, which detects changes in magnetic field, usually a Gradient Echo EPI.

fMRI has its own limitations both in spatial resolution and temporal resolution. In terms of spatial resolution, although for a standard image the voxel volume is, approximately, $3 \times 3 \times 5$ mm^3, it is theoretically possible to go down to $0.5 \times 0.5 \times 1$ mm^3. The temporal resolution is limited by the hemodynamic lag of 4 to 8 sec in the response to the neuronal electrical activity and the speed of the scanner hardware, presently of the order of 10 frames per sec. To achieve optimal functional imaging it is important to have the highest possible magnetic field, powerful and fast gradients and a powerful computer with adequate software to manipulate the image.

The pulse sequence used will look for small variations in the signal of the T2* weighted image. Since these variations are very small it is necessary to obtain a large number of images as the activation paradigm is performed. The paradigms normally consist of blocks of 30 sec of rest followed by 30 sec of activation. During rest no activity is maintained. During the activation period the task being studied is performed. The activation can be motor, sensory, visual, auditory, language generation and others.

The images obtained directly from the system do not show any visible characteristic to the naked eye. It is necessary to treat the images mathematically by comparing the variation of intensity of the pixels in a certain image as a function of time with the variation that one would expected in the theoretical ideal case of the particular activation paradigm, which corresponds to a square function. This comparison is done statistically, pixel by pixel, and colour coded to indicate if they are more or less correlated with the activation paradigm. The final images are then obtained by superimposing the statistically processed EPI images on the anatomical images obtained for the same slices. It should be noted however that the images obtained with the EPI sequence are very sensitive to changes in magnetic susceptibility and can be heavily distorted. It is normally necessary to correct the EPI images by computer. It is important to stress that there can be false positive signals, due in particular to blood vessels and eye movement (Fig. 9).

There are several clinical applications for fMRI now being tried, like tumor surgery planning, AVMs, epi-

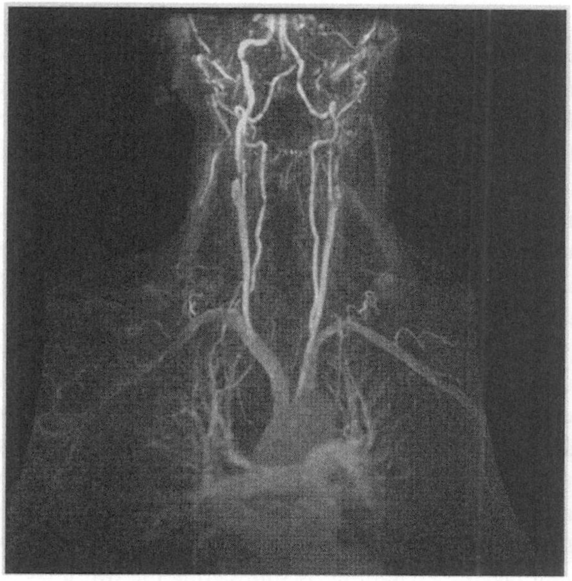

Fig. 8. A CE-MRA image showing the carotids all the way from the aorta

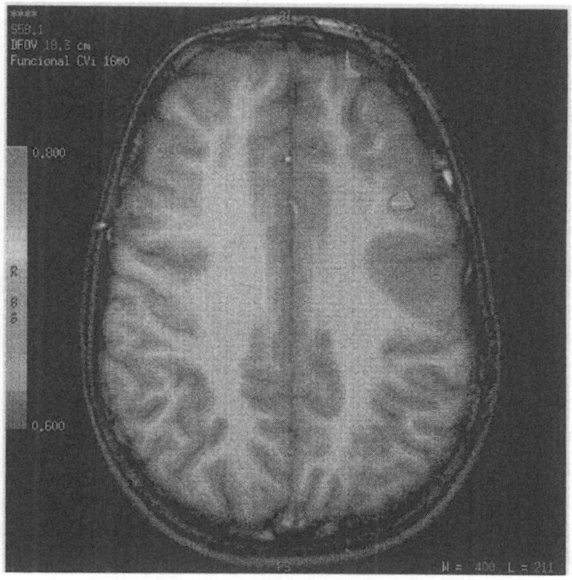

Fig. 9. A functional image showing the language premotor cortex activation for a phonetic language generation paradigm

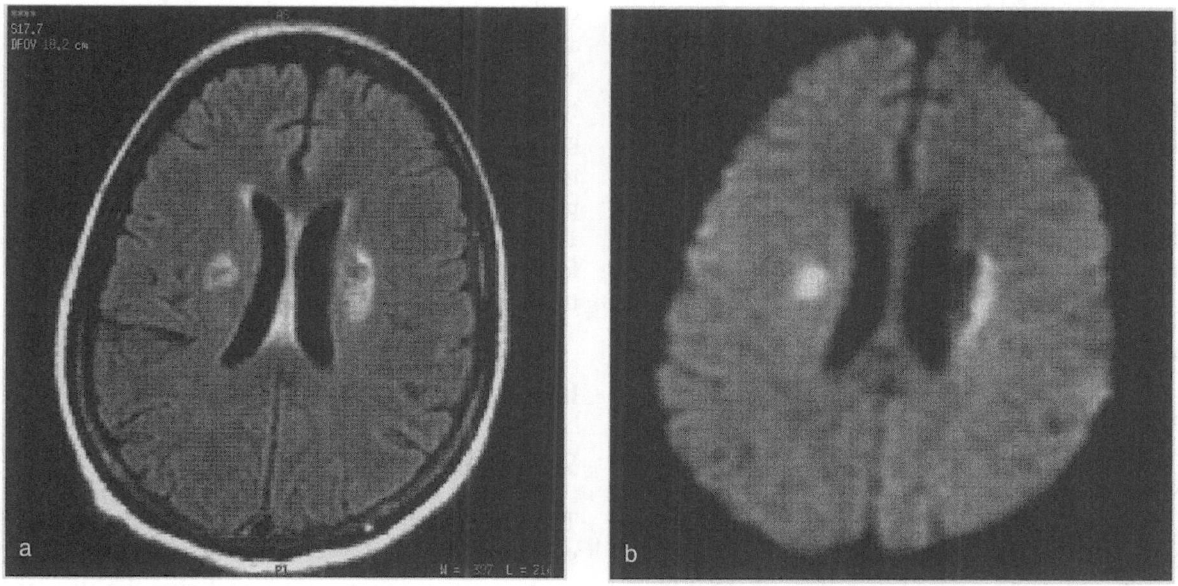

Fig. 10. Two images on the same location of a patient with infarcts: **a)** a FLAIR image, and **b)** a diffusion image

lepsy, addiction, schizophrenia and AIDS. Functional brain activation imaging with MR promises to be clinically useful, but only with a more robust and complete image processing and being very careful with the definition of the paradigm used and the verification of its implementation.

Diffusion

While MR angiography and flow measure the movement of spins from voxel to voxel, MR is capable of measuring microscopic translational motion within each voxel. This motion can be the molecular diffusion of water and the microcirculation of blood in the capillary network, referred to as perfusion. Diffusion is the process by which molecules and other particles mix and migrate due to their random thermal motion.

Diffusion imaging is acquired in a similar way to phase contrast angiography, using a specific bipolar gradient with very high strength and duration of the gradient lobes to detect the slow molecular diffusion in the body. This bipolar gradient will cause a signal loss in the diffusing spins, which depends on the diffusion coefficient and the B value. The B value is determined by the strength and duration of the gradients.

High B values can eliminate the T2 effect and improve the visualisation of the white matter fibres and can be useful to differentiate subacute from chronic infarcts.

It is possible to obtain maps of diffusion in the three different orthogonal directions, or combine the three images into a single map of overall diffusion.

With specific research software it is possible to calculate the diffusion tensor and deduce the actual direction of the diffusion, and even obtain the direction of the neuronal axons (Fig. 10).

Perfusion

Perfusion is the study of the net transport of magnetization into a volume of tissue. It can be performed either with endogenous or exogenous contrast agents.

One way of achieving this is by injecting a bolus of contrast agent, like Gadolinium, and acquiring rapidly, usually with EPI, a series of slices through the region of interest. And then, repeating the multislice acquisition at a rapid rate, of the order of one per second. This repetition is performed from before the injection until about 30 seconds after the arrival of the bolus. The images are then analysed to look for variations in the arrival of the contrast agent between the pathological and normal regions.

The analysis can be made in several ways: the negative enhancement integral (NEI), which produces a map of the susceptibility effect caused by the flow of blood containing contrast; the mean time to enhance (MTE), which produces a map based on the time of arrival of the contrast in different regions; and the positive enhancement integral, which produces a map of the contrast enhancement rates based on the increase in SNR due to increased concentration of contrast material.

Another way to perform perfusion studies is by using tagging methods, which avoid the injection of a contrast agent. The tagging is achieved by applying a saturation pulse to the feeding arteries, which prepares the blood before it enters the slice of interest, and then acquiring an image of the slice. Following that, a second image acquisition is obtained but with a different tagging excitation. By subtracting the two images, signal differences are obtained only in the regions where the tagged blood has reached. Because of the small signal, it is normally necessary to average over a large number of acquisitions.

Sequence comparison table

With the evolution of MRI many manufacturers started developing their own sequences, or their versions of the standard sequences. Some of these sequences although slightly different and called different names are practically equivalent. This diversity of nomenclature can introduce some chaos in the field when dialoguing amongst different platforms, since it is important to know if the same sequences are being used for a particular study. With this in mind, we present in Table 1 sequences comparison where equivalent sequences for four of the major manufacturers are presented on the same line.

Image parameters and contrast

Signal to noise ratio

The main parameter to assess signal quality is the signal to noise ratio, designated by SNR. This is defined as the ratio of the average signal over the standard deviation of noise. The signal comes only from the spins that were excited intentionally when selecting the slice, or volume, of interest and the noise comes from many other sources, the main one being the patient. As mentioned above, only the spin excess between the aligned and anti-aligned spins contribute to the signal, but all the other spins, as they jump up and down from

Table 1. Sequences comparison

	GE	Philips	Picker	Siemens
• SPIN-ECHO SEQUENCES				
Single-Echo Spin Echo	Spin Echo	Spin Echo Modified Spin Echo	Spin-Echo throughput heightened rapid increased flip T2 (THRIFT) Phase symmetrized rapid increased flip spin echo (PRISE)	Single Spin Echo
Multi-Echo Spin Echo	Multiecho Multiplanar (MEMP) Variable Echo Multiplanar (VEMP)	Multiple Spin Echo (MSE)	Multiecho THRIFT	Spin Echo Double Echo
Echo Train Spin Echo	Fast Spin Echo (FSE) Single Shot FSE (SS-FSE)	Turbo Spin Echo (TSE) Ultra Fast Spin Echo (UFSE)	Fast Spin Echo (FSE)	Turbo Spin Echo (TSE) Half Fourier Acquisition Turbo Spin Echo (HASTE)
Hybrid Gradient Echo Spin Echo	Gradient-Spin Echo (GRASE)	GRASE	GRASE	Turbo Gradient Spin Echo (TGSE)
• INVERSION-RECOVERY SEQUENCES				
Standard Inversion Recovery	Multiplanar IR (MPIR)	IR	IR	IR
Echo Train Inversion Recovery	Fast Multiplanar IR (FMPIR)	IR-Turbo Spin Echo (IR-TSE)	IR	Turbo IR
Interleaved Excitation	Nonsequential	—	Slice Interleaved	Interleaved
Magnitude Reconstruction	—	Modulus	—	Absolute Value Magnitude
Phase Sensitive Reconstruction	—	Real	—	True IR
• GRADIENT-ECHO SEQUENCES				
Spoiled	Spoiled GRASS (SPGR) Fast Spoiled GRASS (FSPGR) Multiplanar Spoiled GRASS (MPSPGR) Fast Multiplanar Spoiled GRASS (FMPSPGR)	T1 Contrast-Enhanced Fast Field Echo (T1 CE-FFE)	Radio-Frequency Spoiled Fourier-Acquired Steady State (RF-FAST)	Fast Low Angle Shot (FLASH)
Refocused, Postexcitation	Gradient Acquisition in the Steady State (GRASS) Fast GRASS Multiplanar GRASS (MPGR) Fast Multiplanar GRASS (FMPGR)	Fast Field Echo (FFE)	FAST	Fast Imaging with Steady-State Precession (FISP)
Refocused, Preexcitation	Steady State Free Precession (SSFP)	T2 Contrast-Enhanced FFE (T2 CE-FFE)	Contrast-Enhanced FAST (CE-FAST)	Reversed FISP (PSIF)
Magnetization Prepared	IR-Prepared Fast GRASS Driven Equilibrium (DE)-Prepared Fast GRASS	Turbo Field Echo (TFE)	Rapid Acquisition Magnetization-Prepared FAST (RAM-FAST)	Turbo FLASH Magnetization-Prepared Rapid Acquisition Gradient Echo (MP-RAGE)
• ECHO-PLANAR IMAGING				
Single Shot	Single Shot	Single Shot	Single Shot	Single Shot
Multishot	Multishot	Multishot	Multishot	Multishot

(continued)	GE	Philips	Picker	Siemens
Blipped Phase Encoding	Blipped Phase Encoding	Blipped Phase Encoding	Blipped Phase Encoding	Blipped Phase Encoding
Constant Phase Encoding	Constant Phase Encoding	Constant Phase Encoding	Constant Phase Encoding	Constant Phase Encoding
Spiral Phase Encoding	Spiral Phase Encoding	Spiral Phase Encoding	Spiral Phase Encoding	Spiral Phase Encoding
• SPECTROSCOPY				
Point-Resolved Spectroscopy	PRESS	PRESS	PRESS	PRESS
Stimulated Echo Acquisitiom	STEAM	STEAM	STEAM	STEAM
Single Voxel Spectroscopy	SVS	SVS	SVS	SVS
Two-Dimensional Chemical Shift Imaging	2D-CSI	2D-CSI	2D-CSI	2D-CSI
2D MR Spectroscopic Imaging	2D-MRSI	2D-MRSI	2D-MRSI	2D-MRSI
• OTHER TECHNIQUES				
Spatial Presaturation	Saturation (Sat)	Regional Saturation Technique (REST)	Presaturation (PS)	Presaturation (Presat)
Short Inversion Time Inversion Recovery	STIR	STIR	STIR	STIR
Fat Saturation	Fatsat Chemsat Spectral Presaturation	Spectral Presaturation with Inversion Recovery (SPIR)	Chemical Shift Imaging -Dixon Method	Fatsat
Magnetization Transfer	MT	Magnetization Transfer Contrast (MTC)	MT	MT
Angiography Nonuniform Excitation	Ramped Excitation	Tilted Optimized Nonuniform Excitation (TONE)	Ramped excitation	TONE
Motion Compensation	Flow Compensation (FLOW COMP)	Flow Adjustable Gradients (FLAG)	Motion Artifact Suppression Technique (MAST)	Gradient Motion Rephasing (GMR)

the two energy levels, can emit a random radio frequency photon, which contributes to the background noise. Obviously if the sensitive volume of the coil is large it will detect a large number of these random transitions, producing a large noise. Therefore, the smaller the coil the closer it will be to the excited spins, producing a larger signal, and detecting fewer random transitions, consequently producing less noise; that is, a smaller coil will have a higher SNR. For example, a head coil has a higher SNR than a body coil, and a small surface coil placed close to the anatomy of interest will have an even higher SNR than the head coil.

There are several factors affecting SNR, like voxel size, number of excitations and bandwidth. The voxel size affects the signal and the other parameters affect the noise. The larger the voxel size the larger the number of spins inside it, so the signal is directly proportional to the voxel size within a tissue of uniform spin density. The SNR also depends on the number of excitations, but not in a linear way. In fact the SNR is proportional to the square root of the NEX, for example, going from 1 to 4 NEX only improves the SNR by a factor of 2. Another factor that influences the SNR is the receiver bandwidth, but here the dependence is inverted, the SNR is inversely proportional to the square root of the bandwidth, that is, if the bandwidth is increased by a factor of 4 the SNR is decreased by a factor of 2.

Contrast

Contrast is the relative difference between the signals of adjacent voxels and can be defined as the difference of signal intensities divided by the average signal intensity in two adjacent regions. To differentiate one tissue from another it is very important to increase the contrast between them.

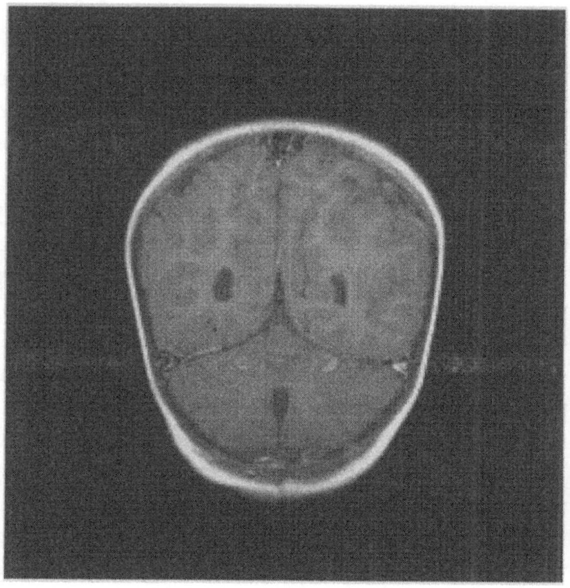

Fig. 11. An image showing both a motion artifact on the left and right and a blood flow artifact running from left to right

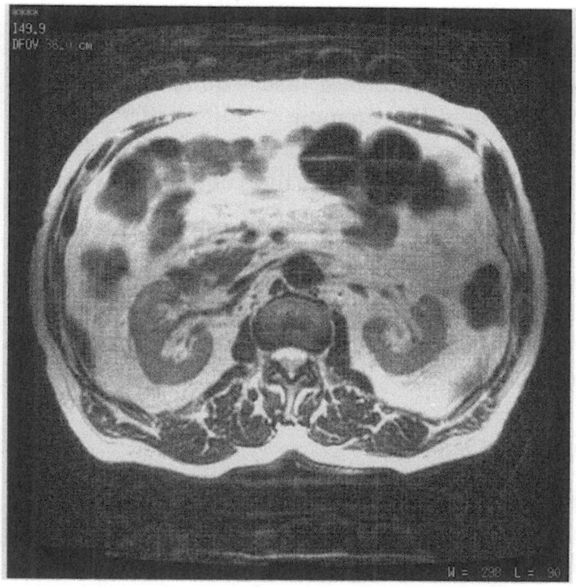

Fig. 12. An image showing a respiratory artifact, from top to bottom

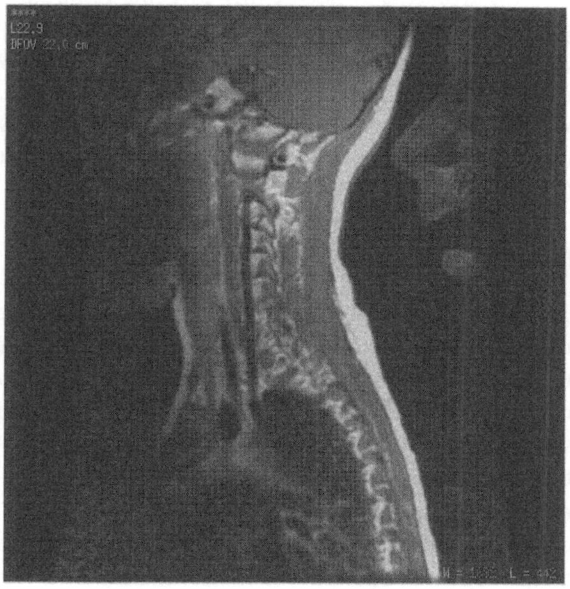

Fig. 13. An image showing a wrap around artifact, with the nose and mouth appearing at the back of the head

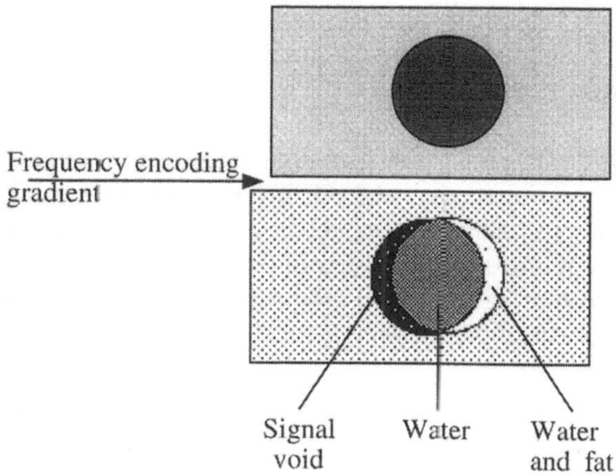

Fig. 14. Diagram showing the original object and the image obtained due to chemical shift

The main factor determining the tissue contrast is the choice of sequence and its parameters, since for a particular sequence each tissue will have a particular signal height. For example, to differentiate a tumor from the surrounding tissue it is essential to choose a sequence that maximizes the contrast between the tumor and the surroundings.

Once the image is acquired it is possible to improve the contrast by manipulating the image in the post-processing stage, but this should be done very carefully because it can mask the original information obtained.

Spatial resolution

The spatial resolution of an image is determined by the number and size of points composing it and it will determine the smallest anatomical structure that can be resolved. It is inversely related to voxel size: the higher the spatial resolution, the smaller the voxel size. The two main factors determining spatial resolution are the field of view (FOV) and the matrix size. If you diminish the FOV maintaining the same matrix the voxel size goes down, and if you maintain the FOV but increase the matrix the voxel size goes down as well. We should bear in mind that if the matrix size is doubled, for instance from 256 x 256 to 512 x 512, the number of voxels goes up by a factor of four and their volumes go down by a factor of 4. Therefore, as the resolution goes up the voxel size goes down and this implies that the SNR goes down as well. As always in MRI there is a price to pay for a particular improvement.

Image artifacts

MRI has a multitude of factors that affect the appearance of the image and this makes it very interesting and rich. However, one of the problems with this is that it is possible that a few of these factors will go out of control producing an image that does not reflect the real state of the anatomy, in other words, the image will have artifacts.

Although sometimes artifacts may destroy the quality of the image, with the possibility of making it useless, some other times they can just be ignored, but the major danger of some of the artifacts is that they can be confused with pathology, leading to misdiagnosis.

Therefore, it is important to understand the aspects and the causes of the major artifacts and how to deal with them. All manufacturers have a series of standard techniques to eliminate or compensate for the major artifacts.

Motion and ghosting

The most common cause of image artifacts is patient motion. Random motion will just produce a blurred image and is avoided by asking the patient to be still or sedating in extreme cases.

However, not all movement can be controlled by the patient, for instance, the blood keeps on pulsating. And any motion that occurs regularly in a repeating pattern will not cause a blurred image, but will produce one or more "ghosts" in the phase encoding (PE) direction. If the "ghost" occurs inside the image it can produce either a darker or brighter area in the surrounding tissue that can be confused with pathology. The way to avoid this artifact or at least reduce it is by using cardiac gating or a spatial presaturation pulse on the side of the incoming blood. Sometimes, if it is unavoidable, changing the order of the phase encoding and the frequency encoding gradients shifts the "ghosting" to an area that is not important and leaves the area of interest clear (Fig. 11).

Respiratory artifact

Another source of periodic motion is respiration and this can also cause ghosting, but since it is the whole chest and abdomen that moves, the artifact appears above, below and throughout the body.

This artifact is avoided either by using respiratory gating, improving the quality of the image significantly (but doubling, or more, the acquisition time), or by using a method known as respiratory compensation, which will clear the coherent "ghosts" above and below the image though producing a slight blur all across it (Fig. 12).

Magnetic susceptibility

To obtain good MRI images it is essential that the magnetic field is as homogeneous as possible, because the spatial encoding is based on a precise distribution of the linear field gradients. Unfortunately, there are several things that can distort the magnetic field affecting adversely the quality of the image. The main parameter affecting the homogeneity is the magnetic susceptibility, which says how much a substance will be magnetized when placed in a magnetic field. This can be a problem not just due to the presence of foreign objects within the body but also due to differences in magnetic susceptibilities of adjacent tissues.

If the magnetic susceptibility of the region being imaged is fairly homogeneous there will be no major changes in the magnetic field. However, if the magnetic susceptibility of adjacent tissues differs much, like the transition between air and tissues (lungs and sinuses), there will appear an artifact at the separating edges.

Because of their high magnetic susceptibility, the presence of metallic objects (like dental implants, clips and shunts) within the volume to be imaged will distort drastically the local magnetic field and it is possible to have a total loss of signal in a particular region, surrounded by a strong distortion of the signal.

Partial voluming

Normally the width of the voxels, that determines the resolution in the imaging plane, is smaller than their depth, except in the case of 3D volume acquisitions. This depth, called the slice thickness, is normally of the order of 5 mm. If a group of voxels incorporates the edge of a structure there will be a blurring of the image around these voxels because their volume averages the signal from different tissues. This process is called partial volume. The way to diminish this problem is to reduce the slice thickness.

Wrap around

The phase encoding (PE) gradient produces a phase shift in the spins that varies between 180° and ñ180°. However, if the field of view is too small there will be excited tissue outside the FOV, producing phase shifts above 180° and below –180°. Because the equipment cannot measure phases outside the 180° and –180° range, values above 180° will be confused with those near the –180°, and values below –180° will be confused with those near the 180°. As an example, some tissue outside the FOV assigned a phase shift of 190° will be confused, in the reconstruction, with the tissue inside the FOV assigned a phase shift of –170°. This will produce what is called a wrap around in the PE direction. In a sagittal slice of the head this can cause the nose to appear on the back of the head (Fig. 13).

Chemical shift

The different chemical environment of the nuclei of hydrogen in water and fat produces a slightly different magnetic field around them, that causes the protons in fat to resonate at a frequency lower than the protons of water. This difference is 3.5 ppm, which at 1.5 T corresponds approximately to 224 Hz.

One consequence of this is the appearance of white and dark bands at fat/tissue boundaries, as exemplified in Fig. 14.

The image reconstruction computer assumes that a spin with a particular frequency comes from a particular point in space. However, because the fat protons have a resonant frequency lower than that for the water protons, along the frequency encoding gradient all their spins will be considered to come from a slightly displaced position when compared with the water protons. This can have two effects. One is producing a dark band, or signal void, in an area where the fat signal should be, but no water exists. The other is the appearance of a white band due to the existence of two different signals, from water and fat, with exactly the same frequency which add up.

By exchanging the frequency and phase encoding gradient directions these artifacts can disappear from some of the boundaries.

Note

All the images shown were obtained on a 1.5 T GE CVi system at Resonancia Magnètica de Caselas, Lisbon, Portugal.

Selected references

1. New Aiver M (1997) All you really need to know about MRI physics. Simply Physics, Baltimore
2. Finn JP (1999) Physics of MR Imaging. Magnetic Resonance Imaging Clinics of North America 7(4), WB Saunders, Philadelphia
3. Haacke M, Brown RW, Thompson MR, Venkatesan R (1999) Magnetic Resonance Imaging. Physical Principles and Sequence Design. Wiley-Liss, New York
4. Brown MA, Semelka RC (1999) MR Imaging Abbreviations, Definitions and Descriptions: a review. Radiology 647

In vivo Magnetic Resonance Spectroscopy in neurosciences

B. Soussi

Aim of chapter

The aim of this chapter is to provide a comprehensive introduction to the new possibilities that Magnetic Resonance Spectroscopy (MRS) offers in clinical neurosciences. Focus will be on what MRS *can do* rather than what MRS *is*. For simplicity, basic physical and chemical principles will not be much explored and are referred to elsewhere.

Introduction

For over half a century, interest in Nuclear Magnetic Resonance (NMR) has been continuously increasing. From structural analysis in smaller organic molecules, to biochemical macromolecules, tissue extracts, isolated intact organs and *in vivo* studies in animals and humans.

For almost two decades, *in vivo* MRS has been a revolutionary technique in biomedical research. Today, it is a powerful tool in neurosciences giving noninvasive access to the chemistry of the human brain in health and in disease.

Nuclei like ^{31}P, ^{1}H, ^{13}C, ^{19}F and ^{23}Na have been studied in various organs. However, early applications of *in vivo* MRS began with the measurements of ^{31}P metabolites in isolated organs and surface regions like skeletal muscles from intact animals.

Historically, ^{31}P has been the most studied nucleus. However, MRS of the brain today relies mostly on ^{1}H examination due to its relative ease, i.e. high natural abundance (99.9%) and sensitivity (100%). Numerous studies have shown that MRS can detect pathophysiological changes in the brain tissue in a number of diseases. Therefore, this chemically specific technique with its ability to examine the mechanisms of disease is continuously gaining attention from clinicians.

In vivo MRS should be seen as complementary to the well established clinical MRI, providing quantitative nondestructive analysis of the biochemistry of the brain cells without the use of radioactive tracers.

It is possible to integrate spectroscopy with conventional MRI equipment of 1.5 T or higher magnetic field by adding appropriate hardware and software available from MR manufacturers.

Theoretical background

MR theory is described elsewhere. For more detailed physical and chemical aspects of the technique see references.

The basic principles for MRS are the same as for MRI. It is suitable, however, to briefly mention some aspects that are related to spectroscopy, and put them in the simplest form:

- the interaction between atomic nulclei (possessing a spin that gives a magnetic moment) and radio waves when an external static magnetic field is applied gives rise to a electromagnetic signal;
- the electromagnetic signal obtained after the application of a 90° radiofrequency pulse is called free induction decay (FID);
- at the same time, each nucleus is characterized by the time constants T_1 (longitudinal relaxation) and T_2 (transversal relaxation);
- the decaying signal is the result of the relaxation of the nuclei from their excited state to their relaxed state;
- the FID is then converted to a spectrum by a Fourier transformation (mathematical algorithm);
- the "spectral" chemical shift (σ) is measured in parts per million (ppm) and is a characteristic of the variation in resonance frequency. Its specific dependency on the chemical environment of a particular nuclei makes it like a "finger print" of the analyzed substance. Figure 1 shows the conversion of a FID to a spectrum by Fourier transformation.

Localization

Image guided spectroscopy is shown in Fig. 2, which illustrates the selection of a volume of interest (VOI) based on a topographical MR image in order to acquire a proton MR spectrum.

The same strategy is used in the example in Fig. 3 to get a ^{31}P MRS localization based on a topographical MRI.

Localization methods

Early localization methods started with surface coil localization which is based on RF pulses and the use of surface coils for spatial localization.

A disadvantage of this procedure is surface tissue contamination of the spectra.

Multi-shots methods

ISIS

Image-selected *in vivo* spectroscopy (ISIS) uses a combination of 8 pulses. The VOI is preselected, based on MRI scan and is repeatedly excited. The ISIS method has been applied to both ^{31}P and ^{1}H. One advantage of this method is that it can be used without T_2 weighing. However, the eight phase cycles used in localization might make shimming difficult.

Single-shot methods

Two methods are widely used and basically similar.

STEAM

Stimulated Echo Aquisition Mode (STEAM) uses a stimulated echo generated by three 90° pulses (90°-90°-90°). It is mostly used in ^{1}H spectroscopy. Signal loss due to motion sensitivity at long echo times is a disadvantage. This method is suitable for short TE acquisitions.

PRESS

Point Resolved Spectroscopy (PRESS) involves a double spin echo scheme (90°-180°-180°) which theoretically gives improved S/N. This method is most suitable to ^{1}H spectroscopy where small volumes and/or metabolites with long relaxation times T2 are of interest.

Characteristic patterns seen in STEAM and PRESS spectra in patients with acute brain injury are shown in Fig. 6.

Spectroscopic imaging

Spectroscopic imaging is the simultaneous acquisition of spectra from many volumes using phase encoding. It is suitable for both ^{1}H and ^{31}P. This method offers the advantage of investigating many slices simultaneously. However, the S/N is lower than in single-voxel techniques.

Water and lipid suppression

The ^{1}H peak from brain water is dominant as well as the resonance from precranial lipids. Since most ^{1}H signals from brain metabolite are present at concentrations less than 10 mM, water and lipid suppression techniques are essential in ^{1}H MRS. Water suppression can be done using Gaussian chemical shift selective pulses (CHESS). The water signal is presaturated by using frequency selective 90° pulses.

Outer volume selective pulses may be applied to presaturate the lipid resonance. However, by using localization technique such as PRESS and STEAM, lipid areas can be kept outside the VOI.

Sensitivity

The analytical limit is around 1 mM. MRS is thus not a very sensitive technique. However, many of the 100% naturally abundant ^{31}P and ^{1}H metabolites are present in cellular concentrations in the mM range.

In localized *in vivo* spectroscopy, theoretical minimum resolution is around 1 ml for ^{1}H and 15 ml for ^{31}P. Generally, volumes for brain ^{1}H MRS vary from 4 to 30 ml at 1.5 T, and typically used VOI is around 8 ml. Resolution can be improved at longer aquisition times and with increasing magnetic field strength.

Several factors can influence the sensitivity during an MR examination. For example, the presence of paramagnetic species, or the slow exchange between bound and unbound forms of molecules, can cause signal broadening.

Changes in viscosity, inhomogeneity of magnetic field and many exchange processes could also affect the line shape of a resonance.

However, despite this relative insensitivity, no other method can do today what MRS can.

Field strength

Most clinical MRS are performed at 1.5 T to this date. Higher field strength permits better resolution of overlapping peaks. Field strengths of 3 and 4 T for clinical research have been available for a few years. Today, *in vivo* magnets of 9 T for experimental research are commercially available. A comparison illustrating improvement in resolution with increased magnetic field strength is shown in Fig. 4.

Spectral quantitation

For calculation of *in vivo* metabolite concentrations it is important to apply quantitation methods using internal and/or external standards.

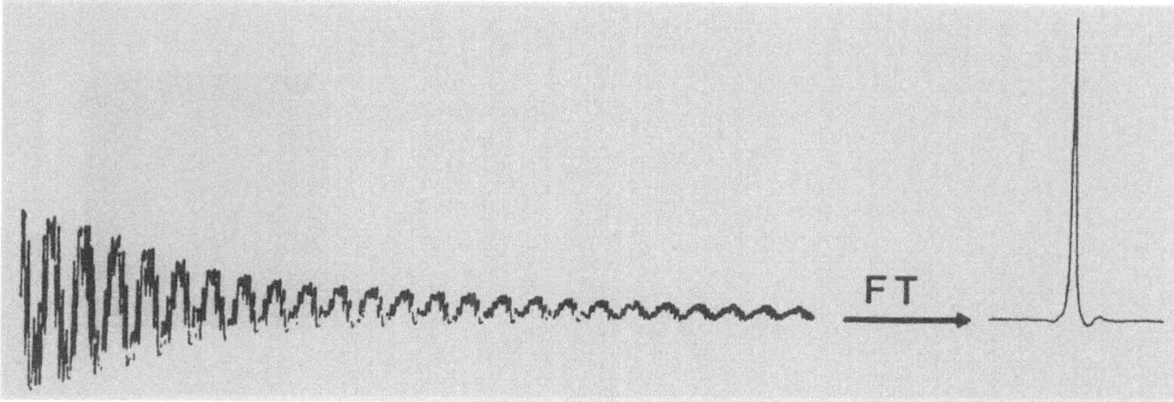

Fig. 1. The free induction decay is converted to a spectrum by a Fourier transformation. The FID signal (= amplitude *vs* time) is converted to a spectrum (= amplitude *vs* frequency)

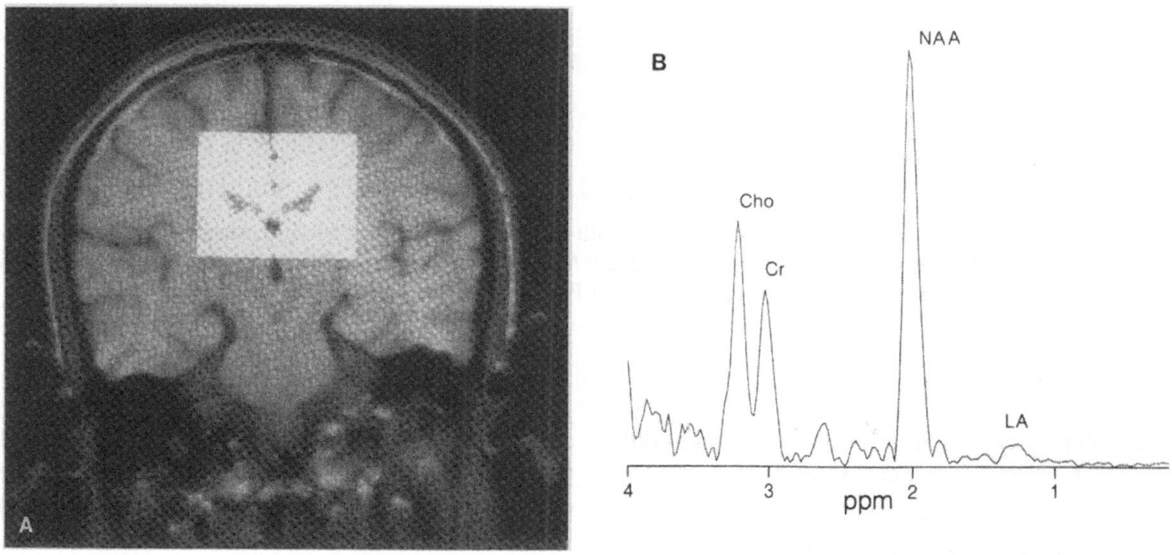

Fig. 2. Volume selection and spectral acquisition: **A**) MRI of normal human brain, illustrating the VOI = 50 × 40 × 50 mm; **B**) proton MRS spectrum of the selected volume showing the major proton metabolites (reproduced from [47])

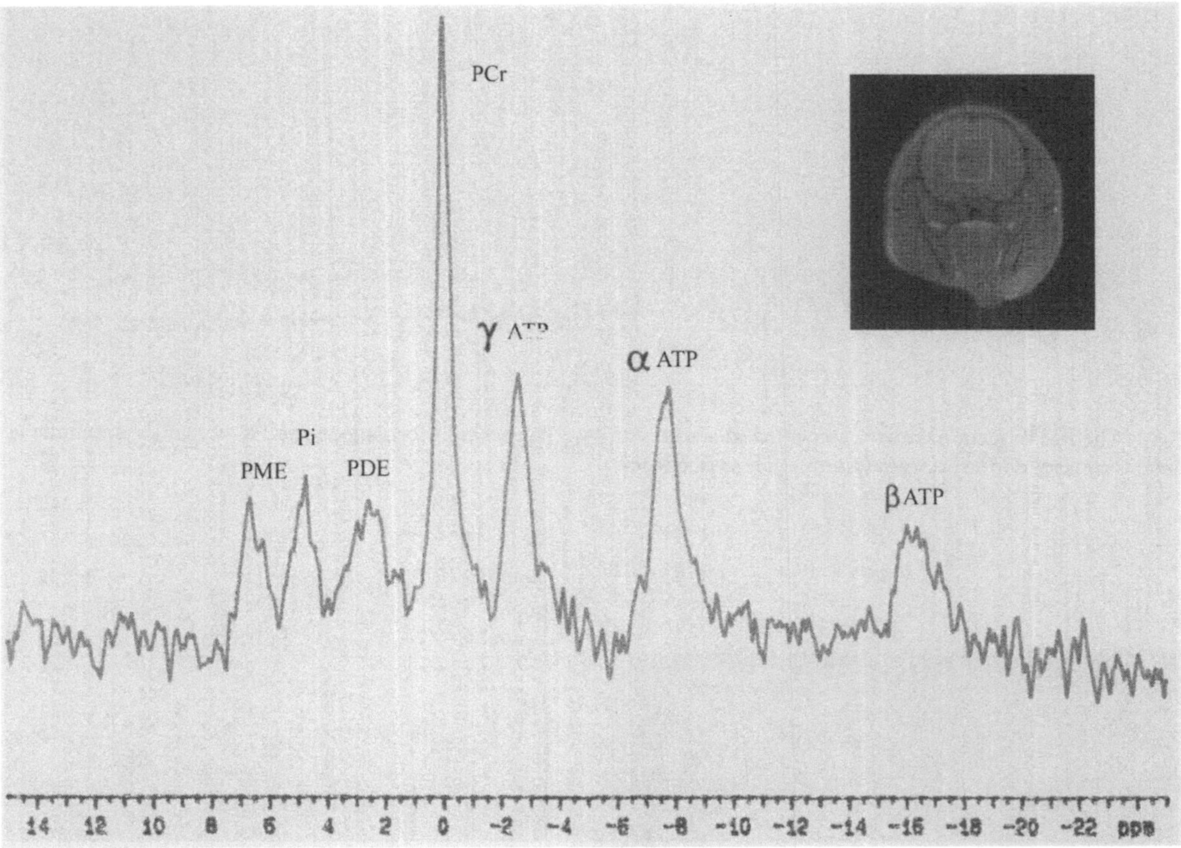

Fig. 3. *In vivo* ^{31}P MR spectrum localised from rat brain with ISIS (VOI = 10 × 10 × 10 mm). Magnetic field strength = 2.35 T. The peaks of the adenosine triphosphates α-, β-, and γ-ATP, the phosphocreatine PCr, the inorganic phosphate Pi as well as the phosphomonoesters PME and the phosphodiesters PDE are assigned

Absolute quantitation is possible but remains difficult. Relative concentrations and areas of peak ratios are also useful and widely used.

Problems associated with spectral quantitation

Common technical problems encountered arise from:

- motion artifacts;
- magnetic susceptibility effects;
- partial volume effects.

Motion artifacts may arise from breathing or any other movement. Susceptibility effects may arise from the variety of adjacent tissue to the VOI complicating shimming and affecting field homogeneity.

Partial volume effects are caused by the region surrounding the VOI affecting adequate metabolite quantitation. This is particularly problematic when large volumes (> 8 ml) are selected. Smaller VOI can be chosen at the cost of lower signal/noise ratio. Higher magnetic field might solve this problem.

Additionally, general factors like lower field strength, poor shimming and the low concentration of a particular metabolite may complicate calculation of peak areas due to non-Lorentzian lineshapes, baseline distortions and resonance overlap.

Metabolic information

^{31}P MRS

A representative ^{31}P MRS spectrum of the human brain at 1.5 T is shown in Fig. 5A, where peaks of major metabolites observed are assigned.

The peaks of α, β and γ-ATP, and of PCr and Pi can be clearly identified. Phosphomonoesters such as phosphocholine, phosphoethanolamine and sugar phosphates are under normal conditions, also present on both sides of the Pi resonance, and might partly overlap the Pi peak at lower fields.

The β-ATP peak is the most reliable in analyzing ATP concentrations, while the α and γ resonances contain contributions from NAD and ADP, respectively.

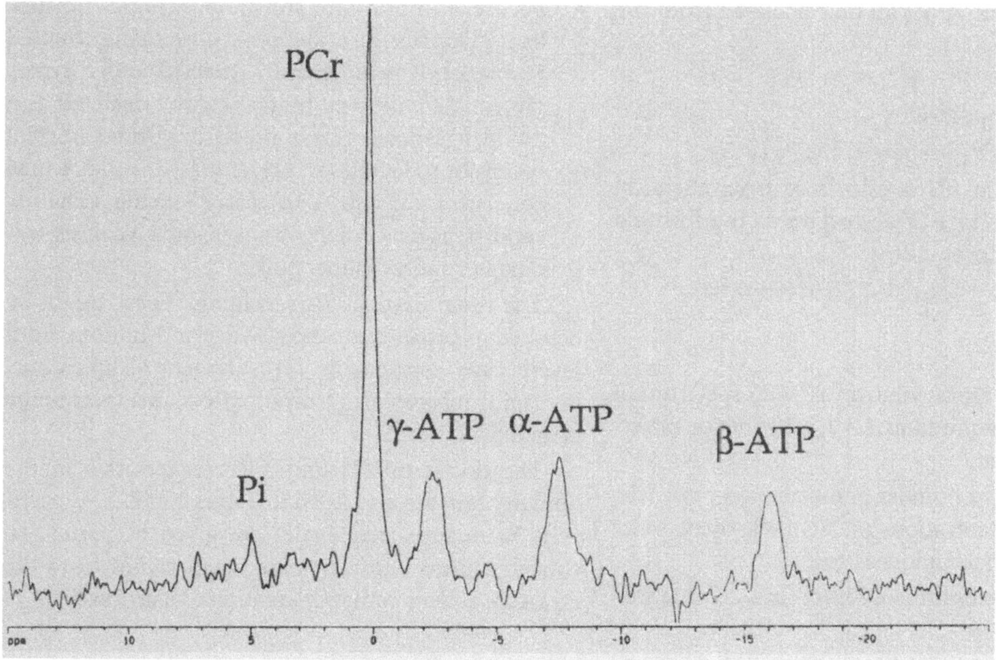

A

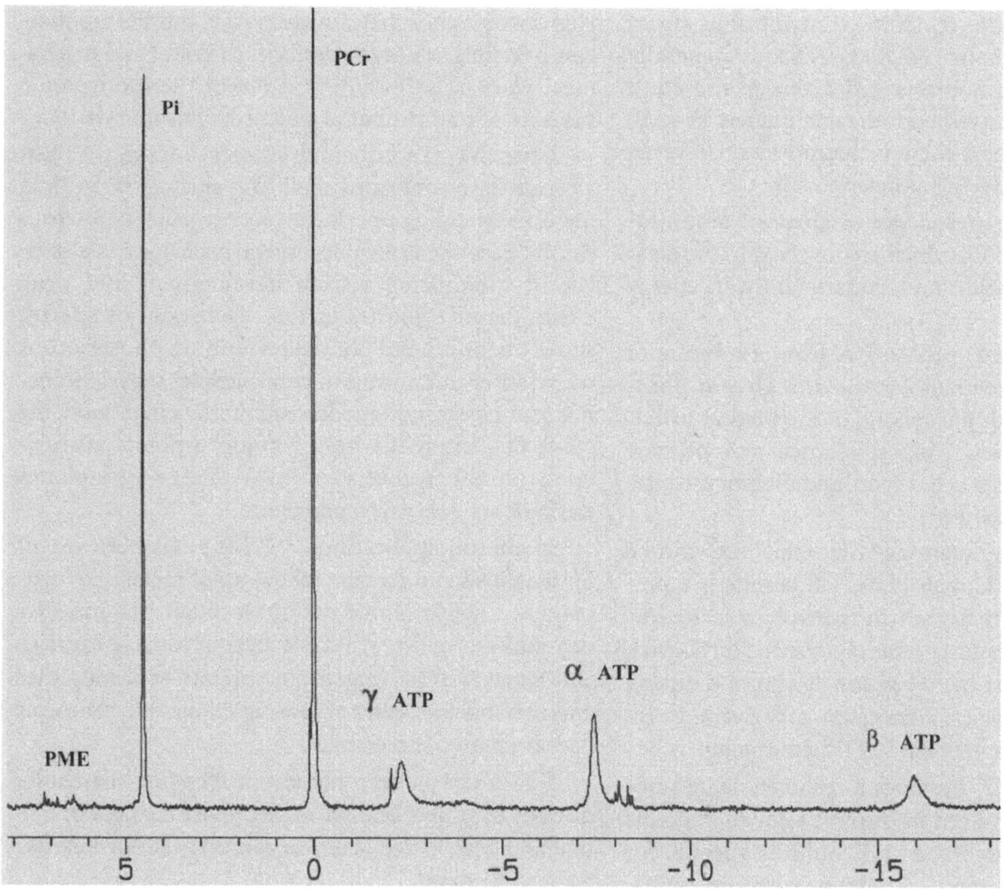

B

Fig. 4. Improved resolution with improved magnetic field strength: **A)** *in vivo* ^{31}P MR spectrum of rat skeletal muscle at 2.35 T; **B)** *in vitro* ^{31}P high resolution NMR spectrum of skeletal muscle extract acquired at 11.74 T. The peaks of the adenosine triphosphates α-, β-, and γ-ATP, the phosphocreatine, PCr, the inorganic phosphate Pi are well resolved. Peaks at 6.3-7.3 ppm are PME including G-6-P at 7.17 ppm. The large Pi peak arises from artifactual degradation of PCr.

Free cytoplasmic ATP can be calculated from the creatine kinase reaction

$$PCr^{2-} + ADP^{-} + H^{+} \longleftrightarrow ATP^{2-} + Cr$$

assumed to be at equilibrium:

$$Keq = [ATP]\ [Cr]/[H+]\ [ATP]\ [PCr]$$

The intracellular pH is calculated from chemical shift of Pi relative to PCr according to the formula where σ is the chemical shift:

$$pH = 6.75 + \log [(\sigma\text{-}3.27)/(5.69\text{-}\sigma)]$$

^{1}H MRS

Figure 5B shows a representative ^{1}H MRS spectrum of the human brain aquired at 1.5 T, with major observable peaks assigned.

^{1}H MRS detects a number of metabolites present in relatively low concentrations (< 10 mM), when water and fat suppression techniques used.

Major ^{1}H metabolites observed are commented below:

- N-acetyl-aspartate (NAA) produces a large resonance in a H^2O suppressed ^{1}H spectrum. The peak may contain up to 20% contributions from Aspartyl-glutamate (NAAG). NAA is generally associated with neurons and axons in the adult brain. It has received considerable interest in several disorders where there is neuron loss. However, its function is largely unknown.
- The Creatine (Cr) resonance originates from intracellular Cr and PCr, which are involved in the creatine kinase reaction and, consequently, in energy metabolism.
- The Choline (Cho) peak arises from a mixture of glycero-phosphoethanolamine and glycero-phosphocholine. Both phospholipids are present in cellular membranes. This resonance can provide information about cell density and membrane integrity (or peroxidation).
- A glutamate and glutamine (Glu, Gln) peak can be detected in the human brain. Glutamine is a precursor of glutamate, which is involved in neurotransmission. Gamma-aminobutyric acid (GABA), also present but in lower concentrations during normal physiological conditions, may overlap with the Glu, Gln resonance at 1.5 T field strength.
- Myo-Inositol (MI) provides a relatively large resonance and is involved in osmotic regulation across the cellular membrane and could be specific for glial cells. The amino acid glycine may also contribute to the myo-inositol resonance.
- Scyllo-inositol, an isomer of inositol appears also as a singlet peak more downfield. Taurine resonates close to the scyllo-inositol region.
- Glucose, an important substrate in brain metabolism gives rise to a weak but observable coupled resonance. It is more easily detected under hyperglycemic conditions. Lactate can be detected as a doublet resonance in brain tissue. Under normal conditions, lactate is present at around 1 mM concentration and is increased during ischemic conditions as a result of anaerobic glycolysis leading to a more distinct peak.

The brain tissue is rich in lipids. These might be detected as broad resonances with contributions from several fatty acyd chains. Measurement of lipids may be useful in evaluating myelination and membrane breakdown.

The dominant ^{1}H and ^{31}P biochemicals in the human brain are also listed in Tables 1 and 2, respectively. Resonance frequencies are given in ppm. The concentrations and ratios are mean values from the literature and are rather orientational than absolute.

MRS and bioenergetics

High energy phosphates such as ATP and PC are markers of cellular ability to perform chemical and mechanical work. The PCr/Pi is a direct thermodynamic measure of mitochondrial oxidative phosphorylation.

Extensive experimental studies during the past 15 years have confirmed the high value of ^{31}P MRS in the understanding of cellular bioenergetics. Numerous studies have used the bioenergetic behaviour as a marker in monitoring disease development and drug effect. Figure 6 illustrates this. The series of spectra show on one hand the behaviour of phosphorous metabolites in an experimental skeletal muscle ischemia and reperfusion model; and on the other hand, the effect of treatment with ascorbate, a potent antioxidant, on the recovery of high energy phosphates during post-ischemic reperfusion.

In clinical applications, ^{31}P MRS has been useful for diagnosis and therapy follow-up of metabolic myopathies. Calculation of the intracellular pH and PCr degradation and resynthesis during muscle exercise and recovery from exercise in patients with muscular and metabolic diseases according to suitable protocols have been used successfully.

^{31}P MRS has been helpful in studying metabolic diseases of mitochondrial origin where changes in lactate and PCr/Pi are taken as markers like in Kearns-Sayre syndrome.

Aerobic oxidation of glucose provides the human brain cells with energy.

^{31}P MRS can register metabolic changes during brain hypoxia where a reduction in oxygen and sub-

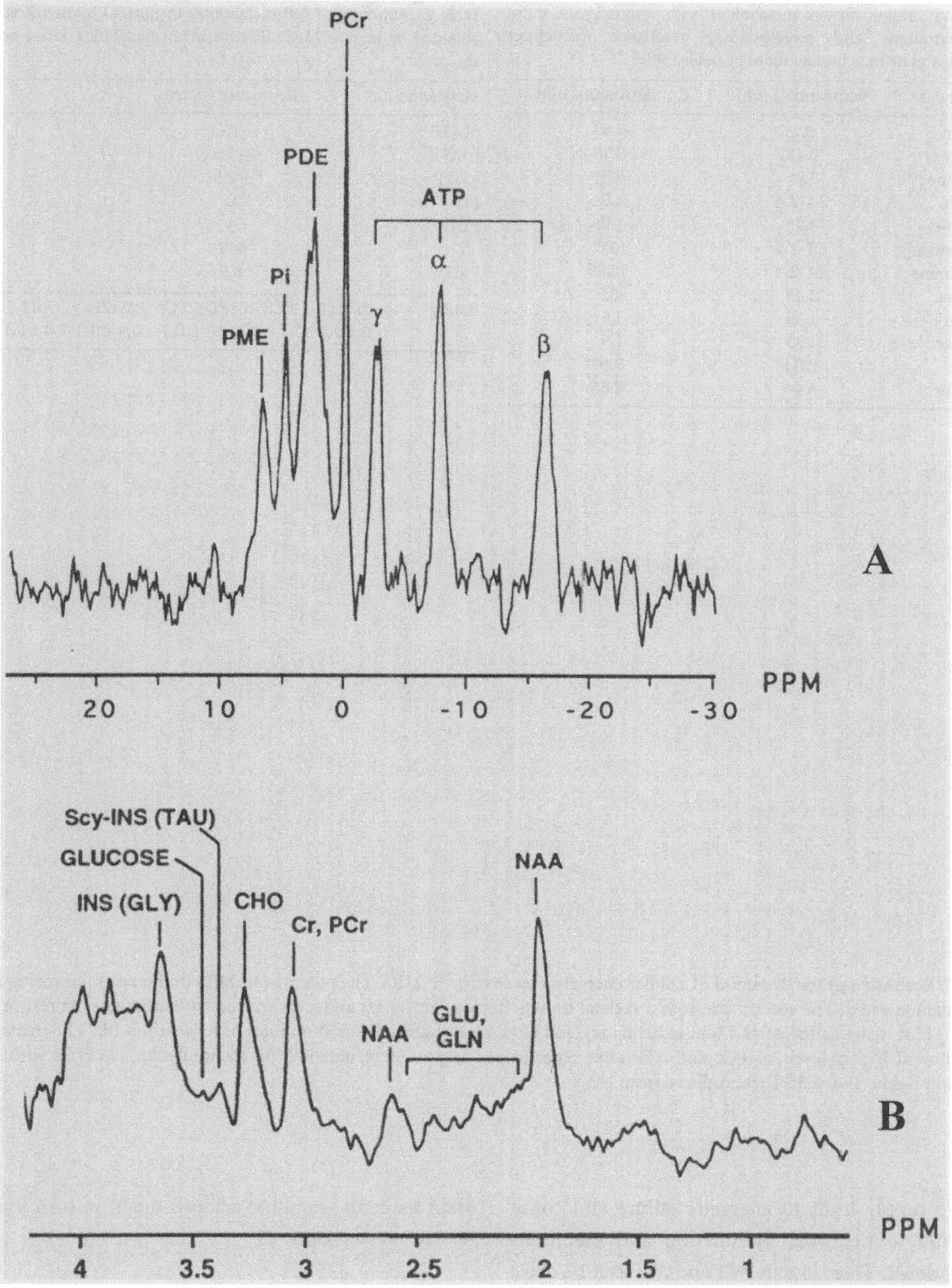

Fig. 5. Localized MRS spectra of normal human brain illustrating the major metabolites observed. **A)** is an ISIS ^{31}P MR spectrum obtained at 2 T (VOI = 100 ml). The adenosine triphosphates α-, β-, and γ-ATP, the phosphocreatine PCr, the inorganic phosphate Pi as well as the phosphomonoesters PME and the phosphodiesters PDE are well resolved; **B)** is a proton MR spectrum at 1.5 T obtained with STEAM combined with CHESS to suppress the water signal (VOI = 8 ml). The assigned proton metabolites are: N-acetylaspartate NAA, glutamate and glutamine GLU-GLN, creatine and phosphocreatine Cr-PCr, choline CHO, inositol INS, scyllo-inositol Scy-INS, taurine TAU, glycine GLY and glucose (reproduced from [3])

Table 1. Major proton metabolites with approximate mean concentrations and corresponding resonance frequencies detected in normal human brain by *in vivo* MRS

Metabolite	Resonance (ppm)	Concentration (mM)
Alanine	3.2	1.85
Cholines	3.24	3.19
Creatine	3.02	9.37
GABA	2.1-2.4	4.79
Glucose	5.23	1.76
Glutamate	2.1-2.4	10.05
Glutamine	2.1-2.4	12.89
Glycine	3.45	0.5
Inositol	1.33	1.91
Lactate	1.33	1.91
NAA	2.02	9.69
Taurine	3.44	2.88

Table 2. Summary of ^{31}P metabolites in normal human brain obtained by *in vivo* MRS. Relative mean metabolite ratios are also given

Metabolite	Resonance (ppm)				
β-ATP	– 16.1				
α-ATP	– 7.6				
γ-ATP	– 2.5				
Pcr	0				
PDE	4				
Pi	4.8				
PME	6.9				
Ratios	PCr/ATP	PCr/Pi	PDE/PCr	PME/PCr	pH
	1.2 ± 0.3	7.7 ± 2.3	1.1 ± 0.1	0.5 ± 0.2	7.0 ± 0.1

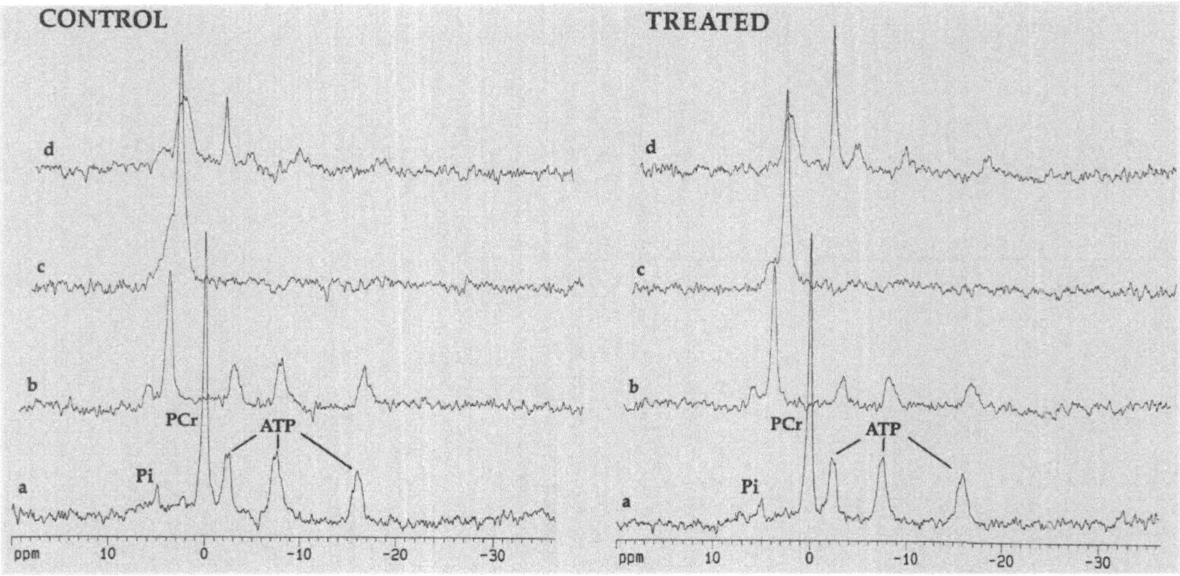

Fig. 6. llustration of the dynamics of cellular energetics by *in vivo* ^{31}P MRS. The potential of MRS in therapy monitoring is also demontrated. The spectra are from a skeletal muscle from a control rat and a rat treated with ascorbate. At rest (**a**); after 2 h of ischemia (**b**), after 4 h of ischemia (**c**), and after 4 h of ischemia + 150 minutes of reperfusion (**d**). The treated rat showed higher levels of PCr and ATP after reperfusion. Spectra were obtained by accumulating 128 FIDs with a repetition time 1s at 2.35 T (reproduced from [13])

strate supply leads to energetic failure and consequently to neuronal dysfunction and membrane breakdown. Thus, loss in PCr and ATP can be detected as well as decreases in intracellular pH. Possible structural membrane changes can be demonstrated from changes in PDE and PME. Intracellular pH and/or lactate are useful markers of low oxygen availability in the cell. It is well known that anaerobic metabolism leads to lactate accumulation, and in the brain tissue the resulting acidosis might in turn lead to neuronal damage.

Metabolic encephalomyopathies

Brain ischemia and hyoxic/ischemic disease in newborns in which cerebral energetics can be monitored to study oxidative and glycolytic metabolism where parameters like pH, Pi/ATP has proven to be good markers.

Anaerobic glycolysis in brain is an indication of impairment in mitochondrial function. Decreased PCr/Pi and elevated lactate levels are indications that could help the diagnosis of that metabolic disorders.

In cases of hepatic encephalomyopathy, Kearns-Sayre syndrome and pyrovate dehydrogenase deficiency, MRS is used to monitor therapy.

Brain trauma

Post-traumatic brain injuries might affect cerebral energy metabolism. Decreases in ATP and in intracellular pH were shown by ^{31}P MRS. Elevated lactate probably due to increased anaerobic glycolysis and diminished NAA were also reported from ^{1}H MRS examinations. In neonates with acute brain injury ^{1}H MRS examination was able to predict outcome through variations in NAA, Glu/Gln and lactate, as illustrated in Fig. 7.

Stroke

Stroke is associated with degradation of high-energy phosphates (ATP, PCr), and increase in inorganic phosphate (Pi) and intracellular acidocis, as documented from early ^{31}P MRS investigations. Additionally, typical ^{1}H MRS of patients with stroke show elevated lactate and reduced NAA. Follow-up after the acute

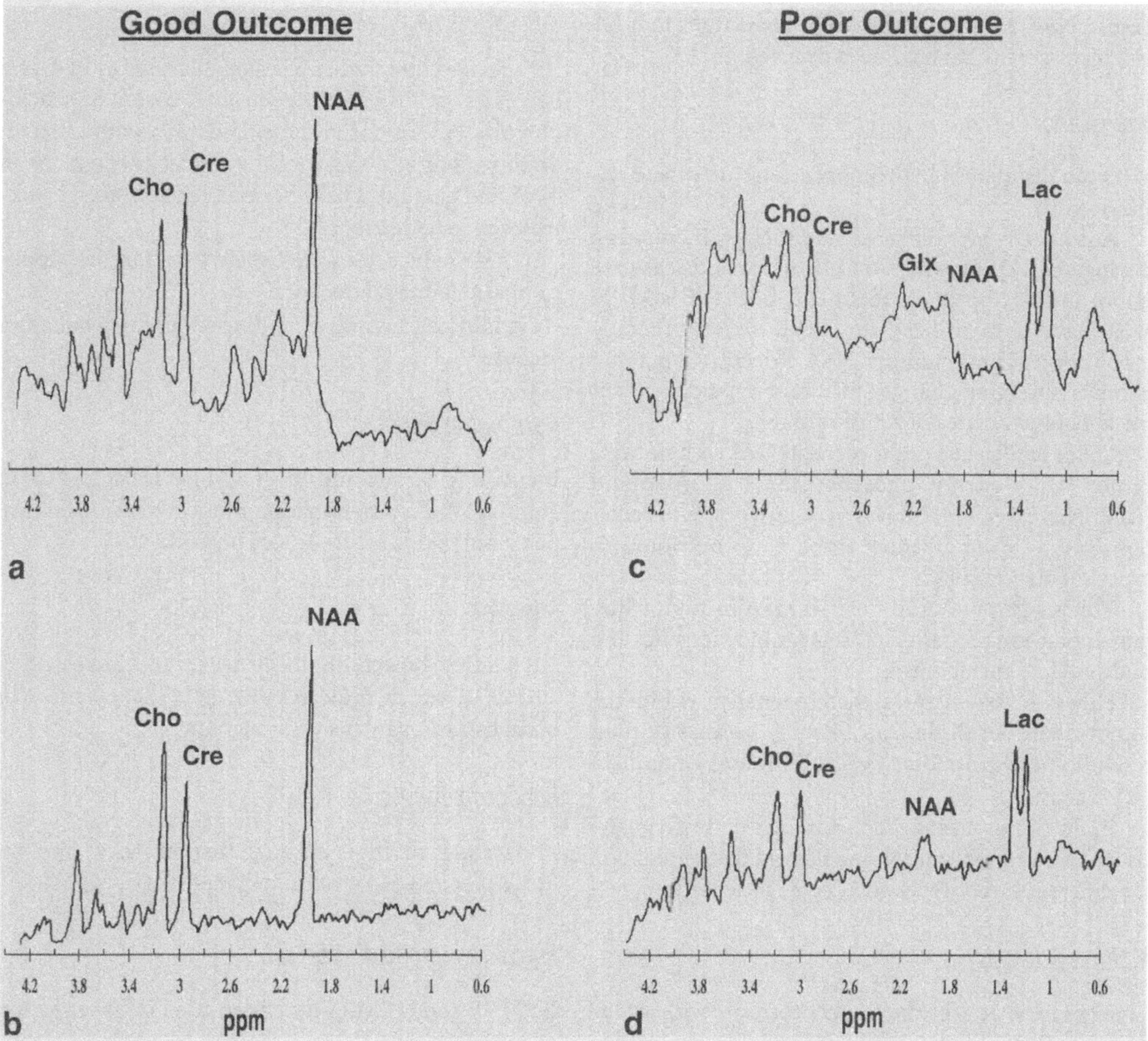

Fig. 7. ^{1}H MRS illustrating patterns seen in STEAM spectra (**a, c**) and in PRESS spectra (**b, d**) from the brain of two children with after birth brain injury. Spectra (**a, b**) are from a patient with a mild brain injury and show good outcome. Spectra (**c, d**) are from a patient with a traumatic brain injury and show poor outcome (note the low NAA signal and the elevated lactate signal)

infarction period might reveal continued loss in NAA as well as acidocis in the ischemic regions of the brain. These parameters are certainly useful in monitoring the effect of medication.

Alzheimer disease

^{1}H MRS using short TE STEAM revealed that myo-inositol is increased in AD. NAA is also decreased in the brain indicating diminished number of healthy neurons.

Figure 8 illustrates abnormalities in ^{1}H MRS spectrum in a patient with AD.

AIDS

Neurologic disorders such as AIDS encephalitis and AIDS dementia resulting from HIV infection have been successfully studied by MRS. Reductions in NAA and increases in Cho have been detected.

Brain tumor

MRS can distinguish between recurrent tumor and tissue necrosis.

Adequate tumor diagnosis and therapy monitoring during the various stages of a tumorous disease are important for optimal treatment. Both ^{31}P and ^{1}H MRS have been utilized for diagnosis and therapy monitoring of brain tumors. NAA is decreased in brain gliomas. Studying changes in tumor-type dependent metabolites is an area of active research.

Lipids and lactate peaks correlate well with necrotic tumor. High-energy phosphate and phospholipid (ATP, PCr, PDE, PME) levels vary in reponse to radiation therapy, chemotherapy and even to nutrition (in experimental cancer).

This suggests to utility of ^{31}P MRS in tumor therapy monitoring focusing on cellular bioenergetics and phospholipid metabolism.

However, biochemical heterogeneity within the tumor tissue is still difficult to study because of poor resolution on commonly available clinical equipment (1.5 T).

Brain tumor classification through network analysis and pattern recognition might shed further light on the different tumor types and degree of activity.

Multiple Sclerosis

Changes in NAA, Cho and lactate correlate with axonal damages, demyelination and inflammation observed in MS patients during various stages of the disease. These metabolites can be monitored to study the outcome of new treatment.

Epilepsy

^{31}P MRS showed that the PCr/Pi is dramatically decreased during seizures and normalized after seizure discharge.

The glutamine and glutamate peak is elevated in the hippocampus, while NAA is diminished in patients with chronic epilepsy.

Changes in GABA have been correlated with drugs affecting GABA metabolism.

An increase in lactate has also been reported in focal epilepsy of extratemporal origin.

These biochemical changes in epileptogenic region of the brain indicate that ^{1}H MRS can be clinically useful in the diagnosis of this disease as a complement to MRI.

Schizophrenia

^{31}P spectroscopy studies revealed increases in PDE and decreases in PME in the prefrontal cortex of schizophrenics. Alterations in these lipids vary with different brain regions and stages of the disease. Reductions in NAA and glutamate have been reported from ^{1}H spectroscopy investigations.

These reductions were largely found in the hippocampal area/mesial temporal lobe.

Additional neurological diseases under evaluation include:

Huntington disease

Increase in lactate and in Pi and decrease in PCr in Huntington disease implicate mitochondrial oxidative phosphorylation in the disease process.

Migraine

^{31}P studies showed diminished PCr and increased Pi and ADP, which indicates energetic disturbances in brain tissue in patients with migraine.

Parkinson disease

A decrease in the neuronal marker NAA and an increase lactate/NAA ratio were reported by ^{1}H MRS.

Psychiatry (mood disorders)

Both ^{31}P and ^{1}H MRS have been used in investigation of mood disorders where changes in energy metabolism, lipids and Cho were observed. This indicates the potential of MRS in monitoring the effect of psychopharmacological drugs.

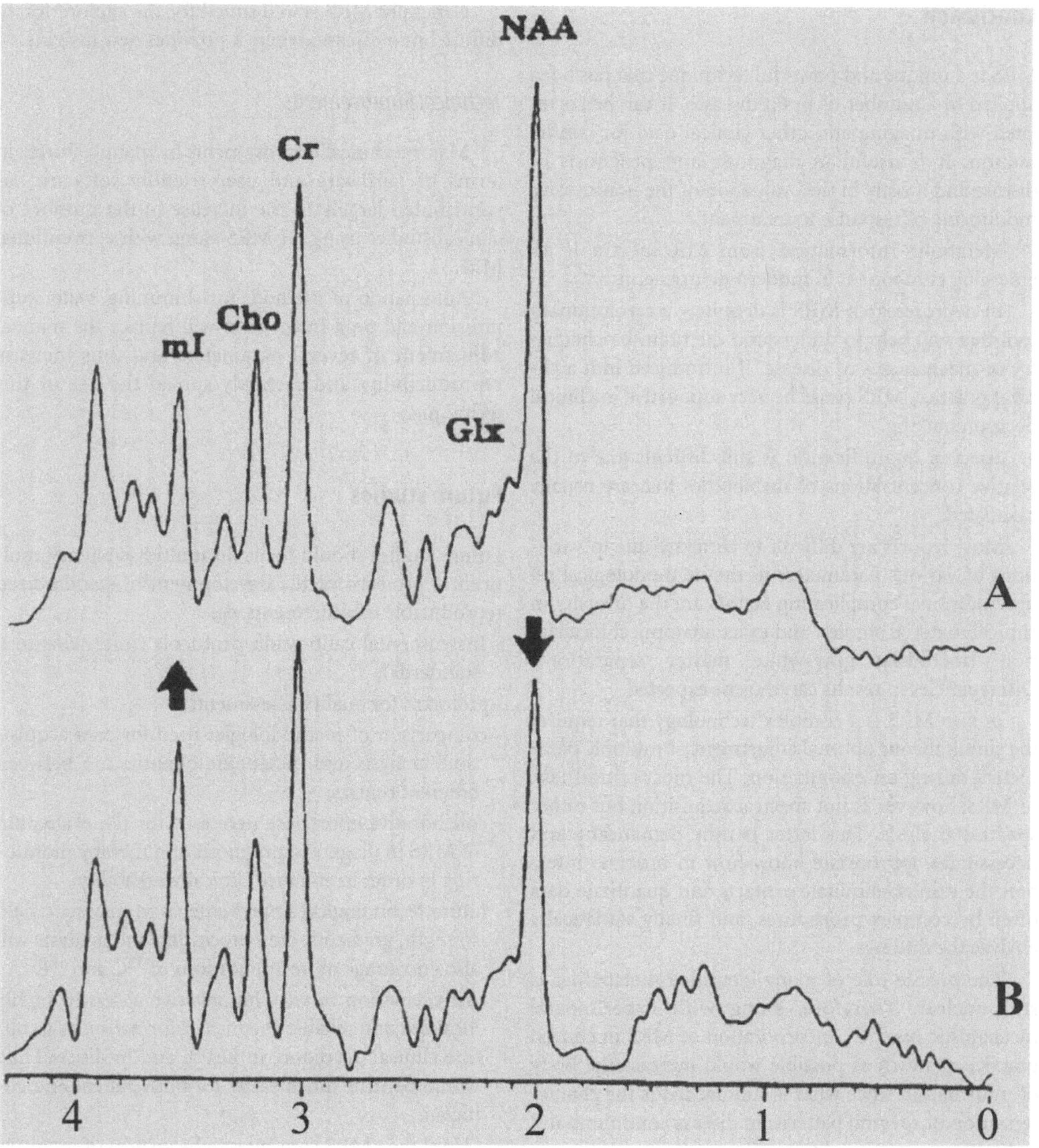

Fig. 8. A proton MRS spectrum from the brain of a normal patient (**A**) compared with a spectrum of a patient with Alzheimer's disease (**B**) (reproduced from [58])

Conclusion

MRS is a unique and powerful technique that has been applied to a number of brain diseases. It can be correlated with imaging and other clinical data for confirmation. It is useful in diagnosis and prognosis of disease and mostly in the evaluation of the noninvasive monitoring of response to treatment.

Metabolic information from MR spectra is an emerging component in modern neurochemistry.

In neuroresearch MRS is definitely a revolutionary tool that will help to understand the brain biochemistry of mechanisms of disease. If introduced into a clinical practice, MRS could be very supportive in clinical decision making.

Spectral quantification is still difficult due to the relative concentrations of metabolites that are usually calculated.

Most reports are difficult to compare due to variations of various parameters in the methodological set up. Additional complicating factors are the diversity in clinical material studied and exact anatomical localization (including gray-white matter separation). Discrepancies in results can thus be expected.

In vivo MRS is a complex technology that requires the simultaneous optimal adjustment of multiple parameters during an examination. The most critical task in MRS, however, is not spectral acquisition but rather spectral analysis. This latter is time demanding and necessitates appropriate know-how in order to interpret the results, eliminate artifacts and quantitate data often by complex procedures, and finally statistically analyze the findings.

The precise role of many identified metabolites is still unclear. Therefore, along with experimental mechanistic research, incorporation of MRS in clinical practice as much as possible would increase the body of information since what is still needed is the characterization of spectral patterns in disease conditions and in healthy control conditions.

Major benefits

High chemical specificity in studying:

- energy metabolism
- lipid metabolism
- amino acid and intermediary metabolism
- noninvasive regional serial measurements of metabolite in patients and controls subjects
- therapy response
- mechanistic studies of inherited and acquired brain metabolic diseases.

Generally, MRS is well suited for the exploration of diffuse brain diseases where it provides new insights.

Technical improvements

Major technical improvements by manufacturers in terms of hardware and user friendly software has contributed largely to the increase in the number of clinical studies using ^{1}H MRS along with conventional MRI.

Automation of methods for shimming, water suppression and peak integration will replace the manual adjustment of several parameters and thus increase reproducibility and certainly spread the use of this technique.

Future studies

Future studies should focus on multidisciplinary multicentre projects for the development of standardized reproducible measurements, eg:

- Instrumental calibration protocols (internal/extern standards);
- protocols for quality assessment;
- comparison of methodologies used for data acquisition, analysis and metabolite quantitation between different centres;
- collaborative efforts are necessary for the evaluation of MRS in diagnosis, prognosis and therapy monitoring in order to enhance clinical workability;
- future technological improvements in magnetic field strength, gradients, data processing and analysis will also encourage more applications of ^{13}C and ^{19}F.
- Lastly, envision *in vivo*, noninvasive access to highly localized and reliable chemical information as a routine clinical procedure in health and in disease. Life would become much easier for both patient and clinician.

Until then MRS continues to be an area of intensive investigation.

References and suggestions for further reading

General reviews

1. Novotny E, Ashwal S, Shevell M (1998) Proton magnetic resonance spectroscopy: an emerging technology in pediatric neurology research. Pediatr Res 44(1): 1-10
2. Rudkin TM, Arnold DL (1999) Proton magnetic resonance spectroscopy for the diagnosis and management of cerebral disorders. Arch Neurol 56(8): 919-926

3. Vion-Dury J, Meyerhoff DJ, Cozzone PJ, Weiner MW (1994) What might be the impact on neurology of the analysis of brain metabolism by in vivo magnetic resonance spectroscopy ? [ed] J Neurol 241(6): 354-371
4. Iles RA, Burns SP (1995) In vitro NMR spectroscopy in the diagnosis and investigation of metabolic disorders. Quart. Magn Res Biol Med 2(1): 35-42
5. Radda GK (1986) The use of NMR spectroscopy for the understanding of disease. Science 233(4764): 640-645
6. Gruetter R, Weisdorf SA, Rajanayagan V, Terpstra M, Merkle H, Truwit CL, Garwood M, Nyberg SL, Ugurbil K (1998) Resolution improvements in in vivo 1H NMR spectra with increased magnetic field strength. J Magn Reson 135(1): 260-264

MR principle

7. Andrew ER (1990) An introduction to nuclear magnetic resonance in biomedicine. Can Assoc Radiol J 41 (1): 2-7
8. Hoffenberg EF, Kozlowski P, Salerno TA, Deslauriers R (1996) Evaluation of cardiac 31P magnetic resonance spectroscopy: reviewing NMR principles. J Surg Res 62(1): 135-143
9. Woodward P (2000) MRI for technologists. McGraw-Hill, p 432
10. Mitchell DG (1998) MRI Principles. Saunders, p 288
11. Longmore DB (1989) The principles of magnetic resonance. Br Med Bull 45(4): 848-880
12. Wehrli WF, Shaw D, Kneeland JB (1988) Biomedical magnetic resonance imaging. VCH Publisher, New York, p 601
13. Gadian DG (1982) Nuclear Magnetic Resonance and its application to living systems. Clarendon, Press Oxford
14. Slichter CP (1978) Principles of Magnetic Resonance Spectroscopy. Second ed, Springer-Verlag, Berlin
15. Farrar TC, Becker ED (1971) Pulse and Fourier Transform NMR. Academic Press, New York

Experimental applications

16. Bollano E, Omerovic E, Bohlooly y M, Kujacic V, Madhu B, Tornell J, Isaksson O, Soussi B, Schulze W, Fu ML, Matejka G, Waagstein F, Isgaard J (2000) Impairment of cardiac function and bioenergetics in adult transgenic mice overexpressing the bovine growth hormone gene. Endocrinology 141(6): 2229-2235
17. Lagerwall K, Daneryd P, Schersten T, Soussi B (1995) In vivo 31P nuclear magnetic resonance evidence of the salvage effect of ascorbate on the postischemic reperfused rat skeletal muscle. Life Sci 56(6): 389-397
18. Lagerwall K, Madhu B, Daneryd P, Schersten T, Soussi B (1997) Purine nucleotides and phospholipids in ischemic and reperfused rat skeletal muscle: effect of ascorbate. Am J Physiol 272(1 Pt 2): H83-90
19. Lukes D, Madhou B, Arvidsson N, Gustavsson M, Mjornstedt L, Soussi B, Olausson M (1997) In vivo 31P MRS evaluation of the rejection process and differences in anesthetic procedures in a concordant xenotransplantation: mouse heart to rat model. Transplant Proc 29(7): 3159-3160
20. Madhu B, Lagerwall K, Soussi B (1996) Phosphorus metabolites in different muscles of the rat leg by 31P image – selected in vivo spectroscopy [published erratum appears in NMR Biomed (1997) 10(6): 302]. NMR Biomed 9(8): 327-332
21. Omerovic E, Basetti M, Bollano E, Bohlooly M, Tornell J, Isgaard J, Hjalmarson A, Soussi B, Waagstein F (2000) In vivo metabolic imaging of cardiac bioenergetics in transgenic mice. Biochem Biophys Res Commun 271(1): 222-228
22. Omerovic E, Bollano E, Basetti M, Kujacic V, Waagstein L, Hjalmarson A, Waagstein F, Soussi B (1999) Bioenergetic, functional and morphological consequences of postinfarct cardiac remodeling in the rat. J Mol Cell Cardiol 31(9): 1685-1695
23. Sorensen V, Jonsson O, Pettersson S, Schersten T, Soussi B (1998) In vivo 31P NMR OSIRIS of bioenergetic changes in rabbit kidneys during and after ischaemia: effect of pretreatment with an indeno-indole compound. Acta Physiol Scand 162(4): 495-500
24. Soussi B, Idstrom JP, Bylund-Fellenius AC, Schersten T (1990) Dynamics of skeletal muscle energetics during ischemia and reperfusion assessed by in vivo 31P NMR. NMR Biomed 3(2): 71-77
25. Soussi B, Lagerwall K, Idstrom JP, Schersten T (1993) Purine metabolic pathways in rat hindlimb perfusion model during ischemia and reperfusion. Am J Physiol 265(4 Pt 2): H1074-1081
26. Hajek M, Burian M, Dezortova M (2000) Application of LCModel for quality control and quantitative in vivo 1H MR spectroscopy by short echo time STEAM sequence. Magma 10(1): 6-17
27. Podo F, Henriksen O, Bovee WM, Leach MO, Leibfritz D, de Certaines JD (1998) Absolute metabolite quantification by in vivo NMR spectroscopy: I. Introduction, objectives and activities of a concerted action in biomedical research. Magn Reson Imaging 16(9): 1085-1092
28. Chen W, Adriany G, Zhu XH, Gruetter R, Ugurbil K (1998) Detecting natural abundance carbon signal of NAA metabolite within 12–cm^3 localized volume of human brain using 1H-[13C] NMR spectroscopy. Magn Reson Med 40(2): 180-184

Epilepsy

29. Duc CO, Trabesinger AH, Weber OM, Meier D, Walder M, Wieser HG, Boesiger P (1998) Quantitative 1H MRS in the evaluation of mesial temporal lobe epilepsy in vivo. Magn Reson Imaging 16(8): 969-979
30. Hajek M, Dezortova M, Komarek V (1998) 1H MR spectroscopy in patients with mesial temporal epilepsy. Magma 7(2): 95-114
31. Rothman DL, Behar KL, Prichard JW, Petroff OA (1997) Homocarnosine and the measurement of neuronal pH in patients with epilepsy. Magn Reson Med 38(6): 924-929
32. van der Grond J, Gerson JR, Laxer KD, Hugg JW, Matson GB, Weiner MW (1998) Regional distribution of interictal 31P metabolic changes in patients with temporal lobe epilepsy. Epilepsia 39(5): 527-536
33. Vainio P, Usenius JP, Vapalahti M, Partanen K, Kalviainen R, Rinne J, Kauppinen RA (1994) Reduced N-acetylaspartate concentration in temporal lobe epilepsy by quantitative 1H MRS in vivo. Neuroreport 5(14): 1733-1736

Stroke and acute brain injuries

34. Blamire AM, Graham GD, Rothman DL, Prichard JW (1994) Proton spectroscopy of human stroke: assessment of transverse relaxation times and partial volume effects in single volume steam MRS. Magn Reson Imaging 12(8): 1227-1235
35. Ford CC, Griffey RH, Matwiyoff NA, Rosenberg GA (1992) Multivoxel 1H-MRS of stroke. Neurology 42(7): 1408-1412
36. Friedman SD, Brooks WM, Jung RE, Chiulli SJ, Sloan JH, Montoya BT, Hart BL, Yeo RA (1999) Quantitative proton MRS predicts outcome after traumatic brain injury. Neurology 52(7): 1384-1391
37. Gideon P, Rosenbaum S, Sperling B, Petersen P (1999) MR-visible brain water content in human acute stroke. Magn Reson Imaging 17(2): 301-304
38. Holshouser BA, Ashwal S, Shu S, Hinshaw DB Jr (2000) Proton MR spectroscopy in children with acute brain injury: comparison of short and long echo time acquisitions. J Magn Reson Imaging 11(1): 9-19
39. Ross BD, Ernst T, Kreis R, Haseler LJ, Bayer S, Danielsen E, Bluml S, Shonk T, Mandigo JC, Caton W, Clark C, Jensen SW, Lehman NL, Arcinue E, Pudenz R, Shelden CH (1998) 1H MRS in acute traumatic brain injury. J Magn Reson Imaging 8(4): 829-840
40. Wardlaw JM, Marshall I, Wild J, Dennis MS, Cannon J, Lewis SC (1998) Studies of acute ischemic stroke with proton magnetic resonance spectroscopy: relation between time from onset, neurological deficit, metabolite abnormalities in the infarct, blood flow, and clinical outcome. Stroke 29(8): 1618-1624

Schizophrenia

41. Keshavan MS, Sanders RD, Pettegrew JW, Dombrowsky SM, Panchalingam KS (1993) Frontal lobe metabolism and cerebral morphology in schizophrenia: 31P MRS and MRI studies. Schizophr Res 10(3): 241-246
42. Nasrallah HA, Skinner TE, Schmalbrock P, Robitaille PM (1994) Proton magnetic resonance spectroscopy (1H MRS) of the hippocampal formation in schizophrenia: a pilot study. Br J Psychiatry 165(4): 481-485
43. Riehemann S, Volz HP, Smesny S, Hubner G, Wenda B, Rossger G, Sauer H (2000) Phosphorus 31 magnetic resonance spectroscopy in schizophrenia research. Pathophysiology of cerebral metabolism of high-energy phosphate and membrane phospholipids. Nervenarzt 71(5): 354-363
44. Sigmundsson T, Maier M, Williams SCR, Simmons A, Greenwood K, Ron MA (1997) Frontal lobe in vivo proton magnetic resonance spectroscopy in schizophrenic patients with negative symptoms. Schizophrenia Research 24(1-2): 182

Cancer

45. Castillo M, Kwock L (1999) Clinical applications of proton magnetic resonance spectroscopy in the evaluation of common intracranial tumors. Top Magn Reson Imaging 10(2): 104-113
46. Kim SH, Chang KH, Song IC, Han MH, Kim HC, Kang HS, Han MC (1997) Brain abscess and brain tumor: discrimination with in vivo H-1 MR spectroscopy [see comments]. Radiology 204(1): 239-245
47. Leach MO (1996) Introduction to in vivo MRS of cancer: new perspectives and open problems. Anticancer Res 16(3B): 1503-1514
48. Negendank W, Li CW, Padavic-Shaller K, Murphy-Boesch J, Brown TR (1996) Phospholipid metabolites in 1H-decoupled 31P MRS in vivo in human cancer: implications for experimental models and clinical studies. Anticancer Res 16(3B): 1539-1544
49. Sijens PE, Levendag PC, Vecht CJ, van Dijk P, Oudkerk M (1996) 1H MR spectroscopy detection of lipids and lactate in metastatic brain tumors. NMR Biomed 9(2): 65-71

Multiple sclerosis and ALS

50. Arnold DL, Matthews PM, Francis G, Antel J (1990) Proton magnetic resonance spectroscopy of human brain in vivo in the evaluation of multiple sclerosis: assessment of the load of disease. Magn Reson Med 14(1): 154-159
51. Block W, Karitzky J, Traber F, Pohl C, Keller E, Mundegar RR, Lamerichs R, Rink H, Ries F, Schild HH, Jerusalem F (1998) Proton magnetic resonance spectroscopy of the primary motor cortex in patients with motor neuron disease: subgroup analysis and follow-up measurements. Arch Neurol 55(7): 931-936
52. Leary SM, Brex PA, MacManus DG, Parker GJ, Barker GJ, Miller DH, Thompson AJ (2000) A (1)H magnetic resonance spectroscopy study of aging in parietal white matter: implications for trials in multiple sclerosis. Magn Reson Imaging 18(4): 455-459
53. Rooney WD, Miller RG, Gelinas D, Schuff N, Maudsley AA, Weiner MW (1998) Decreased N-acetylaspartate in motor cortex and corticospinal tract in ALS. Neurology 50(6): 1800-1805
54. Sarchielli P, Presciutti O, Tarducci R, Gobbi G, Alberti A, Pelliccioli GP, Orlacchio A, Gallai V (1998) 1H-MRS in patients with multiple sclerosis undergoing treatment with interferon beta-1a: results of a preliminary study. J Neurol Neurosurg Psychiatry 64(2): 204-212

Encephalommyopathy and metabolic diseases

55. Soussi B, Schersten T, Waldenstrom A, Ronquist G (1993) Phosphocreatine turnover and pH balance in forearm muscle of patients with syndrome X [letter]. Lancet 341 (8848): 829-830
56. Ronquist G, Soussi B, Frithz G, Schersten T, Waldenstrom A (1995) Disturbed energy balance in skeletal muscle of patients with untreated primary hypertension. J Intern Med 238(2): 167-174
57. Kuhl CK, Layer G, Traber F, Zierz S, Block W, Reiser M (1994) Mitochondrial encephalomyopathy: correlation of P-31 exercise MR spectroscopy with clinical findings. Radiology 192(1): 223-230
58. Mathews PM, Andermann F, Silver K, Karpati G, Arnold DL (1993) Proton MR spectroscopic characterization of differences in regional brain metabolic abnormalities in mitochondrial encephalomyopathies. Neurology 43(12): 2484-2490

Alzheimer´s disease and dementia

59. Cuenod CA, Kaplan DB, Michot JL, Jehenson P, Leroy-Willig A, Forette F, Syrota A, Boller F (1995) Phospholipid

abnormalities in early Alzheimer's disease. In vivo phosphorus 31 magnetic resonance spectroscopy. Arch Neurol 52(1): 89-94

60. Moats RA, Ernst T, Shonk TK, Ross BD (1994) Abnormal cerebral metabolite concentrations in patients with probable Alzheimer disease. Magn Reson Med 32(1): 110-115

61. Ross BD, Bluml S, Cowan R, Danielsen E, Farrow N, Gruetter R (1997) In vivo magnetic resonance spectroscopy of human brain: the biophysical basis of dementia. Biophys Chem 68(1-3): 161-172

62. Schofield PW, Mosesson RE, Stern Y, Mayeux R (1995) The age at onset of Alzheimer's disease and an intracranial area measurement. A relationship. Arch Neurol 52(1): 95-98

63. Schuff N, Amend D, Ezekiel F, Steinman SK, Tanabe J, Norman D, Jagust W, Kramer JH, Mastrianni JA, Fein G, Weiner MW (1997) Changes of hippocampal N-acetyl aspartate and volume in Alzheimer's disease. A proton MR spectroscopic imaging and MRI study. Neurology 49(6): 1513-1521

Part I

Brain and skull

Imaging technique and normal anatomy

1 - Tumoral conditions

2 - Vascular diseases

3 - Inflammatory and infectious diseases

4 - Traumatic conditions

5 - Neurogenetic and developmental disorders

Imaging technique and normal anatomy

It is well established that the anatomical correlations between brain and skull are very poor and so variable as to made precise topometry difficult and doubtful. Thus, all exact topographical measurement of brain structures requires a careful choice of reliable anatomical reference points. Therefore, any precise topographical analysis of brain structures ought to be performed according to judicious and definite cerebral reference lines based on cutaneous, cranial as well as brain landmarks (Tamraz and Comair).

The intercommissural plane of Talairach, which is very close to the orbito-meatal line (1.4°), defined as the plane through the line joining the upper border of the anterior commissure (CA) to the lower border of the posterior commissure (PC), has great validity and continues to be used by neuroradiologists performing CT, despite its great obliquity, considering its stereotactic validation. The ability to perform, using MRI, multiplanar computer reformations permits the evaluation of the anatomical structures in the most accurate orientation. But whatever is the improvement of the technique, direct scanning parallel or orthogonal to the long axis of the structure studied, will give the better spatial resolution.

In order to facilitate the neuro-anatomical approach and optimize biometrical studies using MRI, two recent brain reference lines named the "chiasmato-commissural line" (CH-PC line), defining a chiasmato-mamillo-post-commissural plane, and the "commissural-obex line" (PC-OB line) perpendicular to the previous, are proposed.

The chiasmato-commissural and the commissural-obex reference planes

These orthogonal reference planes, suitable for multimodality imaging, may be used routinely in brain imaging with highly reproducible anatomical results. The anatomical landmarks defining these reference lines are very easily shown on a mid-sagittal MR scoutview, and are present in all vertebrates. The accuracy of these brain midline structures need not to be demonstrated here either from an anatomical or from an ontogenetic and phylogenetic point of view, being located at the mesodiencephalic junction.

Constant angular variations, statistically proven, demonstrate the *in vivo* as well as in cadaver validity of these cephalic orientations:

- the angular relationship between CH-PC line and the bicommissural line AC-PC and named the commissural CH-PC-AC angle averages 24° +/- 2.3°;
- the angular relationship between CH-PC line and PC-Obex brainstem vertical axis joining the posterior commissure to the obex and named CH-PC-Obex troncal angle averages 93° +/- 3.4°.

The chiasmato-commissural (CH-PC) plane

This horizontal reference line, named the chiasmatico-commissural line (CH-PC), runs tangential to the superior border of the chiasmatic point (CH) anteriorly, to join the inferior border of the posterior commissure posteriorly (PC). The consistency of the angle between this line and the bicommissural line, as demonstrated, serves to validate the choice of this pivotal line, situated as it is at the mesodiencephalic junction. This horizontal referential suitable for *in vivo* MR imaging of the brain according to a sylvian referential allows as well the study of fixed isolated brain in primates and can be used in the field of comparative anatomy of the vertebrates brains.

This plane has been shown to be truly horizontal in that it is perpendicular to the great axis of the brainstem materialized by the PC-Obex line. Two main results derive from the morphological analysis of the contiguous MR sagittal cuts of the cerebral hemisphere.

The first concerns the close parallelism of the CH-PC plane with the plane defined by the posterior ramus of the lateral sulcus. In fact, the lateral projection of the plane on successive cuts, oriented parallel to CH-PC referential, shows a close parallelism with the superior temporal sulcus. The value of the commissural CH-PC-AC angle (24°) corresponds to that formed between AC-PC and the lateral sulcus averaging 23° according to Szikla. Thus, the projection onto the insular triangle of the parallel to CH-PC proves to be very close to the projection of the lateral sulcus. The "Sylvian plane" can therefore be projected onto the median plane.

The second finding is that CH-PC plane is almost perpendicular to the main axis of the brainstem as accurately demonstrated by PC-OB line. Anatomoclinical correlations in the coronal and the axial planes of fine structures in the brainstem are consequently made possible with reproducible results. Easier references to the structures within the floor of the fourth ventricle such as the underlying gray nuclei and the nerve fibers bundles, are reliably obtained.

The commissural-obex (PC-OB) plane

This reference line is defined as the line tangential to the anterior border of the posterior commissure (PC) and to the posterior border of the obex (OB) at the caudal extremity of the floor of the fourth ventricle. This line materializes the great vertical axis of the brainstem and appears perpendicular to CH-PC horizontal line.

Coronal cuts are best obtained with the PC-Obex reference plane, due to its perpendicularity to CH-PC plane and thus to the anterior-posterior axis of the first temporal and the lateral sulci, becoming most suitable for the coronal study of the temporal lobes, the inferior frontal gyri and the perisylvian speech areas.

Anatomical correlations applied to brain imaging

The immediate corollary resulting from the close parallelism of the CH-PC plane to the Sylvian plane is the close orientation to the temporal horns of the lateral ventricles and to the great axis of the hippocampi.

CH-PC plane separates within the brain, the cerebellum and brainstem from the cerebral hemispheres, passing through the cisterna ambiens beneath the splenium of the corpus callosum, with an exception for the occipital lobes whose descent is a function of the cranial index and the cranial typology.

The accuracy of these axial and coronal reference planes for imaging the brainstem and the posterior fossa structures, is obvious whenever anatomoclinical correlations are needed. The reproducibility of the cuts oriented according to these referencials is also very helpful for follow-up of discrete lesions, as observed most interestingly in case of ocular motor disturbances.

CH-PC horizontal plane, also tangent to the superior border of the lateral geniculate bodies, shows constant relationships at the mesodiencephalic junction, passing almost invariably through the ambient cistern between the inferior border of the splenium and the upper limit of the culmen cerebelli to intersect most commonly the parietooccipital sulcus or its common stem with the calcarine sulcus. Most extent of the calcarine fissure is therefore reliably found on the lower axial cuts.

CH-PC plane seems well to separate the telencephalon and diencephalon above, from the mesencephalon and rhombencephalon below, behaving consequently a real embryologic and phylogenic significance.

The orthogonality of CH-PC line to the great axis of the brainstem reference line explains that coronal cuts performed parallel to the commissural-obex line, display very nicely the geniculocalcarine tracts and particularly their temporal component, as well as the lateral geniculate bodies found beneath and lateral to the thalami within the slice contiguous and anterior to PC-Obex plane. The cisternal optic tracts and the cisternal and intracanalicular optic nerves are well delineated and easily depicted in such perpendicular orientation, avoiding misdiagnoses due to partial volume effect phenomenon.

The chiasm may benefit also of this cephalic orientation despite its well known angular variations, being roughly oriented in the adult along the CH-PC plane ("X" -shape appearance representing the cisternal optic nerves, the chiasm and the cisternal optic tracts). This X-shape aspect helps to differentiate the CH-PC plane from the N.O.P. plane in most circumstances, due to the higher obliquity of the former with respect to the anthropological base line, closer to the direction of the optic tracts and orbital floor.

The chiasmatico-commissural line is validated from the anatomical, anthropological and phylogenetical points of view. The constancy of its angulation as demonstrated with respect to the bicommissural line of stereotactic neurosurgery, its close parallelism to the parallel sulcus and the lateral fissure, its orthogonality as to the brainstem vertical axis, facilitate morphological approaches to brain anatomy as well as biometrical comparative studies of "sylvian oriented brain", *in vivo* as in cadaver.

Normal anatomy

Plate I: Major brain reference lines (used in neuroimaging)

Fig. a. Horizontal planes / lines:

1 *Bicommissural plane (Talairach, 1952)*
2 *Cephalic plane (Delmas, 1959)*
3 *Neuro-ocular plane (Cabanis, 1977)*
4 *Callosal plane (Olivier, 1985)*
5 *Chiasmato-commissural plane (Tamraz, 1989)*

Fig. b. Vertical planes / lines:

1 *Commissuro-mamillary plane (Guiot, 1959)*
2 *Plane perpendicular to NOP (Cabanis, 1983)*
3 *Commissuro-mamillary plane (Baulac, 1990)*
4 *Commissural-obex plane (Tamraz, 1989)*

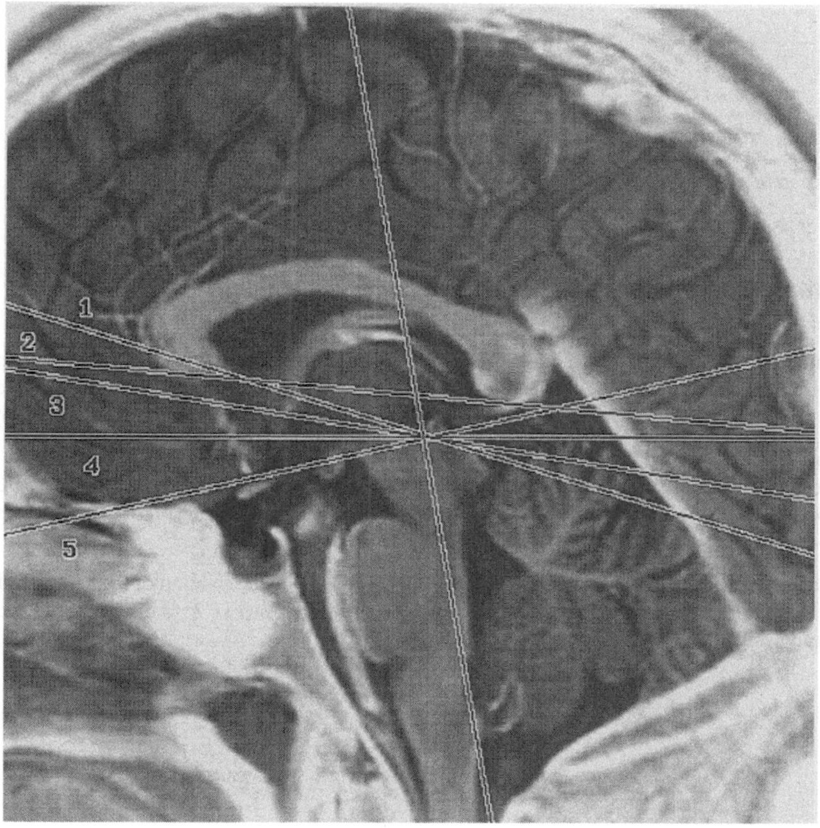

a

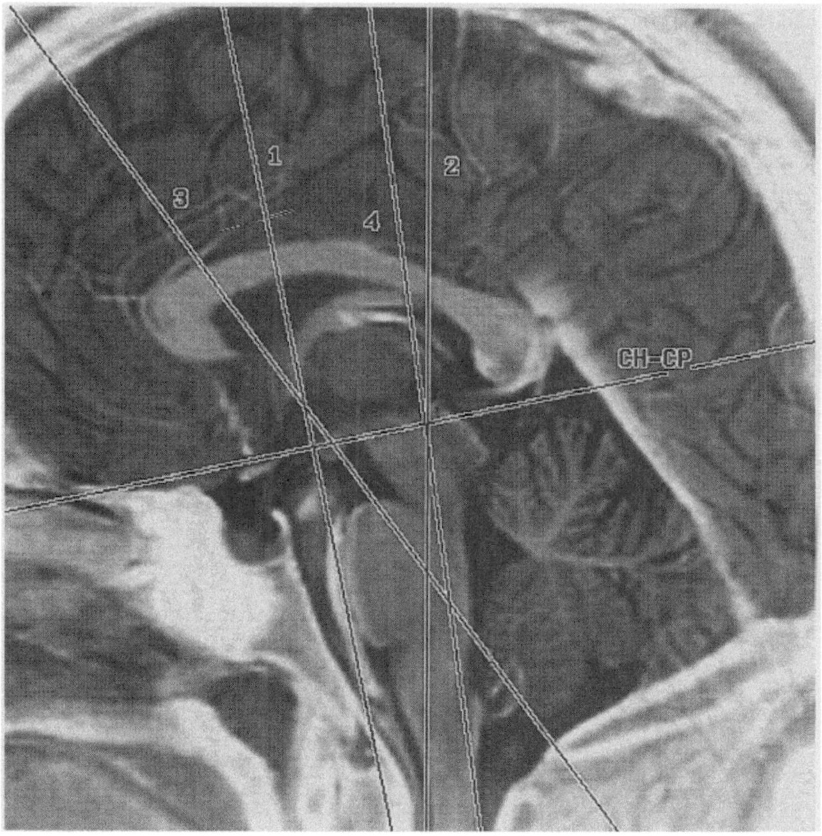

b

Plate II: Chiasmato-commissural plane (CH-PC)

Fig. a. 3D MR volumetric reformation of the brain showing the relation of the CH-PC plane with the mid-sagittal cut

Fig. b. Anatomical landmarks of the chiasmato-commissural plane (CH = chiasmal notch; PC = posterior commissure; MDJ = meso-diencephalic junction; OTC = cisternal optic tract; MB = mamillary bodies; ONC = cisternal optic nerve; TL = temporal lobe

Fig. c. Cutaneous landmarks of the CH-PC plane joining the subnasal point to the superior otobasion, projected on a surface rendering of the head using 3D-MRI

Fig. d. Projections of CH-PC and CP-Obex reference lines on the mid-sagittal cut (brain scout view) and showing the close parallelism to the posthion-lateral sinus groove line (skeletal landmarks)

Fig. e. Troncal angle intercepted between CH-PC and CP-Obex lines (93°) and commissural angle intercepted between CH-PC and AC-PC (24°)

Fig. f. 3D representation of the neuro-ocular plane, plane of the visual pathways (+ 7° to the anthropological baseline)

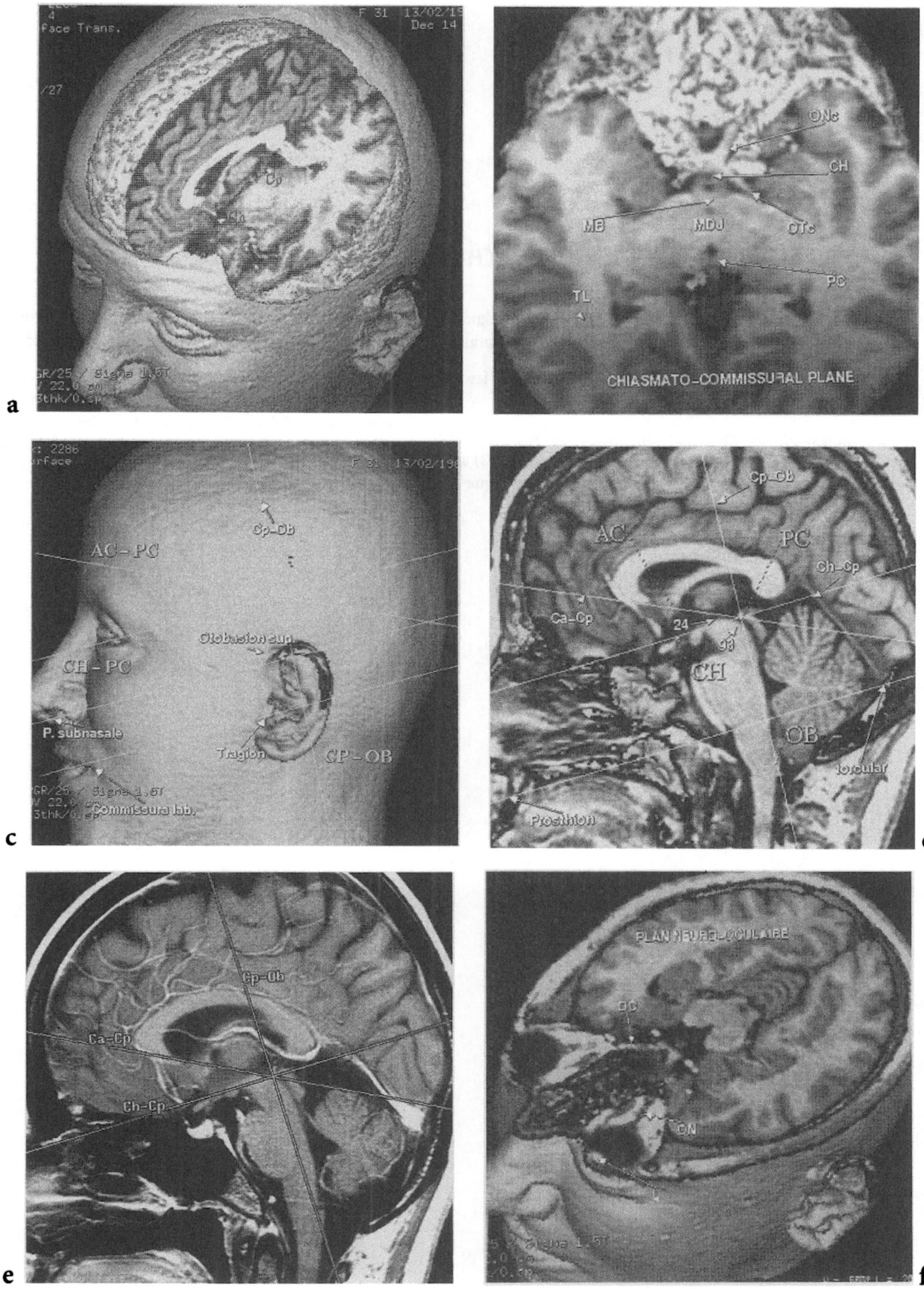
a
Cp
b
ONc
CH
MB
MDJ
OTc
PC
TL
CHIASMATO-COMMISSURAL PLANE
c
Cp-Ob
AC-PC
CH-PC
Otobasion sup.
P. subnasale
Tragion
CP-OB
Commissura lab.
d
Cp-Ob
AC
PC
Ch-Cp
Ca-Cp
24
CH
OB
torcular
Prosthion
e
Cp-Ob
Ca-Cp
Ch-Cp
f
PLAN NEURO-OCULAIRE
OC
ON

Plate III: Chiasmato-commissural plane (CH-PC)

Fig. a. Projections of the CH-CP line as well as its correlated cutaneous reference line, on a 3D MR of the head and brain

Fig. b. Projection of CH-PC plane on the mid-sagittal view, using 3D-MR

Figs. c and d. 3D volume rendering of the brain showing the projection of CH-PC plane on the parallel sulcus (sT-T2) and its close parallelism to the sylvian fissure. Note the oblique direction of the AC-PC plane

Fig. e. MR angiography showing on the axial projection the circle of Willis and the sylvian arteries along their insular route, best visualized in the CH-PC orientation

Fig. f. Cross reference slab showing the successive coronal slices oriented parallel to CP-Obex line as performed for studying the temporal lobes from the anterior pole to the splenium of the corpus callosum

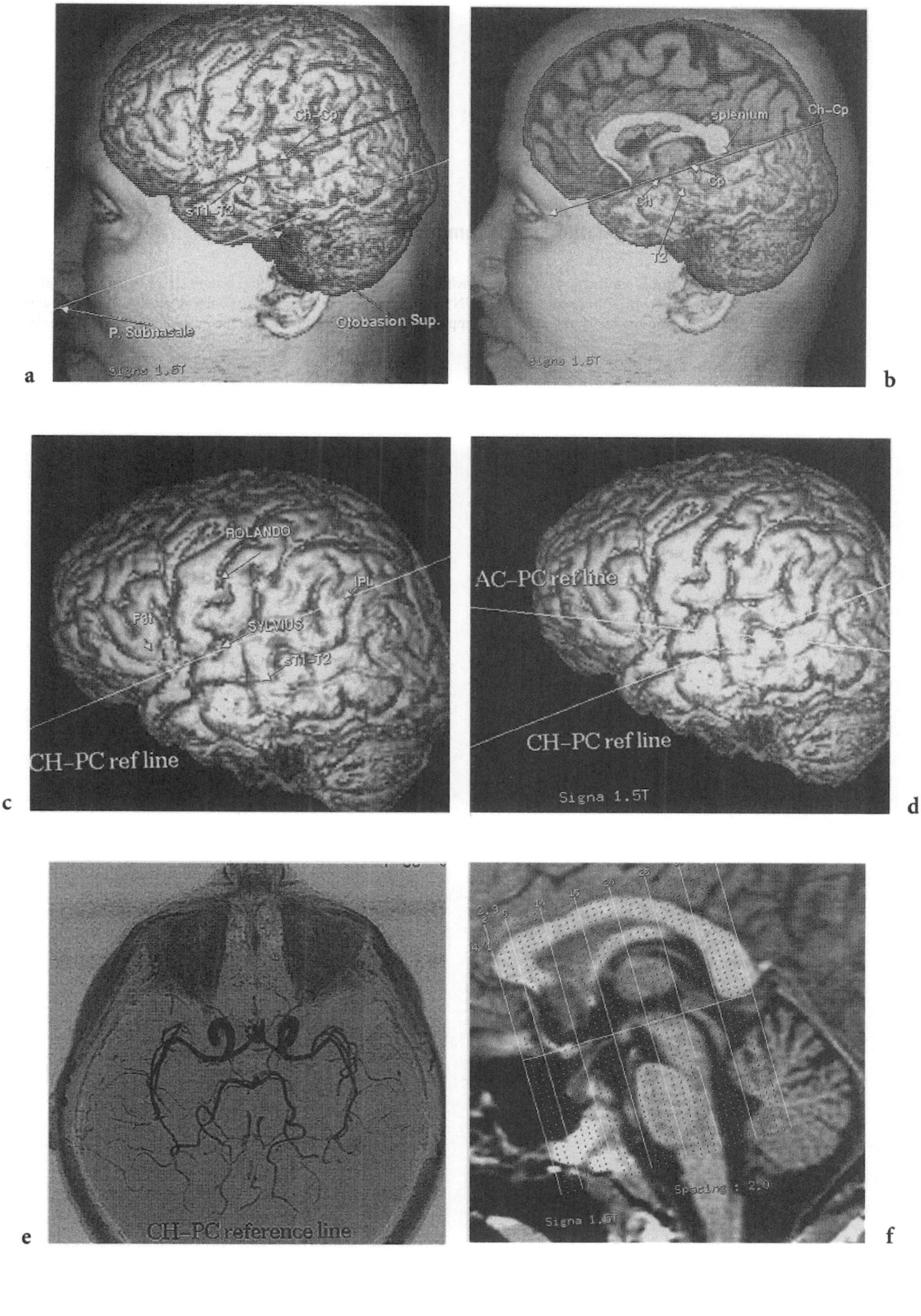
Ch-Cp
sT1-T2
P. Subnasale
Otobasion Sup.
Signa 1.5T
a
Splenium
Ch-Cp
Cp
Ch
T2
Signa 1.5T
b
ROLANDO
IPL
F3t
SYLVIUS
sT1-T2
CH-PC ref line
c
AC-PC ref line
CH-PC ref line
Signa 1.5T
d
CH-PC reference line
e
Spacing : 2.0
Signa 1.5T
f

Plate IV: Anatomical correlations in the commissural-obex plane (PC-OB)

Figs. a and b. 3D MR coronal reformations of the brain with slices oriented parallel to the PC-Obex plane, using a SUN* workstation, demonstrating symmetrically the mesial temporal region (the amygdaloid complex, hippocampal formation and para-hippocampal gyrus), all contained between the anterior commissural plane and the retrosplenial plane of the slab

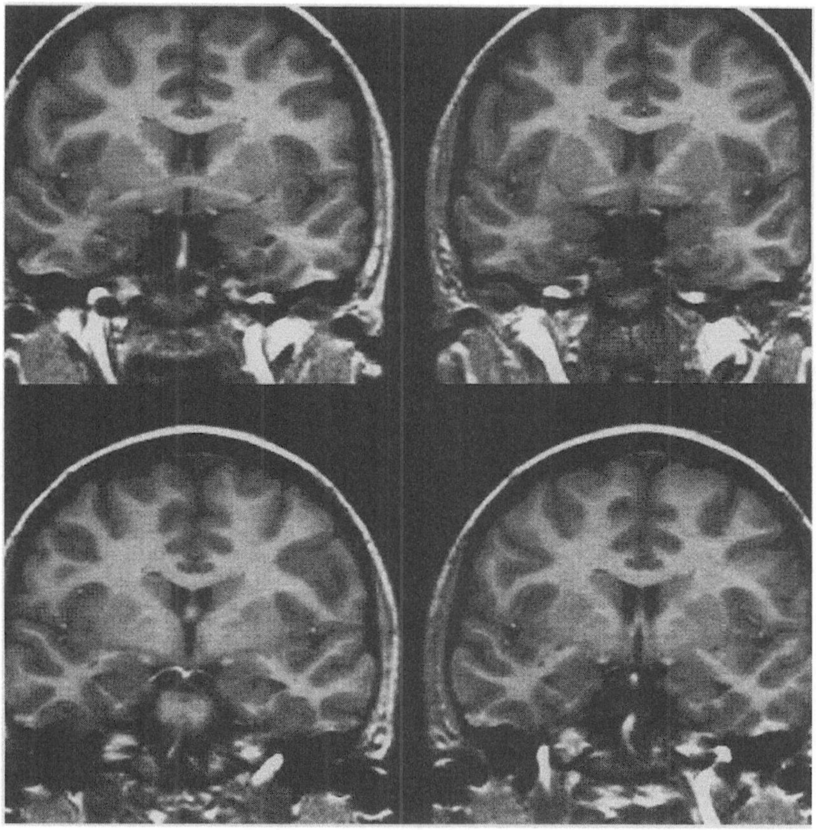

a

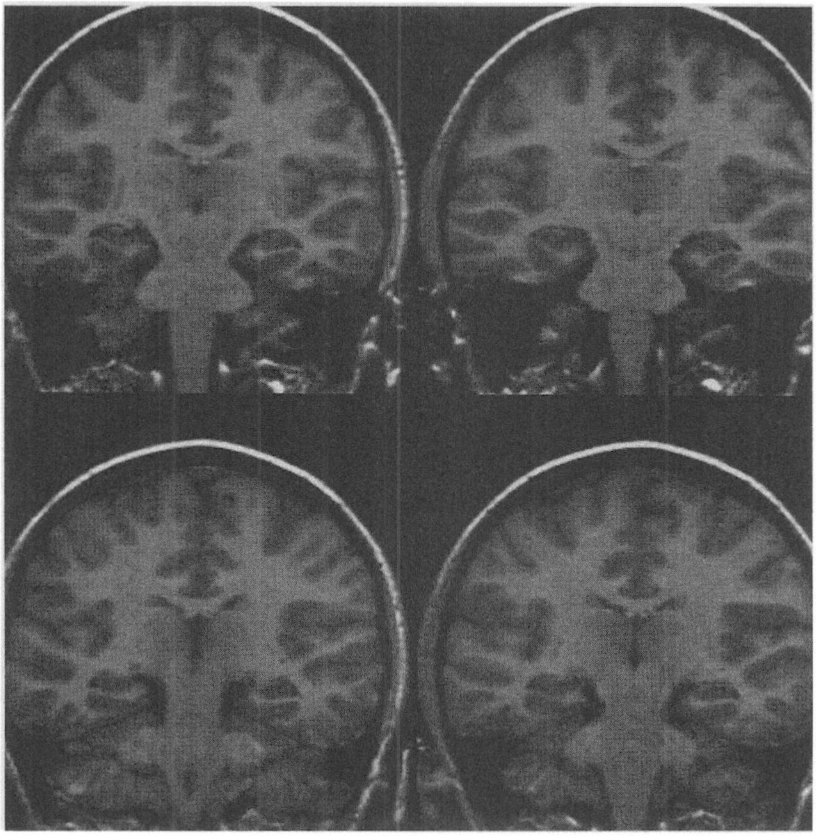

b

Plate V: CH-PC anatomical correlations in the sagittal plane

Figs. a - f. Close parallelism of CH-PC as projected on the sagittal views:

a *to the anterior callosal plane*
b *to the parieto-occipital fissure at its junction with the calcarine fissure*
c *to the tangent to the superior border of the amygdala and the hippocampal tail*
d *to the temporal horns antero-posterior axis*
e *as the bisector of the sylvian triangle*
f *to the parallel sulcus (sT1-T2), as the lateral projection of CH-PC reference line*

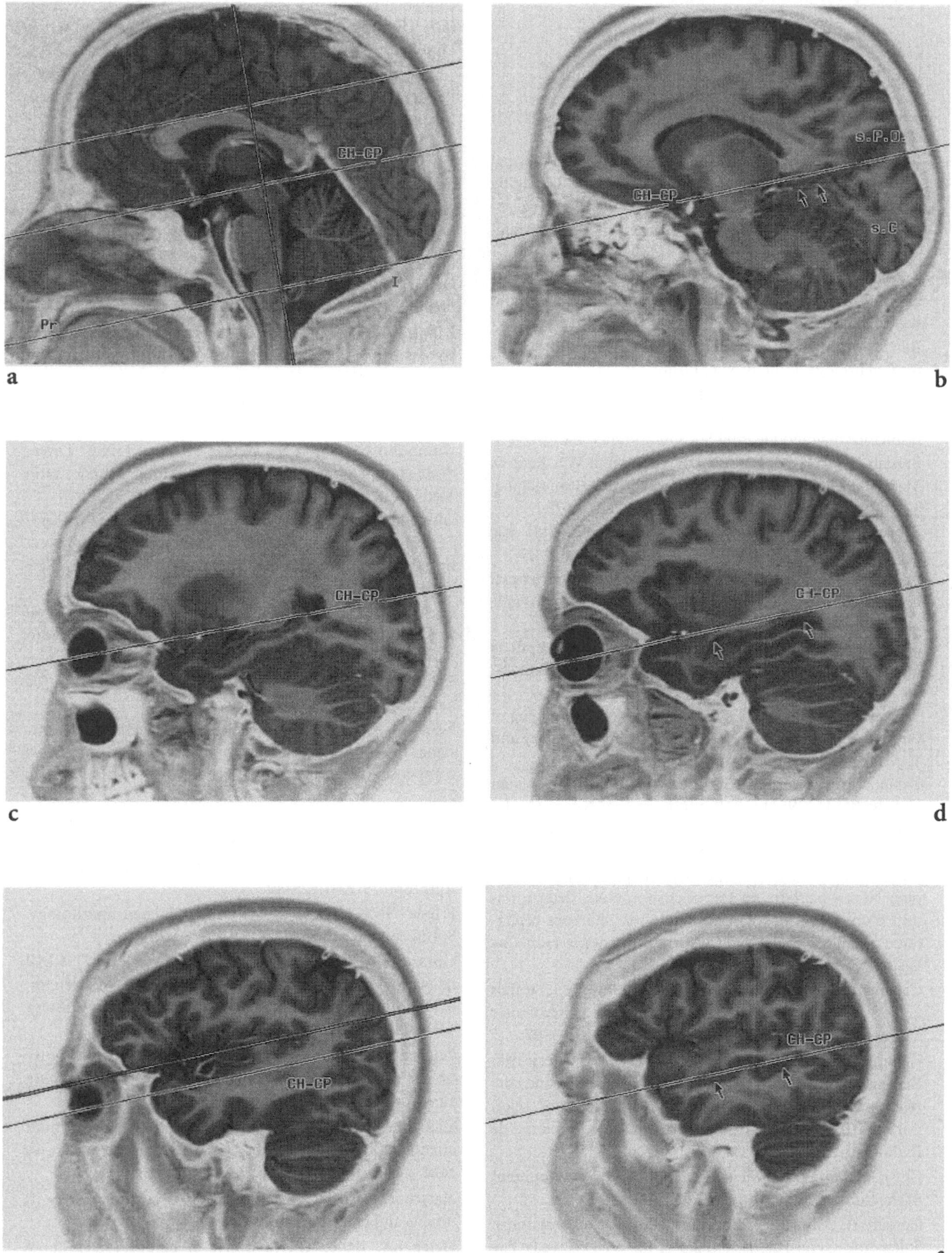
CH-CP
Pr
I
a
s.P.O.
CH-CP
s.C
b
CH-CP
c
CH-CP
d
CH-CP
e
CH-CP
f

Selected References

- Ariens Kappers CU (1947) Anatomie comparée du système nerveux, Masson ed, Paris
- Barkovich AJ, Kjos BO, Jackson DE Jr et al (1988) Normal maturation of the neonatal and infant brain: MR imaging at 1.5T. Radiology 166: 173
- Barkovich JA (1998) MR of the normal neonatal brain: assessment of deep structures. AJNR 19: 1397-1403
- Bradley WG, Schmidt PG (1985) The effect of methemoglobin formation on the MR appearance of subarachnoid hemorrhage. Radiology 156: 99
- Brazis PW, Masdeu JC, Biller J (1990) Localization in clinical neurology. Little Brown, Boston, 76-78
- Bregeat P, Cabanis EA, Iba-Zizen MT, Tamraz J, Stoffels C, Alfonso JM, Lopez A, Le Bihan D (1986) Nerf optique et résonance magnétique. Bull Soc Opht France 5: 627-631
- Breslau J, Jarvik JG, Haynor DR, Longstreth WT, Kent D, Maravilla KR (1999) MR contrast media in neuroimaging: a critical review of the literature. AJNR 20: 670-675
- Broca P (1862) Les projections de la tête. Bull Soc Anthropol Paris 1-3: 514-528
- Broca P (1873) Nouvelles recherches sur le plan horizontal de la tête et sur le degrè d'inclinaison des divers plans crâniens. Bull Soc Anthropol Paris b8: 542-563
- Broca P (1873) Sur le plan horizontal de la tête et sur la méthode trigonométrique. Bull Soc Anthropol Paris 8: 48-96
- Brooks RA, DeChiro G, Patronas N (1989) MR imaging of cerebral hematomas at different field strengths: theory and applications. J Comput Assist Tomogr 13: 194
- Cabanis EA, Pineau H, Iba-Zizen MT, Coin JL, Newman N, Salvolini U (1981) C.T. Scanning in the "Neuro-Ocular Plane": the optic pathways as a "new" cephalic plane. Neuro-Ophtalmology 1(4): 237-252
- Cabanis EA, Iba-Zizen MT, Pineau H, Tamraz J (1982) Le plan Neuro-oculaire en tomodensitométrie, détermination d'un "nouveau" plan horizontal de référence céphalique orienté selon les voies visuelles. Biométrie Humaine Paris 17: 21-48
- Cabanis EA, Laugier EA, Iba-Zizen MT, Tamraz J, Stoffels C (1985) Anatomie de la tête en Imagerie par Résonance Magnétique. Le concours Médical 107(34): 3191-3197
- Cabanis EA, Iba-Zizen MT et al (1986) Imagerie par Résonance Magnétique de la tête à 0.5 Tesla. Atlas de corrélations anatomiques normales dans 3 dimensions, selon l'orientation du plan neuro-oculaire. Feuillets de Radiologie 26(5): 309-416
- Chi JG, Dooling EC, Gilles FH (1997) Gyral development of the human brain. Ann Neurol 1: 86
- Comair YG, Choi HY, Tamraz J (1996) Cortical Anatomy. Sulcal and Gyral Patterns. Chap 1. In: Wyllie E (ed) The treatment of epilepsy: principles and practice. Lea and Febiger, 2nd ed
- Comair J, Tamraz J (2000) Cortical Anatomy: normal and abnormal subcal and gryal patterns. Chapt 6. In: Wyllie E (ed) The treatment of epilepsy: principles and practice. Lippincott Williams and Wilkins, 3rd ed
- Coussement A, Faure C, Coussement-Beylard N (1990) Repères et mesures en imagerie médicale. 4th ed, L'expansion Scientifique Française
- DeCoene B, Hajnal JV, Gatehouse P et al (1992) MR of the brain using fluid-attenuated in version recovery (FALIR) pulse sequences. Am J Neuroradiol 13: 1555
- DeCoene B, Hajnal JV, Pennock JM et al (1993) MRI of the brainstem using fluid attenuated inversion recovery pulse sequences. Neuroradiology 35: 327
- Delmas A, Pertuiset B (1959) Topométrie Crânio-encéphalique chez l'homme. Masson, Paris
- Dillon W (1997) Cryptic vascular malformations: controversies in terminology, diagnosis, pathophysiology, and treatment. AJNR: 18 Controversies 1839
- Dumoulin CL, Hart HR (1986) Magnetic resonance angiography. Radiology 161: 717
- Dumoulin CL, Souza SP, Walker MF et al (1989) Three-dimensional phase contrast angiography. Magn Reson Med 9: 139
- Duvernoy H, Cabanis EA, Iba-Zizen MT, Tamraz J, Guyot J (1991) The Human Brain Surface three-dimensional sectional anatomy and MRI. Springer-Verlag, Wien, New York
- Ehricke HH, Schad LR, Gademann G et al (1992) Use of MR angiography for stereotactic planning. J Comput Assist Tomogr 16: 35
- Escouroulle R, Poirier J (1997) Manuel élémentaire de neuropathologie. Masson, Paris
- Evans AJ, Richardson DB, Tien R et al (1993) Poststenotic signal loss in MR angiography: effects of echo time, flow compensation, and fractional echo. Am J Neuroradiol 14: 721
- Fenart R, Vincent H, Cabanis EA (1982) Le plan orbitaire chez l'adulte jeune, sa position relative à d'autres éléments architecturaux de la tête. Etude vestibulaire. Bull Mém Soc Anthropol; 9(XIII): 29-40
- Friede RL (1989) Developmental neuropathology. Springer-Verlag, Berlin, 2nd ed
- Gomori JM, Grossman RI, Yu-Ip C et al (1987) NMR relaxation times of blood: dependence on field strength, oxidation state, and cell integrity. J Comput Assist Tomogr 11: 684
- Gouaze A, Salamon G (1988) Brain Anatomy and magnetic resonance imaging. Springer-Verlag, Berlin Heidelberg New York
- Hajnal JV, Bryant DJ, Kasuboski L et al (1992) Use of fluid attenuated inversion recovery (FLAIR) pulse sequences in MRI of the brain. J Comput Assist Tomogr 16(6): 841
- Harwood-Nash DC, Fitz CR (1976): Neuroradiology in infants and children. Mosby St Louis
- Heier LA, Bauer CJ, Schwartz L et al (1989) Large Virchow-Robin spaces: MR-clinical correlation. AJNR 10: 929

- Hitmair K, Winberger D, Rand T et al (1994) MR assessment of brain maturation: comparison of sequences. Am J Neuroradiol 15: 425
- Hume Adams J, Corsellis JAN, Duchen LW (1984) Greenfield's Neuropathology. Edward Arnold, London
- Jungresis CA, Kanal E, Hirsch WL (1988) Normal perivascular spaces mimicking lacumar infraction: MR imaging. Radiology 169:101
- Kandel ER, Schwartz JH, Jessel TM (1991) Principles of Neural Science. Elsevier, New York, 3rd ed
- Kemp SS, Zimmerman RA, Bilaniuk LT et al (1987) Magnetic resonance imaging of the cerebral aqueduct. Neuroradiology 29: 430
- Kier EL, Truwit CL (1997) The lamina rostralis: modification of concepts concerning the anatomy, embryologic, and MR appearance of the rostrum of the corpus callosum. Am J Neuroradiol 18: 715
- Kier EL, Kim JH, Fulbright RK, Bronen RA (1997) Embryology of the human fetal hippocampus: MR imaging, anatomy, and histology. AJNR 18: 525–532
- Kucharczyk W, Bernstein M (1997) Do the benefits of image guidance in Neurosurgery justify the costs? From stereotaxy to intraoperative MR. AJNR 18, Controversies 1855
- Kuhn MJ, Mikulis DJ, Ayoub DM et al (1989) Wallerian degeneration after cerebral infarction: evaluation with sequential MR imaging. Radiology 172: 179
- Kurtzke JF (1955) A new scale for evaluating disability in multiple sclerosis. Neurology 5: 580
- Kurtzke JF (1983) Rating neurologic impairment in multiple sclerosis: an expanded disability status scale (EDSS). Neurology 34: 1444
- Lasjaunias PL (1981) Craniofacial and upper cervical arteries. William and Wilkins, Baltimore
- Lazorthes G, Gouaze A, Salamon G (1976) Vascularisation et circulation de l'encéphale. Tome I. Masson, Paris
- Le Bihan D, Breton E, Lallemand D et al (1986) MR imaging of intravoxel incoherent motions: application to diffusion and perfusion in neurologic disorders. Radiology 161: 401
- Lemire RJ, Loeser JD, Leech RW et al (1975) Normal and abnormal development of the human nervous system. Harper & Row New York
- Masaryk AM, Ross JS, DiCello M et al (1991) 3DFT magnetic resonance angiography of the cervical carotid arteries: potential and limitations as a screening exam for arteriosclerotic ischemic cerebrovascular disease. Radiology 179: 797
- Mattle H, Edelman RR, Wentz KV et al (1991) Middle cerebral artery: determination of flow velocities with MR angiography. Radiology 181: 527
- Merran S, Cabanis EA, Chevrot A, Cosnard G, Godefroy G, Iba-Zizen MT, Jardin C, Pharaboz C, Pradel J, Pujol J, Tamraz J, Hureau J (1989) Radio-anatomie en coupes TDM-IRM. Pradel, Paris
- Miller DH, Barkhof F, Berry I et al (1991) Magnetic resonance imaging in monitoring the treatment of multiple sclerosis: concerted action guidelines. J Neurol Neurosurg Psych 54: 683-688
- Murakami JH, Weinberger E, Shaw DW (1999) Normal Myelination of the Pediatric Brain imaged with fluid-attenuated inversion-recovery (FLAIR) MR imaging. AJNR 20: 1406-1411
- Nakagawa H, Iwasaki S, Kichikawa K, Fukusumi A, Taoka T, Ohishi H, Uchida H (1998) Normal myelination of anatomic nerve fiber bundles: MR analysis. AJNR 19: 1129-1136
- Nesbit GM, Forbes GS, Scheaithauer BW et al (1991) Multiple sclerosis histopathologic and MR and/or CT correlation in 37 cases at biopsy and three cases at autopsy. Radiology 180: 467-474
- Ono M, Kubik S, Abernathey C (1990) Atlas of the cerebral Sulci. Georg Thieme Verlag, Stuttgart, New York
- Poser CM, Paty W, Scheinberg L et al (1983) New diagnostic criteria for multiple sclerosis: guidelines for research protocols. Ann Neurol 13: 227
- Poser CM, Paty DW, Scheinberg L et al (1983) The diagnostic criteria for multiple sclerosis: guidelines for research protocols. Ann Neurol 33: 611-617
- Riles TS, Eidelman EM, Litt Aw et al (1992) Comparison of magnetic resonance angiography, conventional angiography, and duplex scanning. Stroke 23: 341
- Rosen BR, Bellivean JW, Verea JM et al (1990) Perfusion imaging with NMR contrast agents. Magn Reson Med 14: 249
- Saban R (1980) Les plans d'orientation de la tête. Cahiers d'Anthropol 3-4: 285-313
- Salamon G et al (1971) Atlas de la vascularisation artérielle du cerveau chez l'homme. Sandoz, Paris
- Salamon G, Lecaque G (1978) Choice of the plane of incidence for computed tomography of the cerebral cortex. J Comput Assist Tomogr 2: 93-97
- Sheldon JJ, Siddharthan R, Tbias J et al (1984) MR imaging of multiple sclerosis: comparison with clinical examinations in 74 patients. AJNR 6: 683-690
- Schrodt JF, Edwards MR, Bognanno JR et al (1988) MR of brain stem corticospinal tract Wallerian degeneration. Am J Neuroradiol 9: 1025
- Szikla G, Bouvier G, Hori T, Petro V, Cabanis EA, Farnarier P, Iba-Zizen MT (1997) Angiography of the human brain cortex. Springer-Verlag, Berlin
- Talairach J, Szilka G, Fournier P (1967) Atlas d'anatomie stéréotaxique du télencéphale. Etudes anatomoradiologiques. Masson, Paris
- Talairach J, Tournoux P (1988) Co-Planar stereotaxic. Atlas of the Human Brain. 3-dimensional proportional system: an approach to cerebral imaging. Georg Thieme Verlag, Stuttgart
- Tamraz J, Iba-Zizen MT, Cabanis EA (1984) Atlas d'anatomie céphalique dans le plan Neuro-Oculaire. J Fr Ophtalmol 7(5): 371-379
- Tamraz J (1983) Atlas d'anatomie céphalique dans le plan neuro-oculaire. Thèse Médecine, Schering Paris
- Tamraz J, Saban R, Reperant J, Cabanis EA (1990) Définition d'un plan de référence céphalique en IRM: le plan chiasmato-commissural. CR Académie des Sciences, Paris, 311 (III): 115-121

- Tamraz J (1991) Morphométrie de l'encéphale par résonance magnétique: Applications à la pathologie chromosomique humaine à l'anatomie comparée et à la tératologie. Université René Descartes, Thèse de Sciences, Paris
- Tamraz J, Saban R, Reperant J, Cabanis EA (1991) A new cephalic reference plane for use with magnetic resonance imaging: the chiasmato-commisural plane. Surg Radiol Anat; 13: 197-201
- Tamraz J, Comair Y (2000) Atlas of regional anatomy of the brain using MRI. Springer-Verlag, Berlin Heidelberg New York
- Tamraz J, Outin C, Fabre V (2000) Imagerie par Résonance Magnétique de la tete et du Rachis. Atlas didactique et guide d'interprétation. Springer Verlag, Paris Berlin Heidelberg, New York
- Tamraz J (2001) Brain Anatomy. In: Structural Neuroimaging. Part IV Luder HO, Comair Y(eds) 2nd ed, Parke-Davis (in press)
- Thorpe JW, Halpin S, MacManus DG et al (1994) A comparison between fast spin echo and conventional spin echo in the detection of multiple sclerosis lesions. Neuroradiology 36: 388
- Tomiak MM, Rosenblum JD, Prager JM et al (1994) Magnetization transfer: a potential method to determine the age of multiple sclerosis lesions, Am J Neuroradiol 15: 1569
- Utsunomiya H, Takano K, Okazaki M, Mitsudome A (1999) Development of the temporal lobe in infants and children: analysis by MR-based volumetry. AJNR 20: 717–723
- Van der Knaap MS, van Wezel-Meijler G, Calk J et al (1996) Normal gyration and sulcation in preterm and term neonates: appearance on MR images. Radiology 200: 389
- Van der Knapp MS, Valk J (1990) MR imaging of the various stages of normal myelination during the first year of life. Neuroradiology 31: 459-470
- Van der Zwan A, Hillen B, Tulleken CA (1992) Variability of the territories of the major cerebral arteries. J Neurorsurg 77: 927-940
- Vanier M, Roch Lecours A, Ethier R, Habib M, Poncet M, Milette PC, Salamon G (1985) Proportional localization system for anatomical interpretation of cerebral computed tomogram. J Comput Assist Tomogr 9: 715-724
- WFN World Federation of Neurology (1962) Problem Commission of Neuroradiology: study meeting on projections and nomenclature. Brit J Radiol 35: 501-503
- Zigmond MJ, Bloom FE, Landis SC, Roberts JL, Squire LR (1999) Fundamental Neuroscience. Academic Press, San Diego London Boston New York
- Zimmerman RA, Bilaniuk LT (1982) Age-related incidence of pineal calcification detected by computed tomography. Radiology 142: 659-662

Part I

Brain and skull

1 - Tumoral conditions

a - Primary supratentorial tumors
b - Incisural and infratentorial tumors
c - Leptomeningeal and Skull voult Tumors
d - Metastatic Tumors

Primary supratentorial tumors

Cases 1- 12: cerebral parenchymal gliomas

Clinical and pathological features

- Overall incidence of intracranial tumors is estimated to reach 5 individuals per 100 000 population per year. The rate of brain tumors is much higher in older patients reaching roughly 40% per year per 100 000 population after age 80.
- Primary brain tumors represented by the gliomas correspond to almost half of all primary intracranial tumors in adults and roughly two thirds of these are found in children less than 15 years old. Metastatic tumors represent in adults (rare in childhood) roughly 1/4 of all intracranial tumors in autopsy series.
- There is a topographic difference between intracranial tumors found in adults and children. Almost 70% of brain tumors are supratentorial in location in adults whereas roughly 70% are found in the posterior fossa in children.
- **Histopathology and classifications:** gliomas may be divided into astrocytomas, oligodendrogliomas and ependymomas corresponding respectively to the three types of glial cells, the astrocytes, oligodendrocytes and the ependymal cells. Almost 5% of these neuroglial tumors are of a mixed type or "mixed gliomas".
- **Astrocytic tumors** ("astrocytomas") may be subdivided into 5 clinicopathological categories: 1- diffuse astrocytomas (more than 60% of all primary brain tumors), subdivided into 3 groups: low grade diffuse astrocytomas (WHO grade II), anaplastic astrocytomas (WHO grade III) and glioblastomas multiformis (WHO grade IV); 2- pilocytic astrocytoma (WHO grade I, found in childhood, usually circumscribed); 3- pleiomorphic xanthoastrocytoma (WHO grade II, but showing some "malignant" features on conventional criteria), usually circumscribed; desmoplastic astrocytoma (commonly mixed neuronal-glial tumor, found in infancy); subependymal giant cell astrocytoma (tuberous sclerosis, usually circumscribed).
- **Oligodendroglial tumors** ("oligodendrogliomas") are subdivided (WHO) into: 1- oligodendrogliomas (WHO grade II), and 2- anaplastic oligodendrogliomas (WHO grade III).
- Mixed gliomas correspond to tumors showing a mixture of neoplastic astrocytic and oligodendroglial components. This pathological entity may be subdivided into 2 groups: oligoastrocytomas (WHO grade II), most frequent of both varieties, and anaplastic oligoastrocytomas (WHO grade III).
- **Ependymal tumors** ("ependymomas") are subdivided into 4 clinicopathological entities: 1- ependymoma (WHO grade II), most and usually intraventricular; 2- anaplatic ependymoma (WHO grade III); 3- subependymoma (WHO grade I) benign in nature. The fourth entity occurring almost always in the conus medullaris and the filum, but which may occur in the lateral ventricle or the brain parenchyma, is the myxopapillary ependymoma (WHO grade I).
- *Gliomatosis cerebri*: very rare, diffuse glial tumor (WHO grade III or IV), infiltrating the cerebral hemispheres (2/3 of the cases) uni or bilaterally and extending over more than one lobe, with associated enlargement of the affected anatomical structure. The centrum semiovale is preferentially infiltrated in over 2/3 of the cases and the cortex in less than 20%, followed by the brainstem in about half the cases, the thalami in 40% and the basal ganglia in 30% of the cases. Leptomeningeal spread occur in about 1/5 of the cases. Age at diagnosis shows a peak incidence between 40-50 years old.
- *Signs and symptoms*: patients with brain tumors may present variable neurological syndromes: 1- a generalized impairement of intellectual function with variable mental changes comprising emotional lability, inertia, psychomotor reduction, disturbance of memory, spatial disorientation and behavioral disorders; 2- a syndrome of increased intracranial hypertension with headaches usually mild, nocturnal and/or worse in the morning and of the pulsatile type, commonly bifrontal and bioccipital, accompanied by vomiting in almost 1/3 of the cases; 3- generalized or focal seizures found in 20-50% of the cases (the presence of an aura being suggestive) and occurring most frequently in more than 2/3 of the cases in slow growing tumors as compared with fast growing tumors (30% of the cases); 4- specific neurological syndromes suggestive of specific tumors and localizations (also developed under the appropriate pathology observations in the following pages of the book).

Cases 1-3: low grade diffuse astrocytomas

Clinical and pathological features

- These primary brain tumors represent 25-30% of all cerebral gliomas.
- They occur in young adults (mean age: 34 years old) between ages of 20 to 45 in 60% of the cases, after 45 years old in about 30% of the cases, and before 20 in about 10%.
- A male predilection is observed with a ratio M:F of 1.8:1.
- These are slowly growing infiltrating tumors occurring anywhere in the central nervous system but with a predilection to the cerebral hemispheres (adults) and the brainstem (children).
- These tumors are well differentiated ("benign") but show a great tendency for malignant transformation (anaplasia). Many of these tumors show areas of different degrees of anaplasia.
- *Histopathology*: formed of neoplastic astrocytes, which are increased in number and in size. Three major variants may be distinguished with respect to the related neoplastic cell type: 1- fibrillary astrocytomas, the most frequent variety, formed of fibrillary astrocytes (elongated cells with little visible cytoplasm, expressing GFAP); 2- gemistocytic astrocytomas, composed mainly of gemistocytic neoplastic astrocytes (plump cells with eosinophilic abundant cytoplasm, expressing GFAP); 3- protoplasmic astrocytomas, a rare variant (stellate cells with small cell bodies and showing few or no stainable fibrils).
- *Signs*: refer to the introductory section on cerebral parenchymal gliomas.

MR characteristic findings

Morphology

- Low grade diffuse astrocytomas are commonly homogeneous infiltrating tumors, developed in the cerebral hemispheres in adults or in children.
- They are found anywhere in the cerebral hemisphere except in the occipital lobe.
- Ill-defined they do not show clear-cut demarcation from the adjacent normal parenchyma (demonstrated on pathological specimen). It has been shown that the tumoral process, due to its infiltrating nature, may extend beyond the abnormality evidenced on imaging. The tumor tend to expand the anatomical structure involved or may infiltrate these structures without marked morphological changes or distortion of normal anatomy.
- Occasionally invasion of the subarachnoid spaces may be shown (usually observed in case of malignant degeneration).
- These low grade tumors usually lack of peritumoral edema. There are no vascular flow voids within the tumoral process and no evidence of intratumoral hemorrhage or necrosis.
- Calcifications which occur in about 15-20% of the cases are seldom evidenced on MR as compared to CT.

Signal intensity

- GE-T1w: hypointense to white matter, homogeneous SI.
- SE-PDw: hyperintense to cortex, homogeneous.
- SE-T2w: hyperintense to white matter.
- Proton spectroscopy: increase in Cho levels, decrease in NAA levels, mild elevation of Cr, nearly absence of lipid resonance peaks.

Contrast enhancement

- Low grade diffuse fibrillary astrocytomas usually do not show contrast enhancement but may show a heterogeneous pattern of enhancement. Note that enhancement is common in the circumscribed partly cystic pilocytic astrocytomas (enhancing mural nodule) or in the pleiomorphic xanthoastrocytomas (enhancing solid portion).
- Enhancement may be demonstrated during the course of the disease mainly with tumor transformation.
- Enhancement do not correlate adequately with the histological grade of the tumor.

Case 1: Cystic thalamic pilocytic astrocytoma

Fig. a: sagittal cut;	GE - T1w	**Fig. c:** axial cut;	SE - PDw
Fig. b: axial cut;	FLAIR - T2w	**Fig. d:** axial cut;	SE - T2w

Right thalamic tumor, sharply demarcated, presenting a large peripheral cystic component slightly hyperintense to the intraventricular CSF on PDw (**c**) and FLAIR (**b**) sequences. The solid mural tumoral component is slightly hypointense to brain on T1w (**a**, ↑↑) and hyperintense on T2w (**b, c, d**) showing contrast enhancement. Note the mass effect on the lateral ventricles (3, ↑).

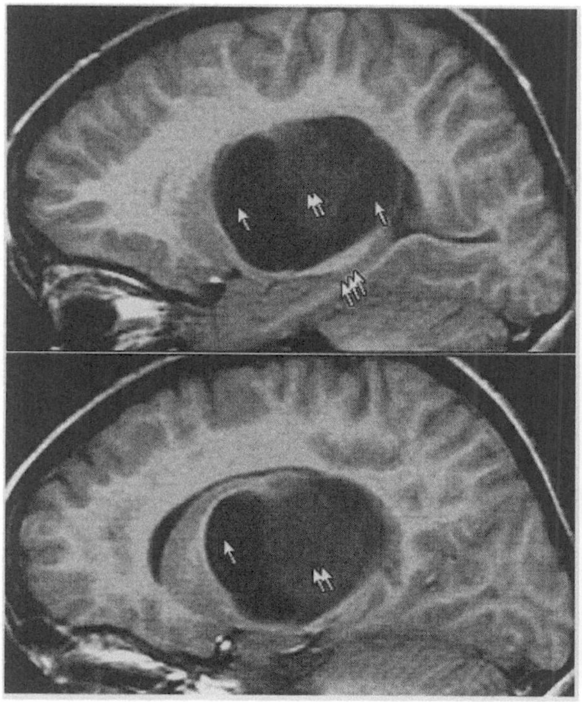

a

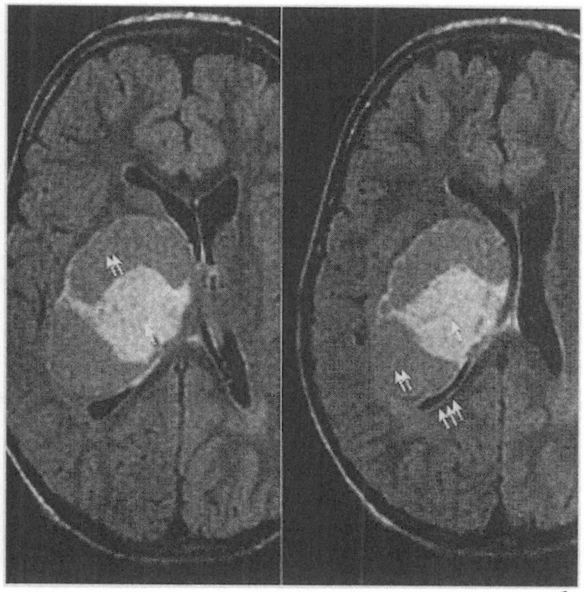

b

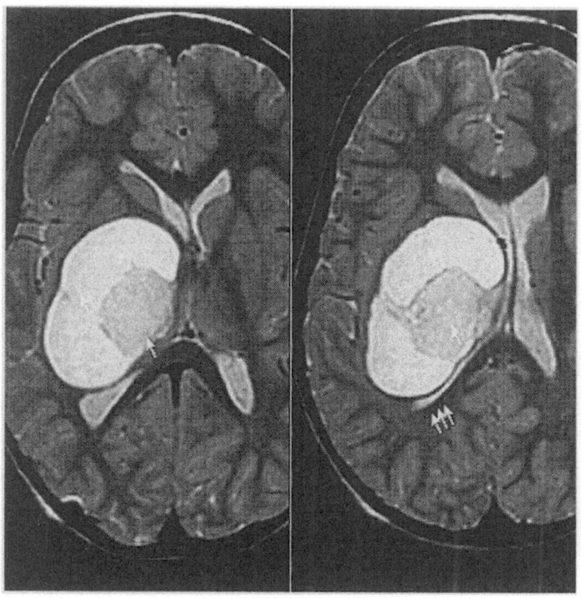

c

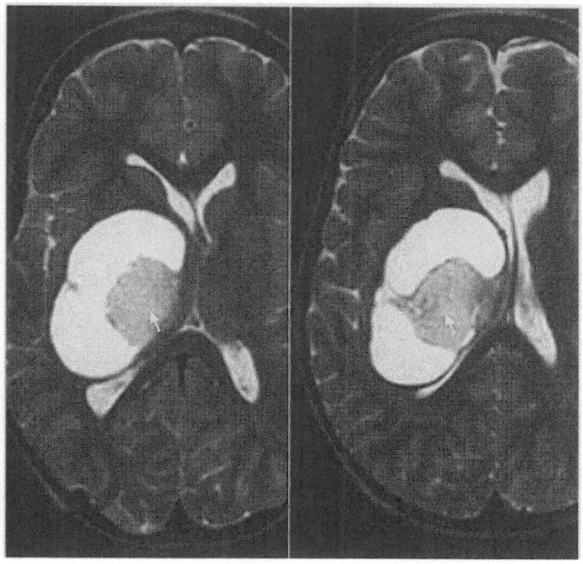

d

Case 2: Gemistocytic astrocytoma, grade II

Fig. a: axial cut; SE - PDw

Fig. b: axial cut; SE - T2w

Temporo-insular well demarcated tumor slightly hyperintense on PDw as well as on T2w sequences, showing areas of intratumoral necrosis (↑) and peritumoral edema (↑↑↑) with mass effect on the atrium of the lateral ventricle (↑↑).

Case 3: Hemorragic astrocytoma, grade II

Fig. c: axial cut; GE (3D) - T1w

Fig. d: axial cut; SE - T2w

Subacute intratumoral hemorrhage with methemoglobin showing a high signal on T1w (↑) and a high signal on T2w, surrounded by edema (↑↑). The lesion shows a peripheral enhancement.

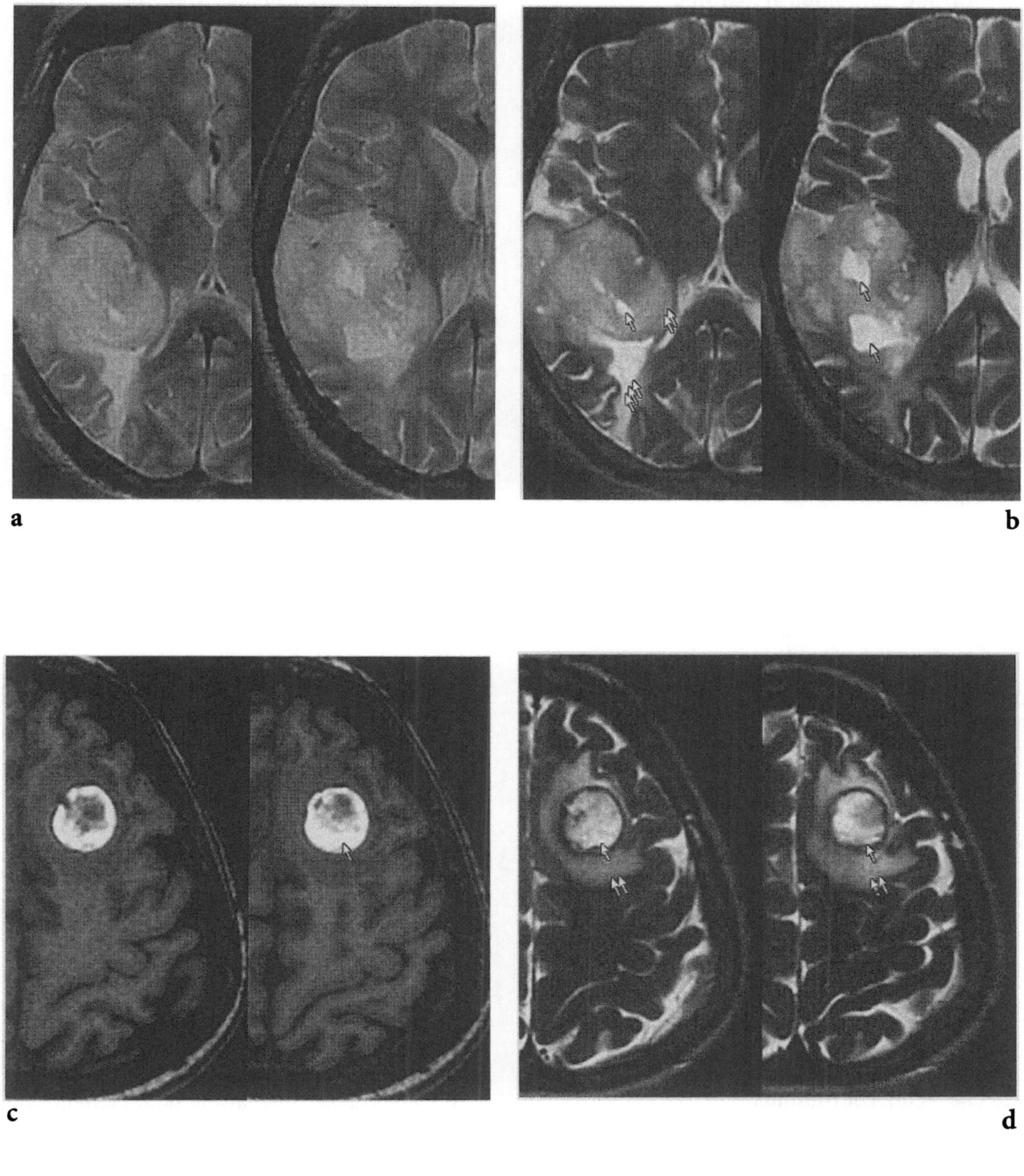

a b c d

Cases 4-8: oligodendrogliomas and mixed gliomas

Clinical and pathological features

- Oligodendrogliomas are infiltrating usually well-defined tumors formed of cells resembling oligodendrocytes.
- They account for about 5 to 7% of primary brain tumors, although highest incidence has been reported.
- They occur in adults in the 5th and 6th decades. About 6% are found in infants or children. A male preponderance is observed (2:1 ratio).
- *Histopathology*: oligodendrogliomas correspond to WHO grade II in case of well differentiated tumors and WHO grade III for anaplastic tumors. The Kernohan and Saint-Anne/Mayo classifications differentiate prognostically dissimilar low grade (1 and 2) from high grade (grade 3 and 4) tumors. Classification is still debated. Oligodendrogliomas, which show a wide variation ranging from well differentiated to highly malignant lesions, are moderately cellular with a characteristic presence of intratumoral microcalcifications (70%) that may also be found in the adjacent cortical area. These tumors develop in the white matter and show propensity to involve the cortex and even invade the leptomeninges with secondary desmoplastic reaction.
- *Signs*: oligodendrogliomas are usually slow growing tumors in the cerebral hemispheres, with a duration between symptoms and surgery estimated as ranging from 2 to 7 years (with tumors near the midline the duration decreases to average 8 months). The presenting sign is most frequently a focal or generalized seizure (50-70%). Signs related to intracranial hypertension with headaches are found in about 1/5 of the cases.

MR characteristic findings

Morphology

- Oligodendrogliomas are supratentorial in location in 90% of the cases of which more than 50% are located in the frontal lobe, followed by the temporal, parietal and occipital lobes. The remaining are found in the third ventricle or the thalami projecting in the 3rd ventricle, or in the lateral ventricles.
- Oligodendrogliomas are commonly fairly well circumscribed tumors with poorly defined boundaries, showing little amount of peritumoral edema. Some may appear heterogeneous due to intralesional presence of small cystic areas of degeneration (20%) and/or hemorrhages (20%) and/or calcifications (50-70%), the latter being best demonstrated on CT (dense or gyriform). Calcifications may be to some extent depicted on MR using gradient echo sequences (magnetic susceptibility), sensitive as well to the presence of iron frequently existing within dystrophic calcifications. Most calcifications are isointense on T1w and may become markedly hypointense on T2w sequences.
- Anaplastic oligodendrogliomas appear as diffusely infiltrating tumors much less demarcated than oligodendrogliomas, with increased associated edema and mass effect. Occuring in adults in the fifties, they are found in the frontal lobes in 60% of the cases, the temporal lobe being involved in roughly 30%. Their estimated incidence is unsettled due to less defined criteria for grading (20-50%). A variable heterogeneous aspect is found owing to the presence of necrotic or cystic areas, hemorrhages and calcifications, similar to what is observed in other anaplastic gliomas entities.
- Mixed gliomas in which is demonstrated two neoplastic components, are most frequently of the oligoastrocytic variety. These WHO grade II oligoastrocytomas (cases 7 and 8), are found in young adults (mean age 45 years), in the frontal and less frequently in the temporal lobes, appear to show no significant imaging differences as compared to WHO grade II gliomas.
- Meningeal gliomatosis is found in almost 10% of oligodendroglial tumors, and a remodeling of the inner table adjacent the skull (scalloping) due to superficial location and slow growth is observed in almost 20% of the cases. Ventricular seeding may be present in 10% of the cases.

Signal intensity

- GE-T1w: hypointense to white matter or isointense to gray matter.

- SE-PDw: hyperintense to cortex (may appear infiltrated).
- SE-T2w: hyperintense to white matter.
- FLAIR: hyperintense to brain.
- In anaplastic oligodendrogliomas, the aspect is more heterogeneous (areas of necrosis) and more infiltrating, with a much less demarcated shape, showing a lower SI on T2w sequences due to hypercellularity.

Contrast enhancement

- Oligodendrogliomas may show contrast enhancement, usually mild, reported in 50 to 70% of the cases.

- Anaplastic oligodendrogliomas may show non specific irregular ring enhancement patterns.

Case 4: Prerolandic oligodendroglioma, grade II

Fig. a: sagittal cut;	GE - T1w	**Fig. c:** coronal cut;	SE - T2w
Fig. b: axial cut;	SE - PDw	**Fig. d:** coronal cut;	GE (3D) - T1w + Gd

Left frontal pre-rolandic, superficial tumor (**a, b, c,** ↑) hypointense on T1w (**a**) with central enhancement (**d**) and surrounding edema (↑↑). The tumor causes mass effect on the adjacent sulci including the sylvian fissure. The lesion is hyperintense on PDw and T2w sequences.

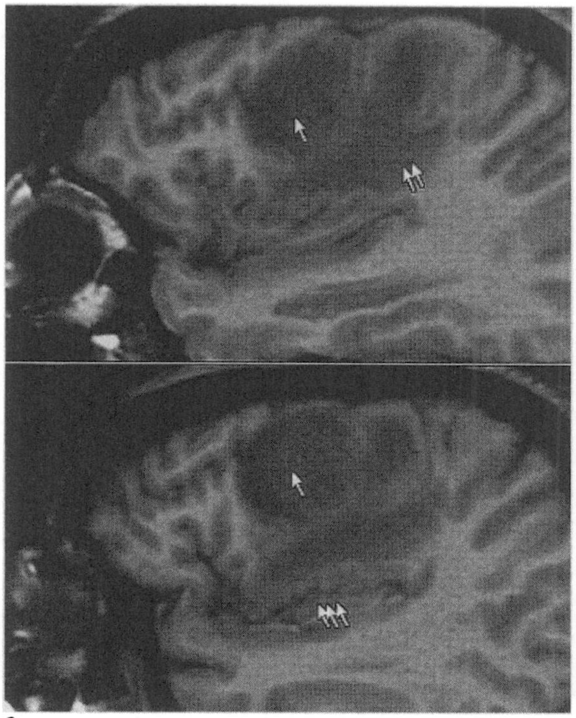

a

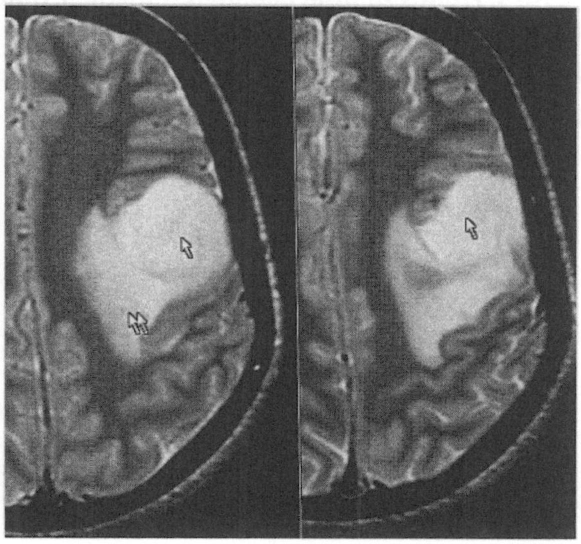

b

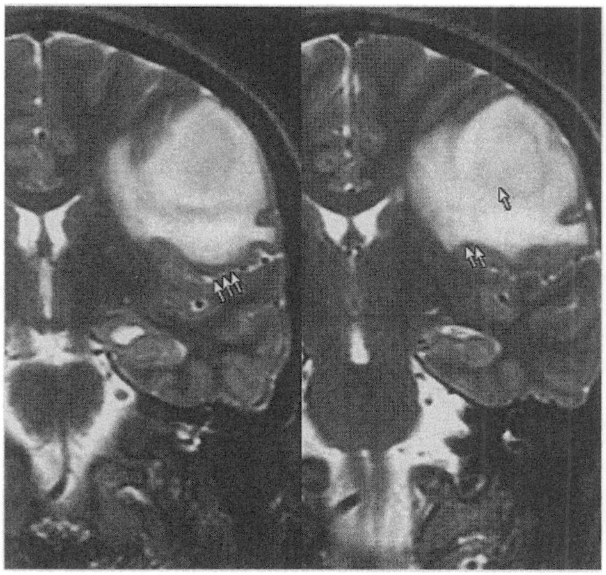

c

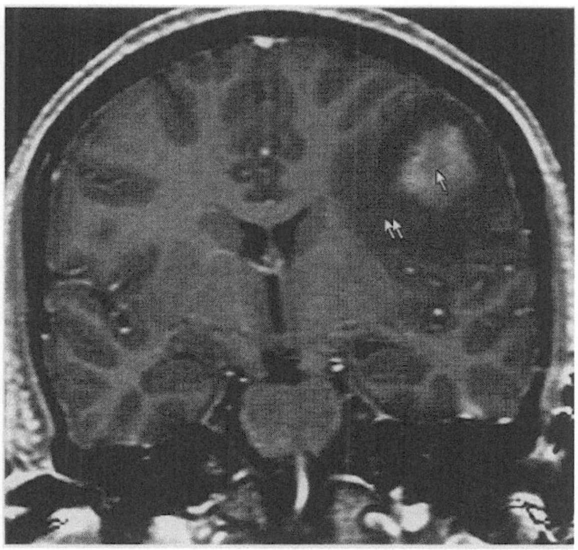

d

Case 5: Fronto-callosal oligodendroglioma, grade II

Fig. a: sagittal cut; GE - T1w

Fig. b: axial cut; SE - T2w

Fronto-callosal poorly demarcated and heterogeneous tumor with areas of necrosis (↑). The lesion crosses the corpus callosum (↑↑) and shows extensive high intensity on T2w, representing tumor plus edema, and is heterogeneously enhancing.

Case 6: Frontal oligodendroglioma, grade II

Fig. c: sagittal cut; SE - T2w

Fig. d: axial cut; GE (3D) - T1w + Gd

Superficial orbito-frontal heterogeneous infiltrating tumor (↑) showing hyposignal on T1w (**d**) sequence, high signal intensity on T2w (**c**) with areas of cystic degeneration (*) and no enhancement. The lesion shows a subpial extension with remodeling of the inner table of the skull (**c**).

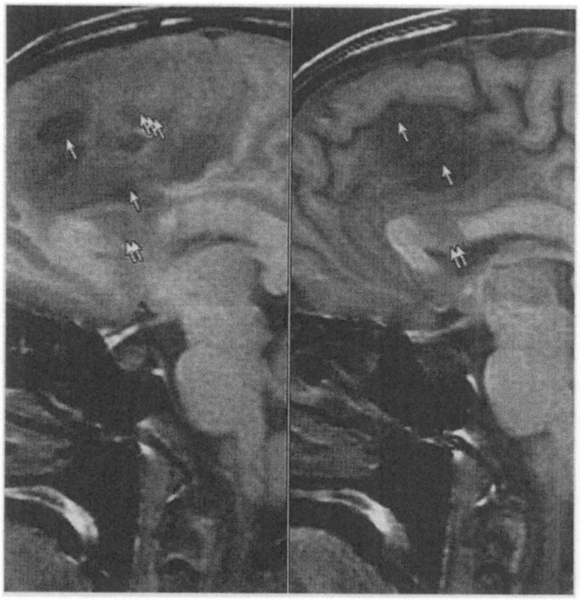

a

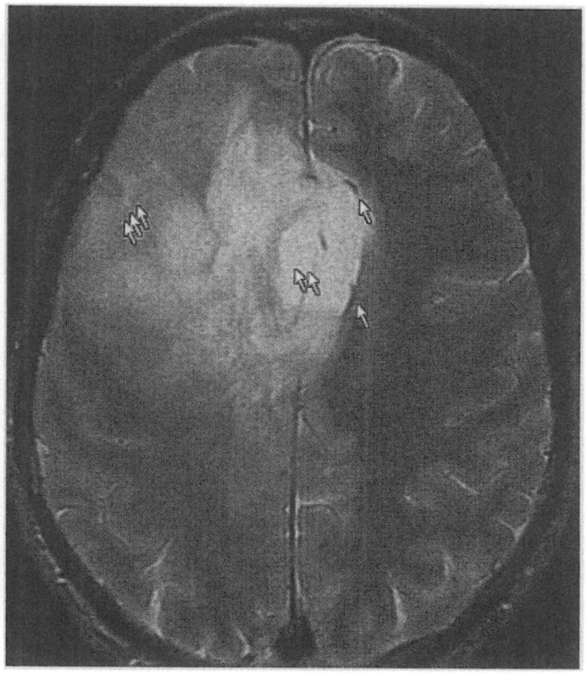

b

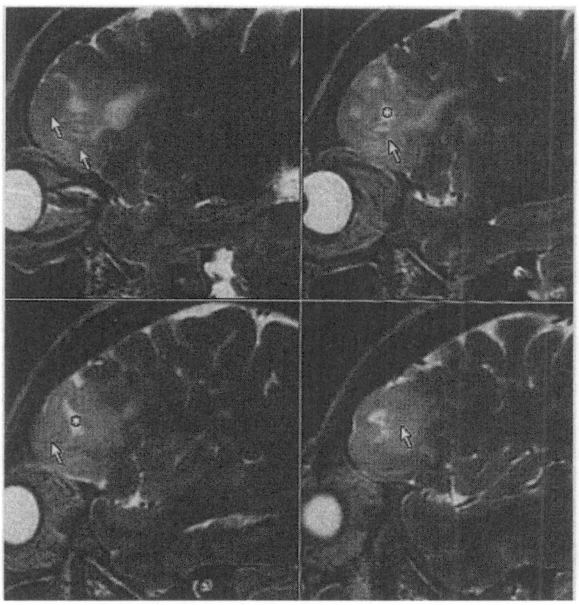

c

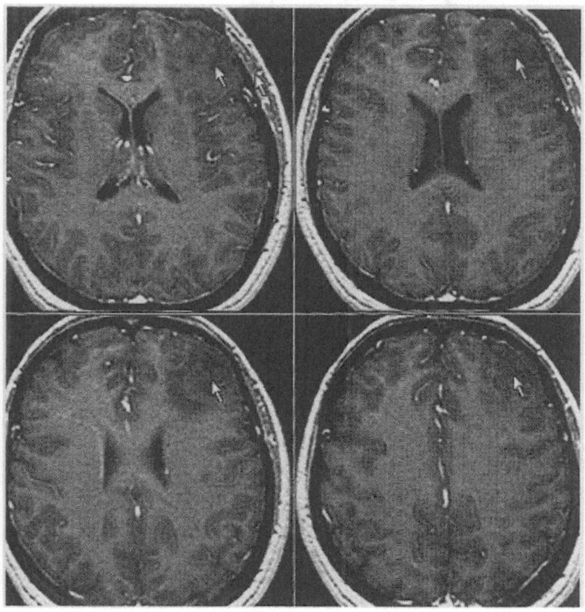

d

Case 7: Oligoastrocytoma, grade II

Fig. a: sagittal cut; GE - T1w + Gd

Fig. b: axial cut; SE - PDw

Left fronto-insular infiltrating and well demarcated tumor, homogeneously hyperintense on PDw sequence (**b**, ↑) and hypointense on T1w without enhancement but with cystic component (**a**, ↑) and no peritumoral edema. Note the mass effect on the lenticular nucleus and the insular cortex.

Case 8: Oligoastrocytoma, grade II

Fig. c: axial cut; SE - PDw and T2

Fig. d: axial cut; FLAIR - T2w

Fronto-insular infiltrating mixed low grade glioma, hyperintense on PDw and T2w (**c**) sequences as well as on FLAIR (**d**). Note the mass effect on the pallidum, hypointense on the FLAIR sequence (↑↑).

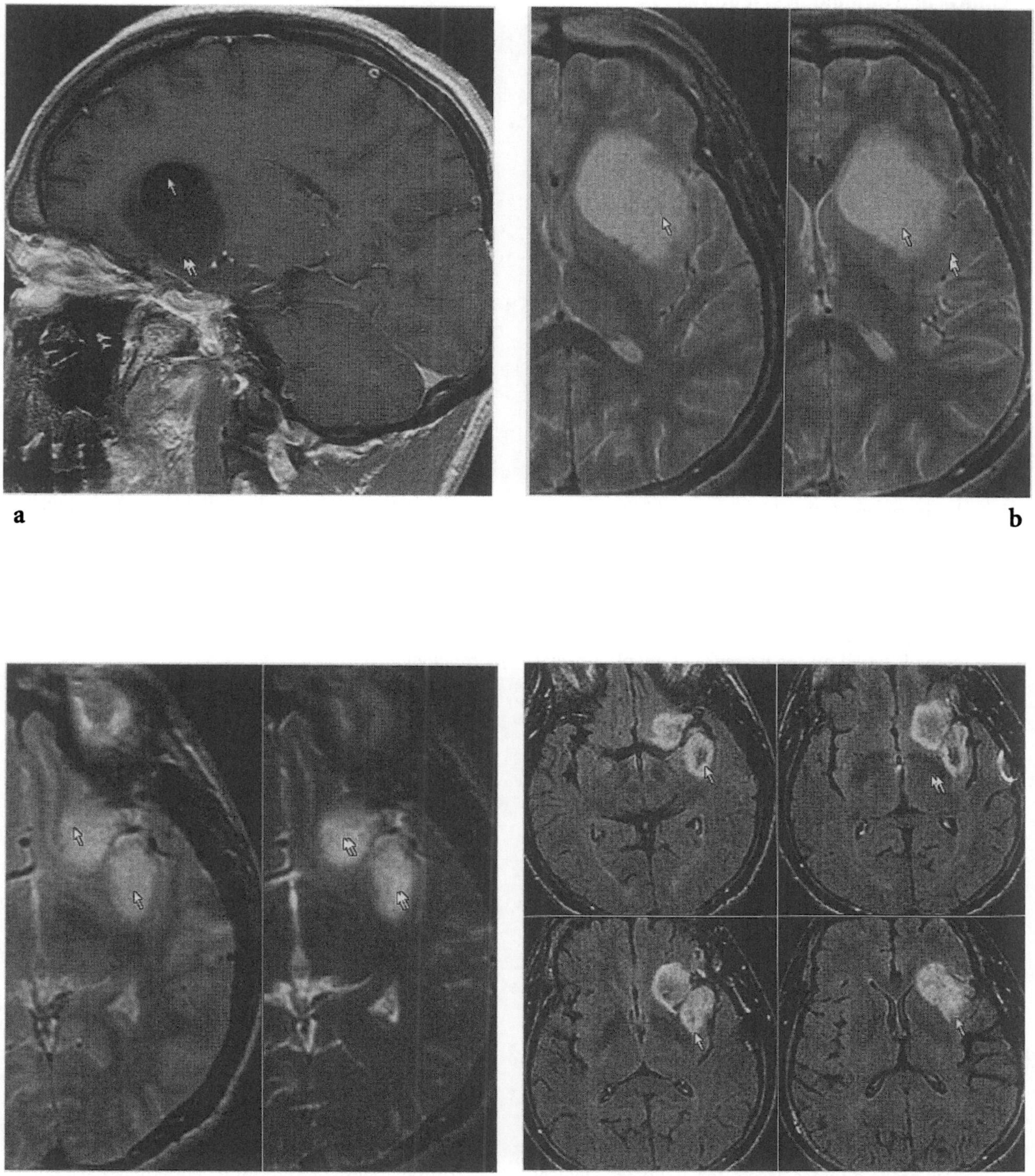
a
b
c
d

Cases 9-12: high grade astrocytomas

Clinical and pathological features

- Malignant astrocytic tumors are represented by the anaplastic astrocytomas (WHO grade III) and the glioblastomas (WHO grade IV).
- Anaplastic astrocytomas are diffuse fibrillary astrocytomas of intermediate form showing on histopathological preparations, focal areas of anaplasia (roughly half the cells) and focal or diffuse increased cellularity (by 50% or more). They typically display mitotic activity. The microcystic changes found in low grade tumors are lacking here. They may arise *de novo* or develop from low grade tumors by malignant transformation.
- Anaplastic astrocytomas show a peak incidence in the 5th decade, with males more frequently affected than females (almost 2:1, male-female ratio).
- Glioblastoma multiformis is the most frequent brain tumor (50-60% of astrocytic tumors in adults) and the most malignant of diffuse astrocytomas. They usually arise as primary tumors but may develop as secondary tumors from low grade diffuse or anaplastic astrocytomas. Histopathologically, the diagnosis of glioblastoma is based on the presence of anaplastic glial cells, frequent mitotic activity, markedly increased vascular proliferation (peripherally or around the necrotic zones), and frequent areas of necrosis (usually in the center of the tumor). Apparent multicentric tumors are reported in as much as 5-15% of the cases, presumably due to the presence of multifocal zones of anaplasia within a low grade tumor. Multiple glioblastomas present as independent tumors in about 2.5 % of the cases.
- Glioblastomas are found at any age, but most commonly in adults between ages 45-70 (mean around 50 years old), and rare in patients under age 30 (less than 3% reported). The incidence in men is more frequent than in women (1.5 to 2: 1). They are found most often in the cerebral hemispheres.
- *Signs and symptoms*: the clinical picture is similar in anaplastic astrocytomas to that presented in low grade gliomas except for a shorter medical neurological history preceding diagnosis and a higher mean age of 41 years (Zurich series). The duration of the clinical history decreases dramatically in glioblastomas developed *de novo*, becoming as short as 3-6 months in almost half the cases, as compared to those developing from preexisting low grade diffuse or anaplastic astrocytomas (4-5 years of duration of the neurological history). In glioblastomas, the symptomatology develops rapidly in a few weeks or months including generalized neurological symptoms and seizures followed by more definite lobar or callosal syndromes according to brain localization. The onset may even be sudden in 3-5%, pseudovascular in progression usually due to intratumoral hemorrhage or cystic formations within the tumor, and leading to the development of focal cerebral signs. Glioblastomas show a very poor prognosis with less than 20% of the patients surviving for 1 year and 10% beyond 2 years after the onset of the clinical symptomatology. Younger patients, less than 45 years old, appear to have a better prognosis than the older, presumably because of the higher frequence of secondary glioblastomas in such younger population. Longer survival would be related to complete resection when possible.

MR characteristic findings

Morphology

- Anaplastic astrocytomas are morphologically often difficult to distinguish from low grade astrocytomas, but appear more heterogeneous. As in diffuse fibrillary astrocytomas, the tumor shows tendency to infiltrate the anatomical structures involved without obvious distortion except for relative enlargement of the corresponding gyri or deep gray nuclei. Vascular proliferation and necrosis are lacking. Occasionally, proliferation of tumor vessels may be observed with flow void foci. Note that anaplastic astrocytomas are found in half the cases of recurrent tumors (malignant transformation). Peritumoral edema may at times be found, but remains a feature of glioblastomas.
- Glioblastomas occur most commonly in the cerebral hemisphere located most frequently in the white matter (epicenter of the tumor) of the frontal lobes, or temporal, and then parietal lobes. Some of these tumors are rather superficial in topography in contact with the leptomeninges and dura (subarachnoid seeding). Cortical infiltration may occur giving rise to a

thickened appearance of gyri surrounding a necrotic area in the subjacent white matter. Brainstem glioblastomas much less frequently occur, mainly in children.

- Glioblastoma commonly appears as an irregular almost well circumscribed mass with highly suggestive central necrosis (as compared to the anaplastic type), surrounded by relatively large areas of peripheral edema. The cellular peripheral zone is highly vascularized, corresponding to zone of "ring" enhancement. The large areas of intratumoral necrosis secondary to destruction of tumoral tissue may correspond to as much as 90% of the total tumoral volume, and appear as characteristic of primary glioblastomas. Hemorrhagic changes are common in the central portion of the tumor. Calcifications are very rare. Mass effect associated to prominent peritumoral edema may be prominent (sulcal effacement or ventricular displacement or collapse).
- Glioblastomas show a high propensity to rapidly expand along the white matter tracts to invade adjacent areas of the cerebral white matter, or spread in the internal capsule or the fornix. The tumor may also typically cross the midline through the callosal fibers bundles to reach the contralateral hemisphere ("butterfly" appearance secondary to ependymal spread, which may also be encountered in lymphomas, but the latter exhibiting a hypointense SI on T2 due to hypercellularity) or through the anterior commissure. Glioblastoma may show subependymal or leptomeningeal spread (best demonstrated after contrast enhancement).

Signal intensity

- Anaplastic astrocytomas: homogeneous SI with slightly heterogeneous areas
  - GE-T1w: hypointense to white matter;
  - SE-PDw: hyperintense to cortex;
  - SE-T2w: hyperintense to white matter;
  - FLAIR: hyperintense to white matter.
- Proton spectroscopy has been reported in anaplastic astrocytomas to show an increase of Cho levels, a decrease of NAA levels and lipid resonance peaks.
- Glioblastoma multiforme: heterogeneous SI
  - GE-T1w: hypointense to white matter with even more hypointense intratumoral necrotic zones, unless presence of hemorrhagic areas (may be hyperintense due to eventual methemoglobin);
  - SE-PDw: hyperintense to white matter, with areas of central necrosis even showing a more hyperintense SI than the solid part (due to high protein content and blood breakdown products such as extracellular methemoglobin) and the extensive peritumoral "vasogenic" edema;
  - SE-T2w: variable SI of the solid portions of the tumor may be mildly hyperintense to brain and/or show iso- to hypointense areas due to tumoral hypercellularity. Serpentine flow void structures corresponding to vessels (tumoral angiogenesis) are also found;
  - FLAIR: mild hyperintensity to normal brain tissue (with variable intensity of the intratumoral necrotic areas) associated to markedly hyperintense SI of peritumoral vasogenic edema. Tumor may show a well circumscribed appearance (similar to enhanced scans), and roughly well separated from the vasogenic edema (case 11).
- Proton spectroscopy has been reported in glioblastomas to show an increase in Cho levels, a decrease in NAA, and the presence of peaks of lactate and lipid. A significant correlation has been observed between higher grade glial tumors and mobile lipid resonance peaks at 1.3 ppm.

Contrast enhancement

- Anaplastic astrocytomas: variable patterns of partial enhancement are observed, ranging from homogeneous to even nodular and occasionally ring-like (more suggestive of a glioblastoma).
- Glioblastoma multiforme: heterogeneous enhancement, with tumoral enhancement more sensitive than with CT, showing a thick and irregular ring-like highly heterogeneous pattern (even more conspicuous with delayed scans), surrounding the areas of necrosis. Contrast enhancement corresponds to the tumoral process but does not limitate the outer tumoral boundary as shown on pathology. The non enhancing surrounding edema has been demonstrated to include microscopic tumor to the edematous changes. The extent of such microscopic adjacent peritumoral zone appears to be most frequently limited to 2 cm, which corresponds to the zone of tumor recurrence as observed in more than 85% of the cases, but may extend to more than 2 cm found as far as 3 or 3.5 cm in 17.1 - 27.3% of the cases (according to CT studies, pathological correlated to imaging studies and stereotactic biopsies).
- The aspect may mimic other malignant tumors such as metastases, anaplastic oligodendrogliomas, lymphomas, or other processes as radiation necrosis, cerebritis and abscesses.

Case 9: Temporo-occipital glioblastoma

Fig. a: axial cut; SE - PDw and T2

Fig. b: axial cut; GE - T1w + Gd

Left temporal highly heterogeneous tumor with central areas of necrosis (↑) associated with hypervascularity. The lesion shows significant mass effect on the left lateral ventricle mainly due to extensive edema in the adjacent white matter (↑↑). Note the heterogeneous peripheral thick and irregular enhancement surrounding the necrotic areas (**b**).

Case 10: Temporal glioblastoma

Fig: c: axial cut; SE - T2w

Fig. d: axial cut; SE - T1w + Gd

Anterior left temporal apparently well delineated tumor, slightly hyperintense on T2w (**c**) with posterior peri-tumoral edema showing a thick ring of enhancement (**d**) surrounding central necrosis (↑↑). Note the mass effect on the uncus and the chiasmal cistern as well as on the left cerebral peduncle (mesial temporal herniation).

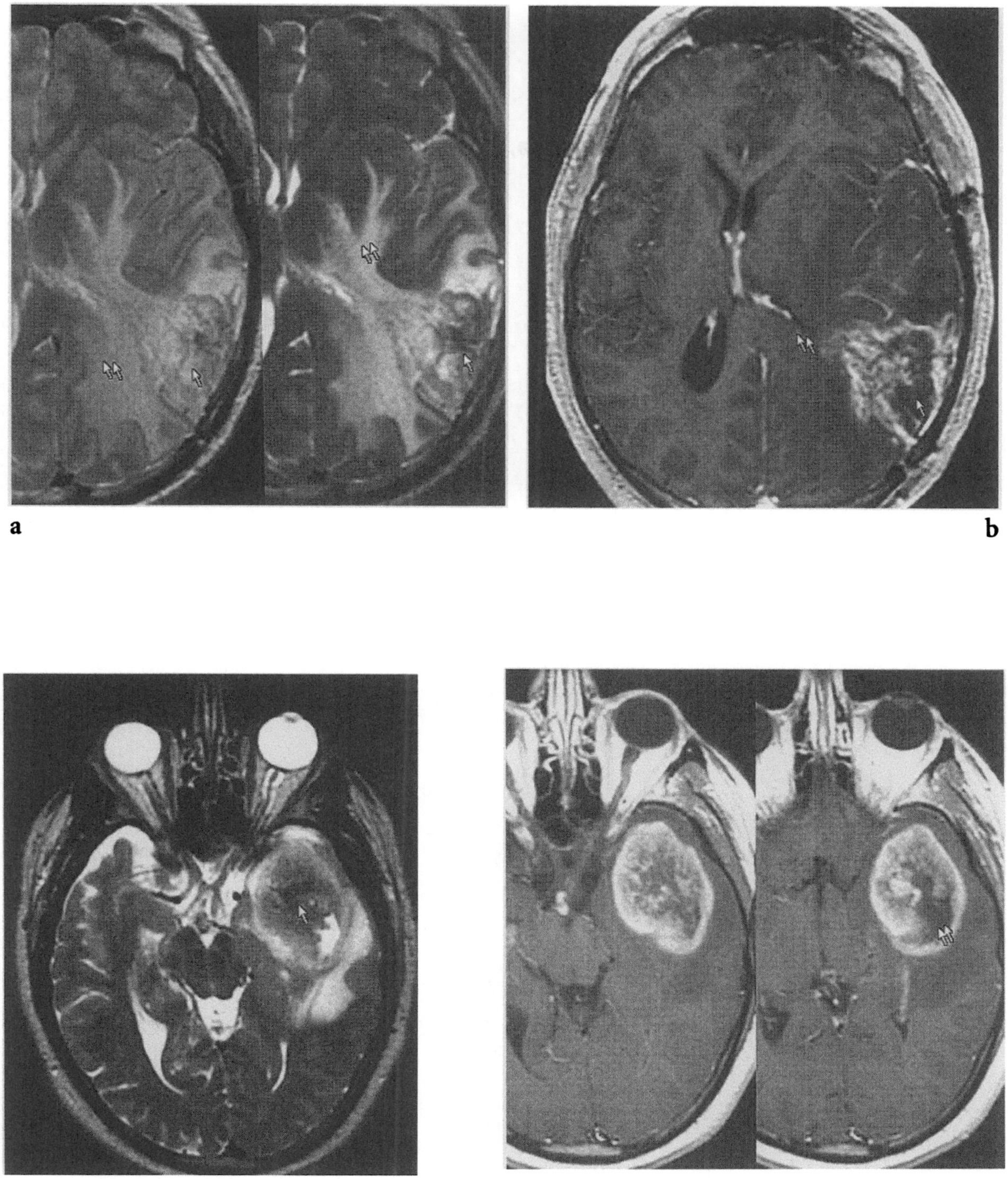

a b c d

Case 11: Temporo-parieto-occipital astrocytoma, grade IV

Fig. a: sagittal cut:	GE - T1w	**Fig. c:** axial cut:	SE - T1w + Gd
Fig. b: axial cut:	SE - T2w	**Fig. d:** axial cut:	FLAIR - T2w

Highly heterogeneous, apparently well-defined temporo-parieto-occipital tumor showing areas of cystic necrosis (**b**, ↑) surrounded by extensive edema, hypointense on T1w and hyperintense on T2w (**b**, ↑↑↑) as well as on FLAIR (↑). The lesion shows irregular enhancement delineating areas of necrosis (**c**).

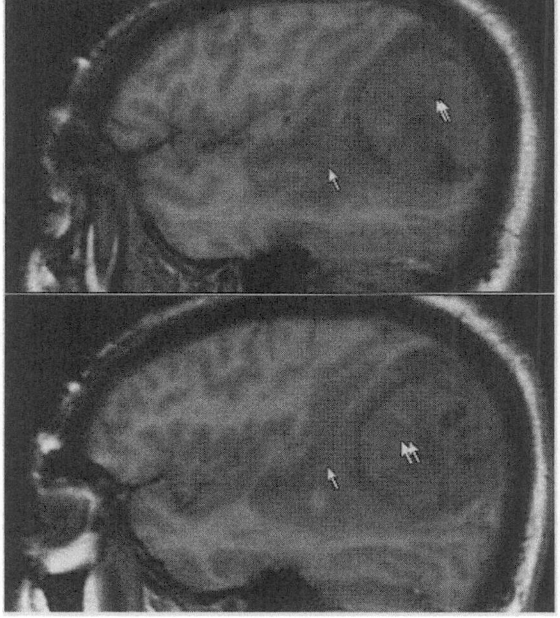

a

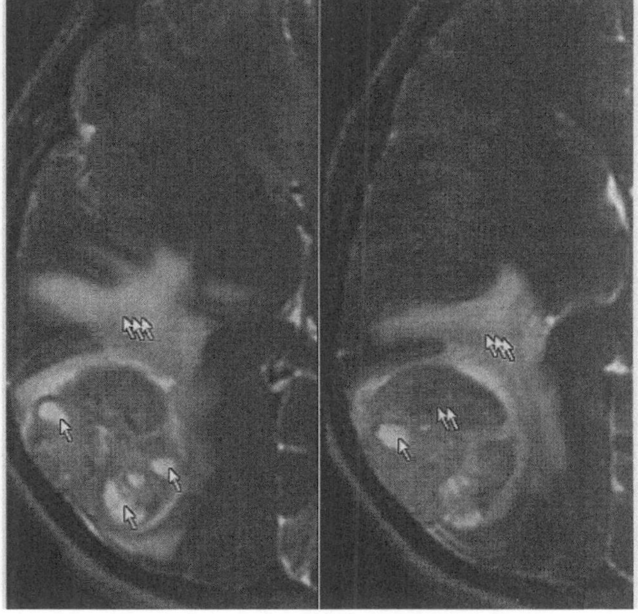

b

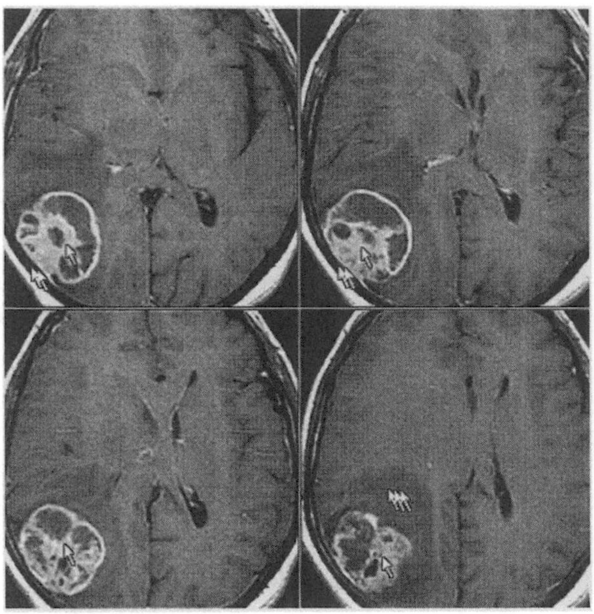

c

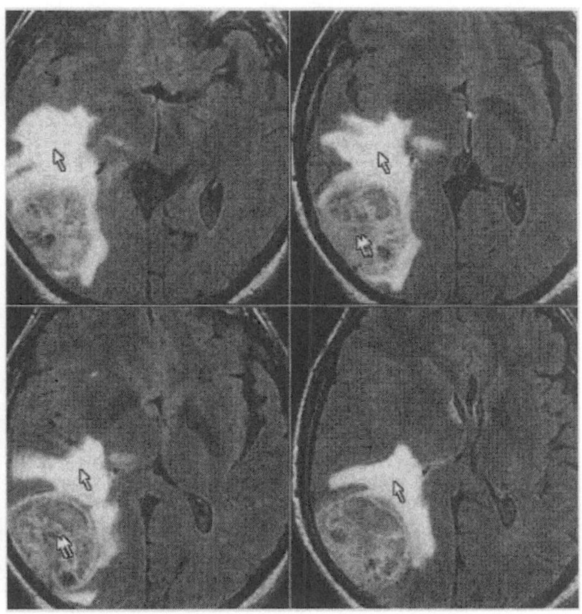

d

Case 12: Temporal glioma, grade III - IV

Fig. a: sagittal cut;	GE - T1w	**Fig. c:** coronal cut (512);	SE - T2w
Fig. b: axial cut;	SE - T2w	**Fig. d:** coronal cut;	SE - T1w + Gd

Huge right temporal tumor showing hypointense signal on T1w sequence (**a**) and hyperintense signal on PD w (**b**, left) as well as on T2w (**b**, right and **c**) sequences. The lesion shows irregular thick peritumoral enhancement (**d**) allowing to depict a subependymal extension of the process involving the lateral ventricle (*) as well as the septum-pellucidum (↑↑ **d** and **c**). Note the mass effect on the external temporal sulci (↑↑↑ **c**).

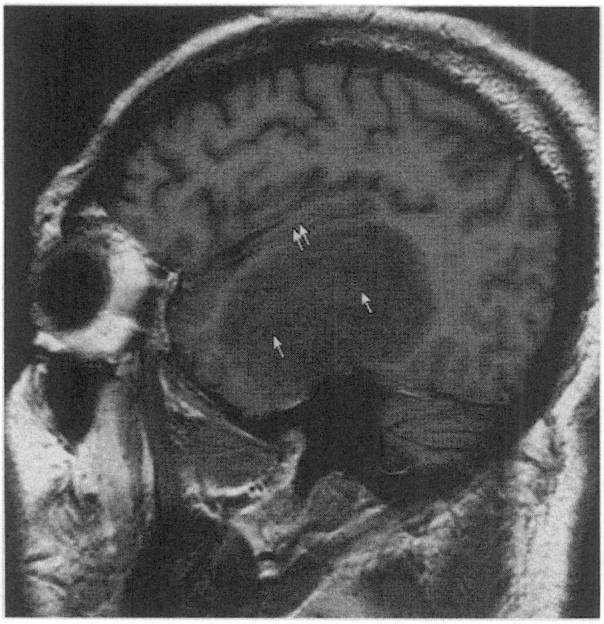

a

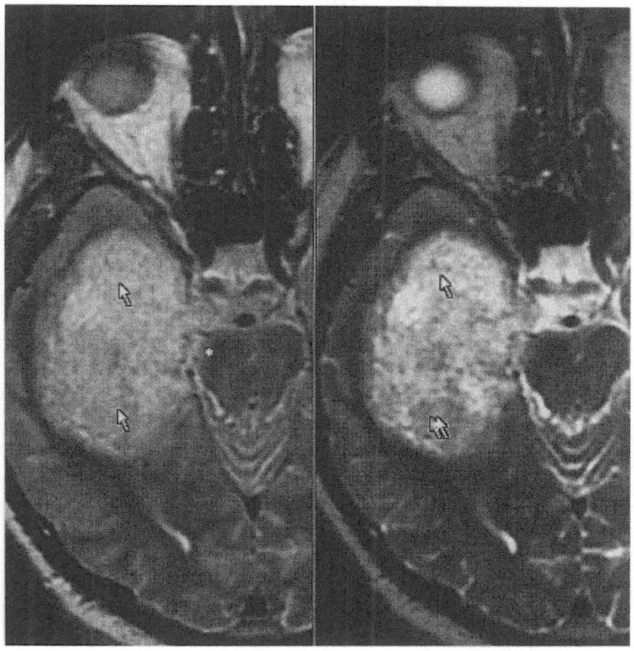

b

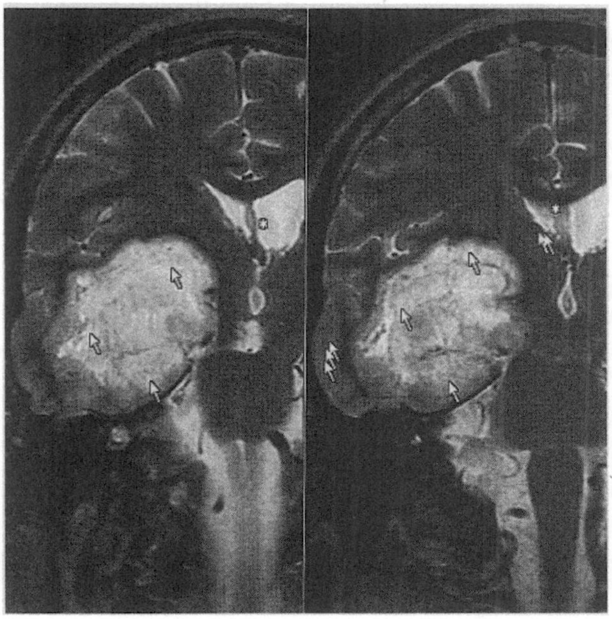

c

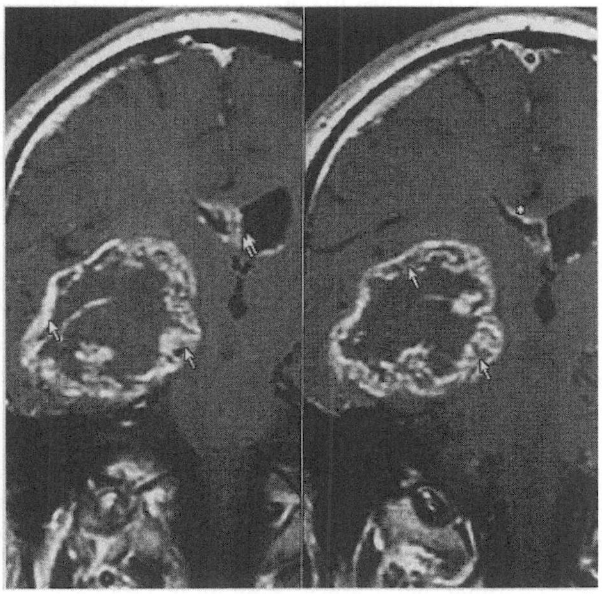

d

Case 13-15: intraventricular and subependymal tumors

Clinical and pathological features

- Intraventricular tumors are of variable types comprising: ependymoma, oligodendroglioma, astrocytoma, subependymal giant cell astrocytoma, and subependymoma, central neurocytoma, meningioma, choroids plexus papilloma, epidermoid cyst, and arachnoid cyst.
- *Central neurocytoma* (WHO grade II) is typically a relatively benign intraventricular neuronal tumor, occurring in young adults between 20-40 years of age (mean age, 29), located in the lateral ventricles and attached to the septum or in the third ventricle. Its incidence is 0.25-0.5% of all intracranial tumors. Histopathologically the lesion mimics an oligodendroglioma. The tumors show a usually benign clinical course and a fairly good prognosis. The presenting signs are those of increased intracranial pressure. At times, endocrine dysfunctions and/or visual defects develop (third ventricular location).
- *Ependymomas* are low grade tumors (WHO grade II) formed of neoplastic ependymal cells occurring mainly in childhood and in young adults, and developing in the ventricular cavities. The most frequent signs observed are those caused by the obstructive hydrocephalus and cerebellar signs. Ependymomas may include necrotic or hemorrhagic foci, calcifications and areas of myxoid degeneration, which account for the heterogeneous aspect (MR). Anaplastic ependymomas (WHO grade III) which may develop secondary to malignant transformation of ependymomas, represent almost 1/4 of all ependymal tumors.
- *Subependymomas* are benign tumors (WHO grade I) incidentally found in middle aged or elderly males, and usually asymptomatic. They are formed of a mixture of astrocytes and ependymal cells.
- *Subependymal giant cell astrocytomas* are WHO grade I tumors, formed of large ganglioid astrocytes, occurring during the first two decades and considered as the most frequent CNS tumor in Tuberous Sclerosis, ranging in frequency between 5 to 15% (as compared to subependymal nodules found in 90-100% of the cases). Histopathologically the tumor is formed of large ganglioid astrocytes, giant pyramidal cells and multinucleated cells. The tumor arises in the lateral ventricle near the foramen of Monro (obstruction of the foramen leading to increased intracranial pressure). The tumor shows a fairly benign course.
- *Choroid plexus papillomas* (WHO grade I) are benign, slow growing tumors developing in the ventricular cavities and arising from the epithelium of the choroid plexus. This tumor accounts for 10-20% of brain tumors found in infants during the first year of life. The other choroid plexus tumor variety is the choroid plexus carcinoma which is a malignant variant with very poor prognosis (WHO grade III) arising in children in 80% of the cases. The overall incidence of choroid plexus tumors is about 0.5% of all brain tumors.

MR characteristic findings

Morphology

- Central neurocytomas are characteristically found in the anterior part (region of Monro foramen) of the lateral ventricle in almost half the cases, mainly on the left side, and bilaterally in 10%. The tumor may develop also in the third ventricle (3%). Extension from the lateral ventricle to the third is observed in 15% of the cases. The tumor is heterogeneous with possible intratumoral calcifications and cystic formations. Differential diagnosis: oligodendrogliomas, ependymomas, subependymal giant cell astrocytoma.
- Ependymomas arise most frequently in the 4th ventricle extending to the cerebellopontine cistern (see posterior fossa tumors) and in the lateral ventricles from which they tend to invade the adjacent brain parenchyma.
- Subependymomas are most frequently found in the fourth ventricle in more than half to almost 2/3 of the cases or in the lateral ventricle. Usually well circumscribed, they may show intratumoral hemorrhage. Of variable size they commonly do not exceed 2 centimeters. The anaplastic ependymomas, which may develop in the intraventricular cavities show a greater propensity than ependymomas to invade the brain and seed the CSF spaces.
- Subependymal giant cell astrocytomas are found at the foramen of Monro, usually well delineated. The tumor does not infiltrate the adjacent

parenchyma nor seeds the CSF spaces. It may cause an obstructive hydrocephalus. The tumor may show at times a massive hemorrhage.

- Choroid plexus papillomas most commonly arise in the lateral ventricles (50%) or the fourth ventricle (40%), and less frequently in the third ventricle. The tumor shows a "cauliflower-like" aspect with lobulated margins, developing within the ventricular cavity (atrium) and fairly well delineated. The tumor may show intratumoral calcifications found in about half the cases and areas of hemorrhages or cystic formations. Enlarged blood vessels may be demonstrated as curvilinear or punctate flow voids. Overproduction of CSF and obstruction of CSF pathways cause frequently associated hydrocephalus. Multifocal tumors may be found in almost 5% of the cases. Choroid plexus carcinomas show a more malignant behavior with regional invasion of the adjacent white matter and associated vasogenic edema. Ventricular and/or leptomeningeal dissemination is more frequent than in papillomas. The tumor may become huge and show intratumoral necrosis and hemorrhages.

Signal intensity

- Central neurocytoma:
  - GE-T1w: isointense to cortex;
  - SE-PDw: hyperintense to cortex;
  - SE-T2w: slightly hyperintense or isointense to cortex.
- Ependymoma: marked heterogeneity
  - GE-T1w: isointense and/or hypointense to brain;
  - SE-PDw: isointense to cortex;
  - SE-T2w: hypointense and/or isointense to white matter.
- Subependymoma:
  - GE-T1w: iso- or hypointense to brain;
  - SE-PDw: hyperintense to gray matter;
  - SE-T2w: hyperintense to brain.
- Subependymal giant cell astrocytoma:
  - GE-T1w: isointense or slightly hyperintense to brain;
  - SE-T2w: slightly hyperintense to brain.
- Choroid plexus papilloma: heterogeneous
  - GE-T1w: iso- or hypointense to gray matter;
  - SE-PDw: isointense to gray matter;
  - SE-T2w: hypointense or isointense to gray matter.

Contrast enhancement

- Central neurocytoma usually shows moderate to intense contrast enhancement.
- Ependymoma usually shows heterogeneous contrast enhancement.
- Subependymoma do not show contrast enhancement.
- Subependymal giant cell astrocytoma shows a marked enhancement.
- Choroid plexus papillomas and choroid plexus carcinomas show intense contrast enhancement.

Case 13: Intraventricular ependymoma

Fig. a: sagittal cut (512); GE - T1w + Gd

Fig. b: coronal cut (512); S.E. - T2w

Fig. c: coronal cut (512); SE - PDw

Fig. d: coronal cut (512); SE - T2w

Left posterior intraventricular mass showing iso-signal on T1w (**a**, ↑) as well as on T2w (**b**, ↑) sequences showing no contrast enhancement (**a**, ↑) but associated to a dilatation of the ipsilateral ventricle with displacement of the septum-pellucidum to the right. The lesion is adherent to the lower aspect of the corpus callosum (**a**,**;**b**, ↑↑).
Huge and heterogeneous intraventricular tumor at the level of the interventricular foramen showing areas of cystic degeneration (↑↑) as well as dense calcifications and intratumoral vessels (↑). The lesion is iso-intense on PDw (**c**) and slightly hyperintense on T2w (**d**) sequences, and invades the third ventricle through the foramen of Monro (**d**, ↑). The mass adheres to the lateral aspect of the lateral ventricle (**d**, ↑↑↑), but does not invade the internal capsule.

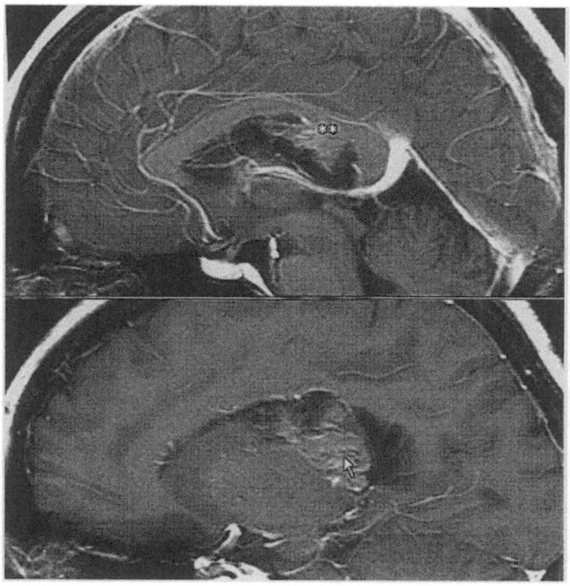

a

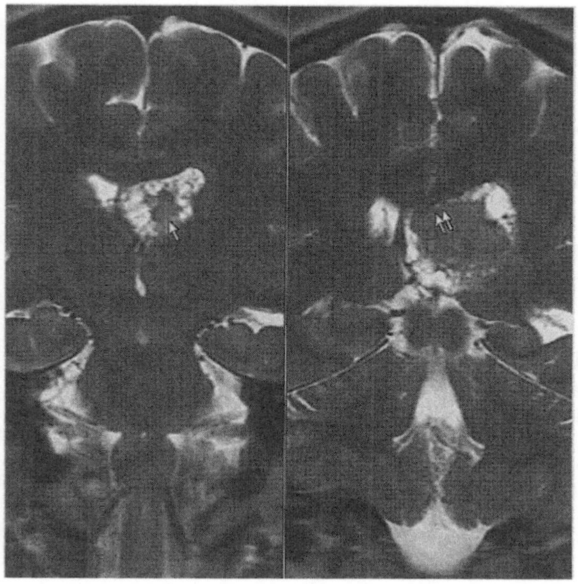

b

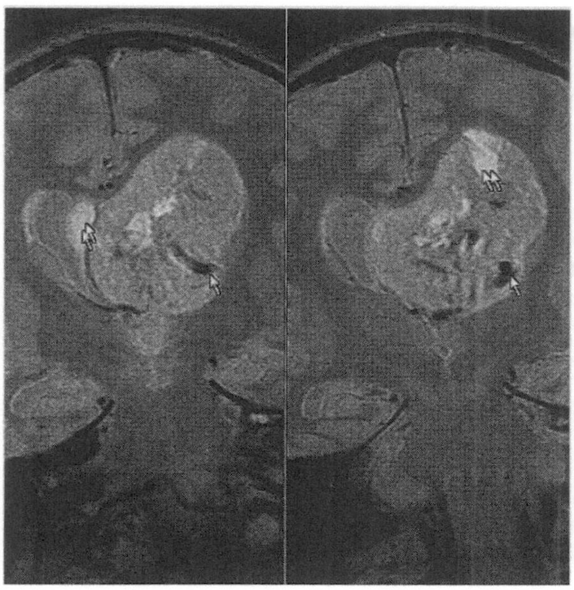

c

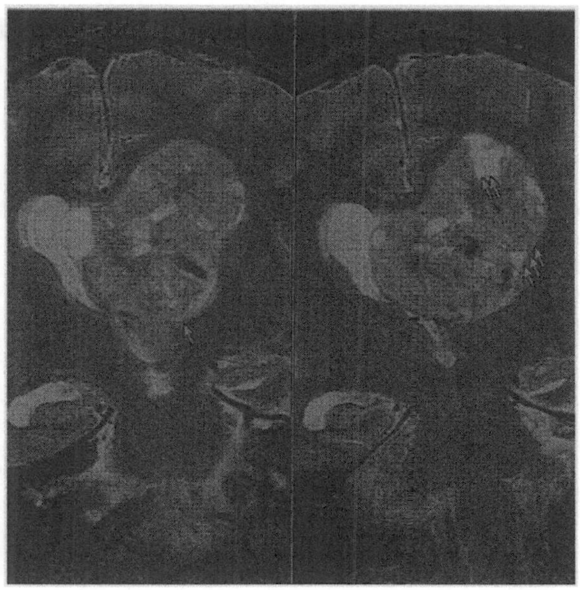

d

Case 14: Intraventricular neurocytoma

Fig. a: sagittal cut (512); GE - T1w + Gd

Fig. b: coronal cut; SE - T2w

Left intraventricular tumor in the frontal horn presenting an iso-signal to brain cortex on T1w (**a**, ↑) and a hyperintense signal on T2w (**b**, ↑). The tumor obstructs the left foramen of Monro with bilateral ventricular dilatation, more prominently on the left side.

Case 15: Subependymal giant cell astrocytoma (Bourneville)

Fig. c: sagittal cut; GE - T1w + Gd

Fig. d: coronal cut; SE - T2w

Small lateral ventricular mass in the region of the foramen of Monro projecting into the right frontal horn, presumably from a subependymal location. The mass is heterogeneous, showing focal areas of necrosis (**d**, ↑). The lesion, slightly hyperintense on T2w (**d**), enhances with IV contrast (**c**, ↑). The frontal horns are dilated and there is a mass effect on the third ventricle and the foramina of Monro (↑↑). The patient presented elsewhere subependymal heterotopic nodules (tubers), some of which are calcified.

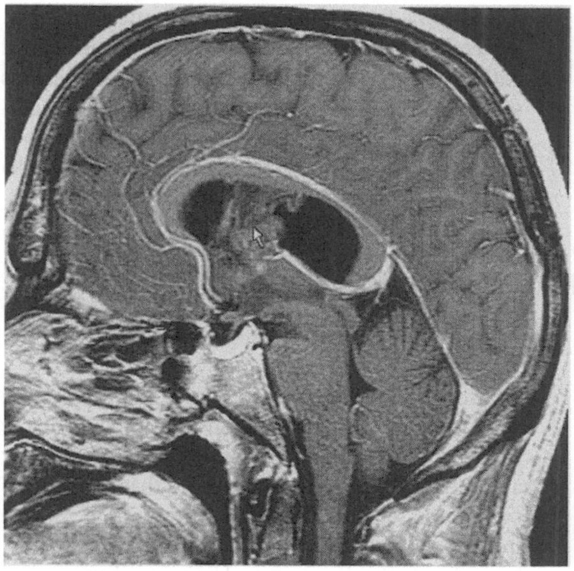

a

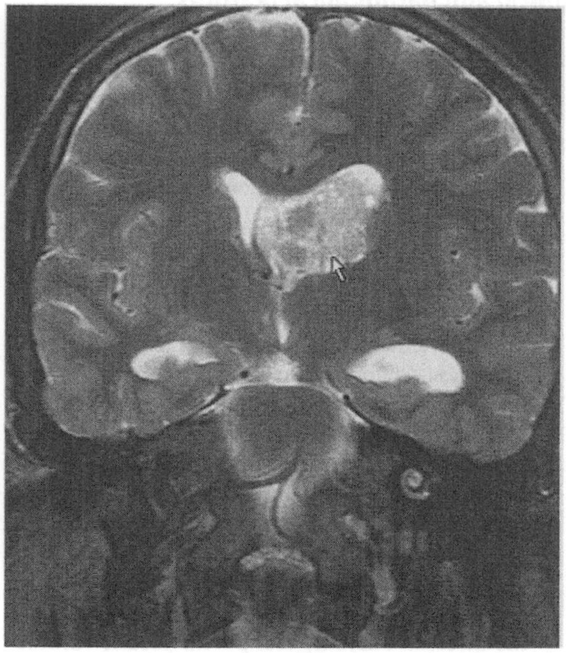

b

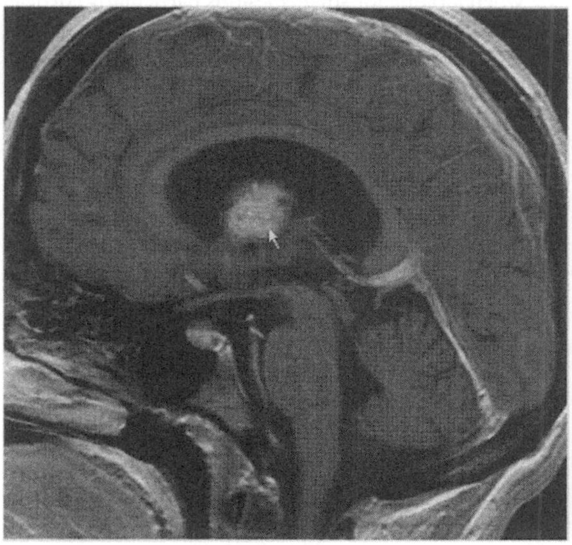

c

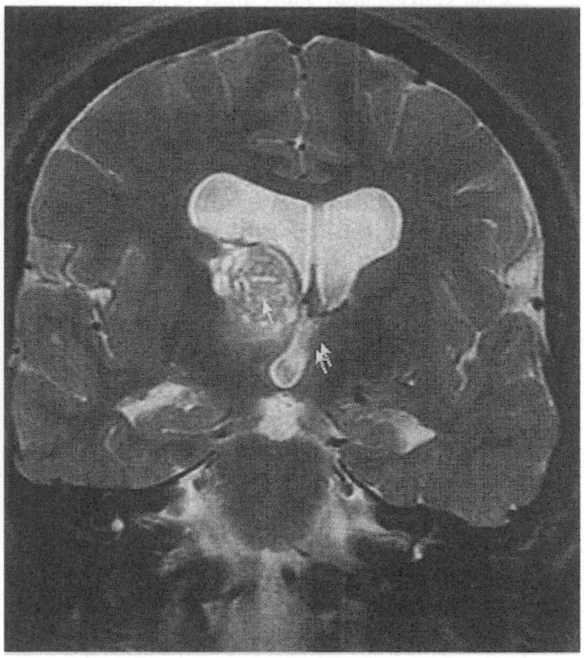

d

Case 16-19: mesial temporal tumors and cortical dysplasias

Clinical and pathological features

- MR is the procedure of choice and the standard of reference in the investigation of patients presenting intractable seizures and particularly those who present with temporal lobe epilepsy. It is an important tool for the clinical evaluation of patients with partial epilepsy.
- In most patients with temporal lobe epilepsy, involvement of limbic or temporal neocortical structures must be rule out.
- Most patients referred with intractable epilepsy have mesial temporal sclerosis. Hippocampal sclerosis is associated with prolonged convulsions in childhood; furthermore, cortical developmental disorders are more frequent than expected. Between 15-30% of patients with mesial temporal sclerosis have associated cortical abnormalities. MR diagnosis of mesial sclerosis necessitates to perform high resolution optimized imaging and to proceed to very careful morphological analysis of the diseased anatomy along with precise volumetric measurements most often to confirm the visually assessed abnormalities.
- Apart from mesial sclerosis, patients with intractable partial epilepsy may show other epileptogenic processes in the temporal lobes such as tumors or vascular (see sections on cavernous angioma and AVM) and developmental (focal cortical dysplasias) lesions. Tumors are represented most frequently by gliomas of the astrocytic, predominantly fibrillary but possibly protoplasmic type, or of the oligodendroglial type, followed less commonly by the mixed neuronal-glial or neuronal tumors such as the gangliogliomas, gangliocytoma and the dysembryoplastic neuroepithelial tumor (DNET).
- The majority of patients with DNETs or gangliogliomas are less than 20 years old. Tumor recurrence after surgery is usually low.
- *Signs and symptoms*: temporal lobe epilepsy may be classified into 5 types according to Wieser: 1- hippocampal-amygdalar or temporobasal; 2- temporal polar or amygdalar-polar; 3- lateral temporal neocortical; 4- opercular-insular, and 5- frontobasal-cingulate. In the first 3 types the onset of seizure is in the temporal lobe whereas in the 2 remaining, seizures arise outside and spread to the temporal lobe. In the most frequent hippocampal-amygdalar type, auras (déjà vu, fear, epigastric sensation without loss of consciousness) are reported in almost 80% of the patients and are most often followed by motor or speech arrest. Seizures are similar in the tempolar-polar type with the adjunction of oroalimentary automatisms, and autonomic disturbances. Complex visual hallucinations and vestibular symptoms are found with lateral temporal neocortical epilepsy. Visceral and auditory auras are frequently associated with the opercular-insular type. Motor automatisms with loss of awareness characterize frontobasal-cingulate seizures. Depth EEG shows that half of temporal lobe seizures arise from the hippocampus, 10% from the amygdala and less than 5% from the neocortex.

MR characteristic findings

Morphology

- In order to identify and recognize the underlying structural abnormality responsible for intractable partial epilepsy, a good knowledge of brain anatomy and the particular lesional processes that may potentially cause seizures, is needed. These lesions may be frequently very subtle in nature and could be overlooked or misdiagnosed. High resolution and contrast coronal MR cuts of the brain and the temporal lobes performed in the accurate cephalic reference plane (PC-OB reference plane) are therefore mandatory. Using surface coils as an additional tool to improve lesional detectability may be of great help when looking for smaller lesions particularly cortical, usually dysplastic. It is obvious that not all structural lesions can be depicted using MR (microdysgenesis).
- *Gangliogliomas and gangliocytomas:*

  - Most of these tumors are found in patients younger than 16 years old, affecting children and adolescents, but have been reported in patients ranging in age from few months to 80 years. Their incidence is slightly higher in males.

  - Gangliocytomas are very rare tumors, formed of neoplastic mature ganglion cells (WHO grade I). Gangliogliomas show the presence of mature ganglion cells with neoplastic glial cells (WHO grade I or II). They represent together 1.3% of all brain tumors.

  - The tumors may occur anywhere but are commonly found in the temporal lobes and tend to spare the lateral neocortex.

  - When superficially located, the tumor may cause thickening of the cortical ribbon and cause a scalloping of the calvarium.

  - Gangliogliomas and gangliocytomas are usually well circumscribed masses, solid or almost cystic. Intratumoral cysts are often multiple.

  - There usually is no obvious mass effect and no peritumoral vasogenic edema.

  - In gangliogliomas, the tumor may involve the leptomeninges and the glial component of this mixed tumor may contribute to give a slight infiltrative appearance of the brain-tumor interface.

- Calcifications are common (1/3), but less frequent than in oligodendrogliomas, and more than in juvenile pilocytic astrocytomas or in pleomorphic xanthoastrocytomas).
- May be associated with mesial temporal sclerosis when located anteriorly.

- *Dysembryoplastic neuroepithelial tumors (DNET):*
  - These tumors are highly associated with long standing intractable epilepsy (85% of patients before age of 15 years). Patients show no associated focal neurological defects or present other congenital neurological abnormality.
  - DNETs are benign, mixed glial-neuronal tumor, predominantly cortical in location. They correspond to WHO grade I lesions. These tumors are relatively rare (more than 300 cases reported).
  - About 50-60% of DNETs are found in the temporal lobes (involving the mesial structures preferentially) and 30% in the frontal lobes. Few were in the parietal or occipital lobes.
  - These tumors are very superficially located, involving mainly the cortex despite a concurrent involvement of the subjacent white matter. The tumor gives rise to a macro-gyriform aspect, rather typically associated with a scalloping of the inner table of the skull vault.
  - These lesions are heterogeneous with a multinodular pseudocystic appearance, but true intratumoral cystic formation is uncommon.
  - Associated edema is lacking and there is no mass effect. Calcifications are found occasionally (CT).
  - DNETs show many of the features of gangliogliomas and oligodendrogliomas.
  - DNETs may be associated with cortical dysplasias and may occasionally occur in neurofibromatosis type I.
- Temporal lobe tumors have to be differentiated from Focal Cortical dysplasia (with balloon cells) particularly when located in the cortical-subcortical zone, non associated with peritumoral edema and lack of contrast enhancement. Cortical dysplasias, more common in the frontal lobes, refer to focal or diffuse developmental abnormalities of brain cortex, due to: 1- abnormal neuronal and glial proliferation (solitary cortical tubers and subependymal hamartomas in tuberous sclerosis; focal transmantle dysplasia; hemimegalencephaly); 2- disorders of neuronal migration (cortical or subcortical heterotopias and subependymal nodular heterotopia); 3- cortical disorganization (focal cortical dysplasia without balloon cells and polymicrogyria). MR signs are blurring of the cortical-white matter interface, slight thickening of the cortical ribbon, often subtle abnormal SI in the subcortical white matter (mild hyperintensity on T2w that tapers as it extends to the lateral ventricles) with indistinct digitations (best appreciated on IR-T1w). There is no mass effect even if the related sulci appear as full and the lesion shows no contrast enhancement (case 19).
- Cortical dysplasia may be associated with primary brain tumor. Actually gangliogliomas and dysembryoplastic neuroepithelial tumors are commonly found with cortical dysplasias. Areas of cortical dysplasia have been demonstrated in association with DNETs in almost 47% of surgical cases. Imaging differentiation between gangliogliomas and dysplastic lesions may be difficult.

Signal intensity

- 3D-GE-T1w: gangliogliomas and DNETs are hypointense to cortex; cystic component of gangliogliomas may be isointense (higher protein content than CSF).
- SE-PDw: solid portion of gangliogliomas is hyperintense to cortex with variable SI of the cystic component which may be even more hyperintense (high protein content); DNETs are hyperintense to adjacent cortex.
- SE-T2w: solid portion of gangliogliomas is mildly hyperintense to brain, less or equally to the cystic component; DNETs are hyperintense to cortex and show no vasogenic edema.
- IR-T1w: due to its high contrast resolution, IR is efficient in the investigation and evaluation of cortical dysplasias, mesial temporal sclerosis and tumoral conditions as DNETs or gangliogliomas/gangliocytomas. Cortical analysis is highlighted.
- FLAIR: very sensitive in imaging of focal cortical dysplasia in which it may show the subtle hyperintensity in the subcortical white matter and occasionally in the overlying cortex itself.
- Diffusion WI: recent studies reported decrease in apparent diffusion coefficient (ADC) in patients with focal status epilepticus. The abnormalities shown, resolved after cessation of the seizures even if little is presently known about the time course of ADC changes post-ictally, may persist for hours or several days after sustained seizures. A concordance of the area of peri-ictal DWI signal changes and the area of focal electrocorticographic seizures has been recently demonstrated in patients. Peri-ictal DW imaging may become a useful non invasive tool for the identification and eventually delineation of epileptogenic foci.
- MR spectroscopy: shows potential clinical utility in providing reliable data with respect to localization and lateralization of the epileptogenic area in patients with temporal lobe epilepsy.

Contrast enhancement

- Enhancement of at least the solid portion of gangliogliomas is variable and irregular, and found in roughly half the cases. The wall of the cystic component may show contrast enhancement.
- DNETs usually do not show contrast enhancement (2/3 of the cases).

Case 16: Amygdalo-hippocampal low grade glioma

Fig. a: coronal cut (512); SE - PDw

Fig. b: coronal cut; SE - T2w

Non-enhancing mesial temporal infiltrating lesion involving mainly the head and the body of the left hippocampal formation, showing an iso-signal to the normal cortex on PDw (**a**, ↑↑) as well as on T2w (**b**, ↑), but with some heterogeneity. The lesion shcws mass effect on the ipsilateral optic tract, raised (**a**, ↑). The tip of the temporal horn is laterally displaced (**b**).

Case 17: Hippocampal low grade glioma, or post-ictal edema (non-verified)

Fig. c: coronal cut (512); STIR

Fig. d: coronal cut; SE - T2w

Increased signal and size of the left hippocampal formation on the STIR sequence (**c**, ↑) as well as on the T2w sequence (**d**, ↑), comparatively to the right hippocampal body. There is no involvement of the para-hippocampal gyrus, neither enhancement after contrast injection.

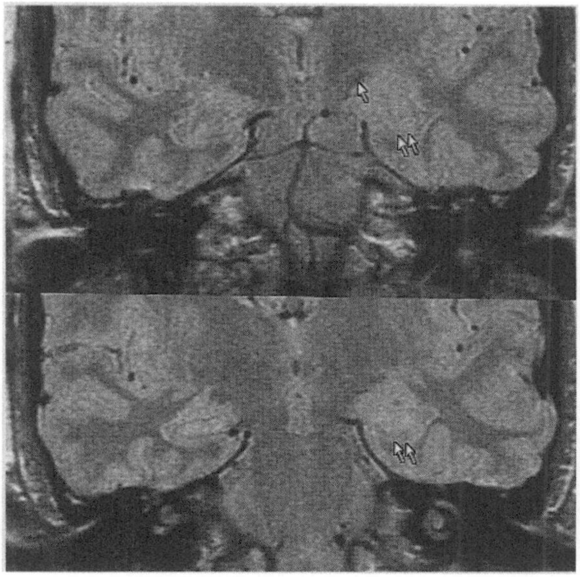

a

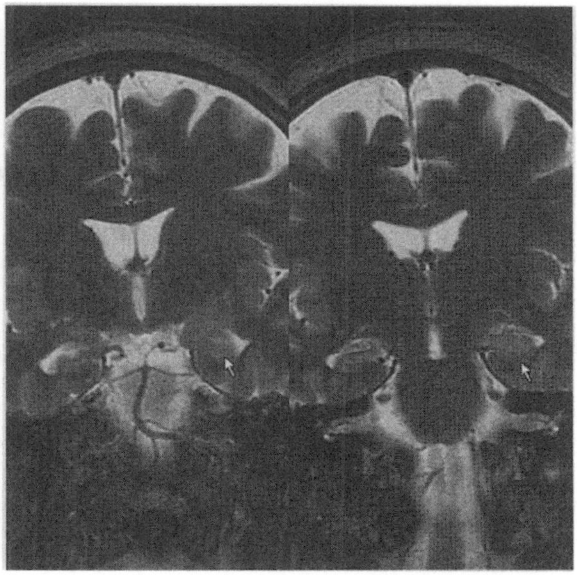

b

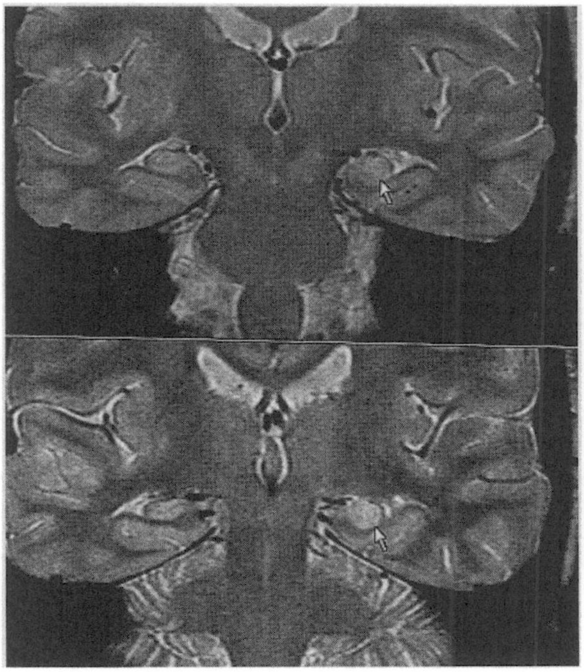

c

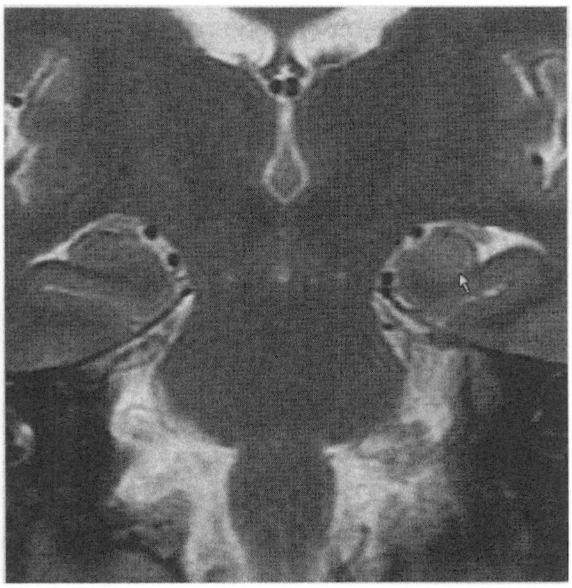

d

Case 18: Hippocampal - parahippocampal cortical dysplasia

Fig. a: coronal cut (512); SE - PDw

Fig. b: coronal cut (512); SE - T2w

There is blurring of the grey-white matter junction in the right para-hippocampal gyrus, extending to the hippocampal formation which is slightly enlarged in comparison to the left side. This apparent thickening of the cortex is slightly hyperintense on the PDw sequence (**a**) remaining slightly hyperintense on T2w (**b**). There is no marked mass effect on the collateral sulcus laterally at that time. The lesion shows no enhancement.

Case 19: Mesial temporal glioma

Fig. c: sagittal cut GE - T1w

Fig. d: coronal cut SE - PDw

Heterogeneous mass, slightly hypointense on T1w (**c**, ↑) and hyperintense on PDw (**d**, ↑), as well as on T2w, occupying the posterior hippocampus and para-hippocampal gyrus with marked mass effect over the collateral eminence and the collateral sulcus.

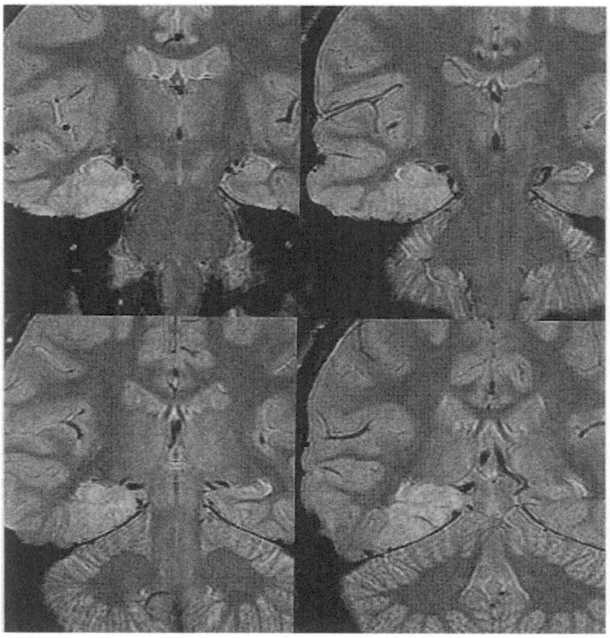

a

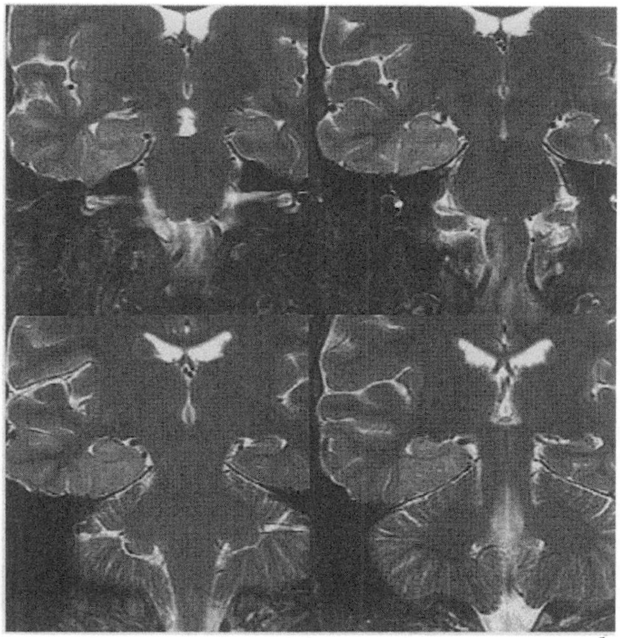

b

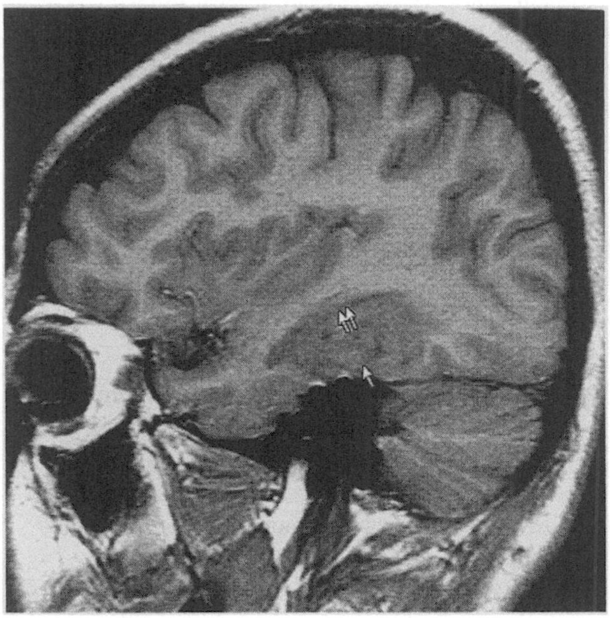

c

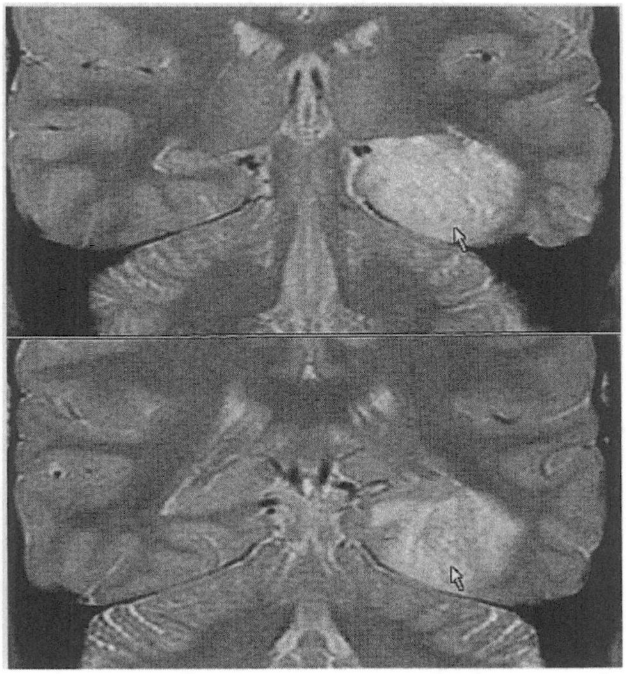

d

Incisural and infratentorial tumors

Cases 20-24: pineal gland and incisural tumors

Clinical and pathological features

- Pineal region tumors are rare intracranial tumors accounting for less than 1-2% reaching 3-8% in children.
- About 15-30% of pineal region tumors are pineal parenchymal tumors. Germ cell tumors account for 0.5-2%.
- Tumors not originating from pinocytes are represented mainly by germ cell tumors. Other tumor-like processes of pineal and incisural region are: gliomas of the tectum or midbrain (case 22), meningiomas, lipomas, pineal cysts, cyst of the cavum septi pellucidi, or arachnoid cysts of the ambient cistern (case 25).
- Pineal parenchymal tumors are: pineocytoma and pineoblastoma or a mixed type.

  - Pineocytomas (almost 45%) are WHO grade II slow growing tumors, occurring at all ages but mainly in adults (25-35 years old), without sex predilection. They are composed histologically by mature cells resembling pinocytes. They usually show a favourable prognosis.

  - Pineoblastomas (slightly more frequent) are malignant embryonal poorly differentiated WHO grade IV tumors, occurring at all ages, but most frequently in childhood during the first and second decades, with a male predilection. These tumors share features in common with PNET tumors (medulloblastoma). A genetic susceptibility is represented by the "trilateral retinoblastoma syndrome" associating a bilateral retinoblastoma and a pineoblastomatous tumor.
- Germ cell tumors are represented mainly by germinomas (most common) and teratomas (mature or immature, or with malignant transformation). Concerning germinomas:

  - Almost 90% occur in patients in the second decade, before age 20.

  - Male predominance is observed (2-2.5:1, male/female ratio). Teratomas (and embryonal carcinomas) occur most often in males.

  - At least 80% of CNS tumors are found in the pineal or suprasellar regions or in both sites (multifocal); thalamus and basal ganglia may be involved bilaterally.

  - Increased risk in Klinefelter syndrome. Reported cases in association with neurofibromatosis type I and Down syndrome.

  - Gonadotrophins could be an aetiological factor.

  - A favourable prognosis due to radiosensitivity characterize localized germinomas (5-years survival in 65-95%). Outcome is poorer in case of hypothalamic infiltration and/or spreading within the leptomeningeal spaces.
- *Signs*: the clinical syndrome is suggestive, comprising signs of increased intracranial pressure and a Parinaud's or dorsal periaqueductal midbrain syndrome: supranuclear paresis of upward gaze with preservation of downward movements, light-near dissociation with mild dilatation of the pupils, convergence paresis with convergence-nystagmus on upward gaze, lid retraction (Collier's sign). Cerebellar ataxia of gait and limbs, precocious puberty in males (10%) and diabetes insipidus.

MR characteristic findings

Morphology

- The pineal region lies in the central portion of the brain, and tumoral conditions developing in this area are related to important deep anatomical structures such as: the tectal plate and midbrain, the third ventricle and the thalami, the splenium of corpus callosum, the deep venous system (vein of Galen and internal cerebral veins), the ambient cistern and the foramen ovale of Pacchioni bounded by the free edges of the tentorium.
- Pineal gland tumors must be suspected in case of a mass larger than 10 mm (AP length of normal gland less than 9 mm) and the presence of

calcifications (CT most sensitive technique) in a child under 6 years of age.

- Cystic lesions must be differentiated from the benign common pineal cysts, incidentally found (40% of brain autopsies) and usually asymptomatic. Normal pineal gland, lacking of blood brain barrier in its capillaries, enhances normally, as the smooth walls of a benign associated cyst. Holocystic enhancement may be observed in case of delayed scanning. Cyst content is usually homogeneous and therefore isointense to CST on all sequences. It is not uncommon to observe a relative mild hyperintense SI to CSF in case of high protein content and/or presence of a sequelar hemorrhage (MR follow-up if questionable aspect). Differential diagnosis: small cystic pineocytoma.
- Pineocytomas are usually well-defined, more or less homogeneous solid masses of less than 3 cm in size, that may commonly show peripheral calcifications and occasionally cystic formations or small areas of hemorrhage. No invasive behavior is observed. Hydrocephalus is a common accompanying feature.
- Pineoblastomas are highly cellular tumors similar to PNETs, which appear as lobulated, ill defined masses. The tumors are commonly hemorrhagic and/or necrotic and tend to infiltrate the adjacent parenchyma and to involve the leptomeningeal spaces. Melanin production may be found.
- Pineal germinomas are usually homogeneous with uncommon cystic changes or calcifications and even hemorrhage which may suggest, if present (more heterogeneous aspect of the tumor), a more virulent behavior or components. Hemorrhages characterize more particularly choriocarcinomas or mixed germ tumors. Germinomas show tendency to metastasize (like pineoblastomas). A suggestive pattern is the encasement of the calcified pineal gland within the tumor as may be interestingly demonstrated with CT (pineal parenchymal tumors showing, a contrario, a dispersion of such calcifications).
- Mature teratomas contain variable tissue elements derived mainly from the ectoderm (skin, choroids plexus) and mesoderm (fat, bone, muscle), but in the majority of these tumors arising in the CNS are found incompletely differentiated components. Teratomas are nearly always hemorrhagic and multicystic accounting for the morphological heterogeneous aspect. Presence of fat, when existing, will explain bright SI on T1w. Calcifications are also common.
- Choriocarcinomas and embryonal carcinomas (rare varieties of germ cell tumors), are of high grade of malignancy and show high propensity to bleed.
- Other rare germ cell tumors are Yolk sac tumor (endodermal sinus), associated with increased level of alpha-fetoprotein.

Signal intensity

- Differential diagnosis of the various pineal gland parenchymal tumors is usually not specific on SI basis using MRI. Pineoblastomas can behave similarly to germinomas on T2w sequences.
- GE-T1w: pineocytomas are hypointense and pineoblastomas and germinomas are hypointense to isointense to brain parenchyma. Fat containing teratoma shows hyperintense SI to brain.
- SE-PDw: pineocytomas are slightly hyperintense to brain, and pineoblastoma and germinomas are isointense to cortex or slightly hyperintense.
- SE-T2w: pineocytomas are moderately hyperintense to brain, and pineoblastomas and germinomas are usually isointense or hypointense to cortex.
- Pineal cysts are usually slightly hyperintense or isointense to CSF on T1w sequences and slightly hyperintense to CST on PDw (high protein content).

Contrast enhancement

- Pineocytomas and pineoblastomas usually show heterogeneous (may be homogeneous in pineocytomas) intense contrast enhancement.
- Germinomas commonly show marked and homogeneous contrast enhancement.
- Contrast enhancement helps to reveal leptomeningeal and adjacent parenchymal spread of pineoblastomas and germinomas.

Case 20: Pinealocytoma

Fig. a. sagittal cut; SE - T2w

Fig. b: sagittal cut; SE - T1w + Gd

Intrinsic pineal mass (**a**, ↑), partly cystic (*), showing homogeneous enhancement (**b**, ↑↑), displacing downward the tectum and the superior colliculi with narrowing of the sylvian aqueduct (**b**, ↑). Note the lack of regional invasion into the adjacent brain parenchyma (**a**).

Case 21: Tectal low grade glioma

Fig. c: sagittal cut; SE - T2w

Fig. d: axial cut; SE - T2w

Homogeneous enlargement of the mid-brain involving mainly the quadrigeminal plate (**c**, ↑, left) extending to the posterior commissure, enlarged (**d**, ↑, right), but respecting the superior medullary velum (**c**, right, ↑↑). The lesion did not enhance, showing a homogeneous hyperintense signal on heavily T2w sequence (**c, d**).

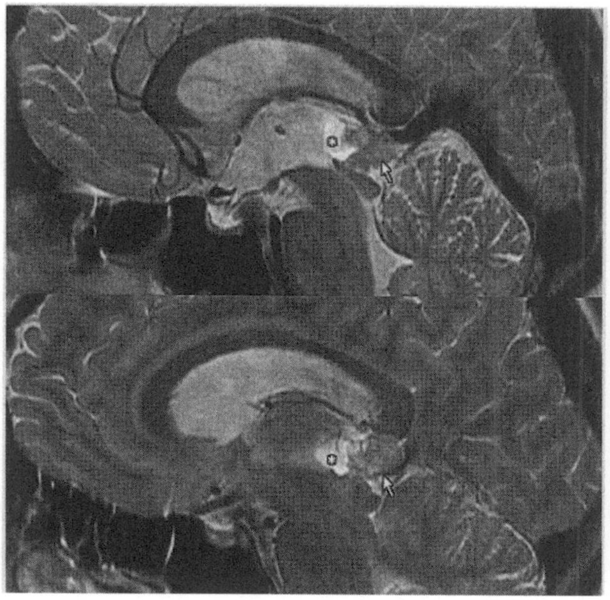

a

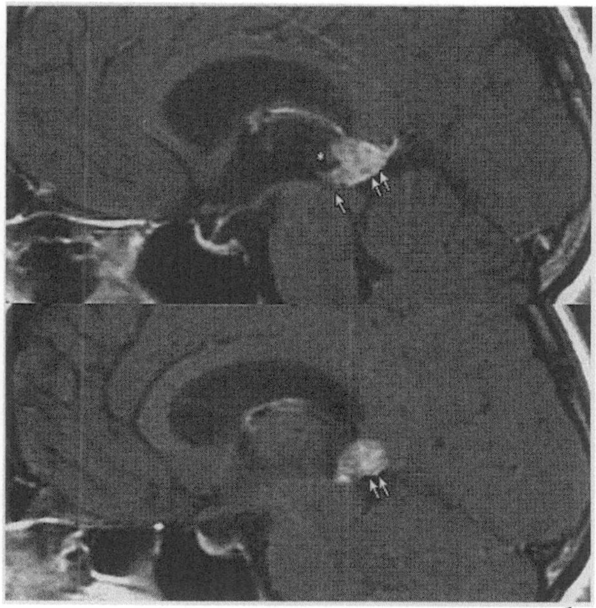

b

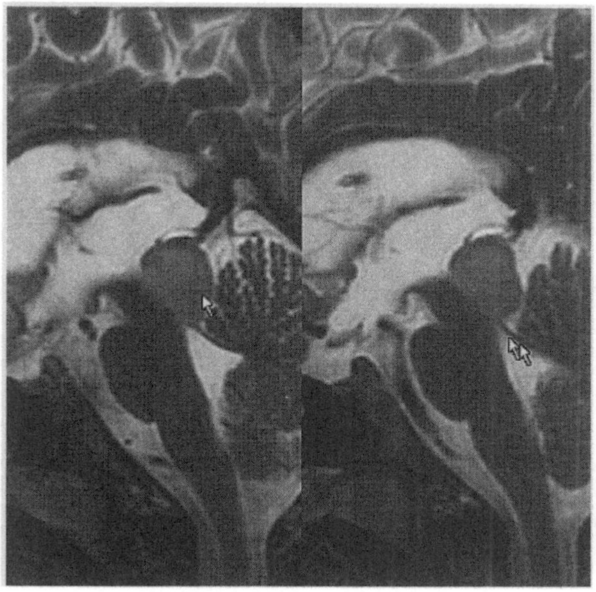

c

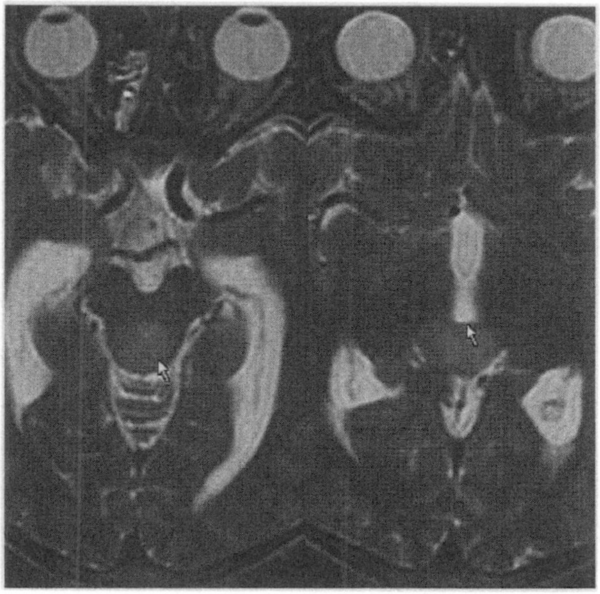

d

Case 22: Mature teratoma of pineal region

Fig. a: sagittal cut (512); GE - T1w **Fig. b:** axial cut; SE - T2w

Highly heterogeneous pineal region tumor showing multiple cystic formations (*) with very low signal on T1w (**a**) and hyperintense CSF-like signal on T2w (**b**), as well as areas of hemorrhage (**a**, up, ↑). The mass invades the posterior 2/3 of the third ventricle, obstructs the foramina of Monro causing hydrocephalus and extends downward toward the fourth ventricle (**a**, down, ↑).

Case 23: Tectal lipoma

Fig. c: sagittal cut; GE - T1w

Very small bright process involving the lower extremity of the tectum on T1w sequence (**c**, ↑). The hypersignal of the lesion close to the signal of the subcutaneous fat was removed on the fat suppression technique, confirming its fatty nature.

Case 24: Arachnoid cyst of the quadrigeminal cistern

Fig. d: sagittal cut; GE - T1w

Sagittal T1w sequence showing a large extracerebral collection in the quadrigeminal cistern which has similar signal intensity to the ventricular CSF. The cyst compresses the posterior aspect of the third ventricle and laminates the tectal plate (↑), displacing the superior vermis inferiorly (↑↑), and the posterior portion corpus callosum and the splenium superiorly (↑↑↑).

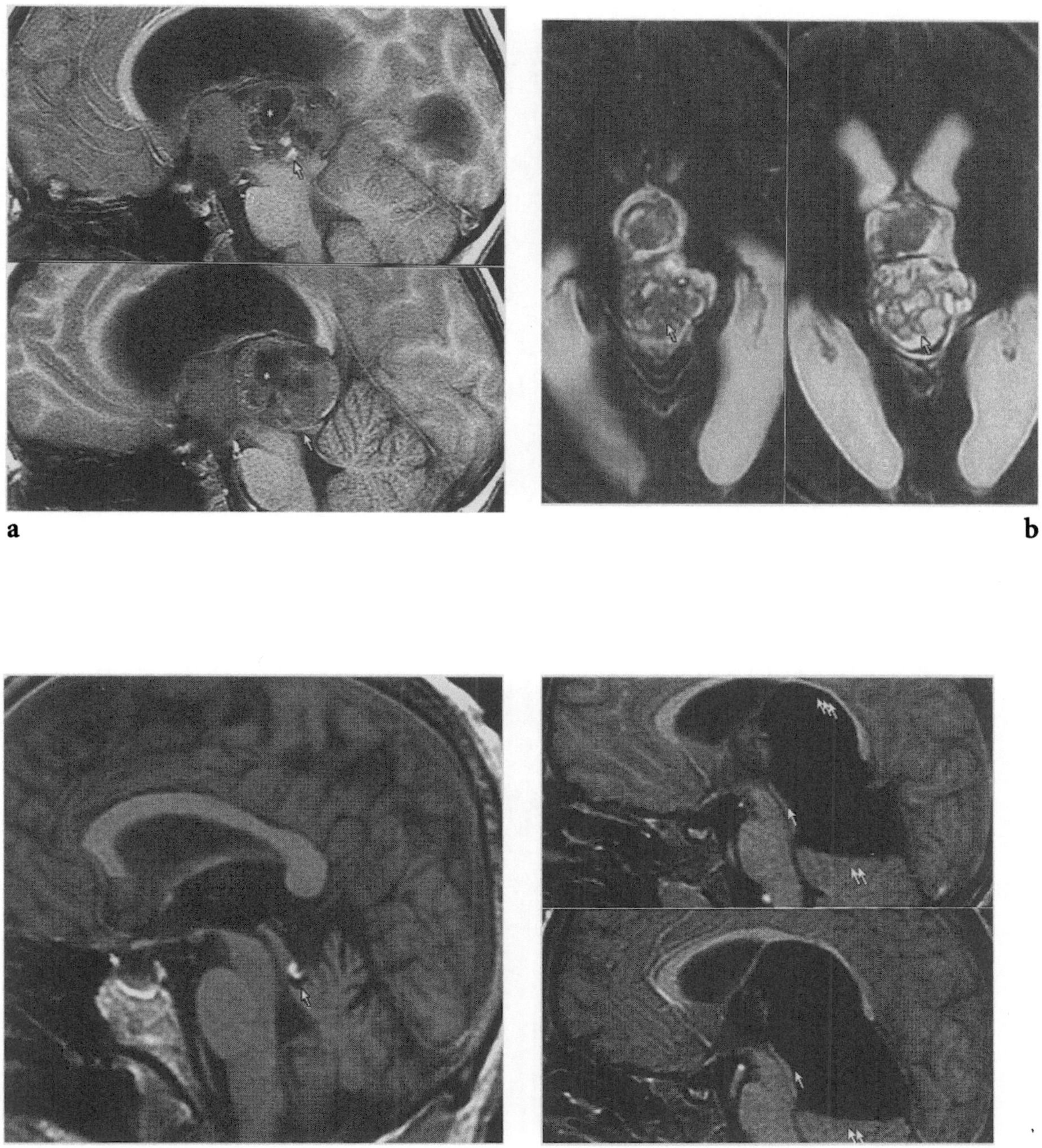

a b c d

Cases 25, 27: cerebellar gliomas

Clinical and pathological features

- Most frequently (30%), posterior fossa tumors arise in childhood and adolescence.
- *Histopathology*: 4/5 of cerebellar tumors are juvenile pilocytic astrocytomas (WHO grade I) and 1/5 are diffuse fibrillary astrocytomas (WHO grade II).
- JPA are present in the first two decades with 2 peaks incidence, at the age of 10 and 20 years, and show no sex predilection. They may be encountered in higher frequency in association with Von Recklinghausen's neurofibromatosis.
- Fibrillary astrocytomas which have to be well separated from the pilocytic astrocytomas, are usually found in young adults (60% between 20-45 years old and 10% below age 20) with a mean age of 34 years and a male predominance ratio (2:1). These tumors show a tendency to malignant progression.
- *Signs*: presenting symptoms of children or adolescents with cerebellar astrocytomas consist most frequently of headaches (about 40%), vomiting (35%), and clumsiness or unsteadiness of gait mainly in case of vermian involvement (15%), partly due to increased intracranial pressure. The duration of symptoms before diagnosis is longer than in medulloblastomas, lasting more than 12 months in more than half the cases and reaching 18 months in about 2/3 of the patients.

MR characteristic findings

Morphology

- The juvenile pilocytic astrocytomas is morphologically a very well delineated cerebellar hemispheric, or less commonly vermian, mass (can be removed totally), presenting most frequently as a large cystic formation containing a mural tumoral nodule in 50-80% of the cases. The cystic component is filled with a protein-rich xanthochromic fluid.
- JPA may also present as a solid mass eventually associated with intratumoral cystic formations or even less commonly may be solid (10%).
- Evidence of previous intratumoral hemorrhage (hemosiderin) and presence of intratumoral necrotic changes or minute calcifications (20%) may be found.
- Characteristically, the tumor does not show invasion of the surrounding parenchyma and exerts a mass effect on the 4th ventricle displaced contralaterally, with or without associated hydrocephalus. Vasogenic edema is found in huge tumors.
- Fibrillary astrocytomas are diffuse homogeneous tumors which tend to widely infiltrate the cerebellar parenchyma with possible secondary extension along the cerebellar peduncles to reach the brainstem. Due to their infiltrative nature they show enlargement and disorganization of the invaded anatomical structures with ill defined boundaries with respect to the adjacent normal tissue.
- Intratumoral microcystic and hemorrhagic changes, or even a single large cyst formation may be uncommonly encountered.

Signal intensity

- GE-T1w: usually hypointense to cerebellar white matter; the cystic portion if present (commonly in JPA) is of intermediate SI between solid portion and CSF due to its high protein content.
- SE-PDw: hyperintense to cerebellar cortex, with a more hyperintense cystic portion (JPA typically).
- SE-T2w: hyperintense SI of both the solid and the cystic components.

Contrast enhancement

- In JPA: intense contrast enhancement of the solid portion or mural nodule due to blood brain barrier breakdown and hypervascularity; the wall of the cystic component may or not show contrast enhancement.
- Fibrillary astrocytomas commonly do no show contrast uptake. Variable enhancement as cystic changes may however occur early in the development, rendering more difficult the differential diagnosis with JPAs (older age group rather than pilocytic tumors).

Case 25: Pilocytic astrocytoma of the cerebellum, low grade

Fig. a: sagittal cut;	SE - T2w	**Fig. c:** coronal cut;	SE - PDw
Fig. b: axial cut;	GE (3D) - T1w	**Fig. d:** coronal cut;	SE - T2w

Homogeneous mass, hypointense on T1w (**b**, ↑), hyperintense on PDw (**c**, ↑) and T2w (**a, d**, ↑) involving the left paravermian area and the postero-lateral aspect of the ponto-cerebellar region, filling the fourth ventricle laminated and displaced to the right (**b**,*). The lesion displaces the brainstem anteriorly (**b**, ↑↑). The right superior cerebellar peduncle is displaced laterally (**d**,*) and the nodule (**c**,*) as well as the tonsils (**d**,**) are displaced downward.

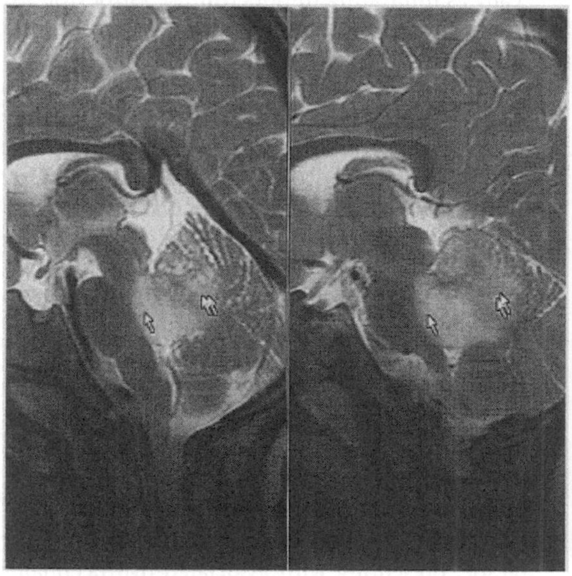

a

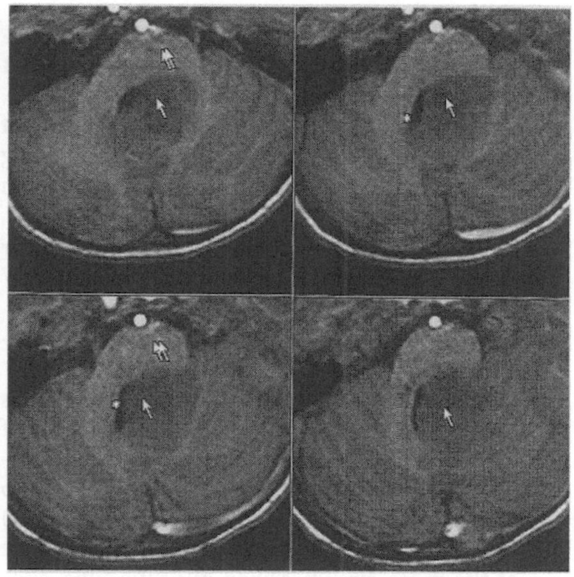

b

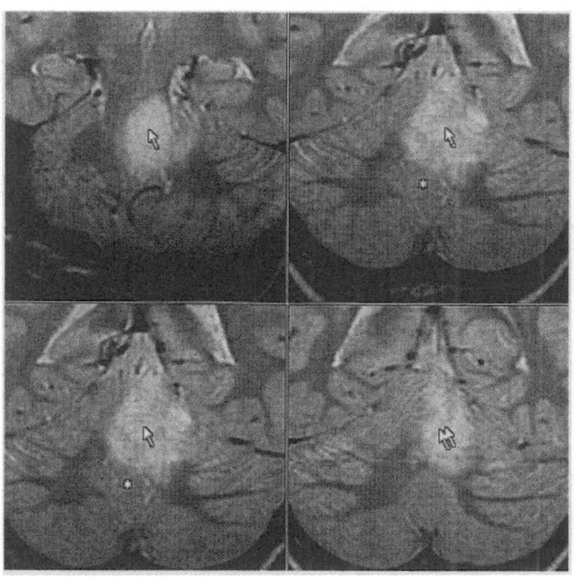

c

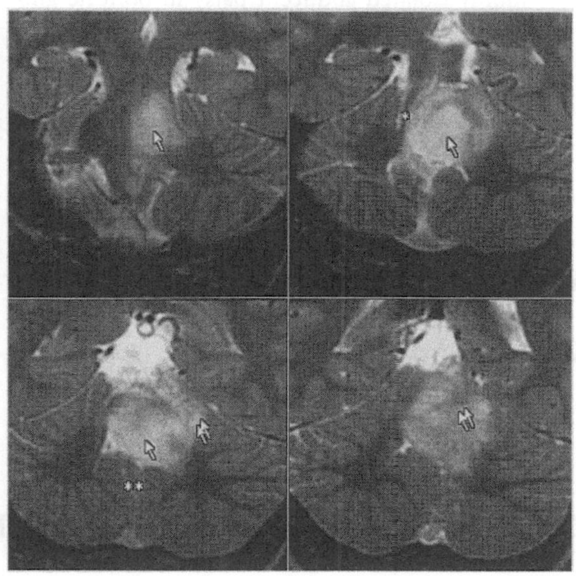

d

Case 26: brainstem gliomas

Clinical and pathological features

- Brainstem gliomas usually of the low grade diffuse type, occur most frequently in children but are also found in young adults.
- The brainstem is the next common localization of these low grade infiltrating tumors, accounting for roughly 1/5 of all infratentorial tumors in childhood.
- More than two thirds of brainstem gliomas occur before age 10 with a peak found typically at 7 years old.
- These tumors show a 2:1 male to female predominance.
- *Histopathology*: low grade diffuse astrocytomas correspond to WHO grade II tumors. Most of these brainstem tumors (4/5) are of the infiltrating fibrillary type with a tendency to progress to a more malignant anaplastic (WHO grade III) phenotype or a glioblastoma. The remaining are of the more circumscribed pilocytic type, showing a better prognosis.
- *Signs*: variable, usually progressive clinical picture with respect to localization in the brainstem; symptoms are generally present 3-5 years before the diagnosis is made. Cranial nerve palsies (commonly isolated abducent palsy in pontine gliomas) are the presenting signs in almost 90% of the patients, followed by long tract signs of the motor and/or sensory types (characteristically crossed with the cranial nerve palsy), disorders of eye conjugate movements and eventually persistent hiccoughs. Course of disease may become more rapidly fatal in case of malignant transformation.

MR characteristic findings

Morphology

- MR is much more sensitive than CT for the diagnosis, the evaluation of extension and follow-up of brainstem tumors.
- Diffuse low grade gliomas appear characteristically as infiltrating tumors (diffuse astrocytomas) that may involve any portion of the brainstem and almost typically the pons (isolated sixth nerve palsy in a child). Exophytic pontine tumors develop in the prepontine cistern and frequently encase the basilar artery. With enlargement of the tumor, the 4th ventricle is displaced posteriorly and almost collapsed (triventricular hydrocephalus). Superior extensions toward the midbrain and as far as the thalamic region (case 27) or inferior extension to the medulla may be progressively observed with time.
- Anaplastic features are not uncommonly found in the diffuse form with presence of necrotic and hemorrhagic areas due to neovascularity (almost half the cases), and may mimic rhombencephalitis. Bleeding is suggestive of malignant transformation. Cystic formations are uncommon.
- Midbrain gliomas are frequently of very low grade, presenting with associated hydrocephalus due to obstruction of the cerebral aqueduct (case 22).
- Exophytic medullary tumors are commonly found and are usually of low grade type, considered to show a better prognosis.

Signal intensity

- GE-T1w: hypointense the white matter.
- SE-PDw and SE-T2w: hyperintense to cerebellar cortex.
- FLAIR: hyperintense to cerebellar cortex.

Contrast enhancement

- Most of the diffuse infiltrating gliomas usually do not enhance at least initially, unless secondary anaplastic transformation.
- The less frequent brainstem pilocytic astrocytomas typically show intense contrast enhancement.

Case 26: Mesodiencephalic astrocytoma, grade II

Fig. a: axial cut;	SE - PDw	**Fig. c:** coronal cut;	SE - T2w
Fig. b: axial cut;	FLAIR - T2w	**Fig. d:** coronal cut;	SE - T1w + Gd

Right mesodiencephalic infiltrating inhomogeneous mass expanding the right portion of the mesencephalon slightly hypointense on T1w, hyperintense on PDw (**a,** ↑) as well as on T2w (**c,** ↑) and FLAIR (**b**). The lesion displaces the aqueduct to the left (**c,** ↑↑), and compresses the posterior third ventricle (**b, d,** *) with invasion of the posterior commissure (**a, b,** ↑↑). The lesion shows small nodular and ring-like partial enhancement (**d,** ↑), involving mainly the pulvinar (**d,** ↑↑) and the lateral quadrigeminal plate (**d,** ↑↑↑).

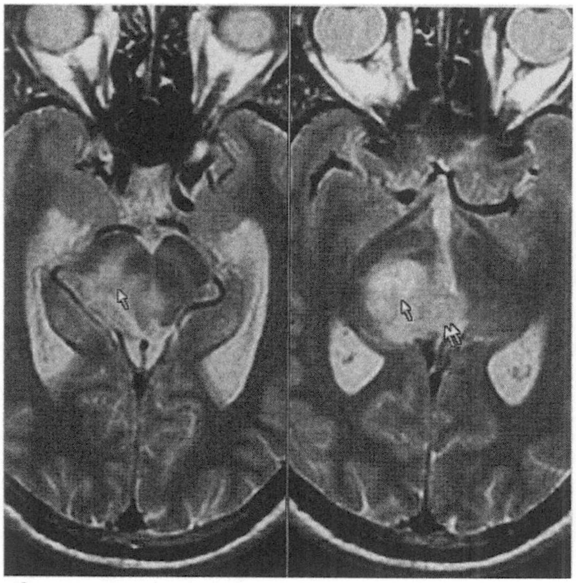

a

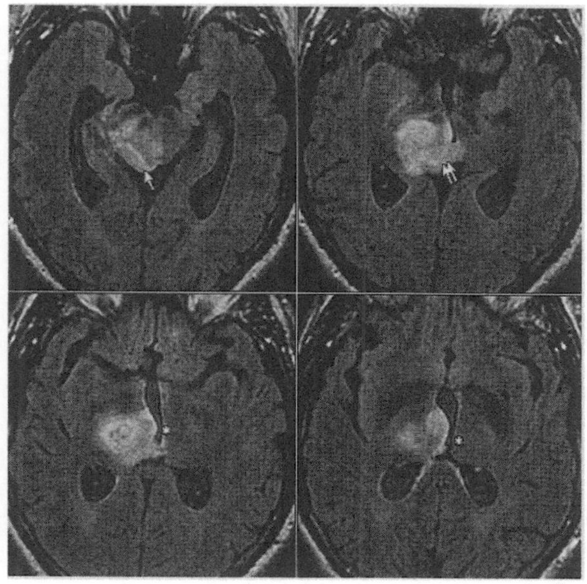

b

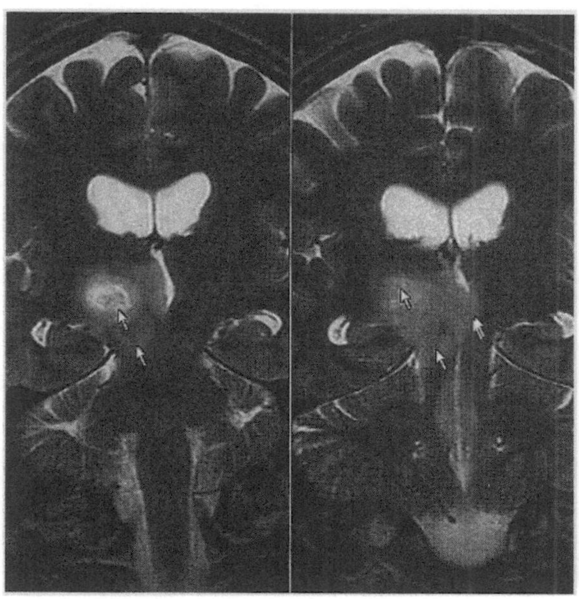

c

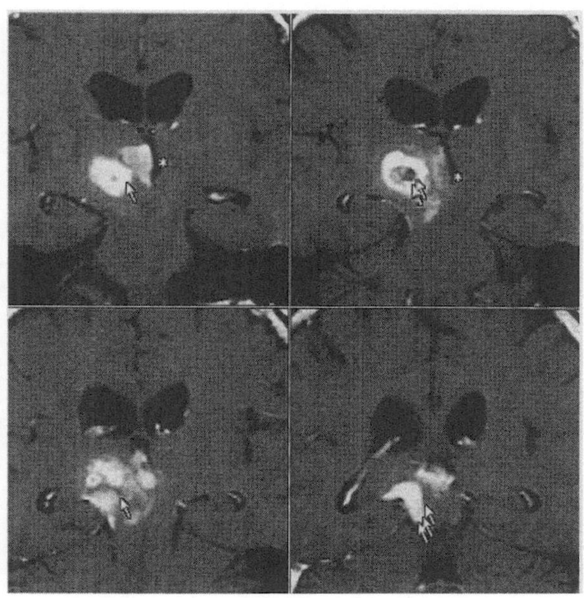

d

Case 27: Oligoastrocytoma of the cerebellum

Fig. a: axial cut;	SE - PDw	**Fig. c:** sagittal cut;	SE - T2w
Fig. b: axial cut;	SE - T2w	**Fig. d:** axial cut;	GE (3D) - T1w + Gd

Extensive ill-defined infiltrating lesion involving the vermis and the left adjacent cerebellar hemisphere showing a slightly hyperintense signal on T1w, non enhancing (**d,** ↑) but with slight mass effect on the postero-lateral aspect of the fourth ventricle (**b, d,** *). The lesion is slightly hyperintense on PDw (**a,** ↑) and hyperintense on T2w (**b, c,** ↑). The mass infiltrates the superior medullary velum (**c,** *).

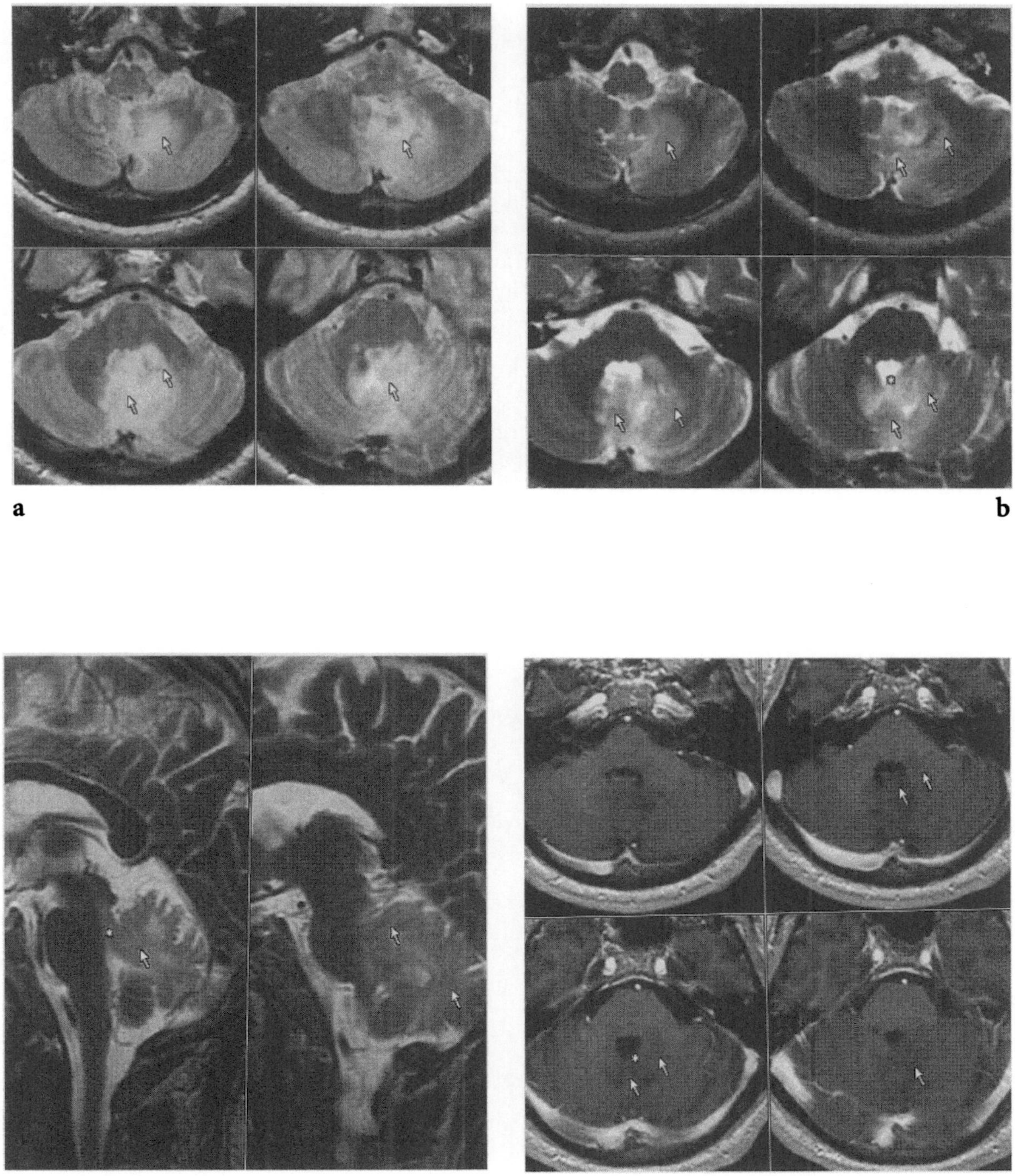
a
b
c
d

Cases 28-30: medulloblastoma and PNET tumors

Clinical and pathological features

- Highly malignant embryonal tumors of the CNS.
- Originate from common stem cells in the subependymal matrix with capacity of differentiation along neuronal, glial and ependymal lines.
- PNET tumors include: the medulloblastoma, the ependymoblastoma, the central neuroblastoma, and the supratentorial PNET. The pineoblastoma (see incisural tumors) shows common features with the medulloblastoma and related PNET.
- These tumors occur preferentially in children. Posterior fossa PNET tumors are second in frequency to astrocytoma of the cerebellum. They account for 5-10% of all supratentorial tumors.
- The annual incidence of medulloblastomas is about 0.5 per 100000 per year. Seventy-five per cent of the tumors occur in patients before age 15 and about half in the first decade. The peak incidence is 7 years old. In young adults about 80% occur between ages of 21-40. A predilection for males is found (2-3:1).
- *Histopathology*: correspond to WHO grade IV tumors. Medulloblastomas are of several types (classic, desmoplastic and a rare large-cell tumor). In the common type, the tumor includes spherical or oval cells densly crowded with ill defined cytoplasms and hyperchromatic nuclei. Vascular proliferation, hemorrhages and calcifications are relatively rare. In the desmoplastic type, an important proliferation of collagen and reticulin is found. The large cell type is similar to the rhabdoid teratoid tumor of the cerebellar area and commonly shows areas of intratumoral necrosis. Most tumors show little if any evidence of differentiation.
- *Signs*: vomiting is the most frequent presenting symptom (30%) of such cerebellar tumors; headaches (10%) diffuse or frontal becoming later suboccipital; these symptoms are usually due to increased intracranial pressure; tilting of the head's child towards the involved hemisphere; truncal ataxia, unsteadiness and gait disturbance due to involvement of vermis (25%); and later, strabismus with slow and coarse oscillations (5%) or dysmetria and adiadochokinesia of the limbs.

MR characteristic findings

Morphology

- The most common presentation mode is found in childhood with about 2/3 of the medulloblastomas arising in the cerebellar vermis, especially from the posterior medullary velum developing in the 4th ventricular cavity, and even extending to the cisterna magna. The mass is usually well circumscribed and homogeneous. It generally infiltrates the floor or the lateral wall of the 4th ventricle, and may also spread over the cerebellar cortex.
- Involvement of the cerebellar hemispheres is more common in adulthood, involving frequently the dorsal surface, and may spread over the cerebellar folia.
- The tumor is characterized by its high propensity to infiltrate the subarachnoid spaces with further seeding throughout the entire CSF pathways, in the spinal canal more frequently than over the cerebral convexity. Leptomeningeal infiltration is of the diffuse and/or nodular type, present in almost 1/3 of the cases at presentation.
- Extracranial matastases are rare (about 5%), usually to bone (vertebral bodies) and even cervical lymph nodes.
- Hemorrhages may be found in almost 10%; but exceptionally, massive hemorrhage may be observed.

Signal intensity

- GE-T1w: hypointense to brain.
- SE-PDw: mildly hyperintense to cerebellar cortex.
- SE-T2w: isointense or hypointense to cortex (due to hypercellularity and minimal intracellular cytoplasm) to cerebellar hemisphere. May show cystic or necrotic hyperintense foci.

Contrast enhancement

- Medulloblastomas show intense contrast enhancement in 95% of the cases.
- Contrast infusion is mandatory to visualize CSF seeding within the leptomeninges and/or on the ventricular surface (intracranially and within the spinal canal), found at presentation in almost 1/3 of the patients.

Case 28: Cerebellar medulloblastoma (adult)

Fig. a: coronal cut; SE - T2w

Fig. b: axial cut; GE (3D) - T1w + Gd

Right cerebellar heterogeneous but well demarcated mass involving the cerebellar cortex and the album cerebelli, slightly hyperintense on T2 w (**a**), partially enhancing (**b**, ↑). The lesion shows areas of cystic necrosis (**a**,*).

Case 29: Cerebellar medulloblastoma (recurrence)

Fig. c:. axial cut; SE - T2w

Fig. d: sagittal cut; GE - T1w + Gd

Highly heterogeneous vermian and paravermian mass showing multiple areas of necrosis (**d**, ↑, **c**, ↑↑). The lesion is iso-signal to the cerebellar cortex on T2w (**c**, ↑), diplaces laterally the left dentate nucleus (**c**, right) and exerts a mass effect on the fourth ventricle (**d**, right, ↑↑ and **c**, right, *). The mass enhances following IV contrast administration and is surrounded by edema, hyperintense on T2w (**c**).

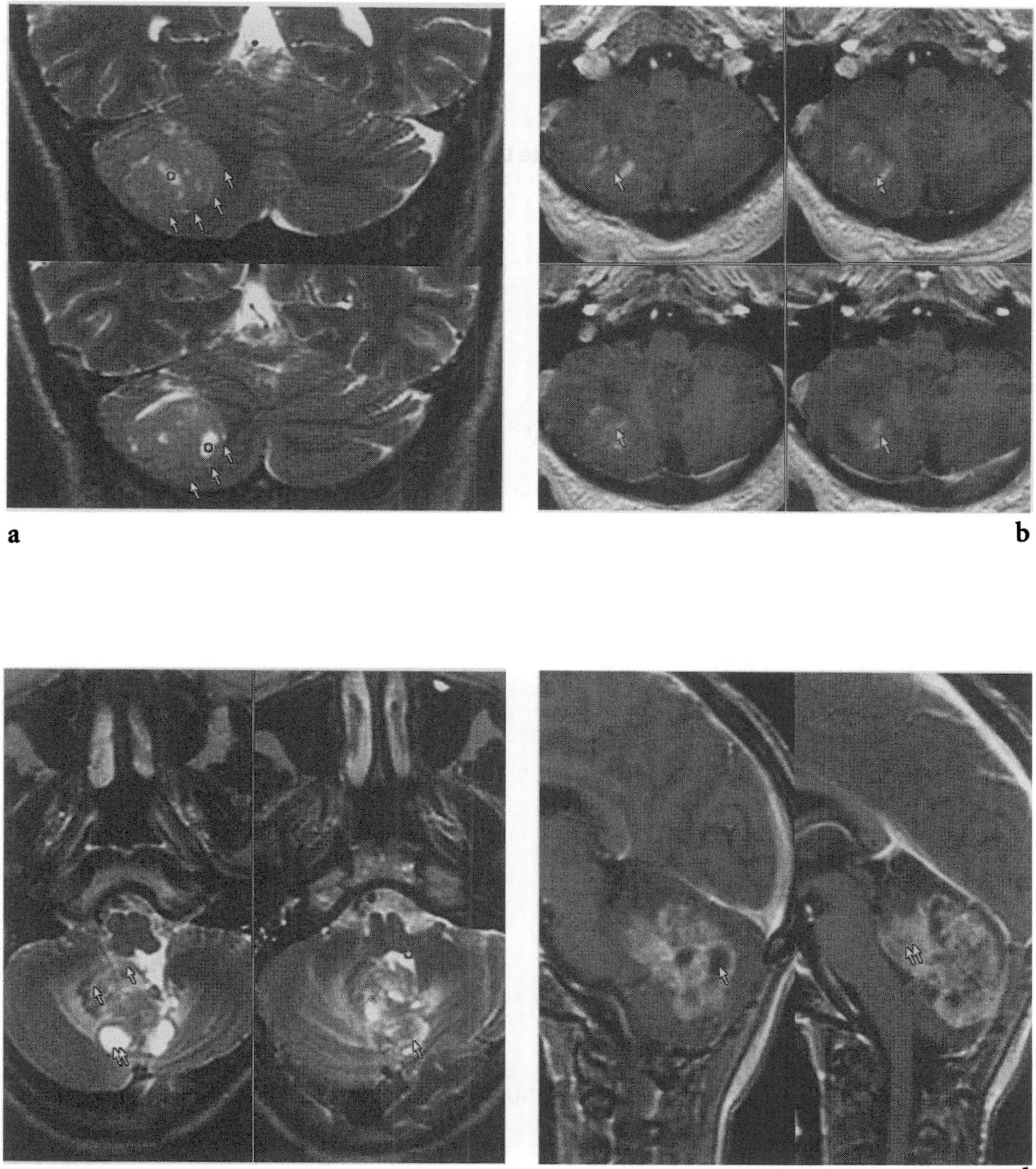

a b c d

Case 30: Primitive neuroectodermal tumor of the tectum

Fig. a: sagittal cut;	GE - T1w	**Fig. c:** axial cut;	SE - PDw
Fig. b: coronal cut;	GE (3D) - T1w + Gd	**Fig. d:** axial cut;	SE - T2w

Enlargement of the posterior aspect of the quadrigeminal plate extending to the anterior-superior part of the vermis, showing a heterogeneous low signal on T1w (**a,** ↑) and a limited dense area of enhancement (**b,** up, ↑). On PDw sequence, the mass is slightly hyperintense to the cerebellar cortex (**c,** ↑) and iso-intense on T2 w (**d**) with areas of high signal corresponding to associated edema and microcystic areas of necrosis (**d,** up, ↑↑). Note the leptomeningeal seeding of the lesion extending laterally to the upper cerebellar cortex (**d,** ↑↑), with some enhancement (**b,** down, ↑↑). Note the hydrocephalus (**a,** ↑↑) and the mass effect on the brainstem (**c,** *).

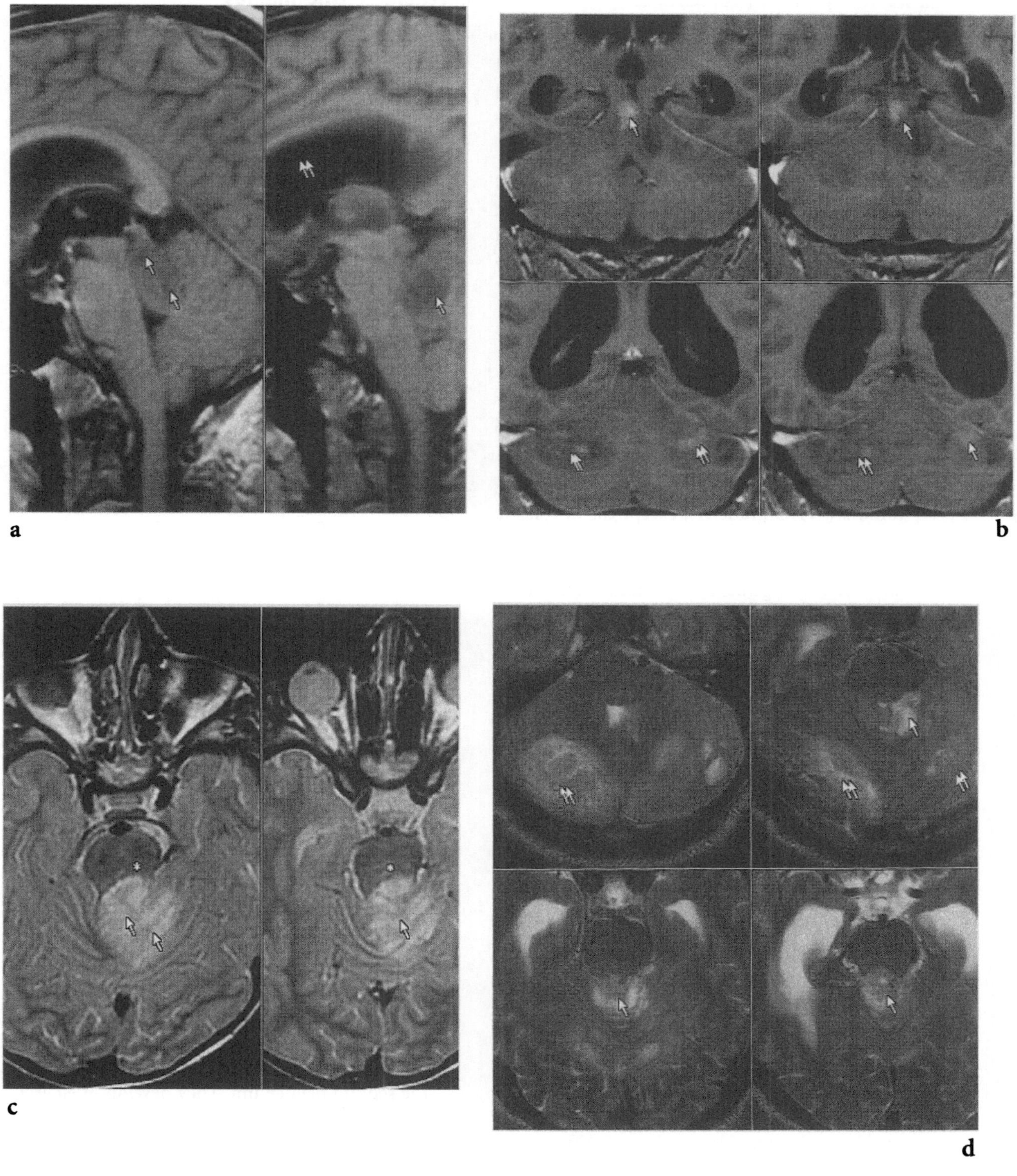
a
b
c
d

Case 31: ependymoma of the fourth ventricle

Clinical and pathological features

- Account for 3-10% of all neuroepithelial tumors, composed of neoplastic ependymal cells, which may occur at any site of the ventricular system.
- Correspond to 8% (6-12%) of intracranial tumors in children and adolescents, with a peak incidence between ages of 10-15 years. One third of the cases are found in children before year 3 of age.
- Ependymomas may be found in all age groups with no sex predilection.
- Peak incidence of spinal tumors is found between ages 30 to 40, characteristically seen along the filum terminale.
- *Histopathology*: ependymomas are low grade tumors (WHO grade II), except the anaplastic type (WHO grade III). Ependymomas are moderately cellular tumors showing a perivascular radial arrangement of the tumors cells. They may contain foci of necrosis, hemorrhages, calcifications and myxoid degeneration. Four subtypes are defined (cellular, papillary, clear cell and tanycytic). The cellular subtype shows a more compact cellular pattern, and a papillary subtype (uncommon) which may mimic a choroid plexus papilloma.
- Most ependymomas show cytogenetic modifications (aberrations of chromosome 22 in 30% of the cases, between others).
- *Signs*: the clinical manifestations are usually similar to those found in posterior fossa tumors in children (see medulloblastoma), caused by increased intracranial pressure and involvement of the cerebellum.
- Prognosis is poorer in children (posterior fossa location), improved with morphologically complete surgical resection. Survival of patients with anaplastic variant as compared to those with low grade ependymoma do not show significant differences. Local relapses are frequent (follow-up MR exams with contrast administration are necessary).

MR characteristic findings

Morphology

- Ependymomas develop most frequently in the posterior fossa, followed by the lateral ventricles, intracranially.
- Localization in the posterior fossa (2/3 of intracranial ependymomas) is dominant in childhood. Tumors arise from the floor of the lower portion of the 4th ventricle, developing within the ventricular cavity and extending (10% of the cases) through the foramina of Luschka toward the cerebellopontine angle or through the foramen of Magendie and foramen magnum to penetrate the upper cervical canal ("tongue-like projection").
- More than 1/3 of the tumors are supratentorial originating from ependymal cell rests and growing in the body of the lateral ventricle (2nd most frequent location) with a greater propensity to expand in the adjacent white matter of the frontoparietal lobes. These masses are usually large with necrotic and hemorrhagic components.
- Calcifications that may be dense, are found in almost half the tumors.
- Hydrocephalus (triventricular in 4th ventricular ependymomas) is a frequently associated feature.
- Seeding of the CSF may be found but much less frequently than with PNET tumors (less than 5% of the cases) and later (at recurrences). Extraneural metastases have been reported (lung).
- A variant (subependymoma, WHO grade I), is often incidentally found in 2/3 of the cases attached to the lower medulla and filling the inferior portion of the 4th ventricle, or in contact with the septum pellucidum. They develop in elderly males (about 60 years old) and are very rarely found in children. The tumor is usually small and homogeneous, and does not show contrast enhancement. It is a slow growing tumor which shows a very good prognosis.

Signal intensity

- GE-T1w: hypointense to brain with areas of marked hypointensities corresponding to cystic or necrotic foci and areas of bright SI in case of hemorrhage.
- SE-PDw: hyperintense to cortex.
- SE-T2w: hyperintense to cortex; may be heterogeneous with areas of hypointense SI due to hypercellularity or recent bleeding, or calcifications or presence of blood vessels, and areas of bright signal to necrotic zones and/or trapped CSF.

Contrast enhancement

- Most ependymomas commonly show mild to moderate contrast enhancement, varying from homogeneous to heterogeneous.
- Subependymomas characteristically do not show contrast enhancement.

Case 31: Ependymoma of the fourth ventricle

Fig. a: sagittal cut; GE - T1w

Fig. b: axial cut; SE - PDw and T2w

Fig. c: coronal cut; SE - PDw

Fig. d: coronal cut; SE - T2w

Intraventricular mass developed in the fourth ventricle showing a hyposignal on T1w (**a,** ↑↑) and extending downward toward the foramen magnum (**c,** ↑, **d**). The lesion is hyperintense on PDw (**c,** ↑↑) as well as on T2w (**d,** ↑↑), occluding the fourth ventricular cavity and displacing laterally the superior cerebellar peduncles (**c,** *) with subsequent hydrocephalus (**c,** **). Note the presence of tumoral cystic formations (**a,** ↑, **d,** ↑).

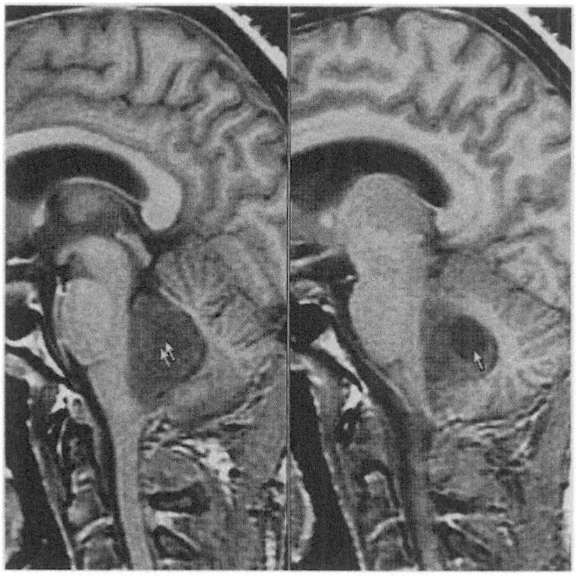

a

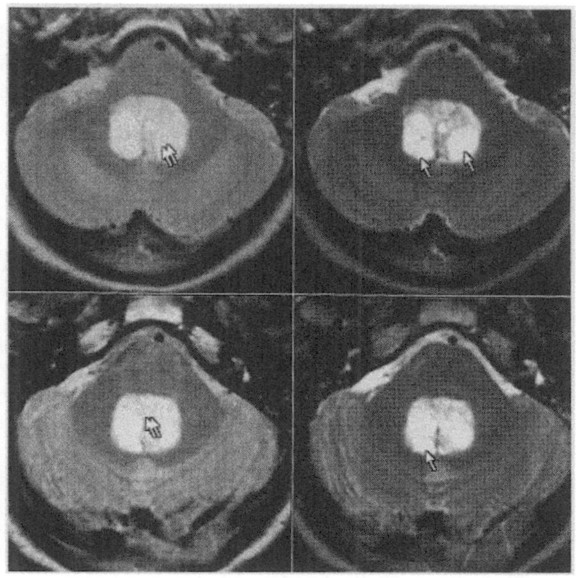

b

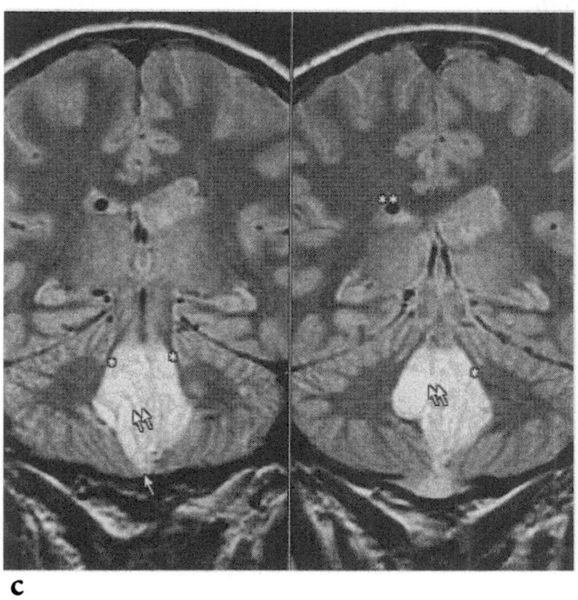

c

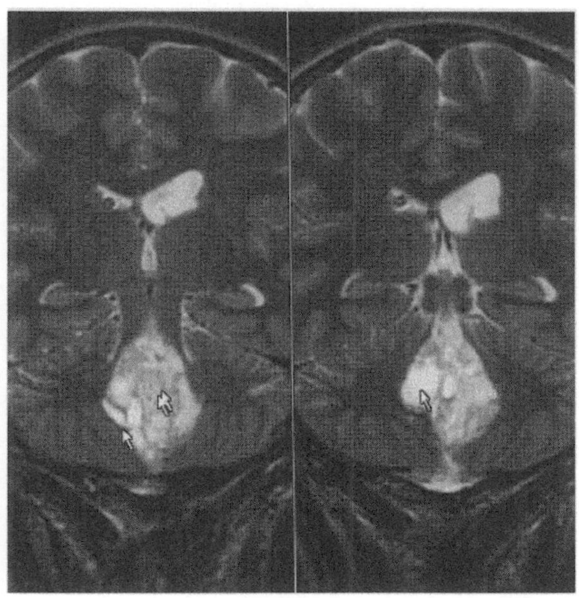

d

Leptomeningeal and skull voult tumors

Cases 32-37: intracranial meningiomas

Clinical and pathological features

- Meningiomas are the most frequent primary intracranial non glial tumors arising from meningothelial arachnoid cells (15% of intracranial tumors).
- Meningiomas affect the adult of 20 to 60 years of age, with a highest incidence in the 7th decade and a female predilection (2:1). Childhood tumors are rare, proved to be more aggressive, and are frequently associated with neurofibromatosis type 2.
- Meningiomas are multiple in about 10% of cases (meningiomatosis).
- They are usually benign, slow growing extra-axial tumors, most frequently found along the brain convexity, or at the skull base (anterior basal) or in the ventricular cavities (glomus of the lateral ventricle). They tend to occur in specific areas in which are usually found arachnoid cell nests in the dura (such as in the vicinity of the venous sinuses). Meningiomas show some propensity to develop in patients irradiated during childhood.
- *Histopathology*: the WHO classifies the meningiomas into a benign typical type, an atypical type and a malignant type (rare). Three different subtypes without any prognostic significance may be identified within the first group including: a meningothelial (or syncytial), a fibroblastic (or fibrous) and a transitional (or mixed subtype, intermediate between the 2 formers). Malignant meningiomas, more frequent in males, show variable degree of anaplasia and may even give rise to extracranial metastases.
- *Signs*: may be an incidental finding on imaging, asymptomatic for long time before coming to medical attention (small meningiomas less than 20 mm are commonly found at autopsy in elderly patients). Meningiomas may reach a huge size before diagnosis is made due to their very slow rate of growth (over 10 to 15 years of duration). This concerns more commonly convexity meningiomas, in which focal seizure is often an early sign. Parasagittal tumors developing in the frontal or parietal interhemispheric fissure may cause progressive unilateral or bilateral spastic paresis or numbness of the inferior limbs. Distinctive clinical syndromes characterize some varieties such as: olfactory grooves, tuberculum sellae and/or lesser sphenoid wing, and petro-clival, cerebellopontine angle or foramen magnum meningiomas (see skull base tumors). The size reached by the tumors before they become symptomatic varies with their anatomical location. At the olfactory grooves: anosmia, mental changes, ipsilateral visual loss or Foster-Kennedy syndrome. At the spheno-cavernous region: progressive exophthalmos, oculomotor palsy, optic atrophy, uncinate fits and hyposmia.

MR characteristic findings

Morphology

- More than 90% of meningiomas are supratentorial, found in the anterior half of the skull in about 2/3 of the cases.
- Meningiomas are more precisely located in the following sites: parasagittal interhemispheric space (25%); lateral convexity (20%) from which more than half are found anterior to the central sulcus or at the coronal suture; sphenoidal ridge (20%) including the inner anterior clinoid variety, the outer pterion, and the in between parasellar variety which tends to develop in the lateral fissure; olfactory grooves (10%); sellar region including tuberculum sellae (10%); posterior fossa (10%) including tentorial leafs, free edges and torcular, cerebellopontine angle, clivus and foramen magnum; the remaining being intraventricular, growing mainly in the glomus of the lateral ventricle (to differentiate from choroid plexus papillomas but usually found in children) or even extracranially (orbit, nasopharynx, scalp).
- Meningiomas are typically well demarcated globular or crescentic or lobulated extra-axial

masses, showing usually a broad dural attachement. They may appear also as pedunculated masses or flat so-called "en plaque meningioma". Absence of dural attachment concerns the intraventricular variety (almost 90% found in the lateral ventricles).

- Calcifications are found in almost 1/4 of the meningiomas (best identified on CT, if needed), tumoral vascularity in 1/3, and central areas of necrosis with cystic degeneration are not infrequent (found particularly in angioblastic meningiomas), and hemorrhagic foci may be depicted in about 5% of the cases. These secondary features contribute to give a heterogeneous intensity pattern to almost 2/3 of the meningiomas. At times, arachnoid cysts may be found at the tumor margin.
- Its extra-axial location is demonstrated by the following morphological features: presence of marginal vessels (flow void curvilinear or punctate structures) and/or CSF clefts at the tumor-brain cortex interface found in almost 2/3 of the cases; and accordion-like pattern of deformation and compression of the adjacent gyri.
- Hyperostosis of adjacent skull bone and or "blistering" (jugum sphenoidal meningiomas) or even bony invasion, are other common accompanying features.
- Almost characteristic is the "tail sign" frequently evidenced adjacent to the tumor's base in particular after contrast infusion. This dural enhancement pattern, found in 1/3 of convexity meningiomas, seems to be related either to dural infiltration by the meningiomatous process or probably simply due to secondary reactive edematous and hyperemic thickening of the adjacent dura.
- Unusual variety may be encountered: the lipomatous meningiomas (secondary metaplastic transformation into adipocytes, case 33).

Signal intensity

- SI pattern may help to differentiate meningiomas.
- SE-T1w or GE-T1w: usually hypointense to white matter or isointense to cortex.
- SE-PDw: isointense or slightly hyperintense to cortex.
- SE-T2w: variable and may add characterization to some extent; meningiomas appear isointense to white matter (hypointense SI in most of the fibroblastic subtype) or hyperintense to brain (in meningothelial/syncitial subtype). Lipomatous variety shows a hypointense SI to cortex on heavily T2w (may show a bright SI on T1w) and a chemical shift artifact at the fat-water interface. Angioblastic tumors may show intratumoral multiple, more or less confluent cystic hyperintense foci. Exceptional cystic tumors show marked hyperintense SI on T2w as on PDw sequences (case 36).
- Spin echo sequences (T1w and T2w) must be used instead of GE sequences in the evaluation of skull base meningiomas (coronal cuts) or in case of skull invasion.
- Calcifications are better seen with CT. Gradient echo sequences may improve their detectability using MR.

Contrast enhancement

- Meningiomas usually show intense contrast enhancement.
- Contrast infusion reveals the typical but not pathognomonic "dural sign".
- Contrast infusion is mandatory to detect and evaluate "en plaque" meningiomas, as well as anterior skull base and perioptic meningiomas.

Case 32: Lipomatous meningioma (metaplastic)

Fig. a: axial cut; SE - PDw

Fig. b: axial cut; SE - T2w

Fig. c: sagittal cut; GE - T1w

Fig. d: coronal cut; SE - T1w + Gd

Extra-axial mass in the left frontal region containing multiple nodular formations. The lesion is iso-intense on PDw (**a,** ↑↑) becoming very hypointense similar to the white matter on T2w (**b,** ↑↑). The mass, hypointense on T1w (**c,** ↑), shows marked enhancement (**d,** ↑↑). On PDw and mainly on T2w (**a, b**), a hyperintense CSF cleft is noticed (↑), surrounding the inner border of the mass, due to its extra-axial location. Its external aspect adheres to the skull vault (**b,** *). Note the enhancing dural tail (**d,** *) and the accordion-like compression of the adjacent cortical convolutions anteriorly and posteriorly (**c,** *).

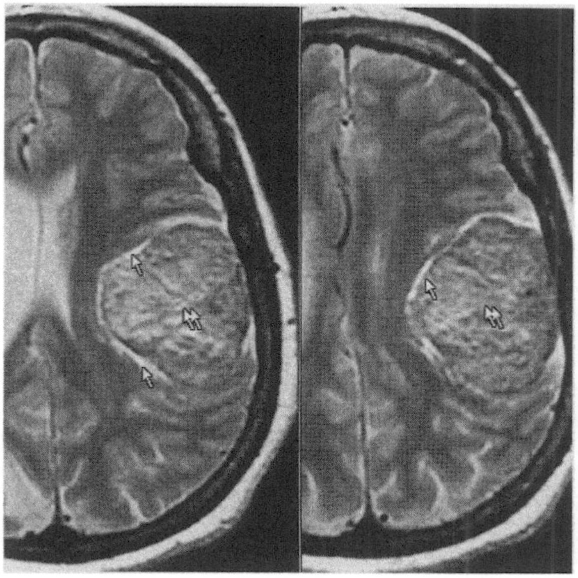

a

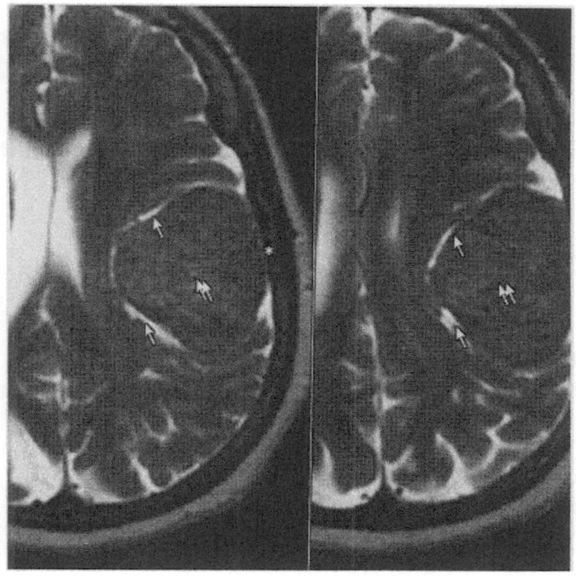

b

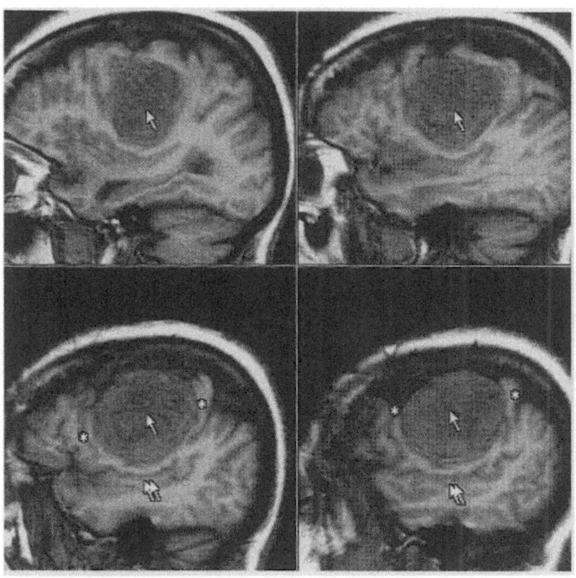

c

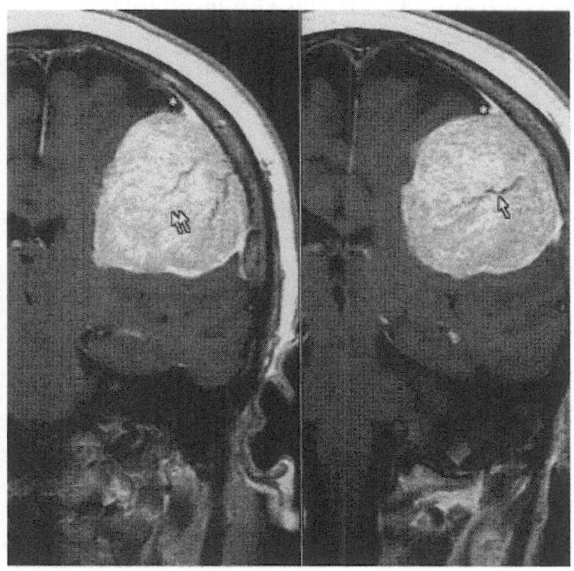

d

Case 33: Jugo-olfactory fibroblastic osteo-meningioma

Fig. a: sagittal cut; SE - T2w

Fig. b: sagittal cut; SE - T2w

Huge mass developed in the anterior skull base, with a broad dural-based margin, inserted on the jugum and the olfactory grooves, with osteoma formation (**a,** ↑↑). The tumor is hypointense to the cortex and white matter on heavily T2w sequences (**a, b**). The mass raises the fronto-orbital gyri (**a,** down, ↑). Note the presence of a CSF interface at the tumor-brain margin (**a,** down, ↑) and a vasogenic edema in the adjacent frontal lobe (**b**). Note also the dilation of the perioptic sub-arachnoid spaces (**b,** ↑↑↑), and the intracanalar portion of the optic nerve (**b,** ↑↑).

Case 34: Sphenoïdal wing meningioma

Fig. c: axial cut (512); SE - T1w + Gd

Fig. d: coronal cut (512); SE - T1w + Gd

Post contrast cuts showing marked homogeneous enhancement of the mass inserted on the sphenoid ridge and developed the temporal polar region, with mass effect on the temporal lobe displaced posteriorly and on the orbital gyri displaced upward. The mass shows intra-tumoral small vessels (curvilinear signal voids) as well as displaced vessels at the tumor bred interface.

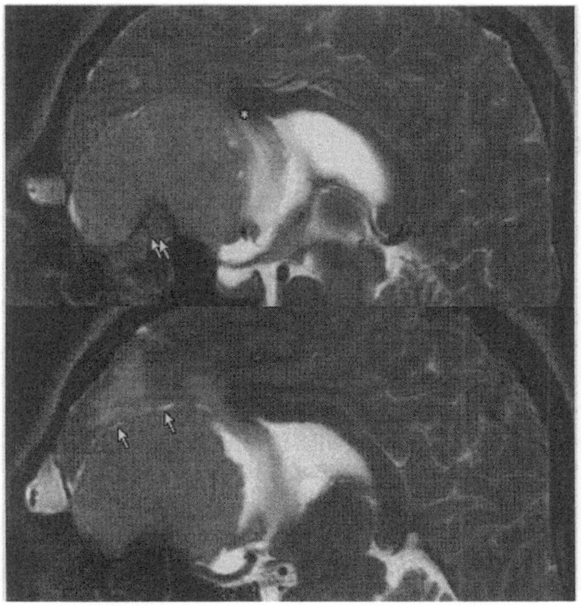

a

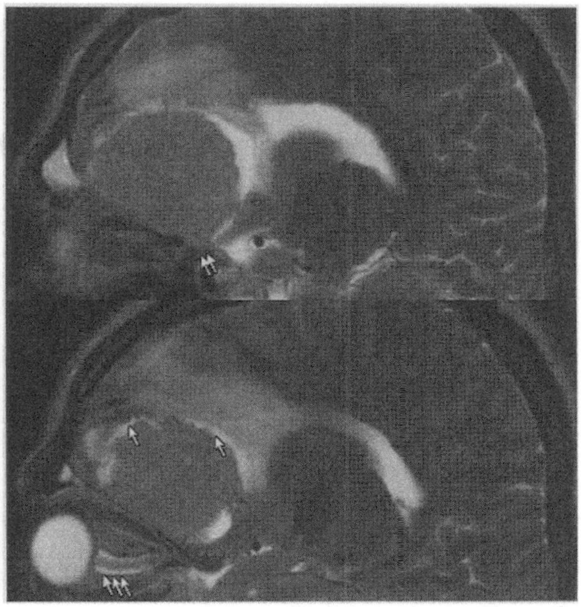

b

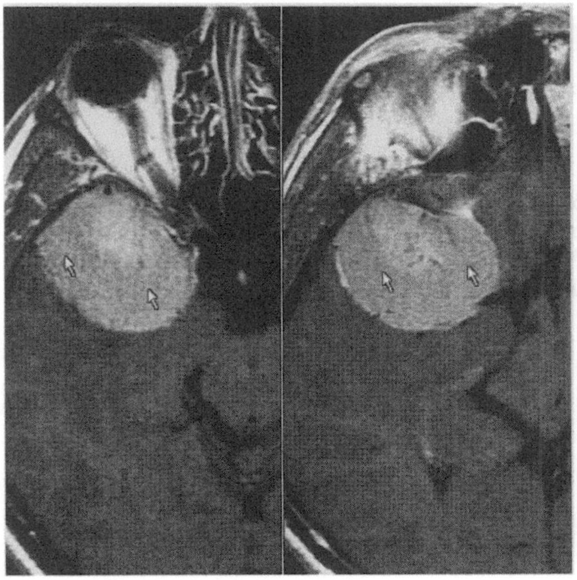

c

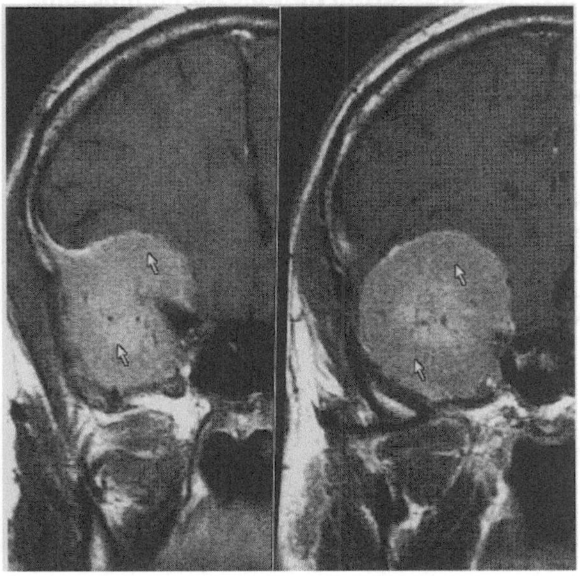

d

Case 35: Cystic hemangioblastic meningioma (child)

Fig. a: axial cut; SE - PDw

Fig. b: axial cut; GE (3D) - T1w – Gd

Exceptional and unusual appearance of a huge left frontal multi-cystic mass, hyperintense on PDw sequence and hypointense on T1 w (**b**) with very thin marginal rim of enhancement. The mass displaces the frontal horns to the right side and the front cortex posteriorly with accordion-like appearance of the cortex. There is no real identification of the various anatomic interfaces at the tumor-brain margin: no CSF cleft nor arterial vascular rim. Note also the absence of dural tail enhancement superficially or remodeling of the contiguous inner table of the skull voult.

Case 36: Parasagittal fibroblastic meningioma

Fig. c: coronal cut; SE - T1w + Gd

Fig. d: sagittal view, MR angio; 2DTOF

Slightly heterogeneous left parasagittal mass, with marked enhancement on T1w sequence (c) presenting a small trans-falcorial extension (**c**, down) and a compression of the superior sagittal sinus shown on the sagittal MIP of the MR-angio sequence (**d**, ↑↑). Note the presence of the small blood vessels, branches of the pericallosal arteries, at the medial aspect of the tumor. Note also the mass effect with displacement of the falx and downward displacement of the corpus callosum and the frontal horns, partly collapsed.

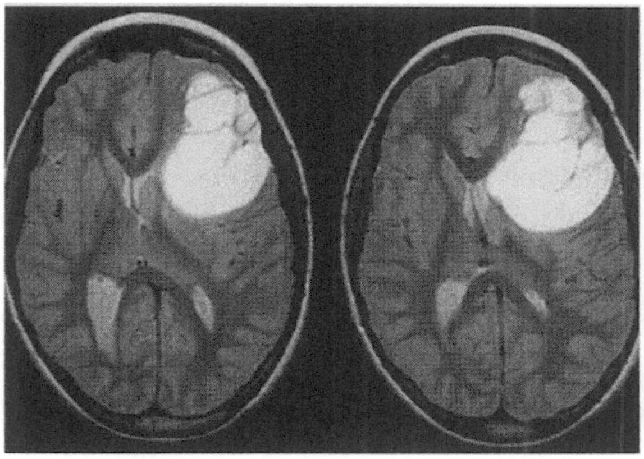

a

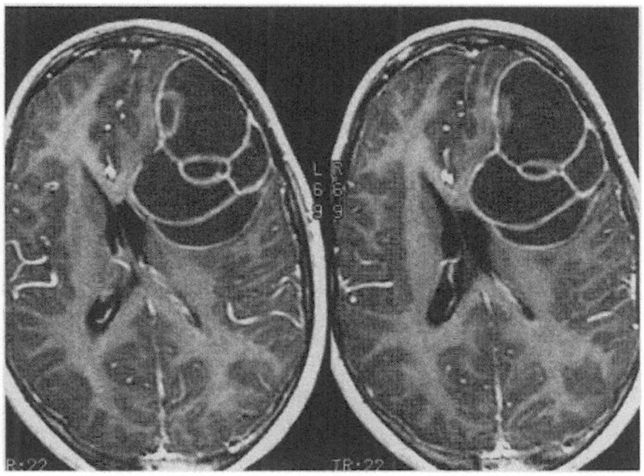

b

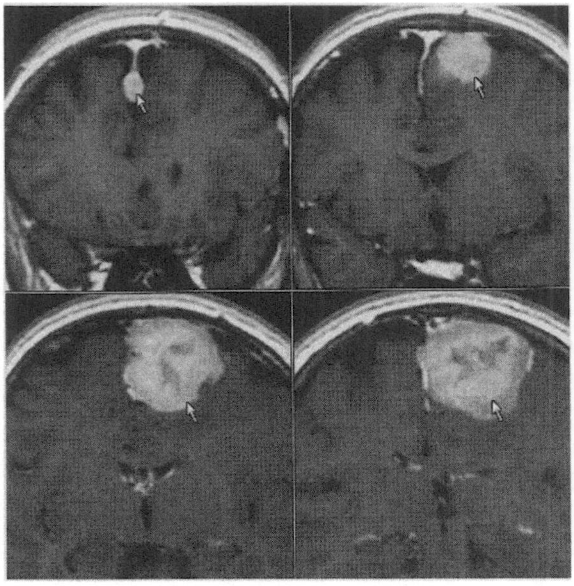

c

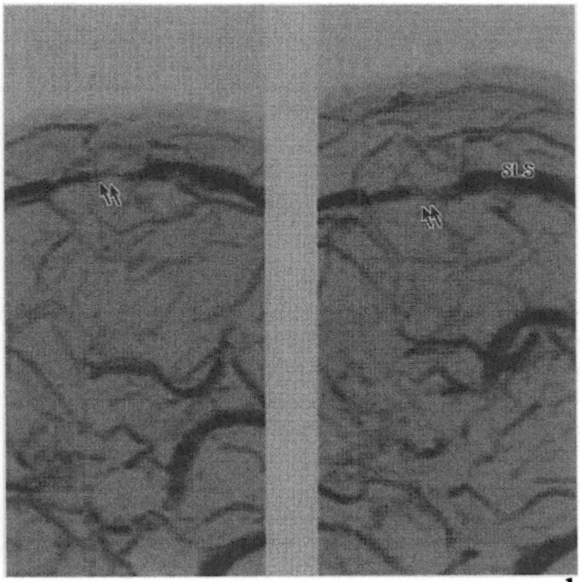

d

Case 37: Anaplastic malignant osteomeningioma of the voult

Fig. a: coronal cut (512); SE - T2w

Very large frontal osteomeningioma of the convexity growing through the skull vault with hyperostotic bony change in the calvarium. The lesion is iso-intense to the cortex on T2w (**a**, right, ↑) and enhances moderately (**b**). Its extracranial component presents as a larger scalp mass (**b**). The outer table of the skull shows increased signal void. This increased thickening and hypointensity of the skull indicates an osteblastic hyperostotic reaction. The intracranial component of the lesion invades and displaces downward the superior sagittal sinus (**a**, left image, ↑).

Case 38: Voult metastasis

Fig. c: coronal cut; SE - T1w + Gd

Fig. d: sagittal cut; GE - T1w + Gd

Left parasagittal frontal convexity highly heterogeneous and lytic tumor presenting as a scalp mass with an extradural major component (**c, d,** ↑↑). The mass is highly heterogeneous with areas of cystic necrosis displacing the leptomeninges downward and with brain compression (**d**). There is invasion of the subcutaneous tissues extra-cranially (**d,** ↑↑↑), but no invasion of the brain. Note another localization of the metastatic disease in the region of the left superior orbital fissure (**c,** ↑).

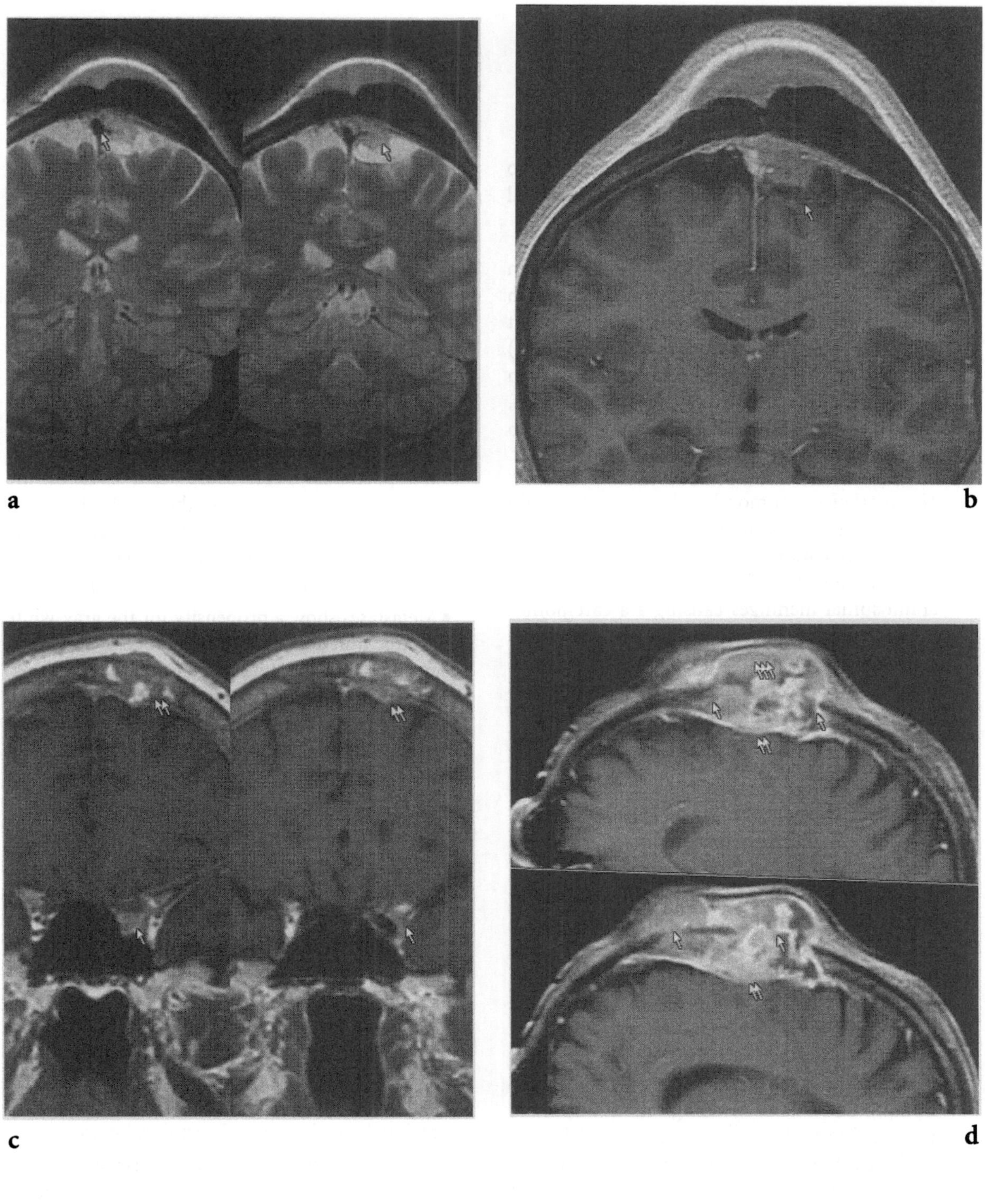
a
b
c
d

Metastatic tumors

Cases 38-50 and 53: metastases to the brain and leptomeninges

Clinical and pathological features

- Cancer can involve the brain by metastasing to its parenchyma, the leptomeninges and the skull voult or base.
- The incidence of brain metastases is high in patients between the fourth and seventh decades. Metastases to the brain are found at least twice more frequently in adults (15-20%) than in children (7%). The incidence is even higher in autopsy series reaching almost 30%. Metastatic brain tumors correspond roughly to 1/5 of all detected brain tumors.
- *Histopathology*: tumoral cells reach the brain most frequently hematogeneously (via systemic arterial circulation), or by direct extension from the skull and dura. The remaining involve the craniospinal meninges causing a a carcinomatous meningitis. Microscopically, the metastasic cells are well demarcated from the adjacent parenchymal tissue contrary to primary brain tumors. This does not exclude spreading of tumoral cells in the adjacent perivascular spaces.
- *Signs*: neurological signs and symptoms from brain metastases lead to the discovery of the primary cancer in nearly half the cases. These are the following, in order of frequency: headache, usually mild; mental changes; seizures (common in melanomas and leptomeningeal lesions); hemiparesis and ataxic gait; signs of intracranial hypertension; visual field defects and papilledema; aphasia and cognitive disorders. The symptomatology is often mild but usually does not differ from that of a glioblastoma multiformis. Metastases to the skull voult are commonly asymptomatic, those invading the skull base are associated to multiple and variable cranial nerve palsies (Garcin's syndrome). Search for the primitive malignancy if unknown, and evaluation of the disease extent clinically and radiologically is mandatory in order to take the accurate therapeutic decision.

MR characteristic findings

Morphology

- Metastases to the brain are most frequently multiple in about 70% of the patients (cases 40, 41). Solitary metastasis is not uncommon, encountered in more than 30% of the cases usually in melanoma, breast and lung carcinomas.
- Localization of metastases varies in frequency with respect to the relative volume of the lobe or the anatomical structure and the blood flow to the particular area. More than 80% of the metastases are found in the cerebral hemispheres, the remaining in the posterior fossa (cerebellum corresponding to 1/10 of volume of the cerebral hemispheres) (cases 49-51).
- Metastases show a propensity for the gray-white matter junction, due to hematogenous spread, mainly from the lung and breast or melanoma, and from the gastrointestinal (colon) and the genito-urinary (kidneys) systems. Prostatic, and oesophageal carcinomas metastasize rarely to the brain parenchyma.
- Skull and dural metastases are focal lytic or blastic masses occuring in tumors which usually metastasize to bone such as: prostate, breast and lung or kidney carcinomas, and in lymphomas and metastatic neuroblastoma. These masses developed in the diploë show destruction of the inner and/or outer calvarial tables, eventually extending to the epidural space. The epidural extension is commonly contained by the dura which may at times be transgressed. Subdural localizations of metastases are presumably due to hematogeneous spreading rather than locoregional invasion from the epidural space.
- Morphologically, metastases appear as well circumscribed nodular masses and well demarcated from brain parenchyma (as compared to primary gliomas particularly high grade), usually solid though they may show central necrosis.

- The metastases evoke little glial reaction but are surrounded by a large amount of vasogenic edema that may follow the white matter tracts but which does not reflect necessarily the size of the tumor (may mimic brain abscesses) (cases 42, 43, 45). This edema does not involve the cortical mantle and usually does not extend through the corpus callosum as a primary infiltrating brain tumor. Small cortical metastatic lesions usually lack of peritumoral edematous changes and may be missed on non enhanced exams.
- Hemorrhagic components are most frequently associated with metastatic melanoma, renal cell carcinoma, chorioepithelioma, and thyroid carcinoma.
- It is not uncommon to find a huge solitary tumor with irregular margins and necrotic center which may be confused with a high grade primary tumor (glioblastoma or high grade gliomas) (cases 46, 47). Thyroid carcinoma and hypernephroma tend to give a solitary metastasis more frequently than other tumors (case 44).
- Carcinomatous meningitis or leptomeningeal carcinomatosis due to widespread dissemination of the tumoral metastatic process throughout the leptomeninges and the ventricular system, are best demonstrated using MR particularly after contrast infusion, and are found in about 5% of the cases. The causal primary malignancies are: the breast, lung and stomachic carcinomas, melanoma, lymphoma and leukemia in children. Numerous focal localizations may also be found along the ependymal wall of the ventricular cavities confirming the subarachnoid seeding. The presence of an associated communicating hydrocephalus is also a highly suggestive associated feature. Note that tumoral seeding within the leptomeningeal spaces cannot be differentiated from other meningeal diseases associated with meninges thickening (pachymeningitis, sarcoidosis with common leptomeningeal and dural sarcoid) and sometimes from subarachnoid seeding from some of the primary brain tumors such as: high grade gliomas (case 54), lymphomas, medulloblastoma or germinoma.

Signal intensity

- GE-T1w or SE-T1w: hypointense to white matter, surrounded by markedly hypointense vasogenic edema. May show areas of hyperintense SI in case of intratumoral hemorrhage (paramagnetic methemoglobin).
- FLAIR or SE-T2w: variable SI of the tumor, frequently iso- or hypointense and more or less heterogeneous (necrosis, hemorrhagic components, hypercellularity, non-brain tissue), surrounded by the hyperintense SI of the vasogenic edema, usually well demarcated from the metastatic lesion (pattern usually not shown in high grade primary brain tumors). Cystic changes secondary to necrosis also show a hyperintense signal.

Contrast enhancement

- The majority of brain metastases show almost immediate intense nodular or thick and irregular ring-like (central necrosis) contrast enhancement (similarly to high grade gliomas). Enhancement pattern helps to demonstrate the clear-cut demarcation of the solid tumor from the wide peritumoral vasogenic edema (almost similarly to FLAIR or heavily weighted FSE-T2 sequences). Disseminated small cortical-subcortical lesions usually lacking of significant peripheral edema, have to be differentiated from infectious granulomas and abscesses such as with tuberculosis or parasitosis.
- Contrast enhanced MR is much more sensitive than T2w sequences and CT with i.v. contrast, for the detection of the metastatic lesions. Contrast infusion is mandatory to demonstrate a carcinomatous meningitis showing enhanced variable thickening of the leptomeninges. MR is the modality of choice in such cases were CT is usually insensitive (particularly in lymphomas or leukemia). Contrast infusion improves also the evaluation of metastases to the diploïc space (SE-T1w and fat suppression after Gd infusion) as those extending to the extradural or subdural spaces (case 39).
- Detection sensitivity of contrast MR is improved in case of administration of a double (0.2 mmol/kg) or even a triple dose (0.3 mmol/kg) of Gadolinium or when using a normal single dose of paramagnetic contrast and performing a T1w sequence with magnetization transfer pulse (probably more cost-efficient).

Case 39: Multiple small nodular metastases (breast)

Fig. a: axial cut; SE - pDw

Fig. b: axial cut; SE - T1w + Gd

Multiple small nodular lesions disseminated in the cerebral hemispheres involving the periventricular white matter, as well as the subcortical (**a,** *) and the CSF cortical junction (**b,** ↑↑). Note the paraseptal intraventricular lesion (**a, b,** ↑). The metastases are iso-intense to the cortex on PDw (**a**) and show contrast enhancement with better identification of the more cortically located foci. Note also the absence of consistent peritumoral vasogenic edema.

Case 40: Breast carcinoma metastases, grade II

Fig. c: coronal cut; SE - T1w + Gd

Fig. d: sagittal cut; SE - T1w + Gd

Multiple metastatic foci mainly visible after double dose contrast infusion, disseminated throughout the brain, especially in the periventricular areas (**c,** ↑), yet involving also the cortical mantle (**d,** ↑↑) as well as the insular cortex (**d,** ↑).

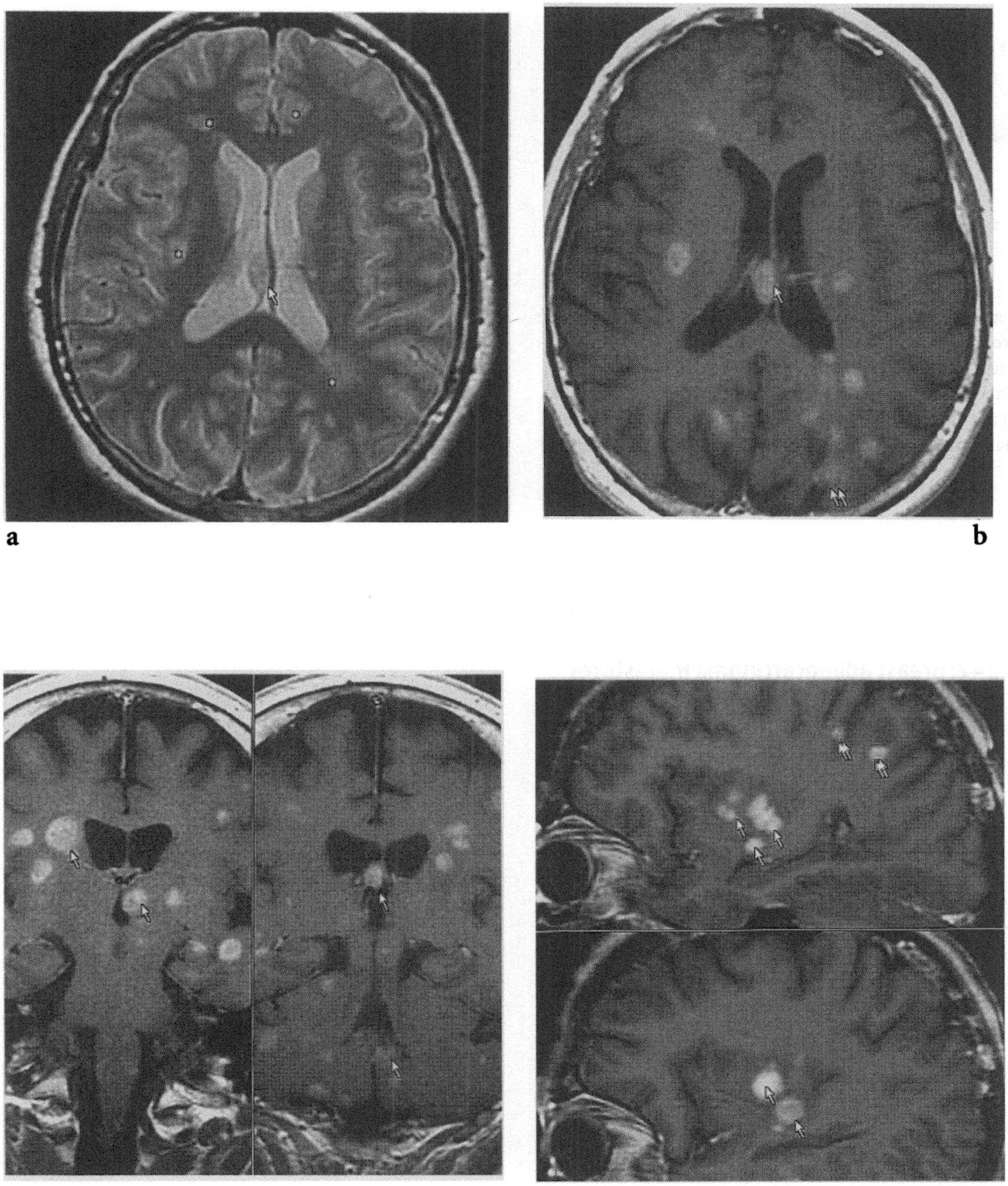

a b

c d

Case 41: Breast metastases

Fig. a: axial cut; FLAIR - T2w

Fig. b: axial cut (512); SE - T1w + Gd

Metastatic tumors of different size showing central necrosis surrounded by massive amounts of edema very well depicted on the FLAIR sequence (**a,** ↑↑↑). The vasogenic edema (**a,** ↑↑↑) delineate nicely the metastatic process (**a,** ↑↑) as compared with the T1w sequence after Gadolinium enhancement, (**b,** ↑↑ and ↑↑↑). This shows two additional periventricular lesions (**b,** ↑) that could be confused with the vasogenic edema on the FLAIR sequence. Note the mass effect of the large frontal metastasis in the lateral ventricle. Therefore, small metastatic lesions may be missed if one relies only on T2w sequences.

Case 42: Breast adenocarcinoma metastases

Fig. c: axial cut; SE - T2w

Fig. d: axial cut (512); SE - T1w + Gd

This case shows cortical metastasis (**c,** ↑) well circumscribed by the extensive vasogenic edema (**c,** ↑↑), spreading along the white matter core of the adjacent cerebral convolutions. The left parasagittal lesion is well shown after enhancement on the T1w sequence (**d,** ↑), surrounded by the vasogenic edema, hypointense on T1w (**d,** ↑↑). The left lesion shows on T2w (**c,** ↑↑) areas of high signal corresponding to non hemorrhagic cystic necrosis, in this case of poorly differenciated breast adenocarcinoma.

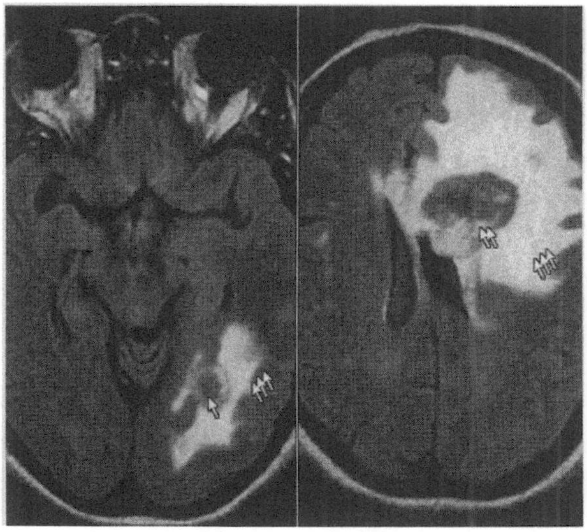

a

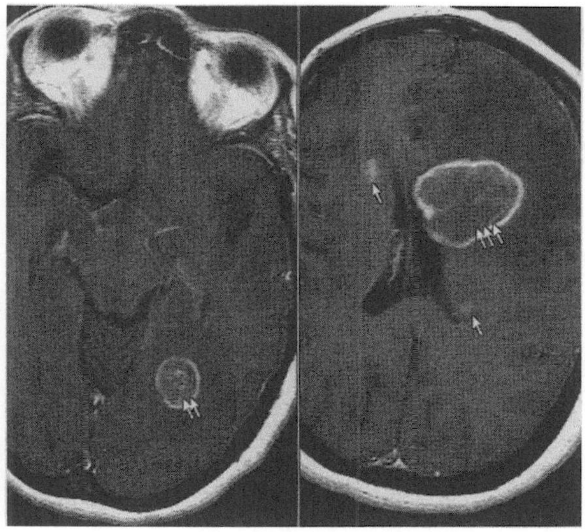

b

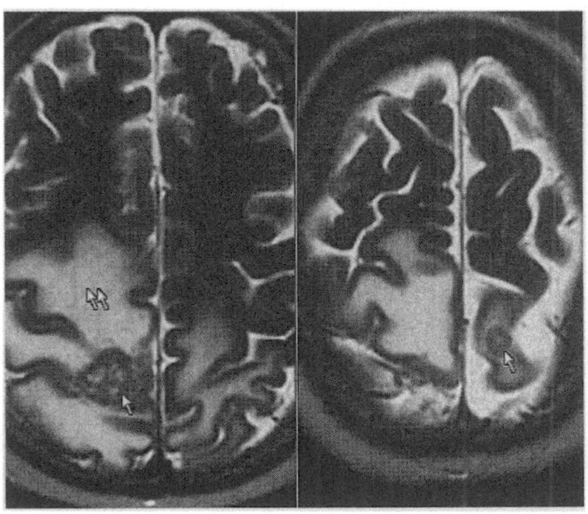

c

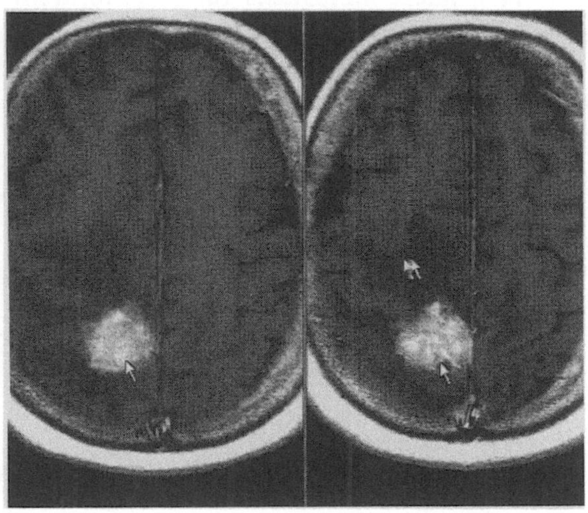

d

Case 43: Single metastasis (thyroid carcinoma)

Fig. a: axial cut; SE - T2w

Fig. b: axial cut; SE - T1w + Gd

Isolated large cystic para-ventricular mass, hyperintense on T2w, (a, ↑↑) surrounded by a small amount of edema (a, ↑). The mass is slightly hypointense on T1w (b, ↑↑) showing a thin rim of contrast enhancement (b, ↑).

Case 44: Isolated metastasis (lung adenocarcinoma)

Fig. c: axial cut; SE - T2w

Fig. d: axial cut; SE - T1w + Gd

Similar case showing an isolated subcortical mass with central necrosis (**c,** ↑↑) and irregular peripheral enhancement on T1w (**d,** ↑), delineating the area of central necrosis, fluid filled (**d,** ↑↑). This metastatic poorly differenciated adenocarcinoma is surrounded by a vasogenic edema readily identified as finger-like projections in the white matter, yet without involvement of the cerebral cortex (**c,** ↑↑↑). This feature may help to distinguish the edematous areas of bright signal on T2w or FLAIR from the primary infiltrative malignancy.

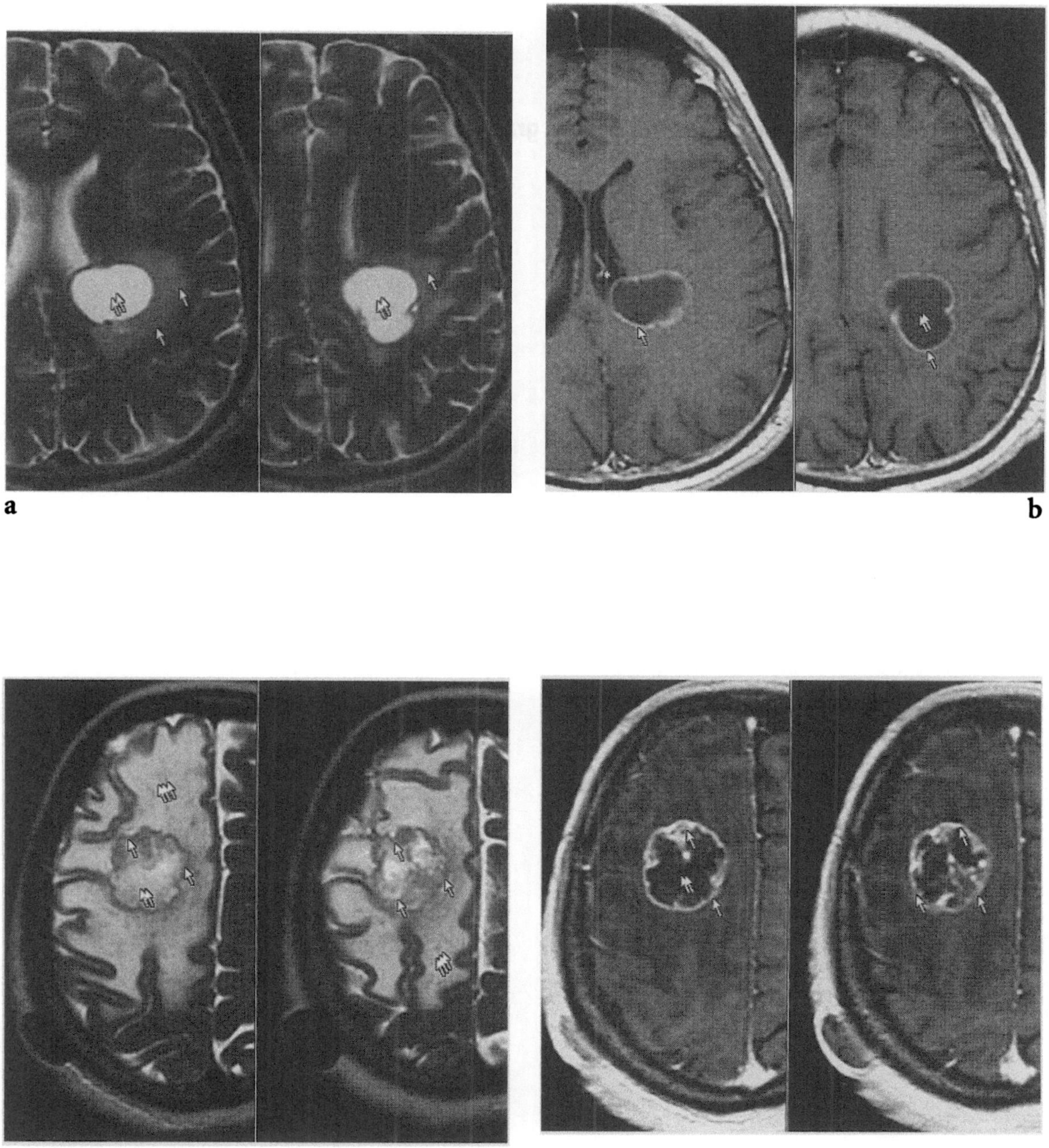

a b c d

Case 45: Isolated metastasis (breast carcinoma, grade III)

Fig. a: axial cut; SE - PDw

Fig. b: axial cut; SE - T1w + Gd

Large, mostly necrotic metastatic lesion involving the left parietal lobe, showing peripherally a solid component which is iso-signal to the cortex on PDw (**a**) enhancing on T1w after Gadolinium administration (**b,** ↑↑). The fluid filling cystic cavity is hypointense on T1w (**b,** ↑) and hyperintense on PDw (**a,** ↑) as well as on T2w, but more hyperintense than the surrounding edema on PDw (**a,** ↑↑). Note the important mass effect with displacement of the left lateral ventricle to the right. Note also that the vasogenic edema does not cross the splenium of the corpus callosum, thus distinguishing this isolated highly malignant lesion from an infiltrating high grade glioma.

Case 46: Isolated metastasis (carcinoma)

Fig. c: sagittal cut; GE - T1w

Fig. d: axial cut; SE - T1w + Gd

Similar case of a big isolated metastatic mass with large cystic necrosis, hypointense on T1w (**c, d,** ↑↑↑), showing a small mural solid portion in the fronto-polar region (**c,** ↑). The lesion shows a thin rim of peripheral enhancement surrounding the non hemorrhagic CSF-like cystic necrosis (**d,** ↑) as well as a heterogeneous enhancement of the para-sagittal mural nodule (**d,** ↑↑). Note the mass effect on the ventricles and the paracallosal region, diplaced posteriorly (**c,** ↑↑).

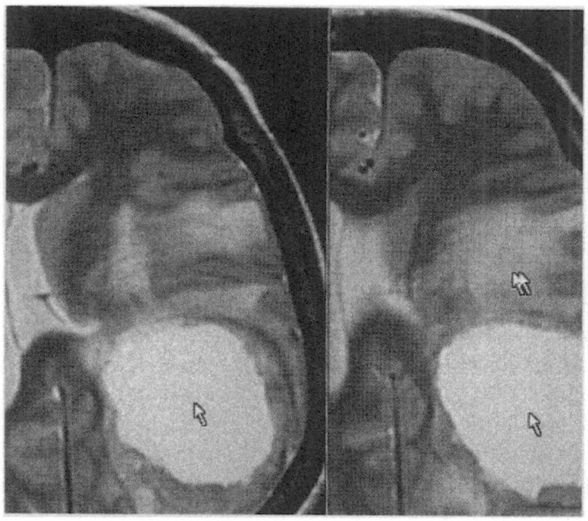

a

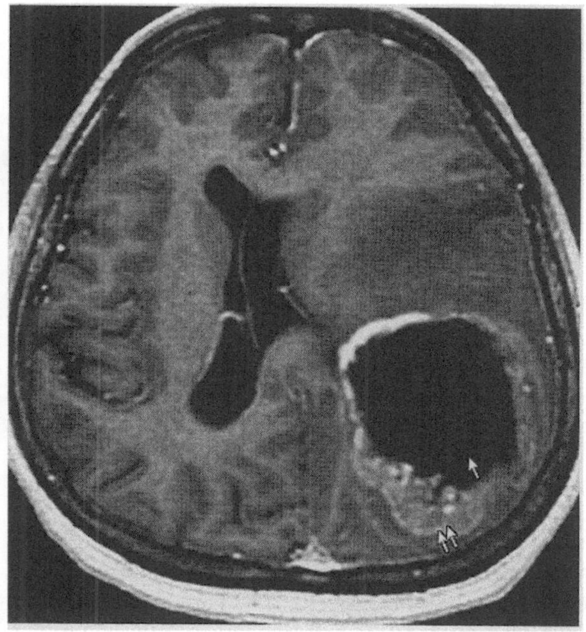

b

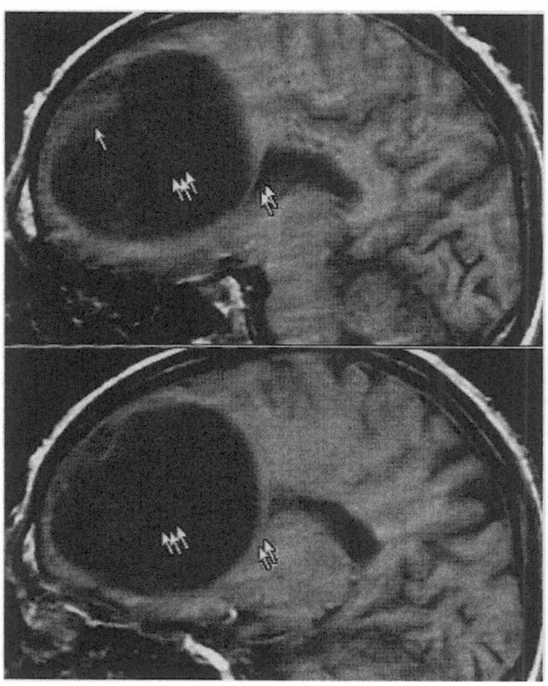

c

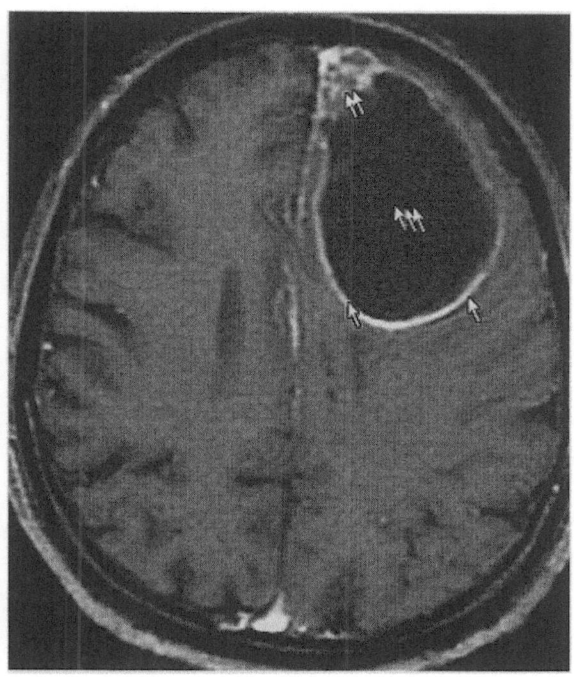

d

Case 47: Multiple metastasis (osteosarcoma of the fibula)

Fig. a: axial cut (512);	GE - T1w	**Fig. c:** coronal cut (512);	SE - T2w
Fig. b: axial cut (512);	SE - T2w	**Fig. d:** coronal cut;	SE - T1w + Gd

Rare metastatic osteosarcoma to the brain and the leptomeninges with a highly heterogeneous and hemorrhagic mass involving the mesial aspect of the right temporal lobe (surrounded by vasogenic edema). The metastasis is hypointense on T1w (**a**) and iso-intense to the cortex on T2w (**b,** ↑, **c,** ↑↑) enhancing after contrast infusion. The tumor is partly cystic due to intratumoral necrosis with very high signal on T2w, iso-signal on T1w with some hemorrhage into the solid portion showing high signal on T1w (**a,** ↑↑) as well as in the cystic portion with a fluid-fluid level showing a very low signal intensity on T2w (**b,** *) due to acute bleeding with deoxyhemoglobin deposition (**b,** *, **a,** ↑). Note the mass effect on the midline structures displaced to the left. Note the vasogenic edema surrounding the frontal parasagittal lesion and the presence of diploïc high signal abnormalities in the skull voult bilaterally. Note the widespread dissemination of the metastatic lesion involving the whole brain with subependymal and leptomeningeal carcinomatosis best seen on T1w (**d**), after contrast infusion.

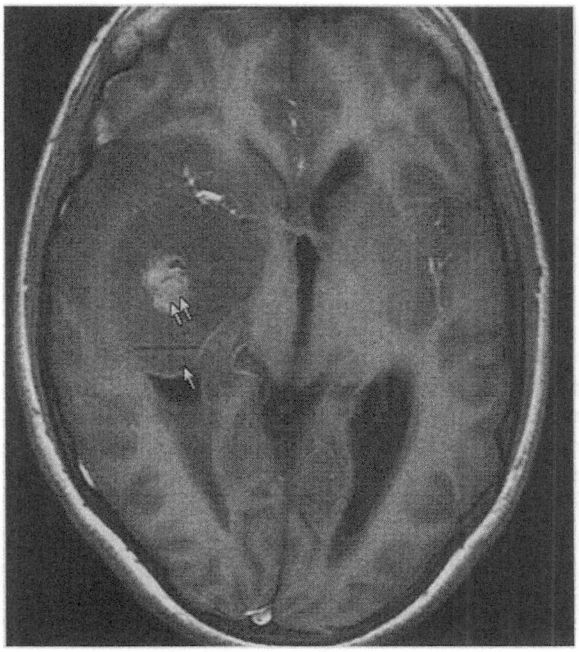

a

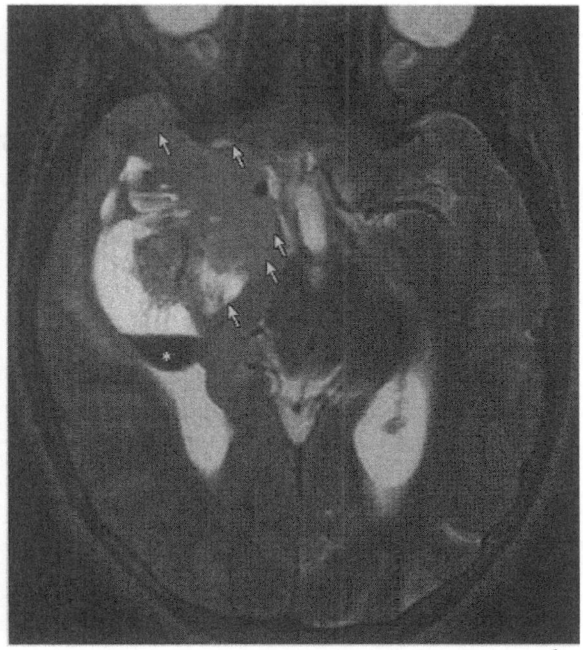

b

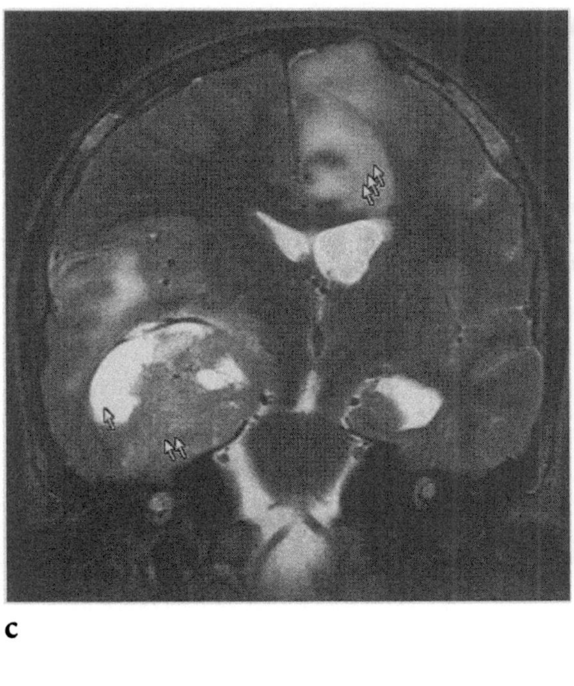

c

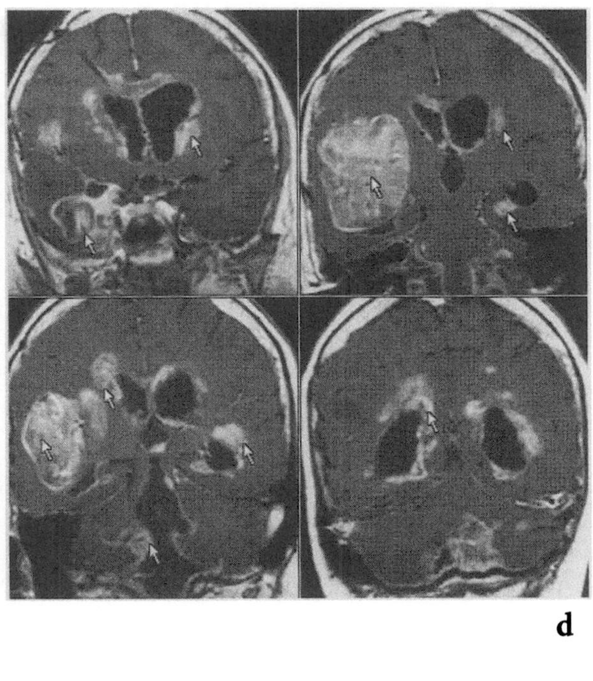

d

Case 48: Single and cystic mesencephalic metastasis (lung)

Fig. a: axial cut; SE - T2w

Fig. b: axial cut; SE - T1w + Gd

Fig. c: axial cut; FLAIR - T2w

Single metastatic lesion of lung carcinoma developped in the upper pons, highly necrotic, showing a hyperintense SI on T2w (**a,** ↑) and surrounded by peritumoral edema (**a,** ↑↑). The necrotic zone shows an isointense signal on FLAIR (**c**), well differenciated pars the hyperintense SI of vasogenic edema which involves diffusely the upper pons and midbrain tegmentum, and dissecting precisely the red nuclei (**c**). After contrast infusion, the small metastasis is enhanced in its periphery (8, ↑) surrounding the central necrotic area (6, ↑↑), wich remains hypointense. Note the moderate mass effect causing an elargement of the left ponto-mesencephalic region, witch associated displacement of the cerebral acqueduct to the right side.

Case 49: Single and cystic mesodiencephalic metastasis (oat cell carcinoma)

Fig. d: axial cut; E.G. (3D) - T1w + Gd

Isolated metastatic lung carcinoma with cystic necrosis developed in the right mesodiencephalic region with mass effect on the third ventricle displaced on the left. The lesion is largely cystic, hypointense on T1w (**d,** ↑↑), behaving like intraventricular CSF (↑↑↑), showing a thin curvilinear enhancement after contrast IV infusion (↑). The mass was unique and very well circumscribed but somehow unusual.

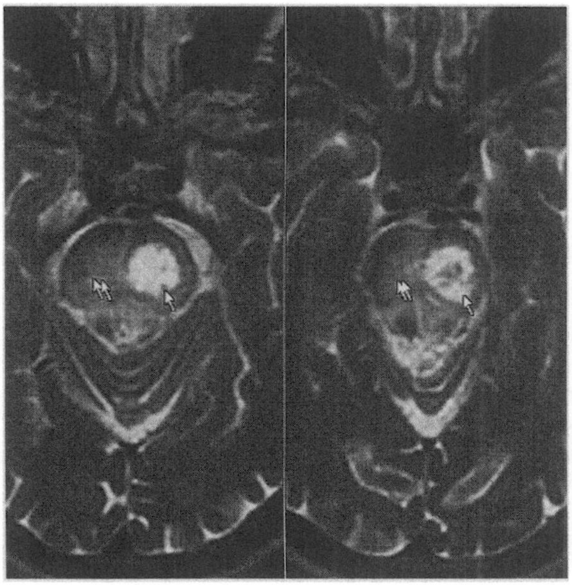

a

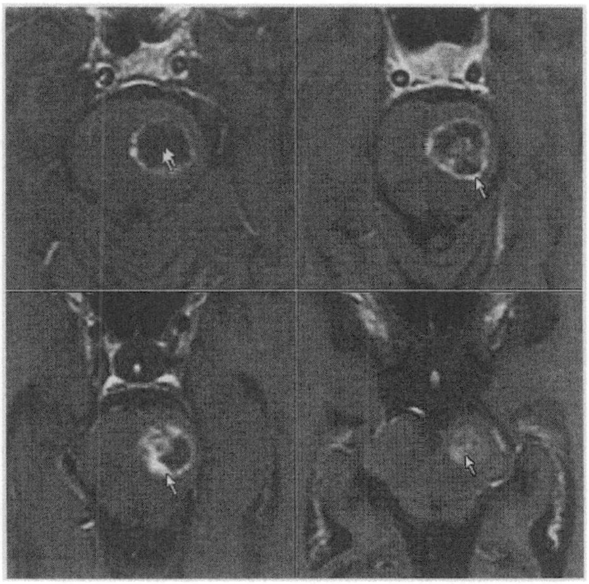

b

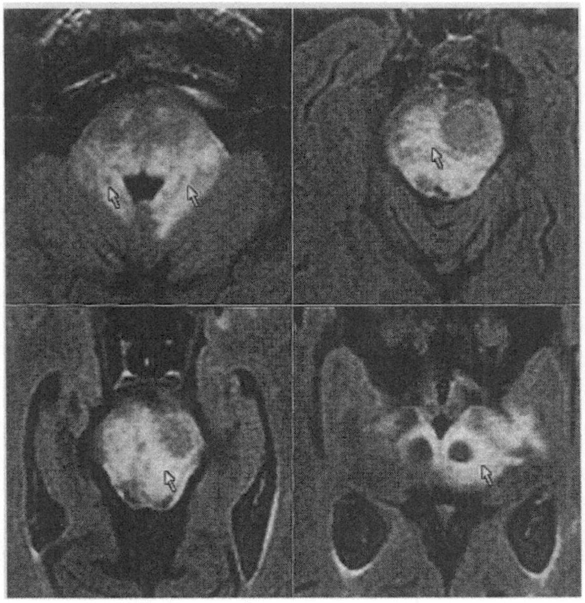

c

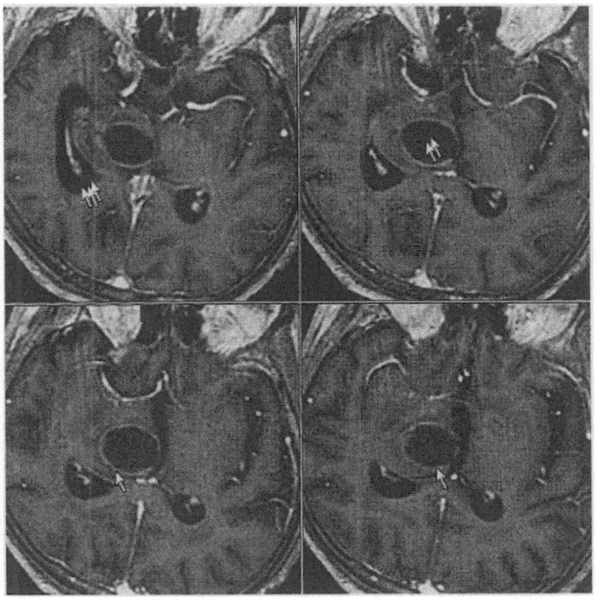

d

Case 50: Single metastasis of the cerebellum (lung)

Fig. a: axial cut; SE - PDw and T2w

Fig. b: coronal cut; SE - T2w

Isolated, well circumscribed and rounded mass involving the inferior aspect of the left cerebellar hemisphere, showing an iso-signal on PDw and T2w sequences (**a,** ↑), surrounded by large amount of vasogenic edema involving totally the album cerebelli (**a,** ↑↑; **b,** ↑↑↑). The vasogenic edema displaces the dentate nucleus upward (**b,** ↑↑), nicely dissected by the hyperintense edema on this heavily T2w sequence (**b**). Note a small area of necrosis in the mass (**b,** down, ↑) and also the lateral displacement of the vermis to the right. Note that metastatic processes are generally well differenciated from primary brain tumors due to the different nature of their constitutive tissue.

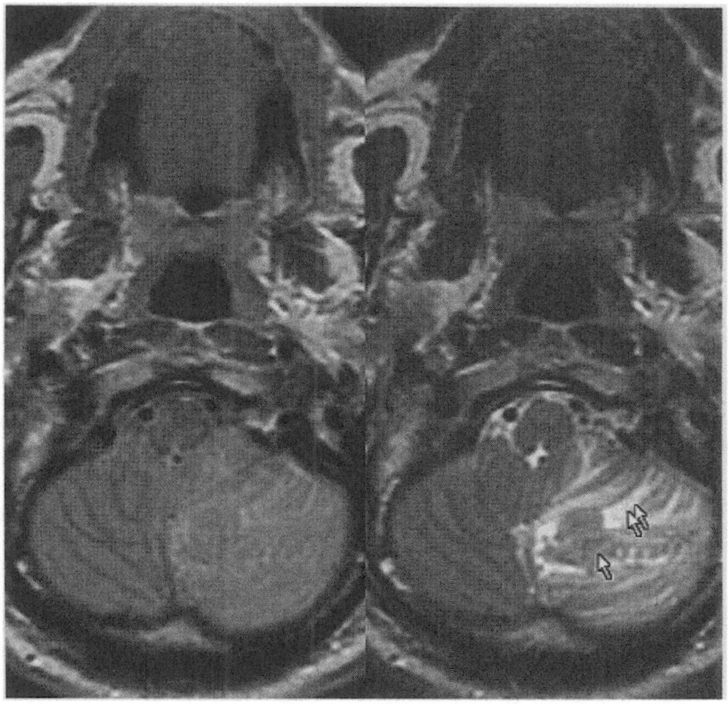

a

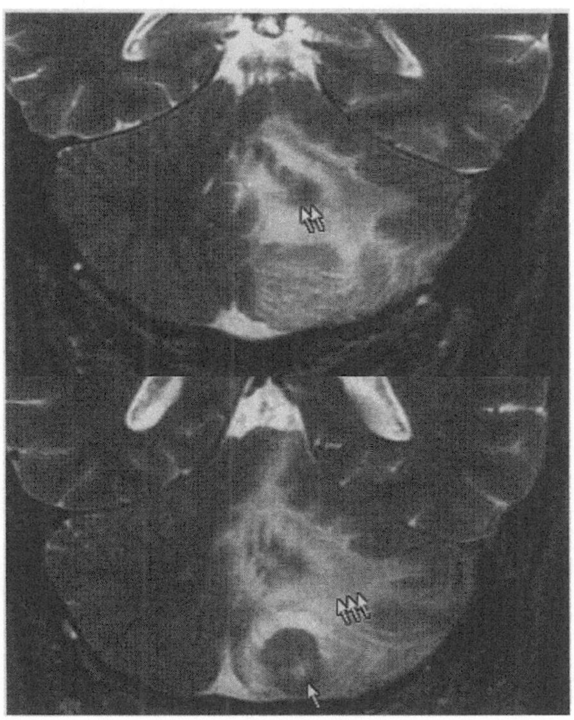

b

Cases 51, 52: lymphomas involving the CNS

Clinical and pathological features

- CNS lymphomas account for 1-3% of CNS tumors, but they are growing in frequency due to increase in number of AIDS and immunosuppressed or other immunodeficient patients in whom the average is 5-10%. It is half as frequent in children. It is believed that the overall frequency will increase.

- The disease affects usually patients in the 5th to 8th decades, with a peak incidence during the 6th decade (excluding the commonly younger AIDS population).

- Lymphomas may be primary or may occur as part of a systemic involvement. Almost half the patients with brain lymphomas show a systemic disease.

- Hematogeneous CNS invasion occurs frequently in case of systemic lymphoma, reaching most commonly the leptomeninges with a secondary infiltration of the brain parenchyma along the Virchow-Robin spaces.

- *Histopathology*: infiltration of brain or cord parenchyma is considered to be invariably the result of an extension from the arachnoid and the perivascular spaces of Virchow-Robin through destruction of the pial-glial membrane. Parenchymal infitration is quite variable. In most cases, the infiltrating process is limited to the area adjacent to the blood vessels. Leptomeningeal infiltration may be found at autopsy in the absence of any previous signs and symptoms.

- *Signs*: the clinical picture of intracranial lymphomas may be divided into 2 major presentation modes: 1- focal neurological signs of an expanding brain (or cord) mass with variable headache, mental changes and confusional states, drowsiness and uncommonly, seizures, which may mimic a high grade glioma or a metastasis; 2- signs of meningeal irritation eventually associated with a syndrome of intracranial hypertension, associated or not with signs of cranial (or spinal) nerve roots involvement. The duration of the clinical symptoms range from 3 to 6 months and the course of disease is more rapid in focal tumors.

MR characteristic findings

Morphology

- Lymphomas of the brain may appear as discrete tumoral processes and/or may show a diffuse infiltrative pattern. The occurrence of multiple focal masses associated to infiltrative growth patterns through the perivascular spaces, is characteristic of the disease process.
- Involvement of brain parenchyma is observed in non Hodgkin's lymphomas.
- Most brain lymphomas involve the cerebral hemispheres, found in 2/3 of the cases in the white matter and the basal ganglia, particularly adjacent to the ventricular walls or the pia-arachnoid.
- Intra-axial masses are frequently multifocal (especially in AIDS patients) in almost half the cases. Though huge, brain lymphomas show mild if any mass effect or associated vasogenic edema.
- Extra-axial subdural and epidural involvement, best evidenced using MR after contrast infusion, are secondary to a systemic disease and ought to be differentiated from "en plaques" leptomeningeal deposits such as meningiomas.

Signal intensity

- GE-T1w: hypointense to white matter.
- SE-PDw: isointense to cortex or slightly hyperintense to brain.
- FLAIR or SE-T2w: variable, isointense to cortex and hyperintense to white matter; but may be hypointense to brain due to its increased cellularity.

Contrast enhancement

- Focal lesions show a homogeneous enhancement in 2/3 of the cases and a peripheral enhancement pattern in 1/3 of the cases.
- Leptomeningeal involvement, usually non visible on CT, is best evidenced on enhanced MR scans, showing a diffuse enhancement pattern and/or focal thickenings due to focal deposits. Identification of trans-arachnoidal extension and seeding along the subjacent cortical perivascular spaces is highly suggestive.

Case 51: High grade lymphoma

Fig. a: axial cut; SE - PDw

Fig. b: axial cut; FLAIR - T2w

Presence of multiple circumscribed lesions involving the subcortical white matter of both hemispheres, showing an isosignal on PDw sequence (**a**) and a hyperintense signal on FLAIR (**b**), in an AIDS patient. The appearance and the signal of the lesions are non specific.

Case 52: Non Hodgkin malignant lymphoma (T cell)

Fig. c: axial cut; SE - PDw and T2w

Fig. d: axial cut; GE (3D) - T1w + Gd

Isolated space occupying lesion in the subcortical white matter of the left hemisphere, well demarcated, showing an isosignal to the cortex on PDw (**c**, left) and T2w (**c**, right) sequences, with a central area of nodular enhancement (**d**, ↑↑). Note the enhancement and the thickening of the leptomeninges (**d**, ↑).

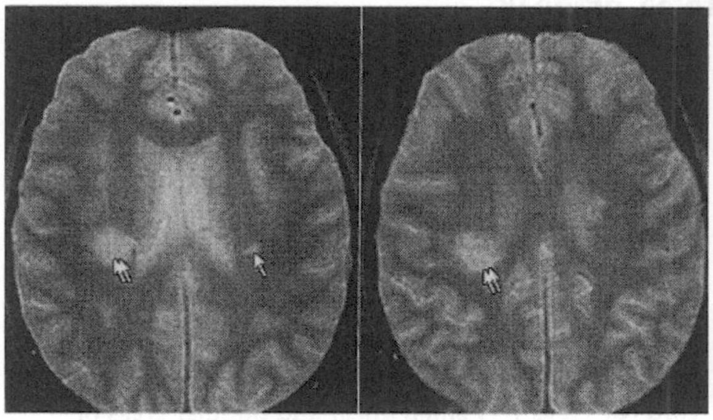

a

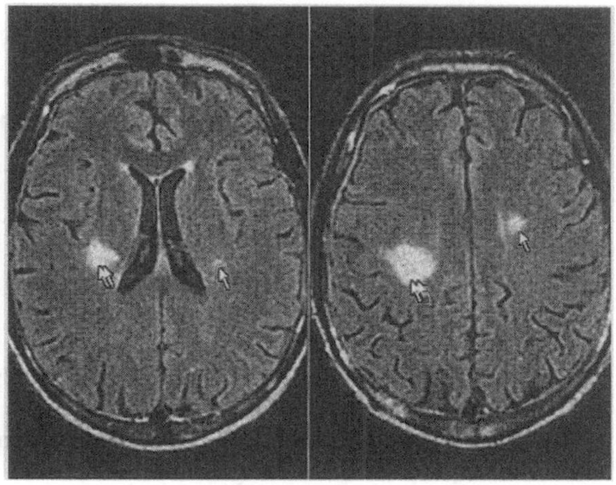

b

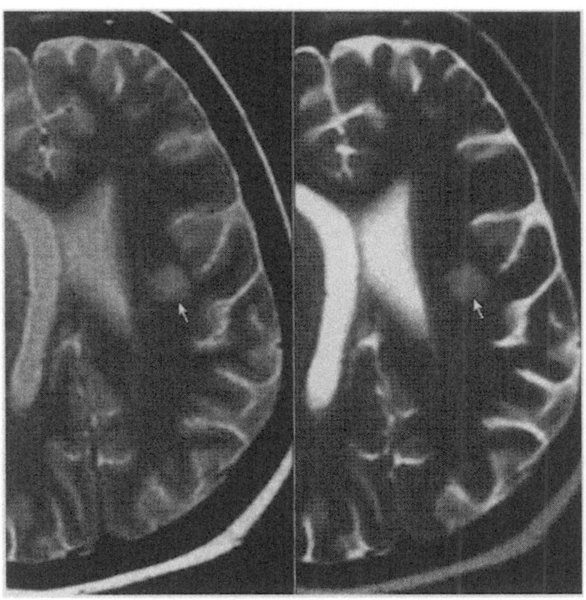

c

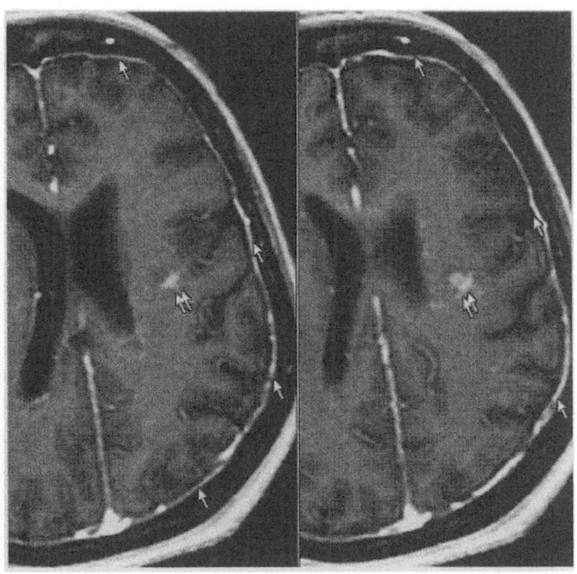

d

Case 53: Metastatic seeding from cervical astrocytoma, grade IV

Fig. a: axial cut; SE - T2w

Fig. b: axial cut; GE (3D) - T1w + Gd

Fig. c: sagittal cut; GE - T1w + Gd

Fig. d: sagittal cut; SE - T1w + Gd

Multiple nodular small space occupying lesions disseminated throughout the leptomeningeal spaces mainly in the posterior fossa, hyperintense on T2w (**a**, ↑), much better depicted and more numerous on T1w sequence after Gadolinium uptake (**b**, **c**), involving mainly the superior and inferior aspects of the fourth ventricle as well as the posterior aspect of the tectal plate, the infundibular recess (**c**, ↑), as well as the cranio-cervical junction (**c**, ↑↑). A subarachnoid seeding of the primitive high grade cervical associed astrocytoma is also noted along the dorsal spinal cord (**d**, ↑), as very small linear and nodular areas of enhancement.

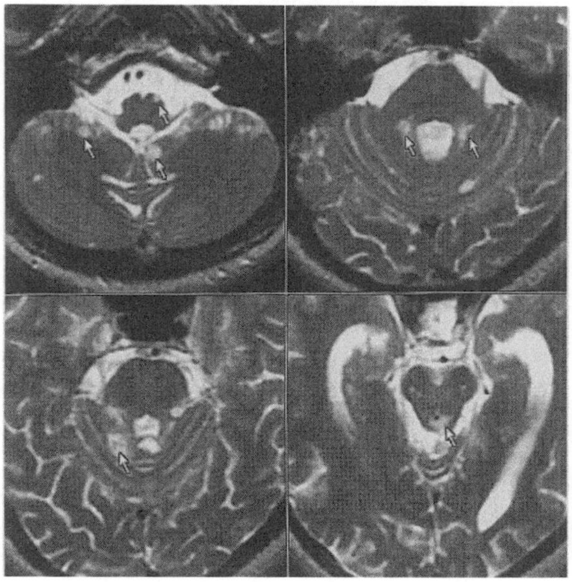

a

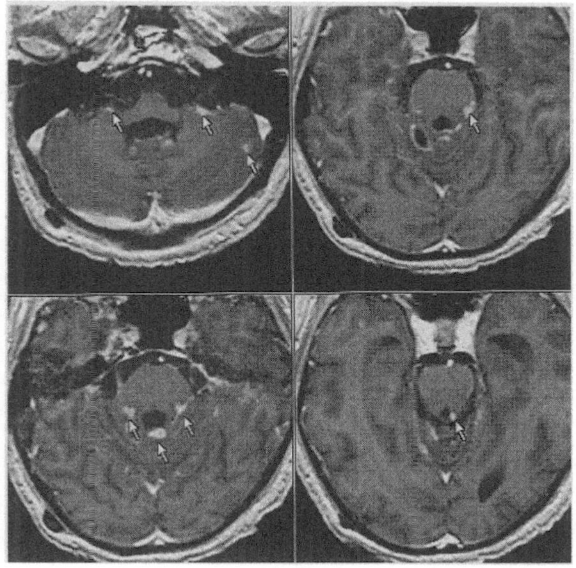

b

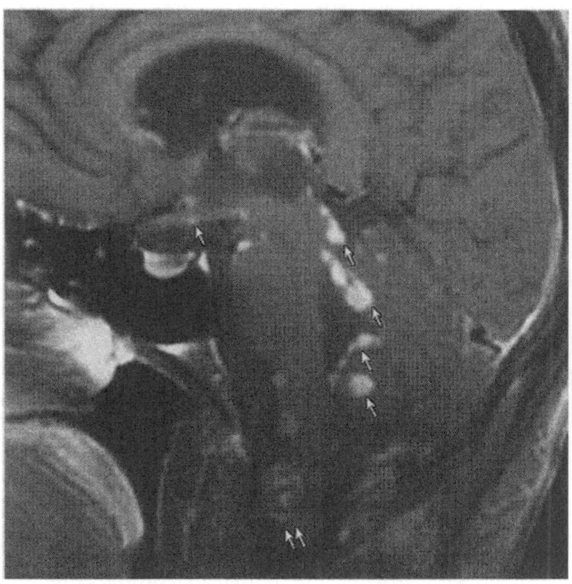

c

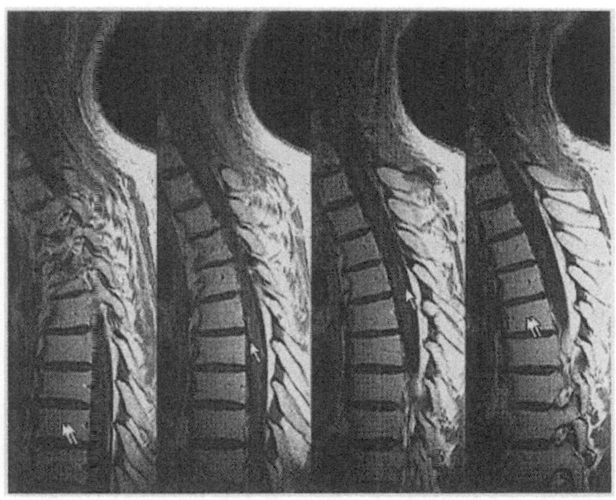

d

Selected References

- Allen RW, Harnsberger HR, Shelton C et al (1996) Low cost high resolution fast spin echo MR of acoustic schwannoma: an alternative to enhanced conventional spin echo MR. AJNR 17:1205-1210
- Andermann F (1997) Brain structure and epilepsy: The impact of modern imaging. AJNR 18:302-306
- Aprile I, Iaiza F, Lavaroni A et al (1999) Analysis of Cystic Intracranial Lesions Performed with Fluid-Attenuated Inversion Recovery MR Imaging. AJNR 20:1259-1267
- Altman NR (1988) MR and CT characteristics of gangliocytoma: rare cause of epilepsy in children. Am J Neuroradiol 9:917
- Anthony DC, Goumnerova L, Tarbell NJ (1998) Tectal Tumors of Childhood: Clinical and Imaging Follow-up. AJNR 19:977–983
- Asari S, Oto T, Tsuchida S et al (1990) MR appearance and cyst content of Rathke cleft cysts. J Comput Assist Tomogr 14:532
- Barnard RO, Geddes JF (1987) The incidence of multifocal gliomas: a histologic study of large hemisphere sections. Cancer 60:1519
- Bjornsson J, Scheithauser BW, Okazaki H et al (1985) Intracranial germ cell tumors: pathobiological and immunohistochemical aspects of 70 cases. J Neuropathol Exp Neurol 44:32-46
- Boring CC, Squires TS, Tong T (1992) Cancer statistics. Cancer J Clin 42:19-38
- Boyko OB, Curnes JT, Oakes WJ et al (1991) Hamartomas of the tuber cinereum: CT, MR and pathologic findings. Am J Neuroradiol 12:309
- Burger PC, Scheithauer BW (1994) Tumors of the central nervous system. Vol 10, Armed Forces Institute of Pathology, Washington DC
- Castillo C, Davis PC, Takei Y et al (1990) Intracranial gangliogliomas: MR, CT, and clinical findings in 18 patients. Am J Neuroradiol 11:109-114
- Cordoliani Y-S, Derosier C, Pharaboz C et al (1992) Primary cerebral lymphoma in patients with AIDS: MR findings: 17 cases. Am J Roentgenol 159:841
- Criscuolo GR, Symon L (1986) Intraventricular meningioma: a review of 10 cases of the National Hospital, Queen Square (1974-1985) with reference to the literature. Acta Neuropathol 83:83
- Daemerel P, Wilms G, Lammeus M et al (1991) Intracranial meningiomas: correlation between MR imaging and histology in fifty patients. J Comput Assist Tomogr 15:45
- Daumas-Duport C, Meder JF, Monsaingeon O et al (1983) Cerebral gliomas: malignancy, limits and spatial configuration. J Neuroradiol 10:51-80
- Daumas-Duport C, Scheithauer BW, Chidkiewicz JP et al (1988) Dysembryoplastic neuroepithelial tumor: a surgically curable tumor of young patients with intractable partial complex seizures. Neurosurgery 23:545
- Daumas-Duport C (1993) Dysembryoplastic neuroepithelial tumors. Brain Pathol 3:283
- David PC, Hudgins PA, Peterman SB et al (1991) Diagnosis of cerebral metastases: double-dose delayed CT vs contrast-enhanced MR imaging. AJNR 12:293-300
- Davis PC, Wichman RD, Takei Y et al (1990) Primary cerebral neuroblastoma: CT and MR findings, 12 cases. Am J Neuroradiol 11:115-120
- Dean BL, Drayer BP, Bird CR et al (1990): Gliomas: classification with MR imaging. Radiology 174:441
- Desprechins B, Stadnik T, Koerts G et al (1999) Case report. Use of diffusion-weighted MR imaging in differential diagnosis between intracerebral necrotic tumors and cerebral abscesses. AJNR 20:1252-1257
- Diepholder HM, Schechheimer K, Mohadjer M et al (1991) A clinicopathologic and immunologic study of 13 cases of ganglioglioma. Cancer 68:2192
- DiPaolo D, Zimmerman RA, Rorke LB et al (1995) Neurofibromatosis type 1: pathologic substrate of high-signal intensity foci in the brain. Radiology 195:721
- Dolinskas CA, Simeone FA (1987) CT characteristics of intraventricular oligodendrogliomas. Am J Neuroradiol 8:1077-1082
- Donati F, Vasella F, Kaiser G et al (1992) Intracranial lipomas. Neuropediatrics 23:32
- Earnest F, Klly PJ, Scheithausser BW et al (1988) Cerebral astrocytomas: histopathologic correlation of MR and CT contrast enhancement with stereotactic biopsy. Radiology 166:823
- Elster AD, Challa VR, Gilders TH et al (1989) Meningiomas: MR and histopathologic features. Radiology 170:857
- Epstein FJ, Farner J-P (1993) Brainstem glioma growth patterns. J Neurosurg 78:408
- Fitzgerald DB, Cosgrove GR, Ronner S, Jiang H et al (1997) Location of language in the cortex: A comparison between functional MR imaging and electrocortical simulation. AJNR 18:1529-1539
- Fujii K, Fijita N, Hirabuki N et al (1992) Neuromas and meningiomas: evaluation of early enhancement with dynamic MR imaging. Am J Neuroradiol 13:1215
- Georgen SK, Gonzales MF, McLean CA (1988) Intraventricular neurocytoma: clinicopathological features of 6 cases. J Neurosurg 68:665
- Glaiser CM, Husain MM, Chadduck W et al (1993) Meningiomas in children. Cancer 67:1223
- Goldsher D, Litt AW, Pino RS et al (1990) Dural "tail" associated with meningiomas on Gd-DTPA-enhanced MR images: characteristics, differential diagnostic value and possible implications of treatment. Radiology 176:547
- Guidetti B, Spallone A (1981) The surgical treatment of choroids plexus papillomas. The results of 27 years experience. Neurosurg Rev 4:129-137
- Harwood-Nash DC (1992) Primary meningeal tumors in children: correlation of clinical and CT findings with histological type and prognosis. Am J Neuroradiol 13:1353

- Hashimoto, M, Fujimoti K, Shinoda S et al (1993) Magnetic resonance imaging of ganglion cell tumors. Neuroradiology 35:181
- Hassoun J, Gambarelli D et al (1993) Central neurocytoma: a synopsis of clinical and histological features. Brain Pathol 3:297
- Hochberg FH, Miller DC (1988) Primary central nervous system lymphoma (review article). J Neurosurg 68: 835-853
- Isiklar I, Leeds NE, Fuller GN et al (1995) Intracranial metastatic melanoma: correlation between MR imaging characteristics and melanin content. AJR 165:1503-1512
- Jack CR, O'Neill BP, Banks PM et al (1988) Central nervous system lymphoma: histologic types and CT appearance. Radiology 167: 211-215
- Jack CR, Reese DF, Scheithauer BW (1986) Radiographic findings in 32 cases of primary CNS lymphoma. AJR 146: 271-276
- Jelenik J, Smirniotopoulos JG, Parisi JE et al (1990) Lateral ventricular neoplasms of the brain: differential diagnosis based on clinical, CT and MR findings. Am J Neuroradiol 11:567
- Johnson PC, Hunt SJ, Drayer BP (1989) Human cerebral gliomas: correlation of postmortem MR imaging and neuropathologic findings. Radiology 170:211
- Johnson BA, Fram EK, Johnson PC, Jacobowitz R (1997) The variable MR appearance of primary lymphoma of the central nervous system: comparison with histopathologic features. AJNR 18:563-572
- Kane AG, Robles HA, Smirniotopoulos JGM et al (1993) Diffuse pontine astrocytoma. Am J Neuroradiol 14: 941
- Kaplan RD, Coon S , Drayer BP et al (1992) Characteristics of meningioma subtypes at 1.5 Tesla. J Comput Assist Tomogr 16:366
- Kaplan AM, Albright AL, Zimmerman RA et al (1996) Brain stem gliomas in children: a children's Cancer Group review of 119 cases. Pediatr Neurosurg 24:185-192
- Kelly PJ, Daumas-Duport C, Scheithauer BW et al (1987) Stereotactic histologic correlations of computed tomography and magnetic resonance imaging defined abnormalities in patients with glial neoplasms. Mayo Clin Proc 62:450-459
- Ken JG, Sobel DF, Copeland B et al (1991) Choroid plexus papillomas of the foramen of Luschka: MR appearance. Am J Neuroradiol 12:1201
- Kim DG, Chi JH, Park SH et al (1992) Intraventricular neurocytoma: clinicopathological analysis of seven cases. J Neurosurg 76:759-765
- Koeller KK, Dillom WP (1993) Dysembryoplastic neuroepithelial tumors. Brain Pathol 3:283
- Krol G, Sze G, Malkin M et al (1988) MR of cranial and spinal meningeal carcinomatosis: comparison with CT and myelography. AJNR 9:709-714
- Kurowa T, Bergey GK, Rothman MI et al (1995) Radiologic appearance of the dysembryoplastic neuroepithelial tumor. Radiology 197:233-238
- Lee YY, Van Tassel PM (1989) Intracranial oligodendrogliomas: imaging findings in 35 untreated cases. AJNR 10: 119-127
- Lee DH, Gao FQ, Rogers JM et al (1998) MR in temporal lobe epilepsy: analysis with pathologic confirmation. AJNR 19:19-27
- Lee C, Ward H, Sharbrough F et al (1999) Assessment of Functional MR Imaging in Neurosurgical Planning. AJNR 20:1511-1519
- Lipper MH, Eberhard DA, Phillips CD et al (1993) Pleomorphic xanthoastrocytoma, a distinctive astroglial tumor: neuroradiologic and pathologic features. Am J Neuroradiol 14:1397-1404
- Luyten PR, Marien AJH, Heindel W et al (1990) Metabolic imaging of patients with intracranial tumors: H-a MR spectroscopic imaging and PET. Radiology 176(3): 791
- Maeder PP, Hotas SL, Basibuyuk LN et al (1990) Colloid cysts of the third ventricle: correlation of MR and CT findings with histology and chemical analysis. AJNR 11: 575-581
- Majos C, Cucurella C, Aguilera C et al (1999) Case report intraventricular meningiomas: MR imaging and MR spectroscopic findings in two cases. AJNR 20:882-885
- Mamourian AC, Cho CH, Saykin AJ, Poppito N (1998) Association between size of the lateral ventricle and asymmetry of the fornix in patients with temporal Lobe Epilepsy. AJNR 19:9-13
- Martin DS, Levy B, Awwad EE et al (1991) Desmoplastic infantile ganglioglioma: CT and MR features. Am J Neuroradiol 12:1195
- Meyerand ME, Pipas JM, Mamourian A et al (1999) Classification of biopsy-confirmed brain tumors using single-voxel MR spectroscopy. AJNR 20:117-123
- Meyers SP, Kemp SS, Tarr RW (1992) MR imaging features of medulloblastomas. AJR 158:859-865
- Milbouw G, Born JD, Martin D et al (1988) Clinical and radiological aspects of dysplastic gangliocytoma (Lhermitte-Duclos disease): a report of two cases and review of the literature. Neurosurgery 22:124-128
- Mirowitz SA, Sartor K, Gado M (1990) High-intensity basal ganglia lesions on T1-weighted MR images in neurofibromatosis. Am J Neuroradiol 10:1159
- Negendank WG, Sauter R, Brown TR et al (1996) Proton magnetic resonance spectroscopy in patients with glial tumors: a multicenter study. J Neurosurg 84:449
- Oppenheim C, Dormont D, Biondi A et al (1998) Loss of digitations of the hippocampal head on high-resolution fast spin-echo MR: A sign of mesial temporal sclerosis. AJNR 19:457-463
- Ostertun B, Wolf HK (1996) Dysembryoplastic neuroepithelial tumors: MR and CT evaluation. AJNR 17:419-430
- Packer RJ, Johnson D et al (1992) Choroid plexus carcinoma of childhood. Cancer 69:580
- Palma L, Guidetti B (1985) Cystic pilocytic astrocytomas of the cerebral hemispheres. J Neurosurg 62:881
- Poussaint TY, Kowal JR, Barnes PD et al (1998) Tectal Tumors of Childhood: Clinical and Imaging Follow-up. AJNR 19:977-983
- Ricci PE, Karis JP, Heiserman JE, Fram EK et al (1998) Differentiating recurrent tumor from radiation necrosis: time for re-evaluation of positron emission tomography? AJNR 19:407-413

- Rippe DJ, Boyko OB, Fuller GN et al (1992) d-DTPA MRI of gliomatosis cerebri: appearance mimicking leptomeningeal tumor dissemination. Am J Neuroradiol 13:1207
- Rippe DJ, Boyko OB, Fuller GN et al (1993) Leptomeningeal gliomatosis: MR findings. J Comput Assist Tomogr 17:317
- Robertson PL, Muraszko KM, Brunberg JA et al (1995) Pediatric midbrain tumors: a benign subgroup of brainstem gliomas. Pediatr Neurosurg 22:65-73
- Ruiz A, Ganz W1, Post MJD et al (1992) Use of thallium201 brain SPECT to differentiate cerebral lymphoma from Toxoplasma encephalitis. Am J Neuroradiol 13:1207
- Schmalbrock P, Chakeres D, Monroe J et al (1999) Assessment of internal auditory canal tumors: A comparison of contrast-enhanced T1-weighted and steady-state T2-Weighted gradient-echo MR imaging. AJNR 20:1207-1213
- Schwaighofer BW, Hesselink JR, Press GA et al (1989) Primary intracranial CNS lymphoma: MR manifestations. Am J Neuroradiol 10:725
- Sevick RJ, Barkovich AJ, Edwards MSB et al (1992) Evolution of white matter lesions in neurofibromatosis type 1: MR findings. Am J Neuroradiol 159:171
- Shaw EG, Scheithauer BW, O'Fallon JR et al (1992) Oligodendrogliomas: the Mayo Clinic experience. J Neurosurg 76:428
- Smirniotopoulos JG, Rushing EJ, Mena H (1992) Pineal region masses: differential diagnosis. Radiographics 12:577
- Smoker GP, Press GA, Hesselink JR et al (1990) Intracranial ependymomas and subependymoma: MR manifestations. Am J Neuroradiol 11:83
- Stroink AR, Hoffman HJ, Hendrick EB et al (1986) Diagnosis and management of pediatric brainstem gliomas. J Neurosurg 65:745
- Tampieri D, Mamdjian R, Melanson D et al (1991) Intracranial gangliogliomas in patients with partial complex seizures: CT and MR findings. Am J Neuroradiol 12:749-755
- Tart RF, Quinsling RG (1991) Curvilinear and tubulonodular varieties of lipoma of the corpus callosum: an MR and CT study. J Comput Assist Tomogr 15:85
- Tien RD, Barkowich AJ, Edwards MSB (1990) MR imaging of pineal tumors. AJNR 11:557-565
- Tien R, Tuori SL, Pulkingham N et al (1992) Ganglioglioma with leptomeningeal and subarachnoid spread: results of CT, MR, and PET imaging. Am J Roentgenol 159:391
- Tokumaru A, Ouchi T, Eguchi T et al (1990) Prominent meningeal enhancement adjacent to meningioma on Gd-DTPA enhanced MR images: histopathologic correlation. Radiology 175:431-433
- Tortori-Donati P, Fondelli MP, Cama A et al (1995) Ependymomas of the posterior cranial fossa: CT and MRI findings. Neuroradiology 37:238-243
- Truwit CL, Barkowich AJ (1990) Pathogenesis of intracranial lipoma: an MR study in 42 patients. AJR 155:855-864
- Vandertop WP, Hoffman HJ, Drake JM et al (1992) Focal midbrain tumors in children. Neurosurgery 31:186
- Vogl T, Bruning R, Schedel H et al (1989) Paragangliomas of the jugular bulb and carotid body. MR imaging with short sequences and Gd-DTPA enhancement. AJNR 10:823-827
- Hwang JH, Egnaczyk GH, Ballard E et al (1998) Proton MR spectroscopic characteristics of pediatric pilocytic astrocytomas. AJNR 19:535-540
- Watanale M, Tanaka R, Takeda N (1992) Magnetic resonance imaging and histopathology of cerebral gliomas. Neuroradiology 35:463
- Weissman JL, Hirsch BE, Fukui MB, Rudy TE (1997) The evolving MR appearance of structures in the internal auditory canal after removal of an acoustic neuroma. AJNR 18:313-323
- Wilms G, Lammenus M, Marchal G et al (1991) Prominent dural enhancement adjacent to non-mengiomatous malignant lesions on contrast-enhanced MR images. Am J Neuroradiol 12:761
- Yoshino MT, Lucio R (1992) Pleomorphic xanthoastrocytoma. AJNR 13:1330-1332
- Yuh WTC, Tali ET, Nguyen HD (1995) Effect of contrast dose, imaging time and lesion size in the MR detection of intracerebral metastasis. AJNR 16:373
- Zagardo MT, Castellani RT, Rees MH et al (1998) Radiologic and pathologic findings of intracerebral schwannoma. AJNR 19:1290-1293
- Zimmermann RA (1990) Central nervous system lymphoma. Radiol Clin North Am 28:697-721
- Zimmerman RA (1996) Neuroimaging of primary brain stem gliomas: diagnosis and course. Pediatr Neurosurg 25:45-53
- Zimmerman RD (1991) MRI of intracranial meningiomas. In: Al-Mefty O (ed). Meningiomas. Raven Press, New York, pp 209-223

Part I

Brain and skull

2 - Vascular diseases

a - Infarct and Ischemia
b - Hematoma and hemorrhage
c - Vascular malformations and anomalies

Infarct and Ischemia

Cases 55, 56: normal and dilated Virchow-Robin spaces

Clinical and pathological features

- The perivascular spaces of Virchow-Robin are normal extensions of the leptomeninges which accompany the perforating arteries into the brain.
- These perivascular spaces may be enlarged and more frequently visualized in the elderly, exceeding 2 mm in diameter but remaining less than 5 mm in size.
- The difficulty is to separate dilated V-R spaces from lacunar infarcts.
- *Histopathology*: perivascular spaces are cavities lined by epithelial leptomeningeal cells containing a small perforating artery.
- Dilated perivascular spaces may be observed in the elderly, in patients with brain atrophy and/or with tortuous, usually atherosclerotic small penetrating arteries. Widening of these perivascular spaces is also found in patients with subcortical arteriosclerotic encephalopathy (hypertension), CADASIL, dementia, mucopolysaccharidosis, meningitis.

MR Characteristic Findings

Morphology

- Virchow-Robin CSF containing spaces are enlarged and become visible on MRI in normal aging or under certain pathological states (type 3 of Poirier et al. classification of lacunae). Particular variants may be found: the "état criblé" (3a), "lacunes de desintegration" (3b), a subputaminal cavity (3c), and enlarged spaces associated with mass effect and gliosis (3d).
- They are commonly found at the base of the brain, involving the inferior part of the striatal lenticulostriate perforating spaces of the anterior perforated substance.
- The may be found also in the centrum semiovale, in its superior part electively, following the path of the deep penetrating cortical arteries.
- They appear as radiate, curvilinear or punctate foci iso-intense to CSF, with well defined and smooth margins, averaging usually 2 to 5 mm in size.
- Remnants of hippocampal sulcus are similar to dilated V-R spaces (1-2 mm) found bilaterally between Ammon's horn and gyrus dentatus.
- They may be confused with lacunar infarcts (type 1) or old hemorrhages (type 2), though they are usually larger than 5 mm, do no follow SI of CSF (hyperintense on FLAIR) and may show presence of hemosiderin (hypointense on T2*) (type 2).

Signal intensity

- Virchow-Robin spaces follow the signal intensity of CSF on all sequences.
- GE-T1w or SE-T1w: hypointense to brain, identical to CSF.
- SE-T2w and SE-PDw: hyperintense to brain and iso-intense to CSF.
- FLAIR: hypointense identical to CSF, easily differentiated from lacunar infarcts.

Contrast enhancement

- No contrast enhancement of dilated perivascular spaces.

Case 56: artherosclerotic leucoencephalopathy

Clinical and pathological features

- Also called subcortical atherosclerotic leucoencephalopathy or leukoaryosis or microinfarcts or periventricular hyperintense white matter lesions.
- Concerns the white matter hyperintense foci commonly found on MRI with advancing age in normal and diseased elderly patients.
- May be found in normal elderly (10-30%) and may appear as frequent as 80% after 75 years old, growing in number with increasing age particularly in patients presenting cerebrovascular risk factors (such a high ischemic score on Hachinski scale) and/or disease susceptible to cause a reduced perfusion pressure.
- *Histopathology*: these foci of hyperintense SI on T2w correspond to areas of atrophic demyelination found around hyalinized arteries or even perivenously.

MR characteristic findings

Morphology

- The hyperintense foci are distributed in the periventricular and deep white matter as well as in the corona radiata and the centrum semiovale, sparing the central part of the corpus callosum. The involvement of the subcortical white matter invariably spares the arcuate U-fibers.
- These foci involve commonly the basal ganglia and usually the central portion of the pons, sparing the medulla, the midbrain and the cerebellum.
- The lesions range in size from punctate to larger patchy white matter foci, usually less than 5-6 mm. The lesions may be smooth or irregular and more or less confluent, in a roughly almost symmetrical distribution.

Signal intensity

- GE-T1w or SE-T1w: slightly hypointense to brain.
- SE-PDw: hyperintense to brain and cortex.
- SE-T2w: hyperintense to brain.
- FLAIR: hyperintense foci (periventricular, deep white matter and basal ganglia best seen; brainstem-pons lesions probably less conspicuously; helps to differentiate punctate foci from dilated Virchow-Robin spaces particularly in the basal ganglia: V-R spaces remaining hypointense similar to CSF).

Contrast enhancement

- Usually no contrast enhancement of the small patchy white matter lesions.
- But small infarcts may show contrast enhancement.

Case 54: Virchow-Robin spaces (normal)

Fig. a: coronal cut (512); SE - PDw

Fig. b: sagittal cut (512); SE - T2w

High resolution PDw (**a**), and T2w (**b**) images, showing linear and punctate foci in the white matter corresponding morphologically and topographically to dilated Virchow-Robin spaces. These VRS are iso-intense to CSF on all sequences and correspond to the pathway of the small penetrating arteries. Note that the VRS follow the course of the perforating cortical arterioles from the brain convexity into the centrum semiovale. In the base of the brain, the VSR follow the route of the lenticulostriate arteries which enter the basal ganglia through the anterior perforated substance. The VRS are < 2 mm in diameter. They are found in all age groups but increase in size and frequency with advancing age. Anatomically, the perivascular spaces are an extension of the subarachnoid space accompanying the small perforating vessels into the brain parenchyma. High resolution images demonstrate them routinely (anatomical variant).

Case 55: "État criblé" of the basal ganglia

Fig. c: coronal cut; SE (512) - T2w

Fig. d: coronal cut; SE (512) - T2w

This cribriform aspect of the basal ganglia called "status cribrosus", and involving mainly the caudate nuclei (**c**, ↑; **d**,*), the lentiform nuclei, and mainly the putamen (**c**, **d**: ↑↑), but also as in this case the thalami (**d**, ↑), correspond to dilated perivascular spaces hyperintense on T2w but iso-intense to CSF. These dilated VRS are well differenciated on high resolution images from the areas of hyperintense signal of the periventricular white matter, corresponding to gliosis and/or demyelination, associated and surrounding the VRS. This is much better shown on FLAIR sequences in which the VRS spaces remain hypointense as the intraventricular CSF contrasting with the hyperintense signal of the periventricular and peri-VRS areas of demyelination and gliosis. Note that the subcortical dilated VRS (**a**,*) associated to the white matter lesions that caracteristically spare the subcortical arcuate "U" fibers are better shown also on FLAIR sequence.

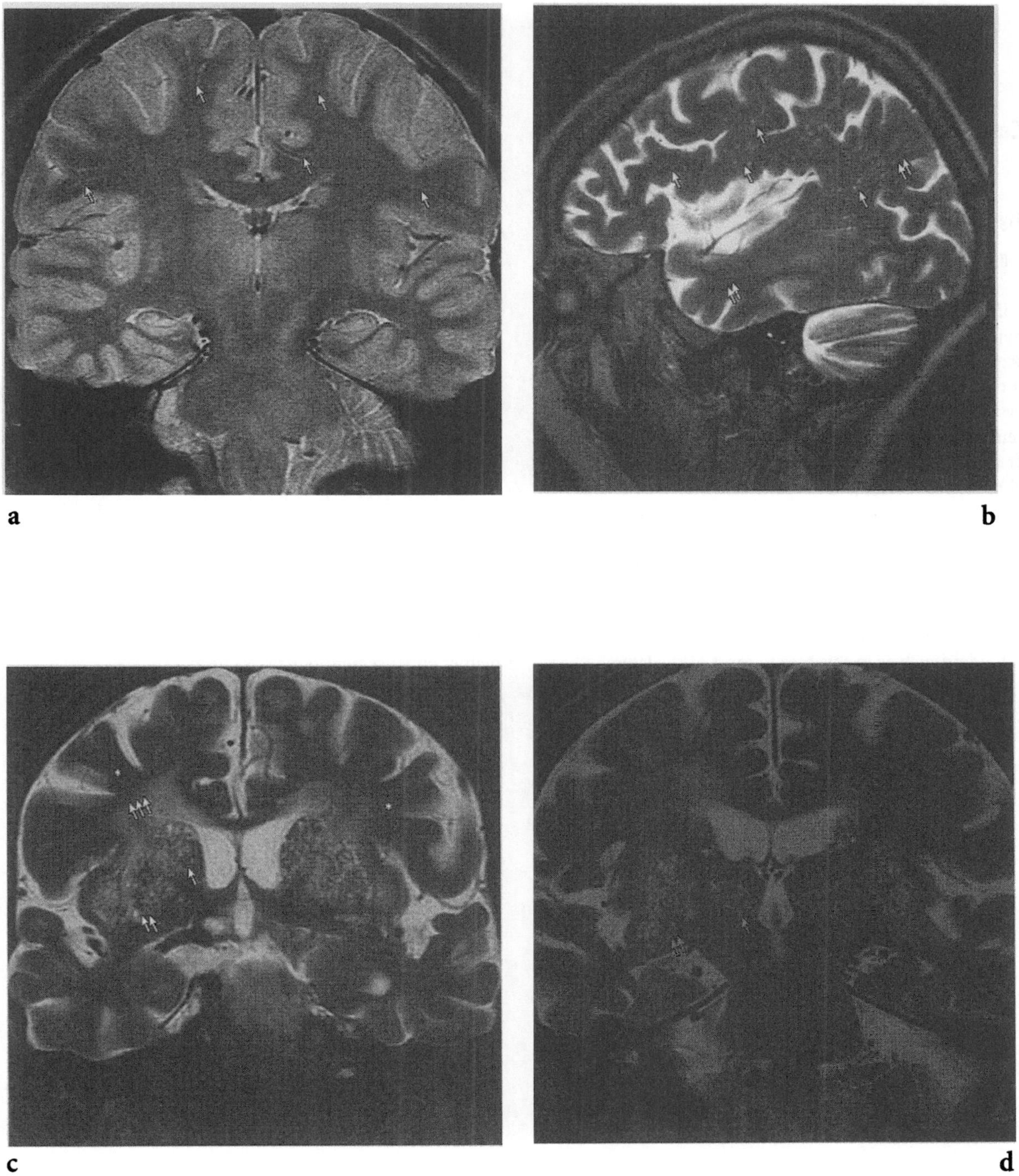
a
b
c
d

Case 56: Arteriosclerotic leucoencephalopathy

Fig a: axial cut;	SE (512) - T2w	**Fig. c:** axial cut;	FLAIR - T2w
Fig. b: axial cut;	FLAIR - T2w	**Fig. d:** axial cut;	FLAIR - T2w

Diffuse subcortical and deep white matter lesion hyperintense to brain on T2w (**a**) as well as on FLAIR sequence (**b**, **c**, **d**), and even much more conspicuous on the latter sequences, showing more particularly the areas of cystic and/or lacunar necrosis (**c**, ↑;**d**, ↑), fluid-filled and remaining hypointense like CSF, and contrasting with the areas of extensive demyelination and/or gliosis (**d**, ↑↑). Note that the VRS spaces in the lenticular nuclei show a hyperintense signal on T2w (**a**, ↑), but remain hypointense like the intraventricular CSF on FLAIR (**b**). Note also that the periventricular hyperintense white matter lesions may be restricted to the margin of the anterior horns of the lateral ventricles called ependymitis granularis (within normal limit variants if isolated, increasing with age), extending such as in this case to most of the lateral ventricular margin as well as to the deep white matter. Note that these white matter abnormalities spare the subcortical arcuate "U" fibers. These extensive abnormalities include presumably a spectrum of underlying insults ranging from perivascular atrophic demyelination to incomplete infarcts and areas of gliosis. This subcortical arteriosclerotic encephalopathy typically involves the centrum semiovale mainly in the upper portions of the lateral ventricles (**c**, **d**) as well as in the posterior junctional regions (**a**, **b**, ↑↑↑) at the posterior extremity of the lateral ventricles and the atria. All these areas are vulnerable to hypoperfusion which is reduced in case of congestive heart failure, hypotension and vascular occlusive disease, like in this patient. Of course, these subcortical and periventricular lesions show neither enhancement nor mass effects.

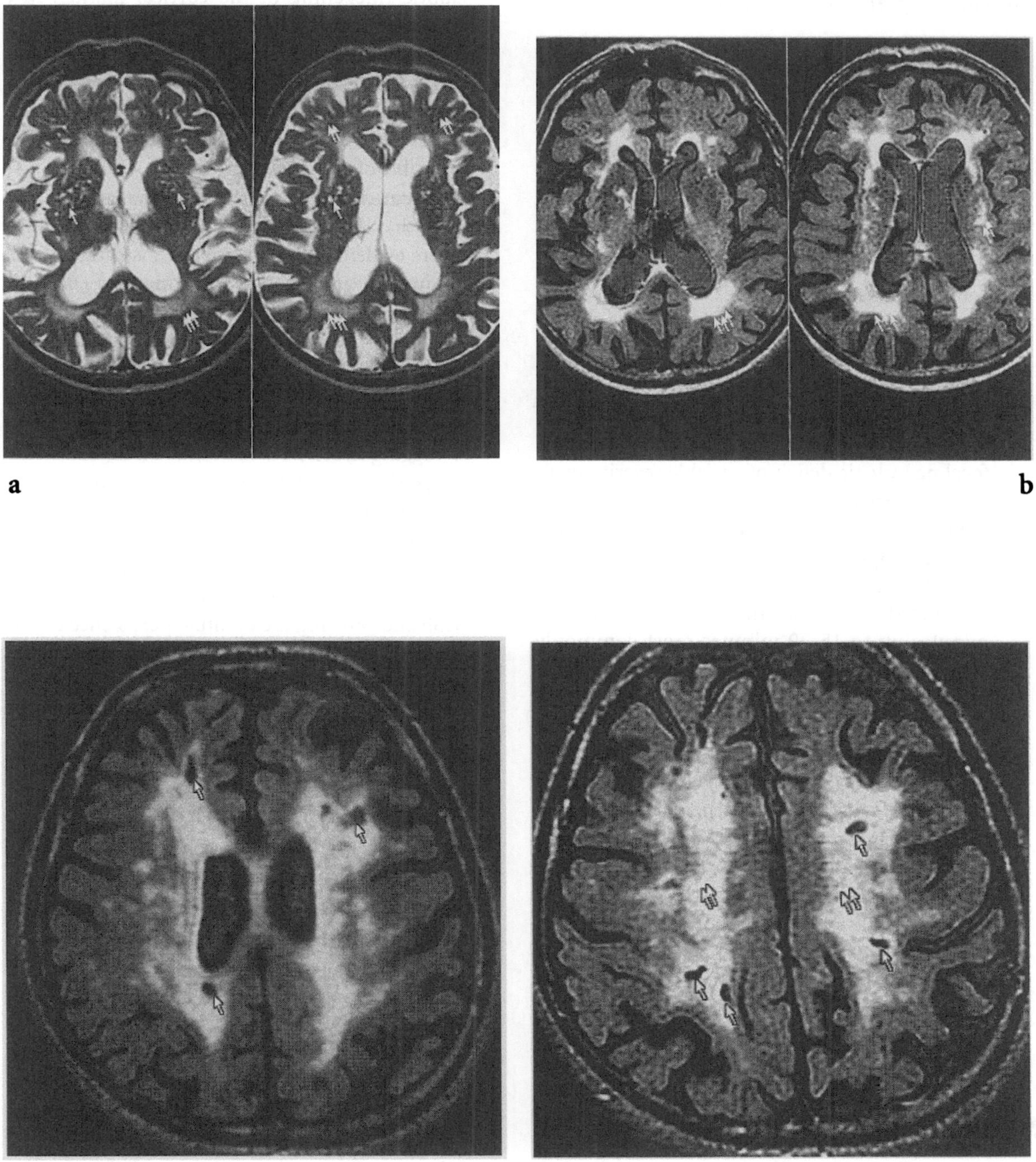

a b

c d

Cases 57-61 and 63-75: cerebral infarcts

Clinical and pathological features

- Stroke is the third cause of mortality.
- Infarcts account for more than 4/5 of all strokes with atherosclerosis being the etiologic factor in 90% of the cases. The other causes of stroke include: embolism, vasculitis, disorders of coagulation, extracerebral arteries dissection or fibromuscular dysplasia, and moyamoya. In children, congenital heart diseases are the most common cause.
- The major risk factors are: hypertension, hyperlipidemia, diabetes mellitus and tobacco smoking.
- *Signs*: the common mode of installation of thrombotic strokes, is characterized by a sudden focal neurologic deficit of variable severity, evolving by steps over several seconds or hours and even days (stroke in evolution), and correlating with the topography and extent of the infarction. Transient ischemic attacks may precede the completed stroke lasting a few seconds or minutes up to 15-30 minutes most commonly and resolving within 24 hours. Half the patients presenting a TIA will experience a stroke within a year. About one third of the patients with transient neurological signs may show sequelar infarcts in imaging. Another clinical presentation mode is the reversible ischemic neurological deficit corresponding to a rapidly regressive stroke of more than 24 hours duration, resolving in less than 20 days. Embolic strokes (1/3 of the cases) show a much more abrupt clinical picture, usually without prodromal symptoms. The embolic material is usually of cardiac origin (mural thrombus on the endocardium) and less frequently the origin is from an arterial atheromatous plaque.

MR characteristic findings

Morphology

- **Acute infarct** (first 24 hours): corresponds to the initial pathological stage of coagulative necrosis, with swelling of the infarcted area, blurring of the cortico-medullary junction and/or effacement of the related sulci and/or subtle thickening of the cortical mantle. MR appears more sensitive to CT within the first 48 hours. During this stage CT discloses abnormalities in about half the cases as compared to conventional MRI (more than 80%), but contribute actually to rule out any hemorrhage responsible of the ictus.
- **Subacute infarct** (starting from the 2nd day to 1 month): the area of brain softening becomes more evident with increase in conspicuity of the white and gray matter abnormalities (SI) and swelling with mass effect on adjacent anatomical structures secondary to edema, prominent in larger infarcts. During the 2nd-3rd weeks MR is more sensitive than CT to detect and evaluate the infarcted area due to its isodense aspect ("fogging effect").
- **Chronic infarct** (after the 4th week): removal of the necrotic infarcted tissue, mainly in the central part of the infarction, leads to the formation of a residual cavity (following the process of phagocytosis and removal of debris) filled with CSF and showing the eventual persistence of few vessels and bounded by a glial scar. An associated ex vacuo cortical and/or subcortical brain atrophy may be observed, with enlargement of the related subarachnoid spaces and corresponding sulci and/or dilatation of the adjacent ventricular cavity.
- The frequency of infarctions occurring in the carotid system territory (80%) as compared to those involving the posterior vertebro-basilar system (15-20%) corresponds to the anatomical vascular supply to the brain (about 90% from the carotid system).
- The topography, the size and extent of the cerebral infarct depend on the site and type of the vascular occlusion, the presence or absence of anastomoses and their relative efficient patency, and the anatomical vascular territory of the involved vessel.
- Three major anastomosis may influence the importance of the residual infarct: the circle of Willis, the ophthalmic system (connecting the external carotid to the internal carotid systems) and the leptomeningeal vascular system comprising the anastomoses between the 3 major cerebral arteries (anterior, middle and posterior cerebral arteries).

- The cortical-subcortical extent of the infracted area shows variable patterns, ranging from cortical with subcortical involvement (commonly) to an almost restricted cortical softening, the so-called laminar necrosis, or may even be confined to the white matter.
- The importance of the damage as observed in the infarcted area may range from complete and permanent (leading to the cavitation, centrally) to incomplete and at times functionally reversible (peripheral "penumbra" with diminished blood flow). The overall morphology is usually that of a wedge-shaped lesion extending deeply toward the deep white matter core.

Signal intensity

- **Acute infarct** (initial 6-24 hours from onset):
  - GE-T1w: iso-intense or slightly hypointense to white matter.
  - SE-PDw: hyperintense to brain, with particular signal enhancement of the cortex and usually better demonstration of involvement of the basal nuclei.
  - SE-T2w: hyperintense to brain, usually seen 8-12 hours after the stroke episode but may be encountered early as 2 hours after.
  - FLAIR: hyperintense to brain, more sensitive than SE-T2w.
  - Diffusion: hyperintense to brain, most sensitive sequence revealing the ischemic cytotoxic edema as early as 90 minutes following the insult (on experimental studies). Signal abnormalities may resolve if the ischemic area is reperfused within two hours.
- **Subacute infarct** (24 hours to 1 month):
  - GE-T1w: hypointense to white matter with eventual association of areas of hyperintense SI due to hemorrhagic component (methemoglobin) involving the cortex at the periphery of the infarcted area.
  - SE-PDw: hyperintense signal to brain.
  - SE-T2w: hyperintense signal to brain.
  - FLAIR: hyperintense SI to brain. Gyriform pattern of cortical involvement and bright SI of the basal ganglia are even better delineated and differentiated from CSF (dark SI) than on SE-T2.
  - Diffusion: area of hyperintense SI may show further extension as well as an increased diffusion, due to vasogenic edema.
- **Chronic infarct** (starting by the end of the 1st month to 3 months or more):
  - GE-T1w: hypointense SI, may become later almost similar to CSF in case of cavitation. CSF-filled secondary to the encephaloclastic process.
  - SE-PDw: hyperintense SI close to CSF.
  - SE-T2w: hyperintense SI close to CSF.
  - FLAIR: hypointense SI similar to CSF (sequelar cavitation) which may be surrounded by high SI of cicatricial gliosis.
- Hemorrhagic areas of infarction may be found in: embolic infarcts (secondary to sudden arterial occlusion with extravasation of red blood cells and/or reperfusion of the infarcted zone) showing blood breakdown products usually in the periphery of the infarct; or in venous occlusions (i.e. occlusion of a dural sinus) in which the hemorrhagic component is found particularly in the subcortical white matter with frequent bilateral distribution (which does not follow an arterial anatomical territory).

Contrast enhancement

- Different, almost characteristic patterns of contrast enhancement may be observed. Gde infusion may improve the visualization of the infarcted area particularly during the acute and subacute stages (particularly when "fogging effect" is present on CT) and may be useful to differentiate recent from old infarcts if diffusion sequence is not available on the MR system used (multiple infarcts on FLAIR).
- During the first 48-72 hours: cortical intravascular enhancement (related to slow flow in involved vessels) in about 2/3 of the cases and less frequently (1/3) meningeal enhancement contiguous to large infarcts (due to irritative process and/or luxury perfusion and/or venous engorgement).
- During the first and second weeks: parenchymal typically gyriform enhancement is observed in the majority of the cases and in virtually all patients when adding magnetization transfer technique or using high-dose contrast infusion. Contrast infusion is useless in chronic stages after 3 weeks (normalized blood brain barrier).

Case 57: Bithalamic lacunar infarct

Fig. a: coronal cut; SE - PDw

Fig. b: coronal cut; SE - T2w

Bilateral small deep lesions involving roughly symmetrically both thalami, showing hyperintense signal on PDw (**a**, ↑) to cortex and hyperintense signal close to CSF on T2w (**b**, ↑). These ischemic foci correspond to bilateral lacunar infarcts in the territory of the paramedian artery of Percheron.

Case 58: Thalamic chronic infarct

Fig. c: sagittal cut; GE - T1w

Fig. d: axial cut; SE - PDw and SE - T2w

T1w MR image showing a hypointense lesion (**c**, ↑) involving the thalamus. The lesion appears hyperintense on PDw sequence close to CSF signal (**d**, left, ↑) and remains hyperintense to the brain on T2w and close to CSF signal, due to its necrotic nature (**d**, right, ↑). The lesion shows no mass effect on the adjacent frontal horn and third ventricle. This lesion is consistent with an infarct in the posterior cerebral artery territory.

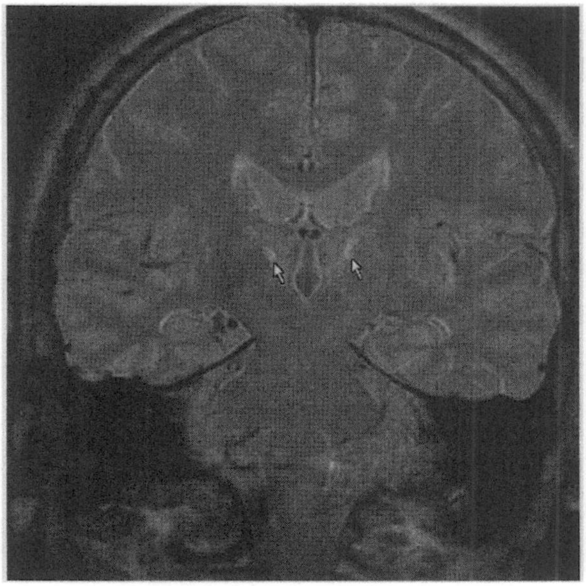

a

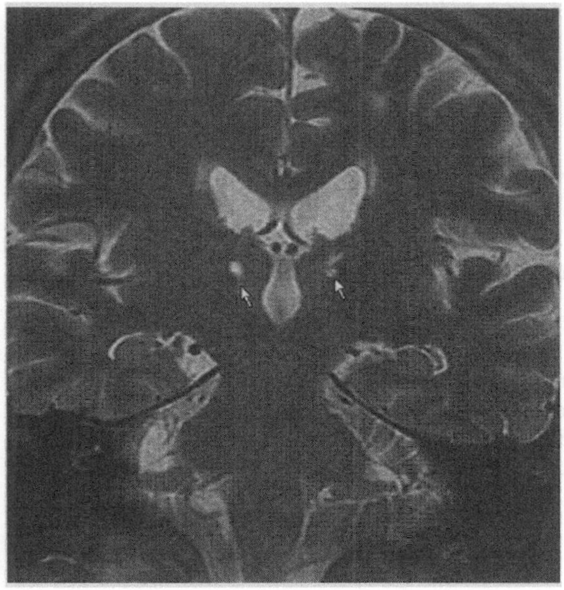

b

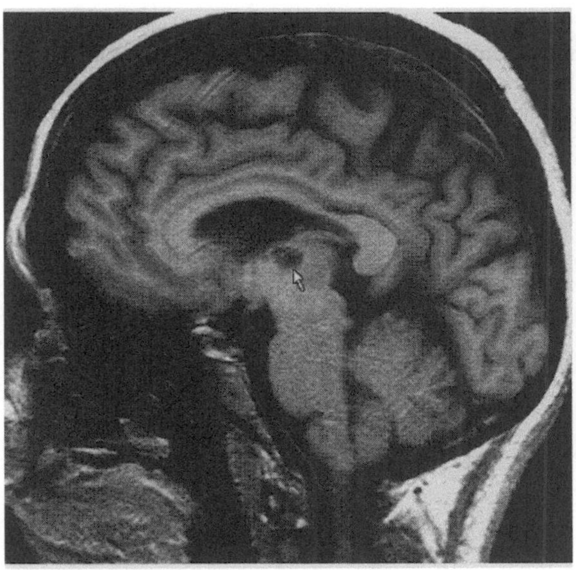

c

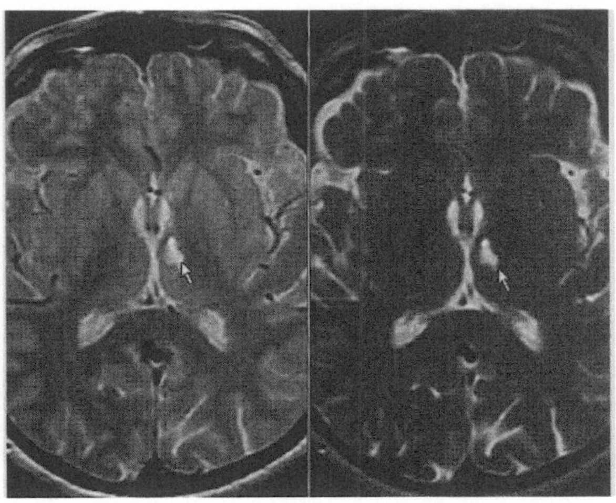

d

Case 59: Putamino-caudate subacute infarct

Fig. a: axial cut; SE - T2w

Fig. b: coronal cut; SE - T2w

Triangular area of bright signal involving the right posterior lenticular nucleus (**a**, ↑) and mainly the putamen (**b**, ↑↑) extending to the body of caudate nucleus (**b**, ↑) and the upper internal capsule. There is a slight mass effect on the ipsilateral frontal horn and a swelling of the posterior putamen due to associated edema (**a**, ↑; **b**, ↑↑). This corresponds to an infarct in the central territory of the right middle cerebral artery.

Case 60: Thalamic acute infarct

Fig. c: axial cut; SE - PDw

Fig. d: coronal cut; SE - T2w

Area of hyperintense signal on PDw (**c**) as well as on T2w (**d**) involving the posterior left thalamus (**c**, **d**, ↑) extending to the subthalamic area with associated lesion in the ipsilateral splenium of the corpus callosum (**a**, ↑↑) and hippocampal formation (**b**). This corresponds to infarct in the territory of the left posterior cerebral artery involving presumably the regions of vascular supply of the posterior choroidal arteries and the dorsal callosal artery.

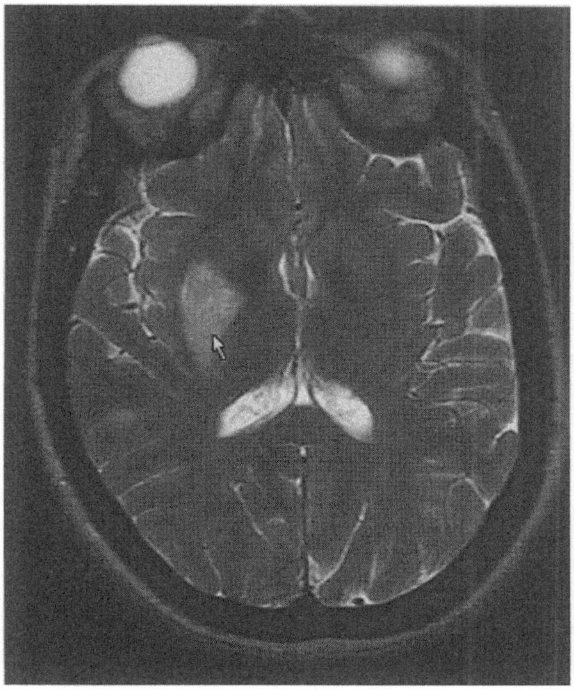

a

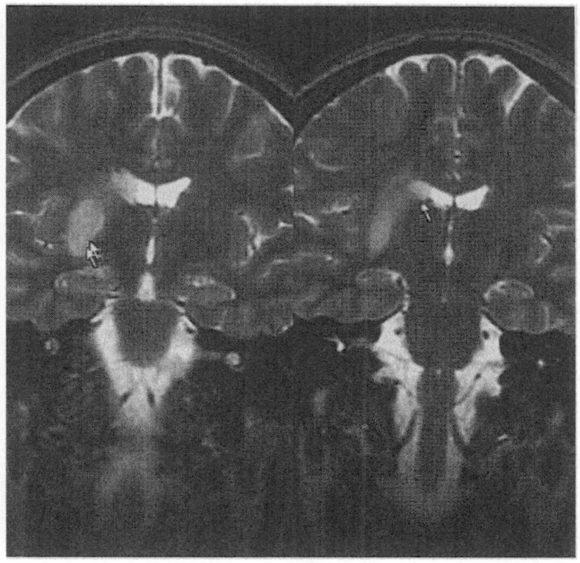

b

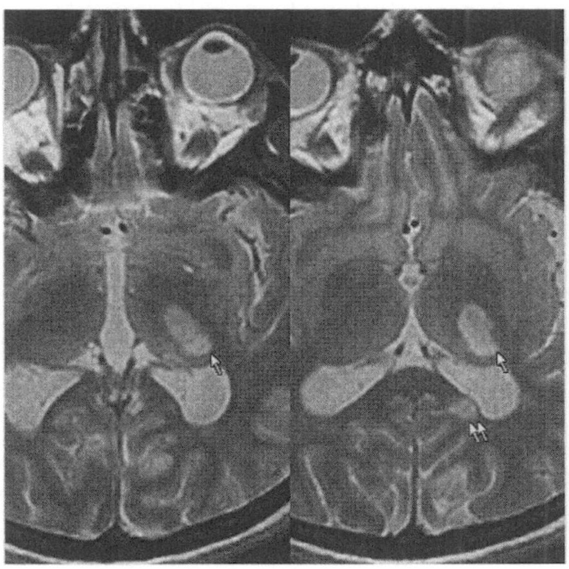

c

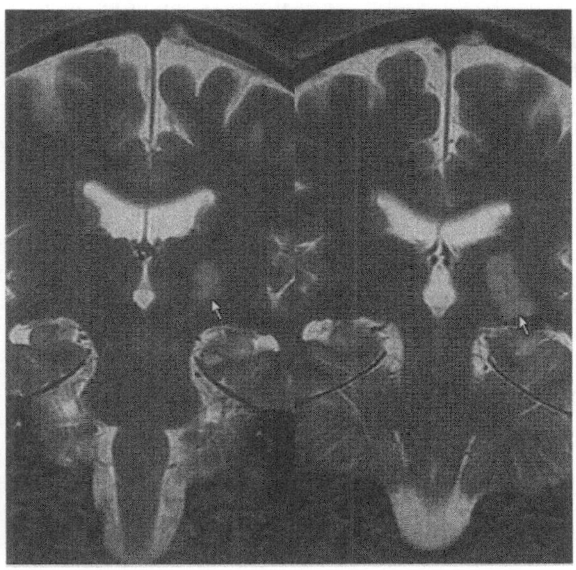

d

Case 61: Putaminal chronic infarct

Fig. a: coronal cut; SE - PDw

Fig. b: coronal cut; SE - T2w

Chronic "slit-like" lesion involving the external aspect of the right putamen associated to an atrophy of the right lenticular nucleus. The lesion is hyperintense to the cortex on PDw (**a**, ↑) and hyperintense on T2w (**b**) with CSF-like signal. This corresponds to a long term MR follow-up examen of a patient presenting a previous infarct in the deep middle cerebral artery territory.

Case 62: Putaminal chronic hematoma

Fig. c: coronal cut; GE - T2*w

Fig. d: axial cut; SE - PDw

Cicatricial chronic right posterior capsulo-lenticular hematoma showing a very small central CSF-like intensity area, surrounded by an extensive peripheral hypointensity, shown on PDw sequence (**d**, ↑) and more obviously and extensively on T2*w sequence (**c**, ↑) due to the cicatricial ferritin/hemosiderin deposition. In fact, the MR exam has been performed several months after the hemorrhagic event. Note that the residual hemosiderin deposit in the brain is much better depicted on T2*w sequence, due to its paramagnetic property and the susceptibility effect.

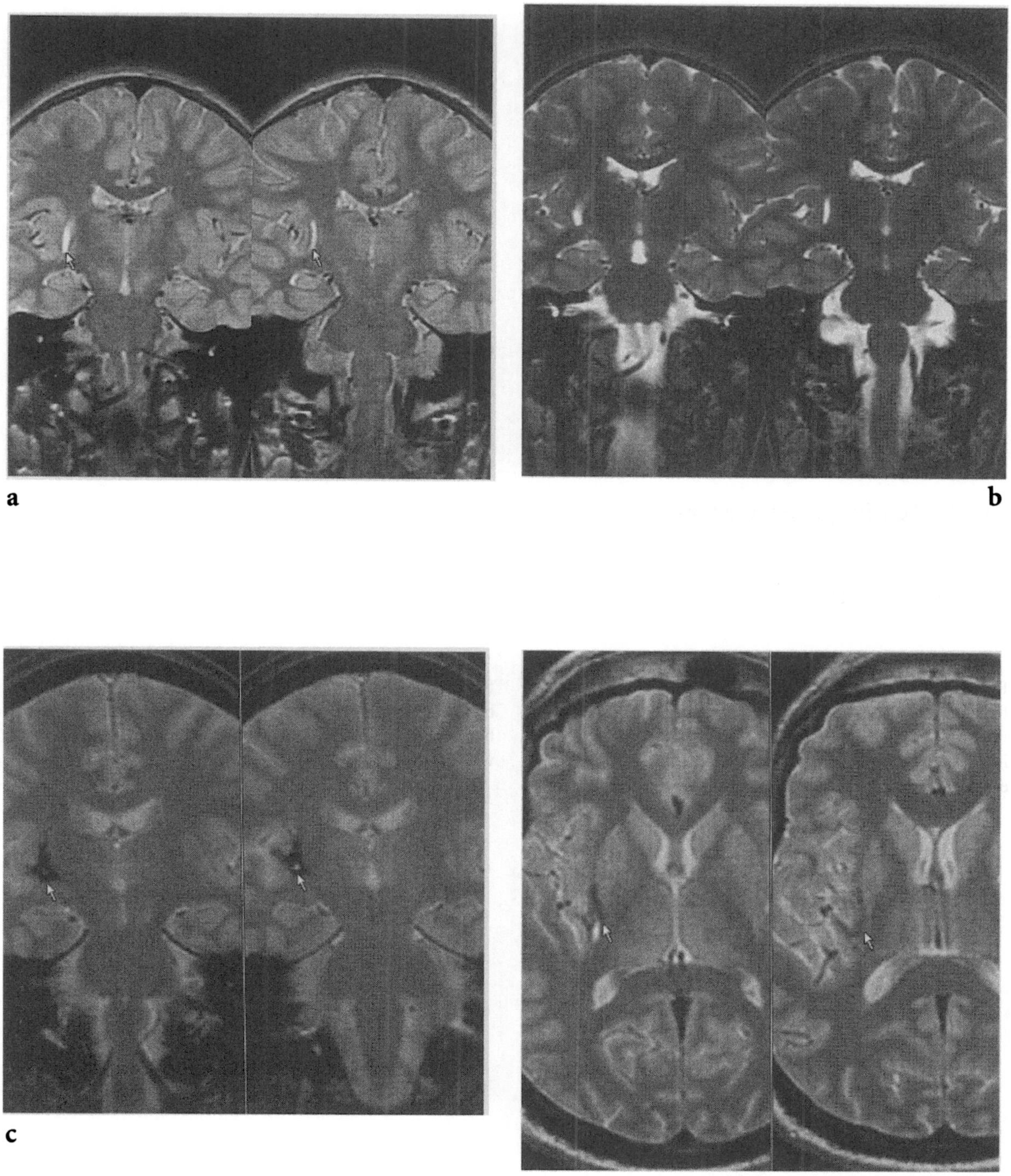

a b c d

Case 63: Acute mesencephalic infarct

Fig. a: sagittal cut; SE - T2w

Fig. b: axial cut; SE - T2w

Ill-defined hyperintense lesion on T2w (**a, b**) sequence involving the left mesencephalic region (↑), with swelling of the left cerebral peduncle and high signal of the left paramedian mesencephalic tegmentum, with small mass effect on the cerebral aqueduct. This lesion was ill-defined on CT which is limited in case of brainstem infarcts due to artifacts emanating from the bony skull base and the relatively poor contrast resolution. Patients with brainstem and cerebellar infarcts are good candidates for MR exams.

Case 64: Pontine paramedian infarct

Fig. c: axial cut; SE - PDw

Fig. d: axial cut; SE - T2w

Roughly triangular area of high signal intensity hyperintense to the cortex on PDw (**c**, ↑↑) sequence and remaining hyperintense on T2w (**d**, ↑↑). The infarct involves the left paramedian vascular territory in the pons entirely, from the anterior aspect of the pons to the floor of the fourth ventricle (↑). There was no occlusion of the basilar artery showing a normal flow void phenomenon on both echos of this T2w dual echos sequence. The angio MR was also normal.

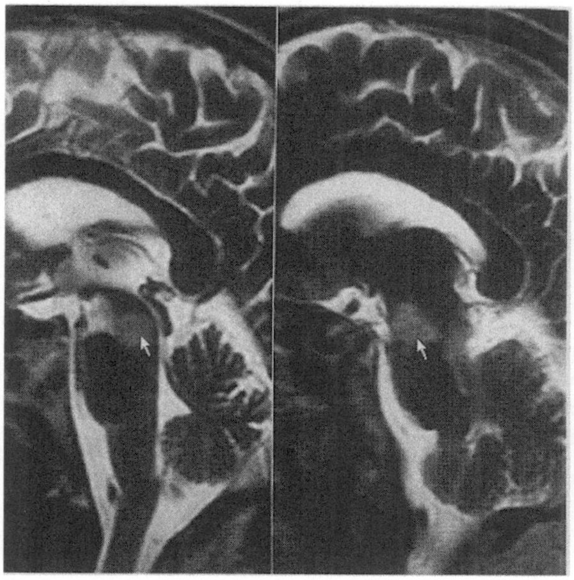

a

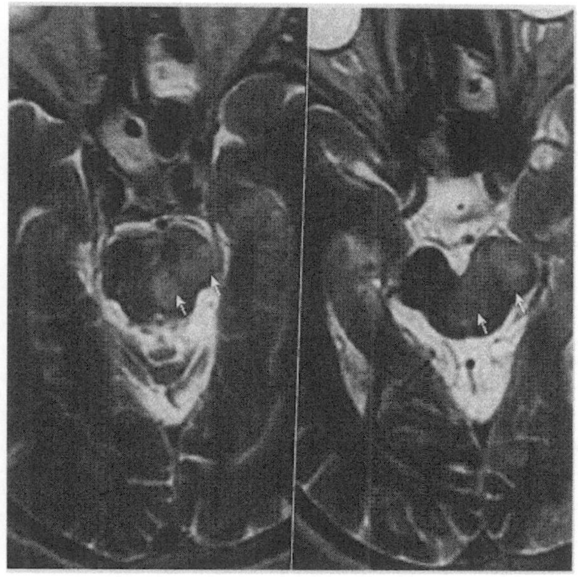

b

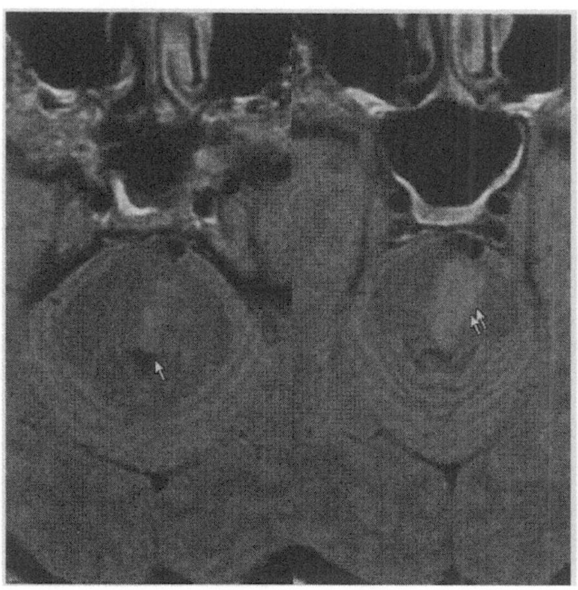

c

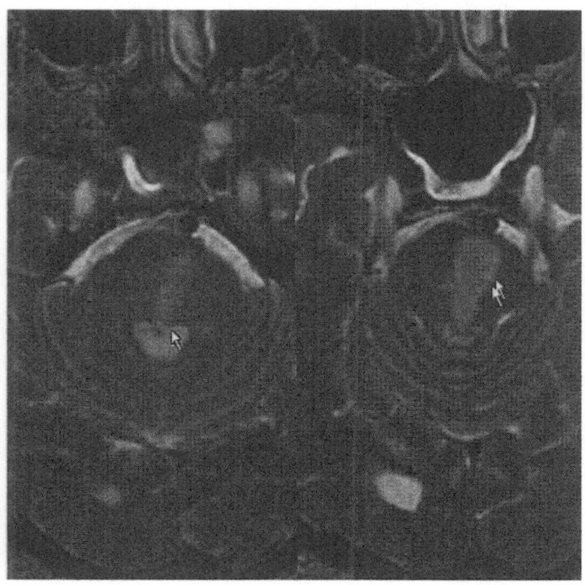

d

Case 65: Pontine paramedian infarct

Fig. a: sagittal cut; SE - T2w

Fig. b: angio, coronal view; 3DTOF

Similar case (case 65) showing an extensive paramedian pontine infarct with an anterior broad base and a posterior extremity reaching the floor of the fourth ventricle (**a**, ↑), well seen on this T2w sequence. The MR-angiography performed (**b**, coronal MIP) showed an irregular atherosclerotic stenosis of the mid-upper portion of the basilar artery (**b**, ↑) demonstrating a severe stenosis, with presumably subsequent occlusion of small paramedian perforating arteries.

Case 66: Lateral medullary infarct (Wallenberg's syndrome)

Fig. c: axial cut; SE - T2w

Fig. d: coronal cut; SE - T2w

Hyperintense lesion involving the left posterolateral region of the medulla oblongata well seen on the T2w sequences in the axial (**c**, ↑↑) as well as in the coronal (**d**, ↑↑) planes. The infarct shows at its upper border a small area of necrosis (**c**, **d**, ↑) presenting a more hyperintense signal similar to the CSF. The patient was referred for a Wallenberg's syndrome.

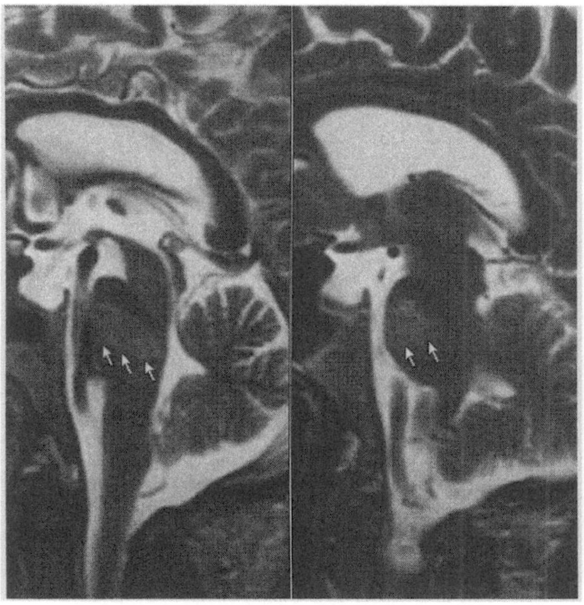

a

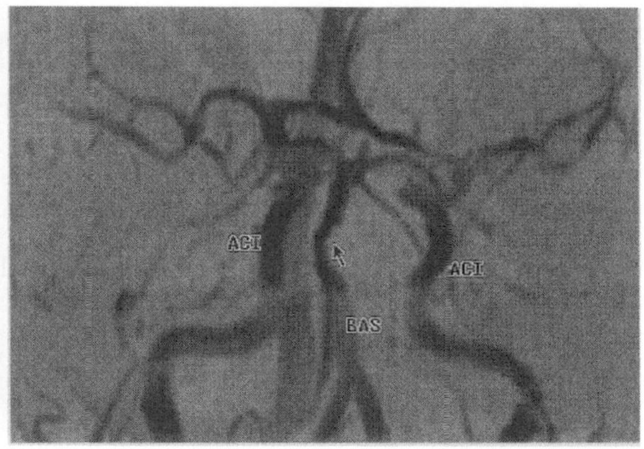

b

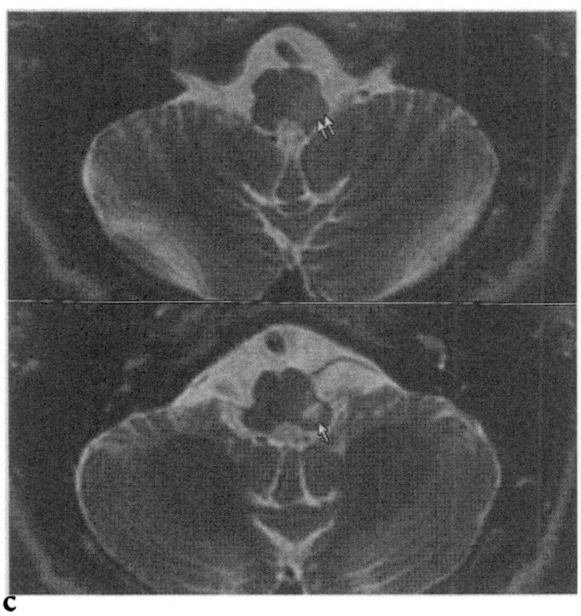

c

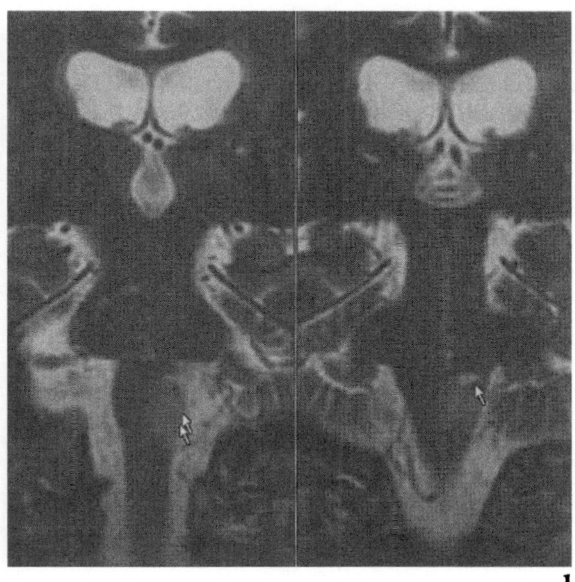

d

Case 67: Inferior lateral bulbar infarct

Fig. a: axial cut; SE - T2w and SE - T2w

Fig. b: axial cut; FLAIR - T2w

Patient presenting a left lateral medullary syndrome. MR shows a small area of high signal intensity slightly hyperintense to the cortex on PDw sequence (**a**, up, ↑) and hyperintense on T2w (**a**, down, ↑), involving the lateral aspect of the inferior medulla. The lesion is very limited, well shown on SE-T2w and even easier on FLAIR sequence (**b**, ↑). The FLAIR sequence appears more sensitive in the depiction of superficial infarcts of the brainstem, and should be performed in case of doubt on SE-T2w sequences due to flow artifacts.

Case 68: Venous infarct (sequelae)

Fig. c: axial cut; SE - PDw and SE - T2w

Fig. d: sagittal cut; GE - T2*w

Large paramedian right posterior frontal-parietal chronic hemorrhagic infarct with a regional atrophy of the cortex and enlargement of the subarachnoid spaces (**c**). The infarct is hyperintense on PD w and T2w (**c**), involving mainly the subcortical white matter (**c**, ↑) with peripheral gliosis and/or edema (**c**, ↑↑). The areas of very low signal mainly objectivated on T2*w sequence (**d**, ↑) are due to the paramagnetic effects of the ferritin-hemosiderin from prior hemorrhagic venous infarct in a patient presenting blood hypercoagulability.

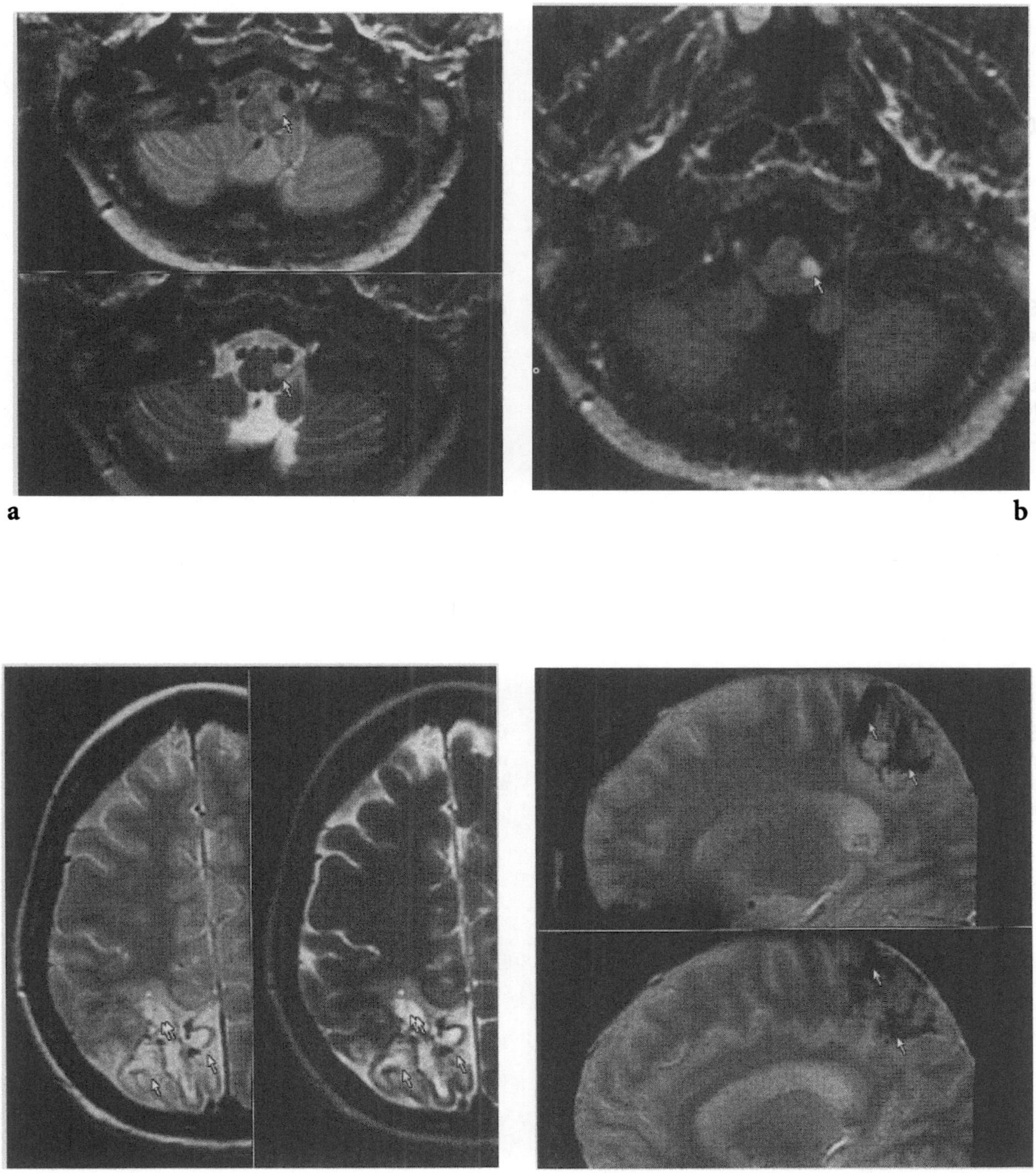

a b c d

Case 69: Subacute hemorrhagic occipital infarct

Fig. a: sagittal cut; GE - T1w

Fig. b: coronal cut; SE - PDw

Subacute right occipital infarct, showing hyperintense signal involving the occipital cortex on T1w (**a**, ↑↑), particularly the lingual gyrus (**a**, ↑). On PDw sequence (**b**), the hemorrhagic infarct is associated with edema showing hyperintense signal (**b**, ↑). The hyperintense signal of the cortex (**a**) on T1w sequence corresponds to methemoglobin deposition but with persistence in the subcortical white matter of some deoxyhemoglobin and/or intracellular methemoglobin, explaining the relative partial hyposignal on PDw (**b**, ↑↑), as well as on T2w sequence.

Case 70: Acute mesial temporal infarct

Fig. c: axial cut; SE - PDw

Fig. d: angio, axial view; 2D - PC

Sudden onset of a right visual field defect. The MR performed showed a gyriform hyperintense signal localized to the internal portion of the left temporal lobe (**c**, ↑) on PDw sequence. MR-angio demonstrated total occlusion of the left posterior cerebral artery (**d**); the proximal latero-mesencephalic segment of the right posterior cerebral artery is shown (**d**, ↑).

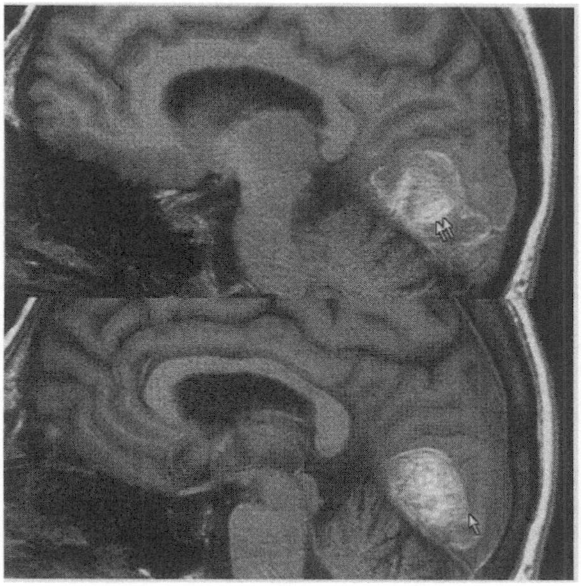

a

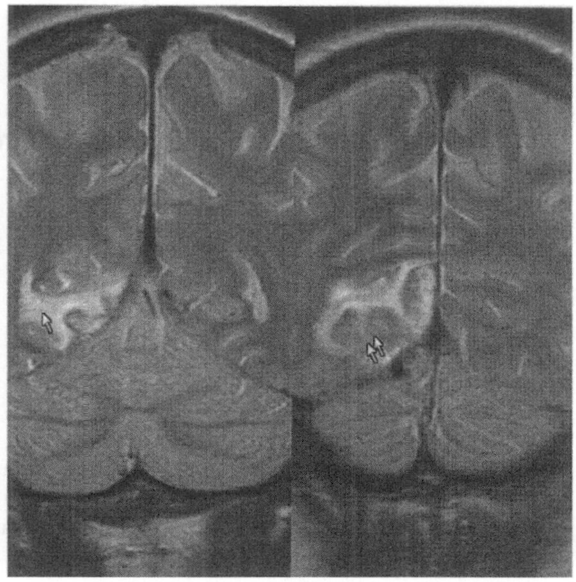

b

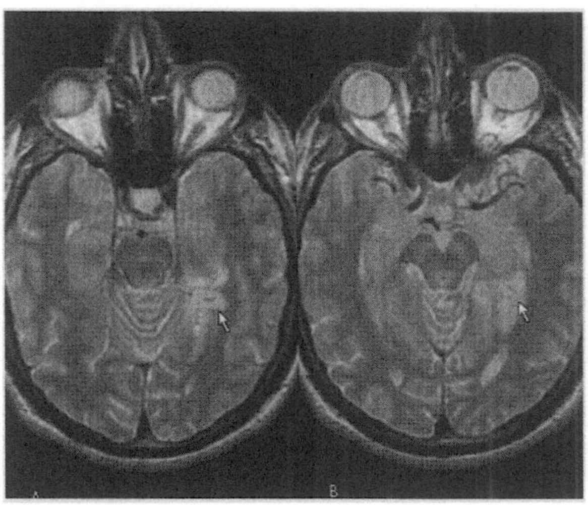

c

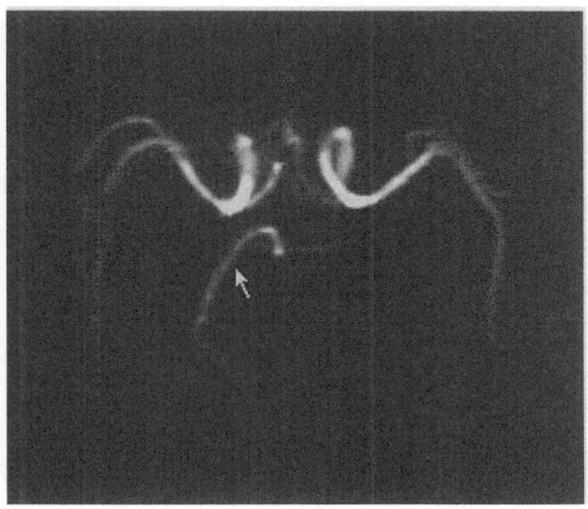

d

Case 71: Acute paramedian frontocallosal infarct

Fig. a: sagittal cut; GE - T1w

Fig. b: sagittal cut; SE - T2w

Fig. c: axial cut; FLAIR - T2w

Fig. d: axial cut; FLAIR - T2w

The MR shows an abnormal signal intensity involving the most anterior part of the corpus callosum as well as the posterior part of its body, extending to the anterior cingulate gyrus and the adjacent anterior part of the internal frontal gyrus with swelling of the corpus callosum (**a**, **b**, up, ↑) and mass effect on the pericallosal and frontal corresponding sulci. The infarct shows a hypointense signal on T1w (**a**, ↑), and a hyperintense signal on T2w (**b**, ↑).

The lesion is very well depicted with hyperintense signal on FLAIR (**c**, **d**: ↑). These abnormalities correspond to an infarct involving the anterior cerebral artery territory. Another small hemorrhagic infarct is also depicted in the posterior cerebral artery territory involving the lingual gyrus (**a**, ↑↑) with hyperintense signal on T1w due to hemorrhagic suffusion with presumably methemoglobin deposition.

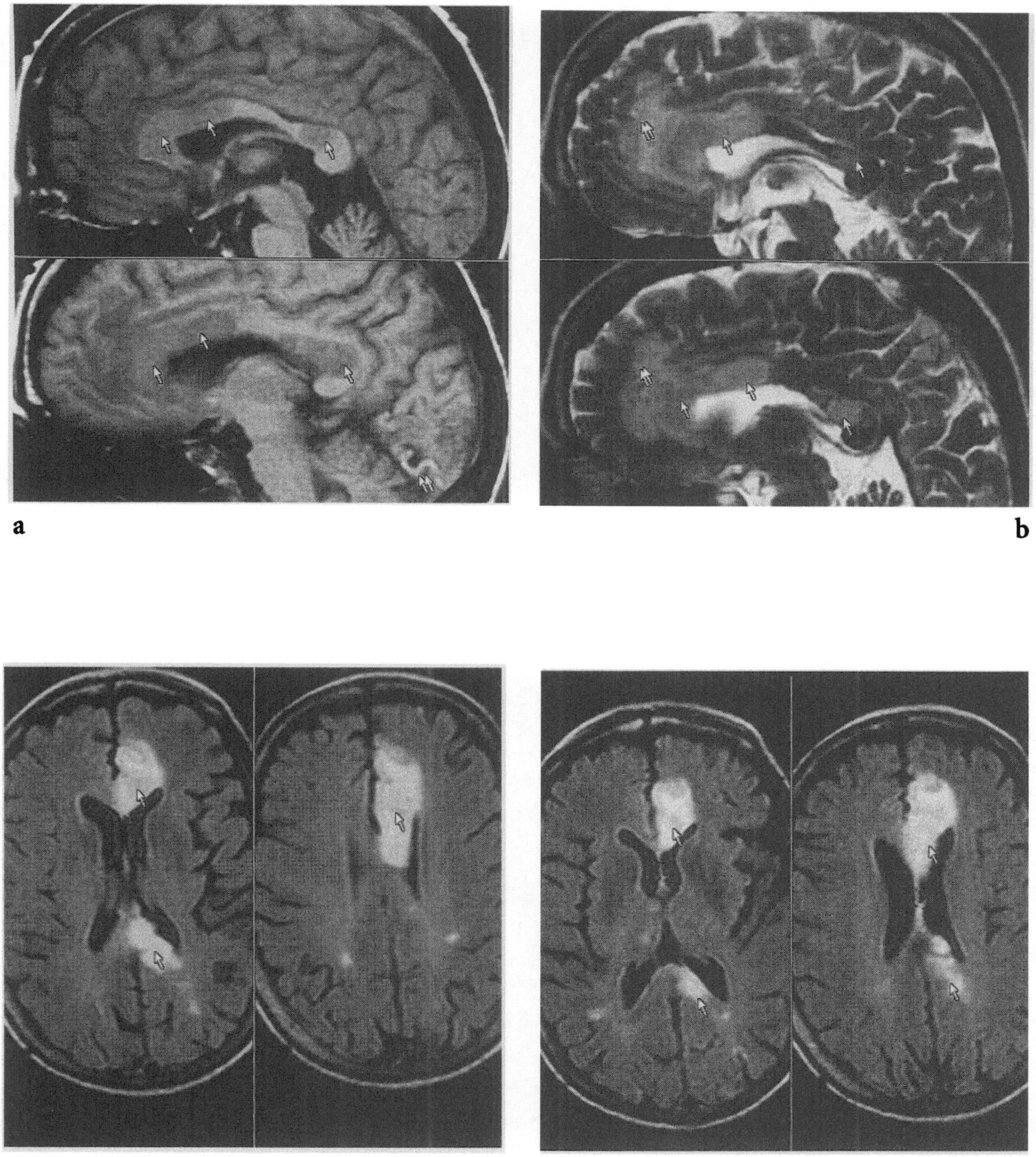

a b

c d

Case 72: Acute sylvian infarct

Fig. a: axial cut; SE - PDw

Fig. b: axial cut; SE - T2w

Cortical infarction involving the fronto-temporo insular region and showing a hyperintense signal on PDw (**a,** ↑) as on TRw (**b,** ↑) sequences, with an apparent thickening of the cortical mantle (**a, b**) and collapse of the subarachnoid espaces within the sulei. The infarct involves the superficial territory of the left MCA, and respects the deep territory (normal lenticulo-striate region).

Case 73: Subacute - chronic sylvian infarct

Fig. c: axial cut; SE - T2w

Fig. d: axial cut; FLAIR - T2w

Focal left frontal infarct limited to the cortex and the subcortical white matter, presenting a hyperintense signal on T2w (**c,** ↑) much better delineated on FLAIR sequence (**d,** ↑), showing the subcortical ischemic demyelination (**d,** ↑↑), as well as the wallerian degeneration.

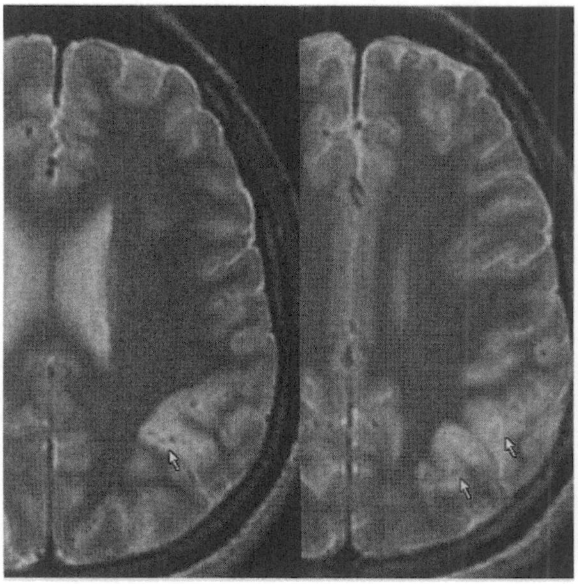

a

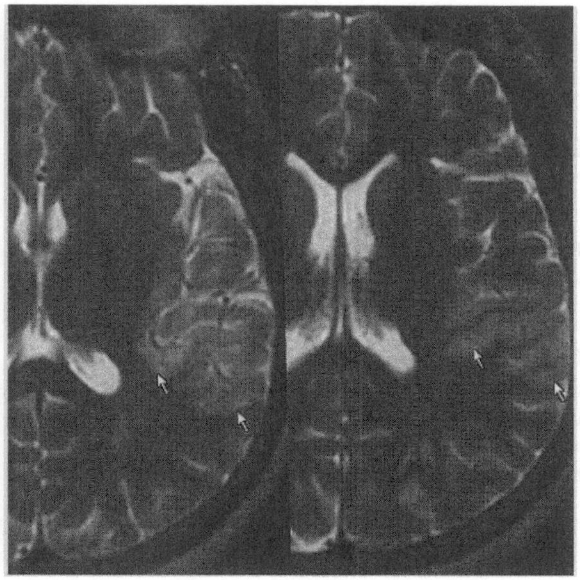

b

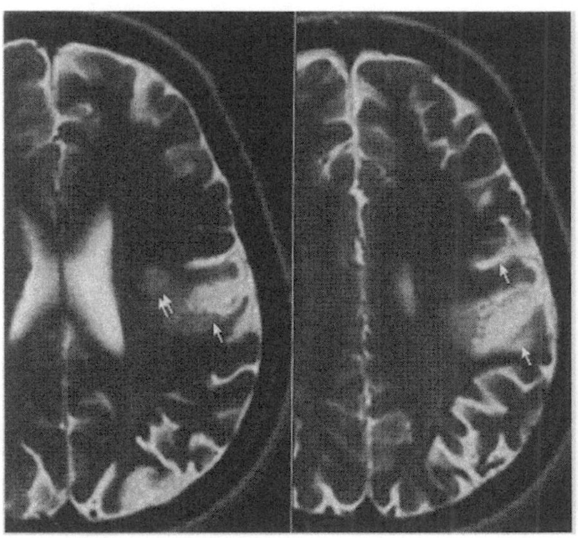

c

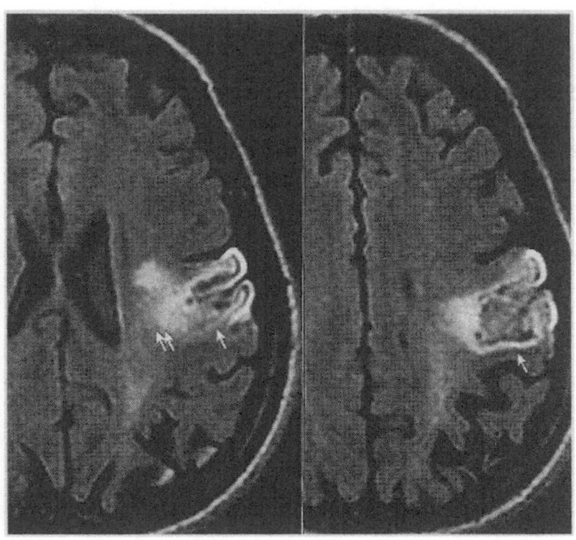

d

Case 74: Chronic sylvian infarct (adult)

Fig. a: axial cut; SE - T2w

Fig. b: axial cut; FLAIR - T2w

Characteristic chronic infarct in the right middle cerebral artery distribution. The signal intensity within the infarcted area is similar to that of CSF, on PDw (**a**, ↑) as well as on FLAIR-T2w sequence (**b**, ↑↑). MR shows also the associated enlargement of the ipsilateral ventricular system indicating subcortical atrophy. The FLAIR sequence (**b**) permits to differentiate between the cicatricial encephaloclastic area and the perilesional extensive gliosis (**b**, ↑).

Case 75: Chronic sylvian infarct (child)

Fig. c: sagittal cut (512); GE - T1w

Fig. d: axial cut; SE - PDw and SE - T2w

Similar case showing the hypointense signal on T1w (**c**, ↑) of the extensive superficial sylvian infarct with atrophy of the adjacent gyri and enlargement of the ipsilateral ventricular atrium (**d**). The infarct is hyperintense on PDw and on T2w sequences (**d**). The ipsilateral lenticular and caudate nuclei are preserved (**d**, ↑) at the inner aspect of the infarct. Note also the underdevelopment of whole ipsilateral hemisphere (**d**).

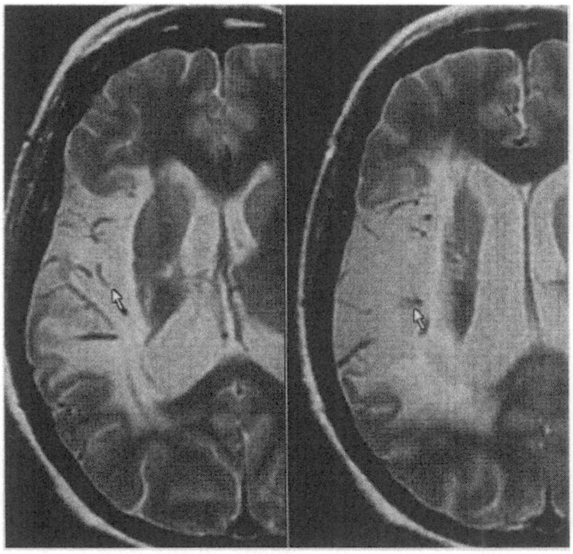

a

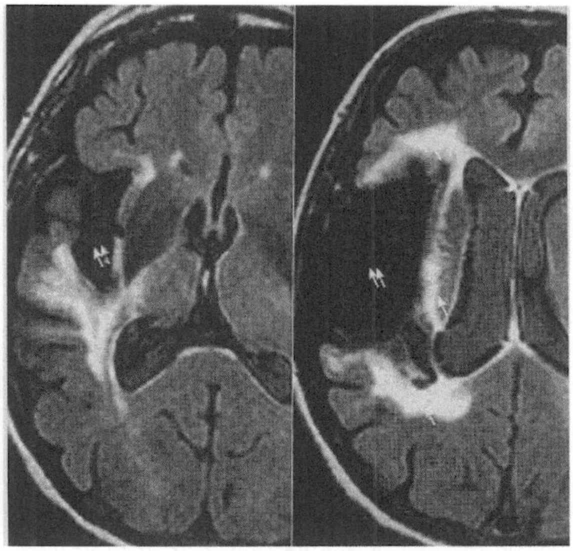

b

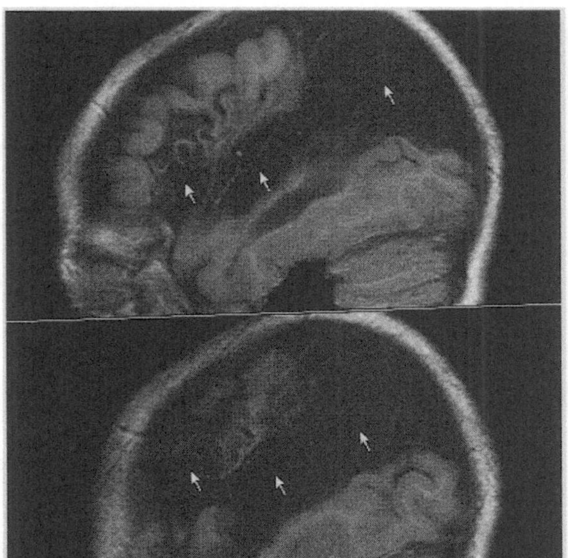

c

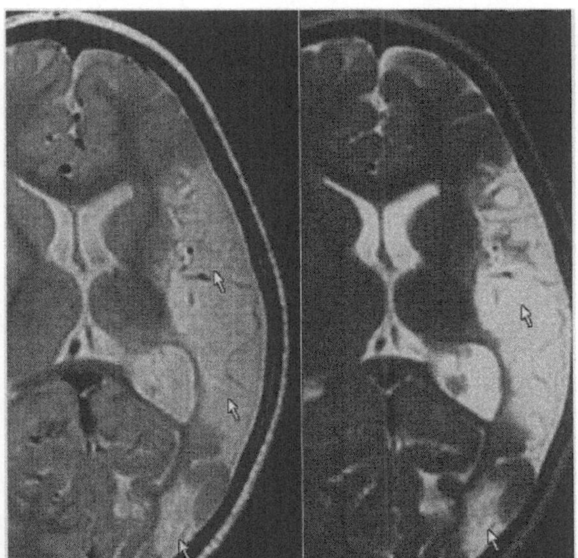

d

Hematoma and hemorrhage

Cases 62 (p. 160-161) and 76-79: intracerebral hemorrhages

Clinical and pathological features

- It is the third most frequent etiology of stroke.
- Due to various causes (excepting trauma) including most commonly: hypertension (about half the cases), rupture of saccular aneurysm or arteriovenous malformation, intratumoral hemorrhage, venous or arterial (embolic) hemorrhagic infarction, hematologic and bleeding disorders such as leukemia, hemophilia, thrombocytopenic purpura, aplastic anemia, anticoagulant therapy, hyperfibrinolysis.
- No sex or age predilection. Primary intracerebral hemorrhage occurs usually at lower age than in thrombotic infarcts.
- *Signs*: sudden onset and rapid evolution (from few minutes or hours to 24 hours) of neurological signs and symptoms. The patient is usually active and in exertion, and exceptionally asleep. Existence of a hypertension is commonly found. Resorption of the intraparenchymal hemorrhage is a slow process lasting weeks or months. Severe headache is a common complaint (may be mild or absent in about half the cases), and may be associated with stiffness of the neck and/or vomiting or focal seizures (10%). Coma may result in case of massive bleeding into the ventricles or in pontine hemorrhage. The presenting signs vary with the localization of the hematoma: lenticulo-capsular (almost half the cases); white matter; thalamic; cerebellar hemispheres; pontine.
- Intracerebral hemorrhage shows a poor prognosis if its volume exceeds an average of 50 cm^3, and about 70% of the patients die in the following month.

MR characteristic findings

Morphology

- Intratumoral hemorrhage (10%): occurs into primary gliomas in 2-5% of cases (glioblastoma, ependymoma, oligodendroglioma) or secondary in about 15% of cases (metastatic melanoma, chorioepithelioma, renal cell and bronchiogenic carcinomas).
- Intraparenchymal hemorrhage may be of different size. It may be massive reaching several centimeters, small averaging 1-2 centimeters, or slit-like corresponding to those occurring at white-gray matter junction. Lenticular-putaminal and cerebellar hemorrhages are usually larger than thalamic or pontine.
- MRI aspect of the intraparenchymal hemorrhage is highly dependent on the field strength and particularly on the stage of the hematoma.
- Considering the evolutive features of a resolving hematoma, it is possible to recognize 5 successive stages: hyperacute (first few minutes to 24 hours); acute (several hours to 3 days); early subacute (from 3 days to 1 week); late subacute (after 1st week), and chronic (3rd week to months or years).
- Blood breakdown products contained within the hematoma and responsible for the SI observed are respectively the following: oxyhemoglobin (hyperacute stage); deoxyhemoglobin (acute); intracellular methemoglobin (early subacute); extracellular methemoglobin (late subacute); chronic (hemosiderin with possible presence of free methemoglobin).

Signal intensity

- SI of blood is particularly related to the presence of iron and the variation of its magnetic properties according to the stage of the hematoma.
- SE-T1w: isointense or slightly hypointense to brain (oxyhemoglobin or deoxyhemoglobin); hyperintense to brain (paramagnetic methemoglobin); isointense or slightly hypointense (hemosiderin/ferritin).
- SE-T2w: slightly hyperintense to brain (diamagnetic oxyhemoglobin); markedly hypointense, particularly on longer TE's (paramagnetic intra-

cellular deoxyhemoglobin and intracellular methemoglobin); markedly hyperintense (extracellular methemoglobin); markedly hypointense particularly on high fields (hemosiderin/ferritin).

- GE-T2*: more sensitive than SE-T2w particularly using high fields, for the detection of hemosiderin and ferritin contained in macrophagic ang glial cells, due to magnetic susceptibility effect. The presence of a continuous hemosiderin rim is one of the features that helps to rule out a neoplastic hemorrhage. The hematoma at the chronic stage may appear as a residual hemosiderin-stained slit-like lesion. The subarachnoid hemorrhage which may have been associated result in a similar leptomeningeal staining called superficial siderosis (hypointense SI).
- Maturation of the hematoma varies in rate with respect to its intracranial location. This maturation process (cell lysis and oxidations) is longer in case of an intraventricular hemorrhage (oxyhemoglobin) than in intraparenchymal, due to the higher oxygen levels within the ventricular spaces.
- Hyperintense peripheral edema on T2w sequence is shown in acute/subacute stage. The presence of a long-standing associated edema with mass effect must raise the possibility of a neoplastic hemorrhage apart a heterogeneous morphological and SI pattern due to an unusual temporal evolution of the resolving blood within the hematoma, and/or the presence of mixed stages of hemorrhage and, of course, in case of presence (enhancing) of a nonhemorrhagic tumoral component. Fluid-fluid levels due to hematocrit effect, within cystic or necrotic components may be found in presence of an underlying tumoral process.
- Hyperintense SI on non enhanced T1w sequence (due to methemoglobin in resolving hemorrhage) may also correspond to fat, intratumoral melanin, dystrophic calcifications (presence of paramagnetic cations), residual lipiodol droplets (trapped in cisterns), hypermyelination, presence of paramagnetic cations (manganese or copper, such as in the basal ganglia of patients with chronic liver disease), mucinous material or high protein concentration usually within tumoral processes (craniopharyngioma, Rathke cyst pouch), or flow effect.

Contrast enhancement

- No need to perform contrast infusion unless an underlying tumoral process is suspected.

Case 76: Acute temporal hemorrhage

Fig. a: sagittal cut; GE - T1w

Fig. b: axial cut; SE - PDw and SE - T2w

Acute-early sub-acute temporal hematoma showing on T1w (**a**) a central iso-intensity area with a peripheral rim of hyperintensity (↑) representing deoxyhemoglobin surrounded by methemoglobin. The central hematoma is iso-signal on PDw (**b**, left, ↑) becoming very hypointense on T2w (**b**, right, ↑↑) due to the presence of intracellular deoxyhemoglobin responsible of T2 shortening. The hematoma is surrounded by high intensity edema (**b**, ↑↑↑) which is hypointense on T1w (**a**, ↑↑).

Case 77: Acute thalamic hemorrhage

Fig. c: axial cut; SE - PDw and SE - T2w

Fig. d: axial cut; FLAIR - T2w

Right thalamic hematoma markedly hypointense on PDw (**c**, left, ↑) as well as on T2w (**c**, right, ↑↑) and FLAIR sequences (**d**, ↑), hypointense to grey matter also on T1w sequence (not shown), due to the presence of intracellular deoxyhemoglobin. The hematoma is surrounded by a hyperintense rim shown on T2w dual echo sequence (**c**,*) with even a better contrast on the FLAIR sequence (**d**,*). The patient also presents a periventricular demyelination (**c**, **d**, ↑↑) and a diffuse brain atrophy with enlargement of the temporo-insular sulci.

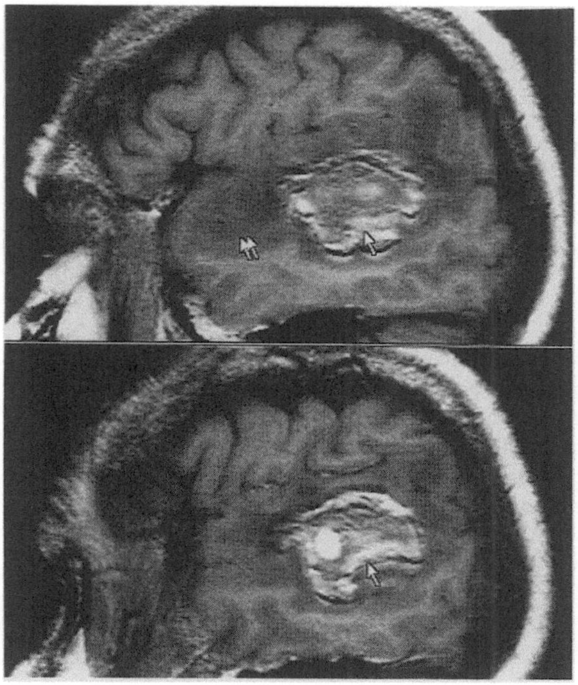

a

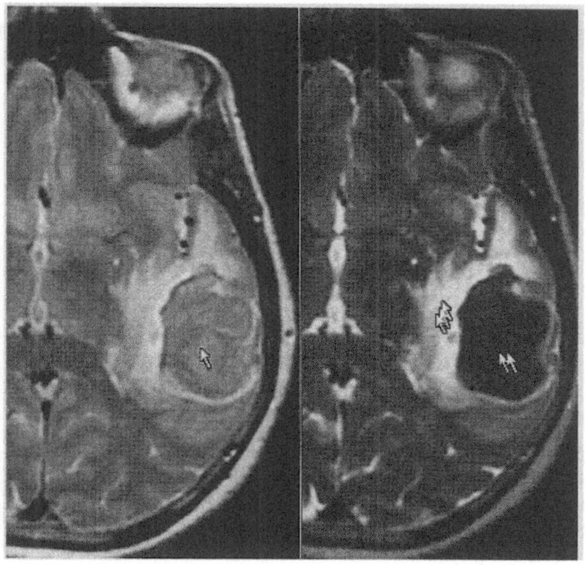

b

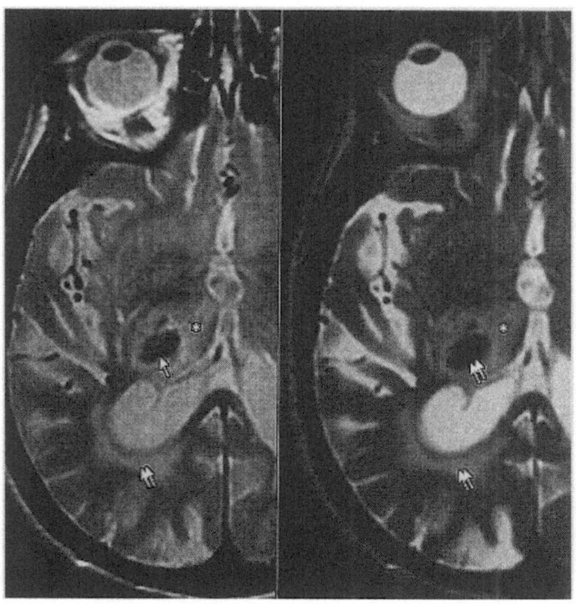

c

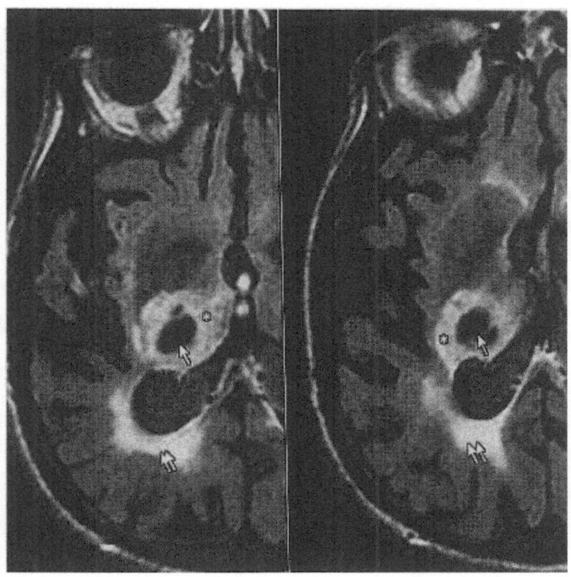

d

Case 78: Subacute frontal parasagittal hemorrhage

Fig. a: axial cut; SE - T1w

Fig. b: coronal cut; SE - T2w

Left frontal parasagittal subacute hematoma, well delineated, presenting on T1w (**a**, ↑) a high signal intensity due to the presence of methemoglobin (oxidized hemoglobin) and a hyperintensity on T2w (**b**, ↑). The center of the hematoma is still iso-hypo signal due to the remaining deoxyhemoglobin surrounded by hyperintense methemoglobin due to the fact that the oxydation begins at the periphery of the hematoma and converges radially inward. Note the existing low-intensity peripheral rim due to the hemosiderin-laden-macrophages noted on T2w sequence (**b**,*). The hematoma is surrounded by vasogenic edema which is hypointense to brain on T1w (**a**, ↑↑) and hyperintense on T2w (**b**, ↑↑), but less hyperintense than the hematic collection which is more hyperintense than edema and CSF mainly or heavily T2w sequences at this stage. Note also the mass effect on the left posterior frontal horn and corpus callosum, displaced downward.

Case 79: Subacute-chronic pontine hemorrhage

Fig. c: sagittal cut; GE - T1w

Fig. d: axial cut; SE - T2w

Right paramedian pontine hemorrhage associated to an infraoccipital subarachnoid hemorrhage (**c**, left, ↑↑). The mass is homogeneously hyperintense on T1w (**c**, ↑) and hyperintense and well limited on T2w (**d**) surrounded by a hypointense rim of hemosiderin on T1w and T2w sequences (**c**, ↑, **d**, ↑). On T2w, a small amount of vasogenic edema persits with slight distortion of the floor of the fourth ventricle (**d**, left, ↑↑).

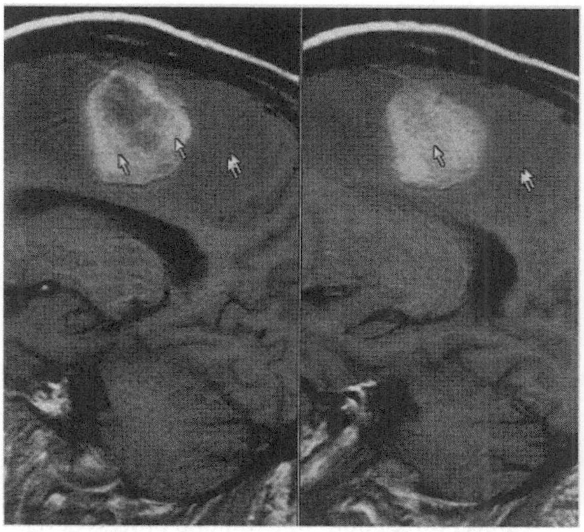

a

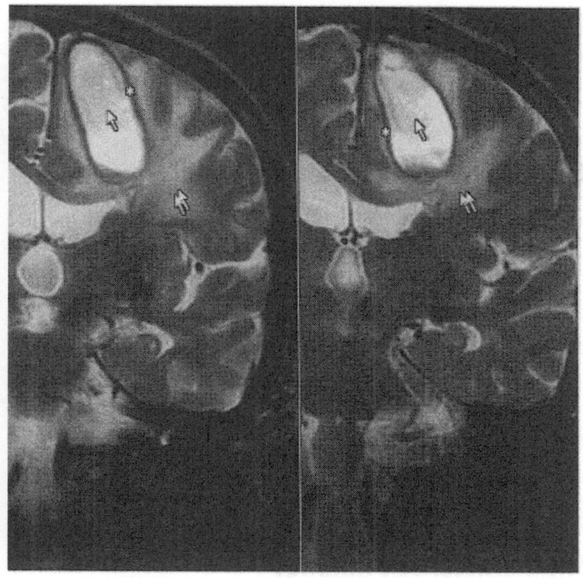

b

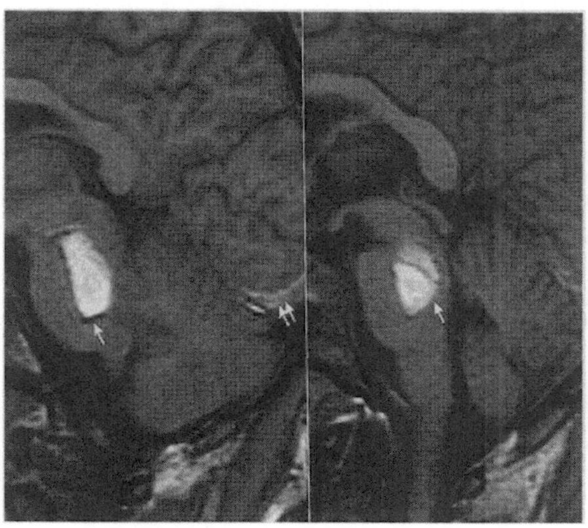

c

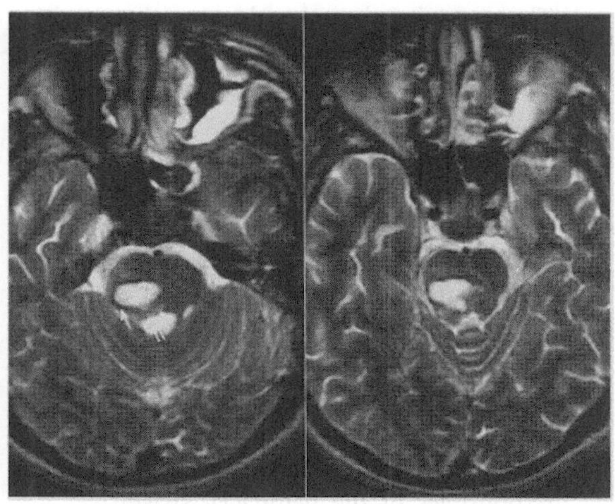

d

Vascular malformations and anomalies

Case 80: basilar dolichoectasia

Clinical and pathological features

- Correspond to a marked elongation and enlargement of a tortuous basilar artery.
- Common cause of vascular compression of the cranial nerves and/or brainstem.
- *Signs*: most commonly hemifacial spasm or trigeminal neuralgia.
- Microvascular decompression may contribute to clinical improvement.

MR characteristic findings

Morphology

- MR shows a dysplastic basilar artery that may involve a vertebral artery.
- The artery is enlarged and tortuous comprising oblique and horizontal segments.
- The lumen is enlarged and usually patent with slow flow.
- The dolichoectatic vessel may show a marked deformation and compression of the brainstem, mainly the pons and its middle cerebellar peduncle and even the medulla. Displacement and compression of the trigeminal (V) or facial (VII) nerve roots entry or exit zones are frequently demonstrated, using axial or coronal cuts, respectively.
- About 60% of the patients show a vascular compression of the facial nerve root and/or the pons, responsibles of the hemifacial spasm. Neurovascular compression or contact may be found in less than 20% of asymptomatic patients.
- Several other vessels (not necessarily dysplastic) may be incriminated, including: the vertebral artery, posterior cerebellar artery (PICA), the anterior cerebellar artery (AICA), and less frequently the basilar artery (midline position) unless tortuous or dolichoectatic.
- Differential diagnosis rules out multiple sclerosis (brainstem plaque involving the intra-axial fibers of the V or VII nerves, common anatomo-clinical correlation) or a tumoral process developed in the cerebellopontine cistern and compressing the cisternal nerve roots.

Signal intensity

- 3D-TOF MR angiography: sequence of choice to visualize the causal vessel responsible of nerve root and/or pontine compression.
- Oblique and coronal reformations, using the native millimetric slices (3D-TOF acquisition), are highly informative to demonstrate the anatomical conflict.

Contrast enhancement

- No need to perform contrast infusion with MR angio.
- Contrast infusion may be needed to depict veins.

Case 81: venous angioma

Clinical and pathological features

- It is a developmental venous anomaly (variant of venous drainage).
- May be considered as the most frequent vascular anomaly of the brain.
- Frequently found since the advent of MR (superior to CT for its detection).
- Etiology is still unknown. It is seldom associated with migrational disorders.
- *Signs*: usually asymptomatic throughout life and found incidentally; in case of hemorrhage an associated usually solitary cavernous angioma must be ruled out.

MR characteristic findings

Morphology

- Venous angiomas show the typical aspect of "medusa head", consisting of radially arranged medullary veins draining into an enlarged collecting vein.
- The sites of predilection are the frontal lobes and cerebellum, but may be found elsewere in the cerebral hemispheres and less commonly in the brainstem or the spinal cord. The intervening brain tissue shows no gliosis nor hemosiderin.
- The collecting vein drains into the superficial (cortical vein or dural sinus) or the deep subependymal venous systems. Cerebellar venous angiomas drain into the cerebellar cortical veins or into the subependymal veins of the 4th ventricle.
- On angiography, the typical "medusa head" anomaly appears in the venous phase after normal arterial and capillary phases.

Signal intensity

- The characteristic "medusa head" is usually shown as linear or curvilinear flow voids (hypointensity) on GE-T1w as well as on SE-T2w.
- GE-T2* sequence may show a marked hypointense area possibly related to the presence of deoxyhemoglobin in the venous channels.
- Vascular nature of the venous angioma may be further confirmed by the eventually associated misregistration artifact, as may be seen on SE-PDw.

Contrast enhancement

- Gd infusion demonstrate the typical aspect of the enhancing venous anomaly.
- MR angiography, if performed, shows the similar angiographic aspect precising the venous drainage.

Case 80: Mega-dolichobasilar artery

Fig. a: coronal cut; SE - T2w

Fig. b: axial cut; SE - T2w

Conventional T2w SE sequence showing in the coronal (**a**) and the axial (**b**) planes a large dysplastic and highly tortuous basilar artery easily depicted on the pre- and lateropontine cisterns (high signal of CSF on T2w), due to the signal void of the flowing blood. Note the mass effect on the anterior aspect of the pons moderately displaced posteriorly as well as on the anterior aspect of the left cerebellar hemisphere. The patient was admitted for intractable facial hemispasm.

Case 81: Venous angioma of the cerebellum

Fig. c: coronal cut (512); SE - T2w

Fig. d: angio, coronal cut; 2D-TOF

Flow-void vermicular and rounded lesion involving the posterior aspect of the left cerebellar cortex, shown on T2w (**c**, ↑), with draining vessel (**c**, ↑↑), without perilesional parenchymal signal abnormality. Note the "spider" aspect of the small vessels of the vascular anomaly (**c**, ↑). The MR-angiogram demonstrated the abnormal cerebellar vein with multiple small venules forming the venous angioma, draining into the left lateral sinus (**d**, ↑↑). The slow flow through the venous angioma is best visualized after administration of Gadolinum. These vascular malformations are incidentally discovered and are more often clinically silent. They correspond to an anomaly in the venous drainage pattern.

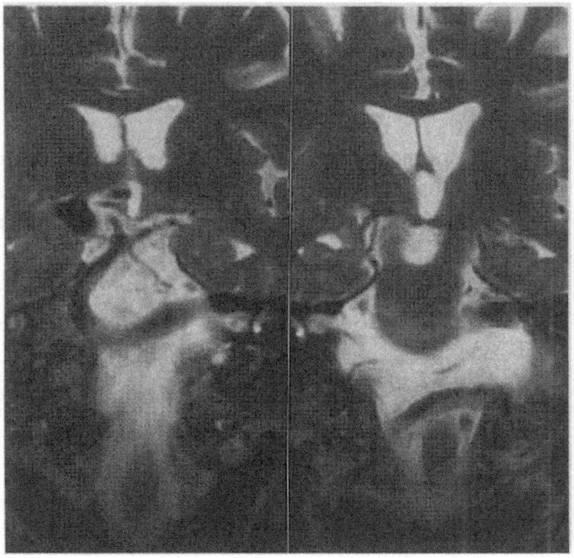
a

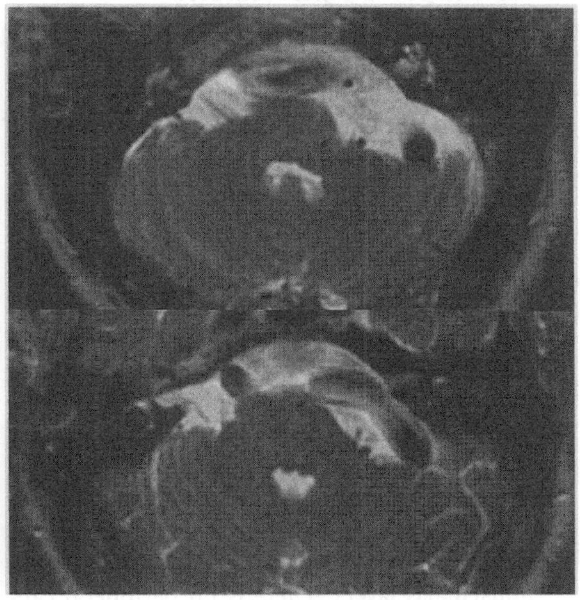
b

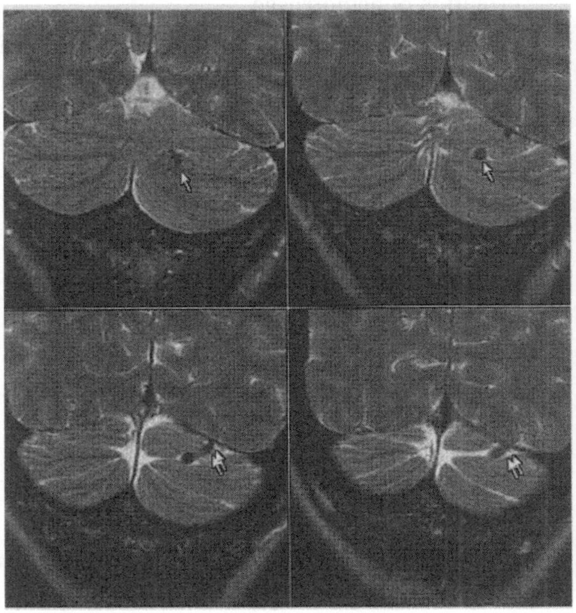
c

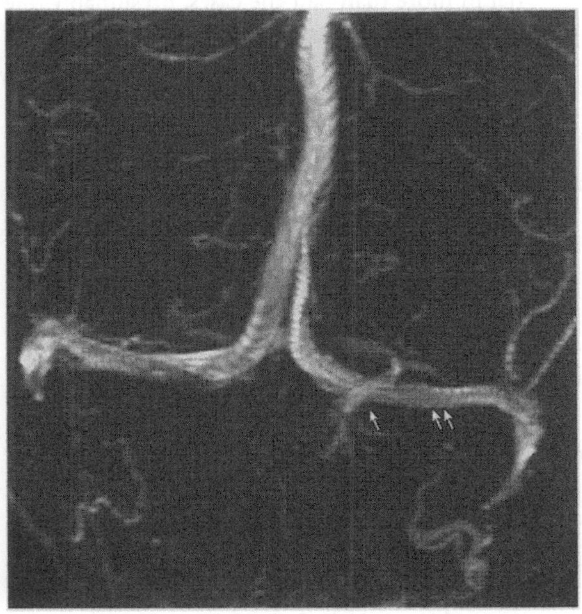
d

Cases 82-85: brain cavernous angiomas

Clinical and pathological features

- Called cavernomas or cryptic vascular malformations (angiographically silent).
- It represents the second most frequent form of vascular malformation.
- Disclosed on MRI in about 0.5% of brain exams.
- May be multiple in 1/4 of the patients and particularly in familial cases (10-15%).
- Males and females are equally involved (female predominance for spine).
- *Histopathology*: endothelial-lined vascular structures containing thrombosed blood of varying stages surrounded by gliosis and hemosiderin containing macrophages. The sinusoidal channels walls are thin or irregularly thick and may calcify. The vessel walls contain no elastica or muscularis layers. There is no peripheral capsule.
- *Signs*: most frequently asymptomatic (common incidental finding on MRI); when symptomatic the presenting signs are: seizures commonly partial in more than half the cases, presumably due to irritative effects to gliosis or hemosiderin deposits from previous hemorrhages; focal progressive neurologic deficits (40-50%) and headaches in a quarter of the cases. Hemorrhage is observed in 10% of the cases mainly during the 2nd or 3rd decades, its risk being evaluated to be about 1% per lesion per year. This risk is even increased after the first bleeding, in young females and for cerebellar or brainstem lesions.

MR characteristic findings

Morphology

- Cavernomas may be found either in the brain or spinal cord.
- Most cavernomas are found in the cerebral hemispheres, particularly close to the ventricular cavities or the leptomeninges. The basal ganglia are involved in 10%.
- Cerebellum and brainstem with predilection to the pons, are involved in 25%.
- Extradural spinal location has been reported.
- Most commonly well demarcated, intraparenchymal cavernomas may be lobulated ("mulberry-like" appearance), containing clotted blood at different stages, surrounded by a peripheral gliosis and hemosiderin rim (pseudocapsule).
- Cavernomas show no intervening normal brain tissue as compared to capillary telangiectasias (consist of dilated capillaries usually in the pons; most often asymptomatic; found almost exclusively on enhanced MR; may be found in association with venous angioma or be part of Rendu-Osler-Weber disease).
- The size of the cavernomas is variable ranging from millimeters to few centimeters. Best disclosed on GE-T2* sequences.
- Possible association between cavernous and venous angioma.

Signal intensity

- SI behavior is characteristic, related to temporal evolution of hemorrhage.
- GE-T1w: heterogeneous SI due to the presence of blood breakdown products (isointense in acute bleeding, hyperintense in subacute/chronic), surrounded by the hypointense rim of hemosiderin (resorption of previous hemorrhages).
- SE-T2w: heterogeneous SI (iso- and hypointense central zone), surrounded by the thin rim of hemosiderin (more hypointense than on T1w).
- GE-T2*: more sensitive than SE-T2w to demonstrate the presence of hemosiderin/ferritin deposits: marked hypointensity with increase in size of the abnormality ("blooming effect"). Note that the susceptibility artifact generated by GE sequence increases the sensitivity of detection of smaller cavernomas.
- Differential diagnosis: presence of small hemorrhagic foci in amyloid angiopathy (subcortical hemorrhagic foci), hemorrhagic metastases (melanoma, chorioepithelioma, etc.), intravascular coagulopathies, petechies due to hypertension, hemorrhage secondary to radiotherapy, treated toxoplasma encephalitis in AIDS.

Contrast enhancement

- Cavernomas are usually not visualized on conventional angiography (occult).
- Unless calcified, cavernomas are iso-dense and show mild enhancement on CT.
- MR shows marked contrast enhancement.
- Differential diagnosis from hemorrhagic tumoral conditions may be difficult. Tumors, mostly high grade, are usually associated with vasogenic edema, show presence of cellular mass, and lack of peripheral hemosiderin deposits due to multiple previous subclinical hemorrhages. Follow-up MR exams with contrast infusion will help to reach the positive diagnosis.

Case 82: Parahippocampal cavernoma

Fig. a: sagittal cut (512); SE - T2w

Fig. b: coronal cut; SE - T2w

Non verified, occult vascular malformation presenting as a round accumulation with heterogeneous content on T2w (**a**, ↑, **b**), comprising areas of hyperintense signal due to the presence of methemoglobin and areas of dense hypointensity corresponding to hemosiderin deposition from old bleeding. The mass is developed in the right lateral occipito-temporal gyrus (**b**, down, ↑↑) and is surrounded by a hyperintense rim of vasogenic edema and/or gliosis. It is bounded laterally by the occipito-temporal sulcus (**b**, up, ↑). The lesion shows no mass effect on the inferior horn of the lateral ventricle (**a**). Follow-up exams are recommended to exclude other ingrowing pathology.

Case 83: Mesial temporal cavernoma

Fig. c: sagittal cut; GE - T1w

Fig. d: axial cut; SE - PDw and SE - T2w

Similar case as the previous one (case 83). The patient presented seizures. MR showed small rounded mass containing heterogeneous material consistent with blood pigments from prior hemorrhages. On T1w the mass contains small collections of hyperintense methemoglobin (**c**, ↑) surrounded by a complete rim of markedly hypointense iron storage products (**c**, ↑, **d**, ↑) with no edema (**d**). Note that these vascular malformations contain essentially stagnant blood and are usually angiographically occult, what may also contribute to the high signal on T1w. Cavernomas are multiple in up to one-third of cases and necessitate extensive brain exploration using the T1w, T2w and mainly T2*w sequences.

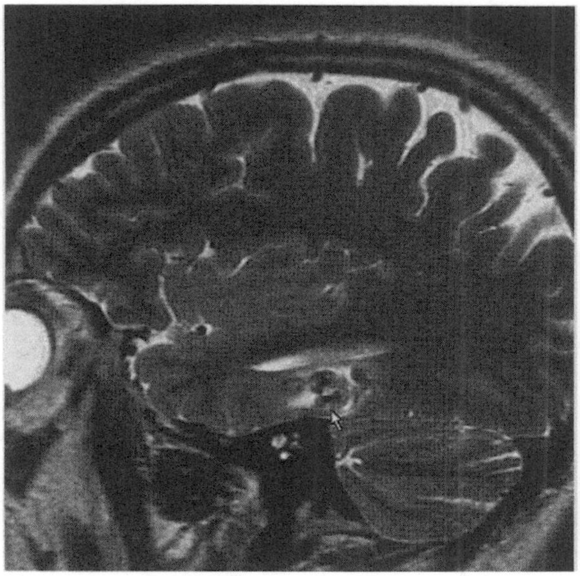

a

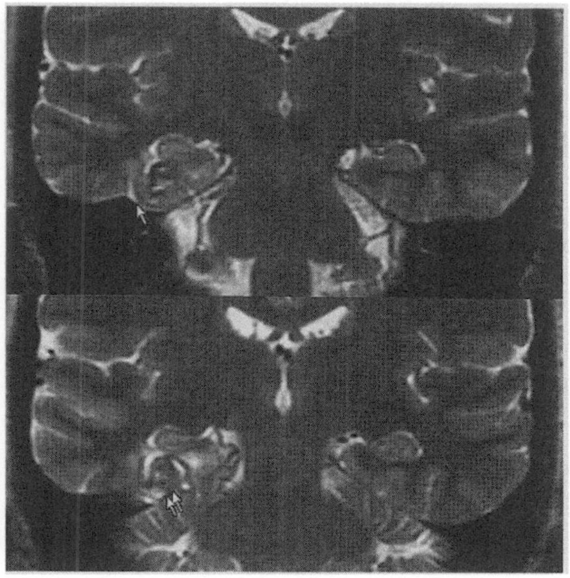

b

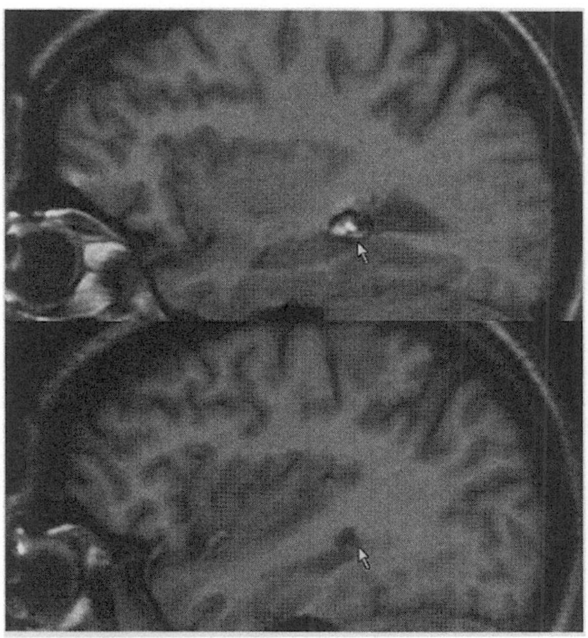

c

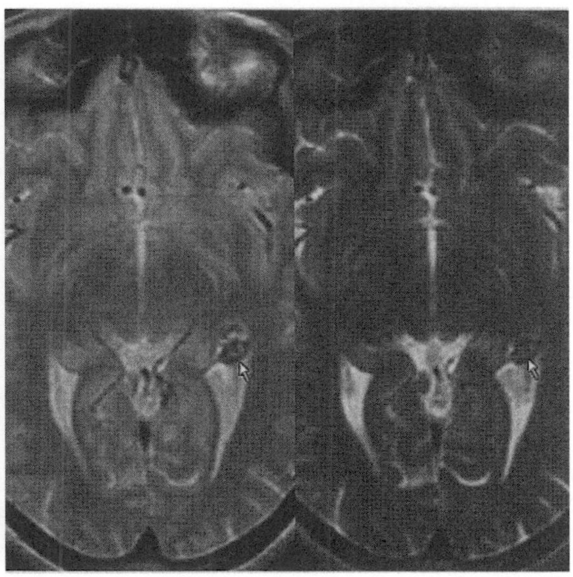

d

Case 84: Frontal cavernoma

Fig. a: coronal cut; SE - PDw

Fig. b: sagittal cut (512); SE - T2w

Subcortical nodular mass involving the frontal lobe, in the post-rolandic area, showing a hyperintense signal to the brain cortex on PDw sequence (**a,** ↑) and a heterogeneous signal intensity on T2w (**b,** ↑) but without any mass effect on the neighbouring sulci and no perilesional edema. The peripheral rim of hypointensity was only shown on T2*w sequence. The mass is well demarcated. The MR is not truly caracteristic. The patient was admitted for Jacksonian fits.

Case 85: Cavernoma of the floor of the fourth ventricle

Fig. c: sagittal cut; GE - T1w

Fig. d: axial cut (512); SE - T2w

Well demarcated round mass visible within the fourth ventricle and developed in the floor of the brainstem, showing a heterogeneous iso- and hyperintense signal on T1w (**c**) and a heterogeneous very low signal on T2w (**d**) due to acute intracavernous hemorrhage. The lesion is surrounded on T2w by an area of high signal involving the pons and the medulla, due to edema and/or gliosis with mass effect on the lateral aspect of the fourth ventricle as well as a symmetric enlargement of the right middle cerebellar peduncle (**d**). The patient presented an acute brainstem syndrome including a lateral gaze palsy.

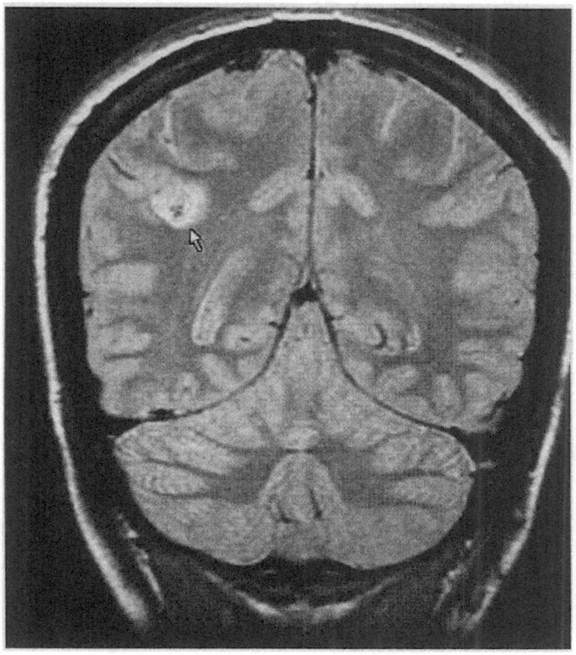
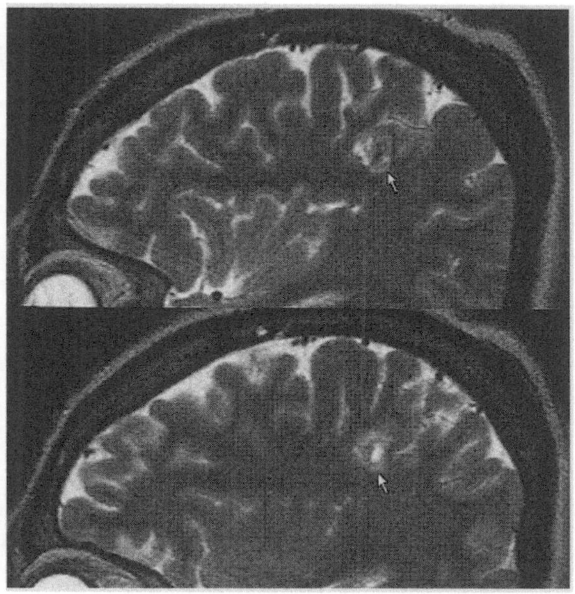

a b

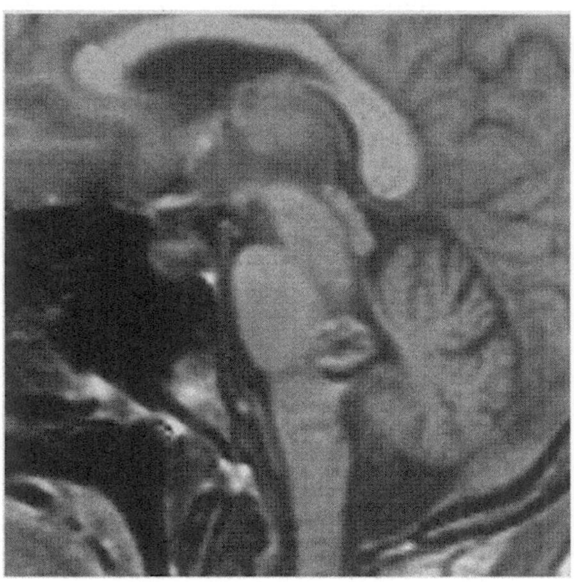
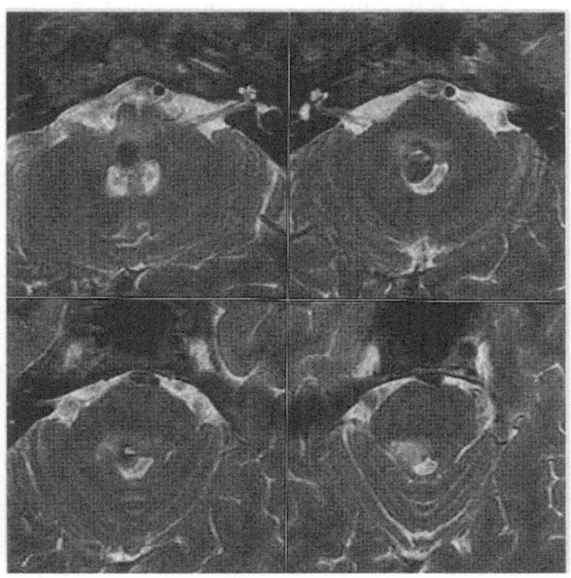

c d

Case 86-87: brain arteriovenous malformations

Clinical and pathological features

- Network of vessels consisting of direct shunts between arterial and venous channels with no intervening capillaries.
- Much less frequent than aneurysms, AVMs are found in about 0.15% of the population, and evidenced in about 30-50% of patients presenting a hemorrhage, most commonly in the 2nd-3rd decades. The frequency of the hemorrhage appears inversely proportional to the size of the AVM.
- The majority of AVMs are found in the cerebral hemispheres (85-90%) and the remaining 10-15% in the cerebellum and brainstem.
- Association with aneurysms is found in about half the cases, with almost all aneurysms occurring on hemodynamically related arteries.
- Multiple AVMs are rarely found in association with Rendu-Osler-Weber autosomal dominant disease and Wyburn-Mason mesencephalo-oculo-facial angiomatosis syndrome.
- *Histopathology*: cluster of abnormal vessels fed by arterial pedicles and drained by dilated collecting and draining veins constituting an arteriovenous shunt. The vessel walls lack of muscularis and internal elastic lamina. The nidus of the AVM is formed of venous channels without intervening brain parenchyma. Focal segments of wall thickening or thinning are observed in arterial or venous vessels.
- Specific arterial feeding arteries and draining veins are best evaluated using conventional catheter angiography which remains superior to MRA, despite the occurrence of roughly 4% neurological events comprising about 1% of neurological deficits.
- *Signs*: most commonly due to intracranial associated hemorrhage, in half the cases, usually intraparenchymal, but possibly intraventricular or subarachnoidal. Risk of bleeding is estimated to be of 2-3 or 4% annually with a rebleeding rate of 6% during the first year. The mortality risk is of 10-15% becoming as high as 25% at the 3rd hemorrhage. Seizures is the most common sign usually associated with large AVMs and observed in half the patients under 30 years of age. Progressive neurological deficit (large AVMs and/or AVMs located in eloquent cortical regions) and headaches (50%) are other frequent findings. Subjective rather than objective intracranial bruit may be reported.
- Complete surgical excision of the AVM nidus is made possible in about 80% of the cases. Radiosurgical procedures are used in non surgical cases, in AVMs measuring less than 3.5 cm. Interventional intravascular embolization contributes to reduce the nidus size prior to the other procedures, or may even cure the malformation in about 10% of the cases.

MR characteristic findings

Morphology

- Diagnosis of AVM is usually evident on MRI.
- Most frequently found over the cerebral convexity (middle third part) in the distribution of the sylvian arteries. May also be located deeply within the cerebral hemispheres, and in the posterior fossa in about 10% of the cases.
- Most frequently unique, AVM typically appears as a wedge-shaped cluster of dilated vessels (linear or vermiculated "signal void" areas on SE due to rapidly flowing blood), with a cortical-based surface and an apex pointing toward the ventricle. The nidus is of variable size, and its anatomical relationships are best evaluated on MR (superior to CT).
- No or minimal mass effect of the nidus on adjacent structures. Atrophic changes may be found in AVMs with associated steal phenomena.
- MR is superior to CT for the depiction of hemorrhagic complications (subacute or chronic blood or cicatricial hemosiderin deposits), which may modify the prognosis as well as the therapeutic decision. MR is also very sensitive for the detection of gliosis and/or edematous changes.
- MR is superior to CT for the evaluation of nidus (operative risk increasing with nidus size) and its obliteration after embolization. Surgical grading of AVMs based on the size, localization with respect to eloquent areas and drainage patterns of the AVM nidus.

- Hemorrhage may obscure the causal AVM which might not be excluded.
- CT is more performant for the depiction of calcified AVMs or the diagnosis of an acute hemorrhage.
- Radiosurgical treatment follow-up: MR is the modality of choice, showing peripheral edema and radiation necrosis with associated mass effect (3 months after), reduction in nidus size, decrease in flow void (consistent after 1 year). MRA is performed (even more than conventional angiography) to differentiate thrombosed from persistant patent vascular channels.

Signal intensity

- SE-T1w and SE-PDw and SE-T2w: network of vermicular signal voids (patent vessels with high velocity blood flow). Hyperintense SI of associated gliosis, demyelination or edema. In case of hemorrhage SI depends on the stage of blood breakdown products.
- GE-T2*: more sensitive than SE-T2 (T2 shortening) to depict the presence of hemosiderin (marked hypointense SI) due to old bleeding (characteristic blooming effect) and even differenciate it from calcification.
- Thrombosed AVM shows a mixed SI, roughly isointense to brain.
- MRA using TOF sequence: feeding arteries are best evaluated using 3D-TOF and dilated and tortuous draining veins best demonstrated using 2D-TOF. MRA brings accurate estimation of the nidus size. Angiogram may be obscured in case of associated hemorrhage (presence of methemoglobin). Phase contrast MRA with accurate velocity encoding must be used.

Contrast enhancement

- Heterogeneous enhancement pattern may be observed involving the draining dilated slow flowing veins and, partly, the nidus of the AVM.

Case 86: Rolandic arteriovenous malformation (child)

Fig. a: axial cut; SE - PDw

Fig. b: angio, axial view; 3D - TOF

Left rolandic subcortical arteriovenous malformation showing multiple large vessels with signal void specific for high flow states on the PDw sequence (**a**, ↑), surrounding the AVM nidus recognizable as wedge-shaped clusters of small vessels with the apex of the wedge directed toward the ventricular surface and the base cortically located (**a**, ↑↑). The MR-angiography without contrast injection shows the AVM unambiguously with its vascular nidus (**b**, ↑↑) in which intervening brain parenchyma is not found, associated to the dilated vascular channels containing relatively rapid flowing blood. The PDw (**a**) and the T2w sequences did not exhibit any peri-AVM gliosis or edema. Note the slight enlargement of the leptomeninges at the level of the enlarged superficial vessels.

Case 87: Thalamic arteriovenous malformation (adult)

Fig. c: axial cut; SE - PDw

Fig. d: axial cut; SE - PDw

Huge right diencephalic arteriovenous malformation involving the thalamo-capsular region. The nidus of the AVM involves the right thalamus entirely (**c**, ↑↑; **d**, ↑↑), with hypointense signal to the normal brain parenchyma on PDw (**c**). There is no associated edema or gliosis and no signs of recent bleeding. Posteriorly and in the midline, MR identifies the large draining vessels with signal void specific for high flow (**d**, ↑). Note that MRI is superior to both CT and conventional angiography in showing the anatomic relationship of the nidus, the feeding arteries and the draining veins of an arteriovenous malformation.

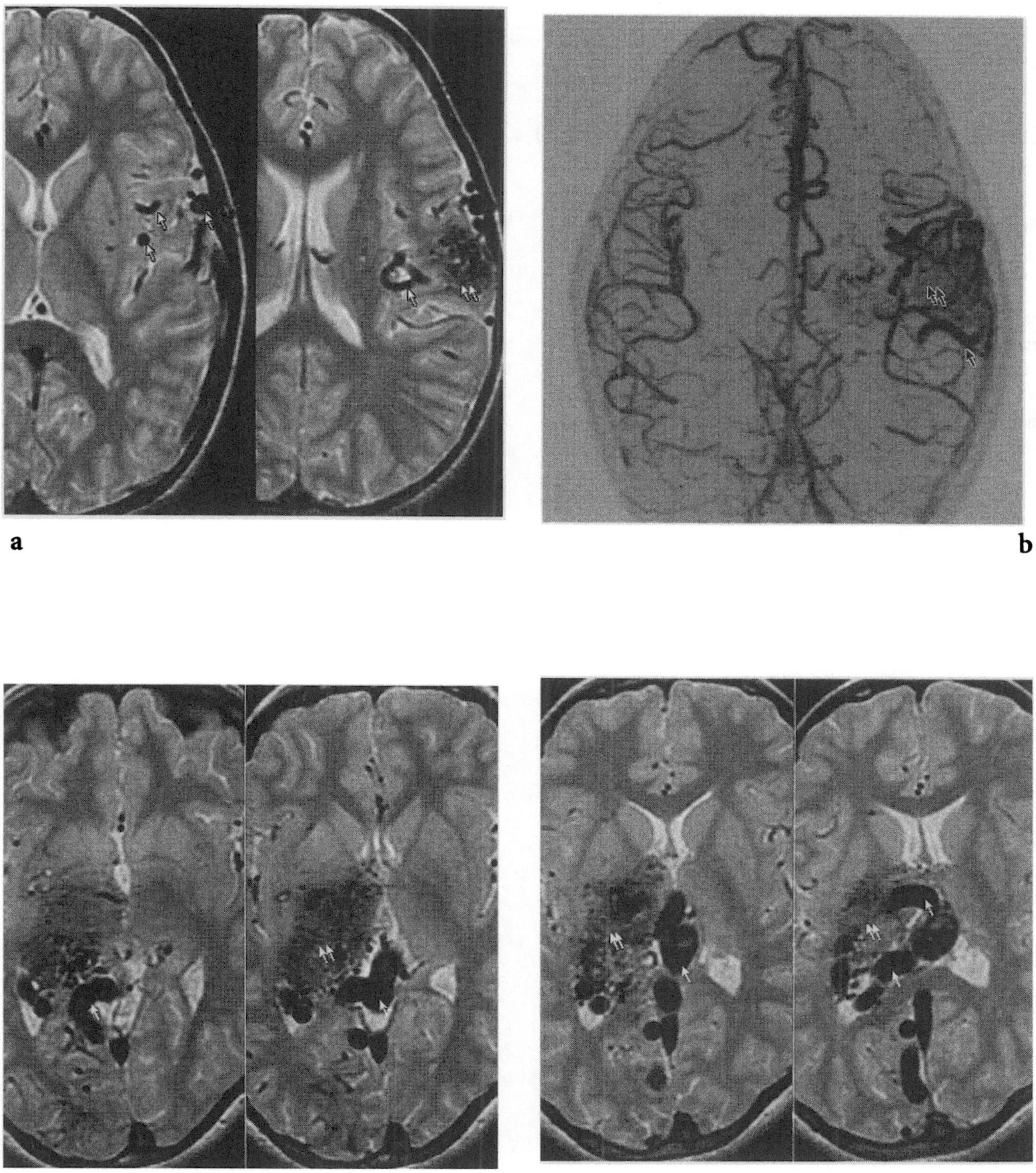

a

b

c

d

Case 88: MR-angio of the circle of Willis (normal)

Fig. a: angio, axial view; 3D-TOF

Axial MIP projection of the circle of Willis (**a**) showing its main arterial branches. **ACA:** anterior cerebral artery; **ACM:** middle cerebral artery; **ACP:** posterior cerebral artery; **BA:** basilar artery; **OPH:** ophthalmic artery; **INS:** insular branches of the ACM; **ACS:** superior cerebellar artery.

Fig. b: coronal cut; SE - T2w

(**b**) Frontal cut passing through the prepontine cistern showing the basilar artery and its branches at the level of the mamillary bodies (CM) and the interventricular foramina (F: fornix); 1: posterior cerebral artery; 2: superior cerebellar artery; 3: middle cerebellar artery; 4: posterior inferior cerebellar artery; 5: vertebral artery.

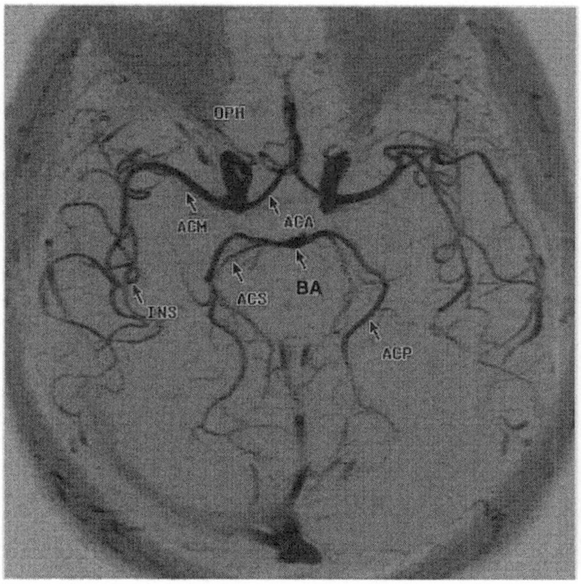

a

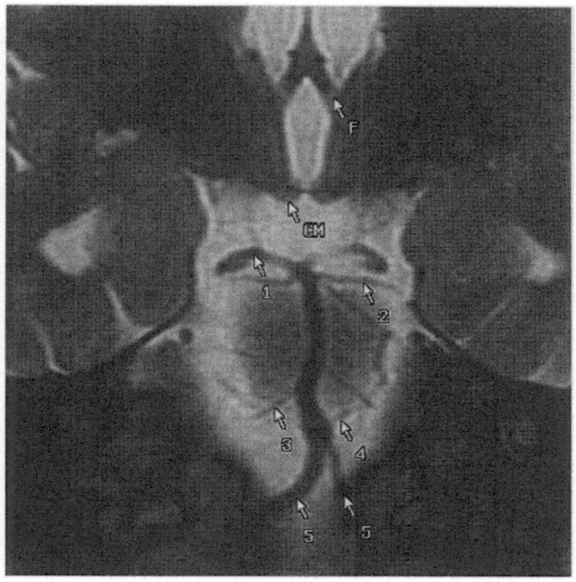

b

Cases 89-92: dissection of the internal carotid artery

Clinical and pathological features

- Internal carotid artery dissection represents 0.5-2.5% of strokes in the general population, with an increasing rate in youth, becoming 10 times more frequent.
- Dissections of cervical arteries may be spontaneous or post-traumatic, involving the cervical ICA distal to the bifurcation or less commonly the supraclinoid segment intracranially. The vertebral arteries are usually involved at the C1-C2 level. Multiple cervical arteries may be involved in about 1/3 of the cases.
- *Histopathology*: dissection results from the presence of a hematoma within the arterial wall usually interposed between the intima and the media; it may rarely be subadventitial. About 15% of the spontaneous dissections appear to be related to fibromuscular diseases: fibromuscular dysplasia, cystic medial necrosis, Marfan and Ehlers-Danlos syndrome.
- *Fibromuscular dysplasia*: rare arterial disease of unknown etiology, consisting of segmental tubular narrowings alternating with adjacent dilatations involving the extracranial ICA (involved in more than 90% of the cases), occurring most frequently on both sides (about 2/3) and mainly in women. Intracranial associated aneurysms are found in 21-51%. The disease process, including a degeneration of the elastic fibers, is of 3 histopathological types: medial (90%), intimal and subadventitial. The disease predispose arteries to dissection.
- *Signs*: headache or cervical pain and/or retrorbital pain on the homolateral side are the most frequent complaints occurring in about 2/3 of the cases; neurological signs and deficits due to an ischemic event may be associated; a unilateral Claude-Bernard-Horner's syndrome is found in half the cases.

MR characteristic findings

Morphology

- MRA is the noninvasive modality of choice for the diagnosis and the follow-up of cervical arteries dissection. More efficient of MRI, it shows a sensitivity of about 95% for the ICA and 20% for the vertebral artery, and a specificity of nearly 100% for both vessels. MRI shows a better sensitivity in the evaluation of the vertebral arteries.
- MRI using SE-T2w or FLAIR sequences is important to screen for a possible associated brain infarct. The intra-arterial wall hematoma is well demonstrated on SE-T1w sequences (SI depending on the stage of the mural blood).

Signal intensity

- SE-T1w sequences (2D-MRI): may depict the intraparietal dissecting hematoma (bright crescent-shaped high SI due to methemoglobin deposit) with an eccentric patent narrowed lumen (hypointense SI on SE-T1w).
- TOF-MRA (2D or 3D): similar aspect on native slices showing the normal hyperintense flow SI surrounded by a slightly hyperintense crescentic rim corresponding to methemoglobin stained mural hematoma.
- MIP MRA: shows a typical tapering of the residual intraluminal flow within the ICA usually starting distal to the bifurcation and extending to the skull base level. Follow-up exams are best performed using TOF sequences.

Contrast enhancement

- No need to perform contrast infusion.

Case 89: MR-angio of the extracranial arteries (neck coil)

Fig. c: angio, coronal view; 2D-TOF

The MR-angiography of the extracranial arteries performed (**c**), without contrast infusion, helps to evaluate the common carotid arteries (**ACC**) as well as their bifurcation into internal carotid (**ACI**) and external carotid artery easily recognized due to its branching in the neck. Both vertebral arteries (**AV**) are also evaluated. Best results are obtained using anterior neck coil, and even better using the neurovascular coil which allows to study the vessels from the aortic arch to the intracranial level, showing also the circle of Willis and its branches.

Case 90: MR-angio: dissection of internal carotid artery (head coil)

Fig. d: angio, coronal view; 2D-TOF

The MR-angio performed using the head coil and focused on the carotid bifurcation region in the upper neck, shows a right internal carotid artery dissection, with decreased caliber of the artery distal to the carotid bulb (**d**, ↑). There were no intramural hematoma on the conventional MR sequences mainly SE-T1w. The contralateral carotid bifurcation is normal.

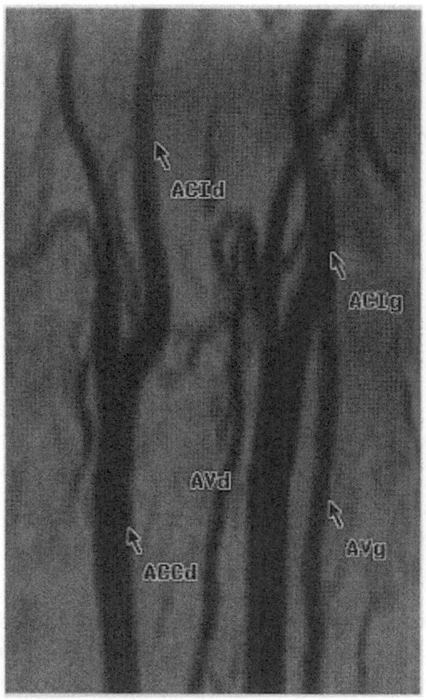

c

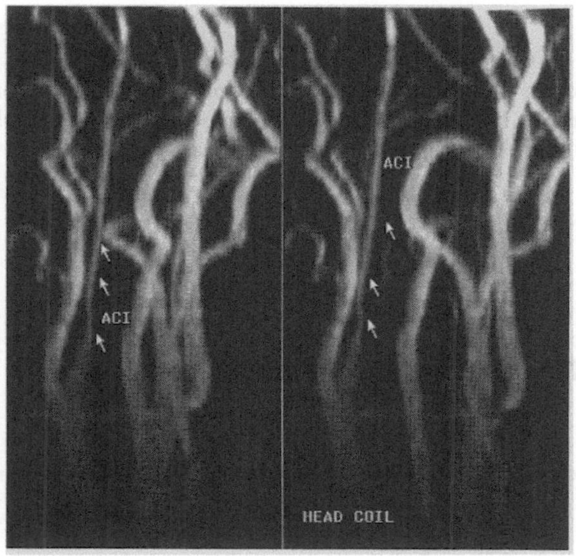

d

Case 91: Occlusion of internal carotid artery

Fig. a: axial cut; SE - T1w

Fig. b: axial cut; SE - T1w

Occlusion of the right carotid artery identified on axial SE T1w sequence (**a, b**) with iso-signal intensity associated to an area of hyperintense signal due presumably to a parietal hematoma (↑), as compared to the left patent carotid artery showing the flow-void signal.

Case 92: MR-angio: occlusion of internal carotid artery

Fig. c: axial cut, angio; 3D-TOF (native cuts)

Fig. d: angio, coronal view; 3D-TOF

Another similar case (**c, d**) performed with MR-angiography shows the absence of the hyperintense signal of the flowing blood in the left internal carotid artery with absence of visualization of its petrous, cavernous and supraclinoid segments, on the source slices in the axial plane (**c**) as well as on the coronal MIP (**d**). The right carotid artery and its terminal bifurcation appear normal as well as the basilar (**TB**) and vertebral arteries (**AV**).

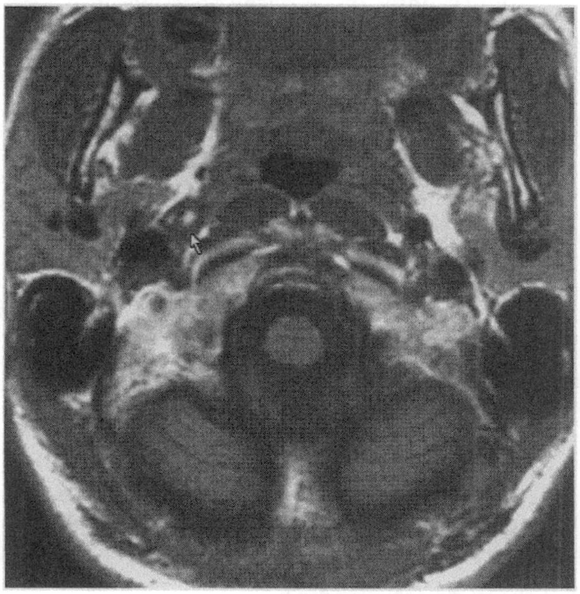

a

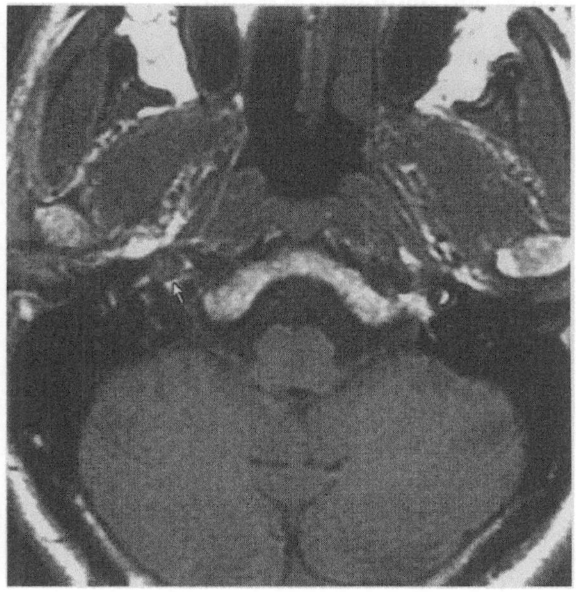

b

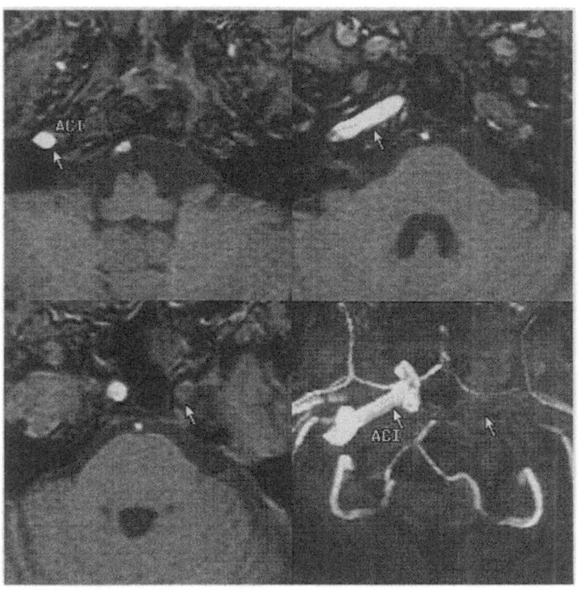

c

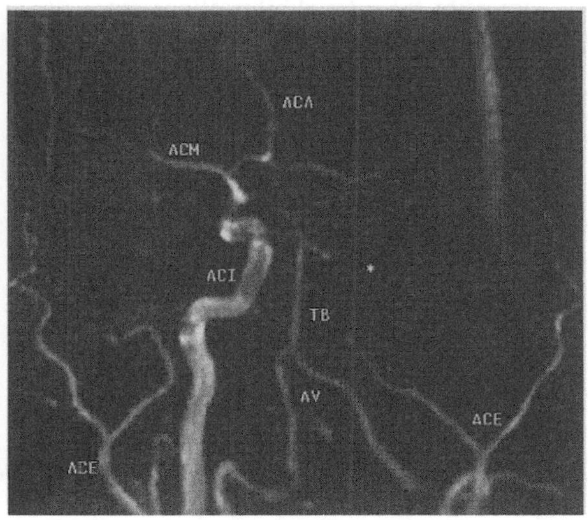

d

Cases 88 (p. 196-197) and 93-97: intracranial aneurysms

Clinical and pathological features

- Berry or saccular outpouchings found in 1% of an angiographic population.
- Aneurysms are rare in children and increase in frequency to reach the maximum between 30 and 70 years of age (90% of ruptured aneurysms vs 5% before age 30). A female preponderance is found (3:2). About 20-25% are multiple.
- Population at risk for intracranial aneurysms are: connective tissue diseases (Marfan's and Ehlers-Danlos syndromes), polycystic kidneys (11-40% risk in autosomal dominant form), fibromuscular dysplasia, coarctation of aorta, phacomatoses (neurofibromatosis and tuberous sclerosis), anomalous arteries (persistent primitive trigeminal, hypoglossal, azygous arteries), moyamoya disease, association with intracranial AVM, and familial form of aneurysms.
- *Pathogenesis*: still debated and not fully elucidated. Thought to be due to a developmental defect (absence of media and elastica in the wall of the vessel), it appears to be caused by hemodynamic forces involving preferentially the apices of the arterial bifurcations especially in the anterior half of the circle of Willis, and are aggravated by hypertension or collagenosis.
- *Histopathology*: the aneurysmal sac shows discontinuity in the media considered as the primary defect and a thickening of the adventitia and the intima. Generally, the lumen is partially obliterated by a mural thrombus due to atherosclerosis. The dome of the aneurysm is the most common site of rupture. Fusiform aneurysms in elderly are secondary to atherosclerosis, usually asymptomatic and seldom bleed. Mycotic aneurysms accounting for 3-4% of all intracranial aneurysms result from infectious bacterial endocarditis, and are preferentially located distally on the parent artery showing a high tendancy to bleed.
- *Signs:* most commonly asymptomatic prior to rupture, unless "giant" cause mass effects on the cavernous sinus and its cranial nerves content or the sella region, rather than subarachnoid bleeding. Subarachnoid hemorrhage due to rupture of the aneurysm is the most common clinical presentation, accounting for 75-90% of non traumatic subarachnoid hemorrhages, and fatal in 30% of the cases at the first hemorrhage. The major clinical manifestation is a generalized, severe and sudden onset of headache (possible ipsilateral preponderance and/or preretinal hemorrhage) associated or not with disorders of consciousness (ranging from drowsiness to deep coma). Most frequently there is no prodrome unless mild leakage occurs. Lateralizing signs are uncommon, present in case of eventually associated meningocerebral hemorrhage or ischemia in the territory of the parent artery. Delayed focal signs (3-5 days) after rupture may be attributed to a vasospasm occurring in about half the cases. A subacute hydrocephalus secondary to blockage of CSF pathways may even develop 2-4 weeks later.
- Postoperatively, it is highly recommended not to scan a patient without precise information on the clip's behavior when submitted to the magnetic field.

MR characteristic findings

Morphology

- Distribution of berry aneurysms in the anterior circulation (90%): junction between anterior communicating-anterior cerebral arteries (1/3); junction of ICA and posterior communicating artery and branches of ICA (1/3); bifurcation or trifurcation of middle cerebral artery (1/3). In the posterior circulation (10%): tip of the basilar artery and junction of the vertebral artery and the posterior inferior cerebellar artery, and superior cerebellar artery. The distribution of giant aneurysms involves the carotid system in 80% of the cases (60% involving the ICA, 10% the anterior communicating and 10% the middle cerebral artery), and the vertebral-basilar system in 20% of the cases.
- Sizes of the saccular aneurysms vary from 1-2 mm to reach 3-4 cm or more, averaging 8-10 mm. Those exceeding 2.5 cm are called "giant" aneurysms (5%), most of them occurring in the cavernous sinus arising from the ICA (cases 97, 98), or involve the middle cerebral artery or the tip of the basilar artery. The berry aneurysms vary in form, being round and connected to the parent vessel by a narrow neck (pedunculated) or broad-based (sessile).
- The fusiform aneurysms secondary to atherosclerosis may be considered as increased type of dolichoectatic arteries, affecting more frequently

the basilar system and less commonly the internal carotid artery. They often contain thrombotic material.

- MR shape of the berry aneurysms is variable depending on its size and the pulse sequence used. The patent lumen shows flow-related signal and the associated thrombus (particularly in giant aneurysms) will show a laminated pattern due to different stages (older in the periphery containing hemosiderin and recent clot near the center which may contain methemoglobin). Eventual calcifications of the aneurismal wall are not detected on MR unlike CT.
- Subarachnoid hemorrhage: in case of bleeding, and particularly in patients with multiple vascular ectasias, location of the cisternal subarachnoid blood may help to detect the ruptured aneurysm: sylvian vallecula and fissure (MCA branching), anterior interhemispheric fissure (anterior communicating artery), suprasellar cistern (supraclinoid ICA, posterior cerebral artery, tip of the basilar artery) and in the fourth ventricle (PICA or basilar artery). Note that sensitivity of CT is similar to CSF analysis during the first 12 hours following the bleeding.
- Aneurysms smaller than 3 mm are very difficult to detect using helical CT angiography, its detection sensitivity ranging from 70-95%. MR angiography (no contrast needed) with accurate postprocessing and reduction of background signal, helps to detect aneurysms more than 3 mm in about 90% of the cases, and may be used therefore as a reliable noninvasive screening technique for intracranial aneurysms, particularly in population with higher risk. Conventional DS angiography remains the gold standard procedure to perform, at least in the acute stage (detection of aneurysm and search for vasospasm). It is maybe non contributive in as much as 20% of the cases. Note that the eventual associated vasospasm may be monitored by transcranial doppler ultrasound.
- Improvements in image resolution and postprocessing algorithms are expected.

Signal intensity

- SE-T1w: hypointense SI "flow void" related to the patent aneurysmal lumen, or hyperintense SI representing methemoglobin in occluded aneurysm.
- SE-PDw: "flow void" (marked hypointense SI of the patent sac).
- SE-T2w: "flow void" (marked hypointense SI of the patent lumen). In giant aneurysms, possible mixed SI with additional high SI of the central portion of the lumen due to refocusing of slow flow by gradient moment nulling.
- GE-T1w: flow-related enhancement (bright SI of the patent lumen).
- GE-T2* and SE-T2w: marked peripheral hypointense SI of the partially thrombosed aneurysm (hemosiderin deposit).
- FLAIR-T2w: there are increasing arguments and data showing that FLAIR sequence, using a fluid attenuated inversion recovery, is sensitive in evaluating diffuse or collected (case 94) subarachnoid hemorrhage (hyperintense SI contrasting with hypointense CSF). This is particularly true for the evaluation of posterior fossa cisterns which are very poorly analysed by CT. CT still remains the modality of choice for the diagnosis of subarachnoid hemorrhage in the early acute stages (first 24 hours). Improvement of fast FLAIR sequence may modify in the future the diagnostic strategy.
- Improvement of image contrast and detectability of smaller aneurysms (2-3 mm) using magnetization transfer technique.
- Presence of ghost images along the phase-encoding axis helps to demonstrate the vascular nature of the malformation.
- 3D-TOF acquisition and MIP algorithm: best for detection and evaluation of smaller aneurysms; analyses of source images are also mandatory. Giant aneurysms are poorly revealed due to saturation effect of slow intraluminal flow (need to perform contrast infusion which will enhance slowly filling lumen or rely on conventional SE acquisitions).
- 3D-PC with a VENC of 30 cm/s: more efficient than 3D-TOF for the evaluation of larger (1.5 cm and more) or giant aneurysms; unlike 3D-TOF, PC-MRA helps to differentiate the patent residual lumen from the adjacent subacute methemoglobin containing thrombus, that may show a high SI which may lead to overestimate the size of the patent lumen. PC is more sensitive than TOF to slow flow and shows a superior background suppression, but a longer acquisition time.
- MRI and MRA must be both performed and their complementary results correlated when screening a patient for the presence of an aneurysm.

Contrast enhancement

- Contrast enhancement usually not needed.
- May be of help for the characterization of partially thrombosed giant aneurysms.

Case 93: Prepontine subarachnoid hemorrhage

Fig. a: sagittal cut; GE - T1w

Fig. b: axial cut; SE - PDw

Fig. c: axial cut; FLAIR - T2w

Fig. d: axial cut; SE - PDw

MR shows a T1 shortening in the chiasmal cistern (**a,** ↑) associated to a similar hyperintense signal of the subarachnoid space in the left pontocerebellar cistern (**c,** ↑) delineating the cochleo-vestibular nerve (**c,** up). The hyperintense signal of the blood shown on T1w (**a, c**) is indicative of methemoglobin related to a subacute hemorrhage. On PDw sequence (**b, d**) the blood remains hyperintense occupying the chiasmal, the prepontine (**b,** ↑) and the left pontocerebellar cisterns (**d**) dissecting at this level the trigeminal nerve (**d,** left, ↑). Note that the subarachnoid hemorrhage is more hyperintense than CSF on PDw (**b, d**). The angiogram performed was normal as well as the MR-angio.

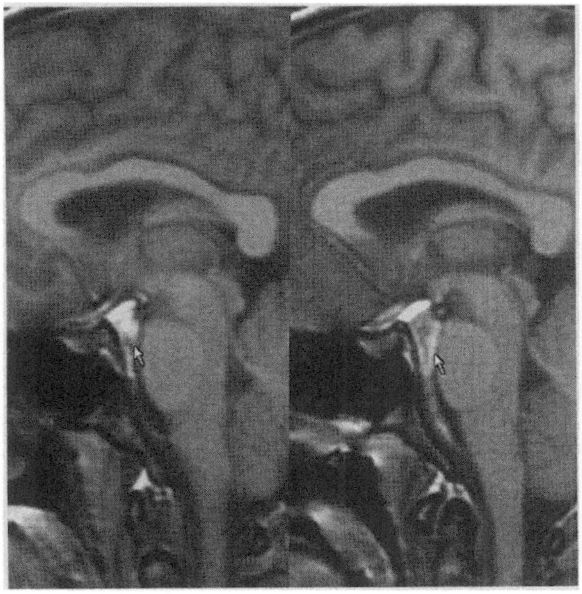

a

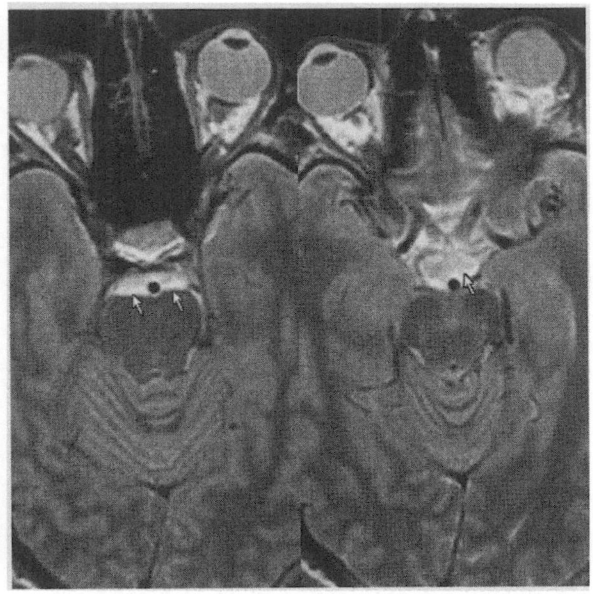

b

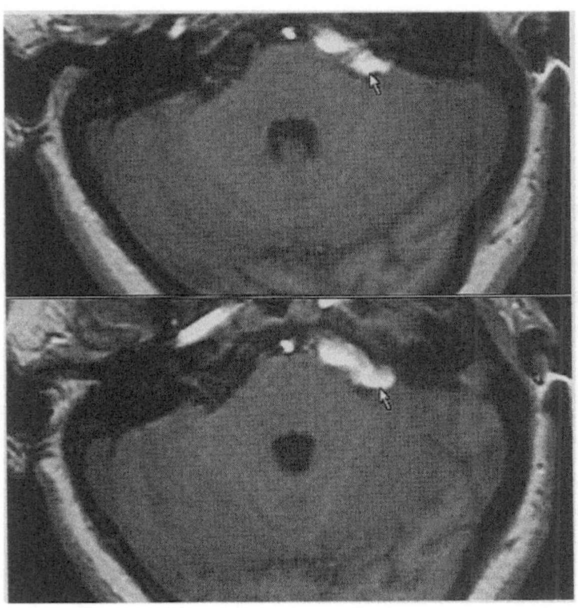

c

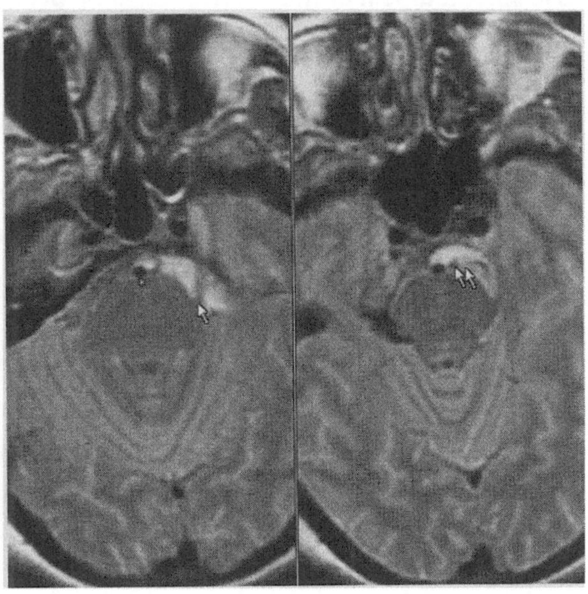

d

Case 94: Aneurysm of anterior communicating artery

Fig. a: axial cut; SE - PDw

Fig. b: angio, coronal cut; 3D-TOF

Subacute intracranial bleeding in the subarachnoid space on the left side (**c, d,** ↑↑↑), associated with an intraventricular hemorrhage with fluid-fluid level (**c, d,** ↑↑) with high signal shown at the posterior extremity of the lateral ventricles and corresponding to methemoglobin deposition. Note also the mass effect on the frontal horns (**c,** ↑) due to bleeding within the anterior interhemispheric fissure and in the frontal lobes (**c**,*), a pattern caracteristic of a ruptured anterior communicating artery aneurysm.

MR-angio (**d**) shows the hyperintense signal of the blood in the subarachnoid spaces and in the ventricular system, and visualizes on the axial projection an ectatic anterior communicating artery (**d,** ↑).

Case 95: Aneurysm of the anterior cerebral artery

Fig. c: axial cut; SE - PDw

Fig. d: angio, coronal view; 3D-TOF

Incidental finding on axial PDw sequence showing a flow void small mass adjacent to the pericallosal arteries (**a,** ↑) and corresponding to an aneurysm. The MR-angiography performed using a 3D-TOF technique without contrast injection shows the aneurysm arising from the left anterior cerebral artery at its bifurcation into two pericallosal arteries (**b,** ↑). MR-angio confirms the flow-void, present centrally within the aneurysmal lumen, on the SE-T2w sequence.

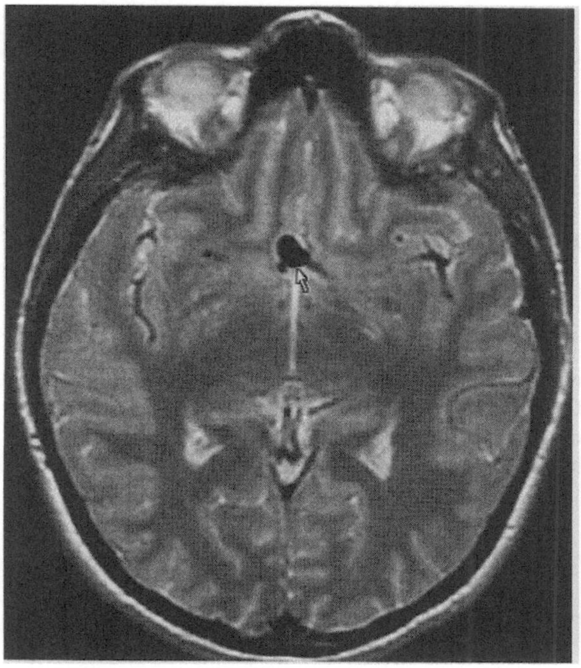

a

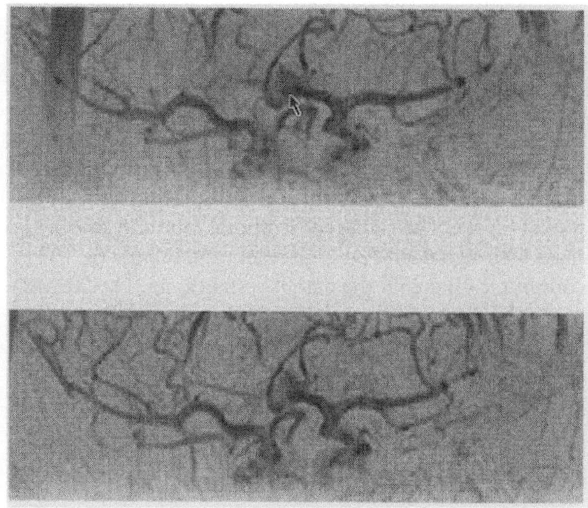

b

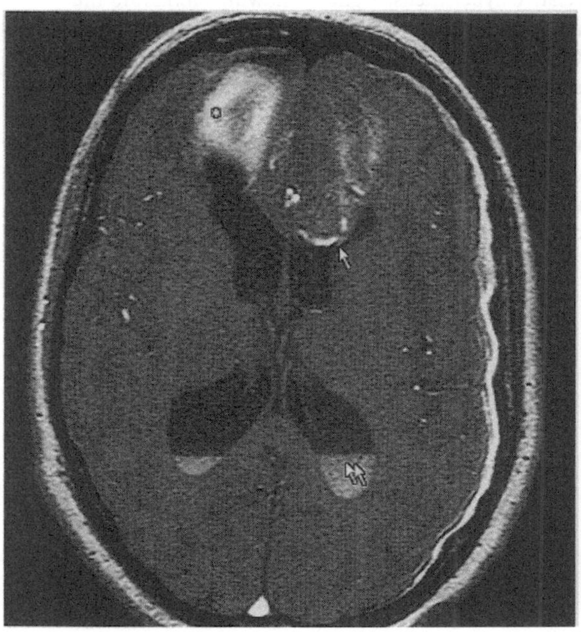

c

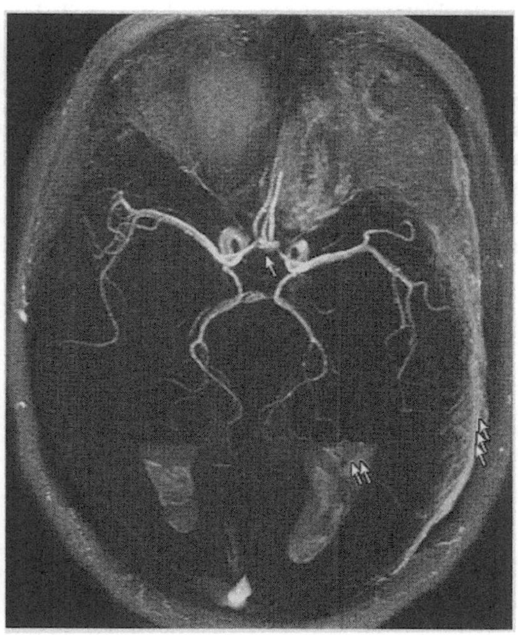

d

Case 96: Bilateral intracavernous aneurysm

Fig. a: coronal cut; SE - T2w

Fig. b: angio, axial view; 3D-TOF

Giant left intracavernous aneurysm displacing laterally the mesial aspect of the adjacent temporal lobe and shown as a mass with signal void on high resolution images (**a**, ↑↑) communicating with the parent vessel (**a**, up, ↑). The MR-angio (**b**) shows on the axial projection the aneurysm (**b**, ↑↑), as well as an anterior hyperintense rim along the anterior aspect of the aneurysm (**b**, ↑) due to incoming flow from the parent vessel. The MR exam shows a dysplastic right internal carotid artery with a small ectasia in the right intracavernous sinus (**a**, **b**, *).

Case 97: Bilateral intracavernous aneurysm

Fig. c: axial cut; SE - T1w

Fig. d: angio, axial view; 3D-TOF

Giant bilateral intracavernous aneurysm shown on SE-T1w sequence (**c**). The left aneurysm shows a hyperintense signal on T1w (**c**, ↑↑) due to thrombosis and methemoglobin deposit, confirmed on MR-angio (**d**) as well as on conventional angiogram. On the right side, the giant aneurysm shows heterogeneous intraluminal low signal intensity on T1w (**c**, left), with flow artifact as ghost images along the phase-encoding axis proving the vascular nature of the mass and its patency. MR angiography shows relative hyperintense signal of flowing blood into the aneurysmal lumen (**d**) with a more hyperintense rim along the anterior border of the aneurysm. Note that on MRA source images large aneurysms appear iso-intense to brain due to slow flow and intravoxel dephasing related to turbulent flow within the aneurysmal lumen, and hence could be poorly visualized. Intravenous contrast administration helps with the visualization of giant aneurysms.

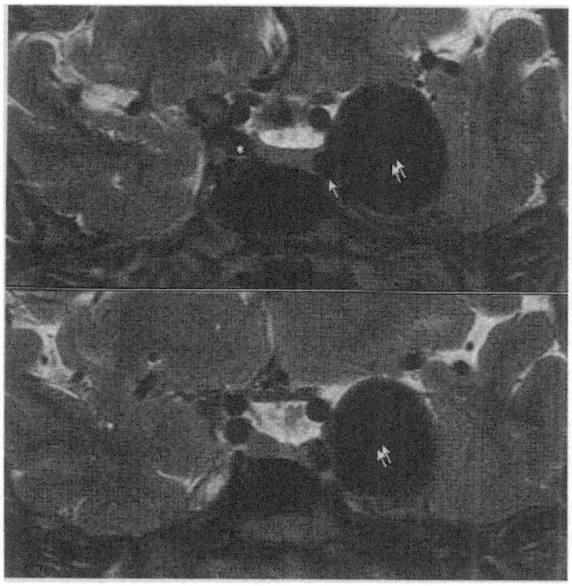

a

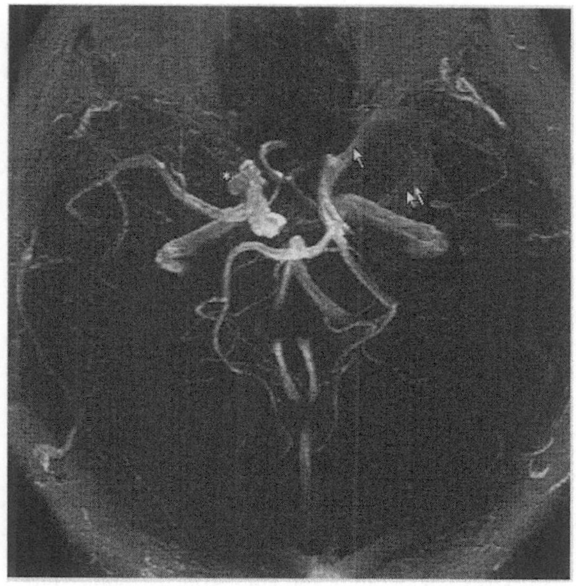

b

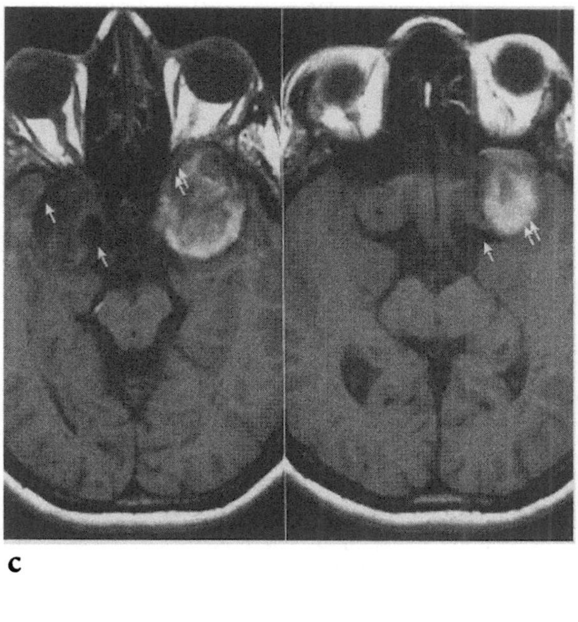

c

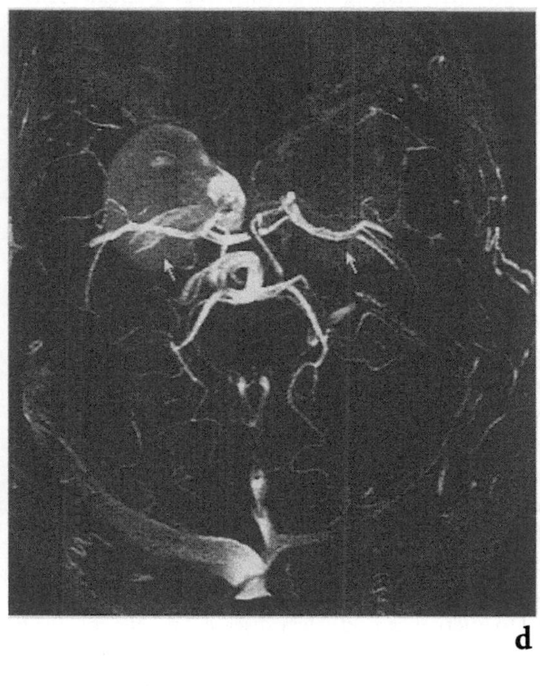

d

Selected References

- Abe T, Singer RT, Marks MP et al (1998) Coexistence of occult vascular malformations and developmental venous anomalies in the central nervous system: MR Evaluation. AJNR 19:51-57
- Aiba T, Tanaka R, Koike T et al (1995) Natural history of intracranial cavernous malformations. J Neurosurg 83: 56-59
- Alexander JA , Sheppard S, Davis PC (1996) Adult cerebrovascular disease: role of modified FLAIR pulse sequences. AJNR 17:1507-1513
- Ameri A, Bousser MG (1992) Cerebral venous thrombosis. In: Barnett HJM, Hachinski VC (eds): Cerebral ischemia: treatment and prevention. Neurol Clin 10:87
- Anal E, Shellock FG, Lewin JS (1996) Aneurysm clip testing for ferromagnetic properties: clip variability issues. Radiology 200:576
- Anderson CM, Saloner D, Lee RE et al (1992) Assessment of carotid artery stenosis by MR angiography: comparison with x-ray angiography and color-coded Doppler ultrasound. Am J Neuroradiol 13:989
- Anderson SC, Shah CP, Murtagh FR (1987) Congested deep cortical veins as a sign of dural venous thrombosis: MR and CT correlations. J Comput Assist Tomogr 11:1059
- Arbelaez A, Castillo M, Mukherji SK (1999) Diffusion-weighted MR imaging of global cerebral anoxia. AJNR 20:999-1007
- Asato R, Okumura R, Konishi J (1991) "Fogging effect" in MR of cerebral infarct. J Comput Assist Tomog 15(1): 160
- Atlas S, Thulborn K (1998) MR Detection of Hyperacute Parenchymal Hemorrhage of the Brain. AJNR 19:1471-1507
- Barrow D (1997) Classification and natural history of cerebral vascular malformations: arteriovenous, cavernous, and venous. J Stroke Cerebrovasc Dis 6:264-267
- Barry Oliver T, Alistair Lammie G, Wright AR et al (1999) Atherosclerotic Plaque at the Carotid Bifurcation: CT Angiographic Appearance with Histopathologic Correlation. AJNR 20:897-901
- Berenstein A, Lasjaunias P (1992) Classification of brain arteriovenous malformations. In: Surgical neuroangiography. Springer-Verlag, Berlin
- Black WC (1994) Intracranial aneurysms in adult polycystic kidney disease: is screening with MR angiography indicated? Radiology 191:18
- Bogousslavsky J, Regli F (1986) Borderszone infarctions distal to internal carotid artery disease. Ann Neurol 20:346
- Bogousslavsky J, Regli F (1987) Ischemic strokes in adults younger than 30 years of age: cause and prognosis. Arch Neurol 44:479
- Bogousslavsky J, Despland PA, Regli F (1987) Spontaneous carotid dissection with acute stroke. Arch Neurol 44:137
- Bogousslavsky J, Regli F, Maeder P et al (1993) The etiology of posterior circulation infarcts: a prospective study using magnetic resonance imaging and magnetic resonance angiography. Neurology 43:1528
- Bousser M, Baron J, Chiras J (1985) Ischemic strokes and migraine. Neuroradiology 27:583-587
- Bradley WG, Shimidt PC (1985) Effect of methemoglobin formation on the MR appearance of subarachnoid appearance. Radiology 1156:99
- Bradley WG, Waluch V, Brant-Zawadski M et al (1984) Patchy periventricular white matter lesions in the elderly: a common observation during NMR imaging. Noninvasive Med Imaging 1:35
- Braffman BH, Zimmerman RA, Trojanowski JQ et al (1988) Brain MR: pathologic correlation with gross and histopathology. I. Lacunar infraction and Virchow-Robin spaces. ANJR 9:621
- Brant-Zadadzki M, Fein G, Van Dyke C et al (1985) MR imaging of the aging brain: patchy white-matter lesions and dementia. AJNR 6:675
- Brant-Zadadzki MN, Atkinson D, Detrick M et al (1996) Fluid attenuated inversion recovery (FLAIR) for assessment of cerebral infarction: initial clinical experience in 50 cases. Stroke 27:187
- Brooks RA, Di Chiro G, Patronas N (1989) MR imaging of cerebral hematomas at different field strengths: theory and applications. J Comput Assist Tomogr 13:194
- Brown RD Jr, Wiebers DO, Forbes GS (1990) Unruptured intracranial aneurysms and arteriovenous malformations: frequency of intracranial hemorrhage and relationship of lesions. J Neurosurg 73:859
- Brown RD Jr, Wiebers DO, Forbes G et al (1988) The natural history of unruptured intracranial arteriovenous malformations. J Neurosurg 68:352-357
- Brown WD, Graves VB, Chun RW (1989) Moyamoya disease: MR findings. J Comput Assist Tomog 13(4):720
- Brugières P, Castrec-Cargo A, Héran et al (1989) Magnetic resonance imaging in the exploration of dissection of the internal carotid artery. J Neuroradiol 16:1
- Bryan RN, Levy LM, Whitlow WD et al (1991) Diagnosis of acute cerebral infarction: comparison of CT and MR imaging. AJNR 12: 611
- Bryan RN (1990) Imaging of acute stroke. Radiology 177:615
- Castaigne P, Lhermitte F, Gautier JC et al (1970) Internal carotid artery occlusion: a study of 61 instances in 50 patients with post-mortem data. Brain 93:231
- Chakeres DW, Bryan RN (1987) Acute subarachnoid hemorrhage: in vivo comparison of magnetic resonance and computed tomography. ANJR 7:223
- Chakeres DW, Schmalbrock P, Brogan M et al (1990) Normal venous anatomy of the brain: demonstration with gadopentetate dimeglumine in enhanced 3-D MR angiography. AM J Neuroradiol 11:1107
- Chapman AB, Rubinstein D, Hughes R et al (1992) Intracranial aneurysms in autosommal dominant polycystic kidney disease. N Engl J Med 327:916
- Chiras J, Dubs M, Bories J (1985) Venous infarctions. Neuroradiology 27:593-600

- Chong J, Lu D, Aragao F et al (1998) Diffusion-weighted MR of acute cerebral infarction: comparison of data processing methods. AJNR 19:1733-1739
- Clark RA, Watanabe AT, Bradley WG Jr et al (1990) Acute hematomas: effects of deoxyhemoglobin, hematocrit, and fibrin-clot formation and retraction on T2 shortening. Radiology 174:201
- Cognard C, Gobin YP, Pierot L et al (1995) Cerebral dural arteriovenous fistulas: clinical and angiographic correlation with a revised classification of venous drainage. Radiology 194:671-680
- Cordobes F, Lobato RD, Rivas JJ et al (1981) Observations on 82 patients with extradural hematoma. J Neurosurg 54:179-186
- Craig DR, Meguro K, Watridge C et al (1982) Intracranial internal carotid artery stenosis. Stroke 13:825
- Crain MR, Yuh WT, Greene GM et al (1991) Cerebral ischemia: evaluation with contrast-enhanced MR imaging. AJNR 12:631
- Curling OD Jr, Kelly DL, Elster AD, Craven TE (1991) An analysis of the natural history of cavernous malformations: incidence and familial occurrence. N Engl J Med 319: 343-347
- De Coene B, Hajnal JV, Gatehouse P et al (1992) MR of the brain stem using fluid-attenuated inversion recovery (FLAIR) pulse sequences. AJNR 13:1555
- De Coene B, Hajnal JV, Pennock JM et al (1993) MRI of the brain stem using fluid attenuated inversion recovery pulse sequences. Neuroradiology 35:327
- DeMarco K, Dillon WP, Halbach VV et al (1990) Dural arterovenous fistula: evaluation with MR imaging. Radiology 175:191
- De Salles A (1997) Radiosurgery for arteriovenous malformations of the brain. J Stroke Cerebrovasc Dis 6(4):277-281
- Domiano TR, Truwit CL, Dowd CF et al (1994) Posterior fossa venous angiomas with drainage through the brainstem. Am J Neuroradiol 15:643
- Drake CG (1981) Management of cerebral aneurysm. Stroke 12:273
- Duyckaertz C, Hauw J (1985) Pathology and pathophysiology of brain ischemia. Neuroradiology 27:460-467
- Edelman RR, Heinrich PM, O'Reilly GV et al (1990) Magnetic resonance imaging of flow dynamics in the circle of Willis. Stroke 21:56
- Edwards T, Murphy M, Geraghty J et al (1999) Intra-arterial Cerebral Thrombolysis for Acute Ischemic Stroke in a Community Hospital. AJNR 20:1682-1687
- Elster AD, Chen MYM (1990) MR imaging of Sturge-Weber syndrome: role of gadopentetate dimeglumine and gradient-echo techniques. Am J Neuroradiol 11:685
- Eskesen V, Rosenorn J, Schmidt K et al (1987) Clinical features and outcome in 48 patients with unruptured intracranial saccular aneurysms: a prospective consecutive study. Br J Neurosurg 1:47
- Fazekas F, Kleinert R, Roob G et al (1999) Histopathologic Analysis of Foci of Signal Loss on Gradient-Echo T2*-Weighted MR Images in Patients with Spontaneous Intracerebral Hemorrhage: Evidence of Microangiopathy-Related Microbleeds. AJNR 20:637-642
- Fisher CM, Ojemann RG, Roberson GH (1978) Spontaneous dissection of cervicocerebral arteries. Can J Neurol Sci 5:9
- Fisher CM (1982) Lacunar, strokes and infarcts: a review. Neurology 32:871-876
- Fisher CM (1979) Capsular infarcts, the underlying vascular lesions. Arch Neurol 36:65
- Fobben E, Grossman RJ, Atlas SW et al (1989) MR characteristics of subdural hematomas and hygromas at 1.5 T. AJNR 10:687
- Frizzel RT, Fisher WS (1995) Cure, morbidity, mortality associated with embolization of brain arteriovenous malformations: a review of 1246 patients in 32 series over a 35 year period. Neurosurgery 37:1031-1040
- Garner TB, Curling O Jr, Kelly DL, Laster DW (1991) The natural history of intracranial venous angiomas. J Neurosurgery 75:715-722
- Gelbert F, Assouline E, Hodes JE et al (1991) MRI in spontaneous dissection of vertebral and carotid arteries. Neuroradiology 33:111
- Gillard JH, Oliverio PJ, Barker PB, Oppenheimer SM, Bryan NR (1997) MR Angiography in Acute Cerebral Ischemia of the Anterior Circulation: A Preliminary Report. AJNR 18:343-350
- Glusker P, Horoupian D, Lane B (1998) Familial arteriopathic leukoencephalopathy: imaging and neuropathologic findings. AJNR 19:469-475
- Gomori J, Grossman RI, Bilaniuk LT et al (1985) High field MR imaging of superficial siderosis of the central nervous system. J Comput Assist Tomogr 9:972
- Gorelik PB, Chatterjee A, Patel D et al (1992) Cranial computed tomographic observations in multi-infarct dementia: a controlled study. Stroke 28:804
- Goulao A, Alvarez H, Garcia Monow R et al (1990) Venous anomalies and abnormalities of the posterior fossa. Neuroradiology 31:476
- Gouliamos A, Gotsis E, Vlahos L et al (1992) Magnetic resonance angiography compared to intra-arterial digital subtraction angiography in patients with subarachnoid hemorrhage. Neuroradiology 35:46
- Graf CJ, Perret GE, Torner JC (1983) Bleeding from cerebral arteriovenous malformations as part of their natural history. J Neurosurg 58:331
- Grindal AB, Cohen RJ, Saul RF et al (1978) Cerebral infarction in young adults. Stroke 9:39
- Grosgogeat Y (1985) Cerebral ischemic accidents of cardiac origin. Neuroradiology 27:579-582
- Gurst G, Steinmetz H, Fischer H et al (1993) Selective MR angiography and intracranial collateral blood flow. J Comput Assist Tomogr 17:178
- Halbach VV, Higashida RT, Hieshima GB et al (1987) Dural fistulas involving the transverse and sigmoid sinuses: results in 28 patients. Radiology 163:443
- Halbach VV, Higashida RT, Hieshima GB et al (1987) Dural fistulas involving the cavernous sinus: results of treatment in 30 patients. Radiology 163:437-442

- Hardy PA, Kucharczyk W, Henkelman RM (1990) Cause of signal loss in MR images of old hemorrhagic lesions. Radiology 174:549
- Harris KG, Tran DD, Sickels WJ et al (1994) Diagnosing intracranial vasculitis: the role of MR and geography. AJNR 15: 317
- Hayashi S, Arimoto T, Itakura T et al (1981) Association of intracranial aneurysms and arteriovenous malformation of the brain. J Neurosurg 55:971
- Hayman LA, Pagan JJ, Kirkpatrick JB et al (1989) Pathophysiology of acute intracerebral and subarachnoid hemorrhage: applications to MR imaging. ANJR 10:457
- Healton EB (1986) Fibromuscular dysplasia. In: Barnett HJM (ed) Stroke. Churchill Livingston, London
- Heier L (1992) White matter disease in the elderly: vascular etiologies. Stroke 2(3):441
- Heiserman JE, Drayer BP, Fram EK et al (1992) Carotid artery stenosis: clinical efficacy of two-dimensional time-of-flight MR angiography. Radiology 182:761
- Heiserman JE, Drayer BP, Fram EK et al (1992) MR angiography of cervical fibromuscular dysplasia. Am J Neuroradiol 13:1454
- Heiserman JE, Drayer BP, Keller PJ et al (1992) Intracranial vascular stenosis and occlusion: evaluation with three-dimensional time-of-flight MR angiography. Radiology 185:667
- Heiskanen O (1981) Risk of bleeding from unruptured aneurysms in cases with multiple intracranial aneurysms. J Neurosurg 55:524
- Hoffimann MW, Bill PLA, Bhigjee AL et al (1992) The clinicoradiological profile of cerebral venous thrombosis. S Afr Med J 82:341
- Horowitz M, Kondziolka D (1995) Multiple familial cavernous malformations evaluated over three generations with MR. Am J Neuroradiol 16:1353-1355
- Horowitz MB, Jungreis CA, Quisling RG, Pollack I (1994) Vein of Galen aneurysms: a review and current perspective. Am J Neuroradiol 15:1486-1496
- Houtteville J (1989) Les cavernomes intra-craniens. Neuroradiology 35:73
- Humphrey PD, Marchall J (1981) Transient ischemic attacks and strokes with recovery prognosis and investigation. Stroke 12(6):765
- Huston J III, Lewis BD, Wiebers DO et al (1993) Carotid artery: prospective blinded comparison of two-dimensional time-of-flight MR angiography with conventional angiography and duplex US. Radiology 186:339
- Huston J III, Torres VE, Sulvinan PP et al (1993) Value of magnetic resonance angiography for the detection on intracranial aneurysms in autosomal dominant polycystic kidney disease. J Am Soc Nephrol 3:1871
- Huston J, Rufnacht DA, Ehman RL et al (1991) Intracranial aneurysms and vascular malformations: comparison of time-of-flight and phase contrast MR angiography. Radiology 181:721
- Ida M, Kurisu Y, Yamashita M (1997) MR Angiography of Ruptured Aneurysms in Acute Subarachnoid Hemorrhage. AJNR 18:1025-1032
- Jack CR Jr, Rydberg Ch, Krecke KN (1996) Mesial temporal sclerosis: diagnosis with FLAIR versus spin echo MR imaging. Radiology 199:367-373
- Johnson BA, Heiserman JE, Drayer BP et al (1994) Intracranial MR angiography: its role in the integrated approach to brain infarction. Am J Neuroradiol 15:901
- Jorgensen HS, Nakayama H, Raaschou SO, Olsen TS (1995) Intracerebral hemorrhage versus infarction: stroke severity, risk factors and prognosis. Ann Neurol 38:45-50
- Kanal E, Shellock FG (1993) MR imaging of patients with intracranial aneurysm clips. Radiology 187:612
- Kido DK, Panzer RJ, Szumowski J et al (1991) Clinical evaluation of stenoses of the carotid bifurcation with magnetic resonance angiographic techniques. Arch Neurol 48:484
- King WA (1997) Dural arteriovenous malformations: indications for treatment and surgery. J Stroke Cerebrovasc Dis 6:292-294
- Klucznik RP, Carrier DA, Pyka R et al (1993) Placement of a ferromagnetic intracerebral aneurysm clip in a magnetic field with a fatal outcome. Radiology 187:855
- Kondziolka D, Nixon BJ, Lasjaunias P et al (1988) Cerebral arteriovenous malformations with associated arterial aneurysms: hemodynamic and therapeutic considerations. Can J Neurol Sci 15:130
- Korogi Y, Takahashi M, Mabuchi N et al (1994) Intracranial aneurysms: diagnostic accuracy of three-dimensional, Fourier transform, time-of-flight MR angiography. Radiology 193:181
- Kucharczyk J, Mintorovitch J, Asgari H (1988) Diffusion/perfusion MR imaging of acute cerebral ischemia. Magn Reson Med 19:331
- Kugler S, Anderson B, Cross D et al (1993) Abnormal cranial magnetic resonance imaging scans in sickle-cell disease. Arch Neurol 50:629
- Lasjaunias P, Garcia-Monaco R, Rodesch, Ter Brugge K, (1991) Deep venous drainage in great cerebral vein (vein of Galen) absence and malformations. Neuroradiology 33: 234-238
- Laster RE Jr, Acker JD, Halford HH III et al (1993) Assessment of MR angiography versus arteriography for evaluation of cervical carotid bifurcation disease. Am J Neuroradiol 14:681
- Leblanc R, Melanson D, Tampieri D et al (1995) Familial cerebral aneurysms: a study of 13 families. Neurosurgery 37:633
- Leblanc R, Worsley KJ, Melanson D et al (1991) Angiographic screening and elective surgery of familial cerebral aneurysms: a decision analysis. Neurosurgery 35:9
- Lee D, Pennington MA, Kenney CM III (1996) MR evaluation of developmental venous anomalies: medullary venous anatomy of venous angiomas. AJNR 17:61-70
- Lefkowitz D, LaBenz M, Nudo SR et al (1999) Hyperacute ischemic stroke missed by diffusion-weighted imaging. AJNR 20:1871-1875
- Levy C, Laissy JP, Raveau V et al (1994) Carotid and vertebral artery dissections: three-dimensional time-of-flight MR angiography and MR imaging versus conventional angiography. Radiology 190:97

- Liang L, Korogi Y, Sugahara T et al (1999) Detection of Intracranial Hemorrhage with Susceptibility-Weighted MR Sequences. AJNR 20:1527-1534
- Lin W, Haacke EM, Smith AS et al (1992) Gadolinium enhanced high-resolution MR angiography with adaptive vessel tracking: preliminary results in the intracranial circulation. J Magn Reson Imaging 2:227
- Litt AW, Eidelman EM, Pinto RS et al (1991) Diagnosis of carotid artery stenosis: comparison of 2DFT time-of-flight MR angiography with contrast angiography in 50 patients. Am J Neuroradiol 12:149
- Locksley HB (1969) Natural history of subarachnoid hemorrhage, intracranial aneurysms, and arteriovenous malformations: based on 6,368 cases in the cooperative study. In: Sahs AL, Perret GE, Locksley HB et al (eds) Intracranial aneurysms and subarachnoid hemorrhage: a cooperative study, Lippincott, Philadelphia
- Lovblad KO, Laubach HJ, Baird AE et al (1998) Clinical experience with diffusion-weighted MR in patients with acute stroke. AJNR 19:1061-1066
- Maeder P, Gudinchet F, Meuli R, de Tribolet N (1998) Development of a Cavernous Malformation of the Brain. AJNR 19:1141-1145
- Marshall VG, Bradley WG, Marshall CE et al (1988) Deep white matter infarction: correlation of MR imaging and histopathologic findings. Radiology 167:517
- Marti-Bonmati L, Menor F, Poyatos C et al (1992) Diagnosis of Sturge-Weber Syndrome: comparison of the efficacy of CT and MR imaging in 14 cases. Am J Roentgenol 158:867
- Mas JL, Bousser MG, Hasboun D (1987) Extracranial vertebral artery dissections: a review of 13 cases. Stroke 18(6):1037
- McQuinn Bam, O'Leary DH (1987) White matter lucencies on computed tomography subacute arteriosclerotic encephalopathy (Binswanger's disease), and blood pressure. Stroke 18:900
- Mohr JP, Caplan Lr, Melski JW et al (1978) The Harvard cooperative stroke registry: a prospective registry. Neurology 28:754
- Morley TP, Barr HWK (1968) Giant intracranial aneurysms: diagnosis, cause, and management. Clin Neurosurg 16:73
- Moseley ME, Cohen Y, Mintorovitch J et al (1990) Early detection of regional cerebral ischemia in cats: comparison of diffusion and T2-weighted MRI and spectroscopy. Magn Reson Med 14:3300
- Moseley ME, Kucharcyzk J, Mintorovitch J et al (1990) Diffusion-weighted MR imaging of acute stroke: correlation with T2-weighted and magnetic susceptibility enhanced MR imaging in cats. AJNR 11:423
- Moseley M, Wendland MF, Kucharcyzk J (1991) Magnetic resonance imaging of diffusion and perfusion. Top Magn Reson Imaging 3(3):50
- Nakagawa T, Hashi K (1994) The incidence and treatment of asymptomatic unruptured cerebral aneurysm. J Neurosurg 80:217
- Nicolaides P, Appelton R (1996) Stroke in children. Devel Med Child Neurol 38:172-180
- Noguchi K, Ogawa T, Inugami A et al (1995) Acute subarachnoid hemorrhage: MR imaging with FLAIR pulse sequences. Radiology 196:773-777
- Noguchi K, Ogawa T, Inugomi A et al (1995) Acute subarachnoid hemorrhage: MR imaging with fluid-attenuated inversion pulse sequences. Radiology 196:773
- Noguchi K, Ogawa T, Inugomi A (1994) MR of acute subarachnoid hemorrhage: preliminary report of fluid-attenuated inversion pulse sequences. ANJR 15:1940
- Offenbacher H, Fazekas F, Schmidt R et al (1996) MR of cerebral abnormalities concomitant with primary intracerebral hematomas. AJNR 17:573-578
- Olsen WL, Brant-Zawadzki M, Hodes J et al (1987) Giant Intracranial aneurysms: MR imaging. Radiology 163:431
- Padayachee TS, Bingham JB, Graves MJ et al (1991) Dural sinus thrombosis: diagnosis and follow-up by magnetic resonance angiography and imaging. Neuroradiology 33:165
- Patel MR, Kuntz KM, Klufas RA et al (1995) Preoperative assessment of the carotid bifurcation: can magnetic resonance angiography and duplex ultrasonography replace contrast arteriography? Stroke 26:1753
- Polak JF, Bajakian RL, O'Leary DH et al (1992) Detection of internal carotid artery stenosis: comparison of MR angiography, color Doppler sonography and arteriography. Radiology 182:35
- Pomper MG, Miller TJ, Stone JH et al (1999) CNS Vasculitis in Autoimmune Disease: MR Imaging Findings and Correlation with Angiography. AJNR 20:75-85
- Pozzati E, Giuliani G, Acciarri N et al (1990) Long-term follow-up of occlusive cervical carotid dissection. Stroke 21:528
- Provenzale JM (1995) Dissection of the internal carotid and vertebral arteries: imaging features. Am J Roentgenol 165:1099
- Putman CM, Chaloupka JC, Fulbright RK et al (1996) Exceptional multiplicity of cerebral arteriovenous malformations associated with hereditary hemorrhagic telangiectasia (Osler-Weber-Rendu syndrome). AJNR 17:1733-1742
- Rapacki TFX, Brantley MJ, Furlow TW et al (1985) Heterogeneity of cerebral cavernous hemangiomas diagnosed by MR imaging. J Comput Assist Tomogr 14:18
- Rapoport S, Sostman HD, Pope C et al (1987) Venous clots: evaluation with MR imaging. Radiology 162:527
- Requena I, Arias M, Lopez-Ibor L et al (1991) Cavernomas of the central nervous system: clinical and neuroimaging manifestations in 47 patients. J Neurology Neurosurg Psychiatry 54:590-594
- Ricci PE, Burdette JH, Elster AD, Reboussin DM (1999) A comparison of fast spin-echo, fluid-attenuated inversion-recovery, and diffusion-weighted MR imaging in the first 10 days after cerebral infarction. AJNR 20:1535-1542
- Rigamonti D, Spetzler RF (1988). The association of venous and cavernous malformations: report of four cases and discussion of the pathophysiological, diagnostic, and therapeutic implications. Acta Neurochir 92:100

- Rigamonti D, Drayer BP, Johnson PC et al (1987) The MRI appearance of cavernous malformations (angiomas). J Neurosurg 67:518
- Rigamonti D, Hadley MN, Drayer BP et al (1988) Cerebral cavernous malformations: incidence and familial occurrence. N Engl J Med 319:343
- Rippe DJ, Boyko OB, Spritze CE et al (1990) Demonstration of dural sinus occlusion by the use of MR angiography. Am J Neuroradiol 11:199
- Robinson JR, Awad IA, Little JR (1991) Natural history of the cavernous angioma. J Neurosurg 75:709
- Rokaninen A, Puranen MI, Hernesniemi JA et al (1995) Intracranial aneurysms: MR angiography screening in 400 asymptomatic individuals with increased familial risk. Radiology 195:35
- Ronkainen A, Hernesniemi JA, Tromp G (1995) Special features of familial intracranial aneurysms: report of 215 familial aneurysms. Neurosurgery 37:43
- Ross JS, Masaryk TJ, Modie MT et al (1990) Intracranial aneurysms: evaluation by MR angiography. Am J Neuroradiol 11:449
- Ruggieri PM, Poulas N, Obuchowski N et al (1994) Occult intracranial aneurysms in polycystic kidney disease: screening with MR angiography. Radiology 191:33
- Russo CP, Smoker WRK (1996) Nonatheromatous carotid artery disease. Neuroimaging Clin North Am 6(4): 811-830
- Rutka JT, Brant-Zawadzki M, Wilon CB et al (1988) Familial cavernous malformations: diagnostic potential of magnetic resonance imaging. Surg Neurol 29:467
- Sato A, Takahashi S, Soma Y et al (1991) Cerebral infarction: early detection by means of contrast-enhanced cerebral arteries at MR imaging. Radiology 178:433
- Satoh S, Kadoya S (1988) Magnetic resonance imaging of subarachnoid hemorrhage. Neuroradiology 30:361
- Schievink I, Torres VE, Piepgras DG et al (1992) Saccular intracranial aneurysms in autosomal dominant polycystic kidney disease. J Am Soc Nephrol 3:88
- Schuier FJ, Hossmann KA (1980) Experimental brain infarcts in cats. II. Ischemic brain edema. Stroke 11:593
- Schwartz RB, Jones KM, Kalina P et al (1992) Hypertensive encephalopathy: findings on CT, MR imaging and SPECT imaging in 14 cases. Am J Roentgenol 159:379
- Schwartz RB, Mulkern RV, Gudbjartsson H, Jolesz F (1998) Diffusion-Weighted MR Imaging in Hypertensive Encephalopathy: Clues to Pathogenesis. AJNR 19:859-862
- Sekhar LN, Heros RB (1981) Origin, growth, and rupture of saccular aneurysms: a review. Neurosurgery 8:258
- Shuaib A, Lee D, Pelz D et al (1992) The impact of Magnetic resonance imaging on the management of acute ischemic stroke. Neurology 42:816
- Sigal R, Krief O, Houttevilk JP et al (1990) Occult cerebrovascular malformations: follow-up with MR imaging. Radiology 176:815
- Simard JM, Garcia-Benyochea F, Ballinger WE et al (1986) Cavernous angioma: a review of 126 collected and 12 new clinical cases. Neurosurgery 18:162
- Takahashi S, Higano S, Ishii K et al (1993) Hypoxic brain damage: cortical laminar necrosis and delayed changes in white matter at sequential MR imaging. Radiology 189:449
- Tomlinson FH, Houser OW, Scheithauer BW et al (1994) Angiographically occult vascular malformations: a correlative study of features on magnetic resonance imaging and histological examination. Neurosurgery 34:792-800
- Toro VE, Geyer CA, Sherman JL et al (1988) Cerebral venous angiomas: MR findings. J Comput Assist Tomogr 12:935-940
- Truwit CL, Barkovich AJ, Koch TK et al (1992) Cerebral palsy: MR findings in 40 patients. AJNR 13:67
- Truwit CL (1992) Venous angioma of the brain: history, significance, and imaging findings. Am J Roentgenol 159:1299-1307
- Turjman F, Massoud TF, Vinuela F et al (1994) Aneurysms related to cerebral arteriovenous malformations: superselective angiographic assessment in 58 patients. AJNR 15:1601-1605
- Ueda T, Yuh WTC, Maley JE et al (1999) Outcome of acute ischemic lesions evaluated by diffusion and perfusion MR imaging. AJNR 20:983-989
- Van Bogaert P, Baleriaux D, Chrisophe C et al (1992) MRI of patients with cerebral palsy and normal CT scan. Neuroradiology 34:520
- Vanninen R, Manninen H, Soimakallio S (1995) Imaging of carotid artery stenosis: clinical efficacy and cost-effectiveness. Am J Neuroradiol 16:1875
- Vaquero J, Salsazar J, Martinez R et al (1987) Cavernomas of the central nervous system: clinical syndromes, CT scan diagnosis, and prognosis after surgical treatment in 25 cases. Acta Neurochir (Wien) 85:29
- Virapongse C, Cazenave C, Quisling R et al (1989) The empty delta sign: frequency and significance in 76 cases of dural sinus thrombosis. Radiology 162:779-785
- Wang PYK, Barker PB, Wityk RJ et al (1999) Diffusion-negative stroke: a report of two cases. AJNR 20:1876-1880
- Wichmann W, Von Ammon K, Flink U et al (1997) Aneurysm Clips Made of Titanium: Magnetic Characteristics and Artifacts in MR. AJNR 18:939-944
- Wilkins RH (1985) Natural history of intracranial vascular malformations: a review. Neurosurgery 17:421-430
- Wilms G, Bleus W, Demaerel P et al (1992) Simultaneous occurrence of developmental venous anomalies and cavernous angiomas. AJNR 15:1247-1254
- Yamada I, Matsushima Y, Suzuki S (1982) Moyamoya disease: diagnosis with three-dimensional time-of-flight MR angiography. Radiology 184:773
- Yuh W, Crain M, Loes D (1991) MR imaging of cerebral ischemia: findings in the first 24 hours. AJNR 12:621-629
- Zimmerman RS, Spetzler RF, Lee KS et al (1991) Cavernous malformations of the brain stem. J Neurosurg 75:32

Part I

Brain and skull

3- Inflammatory and infectious diseases

a - Multiple sclerosis and demyelinating diseases
b - Infectious and granulomatous diseases
c - Metabolic and toxic diseases

Multiple sclerosis and demyelinating diseases

Cases 98-100: multiple sclerosis

Clinical and pathological features

- Demyelinating disease: myelinoclastic process affecting normally constituted myelin sheath, sparing the axons (myelino-axonal dissociation).
- It is the most common cause of neurological disability in young adults between ages 20 to 40 (onset in 2/3 of cases) with a peak incidence at 29 years old. Manifestations of the disease become evident before age 20 in 15% of the cases and after 50 in about 10% of the cases. The incidence is very low in children: less than 0.5% in the first decade of life.
- The highest incidence is found in the northern hemisphere (Europe, USA and Canada) reaching 80 per 100,000 inhabitants. The prevalence is estimated to be less than 1 per 100,000 in equatorial regions, 5-15 per 100,000 in southern hemisphere and middle-east area. It is relatively less common in individuals who immigrate after age 15 show a similar risk to that of their birthplace. A first-degree relative of an MS patient presents a risk of developing the disease estimated to be about 10 times greater than the general population.
- Women are more frequently affected than men by a ratio of 3:2.
- Etiology remains uncertain, but many factors have been implicated: a viral-like infection, a postinfection autoimmune reaction, and environmental factors. Considerable evidences favor the fact that MS is an autoimmune disease. The most common precipitating factors with variable incidence on the course of the disease in term of exacerbations of signs, are: trauma, infections and pregnancy.
- *Histopathology*: destruction of myelin with general sparing of axons; intense inflammation consisting in the presence of perivenular monocytic inflammatory reaction (lymphocytes and monocytes), at the acute stages; glial cells modification with astrocytic proliferation and marked decrease of the oligodendroglia. The astrocytic proliferation of the protoplasmic type is maximal at the periphery of the "plaques" surrounding a central region rich in fibrillary astrocytes. Fibrillary gliosis characterizes the old chronic "plaques" in which some axonal damage may be present. Partial destruction of axons with subsequent wallerian degeneration and associated necrosis may be found in hyperacute forms (such as in the Marburg's type) and/or may evolve in a concentric pattern of demyelination (Balo's concentric sclerosis). The relapsing-remitting evolutional pattern of the disease explains the coexistence of "plaques" at different stages in the same specimen.
- Four different types or variants may be individualized in addition to the commonest "classic" Charcot type: the "acute" variety or Marburg type, the "concentric sclerosis" of Balo, the "diffuse cerebral sclerosis" or Schilder's type (children in the second decade), and the "neuromyelitis optica" of Devic.
- *Signs and symptoms*: the most frequent early manifestations are: sensory (40%), visual (35%) mainly a retrobulbar neuritis (in 25%), cerebellar and brainstem (30%), motor (20%), and urinary (5%). Various neurological signs are frequently reported: numbness or tingling of the extremities or bandlike sensations around the limbs or trunk, partial or total loss of vision (scotoma), weakness or incoordination of a limb, spastic or ataxic paraparesis, cerebellar ataxia, diplopia, vertigo, disorders of micturition including urgency or incontinence, Lhermitte's sign, paroxysmal fits, seizures (in 5%) and deep fatigue (frequently preceeding an aggravation of the clinical state or a relapse).
- Considering natural history, up to 1/3 of these patients will show a very benign course (benign relapsing-remitting form, with EDSS < 3 after more than 10 years of duration); over 90% of the patients will present a relapsing-remitting form with fluctuating periods of exacerbations and remissions of the symptoms and with a secondary progressive course in about 60% of the cases, and the remaining 10% will show a primary progressive course without relapses and remissions.

- *Diagnosis*: even if the advent of MRI has dramatically improved the diagnosis of MS due to its high sensitivity over any other modality, it remains fundamentally clinical and based on clinical, laboratory (CSF, evoqued potentials) and imaging features. The Schumaker's clinically based criteria require a dissociation in place (more than one CNS plaque) and in time (more than one episode or more than half a year progression) of the signs and symptoms. Poser's criteria classify the patients into: possible, probable and definite MS.
- Immunochemical techniques show evidence of intrathecal synthesis of oligoclonal IgG bands in the CSF of about 95% of patients with definite MS (similar to the frequency of T2w high SI lesions found on MRI but non specific to MS). Oligoclonal bands in the CSF may be found also in several other disorders: infectious (syphilis, viral encephalitis, subacute sclerosing panencephalitis, borreliosis, fungal meningitis), inflammatory (lupus erythematosus, Behcet's syndrome, polyarteritis nodosa, Sjögren's syndrome), or granulomatous (sarcoidosis).
- Disease staging is evaluated according to Kurtzke expanded disability status scale (EDSS) grading the functional severity (10 points scale).

MR characteristic findings

Morphology

- Morphology of the "plaques" (foci of demyelination) varies from acute to chronic stages on MR as on pathology. Acute or subacute plaques containing prominently edema and inflammatory process appear as ill-defined lesions associated with eventual mild mass effect. Chronic plaques are usually irregular in shape and sharply marginated. Some plaques characterized by incomplete myelin loss that could be due to partial remyelination, named "shadow plaques", show ill-defined boundaries and are found in the periphery of a "classic" lesion or even independently.
- Morphologically, 2 major patterns may be identified: an irregularly scattered, patchy demyelination foci (most frequent presentation and distribution of plaques), and diffuse, almost continuous large patches of demyelination, most commonly bilateral in the parieto-occipital and periatrial regions or frontal, and more or less symmetric ("butterfly appearance"), that often spares the arcuate U-fibers but involve the corpus callosum and the optic path (in Schilder's disease, thought to be a juvenile variant of MS). A combined patchy and diffuse demyelinating pattern (third one) may also be observed.
- Plaques are most frequently ovoid or rounded in shape, most of them smaller than 10 mm in size. Characteristic lesions may be found in the paracallosal and paraventricular regions involving the corpus callosum and the corona radiata, ovoid or finger-like in shape, oriented perpendicular to the lateral ventricular walls (Dawson's fingers), due to their perivenular distribution and their parallelism to white matter bundles orientation. Lesions involving the corpus callosum, particularly its inferior surface, are highly suggestive of MS, found in more than 90% of MS patients and best shown on T2w sequences (SE or FLAIR) performed in the sagittal plane. Such lesions are rarely associated with arteriosclerotic leucoencephalopathy or vasculitis (small infarcts) which involve preferentially long perforating arteries. Some of the lesions may be large, associated with consistent mass effect and peripheral edema, mimicking a tumor (pseudotumoral plaque; case 100).
- In Devic's disease, there is a predilection for the plaques to occur in the optic nerves (simultaneously or successively) and the spinal cord. The medullary plaques are usually of the necrotizing type (microcystic changes and cavitation on pathology), as in the Marburg's type, involving several segments mimicking myelomalacia. Inflammatory reaction is usually mild.
- In the acute form of MS or Marburg's type there are large areas of acute demyelination associated with an intense inflammatory reaction (enhancement).
- In the Balo's type or concentric sclerosis, a characteristic lamellated pattern consisting of large zones of demyelination alternating with normal myelin (which may correspond to remyelinating areas) is observed on MR as on pathology.
- Topography of the lesions as observed in the classic type of MS (patchy lesions disseminated throughout the CNS white matter and shown as high SI on T2w or FLAIR), is the following: the periventricular areas mainly adjacent to the frontal and temporal-occipital horns, more or less confluent; the optic pathways particularly the optic nerves, chiasm (case 101) and tracts; middle (album cerebelli) and superior cerebellar

peduncles and medial longitudinal fasciculi and periventricular (4th ventricle) and periaqueductal regions of the brainstem; spinal cord. More than 2/3 of the "plaques" are confined to the deep white matter and about 15% involve the gray matter.

- The size of the plaques and their number do not seem to correlate with the severity of the disease.
- With prolonged illness, shrinking of chronic plaques and degeneration of white matter is associated with brain atrophy. Increased iron deposit in the basal ganglia may also be found in long-standing disease.
- *Differential diagnosis*: several diseases may mimick MS on MRI such as: subcortical arteriosclerotic encephalopathy, AIDS; mitochondrial encephalopathies including MELAS (mitochondrial encephalomyopathy with lactic acidosis and stroke-like episodes) and MERRF (myoclonic epilepsy with ragged-red fibers); vasculitis (lupus erythematosus, polyarteritis nodosa, Behcet's disease, giant cell arteritis, Sjögren's syndrome); CADASIL (cerebral autosomal arteriopathy with subcortical infarcts and leucoencephalopathy, linked to 19q12); migraine; spirochaetal encephalitides such as Lyme disease (neuroborreliosis) and neurosyphylis; viral infections such as ADEM (acute disseminated encephalomyelitis), subacute sclerosing panencephalitis, HTLV-1 myelopathy; granulomatous diseases such as neurosarcoidosis and tuberculosis; and B12 deficiency syndrome (combined degeneration of the cord).

Signal intensity

- GE-T1w or IR: hypointense to white matter; T1w sequences help to differentiate acute from chronic plaques: acute lesions tend to be isointense or slightly hypointense and poorly demarcated (due to inflammation and edema) as compared to chronic plaques which appear more hypointense and sharply delineated (demyelination and gliosis).
- SE-PDw: hyperintense to white matter (acute and chronic plaques).
- SE-T2w or FLAIR: hyperintense to white matter (no information about the stage of the lesion: acute vs chronic plaques). STIR-T2w sequence is relatively much less sensitive for the detection of periventricular plaques.
- Periventricular plaques and lesions that abut gray matter are best detected on PDw or even better on FLAIR sequences (hyperintense to intraventricular CSF).
- Quantitative measurement of relaxation times may be of interest, showing differences between "normal appearing" white matter in MS patients as compared to normal referential values.
- Magnetization transfer technique: preliminary results show that MS plaques present a reduction in MT.
- *Diffusion imaging*: lack of anisotropy of diffusion in demyelinated plaques as compared to normally myelinated and mature white matter. Diffusion weighted sequence may be able to differentiate acute from chronic lesions.
- *Proton spectroscopy*: is promising and may contribute to the staging of the plaques and the evaluation of normal appearing brain white matter of MS patients. A decrease in NAA levels is shown in chronic lesions as in white matter. Detection of cholesterol breakdown derivates and free lipids have been reported.

Contrast enhancement

- Enhancement of acute plaques (presence of inflammatory reaction): patchy nodular or ring enhancement. Gd enhanced MRI is much more sensitive in the evaluation of active lesions in comparison to double-dose contrast enhanced CT.
- Complete or partial ring enhancement of the large "butterfly-like" demyelinating areas in Schilder's disease. In the acute form of Marburg's type, a ring enhancement pattern is also shown within the demyelinated area or at its margin, corresponding to the slight hyperintense signal as observed on unenhanced T1w images, due presumably to the presence of free radicals.
- Occasionally, some meningeal enhancement has been reported in MS.
- Steroid treatment may decrease the enhancement of acute plaques and may even reduce its size and/or modify its shape (effect on associated edema).

Case 98: Multiple sclerosis (MS)

Fig. a: axial cut; SE - PDw

Fig. b: axial cut; SE - T2w

Young patient presenting a history of neurologic symptoms with relapse and remissions. MR exam shows on SE-T2w multiple hyperintense nodular abnormalities consistent with white matter demyelination. The lesions appear more hyperintense than the cortex on PDw sequence (**a**), most of them remaining hyperintense and some becoming even more hyperintense on T2w (**b**, ↑). The lesions occur mainly in a periventricular distribution and more caracteristically close to lateral ventricles angles with involvement of the corpus callosum (**a**). Some small plaques comprise the subcortical white matter involving the "U" fibers. There was no enhancement in this case that could indicate acute inflammation. Note that the scattered bright foci depicted on T2w do not essentially correspond to demyelination but might reflect acute transient inflammation.

Case 99: Pseudo-tumoral "plaque" (MS)

Fig. c: axial cut; SE - PDw

Fig. d: axial cut; SE - T2w

Unusual isolated large plaque appearing as a mass lesion in the left parietal region involving the posterior periventricular white matter. The lesion shows mass effect on the left lateral ventricle (**c**,*) and an area of central necrosis (↑), surrounded by a hyperintense rim corresponding to edema and/or demyelination (**c**, **d**, ↑↑). The mass was markedly hypointense on T1w. The patient presented several months before an optic neuritis.

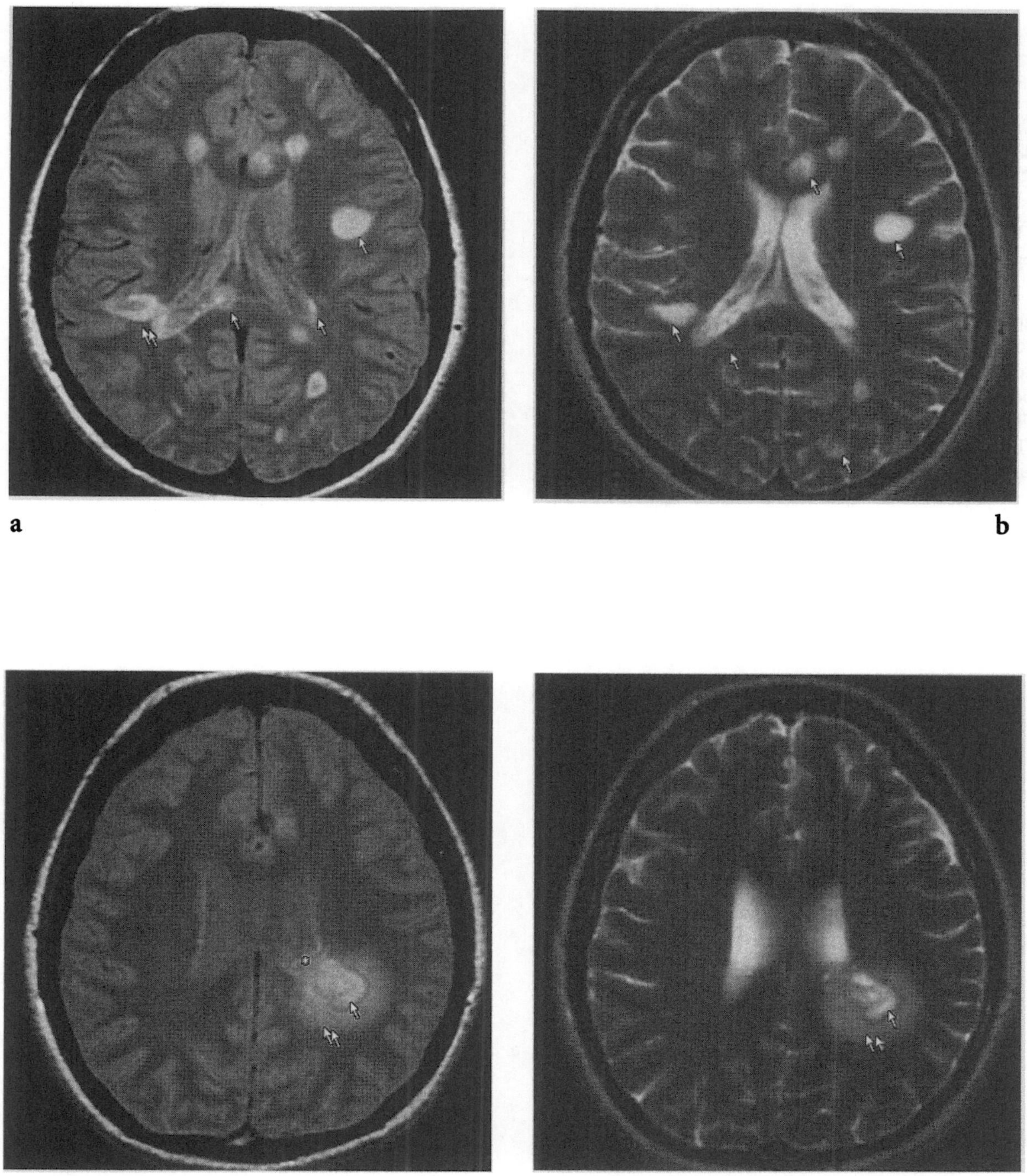

a b

c d

Case 100: Chiasmal neuritis in MS (FLAIR)

Fig. a: axial cut;	FLAIR - T2w
Fig. b: axial cut;	FLAIR - T2w
Fig. c: axial cut;	FLAIR - T2w
Fig. d: axial cut;	FLAIR - T2w

Young patient presenting a relapsing remitting multiple sclerosis with chiasmal syndrome. The MR exam shows on conventional SE-T2w scattered hyperintense foci in the periventricular white matter, much better identified on the FLAIR (**b**, **d**, ↑) some of which presenting the caracteristic "Dawson's" fingers (**b**, ↑). FLAIR sequence shows more easily the periventricular confluent bright white matter lesions (**c**, ↑↑↑) as well as the subcortical (**d**, ↑↑). SE-T2w and PDw sequences missed the extensive plaque involving the chiasm (**a**, ↑↑), corresponding to a chiasmal neuritis explaining the clinical syndrome. In fact, FLAIR sequence is obviously highly sensitive to the depiction of plaques and demyelinating lesions that appear as bright signal areas, best depicted adjacent to the hypointense intraventricular and cisternal or subarachnoid CSF.

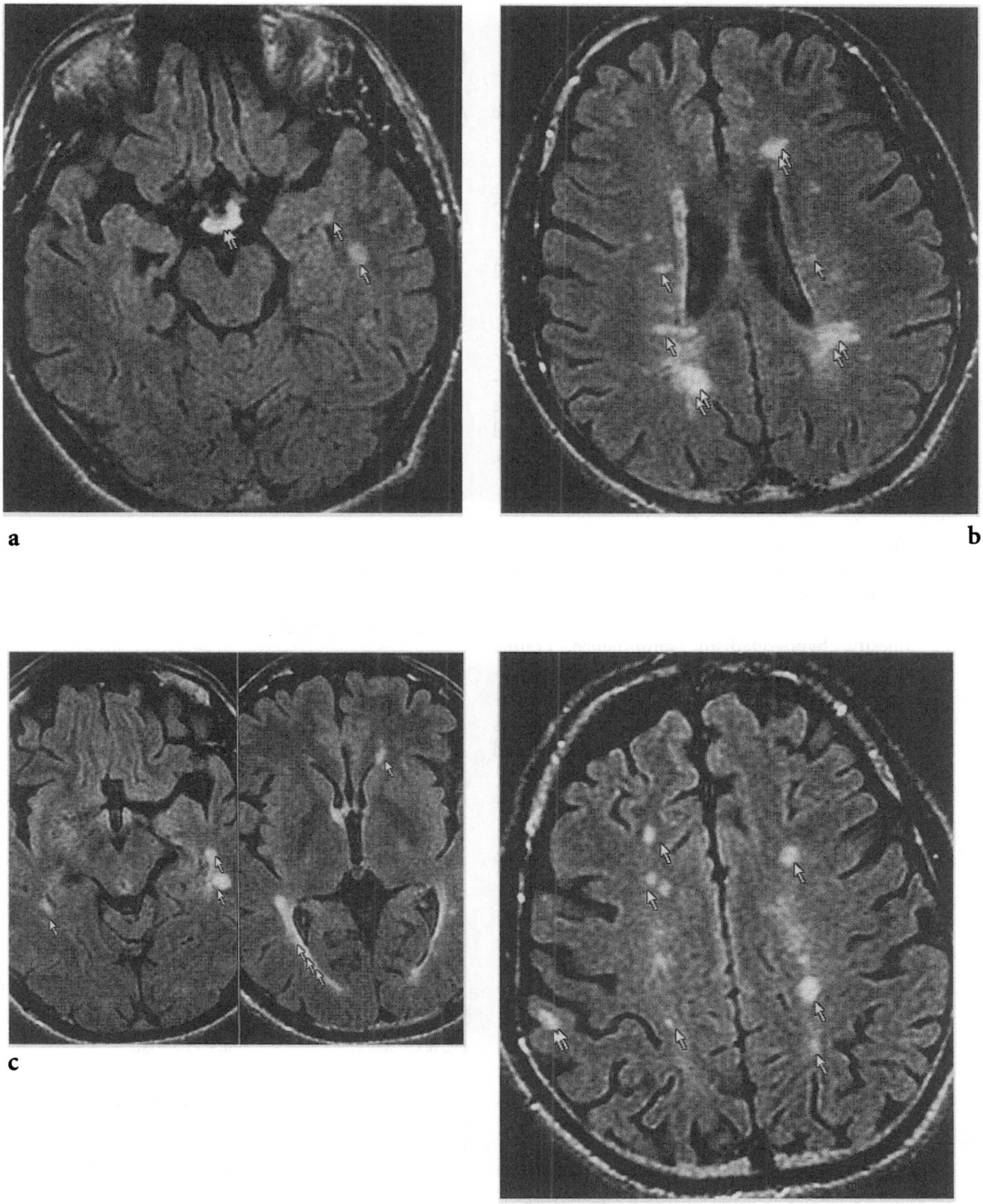

a

b

c

d

Infections and granulomatous diseases

Case 101: Neurosarcoidosis

Clinical and pathological features

- Granulomatous systemic disease involving the nervous system in 15% of cases.
- Not proved to be an infectious disease.
- Close resemblance clinically and pathologically to tuberculosis.
- Affects both sexes, slightly more frequent in females; ten times more in blacks.
- *Pathology*: typical lesion is a non caseating granuloma (sarcoid tubercle), consisting of focal clusters of epithelioid cells surrounded by lymphocytes; giant cells are found. These granulomas are most frequently found in the hilar and peripheral lymph nodes, lung and liver, skin, bones and eyes, parotid glands, and central and peripheral nervous system (about 5-6%).
- *Signs of CNS sarcoidosis*: correspond mainly to leptomeningeal infiltration of the base of the brain: visual loss and atrophy, hypothalamic-pituitary dysfunction with diabetes insipidus, obesity, hypogonadism, somnolence, cranial nerve palsies with the facial nerves more frequently involved within the parotids (uveoparotitis) or peripheral subacute or chronic neuropathy, focal corticospinal and/or cerebellar signs, and hydrocephalus. The course is usually slowly progressive with or without remissions. The diagnosis is made on clinical grounds and on biopsy evidence of non caseating granulomas (skin, lung, uvea).

MR characteristic findings

Morphology

- Leptomeningeal infiltration of the chiasmal cistern and other basal cisterns (even posterior fossa), with involvement of the chiasm and cisternal optic nerves as well as the hypothalamus, the infundibulum and pituitary stalk, and further upward extension to the basal parenchyma via the Virchow-Robin spaces.
- Rarely, involvement of the basal ganglia and upper brainstem may be found will mild associated basal leptomeningeal involvement.
- Rarely, large neoplastic mass in brain parenchyma or spinal adhesive arachnoiditis.

Signal intensity

- SE-T1w: iso-intense to brain.
- SE-T2w: hyperintense to brain.

Contrast enhancement

- Contrast enhancement is usually intense disclosing the leptomeningeal and cranial nerves infiltration and the extension into the adjacent brain parenchyma via Virchow-Robin spaces. Thickening of the infundibular-pituitary axis is suggestive (found in histiocytosis and other rare metastatic diseases). Disseminated intraparenchymal granulomatous lesions (non specific) may be found.

Case 101: Neurosarcoidosis

Fig. a: sagittal cut (512);	GE - T1w + Gd
Fig. b: axial cut;	SE - PDw
Fig. c: coronal cut;	SE - T2w
Fig. d: coronal cut;	SE - T1w + Gd

Middle aged woman presenting a chiasmal and a hypothalamic syndromes. The MR exam of the opto-chiasmatic region showed an infiltrating hyperintense process on T2w (**b**, **c**) involving the third ventricle and mainly its floor, the chiasm and the tuberal region (**c**, 1, 2), extending to the hypothalamic region bilaterally (**c**, 4) and more posteriorly and laterally to involve the cisternal optic tracts (**b**, 1; **c**, 5). The mass stands anteriorly to involve the lamina terminalis and the subcallosal and septal areas. The anterior commissure (**c**, 3), the fornix (**b**, 2) and the mamillo-thalamic tracts (**b**, 3) seem respected and well delineated on T2w. The opto-chiasmatic cistern shows on PDw a hyperintense signal including the latero-mesencephalic cistern and the interpeduncular region (**b**,*). After contrast, the diencephalic mass shows heterogeneous enhancement involving mainly the inferior portion of the third ventricle and the tuberal region (**a**, ↑↑↑; **d**, down, ↑). Adjacent meningeal inflammation with contrast enhancement is also noticed due to associated arachnoiditis.

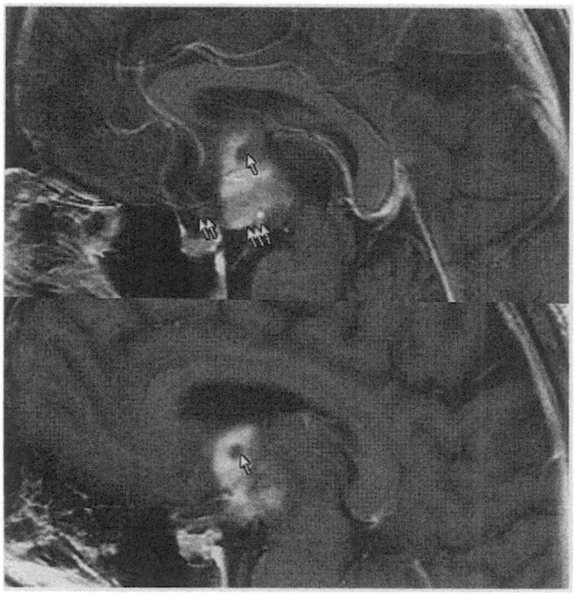

a

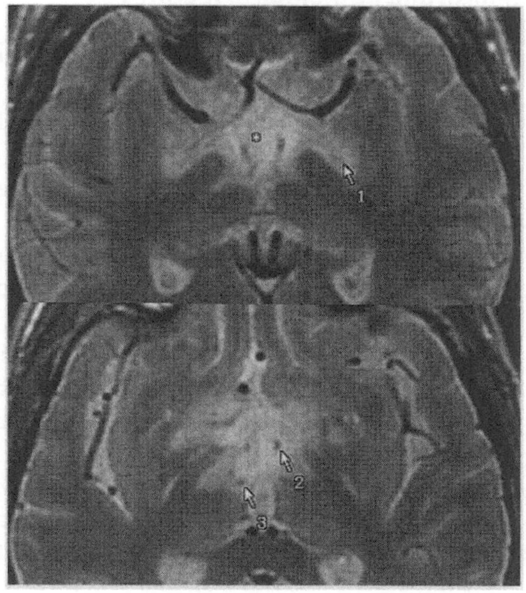

b

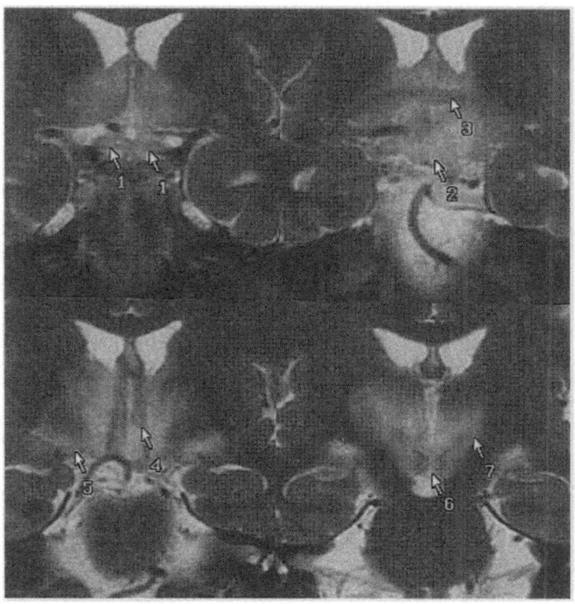

c

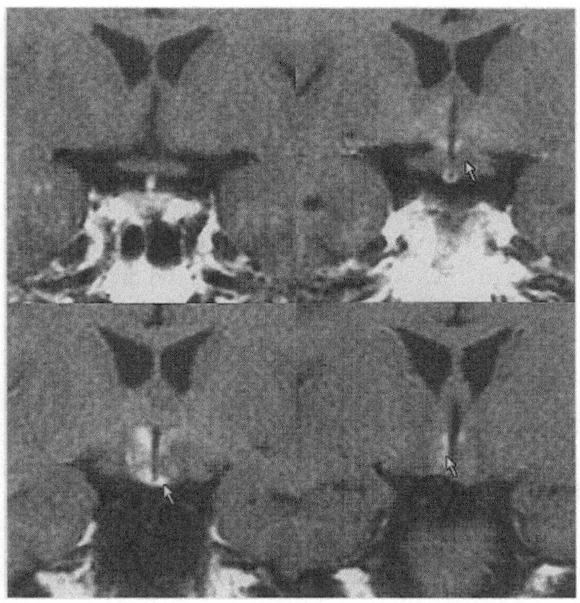

d

Case 102: tuberculous infections

Clinical and pathological features

- Infectious disease caused by mycobacterium tuberculosis.
- The infection shows an increase of its incidence in association with AIDS.
- Central nervous system involvement is observed in about 10% of AIDS patients.
- Occurs in persons of all ages, being more frequent in adults, but is due to the primary infection in most children cases.
- Several clinical pathological forms are individualized: leptomeningitis, tuberculomas of the brain parenchyma and abscesses, epidural infection (due to extension from spine: Pott's disease).
- *Histology*: granuloma with central caseous necrosis surrounded by epithelioid cells, lymphocytes, mononuclear inflammatory cells, some multinucleated giant cells and a capsule of connective tissue. These granulomas are secondary either to spread from the leptomeningeal spaces along the pial vessels toward the underlying brain, or result from hematogenous dissemination. The disease process is therefore a meningoencephalitis. Choroid plexuses and ependyma as well as cisternal cranial nerves are also involved.
- *Signs*: insidious installation of the leptomeningitis over few weeks, with headaches, ocular and cranial nerve palsies, confusional states, fever, irritability in children, signs of raised intracranial pressure, eventual focal neurologic signs.
- CSF study shows a pleocytosis (50-500 white cells /mm^3), a low glucose level of 40/100 ml, an elevated protein level (100-200 mg/100ml) and an increase in adenosine deaminase levels.

MR characteristic findings

Morphology

- *Leptomeningitis*: the exudates is found in the basal cisterns, obliterating the skull base subarachnoid cisterns (chiasmal, interpeduncular and prepontine and infratemporal) with further extension to the floor of the 3rd ventricle, the subthalamic region. Leptomeningeal spaces of the convexity are less prominently involved. It is the most frequent presentation pattern. It may be associated with brain infarcts, often bilateral due to endarteritis (1/3 of cases) involving the perforating arteries branching from the sylvian arteries.
- *Tuberculomas*: multiple in 1/3 of the cases, tuberculous granulomas are found most frequently in the cerebral hemispheres, involving mainly the cortical-subcortical junction and the periventricular white matter. Their aspect varies depending on their size and on the stage of evolution. Edema may be present in the acute stage, less prominent or absent in small lesions and usually relatively less extensive than in pyogenic infections. Calcifications are found in 20% (CT).
- *Complications*: communicating hydrocephalus due to cisternal blockage by the leptomeningeal process and very rarely, large tuberculous abscesses.

Signal intensity

- GE-T1w: isointense to gray matter with eventual mild hyperintense rim.
- SE-T2w: mildly hyperintense or isointense to gray matter; actually variable may become hypointense to brain on late stages due to presence of connective tissue deposits and paramagnetic free radicals. Possible existence of central necrosis may give rise to a hyperintense SI focus.
- GE-T2*: may be useful to disclose calcified tuberculomas (20-25%), hypointense.

Contrast enhancement

- Leptomeningeal basal thickening and infiltration and tuberculous granulomas show intense contrast enhancement.
- Small disseminated tuberculomas show nodular or ring enhancement patterns. May not be associated with leptomeningitis (case 103).

Case 103: cerebritis and brain abscess

Clinical and pathological features

- Abscess is a focal collection of pus secondary to a cerebritis which corresponds to a focal septic encephalitis without tissue destruction.
- Pyogens reach the brain tissue via three routes: hematogeneous spread (about 30%) secondary to a septic embolism or a visceral suppuration; retrograde thrombophlebitis or extension from an adjacent infected structure (mastoiditis, otitis media, frontal or sphenoidal sinusitis) in about 40% of the cases; penetrating wound, compound fracture, or after neurosurgical procedure (10%).
- Brain abscesses are exceptionally secondary to a bacterial meningitis.
- In about 20 % of the cases the origin cannot be found.
- *Etiology*: the most frequently isolated organisms (4/5) are most commonly the streptococci and staphylococci, more rarely the pneumococci and meningococci. Increased frequency of gram-negative organisms is observed. Bacterial species varies with the location of the abscess: anaerobic streptococci are secondary to dissemination from the lungs; enterobacteriaceae are associated with otitis; staphylococcal are secondary due to bacteriemia or penetrating skull fracture.
- *Signs*: headache is the most common complaint (may be localized to the suboccipital and/or posterior auricular regions in cerebellar abscesses), drowsiness, impairement of mental function or confusional states, seizure (may be focal), focal sensorimotor deficits (depending on the localization of the abscess), language disorders (anomia). Fever, usually present in the suppurative cerebritis phase, may resolve as the abscess matures. Unpredictable evolution and worsening of the clinical signs and symptoms particularly in children.

MR characteristic findings

Morphology

- At the acute cerebritis phase (less than 1 week): ill defined area of suppuration including edematous changes and petechial hemorrhages associated to a central zone of necrosis, localized most commonly in subcortical areas of the fronto-parietal lobes (hematogenous spread) or in the temporal lobes (direct spread from otorrhinologic infections), involving the distal territories of the middle cerebral arteries. The cerebellum mainly in its antero-lateral part is involved in less than 15% of cases, most commonly associated to infectious disease of the ear. The mass effect may be limited to a sulcal effacement.
- At the acute abscess phase: pus formation is filling the central cavity, limited by a peripheral zone containing the inflammatory granulation tissue which actually is the ring enhancing zone, and surrounded by a large outer zone of vasogenic edema. At this stage, daughter abscesses or adjacent areas of cerebritis may develop. Involvement of adjacent structures may lead to the formation of an ependymitis or a leptomeningitis.
- At the chronic abscess phase: the abscess is constituted by a liquefied central portion, encapsulated with a collagenous wall and surrounded by a gliosis replacing the peripheral edema and explaining the decrease of the mass effect.

Signal intensity

- GE-T1w or SE-T1w: iso or hyperintense (presence of free radicals) rim to brain surrounding a hypointense central zone of necrosis and liquefaction (slightly more hyperintense to CSF, due to variable increase of protein content).
- SE-T2w: usually non specific hypointense rim surrounding hyperintense central zone. The peripheral vasogenic edema is hyperintense on PDw and T2w.

Contrast enhancement

- In cerebritis enhancement is variable: may be absent but usually is mild, diffuse and heterogeneous and more or less diffuse in early stages, involving the cortex, and later may show a ring-like enhancing pattern.
- In brain abscess: a ring-enhancement of the capsule is evident, usually smooth and thin (less than 5mm) particularly on its medial border, including the smaller "daughter" lesions when present medially. Enhancement as edema decreases with time following treatment as shown on follow-up exams (may persist 2-6 months).
- *Differential diagnosis*: high grade gliomas, metastases, radiation necrosis, malignant lymphomas, hematoma in chonic phase, tumor-like inflammatory processes, which may present as ring-enhancing lesions, but showing focal mural tissular nodule and irregular thickening of the enhancing tissue.

Case 102: Tuberculous encephalitis

Fig. a: axial cut (512); SE - T2w

Fig. b: axial cut (512); SE - T2w

Fig. c: sagittal cut (512); GE - T1w + Gd

Fig. d: axial cut (512); SE - T1w + Gd

Multiple discrete nodular intraparenchymal lesions, showing a hyperintense signal on T2w (**a**, **b**). These disseminated lesions corresponding to tuberculous granulomas presented as small homogeneous areas of hyperintense signal involving both thalami, the genu of the corpus callosum (**b**, ↑) and the subcortical white matter (**b**, right, ↑↑). Some of these tuberculomas show "target" signs on T2w high resolution images surrounded by peripheral edema generally minimal (**a**, ↑↑; **b**, left, ↑↑). This appearance may correspond to the central zone of caseous material surrounded by the reactive granulomatous inflammation, hyperintense on T2w both surrounded by edema. Note that the intraparenchymal granulomas are found at the cortico-medullary junction as well as in the periventricular regions and in this case mainly around the fourth ventricle with mass effect on its floor (**a**). This distribution may be explained by the hematogeneous dissemination of the infection. After contrast, MR shows multiple nodular enhanced foci (**c**, **d**, arrows). Note the more widespread subcortical dissemination of the granulomas giving a "miliary" appearance. There were no associated meningitis with contrast enhancement involving the leptomeninges around the base of the brain.

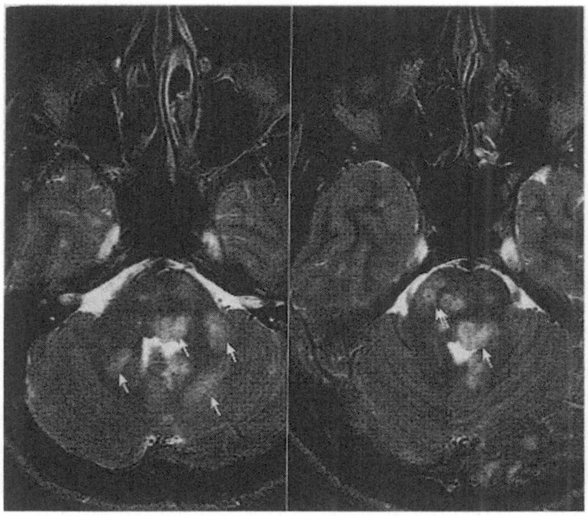

a

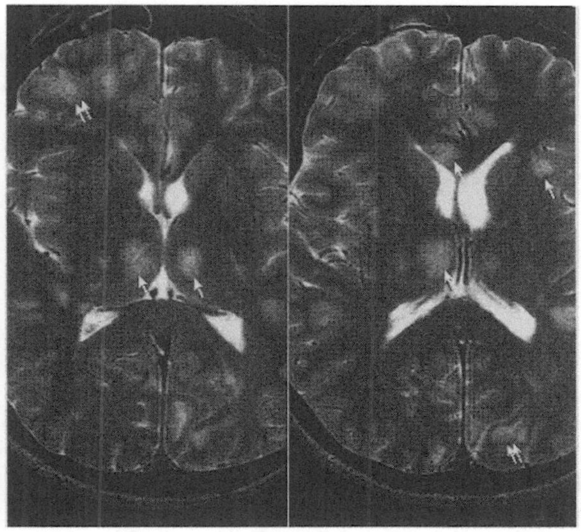

b

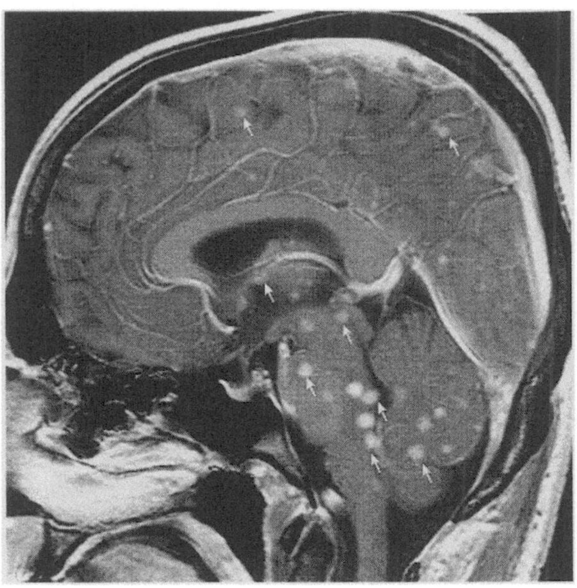

c

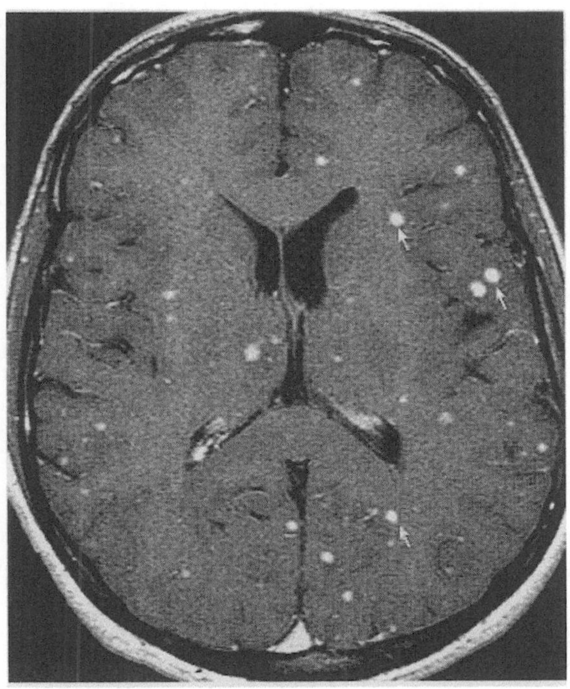

d

Case 103: Basal ganglia abscess

Fig. a: axial cut;	SE - PDw	**Fig. c:** axial cut (512);	GE (3D) - T1w + Gd
Fig. b: axial cut;	SE - T2w	**Fig. d:** coronal cut (512);	GE - T2w

Pyogenic abscess developed in the right basal ganglia region associated to a large amount of vasogenic edema surrounding the abscess, well seen on the PDw (**a,** ↑) as well as on the T2w sequences (**b,** ↑), showing the caracteristic digitations along white matter axis. The edema is responsible of much of the mass effect shown on the third ventricle partly laminated (**b**) with disection of the ipsilateral cerebral peduncle and the posterior arm of the internal capsule (**a, b,** ↑↑). After Gadolinium injection, the T1w sequence (c) showed ring enhancement of the abscess capsule, caracteristically thinner along the medial abscess margin (**c,** ↑↑) surrounding the central necrosis (**c,** ↑). The vasogenic edema much more extensive than the mass itself, is recognized as a peripheral area of hyposignal on T1w (**c,** ↑↑↑) responsible of much of the mass effect on the lateral ventricles and the septum displaced to the left, as well as the swelling of the cortical mantle reducing the subarachnoid spaces. Note improvement of the mass effect and partial reduction of the abscess size (**d,** ↑↑) few weeks after medical treatment. Note in this case the absence of daughter abscesses that generally occur along the weaker thinner wall at the medial margin of the parent mass. Moreover, on the T1w sequence, the abscess capsule was iso-intense to brain parenchyma and did not demonstrate the characteristic hyperintense rim attributed to the presence of paramagnetic free radicals (within the macrophages).

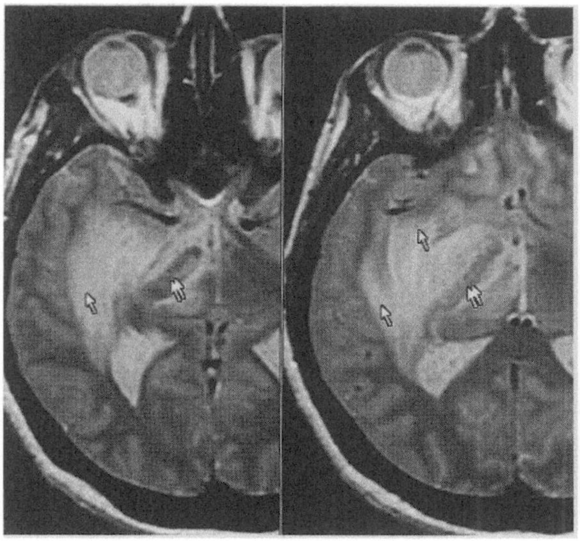

a

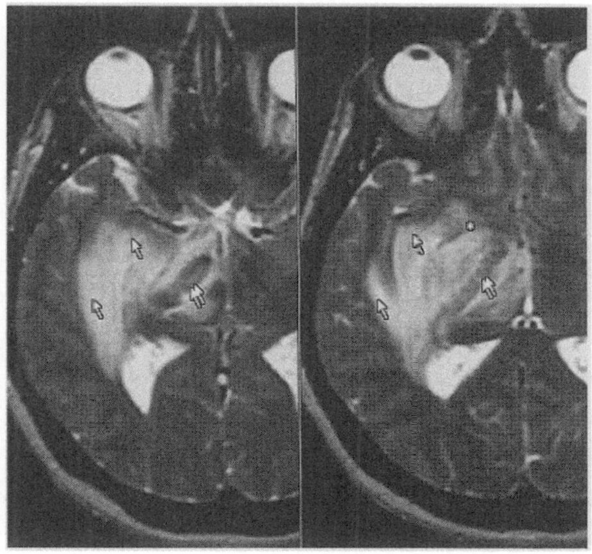

b

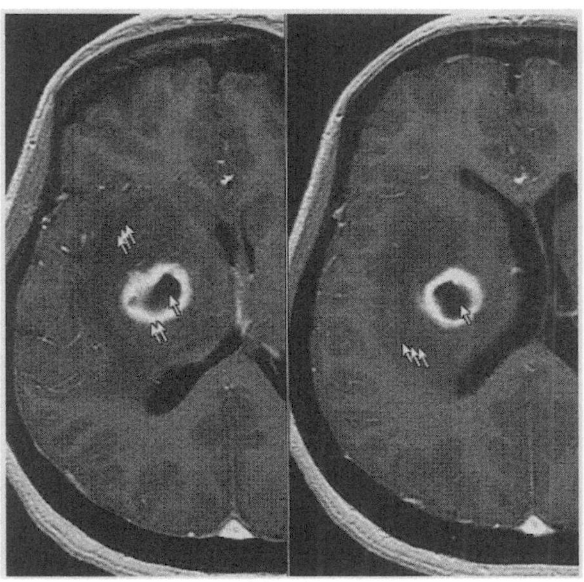

c

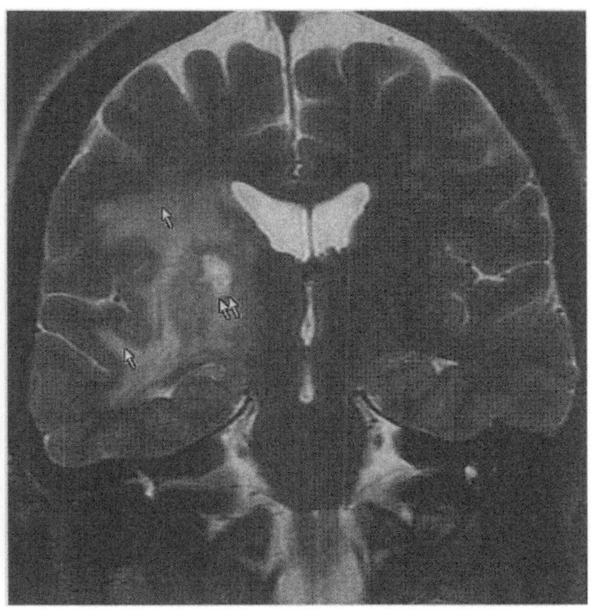

d

Metabolic and toxic diseases

Case 104: pontine and extrapontine demyelination

Clinical and pathological features

- Demyelinating disorder originally described in alcoholics, but found also in association with other diseases accompanied by electrolyte disturbances.
- The exact pathophysiology is still debated and its cause remains unknown, although rapid correction of hyponatremia (osmotic demyelination) is commonly incriminated in the genesis of this disorder.
- This disorder may be found in association with: uremia, diabetic ketoacidosis, inappropriate secretion of antidiuretic hormone, malnutrition, neoplastic diseases, chronic kidney or liver or pulmonary diseases.
- *Histology*: pontine demyelination affects the sheaths of the long descending tracts and pontocerebellar fiber bundles with relative sparing of the axons. Pontine nuclei are involved with sparing of the nerve cells. There is a relative sparing of the descending corticospinal fibers. Most of the axons are usually preserved. Oligodendrocytes are absent or markedly reduced in number and replaced by a reactive fibrillary astrocytosis and macrophagic response. Signs of inflammation are absent. Differentiation from an infarct or multiple sclerosis is easy.
- *Signs* (usually in adults): progressive upper motor neuron tetraparesis or quadriplegia, pseudobulbar syndrome, confusion, lower cranial nerve paresis (involvement of corticospinal tracts), cachexia, disorders of consciousness or locked in syndrome (involvement of central reticular formation), or pseudocoma preceeding death. EEG is rarely abnormal and BAEP shows anomalies confirming the pontine localization.

MR characteristic findings

Morphology

- Central pontine myelinolysis presents at early stages as a single well demarcated triangular or even as trident-shaped demyelinating lesion, becoming wider and almost rounded or oval later, located in the center of the pontes basis (upper or middle pons) and sparing the peripheral pontine region. The lesion may be of few millimeters or occupy the whole pontes basis. It may reach posteriorly the medial lemnisci but only rarely encroach the midbrain and never extends to the medulla. It may be exceptionally unilateral or multiple. Necrosis may even be observed.
- Central pontine lesion must be differentiated from a pontine infarct mainly in the paramedian vascular territories (unilateral and sharply demarcated by the midsagittal plane) or even a multiple sclerosis "plaque".
- Similar extrapontine lesions may be variably found, in about half the cases, in the thalami and lateral geniculate bodies, the striatum, the internal or external or extreme capsules, the anterior commissure, the fornix, the midbrain, the brachium pontis, the cerebral cortex and subcortical cerebral and cerebellar white matter, and in the spinal cord. Extrapontine myelinolysis may be observed alone (case 105) rendering the diagnosis more difficult, the lesional aspect and distribution mimicking encephalitides, acute disseminated encephalomyelitis, or other metabolic encephalopathies.
- MRI may be normal at the onset of the clinical signs and symptoms.
- After clinical recovery, the abnormalities remain visible for long time, although a correlated regression of the demyelinating lesion may be observed with improvement of the clinical status.

Signal intensity

- GE-T1w or IR: hypointense to normal white matter.
- SE-PDw: hyperintense to white matter.
- SE-T2W or FLAIR: hyperintense to white matter.
- SI abnormalities are related to increase in water content within the lesion.

Contrast enhancement

- No enhancement in the central basis pontis or little peripheral ring enhancement in the acute demyelinating phase.

Case 104: Extrapontine myelinolysis

Fig. a: axial cut;	SE - PDw	**Fig. c:** axial cut;	SE - T2w
Fig. b: axial cut;	FLAIR - T2w	**Fig. d:** axial cut;	FLAIR - T2w

Young patient with electrolyte abnormalities presenting symptoms of central pontine myelinolysis with subcomatous state. MR findings failed to demonstrate, on several follow-up exams during 6 months at least, any central pontine demyelination. MR shows hyperintense signal on PDw and T2w (**a**; **c**) involving both striatum with apparent swelling of the putamina (**a**, ↑↑) as well as the left caudate nucleus (**a**), associated to areas of hyperintense signal involving the posterior and medial aspects of both thalami (**a**, ↑). These high signal abnormalities are even more conspicuous on the corresponding images using FLAIR sequence (**b**). MR exam shows also and mainly on the FLAIR sequence (d) as compared to T2w (**c**), cortical and subcortical abnormalities involving both hemispheres, associated with mass effect on the corresponding sulci (**d**, ↑↑). No contrast enhancement was noted after Gadolinium iv injection. The patient improved progressively with betterment and reduction of the hyperintense area.

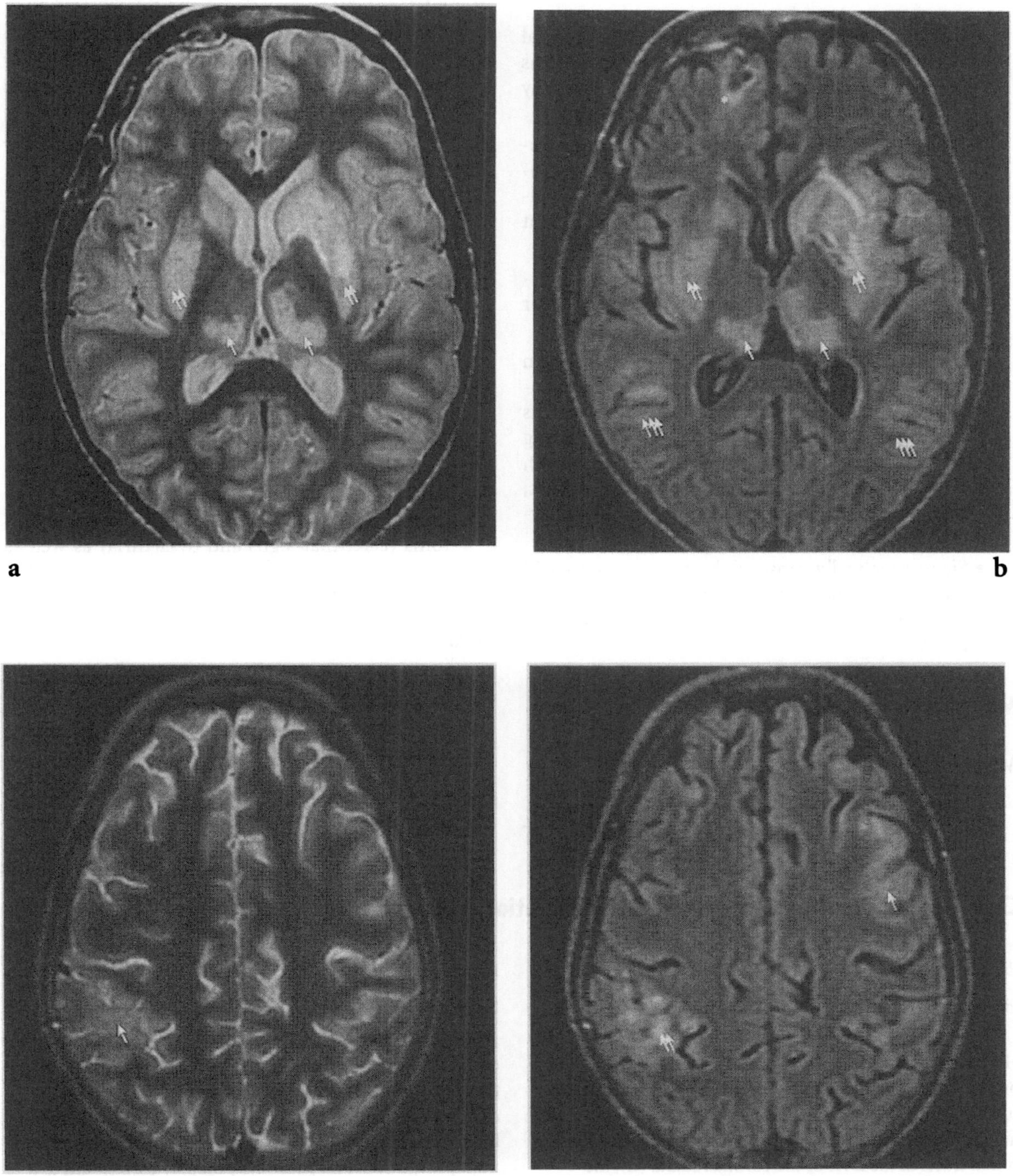

a b

c d

Case 105: toxic encephalopathy

Clinical and pathological features

- Chronic intoxications may lead to neurological syndromes related to interaction of selective areas of the brain with a chemical compound. This may result in a toxic demyelinating encephalopathy.
- Characteristic MR patterns with selective involvement of cerebral structures or anatomo-functional systems may be observed, due to selective vulnerability and affinity of the brain for different toxins.
- Variable groups of toxic materials may be incriminated being of environmental exogenous or biochemical endogenous origin.
- Exogenous toxics that may impair nervous system functions are numerous and cannot be enumerated here, including: industrial solvants, pollutants and volatile substances, insecticides, addicting drugs (opioids: heroin-cocaine sniffing, case 106), alcohol, heavy metals (lead, mercury, arsenic, manganese), chemotherapeutic and other pharmacological drugs.
- *Signs*: practically most of these toxic agents will give rise to: headaches, insomnia, confusional states, seizures, disturbances of consciousness and coma.

MR characteristic findings

Morphology

- MR abnormalities are bilateral and almost symmetrical with a selective regional involvement of the brain or selective involvement of functionally related systems.
- Gray matter structures are known to be more vulnerable to oxygen deprivation (high cellular metabolism) as observed after acute anoxic states, cyanide or carbon monoxide poisoning (insult to globi pallidi) or in Leigh's disease (lenticular and caudate nuclei, periaqueductal gray, dentate cerebellar nuclei).
- Lipophylic toxics (such as organic solvents or toluene or heroin) are known to affect brain white matter and particularly heavily myelinated fiber bundles.
- In toxic encephalopathy due to heroin or even cocaine sniffing, a characteristic pattern of demyelination may be observed (case 106) with preferential initial involvement of the corticospinal pathways from the cortex to the brainstem level as well as related pontocerebellar pathways including middle cerebellar peduncles and cerebellar white matter, bilaterally. Occipital white matter is also demyelinated. The clinical symptoms (cerebellar signs and dysarthria) as well as the white matter abnormalities may recover 1.5-2 years after toxic withdrawal.

Signal intensity

- GE-T1w: hypointense systematized and/or diffuse areas of demyelination.
- SE-T2w or FLAIR: hyperintense areas of demyelination with respect to normal white matter.

Contrast enhancement

- No contrast enhancement.

Case 105: Toxic multisystem wallerian degeneration (opioids)

Fig. a: axial cut (512); SE - T2w

Fig. b: axial cut (512); SE - T2w

Fig. c: axial cut (512); SE - T2w

Fig. d: axial cut (512); SE - T2w

High resolution MR imaging of the brain in a patient presenting cerebellar disorder, presumably of toxic origin (opioids) and showing on T2w sequence (**a-d**), symmetric high intensity abnormalities involving the major corticospinal tracts at the level of the corona radiata (**d**), the posterior arm of the internal capsules (**c**) and the cerebral peduncles (**b**) bilaterally. The high signal abnormality is followed down in the pontes basis showed as two central areas of high signal corresponding to the corticospinal tracts (**a**). Hyperintense lesions involve also, symmetrically, both cerebellar white matter (**a**) as well as the superior cerebellar peduncles and their decussation at the mesencephalic level (**b**). Note the involvement of the parieto-occipital white matter as well as the splenium of the corpus callosum (**c, d**). Note also in the posterior part of the tegmentum pontis the two linear areas of high signal corresponding presumably to the medial lemniscus. These high signal lesions involving bilaterally and symetrically multiple intracerebral pathways probably correspond to a multisystemic wallerian degeneration involving mainly the corticospinal and cerebellar pathways (observed in two cases, similarly). The follow-up MR exams as the clinical follow-up of the patient (2 years), showed improvement of the lesions from the frontal cortex to the cerebellum paralleling the improvement of the clinical status and following the suspension of the toxic uptake. The evolution points out to wallerian regeneration.

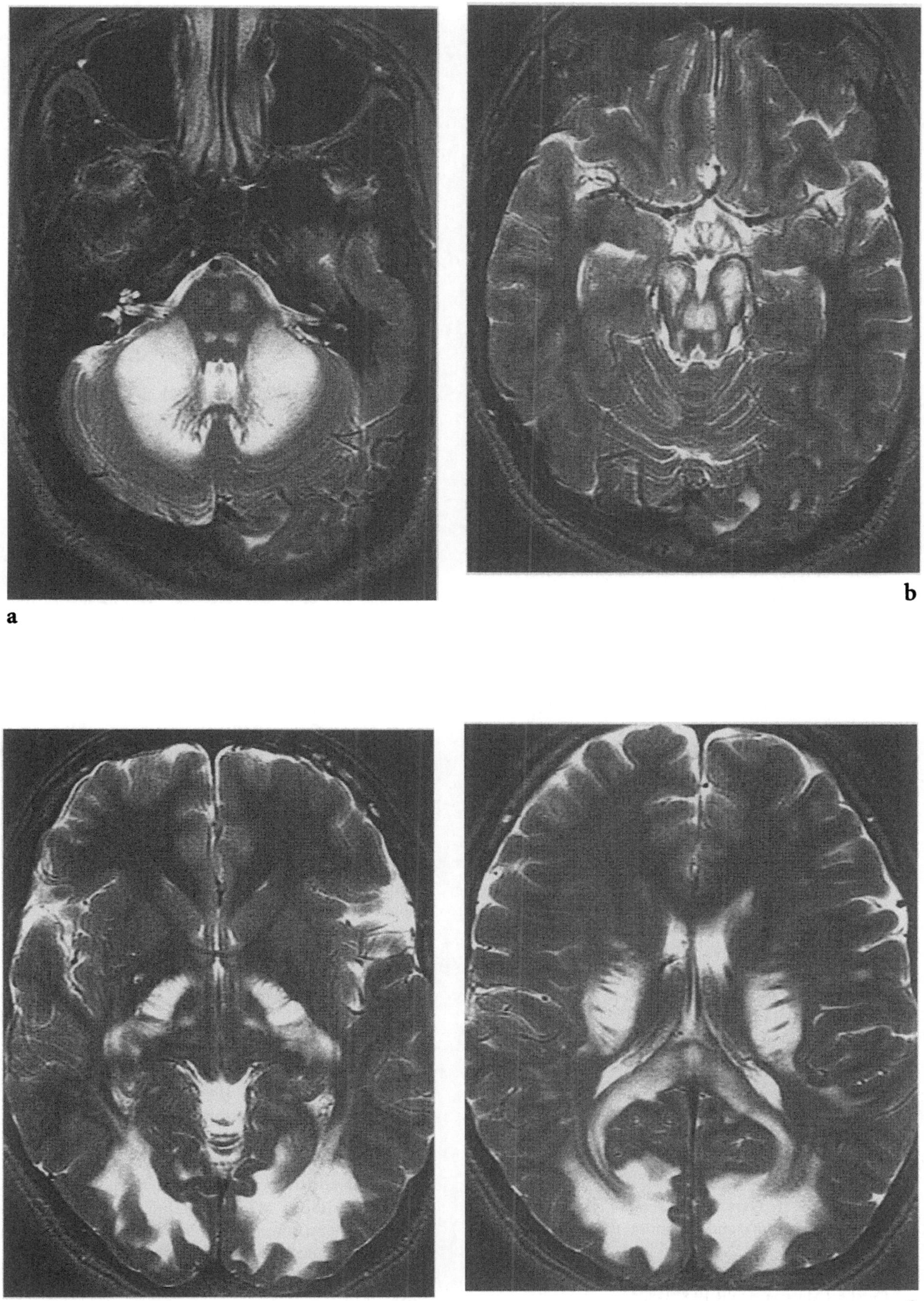

a b

c d

Selected References

- Abbott RJ, Howe JG, Currie S et al (1982) Multiple sclerosis plaque mimicking tumour on computed tomography. Br Med J 285:1616
- Aimes AB, Zimmerman RD, Morgello S et al (1989) MR imaging of brain abscesses. Am J Neuroradiol 10:279
- Bagley LJ, Grossman R, Galetta S et al (1999) Characterization of White Matter Lesions in Multiple Sclerosis and Traumatic Brain Injury as Revealed by Magnetization Transfer Contour Plots. AJNR 20:977-981
- Barkovich AJ, Lindan CE (1994) Congenital cytomegalovirus infection of the brain: imaging analysis and embryologic considerations. AJNR 15:703-715
- Barkovich AJ, Peck WW (1997) Zellweger syndrome. AJNR 18:1163-1170
- Bastianello S, Bozzao A, Paolillo A et al (1997) Fast spin-echo and fast fluid-attenuated inversion-recovery versus conventional spin-echo sequences for MR quantification of Multiple Sclerosis lesions. AJNR 18:699-704
- Baum HM, Rothschild BB (1981) The incidence and prevalence of reported multiple sclerosis. Ann Neurol 10:420
- Baum PA, Barkovich AJ, Koch TK, Berg BO (1994) Deep gray matter involvement in children with acute disseminated encephalomyelitis. AJNR 15:1275-1283
- Beck RW, Cleary PA, Trobe JD et al (1993) The effect of corticosteroids for acute optic neuritis on the subsequent development of multiple scelerosis. N Engl J Med 329:1764
- Beer S, Rosler KM, Hess CW (1995) Diagnostic value of paraclinical tests in multiple sclerosis: relative sensitivities and specificities for reclassification according to the Poser committee criteria. J Neurol Neurosurg Psychiatry 59:152
- Borzone M, Capuzzo T, Rivano C et al (1980) Subdural empyema: fourteen cases surgically treated. Surg Neurol 13:449
- Bourgouin PM, Chalk C, Richardson J et al (1995) Subcortical white matter lesions in osmotic demyelination syndrome. AJNR 16:1495-1497
- Brismar J, Ageel A, Brismar G et al (1990) Maple syrup urine disease: findings on CT and MR imaging of the brain in 10 infants. AJNR 11:1219-1228
- Brownell B, Hughes JT (1962) The distribution of plaques in the cerebrum in multiple sclerosis. Am J Neurosurg Psychiatry 25:315
- Bydder GM, Steiner RE, Young IR et al (1982) Clinical NMR imaging of the brain: 140 cases. Am J Neuroradiol 3:459
- Christoforidis GA, Spickler EM, Recio MV et al (1999) MR of CNS sarcoidosis: correlation of imaging features to clinical symptoms and response to treatment. AJNR 20:655-669
- Cordoliani YS, Sarrazin JL, Felten D et al (1998) MR of Cerebral Malaria. AJNR 19:871-874
- Dagher AP, Smirnitopoulos J (1996) Tumefactive demyelinating lesions. Neuroradiology 38:560-565
- Davidson HD, Steiner RE (1985) Magnetic resonance imaging in infections of the central nervous system. Am J Neuroradiol 6:499
- Dousset V, Grossman RI, Ramer KN et al (1992) Experimental allergic encephalomyelitis and multiple sclerosis: lesion characterization with magnetization transfer imaging. Radiology 182:482
- Dousset V, Armand JP, Lacoste D et al (1997) Magnetization transfer study of HIV encephalitis and progressive multifocal leukoencephalopathy. AJNR 18: 895-901
- Drayer B, Burger P, Hurwitz B et al (1987) Reduced signal intensity on MR images of thalamus and putamen in multiple sclerosis: increased iron content? Am J Neuroradiol 8:413
- Dunn V, Bale JF Jr, Zimmerman RA et al (1986) MRI in children with postinfectious disseminated encephalomyelitis. Magn Reson Imaging 4:25
- Eblen F, Poremba M, Grodd W et al (1991) Myelinoclastic diffuse sclerosis (Schilder's disease): cliniconeuroradiologic correlations. Neurology 41:589-591
- Endo Y, Oda M , Hara M (1981) Central pontine myelinolysis. A study of 37 cases in 1 000 consecutive autopsies. Acta Neuropathol 53:145-153
- Enzmann DR, Britt RH, Yeager AS (1979) Experimental brain abscess evolution: computed tomographic and neuropathologic correlation. Radiology 133:113
- Fabiani A, Trebini F, Torta R (1980) Brain Hydatidosis report of 2 cases. J Neurosurg Psychiatry 43:91
- Fazekas F, Offenbacher H, Fuch S et al (1988) Criteria for increased specificity of MRI interpretation in elderly subjects with suspected multiple sclerosis. Neurology 38: 1822-1825
- Filippi M, Barker GJ, Horsfield MA et al (1994) Benign and secondary progressive multiple sclerosis: a preliminary quantitative MRI study. J Neurol 241:246
- Filippi M, Horsfield MA, Morissey SP et al (1994) Quantitave brain MRI lesion load predicts the course of clinically isolated syndromes suggestive of multiple sclerosis. Neurology 44:635
- Filippi M, Paty DW, Kappos L et al (1995) Correlations between changes in disability and T2-weighted brain MRI activity in multiple sclerosis: a follow-up study. Neurology 45:255
- Filippi M, Rossi P, Campi A et al (1997) Serial contrast-enhanced MR in patients with Multiple Sclerosis and varying levels of disability. AJNR 18:1549-1556
- Filippi M, Rocca MA, Wiessmann M et al (1999) A comparison of MR imaging with Fast-FLAIR, HASTE-FLAIR, and EPI-FLAIR sequences in the assessment of patients with Multiple Sclerosis. AJNR 20:1931-1938
- Filippi M, Rocca MA, Gasperini C et al (1999) Interscanner variation in brain MR lesion load measurements in Multiple Sclerosis using conventional Spin-Echo, Rapid Relaxation-Enhanced, and Fast-FLAIR Sequences. AJNR 20:133-137

- Fulton JC, Grossman RI, Udupa J et al (1999) MR lesion load and cognitive function in patients with relapsing-remitting Multiple Sclerosis. AJNR 20:1951-1955
- Gean Marton AD, Venzina LG, Marton KI et al (1991) Abnormal corpus callosum: a sensitive and specific indicator of multiple sclerosis. Radiology 180:215
- Gonzalez-Scarano F, Grossman RI, Galetta S et al (1987) Multiple sclerosis disease activity correlates with gadolinium-enhanced magnetic resonance imaging. Ann Neurol 21:330
- Grossman RI, Braffman BH, Brorson JR et al (1988) Multiple sclerosis: serial study of gadolinium-enhanced MR imaging. Radiology 169:117
- Gupta RK, Kathuria MK, Pradhan S (1999) Magnetization transfer MR imaging in CNS tuberculosis. AJNR 20:867-875
- Guttman CRG, Ahn SS, Hsu L et al (1995) The evolution of multiple sclerosis lesions on serial MR. Am J Neuroradiol 16:1481
- Hajnal JV, Bryant DJ, Kasuboski L et al (1992) Use of fluid attenuated inversion recovery (FLAIR) pulse sequences in MRI of the brain. J Comput Assist Tomogr 16:841
- Harris KG, Tran DD, Sickels WJ et al (1994) Diagnosing intracranial vasculitis: the roles of MR and angiography. AJNR 15:317-330
- Hashemi RH, Bradley WG, Chen D-Y et al (1989) Suspected multiple sclerosis: MR-clinical correlation. Am J Neuroradiol 10:929
- Hatta S, Mochizuki H, Kuru Y et al (1994) Serial neuroradiological studies in focal cerebritis. Neuroradiology 36: 285-288
- Hendrie HC, Farlow MR, Austrom MG et al (1989) Foci of increased T2 signal intensity on brain MR scans of healthy elderly subjects. Am J Neuroradiol 10:703
- Hiehle JMJ, Grossman RI, Ramer KN et al (1995) Magnetization transfer effects in MR-detected multiple sclerosis lesions: comparison with gadolinium-enhanced spin echo images and non enhanced T1-weighted images. Am J Neuroradiol 16:69
- Honing LS, Sherremata WA (1989) Magnetic resonance of spinal cord lesions in multiple sclerosis. J Neurol Neurosurg Psychiatry 52:459
- Horowitz AL, Kaplam RE, Grewe G et al (1989) The ovoid lesion: a new MR observation in patients with multiple sclerosis. Am J Neuroradiol 10:303
- Issac C, Li DKB, Genton M et al (1988) Multiple sclerosis: a serial study using MRI in relapsing patients. Neurology 38:1511
- Jacobs L, Kinkel WR, Polachini I et al (1986) Correlation of nuclear magnetic, resonance imaging, computerized tomography, and clinical profiles in multiple sclerosis. Neurology 36:27
- Jacobs L, Munschauer FE, Kuba SE (1991) Clinical and magnetic resonance imaging in optic neuritis. Neurology 41:15-19
- Jinkins JR, Siqueira E, Al-Kawi MZ (1987) Cranial manifestations of aspergillosis. Neuroradiology 29:181
- Just M, Kramer G, Higer HP et al (1987) MRI of Listeria rhombencephalitis. Neuroradiology 29:401
- Kendall BE (1992) Disorders of lysosomes, peroxisomes and mitochondria. AJNR 13:621-653
- Kermode AG, Tofts PS, Thompson AJ et al (1990) Heterogeneity of blood-brain barrier changes in multiple sclerosis: an MRI study with gadolinium-DTPA enhancement. Neurology 40:229-235
- Kesselring J, Miller DH, Robb SA et al (1990) Acute disseminated encephalomyelitis: MRI findings and the distinction from multiple sclerosis. Brain 11:291-302
- Kioumehr F, Dadsetan MR, Rooholamini SA, Au A (1994) Central nervous system tuberculosis: MRI. Neuroradiology 36:93-96
- Kocer N, Islak C, Siva A et al (1999) CNS Involvement in Neuro-Syndrome: An MR Study. AJNR 20:1015-1024
- Koudriavtseva T, Pozzilli C, Biasi C et al (1996) High dose contrast enhanced MRI in multiple sclerosis. Neuroradiology 38:S5-S9
- Kovanen J, Erkinjuntti T, Livanainen M et al (1985) Cerebral MR and CT imaging in Creutzfeldt-Jakob disease. J Comput Assist Tomogr 9:125
- Lai PH, Lin SM, Pan HB, Yang CF (1997) Disseminated Miliary Cerebral Candidiasis. AJNR 18:1303-1306
- Lassman H (1983) Comparative neuropathology of chronic experimental allergic encephalomyelitis and multiple sclerosis. Springer-Verlag, Berlin
- Lee KH, Hashimoto SA, Hooge JP et al (1991) MR imaging of the head in the diagnosis of MS: a prospective 2 years follow up with comparison of clinical evaluation, evoked potentials, oligoclonal banding and CT. Neurology 41:657-660
- Lexa FJ (1994) Neuroradiological manifestations of acquired immunodeficiency syndrome. Sem Roentgenol 29: 288-302
- MacDonald JT, Stauffer AE, Heitoff K (1984) Adrenoleukodystrophy: early frontal lobe involvement on computed tomography. JCAT 8:128-130
- Martizez HR, Rangel-Guerra R, Elizondo G et al (1989) MR imaging in neurocysticercosis: study of 56 cases. Am J Neuroradiol 10:1011
- Mastronardo G, Rocca MA, Iannucci G, Pereira C, Filippi M (1999) A Longitudinal MR study of the presymptomatic phase in a patient with clinically definite Multiple Sclerosis. AJNR 20:1268-1272
- McDonald WI (1989) Diagnosis of multiple sclerosis. Br Med J 299:635
- Mehler MF, Rabinowich L (1989) Inflammatory myelinoclastic diffuse sclerosis (Schilder's disease): neuroradiologic findings. AJNR 10:176-180
- Mehta RC, Pike GB, Enzmann DR (1995) Improved detection of enhancing and non enhancing lesions of multiple sclerosis with magnetization transfer. AJNR 16:1771-1778
- Miller DH, Barkhof F, Nauta JJP (1993) Gadolinium enhancement increases the sensitivity of MRI in detecting disease activity. Brain 116:1077
- Miller DH, Rudge P, Johnson G et al (1988) Serial gadolinium enhanced magnetic resonance imaging in multiple sclerosis. Brain 111:927
- Moody DM, Bell MA, Challa VR (1988) The corpus callosum, a unique white matter tract, anatomic features that

may explain sparing in Binswanger disease and resistance to flow of fluid masses. Am J Neuroradiol 9:1051
- Moriarty DM, Blackshaw AJ, Talbot PR et al (1999) Memory dysfunction in Multiple Sclerosis corresponds to juxtacortical lesion load on fast fluid-attenuated inversion-recovery MR images. AJNR 20:1956-1962
- Morriss MC, Rutstein RM, Rudy B, Desrochers C, Hunter JV, Zimmerman RA (1997) Progressive multifocal encephalopathy in an HIV-infected child. Neuroradiology 39:142-144
- Murata R, Nakajima S, Tanaka A et al (1989) MR imaging of the brain in patients with mucopolysaccharidosis. AJNR 10:1165-1170
- Nelson MD Jr, Wolff JA, Cross CA et al (1992) Galactosemia: evaluation with MR imaging. Radiology 184:255-261
- Niebler G, Harris T, Davis T et al (1992) Fulminant multiple sclerosis. Am J Neuroradiol 13:1547
- Ormerod IEC, Miller DH, McDonald WI et al (1987) The role of NMR imaging in the assessment of multiple sclerosis and isolated neurological lesions. A quantitative study. Brain 110:1579-1616
- Paty DW, Oger JJF, Kastrukoff LF et al (1988) MRI in the diagnosis of MS: a prospective study with comparison of clinical evaluation, evoked potentials, oligoclonal banding, and CT. Neurology 38:180
- Phillips MD, Grossman RI, Miki Y et al (1998) Comparison of T2 lesion volume and magnetization transfer ratio histogram. Analysis of atrophy and measures of lesion burden in patients with Multiple Sclerosis. AJNR 19:1055-1060
- Rangel-Guerra R, Martinez HR, Saenz C (1985) Mucormycosis: report of 11 cases. Arch Neurol 42:578
- Riccio TJ, Hasselink JR (1989) Gd-DTPA enhancement MR of multiple cryptococcal brain abscesses. Am J Neuroradiol 510:65
- Rocca MA, Mastronardo G, Rodegher M et al (1999) Long-term changes of magnetization transfer-derived measures from patients with relapsing-remitting and secondary progressive Multiple Sclerosis. AJNR 20:821-827
- Roman-Campos G, Toro G (1980) Herpetic brain stem encephalitis. Neurology 30:981
- Rovira A, Alonso J, Cucurella G, Nos C et al (1999) Evolution of multiple sclerosis lesions on serial contrast-enhanced T1-weighted and magnetization-transfer MR images. AJNR 20:1939-1945
- Runge VM, Clanton JA, Price AC et al (1985) Evaluation of contrast-enhanced MR imaging in a brain-abscess model. Am J Neuroradiol 6:139
- Schaumburg HH, Powers JM, Raine CS et al (1975) Adrenoleukodystrophy: a clinical and pathological study of 17 cases. Arch Neurol 32:577-591
- Shaw DWW, Maravilla KR, Weinberger E et al (1991) MR imaging of phenylketonuria. AJNR 12:403-406
- Heldon JJ, Siddharthan R, Tobias J et al (1985) MR imaging of multiple sclerosis: comparison with clinical and CT examinations in 74 patients. Am J Neuroradiol 6:683
- Soges LJ, Cacayorin ED, Petro GR et al (1988) Migraine: evaluation by MR. Am J Neuroradiol 9:425
- Sze G, De Armond SJ, Brant-Zawakzki M et al (1986) Foci of MRI signal (pseudolesions) anterior to the frontal horns: histologic correlations of a normal finding. Am J Neuroradiol 7:381
- Teitelbaum GP, Otto RJ, Lin M et al (1989) MR imaging of neurocysticercosis. Am J Neuroradiol 10:709
- Thomas DJ, Pennock JM, Hajnal JV et al (1993) Magnetic resonance imaging of spinal cord in multiple sclerosis by fluid-attenuated inversion recovery. Lancet 341:593
- Thompson AJ, Kermode AG, Wicks D et al (1991) Major differences in the dynamics of primary and secondary progressive multiple sclerosis. Ann Neurol 29:53
- Tsuchiya K, Inaoka S, Mizutani Y et al (1997) Fast Fluid-Attenuated Inversion-Recovery MR of intracranial infections. AJNR 18:909-913
- Uldry P-A, Regli F, Uske A (1993) Magnetic resonance imaging in patients with multiple sclerosis and spinal cord involvement: 28 cases. J Neurol 240:41
- Ulner JL, Elster AD (1991) Sarcoidosis of the central nervous system. Neuroim Clin N Am 110:141-158
- Van der Knaap MS, Valk J (1989) Magnetic resonance of myelin, myelination and myelin disorders. 2nd Ed, Springer-Verlag
- Van der Knapp MS, Valk J (1991) The MR spectrum of peroxisomal disorders. Neuroradiology 33:30-37
- Van Waesberghe JHTM, Castelijns JA, Roser W et al (1997) Single-dose gadolinium with magnetization transfer versus triple-dose gadolinium in the MR detection of Multiple Sclerosis lesions. AJNR 18:1279-1285
- Vion-Dury J, Jiddane M, Bouchacour M et al (1987) Mid- and long-term MRI study of sequelae of acute necrotizing viral encephalitis. J Neuroradiol 14:53
- Wallace CJ, Seland TP, Fong TC (1992) Multiple sclerosis: the impact of MR imaging. Am J Neuroradiol 158:849
- Young IR, Hall AS, Pallis CA et al (1981) Nuclear magnetic resonance imaging of the brain in multiple sclerosis. Lancet 2:1063
- Yousry TA, Filippi M, Becker C et al (1997) Comparison of MR pulse sequences in the detection of Multiple Sclerosis lesions. AJNR 18:959-963
- Zimmerman RD, Weingarten K (1991) Neuroimaging of cerebral abscesses. Neuroim Clin N Am 1(1):1-16

Part I

Brain and skull

Cases 106-109 : subdural hematomas (SDH)

Clinical and pathological features

- Extra-axial blood collections accumulating inside the virtual space between the inner dura mater and the arachnoid.
- They result from traumatic tearing of stretched bridging veins connecting the cerebral cortex to the dural sinuses. Unlike epidural arterial hematomas, bleeding is finally stopped by the rising of intracranial pressure.
- Subdural hemorrhages are associated with head trauma of variable severity, presenting with or commonly without skull fractures. The trauma is frequently very mild in elderly patients or alcoholics prone to repeated falls, in whom the atrophic brain becomes more mobile (increased space between brain and skull). It may even result in abused children from whiplash shaking injuries.
- Subdural hematomas may be seen in patients without any history of trauma, or known to suffer from blood dyscrasias or treated by anticoagulant drugs. Spontaneous, so-called idiopathic cases may also be seen (most probably secondary to a minor forgotten trauma).
- *Signs*: they are variable, in acute or chronic SDH. In acute SDH the latent interval is of several days to 1-2 weeks before the symptoms appear: headaches, disturbance of consciousness, mental slowness, even confusion or later some focal neurological signs (subjacent brain contusion). In chronic SDH, the trauma is often trivial or forgotten, the symptoms occuring insidiously several weeks later: headaches, drowsiness, slowness and incoherence of thought, apathy or confusion, focal neurological signs such hemiparesis (contralateral or ipsilateral) or aphasia and eventual seizure. EEG is usually abnormal on both sides, and skull X-rays if performed prior to MR or CT are usually negative, showing no fractures.

MR characteristic findings

Morphology

- SDH are extra-axial crescentic shaped collections of blood, most commonly located over the middle third convexity, near the vertex. They are not limited by skull sutures as the epidural type.
- Less frequent locations: subfrontal or in the posterior fossa, and interhemispheric or tentorial (mainly in abused children).
- SDH are convex-concave in shape becoming biconvex later (mainly in elderly). They are bilateral in about 15% of the cases. They are easily differentiated from epidural hematomas with MR by the direct visualization of the low SI dura mater separating the SDH from the cerebral cortex.
- MR due to its multiplanar capability (coronal and axial) and its high contrast resolution, permits an accurate appreciation of SDH thickness as well as the severity of the mass effect (shift of midline structures and ventricular compression) which is a more reliable prognostic factor. The mass effect particularly in acute SDH may be apparently disproportionate to its size in case of subjacent brain contusion (identified with MR on coronal T2w and T2*w).
- MR is much superior to CT in differentiating hyperacute, acute, subacute or chronic SDH and in differentiating SDH from SD hygroma. SDH of 1-3 mm in thickness are usually missed with CT.
- MR is helpful to show the fibrous septations within subacute and mainly chronic SDH (on T2w) and may be useful in case of rebleedings to guide drainage.
- MR differentiates SDH from SD hygroma which is an extra-axial subdural collection resulting from a tear in the arachnoid causing a CSF leak from the subarachnoid to the subdural space. Hygromas occur bilaterally in 50% of the cases after a trauma.

Signal intensity

- SDH are always hyperintense to the adjacent inner table of the skull voult.
- SI of SDH with respect to brain varies with the age of the hematoma.
- hyperacute stage (24-48 h): hypointense on SE-T1w; hyperintense on SE-T2w;
- acute stage (1st week): isointense on SE-T1w; hypointense on SE-T2w (deoxyhemoglobin);
- subacute stage (1-3 weeks): hyperintense on SE-T1w; hypointense on early subacute (intracellular methemoglobin) or hyperintense (late subacute) on SE-T2w; possible mixed “hematocrit effect”

aspect combining a hypointense dependent part containing cellular debris possibly due to a rebleeding, separated from a hyperintense supernatant fluid;

- chronic stage: iso-intense or hypointense on SE-T1w; hyperintense on SE-PDw, SE-T2w or FLAIR.

- SI of SD hygromas parallels CSF on all pulse sequences. Chronic SDH are hyperintense to CSF on all sequences due to a higher protein content.

Contrast enhancement

- No need to contrast infusion for the diagnosis of SDH using MR. Iodine contrast infusion may be needed with CT in case of isodense SDH in order to opacify the medially shifted veins (on MR isodense SDH are easily shown being hyperintense on all pulse sequences).
- Bridging veins are enhanced after Gd uptake and differentiated from internal chronic SDH septations (dural pseudomembranes).

Case 106: Acute subdural hematoma

Fig. a: coronal cut; SE - T1w

Fig. b: coronal cut; SE - T2w

Extra-axial blood collection easily identified by SE MR images showing hypersignal on T1w (**a**, ↑) due to T1 shortening, related to the presence of extracellular methemoglobin. But the hematoma is still in the acute phase with deoxyhemoglobine and intracellular methemoglobin explaining the T2 shortening with iso-hypo signal on T2 w (**b**,*). Note the mass effect on the cerebral cortex and the displacement of the subarachnoid space shown as a CSF cleft on the inner aspect of the hematoma (**b**, ↑).

Case 107: Bilateral subdural hematoma

Fig. c: axial cut (512); GE - T1w

Fig. d: axial cut; SE - T2w

Bilateral hemorragic subdural collections (↑) with hypersignal on T1w (**c**) and hypersignal on T2w (**d**) due to oxidization of deoxyhemoglobin to methemoglobin. Note the presence of a rim of hyposignal at the inner aspect of the hematoma on T2w (**d**) that would match with hemosiderin deposition. Note the presence of a ventricle of Varga (**c**, ↑↑), anatomical variant.

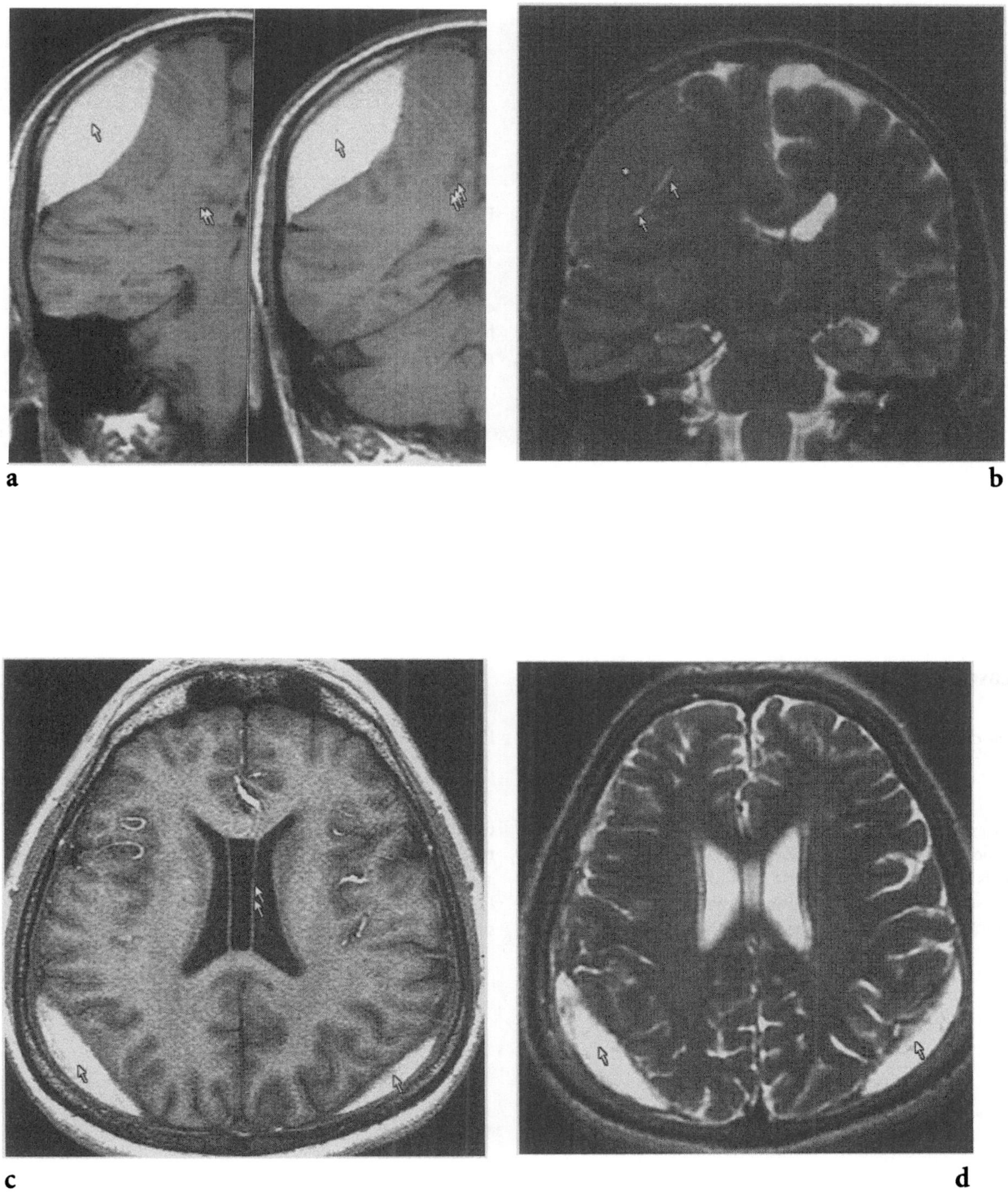
a
b
c
d

Case 108: Chronic subdural hematoma

Fig. a: axial cut; SE - PDw

Fig. b: axial cut; SE - T2w

Extended chronic right frontal subdural hematoma, hyperintense on PDw (**a**), as well as on T2w (**b**, ↑), containing intervening hypointense areas (**b**, ↑↑), suggesting recent rebleeding. Note the mass effect on the cerebral cortex, the ipsilateral ventricle and on the anterior hemispheric fissure displaced to the left and collapsed (**a**, right, ↑↑).

Case 109: Chronic subdural collection (FLAIR)

Fig. c: axial cut; FLAIR - T2w

Fig. d: axial cut; FLAIR - T2w

Extensive right hemispheric hemorrhagic collection hyperintense to brain and CSF on FLAIR sequence (**c**, **d**, ↑). This very thin hemorrhagic collection was not identified on CT scan, neither on the routine conventional SE sequences. The FLAIR sequence showed easily the hyperintense signal to subarachnoid and ventricular CSF of the diluted methemoglobin. Presumed chronic subdural hematomas and hygromas may be more efficiently diagnosed using FLAIR sequences as shown in some cases of loculated subarachnoid hemorrhage.

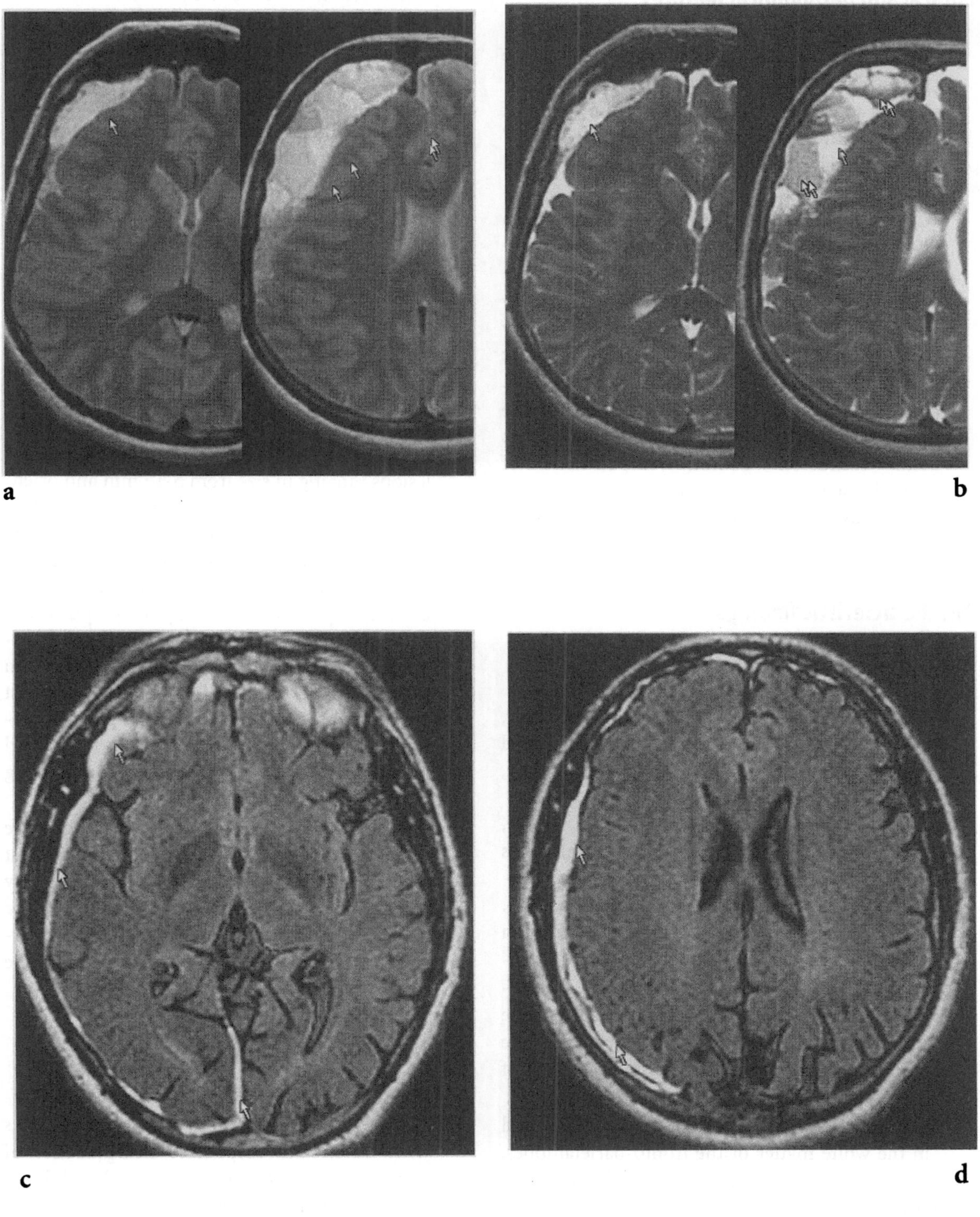
a
b
c
d

Cases 110-111: brain contusions

Clinical and pathological features

- Incidence of head injuries is about 0.5%, more than 1/3 being related to traffic.
- These contusions result from "coup" (site of impact) and more commonly "contre-coup" (opposite site of impact) effects of the causal head traumatism.
- One of the commonest mechanisms involved in non penetrating injuries, is represented by acceleration-deceleration of the skull and brain.
- *Histology*: brain contusion is characterized by petechial hemorrhages secondary to capillary tearing, presumably consecutive to intrinsic brain forces developed by rotational movements.
- *Signs*: variable with respect to severity of the trauma. In mild injuries: transient headaches and confusion without impairment of consciousness. In severe injuries: impairment of consciousness with focal neurologic deficits are present.

MR characteristic findings

Morphology

- MR abnormalities correlate well with pathology. The pattern of predilection of contusional effects of trauma, usually bilateral and multiple, include the following locations:
- inferior aspect of orbitofrontal lobes (30%), temporal polar and inferior temporal lobes (about 40%), less frequently the ventral cerebellar cortex (rule out associated occipital fracture). These sites are explained by their close relation with rough bones of the skull base;
- occipital and parietal lobes (15%) and cerebellum (10%) are the least involved because of the smooth texture of the inner aspect of the parieto-occipital bones and protective effect of the tentorium and the falx against brain motion;
- other lesions and/or petechiae may also be found in the white matter of the fronto-parietal lobes, in the corpus callosum, or in the midbrain.
- Brain contusions more specifically involve the crest of the convolutions and the subjacent white matter core showing a wedge-shaped appearance with the apex directed medially. These lesions are commonly hemorrhagic (over 50%) but may be non hemorrhagic (limited to a focal cortical-subcortical edematous area).
- Most of these acute lesions are to some extent hemorrhagic, this explaining variation of their MR aspect with time. Edema and softening of the injured brain, with similar SI aspect to hemorrhagic infarct on MR, appear in the 4-7 days following the acute insult. Liquefaction of the hemorrhagic lesion follows (2nd-3rd weeks) with associated proliferation of fibrillary astrocytes in the peripheral brain tissue and fibroblasts in the adjacent leptomeningeal space. The resulting chronic scar involves the gyral ridge, eventually associated with enlarged contiguous CSF space or ventricle.
- Diffuse axonal injury lesions are common, usually non hemorrhagic (80%) white matter lesions ranging in size from 5-15 mm and oriented along the direction of the injured fibers tract. They are most commonly depicted in the gray-white matter junctional areas or deeper in the corona radiata or even in the internal and external capsules, followed by the corpus callosum with predilection to the splenium and posterior part of its body, and finally the brainstem mainly at its dorsolateral aspect. They are best depicted on MR. Usually found in patients presenting with severe head injury (sustained impairement of consciousness and progressing focal neurological deficits).
- Isolated true intracerebral hemorrhages are rarely encountered. Delayed hematomas that may occur during the first 3 days in a previously non hemorrhagic site, involve the frontal or temporal lobes and are most frequently observed in patients with low Glasgow Coma Scale scores.

Signal intensity

- MR signal intensity of cortical contusions, diffuse axonal injuries and hematomas, varies with time, with respect to the presence of associated blood and related breakdown products (patient imaged in the acute or subacute or chronic stages).
- GE-T2*: mandatory due to its sensitivity to field inhomogeneities and thus to the detection of

brain hemorrhagic foci in the acute or subacute stages and depiction of hemosiderin deposits (hypointense SI) in the chronic post-traumatic stage (coronal). T2* sequence is most efficient in the evaluation of patients with a remote history of head trauma showing sequelar post-traumatic lesions and scars.

- SE-T2w and SE-PDw: important for the detection of the non hemorrhagic edematous contusions (hyperintense SI) of the cortex as well as the small hyperintense lesions of diffuse axonal injury (during the first 2 weeks).
- Diffusion imaging: may be more sensitive than SE-T2w for the detection of areas of edema associated with contusions explored in the acute stages.

Contrast enhancement

- Subacute hemorrhagic contusions or hemorrhages may show ring enhancement.

Case 110: Post-traumatic contusion

Fig. a: sagittal cut;	GE - T1w
Fig. b: axial cut;	GE - T2*w
Fig. c: axial cut;	SE - PDw
Fig. d: axial cut;	SE - T2w

Patient explored one week after a head trauma. MR showed a heterogeneous cortico/sub-cortical lesion slightly hyperintense on PDw (**c**, ↑) and hyperintense on T2w (**d**, ↑↑), corresponding presumably to edema related to the contusion of the left temporal lobe. The patient also presented an associated subarachnoid hemorrhage shown on T1w GE image (**a**, ↑). The hyperintense signal of the hemorrhagic suffusion along the tentorium cerebelli is due to the presence of methemoglobin. Note the apparent thickening with very low signal of the dura on T2* sequence (**b**, ↑).

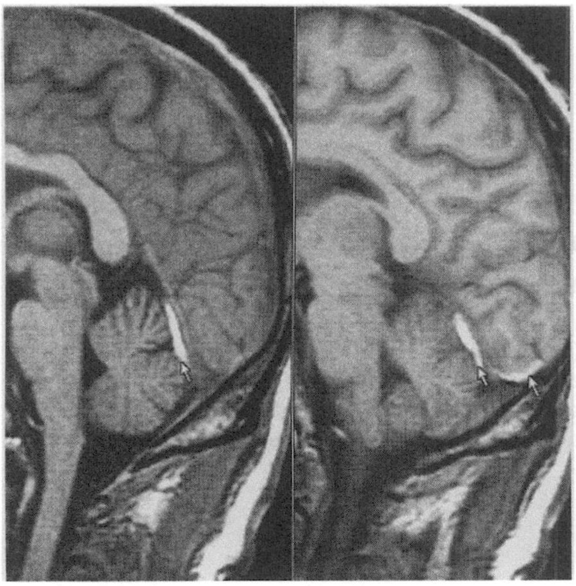

a

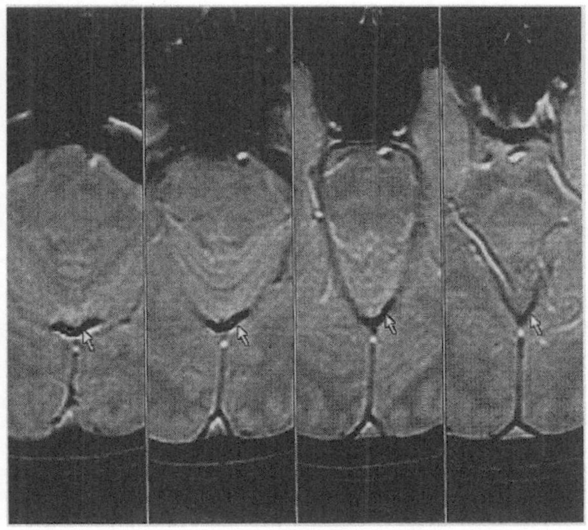

b

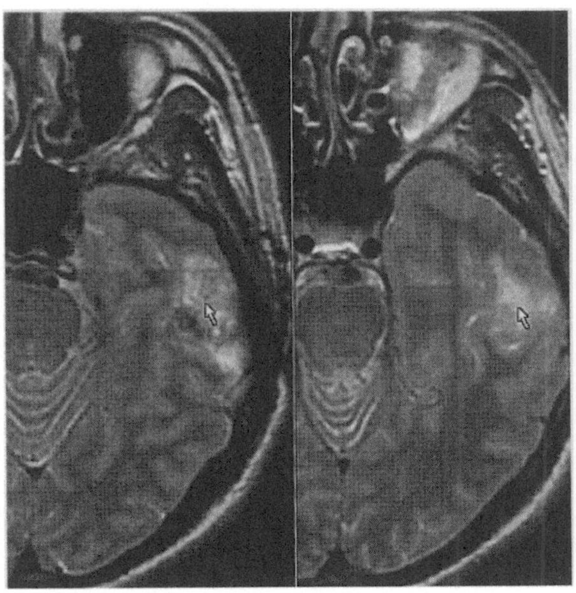

c

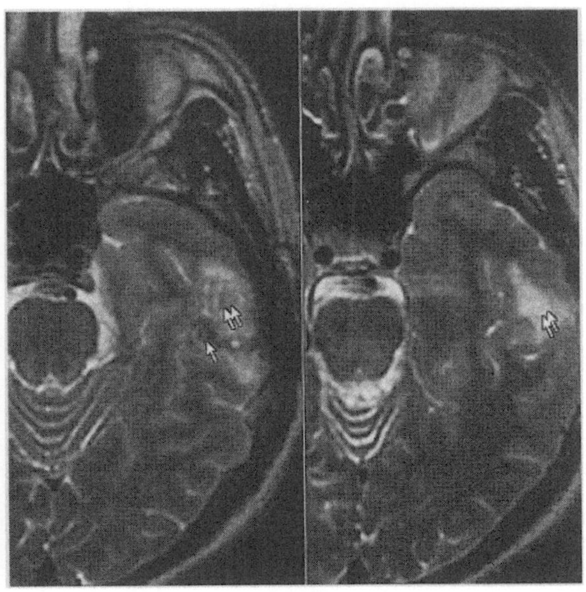

d

Case 111: Traumatic sequelae

Fig. a: sagittal cut;	GE - T1w
Fig. b: coronal cut;	SE - T1w
Fig. c: axial cut;	SE - PDw
Fig. d: axial cut;	SE - T2w

Long term evaluation of an important head trauma with fracture of the anterior skull base and especially the left cribriform plate. MR shows a porencephalic lesion involving the left parasagittal basal frontal lobe CSF filled, and showing CSF-like signal on T1w (**a**, ↑; **b**) and on PDw (**c**) as well as on T2w (**d**) sequences. Note the small mass effect on the right rectus gyrus (**c**, **d**, ↑↑), possibly due to arachnoidal loculation. The encephaloclastic area showed no enhancement after Gadolinium injection (**a**).

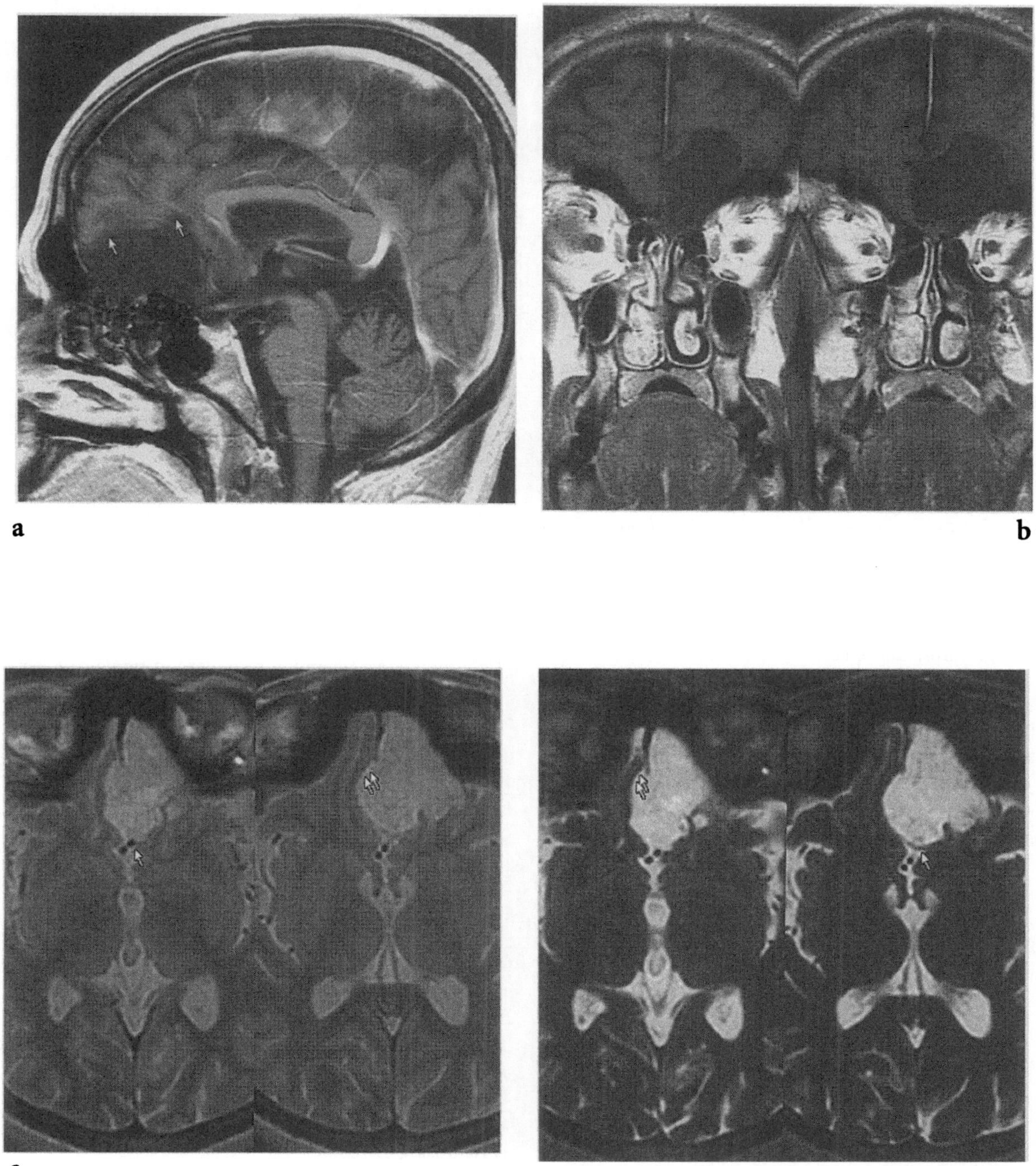

a b c d

Selected References

- Adams JH, Graham DI, Murray LS et al (1982) Diffuse axonal injury due to non-missile head injury in humans: an analysis of 45 cases. Ann Neurol 12:557
- Baykaner K, Alp H, Ceviker N et al (1988) Observation of 95 patients with subdural hematomas and review of the literature. Surg Neurol 30:339
- Bruce DA, Alavi A, Bilaniuk LT et al (1981) Diffuse cerebral swelling following head injuries in children: the syndrome of malignant brain edema. J Neurosurg 54:170
- Bulla WJ, Loes DJ (1994) Trauma to the cerebrovascular system (review). Neuroimaging Clin North Am 4:753-772
- Caplan LR, Zervas NT (1977) Survival with permanent midbrain dysfunction after surgical treatment of traumatic subdural hematoma: the clinical picture of a Duret hemorrhage. Ann Neurol 1:587
- Clark RA, Watanabe AT, Bradley WG et al (1990) Acute hematomas: effects of deoxygenation, hematocrit, and fibrin-clot formation and retraction on T2 shortening. Radiology 175:201
- Cordobes F, Lobato RD, Rivas JJ et al (1981) Observation in eighty-two patients with extradural hematomas. J Neurosurg 54:179
- Fobben ES, Grossman RI, Atlas SW et al (1989) MR characteristics of subdural hematomas and hygromas at 1.5T. Am J Neuroradiol 10:687
- Friedman DP, Flanders AF (1992) Unusual dissection of the proximal artery: description of three cases. Am J Neuroradiol 13:283
- Gallagher JP, Browder J (1968) Extradural hematomas: experience with 167 patients. J Neurosurg 29:1
- Garza-Mercado R (1983) Extradural hematoma of the posterior cranial fossa: report of seven cases with survival. J Neurosurg 59:664
- Gean AD (1994) Imaging of head trauma. Raven Press, New York
- Gentry LR, Thompson B, Godersky JC (1988) Trauma to the corpus callosum: MR features. Am J Neuroradiol 9:1129
- Gomori JM, Grossman RI, Hackney DB et al (1987) Variable appearance of subacute intracranial hematomas on high-field spin-echo MR. Am J Neuroradiol 8:1019
- Gomori JM, Grossman RI, Yu-IP C et al (1987) MR relaxation times of blood: dependence on field strength, oxidation state and cell integrity. J Comput Assist Tomogr 11:84
- Greiner FG, Orrison WW, King JF et al (1991) Vertebral artery injury associated with cervical spine fractures. Am J Neuroradiol 12:1254
- Kaufman HH, Singer JM, Sdahu VK et al (1980) Isodense acute subdural hematoma. J Comput Assist Tomogr 4:557
- Kawano Y, Ochi M, Hayashi K et al (1991) Magnetic resonance imaging of cerebral fat embolism. Neuroradiology 33:72
- Levin AV, Magnussom MR, Rafto SE et al (1989) Shaken baby syndrome diagnosed by magnetic resonance imaging. Pediatr Emerg Care 5:181
- Levy C, Laissy JP, Raveau V et al (1994) Carotid and vertebral artery dissections: three-dimensional time-of-flight MR angiography and MR imaging versus conventional angiography. Radiology 190:97
- Lewin JS, Laub G (1991) Intracranial MR angiography: a direct comparison of three time-of-flight techniques. Am J Neuroradiol 12:1133
- Liu AY, Maldjian JA, Bagley LJ et al (1999) Traumatic brain injury: diffusion-weighted MR imaging findings. AJNR 20:1636-1641
- Mas JL, Bausser MG, Hasboun D et al (1987) Extracranial vertebral artery dissections: a review of 13 cases. Stroke 18:1037
- Mascalchi M, Bianchi MC, Mangiafico S et al (1997) MRI and MR angiography of vertebral artery dissection. Diagn Neuroradiol 39:329
- Mayer CA, Mirvis SE, Wolf Al et al (1991) Acute traumatic midbrain hemorrhage: experimental and clinical observations with CT. Radiology 179:813
- Rosenblum WI, Greenberg RP, Seelig JM et al (1981) Midbrain lesions: frequent and significant prognostic features in closed head injury. Neurosurgery 9:613
- Rothfus WE, Goldberg AL, Tabas JH et al (1987) Callosomarginal infarction secondary to tranfalcine herniation. Am J Neuroradiol l8:1073
- Servadei F, Nanni A, Nasi MT et al (1995) Evolving brain lesions in the first 12 hours after head injury: analysis of 37 comatose patients. Neurosurgery 37:899-906
- Sato Y, Yuh WTL, Smith WL et al (1989) Head injury in child abuse: evaluation with MR imaging. Radiology 173:653
- Strich SJ (1956) Diffuse degeneration of the cerebral white matter in severe dementia following head injury. J Neurol Neurosurg Psychiatry 19:163
- Strich SJ (1961) Shearing of nerve fibers as a cause of brain damage due to head injury, a pathological study of twenty cases. Lancet 1:2443
- Sue DE, Brant-Zawadski M, Chance J (1992) Dissection of cranial arteries in the neck: correlation of MRI and arteriography. Neuroradiology 34:273
- Tamaka T, Skai T, Uemura K et al (1988) MR imaging as predictor of delayed post-traumatic cerebral hemorrhage. J Neurosurg 69:203
- Thulborn K, Sorenson G, Kowall N et al (1990) The role of ferritin and hemosiderin in the MR appearance of cerebral hemorrhage: a histopathologic biochemical study in rats. Am J Neuroradiol 54:291

Part I

Brain and skull

5 - Neurogenetic and developmental disorders

Case 112: corpus callosum anomalies

Clinical and pathological features

- The corpus callosum is the white commissure linking the cerebral hemispheres.
- It normally forms between 11 and 18th weeks of gestation.
- It is divided into 4 parts: rostrum, genu, body and splenium.
- Its development proceeds from anterior to posterior starting with the junctional area between dorsal genu and anterior body at week 13, followed by the body and splenium, and ending at week 18 by the rostrum.
- *Signs*: callosal agenesis is usually asymptomatic unless associated with other congenital anomalies most commonly of the migrational type and giving rise to seizures, moderate to severe mental retardation, hypothalamic disorders, impaired motor, bimanual or visual coordination and may be accompanied by midline craniofacial dysmorphy or macrocephaly.
- Callosal dysgenesis may be part of clinical syndromes such as: Aicardi, acrocallosal or Meckel syndromes.
- Aicardi (X-linked) syndrome includes in females: agenesis of the corpus callosum or hypogenesis associated with an interhemispheric cyst, infantile spasms, chorioretinal lacunae and mental retardation.

MR characteristic findings

Morphology

- Callosal dysgenesis may be classified into 3 forms (Jinkins): **type I**: complete agenesis; **type II**: partial agenesis, involving its posterior part and splenium and the rostrum; **type III**: hypoplasia associated to other developmental cerebral anomalies (heterotopias, callosal lipomas, schizencephalies, sphenoidal encephalocele, cerebellar dysplasias, Dandy-Walker variants, and Chiari II malformation).
- The genu presents in partial agenesis except in semilobar holoprosencephaly in which there is a frontal lobes fusion associated to the presence of the splenium.
- Morphological abnormalities include:
  - on mid-sagittal cut: disorganization of the gyral pattern of the medial aspect of the cerebral hemisphere assuming a radiate arrangement, hypoplasia of the anterior commissure, interhemispheric lipoma (50%) showing frequently a peripheral or central calcification, or eventual "interhemispheric" cyst with parasagittal development, which may or not communicate with the third ventricle;
  - on coronal cuts: presence of two longitudinal callosal bundles of Probst that fail to cross the midline and impress the medial wall of the lateral ventricles, giving a crescentic "steer horn shape", and evertion of cingulated gyri, superior extension of a usually dilated third ventricle into the interhemispheric fissure, malrotation of hippocampal formations with vertically oriented tips of temporal horns and other limbic anomalies;
  - on axial cuts: roughly parallel separated lateral ventricles with marked posterior dilatation of occipital horns and atria of the lateral ventricles (colpocephaly) due to lack of normal development of posterior parieto-occipital white matter (in the absence of splenium).
- May be detected in utero by fetal ultrasound.
- Developmental callosal dysgenesis is most frequently differentiated from a secondary destruction involving the body and/or the genu but respecting the rostrum and splenium portions.
- An enlarged hippocampal commissure connecting the fornices (as demonstrated on coronal cuts) must not be confused with the splenium in a callosal agenesis.

Signal intensity

- SE-T1w: high signal of the lipoma (when present).
- SE-T2w: hypointense signal of the lipoma.
- Chemical shift artifact due to presence of fat (large lipomas).

Contrast enhancement

- No need of contrast infusion.

Case 113: neuronal heterotopias

Clinical and pathological features

- Anomaly of neuronal migration due to arrest of radial migration of neuroblasts.
- Results in accumulation of nerve cells in abnormal locations.
- Found from the subependymal region to the subcortical white matter.
- Divided in 3 major types: subependymal, focal in deep white matter and subcortical.
- *Signs*: represented by seizures and variable developmental and neuropsychological status from normal to mild or severe delay.

MR characteristic findings

Morphology

- MR is the modality of choice to depict these ectopic clusters of neurons.
- Different shape according to location and type.
- Subependymal heterotopia: ovoid nodular masses developing into the ventricular cavity to be differentiated from the tubers of Bourneville's disease.
- *Focal heterotopias*: focal processes found in the deep or subcortical white matter of the centrum semiovale. Regional infolding of the cortex may be found including subarachnoid space and even vessels (as shown on T2).
- *Band heterotopia*: giving rise to a "double cortex" appearance almost symmetrical on both cerebral hemispheres, separated by a thin lamina of white matter. The gray heterotopic band does not follow the infolding pattern of the overlying cortical mantle which may appear as normal, but may be pachygyric (broad, thick and flat convolutions).

Signal intensity

- IR-T1w or GE-T1w and SE-T2w: isointense to normal cortex on all pulse sequences (axial and coronal).
- Subependymal heterotopias must be differentiated from the tubers which are iso- or hypointense to white matter.

Contrast enhancement

- Heterotopic gray matter do not enhance after contrast infusion.
- It may be assumed that heterotopias may show some transient enhancement in the immediate post-ictal phase (?).
- Helps to differentiate subependymal nodules from enhancing subependymal metastases.

Case 112: Callosal agenesis

Fig. a: sagittal cut; GE - T1w

Fig. b: coronal cut; SE - T1w

Child presenting with a developmental delay. The sagittal T1w sequence (**a**) shows a callosal agenesis but the presence of the hippocampal commissure (**a**, ↑↑). Note the radial distribution of some of the internal frontal sulci (**a**, ↑). On the coronal T1w cut (**b**), the frontal horns of the lateral ventricles are small, superiorly pointed and widely separated with concave medial borders. Note the presence of the hippocampal commissure in the coronal plane (**b**, ↑). Note the abnormal shape of the hippocampal formations with patulous temporal horns. Note also the absence of interhemispheric cyst.

Case 113: Periventricular neuronal heterotopias

Fig. c: sagittal cut; GE (3D) - T1w

Fig. d: coronal cut; SE - T2w

Patient with epilepsy. MR exam evidenced on T1w multiple nodular masses lining the lateral ventricles mainly the temporo-occipital horns (**c**, ↑), which show iso-intensity with cortical grey matter on T1w (**c**) and T2w (**d**, ↑). These nodular masses correspond to subependymal heterotopias of grey matter, found in ectopic location. Such heterotopias result from a disturbance in the radial migration of neuroblasts originating at the germinal matrix. Note also areas of abnormal cortex and increased thickness (**c**, ↑↑↑; **d**, ↑↑), due to the presence of cortical heterotopias.

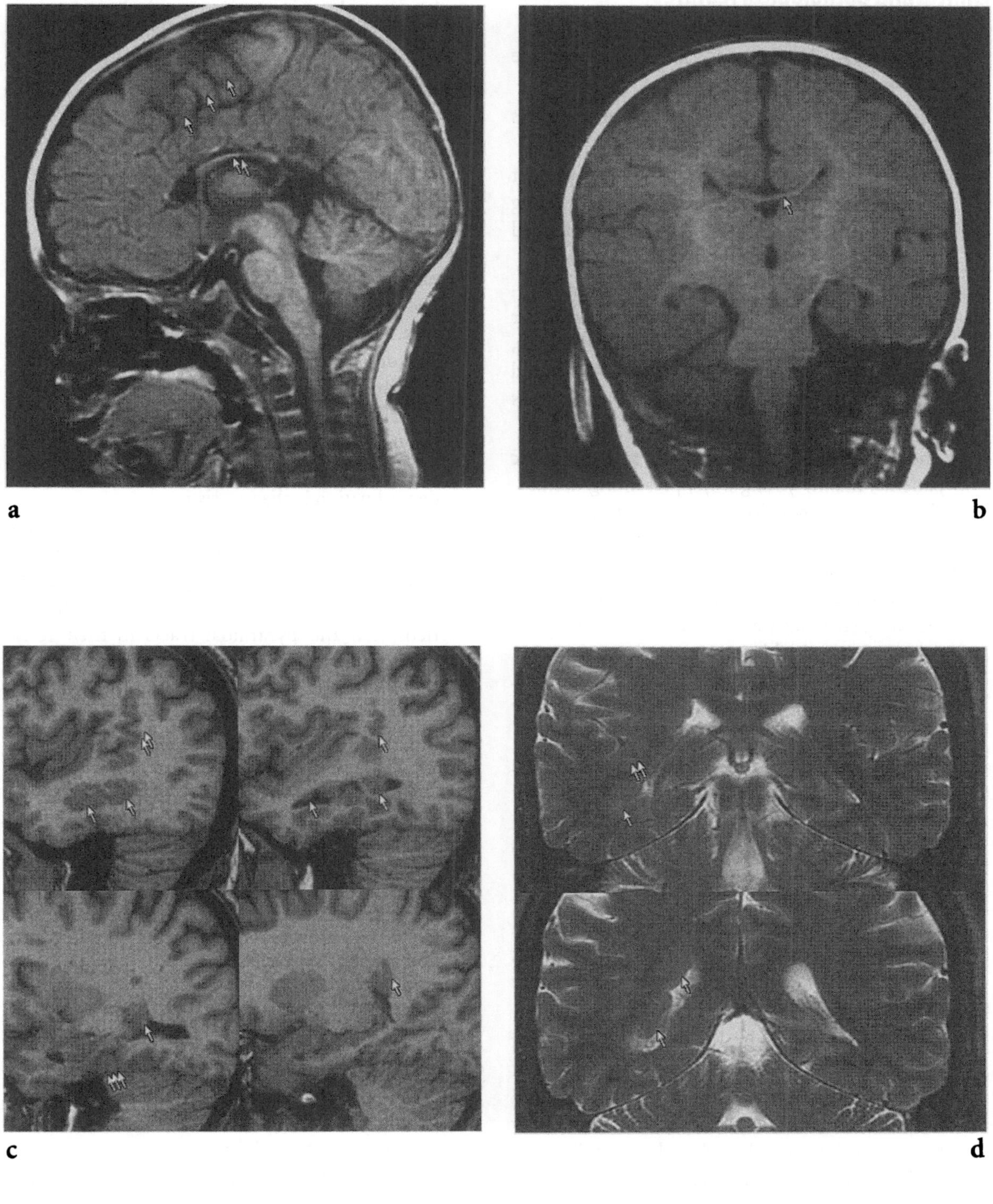

a b c d

Case 114: adrenoleucodystrophies

Clinical and pathological features

- Genetically determined, X-linked recessive disease.
- The most frequent form (50%) occurs in boys during late childhood (5-10 years).
- A similar cerebral form of the disease may present a later onset occurring in adolescents and young adults, but showing a more rapid and prolonged course.
- The other common form (30%), the adrenomyeloneuropathy is an adult variant which may occur in either sex, mainly in the third or fourth decade, showing a more progressive course associated with distal polyneuropathy, spastic paraparesis and cerebellar ataxia.
- The last group corresponds to asymptomatic patients, mainly young boys, presenting the biochemical abnormalities, but remaining free of symptoms.
- Adrenoleucodystrophy is a lipidosis in which there is an accumulation of cholesterol esters with very long chain fatty acids, in all tissues and particularly: cerebral white matter, peripheral nerves, adrenal cortex, due to a deficiency of Acyl CoA synthetase causing lack of oxidation of the very long chain fatty acids.
- *Histopathology*: the active demyelinated region shows a decrease in lipids content with an increase in cholesterol esters, decrease in galactolipids and increase in very long chain fatty acids in sphingomyelin and phosphatidyl choline, associated with increase in water content; in the gliotic region, decrease in galactolipids, increase in very long chain fatty acids content of sulfatides, cerebrosides and phospholipids, in the presence of high water content.
- *Signs* (common cerebral form): average age of onset is 7 years, with progressive deterioration of mental and behavioral functions, poor school intellectual and social performances, followed by neurological signs such as disorders of gait, visual and hearing loss, dysarthria and dysphagia, seizures and cerebellar ataxia, and signs of Addison's disease (variable degree of adrenal insufficiency). At terminal stages within few years of onset (1-5 years), the patient is almost tetraplegic, blind and deaf, and in decorticate posturing before death.

MR characteristic findings

Morphology

- Cerebral abnormalities are typical in the common form in over 2/3 of the cases.
- Extensive, roughly symmetrical areas of demyelination involving the occipital white matter core and including the splenium of the corpus callosum.
- The demyelinative process spares the subcortical U-fibers (best seen on T1w).
- Associated zones of cavitation and even calcification (CT) may be found.
- A forward progression pattern of the area of demyelination is characteristic.
- Other anatomical structures are involved: the geniculate bodies, the optic radiations, the posterior limb of the internal capsules and the auditory radiations, the external capsules and the corticospinal and pyramidal tracts in their route through the brainstem, excluding the frontopontine tracts, and later the cerebellar white matter.
- This usual pattern of demyelination is different in the frontal form (10%, see case 115) in which the demyelinative process starts in the frontal white matter bilaterally, involving the genu and the rostrum of the corpus callosum as well as the anterior limbs of the internal capsule, with subsequent posterior spread of disease process.
- The areas of demyelination may exceptionally develop asymmetrically or unilaterally in the earliest stages of the disease.
- The cortex appears of normal thickness and signal.

Signal intensity

- IR or SE-T1w: hypointense area of demyelination sparing subcortical U-fibers.
- SE-T2w: markedly hyperintense area of demyelination.
- Assessment by MR spectroscopy seems promising and reliable for prognosis.

Contrast enhancement

- The area of demyelination is bordered anteriorly (usual posterior form) or posteriorly (rare frontal form), by a rim of enhancement which separates the severely affected area from the advancing zone of demyelination, and which correspond to the band of active inflammation.
- In fact three zones may be distinguished on enhanced MR paralleling the pathological specimens: an outer edge zone of active demyelination, a middle ribbon-like enhanced zone of active inflammation and demyelination (Schaumberg's), and the largest demyelinated zone.

Case 114: Frontal pseudotumoral adrenoleucodystrophy

Fig. a: sagittal cut (512); GE - T1w

Fig. b: sagittal cut; SE - T2w

Fig. c: coronal cut; SE - PDw

Fig. d: coronal cut; SE - T2w

Young child approximately 5 years old, male, developing behavioral disturbances and developmental delay. MRI shows extensive high intensity changes on PDw (**c**) and on T2w (**b, d**) sequences, being hypointense on T1w (**a**) and involving bilateral frontal white matter through corpus callosum (**b,** ↑↑↑; **d,** ↑↑↑), extending to the corticospinal tracts in the internal capsules (**c, d,** ↑) as well as in the cerebral peduncles (**b,** ↑) and involving the cingulum as well (**d,** ↑↑). These white matter tracts showed marked enhancement after Gadolinium infusion as well as the posterior edges of the bifrontal white matter demyelinating lesions. The contrast enhancement observed at the lateral margin of the zones of demyelination corresponds to areas of active demyelination and inflammation (Schaumburg's zones). Note the mass effect of the white matter lesions on the corpus callosum moderately enlarged (**a, b,** ↑↑). This frontal predominance pattern in unusual and may correspond to a type II variant; the most frequent presentation mode (type I) involves predominantly the parieto-occipital lobes with gradual caudorostral progression and contrast enhancement along the anterior margin of the demyelinated white matter.

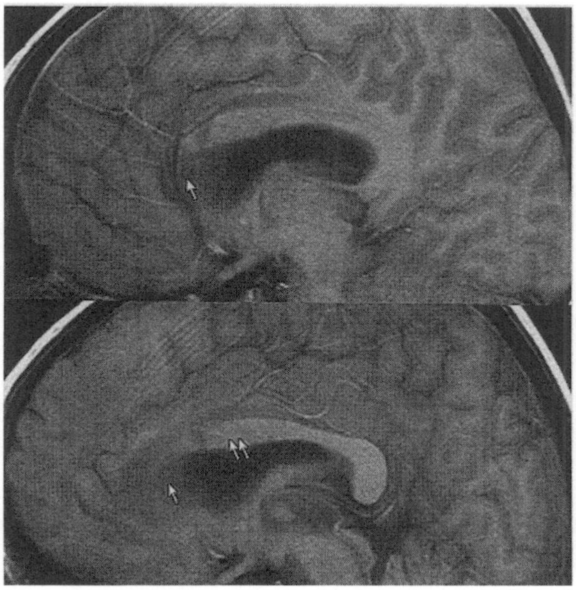

a

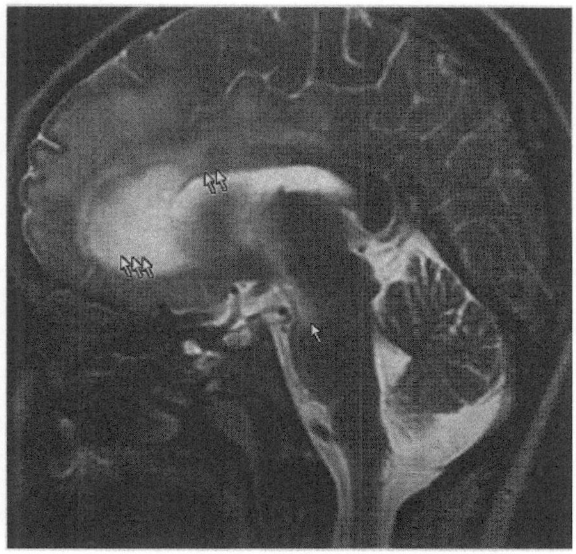

b

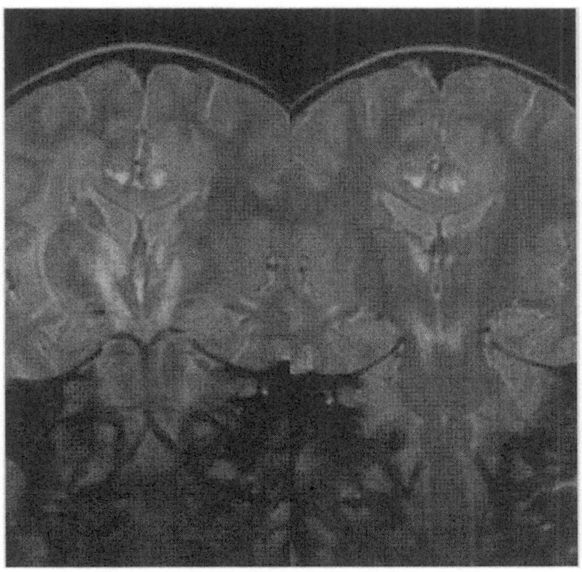

c

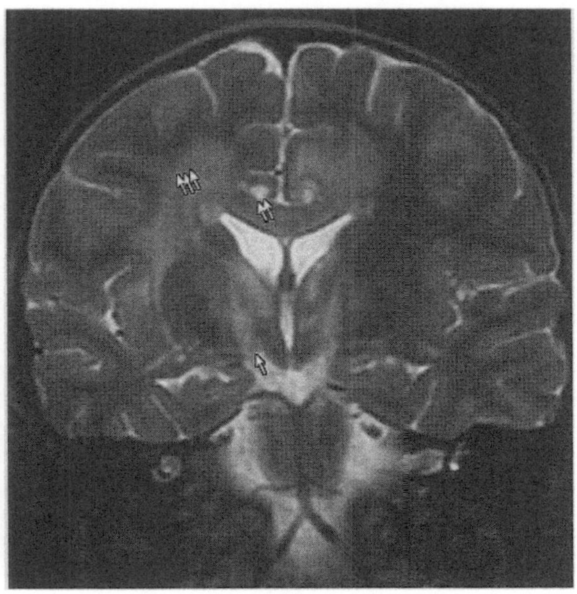

d

Case 115 : mucopolysaccharidosis

Clinical and pathological features

- Family of inherited diseases due to lack of degradation of mucopolysaccharides.
- The glycosaminoglycans are the deficient lysosomal enzymes incriminated.
- The result is the accumulation of MPS in the brain, the viscera and the bones.
- The transmission is autosomal recessive except Hunter's diseases being x-linked.
- The most common types are Hurler's and Hunter's diseases.
- CNS involvement is found in Hurler's (IH), Hurler-Scheie (IH-S), Hunter's (II), Sanfilippo A-D (III), Sly (VII), and Maroteaux-Lamy (VI) syndromes.
- Marked skeletal involvement in Maroteaux-Lamy (VI) and Morquio's A-B (IV) syndromes, and Diferrante's disease (VIII).
- Visceral involvement is found in types I (Hurler, Scheie, Hurler-Scheie), II (Hunter), and VII (Sly), associated with bony and/or CNS deposits.
- Several clinical and biochemical entities are individualized (types I to VII).
- The diagnosis is made on the basis of the clinical picture and the presence of glycosaminoglycans in the urine.
- *Histology*: deposits of MPS in the perivascular spaces of Virchow-Robin.
- *Signs*: variable associating marked (I, II, IV, VI, VII) to moderate (III, V) dwarfism, "gargoyle" appearance, mental retardation (III, IH, II, VII), macrocephaly, corneal clouding (IH, V, VI), skeletal deformations (all types except III), visceral and cardiovascular (valvular) involvement (IH, VI, VII, and II, IV, V later).

MR characteristic findings

Morphology

- MR abnormalities show wide variations in severity.
- Macrocrania, probably resulting from MPS deposits in the calvarium, brain and leptomeninges with eventual associated hydrocephalus.
- Dural and leptomeningeal thickening due to MPS and collagen deposition. Rule out cord compression mainly (IH, IV, V, VI) at atlanto-axial level. Possible associated meningeal cytic formation (suprasellar or other location).
- White matter abnormalities found in nearly all groups: consist of numerous small well defined lesions scattered in the white matter with predominance in the parieto-occipital lobes. Largely cystic in nature, following the signal of the CSF (enlarged perivascular spaces), they show a radial orientation or a rounded shape, more or less confluent due to associated demyelination, infarcts and gliosis.
- Corpus callosum is involved as well. Basal ganglia may be involved (I, II).
- Enlarged ventricular cavities due to concomitant hydrocephalus (impairment of leptomeningeal CSF resorption) or attributed to brain atrophy.
- Atlanto-axial subluxation due to hyperlaxity of the transverse odontoid ligament and/or odontoid dysplasia, hypoplasia or absence to be ruled out (MPS IV, VI).

Signal intensity

- GE or SE or IR-T1w: hypointense SI to white matter and isointense to CSF.
- SE-T2w and SE-PDw: hyperintense SI to white matter and isointense to CSF.
- Punctate cystic and radiate lesions showing SI behavior similar to CSF on all pulse sequences (dilated Virchow-Robin spaces), hypointense on FLAIR.
- Associated demyelination and infarcts; gliosis hyperintense on FLAIR and on T2w but not parallel CSF SI on FLAIR.
- Basal ganglia and thalamus may show a hypointense SI on SE-T2w possibly due to iron deposition (MPS IH and V).
- Proton spectroscopy may add to the diagnosis (decreased NAA/choline ratio and elevation of glutamate/glutamine).

Contrast enhancement

- May be proposed in case of leptomeningeal thickening.

Case 115: Mucopolysaccharidosis

Fig. a: axial cut; SE - T2w

Fig. b: axial cut; SE - T2w

Fig. c: sagittal cut; SE - T1w

Fig. d: axial cut; SE - T1w

Unusual punctate cavitary lesions observed in the white matter showing hypointense signal on T1w (**c**, **d**, ↑) involving here the callosal and para-callosal regions, becoming bright on T2w (**a**, **b**, ↑). The perivascular involvement with mucopolysaccharide accumulation is reflected by this radial as well as punctate bright signal lesions involving the enlarged Virchow-Robin spaces well shown in the subcortical white matter (**a**, ↑↑; **d**, ↑↑) as well as in the periventricular region (**b**, ↑↑). The larger lesions observed may be due to perivascular associated demyelination and/or small infarcts, becoming more widespread as the disease progresses.

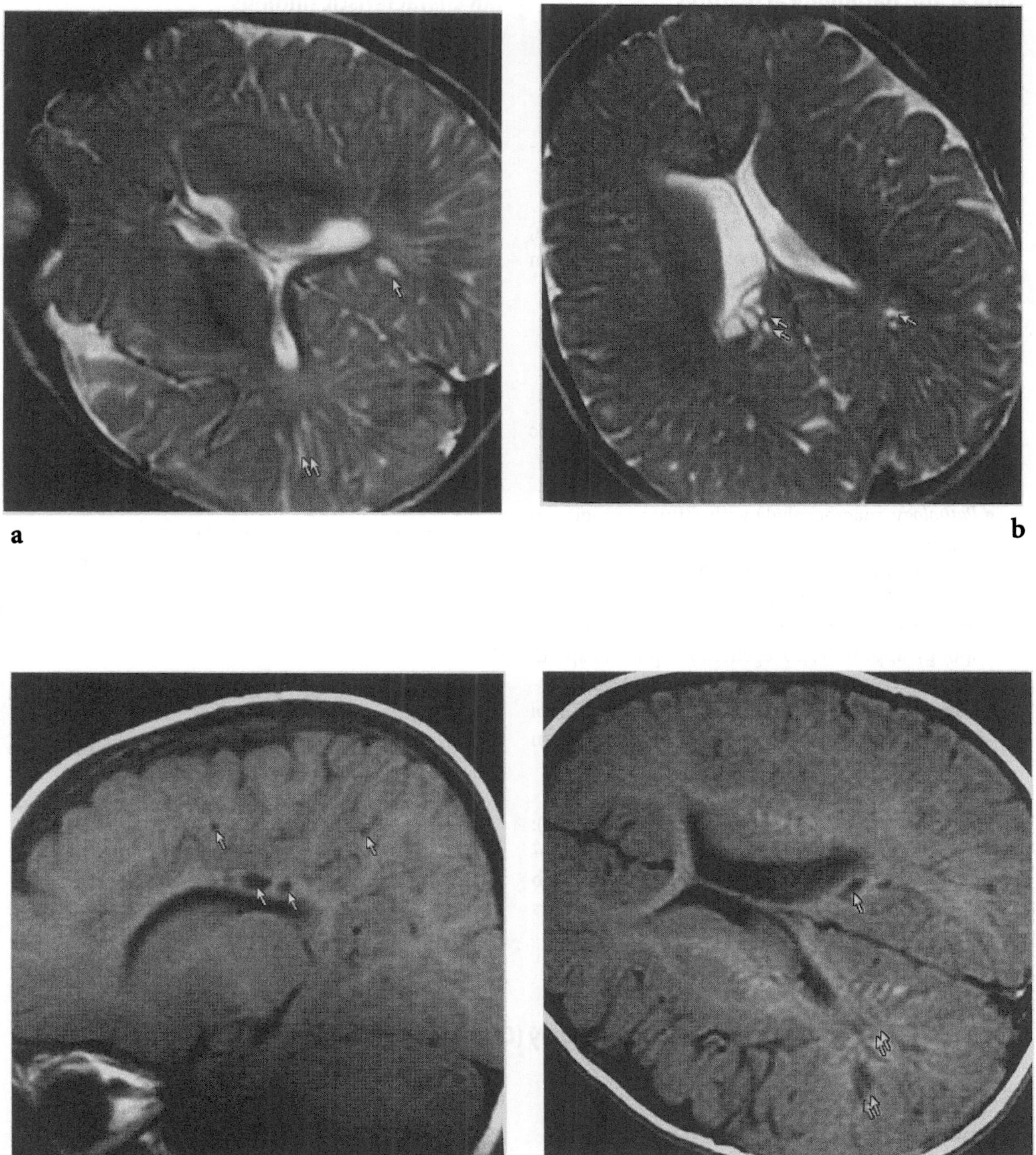

a b c d

Case 116: Leucodystrophy with megalencephaly

Clinical and pathological features

- Rare inherited neurogenetic disease.
- Also called spongy degeneration of Von Bogaert-Canavan.
- Affects white matter of infants and children, mainly in Ashkenazi Jews and Saudi Arabian population.
- Transmitted in the autosomal recessive mode.
- Caused by an enzymatic deficiency of aspartoacyclase, with consequent massive N-acetylaspartic aciduria.
- *Pathology*: macrocephaly with diffuse spongiform degeneration of the cerebral white matter, involving mainly the subcortical white matter, the U-fibers and deep cortical layers, with vacuolation on the histological ground giving rise to the spongy appearance. Degenerative process spares the periventricular white matter, the internal capsules, the corpus callosum and the fornices. Globi pallidi are involved with sparing of the striatum (putamina and caudate nuclei).
- *Signs*: appearing during infancy (first 6 months): hypotonia, macrocephaly, optic atrophy, developmental delay with decreased motor activity, limb spasticity, irritability and seizures. Death occurs after 2-4 years in most cases.

MR characteristic findings

Morphology

- Abnormal symmetrical involvement of the white matter in the centrum semiovale.
- Early in the course of the disease: involvement of the subcortical white matter and U-fibers; differs from metachromatic leucodystrophy and Krabbe's disease involving the deep periventricular white matter.
- Centripetal extension of the degenerative demyelinating process from the subcortical region toward the ventricular cavities.
- Relative sparing of the internal capsules and corpus callosum.
- Involvement of cerebellar white matter.
- Characteristic involvement of the globi pallidi (shown in Mapple Syrup urine disease) and thalami with sparing of the striatum.
- Diffuse cerebral cortico-subcortical atrophy in the late stages.

Signal intensity

- GE or IR or SE-T1w: hypointense to normal white matter.
- SE-T2w or SE-PDw or FLAIR: hyperintense to normal white matter.
- MR spectroscopy shows an elevated level of N-acetylaspartate.

Contrast enhancement

- No enhancement reported.

Case 116: Leucodystrophy with megalencephaly (Canavan)

Fig. a: sagittal cut;	GE - T1w	**Fig. c:** coronal cut;	SE - T2w
Fig. b: axial cut;	SE - PDw	**Fig. d:** coronal cut;	SE - T2w

Infant referred for macrocephaly and deteriorating neurologic function. MR imaging shows on T2w (**c**, **d**) sequences near complete-high signal abnormality in the supratentorial white matter (**c**, **d**, ↑↑) sparing the cerebellar white matter (**d**, ↑). The diffuse increased signal of the white matter is also shown on PD images (**b**, ↑↑). This demonstrates more accurately the relative sparing of both internal capsules (**b**, ↑) showing relative hyposignal due to myelination. Note the associated increase in overall volume of the hemispheric white matter, accounting for the macrocephaly. T1w sequences showed extensive hyposignal of the white matter (**a**, ↑↑), associated to areas of cystic degeneration of the subcortical white matter (**a**, ↑). The most likely diagnosis is spongy degeneration of Von-Bogaert-Canavan. Brain biopsy could not be performed.

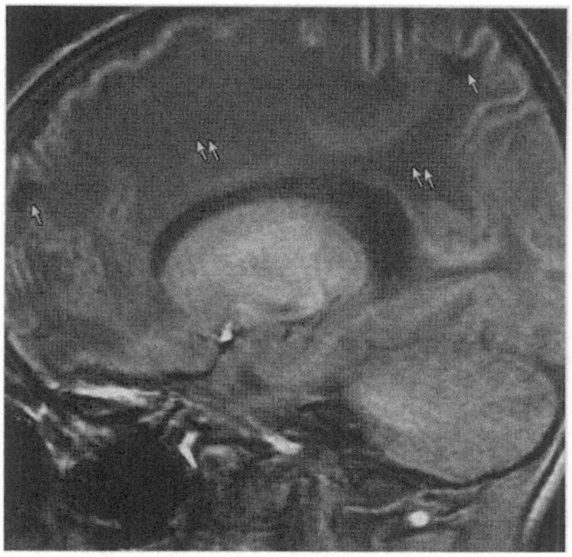

a

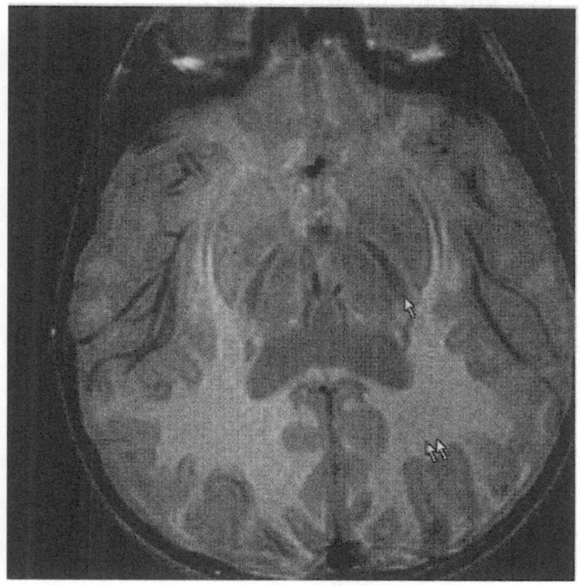

b

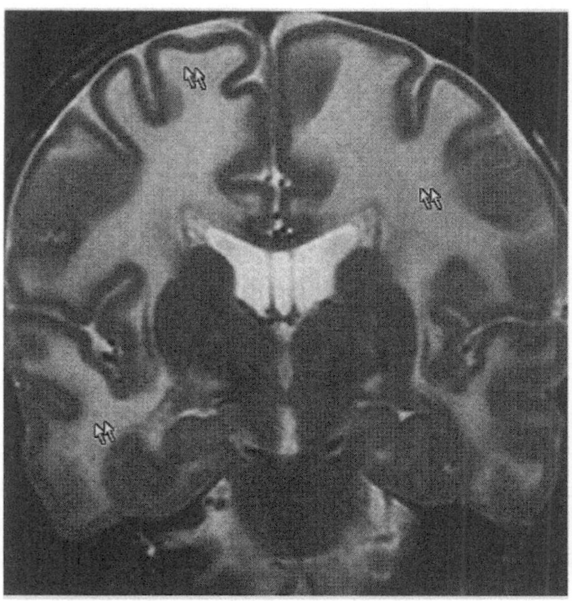

c

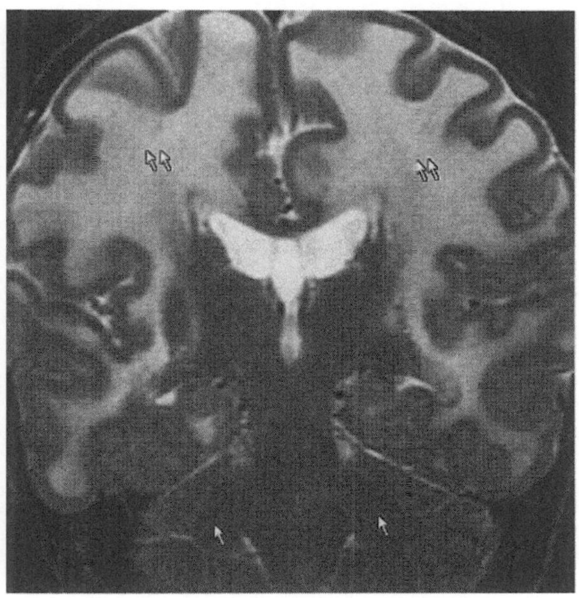

d

Cases 117 - 118: neurofibromatosis type I

Clinical and pathological features

- NF-I is the commonest (90%) form of Von Recklinghausen's disease.
- Its prevalence is around 1: 3000, with no sex predilection.
- Genetically it is an autosomal dominant disorder with high phenotypic expression and pleomorphism between patients.
- High rate of mutations explaining that a family history is found in less than 50%.
- The defectuous gene is located in the long arm of chromosome 17 in NF-I (and on chromosome 22 in NF-II). Production of the neurofibromin protein in the brain, particularly by the NF-I locus, is defective in such patients, causing a proliferation of various neoplastic processes.
- *Signs and diagnostic criteria* (presence of 2 or more of the 7 following features): café-au-lait spots, first manifestations appearing during the 1st year of life, over 6 in number, measuring 5 mm in children and 15 mm in adults; characteristic axillary freckling found in about 2/3 of the cases; 2 or more iris hamartomas (Lisch nodules) appearing in childhood, increasing in number with age and almost always present in adults, being almost diagnostic if bilateral; two neurofibromas or a plexiform neurofibroma; finding of an optic nerve glioma (associated with NF-I in 50% of cases); family history of NF-I; other bony features including: sphenoid wing dysplasia highly characteristic feature found in about 8% of cases (with associated pulsatile exophthalmos), lambdoid suture defect, kyphoscoliosis (50%), enlarged neural foramina or posterior scalloping of the vertebral bodies (non specific of NF-I) due to dural ectasia.
- Optic nerve gliomas is the most frequently found intracranial neoplasm, associated with NF-I in about 50-75% of the cases. These tumors are low grade juvenile pilocytic astrocytomas. Most (2/3) are asymptomatic at time of diagnosis during childhood (5-10 years). If bilateral, these anterior optic paths gliomas are highly suggestive of NF-I. Neuro-ophthalmological and MR follow-up are mandatory and conservative management is the rule, these optic gliomas behaving more as hamartomatous processes than aggressive neoplasms.

MR characteristic findings

Morphology

- Probably more than 50% of NF-I patients will show hamartomatous or neoplastic lesions on brain MR exams including the optic pathways.
- More than 75% of the patients exhibit abnormal foci of high SI on T2w in the brain, involving: the basal ganglia, internal capsules, splenium of corpus callosum, pons, midbrain, bilateral cerebellar white matter. These processes characteristically do not show mass effect neither associated edema, and do not enhance after administration of contrast agent. At times, lesions of the globi pallidi may be associated with a mild mass effect. The nature of these bright foci is still not definitely settled (hamartomas, atypical glial cells, abnormal myelination). They tend to increase in number during childhood but may even regress with advancing age. They are rarely seen after age of 20.
- Optic nerve gliomas, frequently bilateral, involve the orbital and cisternal optic nerves and may extend along the intracranial pathways at time of MR diagnosis. MR shows a more or less tubular nerve thickening with associated kinking and eventual presence of leptomeningeal cysts. Perioptic leptomeningeal spaces may be dilated and/or infiltrated by extra-axial development of the gliomatous process intermingled with an associated arachnoidal hyperplasia. Intracranial extension of the glioma may not be associated with enlargement of the optic canal. Enhancement is variable. High resolution MR imaging needs to be performed in the neuro-ocular plane (axial, SE-T2w) and coronal (SE-T1w and T2w) and completed with sagittal oblique cuts parallel to ON axis and/or optic canals (STIR).
- Pilocytic astrocytomas may involve also the chiasm and cisternal optic nerves and tracts ("mustache-like" appearance) with possible further infiltration of the posterior inferior thalami and basal ganglia. Usually solid infiltrating tumors, they have to be differentiated from hypothalamic gliomas with secondary extension to the chiasm. The latter tumors tend to be more aggressive in behavior, intraventricular, showing

cystic components, and more frequently enhancing.

- Low grade gliomas may be found elsewhere (case 119) in the brain, and particularly in the brainstem and periaqueductal area.
- Orbitocranial neurofibromas and plexiform neurofibromas may be found in NF-I in about 30% of the cases, and may extend through the neural foramina into the cranial cavity.

Signal intensity

- SE-T1w: moderately hyperintense foci (bilateral pallidal hamartomatous processes) or hypointense (in case of low grade glioma).
- SE-T2w or FLAIR: hyperintense foci (usually bilateral processes in globi pallidi, medial cerebellar white matter, optic radiations, may be found in medulla and brainstem or splenium and white matter).

Contrast enhancement

- May help to differentiate the numerous hamartomatous processes (may show in globi pallidi slight mass effect) from true low grade astrocytomas or other neoplastic conditions (mass effect present and eventual vasogenic edema).
- The so-called "hamartomatous" processes (bright SI foci on T2 or FLAIR) in the globi pallidi and cerebellar white matter as in other locations, do not show contrast enhancement. Scattered similar foci found mainly along the optic radiations, may show the same behavioral pattern.
- Pilocytic astrocytomas commonly show variable contrast enhancement.

Case 117: Von Recklinghausen type I neurofibromatosis

Fig. a: sagittal cut; GE - T1w; **Fig. b:** axial cut; SE - PDw

Multiple small foci of hyperintensity on PDw sequence (**b**) involving the right globus pallidus (**b**, ↑↑) outlined anteriorly by the anterior commissure (**b**,*) and involving the postero-lateral aspect of the right thalamus (**b**, ↑) and the antero-medial aspect of the left thalamus (**b**). No real mass effect is observed on the adjacent ventricles and no contrast enhancement is obtained. The globus pallidus high intensity lesions, generally bilateral, are considered as abnormal hamartomatous collections of glial cells, typical for NF I, and do not normally show contrast enhancement. Another manifestation of these lesions involve the splenium of the corpus callosum as a hypointense area on T1w (**a**, ↑).

In this patient, the medulla was enlarged with mass effect on the caudal aspect of the fourth ventricle (**a**, ↑↑) due to the presence of an intramedullary mass, hyperintense on T2w and hypointense on T1w (**a**, ↑). The presence of the mass effect demands close follow-up with Gadolinium infusion. In general, due to the potential of developing of such heterotopias into neoplasms, mainly when mass effect is associated and/or a contrast enhancement is observed, a close follow-up clinically as well as with MRI is highly recommended. In fact, precise histologic correlate of the hyperintense foci is not known in most cases, representing gliosis, hamartomas or even forms "frustes" of gliomas.

Case 118: Neurofibromatosis type I: glioma

Fig. c: coronal cut (512); SE - Pdw; **Fig. d:** sagittal cut (512); GE - T1w + Gd

Patient presenting with seizure activity. The MR exam showed bilateral globus pallidus high intensity hamartomas on T2w slightly hyperintense on T1w, in the typical location for NF I patient. The MR exam also showed a small well circumscribed homogeneous hyperintense lesion on PDw (**c**) cortically located, involving the posterior 3rd of the second right temporal gyrus (**c**, ↑↑) with mass effect on the adjacent cortex (**c**, ↑), but without perilesional edema.

The mass enhances homogeneously after Gadolinium injection indicating abnormality in the blood-brain barrier, suggesting an eventual glioma. A follow-up exam (**d**) 5 months later showed an increase of the mass volume with a posterior cystic formation (**d**, ↑↑) hypointense on T1w, the neural tumoral nodule still enhancing (**d**, ↑). The patient is candidate for surgery.

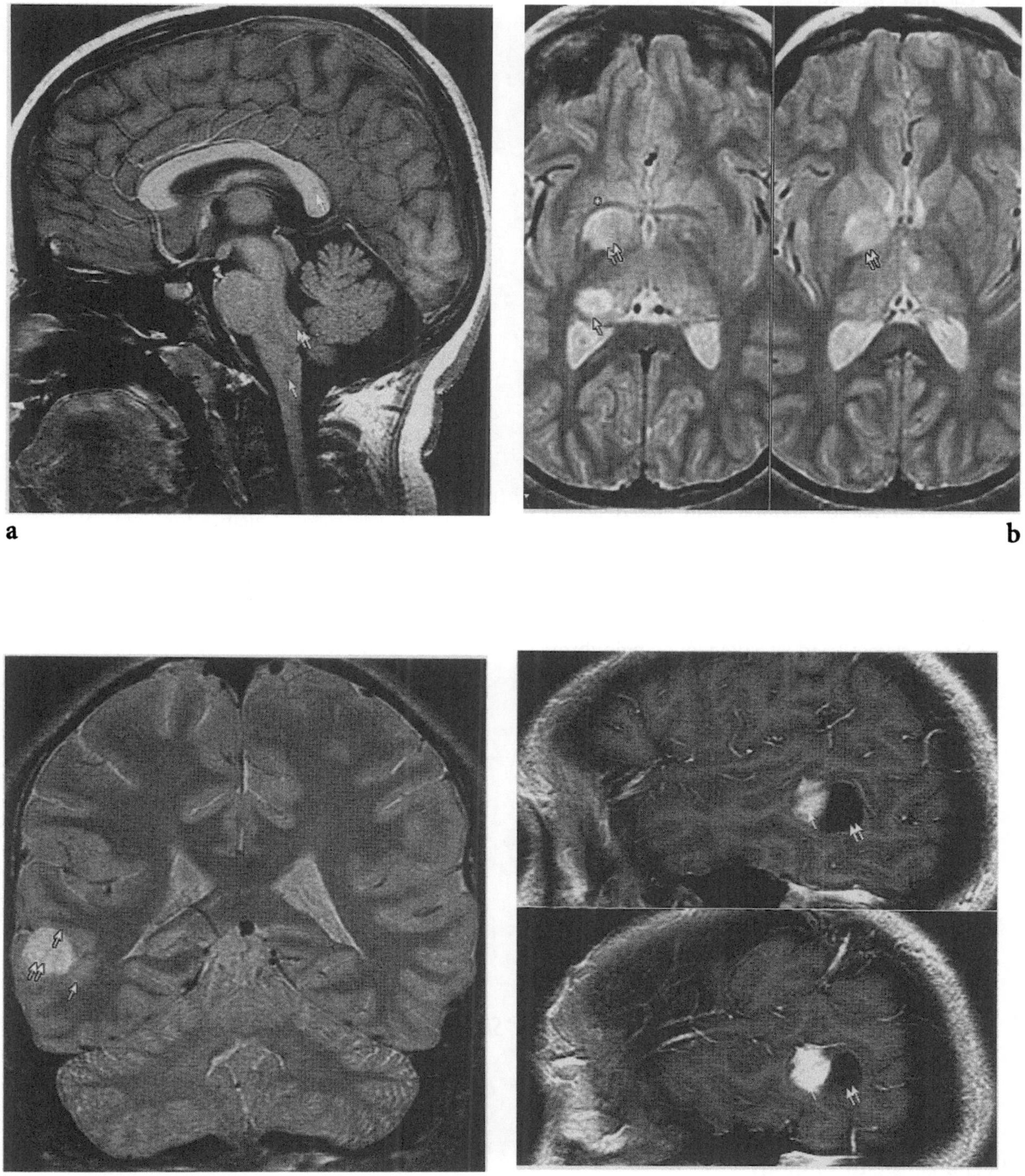

a b

c d

Cases 119, 15 (p. 80-81): tuberous sclerosis

Clinical and pathological features

- Bourneville's disease is a phakomatosis, including cutaneous, visceral and central nervous system manifestations.
- Genetically determined and familial, transmitted in an autosomal dominant mode with low penetrance.
- Results from chromosomal aberrations: most frequently, deletion of short arm of chromosome 16 (16p13), or aberration on long arm of chromosome 9 (9q34) or 11. Spontaneous mutations may occur in over 50%.
- Its incidence is evaluated as being 1: 100-150,000 patients or may even be as high as 1: 10-20,000. No sexual or racial preponderance.
- *Histology*: tubers are characteristic multiple hamartomatous nodules of disorganized cortical tissue containing atypical giant bizarre neurons and astrocytes. The subjacent white matter contains few myelinated fibers and fibrillary gliosis. Affected cortex does not show histologically the normal six-layered lamination.
- *Signs*: classic triad of cutaneous angiofibroma (90%) appearing between 1-5 years old, in the nasolabial folds or extending to the cheeks and the malar region (butterfly appearance); myoclonic seizures and hypsarrhythmia (80%) and mental deficiency (about 60%) of variable severity (mild in 1/3 of the patients). Other cutaneous signs are noted and some fairly characteristic: subungual angiofibromas (25%), hypomelanotic oval ash leaf macules in the trunk and extremities, lumbosacral or intertriginous subepidermal fibrosis (shagreen patches) (25%), non specific café-au-lait spots (25%), highly characteristic punctate areas of depigmentation ("confetti-like") in the axillary region. Retinal phakomas or hamartomas (50%), frequently bilateral, found near the optic cup causing leukocoria.

MR characteristic findings

Morphology

- Cortical-subcortical cerebral tubers (found infrequently in the cerebellum: 20%) are the typical findings. Their number is variable ranging from 2 or 3 to more than 20. Most tubers are found in the frontal lobes. Tubers tend to calcify with advancing age (about 50% in late childhood). They are frequently associated with expansion of the overlying cortex, appearing as enlarged gyri, in infants and neonates. The internal borders of the mass is ill-defined, although its outer limit appears almost clear-cut from the overlying cortex. Cortical tuber exceptionally degenerate. The tubers may appear as linked to the lateral ventricle by a linear or curvilinear band of heterotopic cells and demyelinated fibers with gliosis remnants from abnormal migration pattern.
- Subependymal hamartomas are usually found in the thalamostriate groove between the thalami and caudate nuclei, protruding in the ventricular cavities along the lateral walls ("candle-guttering" masses). Less frequently, these nodules are found in the frontal or temporal horns or even the 3rd or 4th ventricles. Calcifications are most frequently present in the 2nd or 3rd decades (CT).
- Glial nodules in the white matter: correspond to clusters of heterotopic atypical astrocytes that may partially calcify, associated with areas of hypomyelination.
- Giant cell astrocytoma (5-10%): slowly growing tumor (grade I), typically subependymal, tend to occur in the region of the foramen of Monro (obstructive hydrocephalus). The lesion usually does not infiltrate the adjacent parenchyma and does not show subarachnoid seeding. May be rarely an isolated finding (possible "forme fruste" of the disease).
- Miscellaneous: wedge-shaped cerebral or cerebellar calcifications within the cortex (CT), hyperintense cortical areas on T2w.

Signal intensity

- GE or SE-T1w: tubers are centrally hypointense to white matter, or iso-intense, but hyperintense in infants as compared to unmyelinated white matter.
- SE-T2w or FLAIR: hyperintense to white matter, mainly centrally.
- In case of associated calcifications of subependymal nodules or cortical tubers, GE-T2*w sequences are helpful for depicting these calcified portions due to magnetic susceptibility diffe-

rences with adjacent brain tissue. Such calcifications may show a hypointense SI on SE-T2w (subependymal nodules particularly) and may even appear as bright areas on T1w.

Contrast enhancement

- Cortical-subcortical tubers usually do not show contrast enhancement as hamartomatous processes. Mild peripheral enhancement may be observed using magnetization-transfer techniques.
- Subependymal hamartomatous nodules and glial nodules within the white matter do not enhance (unless undergoing exceptional degeneration). Enhancement of subependymal nodules must be followed up with MR in order to rule out an eventual transformation to a giant cell astrocytoma.
- Subependymal giant cell astrocytoma characteristically enhances.

Case 119: Tuberous sclerosis (Bourneville)

Fig. a: sagittal cut;	GE - T1w	**Fig. c:** axial cut;	SE - T2w
Fig. b: axial cut;	GE (3D) - T1w	**Fig. d:** axial cut;	FLAIR - T2w

Young patient presenting with epilepsy and showing typical nasolabial sebaceous adenomas. MR shows multiple small subependymal calcified nodules creating filling defects within the lateral ventricles and presenting hyperintense signal close to the white matter on T1w (**a**, **b**, ↑) and a marked hypointense signal on T2w (**c**, ↑↑) well delineated by the bright signal of the CSF. Note the presence of small subependymal hamartomas located just posterior to the foramina of Monro close to the head of the caudate nucleus, as commonly observed (**b**, left; **c**, *). The other common locations previously pointed out are the lateral borders of the bodies of the lateral ventricles and the anterior aspect of the ventricular atrium. MR shows on T2w (**c**, left, ↑) but more obviously on FLAIR sequence (**d**, ↑) multiple cortical-sub-cortical hyperintense lesions in both hemispheres corresponding to uncalcified cortical hamartomas (tubers). These cortical hamartomas are larger than the subependymal nodules and are best visualized on MRI than on CT, using T2w SE or FLAIR sequences. All these lesions show no contrast enhancement which is typical for the cortical lesions. In case of enhancement of the subependymal hamartomas, mainly those located at the foramina of Monro, a giant cell astrocytoma should be ruled out and followed up with MR until proven and treated. Note that enhancement is occasionally observed in uncomplicated hamartomas.

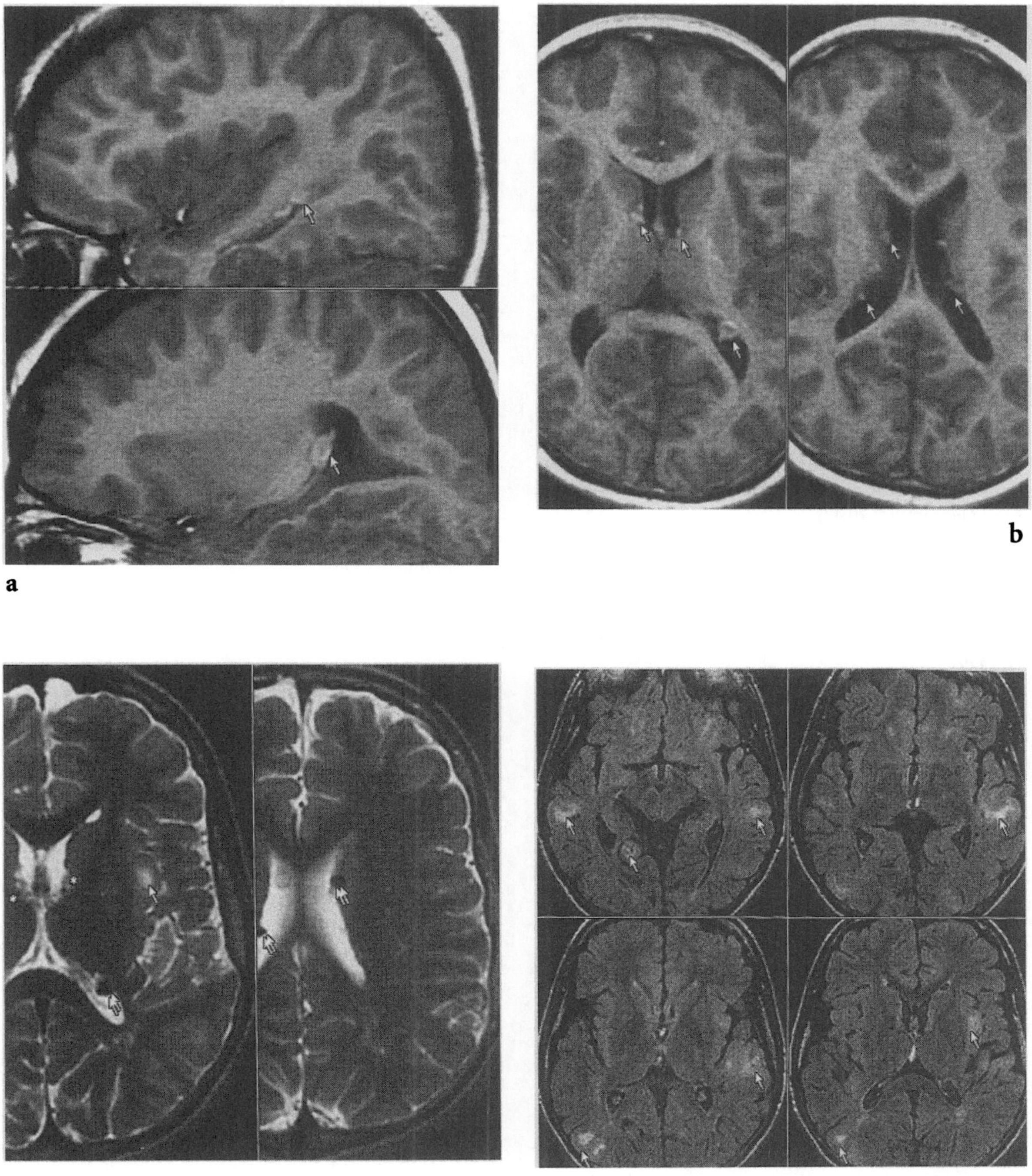

a b c d

Selected References

- Adachi M, Hosoya T, Haku T et al (1999) Evaluation of the Substantia Nigra in Patients with Parkinsonian Syndrome Accomplished Using Multishot Diffusion-Weighted MR Imaging. AJNR 20:1500-1506
- Baenziger O, Martin E, Steilin M et al (1993) Early pattern recognition in severe perinatal asphyxia: a prospective MRI study. Neuroradiology 35:437
- Baron Y, Barkovich A (1999) MR Imaging of Tuberous Sclerosis in Neonates and Young Infants. AJNR 20:907-916
- Barkovich AJ (1995) Pediatric Neuroimaging. 2nd Ed, Raven Press, New York
- Barkovich AJ, Chuang SH (1990) Unilateral megalencephaly: correlation of MR imaging and pathologic characteristics. Am J Neuroradiol 11:523
- Barkovich AJ, Kjos BO (1992) Schizencephaly: correlation of clinical findings with MR characteristics. Am J Neuroradiol 13:85
- Barkovich AJ, Newton TH (1989) MR of aqueductal stenosis: evidence of a broad spectrum of tectal distortion. Am J Neuroradiol 10:471
- Barkovich AJ, Norman D (1988) Absence of the septum pellucidum: a useful sign in the diagnosis of congenital brain malformations. Am J Neuroradiol 9:1107
- Barkovich AJ, Sargent SK (1995) Profound asphyxia in the premature infant: imaging findings. Am J Neuroradiol 16:1837
- Barkovich AJ, Chuang SH, Norman D (1987) MR of neuronal migration anomalies. Am J Neuroradiol 8:1009
- Barkovich AJ, Kjos BO, Norman D et al (1989) Revised classification of posterior fossa cysts ans cystike malformations based on the results of multiplanar MR imaging. Am J Neuroradiol 10:997
- Barkovich AJ, Westmark KD, Partridge JC et al (1995) Perinatal asphyxia: MR findings in the first 10 days. Am J Neuroradiol 16:427
- Barkovich AJ (1988) Abnormal vascular drainage in anomalies of neuronal migration. Am J Neuroradiol 9:939
- Barkovich AJ (1992) MR and CT evaluation of profound neonatal and infantile asphyxia. Am J Neuroradiol 13:959
- Barkovich AJ (1998) Neuroimaging Manifestations and classification of congenital muscular dystrophies. AJNR 19:1389-1396
- Barkovich J, Peck W (1997) MR of Zellweger Syndrome. AJNR 18:1163-1170
- Barkovich AJ, Hevner R, Guerrini R (1999) Syndromes of Bilateral Symmetrical Polymicrogyria. AJNR 20:1814-1821
- Bonawitz C, Castillo M, Chin C et al (1998) Usefulness of contrast material in MR of patients with neurofibromatosis Type 1. AJNR 19:541-546
- Brismar J, Brismar G, Gascon G et al (1990) Canavan disease: CT and MR imaging of the brain. Am J Neuroradiol 11:805
- Bydder GM (1997) Visualization of the germinal matrix with postnatal MR imaging very premature infants. Radiology 205(P): 210
- Cho WH, Seidenwurm D, Barkovich J (1999) Adult-onset neurologic dysfunction associated with cortical malformations. AJNR 20:1037-1043
- Christophe C, Clerex A, Blum D et al (1994) Early MR detection of cortical and subcortical hypoxic-ischemic encephalopathy in full-term-infants. Pediatr Radiol 24:581
- Demaerel P, Lagae L, Casaer P, Baert A (1998) MR of cerebellar cortical dysplasia. AJNR 19:984-986
- Diebler C, Dulac O (1983) Cephalocles: clinical and neuroradiological appearance: associated cerebral malformations. Neuroradiology 25:199
- Dietrich RB, Bradley WG, Zaragova EJ et al (1988) MR evaluation of early myelination patterns in normal and developmentally delayed infants. Am J Neuroradiol 9:69
- Dietrich RB, Vining EP, Taira RK et al (1990) Myelin disorders of childhood: correlation of MR findings and severity of neurological impairment. J Comput Assist Tomogr 14:693
- Farell K, Chuang S, Becker LE (1984) Computed tomography in Alexander's disease. Ann Neurol 15:605
- Fitz CR (1983) Holoprosencephaly and related entities. Neuroradiology 25:225
- Flodmark O, Lupton B, Li D et al (1989) MR imaging of periventricular leukomalacia in childhood. Am J Neuroradiol 10:111
- Friede RL (1989) Developmental neuropathology. 2nd ed, Springer-Verlag, Berlin
- Hager BC, Dyme IZ, Guertin SR et al (1991) Linear nervus sebaceous syndrome: megalencephaly and heterotopic gray matter. Pediatr Neurol 7:45
- Hart MN, Malamed N, Ellis WG (1972) The Dandy-Walker syndrome: a clinicopathological study based on 28 cases. Neurology 22:771
- Joubert M, Eisenring J, Robb JP et al (1969) Familial agenesis of the cerebellar vermis. Neurology 19:813
- Journel H, Roussey M, Gaudon Y et al (1987) Magnetic resonance imaging in Pelizaeus-Merzbacher disease. Neuroradiology 29:403
- Juottonen K, Laakso M, Partanen K, Soininen H (1999) Comparative MR Analysis of the Entorhinal Cortex and Hippocampus in Diagnosing Alzheimer Disease. AJNR 20:139-144
- Kendall BE (1983) Dysgenesis of the corpus callosum. Neuroradiology 25:239
- Kier EL, Truwit CL (1996) The normal and abnormal genu of the corpus callosum: an evolutionary, embryologic, anatomic, and MR analysis. Am J Neuroradiol 17:1631
- Kim TS, Kim IO, Kim WS, Choi YS et al (1997) MR of childhood metachromatic leukodystrophy. AJNR 18:733-738
- Kjos BO, Umansky R, Barkovich AJ (1990) Brain MR imaging in children with developmental retardation of unknown cause: results in 76 cases. Am J Neuroradiol 11:1035
- Kumar AJ, Rosenbaum AE, Naidu S et al (1987) Adrenoleukodystrophy: correlating MR imaging with CT. Radiology 165:497

- Kwan E, Drace J, Enzmann D (1984) Specific CT findings in Krabble disease. Am J Roentgenol 143:665
- MacDonald JT, Stauffer AE, Heiroff K (1984) Adrenoleukodystrophy: early frontal lobe involvement on computed tomography. J Comput Assist Tomogr 8(1):128
- Manefle C, Rochiccioli P (1979) CT of septo-optic dysplasia. Am J Neuroradiol 133:1157
- Manefle C, Sevely A (1982) Neuroradiological study of holoprosencephalies. J Neuroradiol 9:15
- Marler JR, O'Neill BP, Forbes GS et al (1983) Adrenoleukodystrophy (ALD): clinical and CT features of a childhood variant. Neurology 33:1203
- Menkes J, Curran JG (1994) Clinical and magnetic resonance imaging correlates in children with extrapyramidal cerebral palsy. Am J Neuroradiol 15:451
- Moser HW, Moser AE, Singh I et al (1984) Adrenoleukodystrophy: survey of 303 cases: biochemistry, diagnosis, and therapy. Ann Neurol 16:628
- Naidich TP, McLone DG, Fulling KH (1983) The chiari II malformations. IV. The hind-brain deformity. Neuroradilogy 25:179
- Osborn RE, Byrd SE, Naidich TP et al (1988) MR imaging of neuronal migrational disorders. Am J Neuroradiol 9:1101
- Parisi JE, Collins GH, Kim RC et al (1983) Prenatal symmetrical thalamic degeneration with flexion spasticity at birth. Ann Neurol 13:94
- Parrish ML, Roessman U, Levinsohn MW (1979) Agenesis of the corpus callosum: a study of the frequency of associated malformations. Ann Neurol 6:349
- Pasternak JF, Predey TA, Mikhael MA (1991) Neonatal asphyxia: vulnerability of basal ganglia, thalamus, and brainstem. Pediatr Neurol 7:147
- Paul KS, Lye RH, Strang FA et al (1983) Arnold-Chiari malformations: review of 71 cases. J Neurosurg 58:183
- Penner MW, Li KC, Gebarski SS et al (1987) MR imaging of Pelizaeus-Merzbacher disease. J Comput Assist Tomogr 11:591
- Pojunsa K, Williams AL, Daniels DL et al (1984) Syringomyelia and hydromyelia: magnetic resonance evaluation. Radiology 153:579
- Raybaud C (1982) Cystic malformations of the posterior fossa: abnormalities associated with the development of the roof of the fourth ventricule and adjacent meningeal structures. J Neuroradiology 9:103
- Raybaud C (1983) Destructive lesions of the brain. Neuroradiology 25:265
- Rutheford MA, Pennock JM, Schwiesco JE et al (1995) Hypoxic ischaemic encephalopathy: early magnetic resonance imaging findings and their evolution. Neuropediatrics 26:183
- Sawada H, Udaka F, Seiru N et al (1990) MRI demonstration of cortical laminar necrosis and delayed white matter injury in anoxic encephalopathy. Neuroradiology 32:319
- Schouman-Clays E, Henry-Fuegeas MC, Roset F et al (1993) Periventricular leukomalacia: correlation between MR imaging and autopsy findings during the first 2 months of life. Radiology 189:59
- Silverstein AM, Hirsch DK, Trobe JD et al (1990) MR imaging of the brain in five members of a family with Pelizaeus-Merzbacher disease. Am J Neuroradiol 11:495
- Takahashi S, Higano S, Ishii K et al (1993) Hypoxic brain damage: cortical laminar necrosis and delayed changes in white matter at sequential MR imaging. Radiology 189:449
- Takashima S, Tanaka K (1978) Development of the cerebrovascular architecture and its relationship to periventricular leukomalacia. Arch Neurol 35:11
- Valk J, van der Knaap MS (1989) Canavan's disease. In: Valk J, van der Knapp MS (eds) Magnetic resonance of myelin, myelination, and myelin disorders. Springer-Verlag, Berlin
- Valk J, van der Knaap MS (1989) Pelizaeus-Merzbacher disease. In: Valk J, van der Knapp MS (eds) Magnetic resonance of myelin, myelination, and myelin disorders. Springer-Verlag, Berlin
- Van der Knaap MS, Valk J (1990) MR imaging of the various stages of normal myelination during the first year of life. Neuroradiology 31:459
- Van der Knaap MS, Valk J (1989) The reflection of histology in MR imaging of Pelizaeus-Merzbacher disease. Am J Neuroradiol 10:99
- Van der Knaap MS, Valk J, Bakker CJ et al (1991) Myelination as an expression of the functional maturity of the brain. Dev Med Child Neurol 33:849
- Yakovlev PI, Wadsworth RC (1946) Schizencephalies: a study of the congenital clefts in the cerebral mantle. I. Clefts with fused lips. Neuropathol Exp Neurol 5:116

Part II

Skull base and face

Imaging technique and normal anatomy

The neuro-ocular plane (N.O.P.) remains the most accurate cephalic orientation for the investigation and biometrical study as well as for follow-up of diseases of the eyes and the optic nerves in the axial or in the coronal planes. The same applies to the screening and exploration of diseases involving the face and skull base, due to the close parallelism of NOP to the Frankfurt-Virchow anthropological baseline. Several works demonstrated these widely adopted results.

The neuro-ocular plane (NOP)

In his writings on the projections of the head, Paul Broca stated: "*the direction of gaze is the only characteristic of the living by which it may be determined that the head is horizontal. When a man is standing and his visual axis is horizontal, he is in his natural attitude*". Broca's horizontal plane approaches the alveolar-condylar plane defined by him because "*the natural position of the head could not be determined on the skull*".

One century after his death, CT demonstrates in a study of the optic nerve in papilledema the validity of Broca's opinion about horizontality of the visual pathways. The neuro-ocular plane, best materializes this optimal cephalic orientation for scanning of patients with visual complaints. Cabanis defined it as the "*plane passing through the lenses, the optic nerve heads and the optic canals with the patient maintaining primary gaze*" (Cabanis et al; Salvolini et al); Sagittal oblique cuts oriented along the optic nerve axis or the optic canal if needed, may also be of great help in special conditions, such as in tumoral processes arising in or about the intraorbital optic nerves and sheaths, or within the optic canal.

The visual pathway, in their orbito-cranial route, maintain a roughly horizontal orientation extending from the eyes to reach the calcarine fissure. For such a reason, the neuro-ocular plane taken as the cephalic reference plane, may be considered as the most accurate compromise for studying the brain and the visual pathways and the most accurate for studying the orbital optic nerves. In order to visualize the optic nerves, authors proposed reference planes calculated with respect to the widely used orbito-meatal radiological baseline, or according to the anthropologic baseline (Unsold et al). The mean angle between NOP and Frankfurt-Virchow plane (FVP) averages –7°. To remind you, the orbito-meatal plane, which is the classic radiologic reference plane (World Federation of Neurology, 1962) is angled at approximately +10° with respect the FVP plane. Exhaustive works about the relation of the orbital axis plane with several craniofacial reference lines have been reported (Saban et al; Fenart et al), showing the close parallelism of the NOP plane to the alveolo-condylar plane, the palatine plane and the prosthion-opisthion cranial reference line (Fenart et al). This explains the accuracy of this cephalic orientation as the reference plane for screening patients with orbito-maxillofacial diseases.

Anatomical correlations

NOP orientation provides the optimal condition for CT or MR explorations of the intraorbital structures. The partial volume effect on the optic nerves is reduced to the minimum. Visualization of the chiasm in the chiasmal cistern using CT or MR axial cuts in the NOP orientation, is determined by the individual anatomy of this region, the height of the sella turcica and the type and the variable obliquity of the chiasm.

Considering the retrochiasmal visual pathways, they beneficiate also of this cephalic orientation, due to the gross parallelism of NOP with respect to the orientation of the temporal horns of the lateral ventricles. However, parts of the visual pathways depart somewhat from the ideal horizontal plane: the chiasm, for example, due to its variable position in the chiasmal cistern, and the calcarine fissure in its ascending portion.

The angulation of the visual pathways with respect to the base of the skull changes with age due to the well established occipital descent. Once maturation is completed the angle between the visual pathways and the skull remains constant. Therefore, the angular relationship between NOP and FVP becomes constant in the adult.

Radiological correlations

Practically, the NOP is easily determined and allows rapid alignment of the head in the gantry. The line joining the centers of the pupils, which is precisely horizontal in the normal, determines the anterior portion of the plane. The second point is found about 33 mm above the tragion.

External cutaneous landmarks, materialized by the acantho-meatal line, are helpful to orient the patient's head in clinical practice. Bony landmarks materialized by the prosthion-opisthion line are also available and may be used on a scout-view.

The availability of a reference atlas (Tamraz et al; Cabanis et al) of cross-sectional anatomy of the head oriented in the neuro-ocular plane and reporting the major anatomical correlations observed, helps routine interpretation of brain slices. All cuts are annoted using the international nomenclature (Parisiensa nomina anatomica).

Oculo-orbital topometry

The major advantage of NOP orientation, as shown by Cabanis, is the ability to perform biometrical measures which are of great help in the evaluation of various normal, developmental and pathological conditions.

It is interesting to note that NOP provides a meridional cut through the globe and through the horizontal recti muscles, from the annulus of Zinn to their tendinous attachment on the eyeball. Cuts inferior and superior to NOP show the vertical recti muscles.

Several distances are defined and measured, some of which are routinely used:

- the external bicanthal plane (PBCE) constitutes the first reference line joining both external orbital rims in the NOP (average 97,5 mm +/- 4,4);
- the interocular distance (DIO) joining the center of both lenses (average 63,7 mm +/- 3,6);
- the maximal interplanar distance (DIPM) separating the two internal orbital walls (average 28,7 mm +/- 2,6 mm);
- the maximal anterior posterior axial length of the eyeball (LAM) (average 24,1 mm +/- 1, on the right as well as on the left side);
- the pre-PBCE segment of the eyeball (average 16 mm +/- 1,9);
- the retro-PBCE segment of the eyeball, corresponding to the complementary of the last value;
- the transverse diameter of the intraorbital optic nerve (DNO), measured at its middle portion (average 3,5 mm +/- 0,5);

Several indices are also determined and used:

- the neuro-ocular index (INO), corresponding to DNO/LAM (average 14,8 +/- 0,74);
- the oculo-orbital index or index of exophthalmy, corresponding to PBCE segment/LAM (average 65, 44). This means that 65% of the eyeball in the adults is located anterior to the external bicanthal plane in normal conditions;
- interocular distance index (IDIO) corresponding to DIO/DBCE (average 65, 3). This indicates that the interpupillary distance corresponds to the 2/3 of the external intercanthal distance.

Selected References

- Bilaniuk LT (1990) Magnetic resonance imaging: orbital anatomy. In: Newton TH, Bilaniuk LT (eds) Modern Neuroradiology. Vol 4, Radiology of the Eye and Orbit. Raven Press, New York
- Bregeat P, Cabanis EA, Iba-Zizen MT, Tamraz J, Stoffels C, Alfonso JM, Lopez A, Le Bihan D (1986) Nerf optique et résonance magnétique. Bull Soc Opht France 5:627-631
- Breslau J, Dalley RW, Tsuruda JS et al (1995) Phased array surface coil MR of the orbits and optic nerves. Am J Neuroradiol 16:1247
- Cabanis EA, Salvolini U, Rodallec A, Menichelli F, Pasquini U, Bonnin P (1978) Computed tomography of the optic nerve. Part II: Size and shape modifications in papilledema. J Comput Assist Tomogr 2:150-155
- Cabanis EA, Haut J, Iba-Zizen MT, Palacin JJ, Porret G, Laplaud O, Guillaumat L (1980) Exophtalmométrie tomodensitomtrique et biométrie. TDM Oculo-orbitaire. Bull Soc Opht France 1 80:62-66
- Cabanis EA, Pineau H, Iba-Zizen MT, Coin JL, Newman N, Salvolini U (1981) CT Scanning in the "Neuro-Ocular Plane": the optic pathways as a "new" cephalic plane. Neuro-Ophtalmology, 1 4 237-252
- Cabanis EA, Iba-Zizen MT, Pineau H, Tamraz J (1982) Le plan Neuro-oculaire en tomodensitométrie, détermination d'un "nouveau" plan horizontal de référence céphalique orienté selon les voies visuelles. Biométrie Humaine 17:21-48
- Cabanis EA, Iba-Zizen MT, Tamraz J, Muzac S et al (1982) Topométrie Oculo-orbitaire, Aspect dynamique normal et pathologique. Rapport An Soc Opht France, Paris; Bull Soc Opht France 57-95
- Cabanis EA, Carteret M, Pineau JC, Massin M, Abanou A Lopez A, Iba-Zizen MT (1992) IRM orbitaire dynamique: un protocole biométrique de référence. Ophtalmologie 6:174-178
- Cabanis EA, Iba-Zizen MT, Lopez A (1992) Neuro-Imagerie. In: Godde-Jolly D, Dufier JL (eds) Ophtalmologie pédiatrique. Masson, Paris
- Cabanis EA, Bourgeois H, Iba-Zizen MT et coll (1996) L'imagerie en ophtalmologie. Rapport Annuel SFO, Masson, Paris
- Cavezian R, Cabanis EA, Pasquet G, Iba-Zizen MT, Le Bihan D, Tamraz J, Roger B (1986) Imagerie par Résonance Magnétique. Principes physiques, applications biomédicales actuelles, perspectives à l'étage maxillo-facial. Actualités Odonto-Stomatologiques 154:219-232
- Cosnard G, Pasquet G, Tamraz J, Cavezian R, Iba-Zizen MT, Cabanis EA (1987) Imagerie par Résonance Magnétique et massif facial. Encyclopédie Médico-Chirurgicale Stomatologie 6 p 22010 D 50, 2
- Coussement A, Faure C, Coussement-Beylard N (1990) Repères et mesures en imagerie médicale. 4th éd, L'Expansion Scientifique Française
- Davis PC et al (1991) Pituitary adenoma: correlation of half-dose gadolinium-enhanced MR imaging with surgical findings in 26 patients. Radiology 180:779
- DeLano MC, Yun FY, Zinreich SJ (1996) Relationship of the optic nerve to the posterior paranasal sinuses: a CT anatomic study. AJNR 17:669-675
- Doyle A (1990) Optic chiasm position on MR images. AJNR 11:553-555
- Doyon D, Laval-Jeantet M, Halimi PH, Cabanis EA, Frija J (1988) Tomodensitométrie. Masson, Paris
- Doyon D, Cabanis EA, Laval-Jeanter M, Frija J, Pariente D, Idy-Peretti I (1994) Imagerie Par Résonance Magnétique. Masson, Paris
- Duke-Elder S, Wybar KC (1961) The anatomy of the visual system. In: Duke-Elder S (ed) System of ophthalmogy. Vol 2, Kimpton, London
- Han JS, Huss RG, Benson JE et al (1984) MR imaging of the skull base. J Comput Assist Tomogr 8(5):944
- Harnsberger HR, Osborn AG (1991) Differential diagnosis of head and neck lesions based on their space of origin. I. The suprahyoid part of the neck. Am J Roentgenol 157:147
- Helms CA, Kaplan P (1990) Diagnostic imaging of the temporomandibular joint: recommendation for use of the various techniques. AJR 154:319-322
- Higgins CB, Hricak H, Helms CA (1992) Magnetic Resonance Imaging of the body. Raven Press, New York
- Hilal SK, Trokel SL (1997) Computerized tomography of the orbit using thin sections. Semin Roentgenol 12:137-147
- Hoffmann KT, Hosten N, Lemke AJ, Sander B et al (1998) Septum orbitale: high-resolution MR in orbital anatomy. AJNR 19:91-94
- Jones JS, Jakobiec FA (1979) Diseases of the orbit. Harper & Row, New York
- Lang J (1987) Clinical anatomy of the head neurocranium, orbit, cranio-cervical region. Springer Verlag, Berlin, Heidelberg, New York
- Langer BG, Mafee MF, Pollock S et al (1987) MRI of the normal orbit and optic pathway. Radiol Clin North Am 25:429
- Lasjaunias PL (1981) Craniofacial and upper cervical arteries. William and Wilkins, Baltimore
- Lemmerling MM, Stambuk HE, Mancuso AA et al (1997) CT of the normal suspensory ligaments of the ossicles in the middle ear. AJNR 18:471-477
- Lo WWM, Daniels DL, Chakeres DW et al (1997) The endolymphatic duct and sac. AJNR 18:881-887
- Madeline LA, Elster AD (1995) Postnatal development of the central skull base: normal variants. Radiology 196:757
- Mafee MF (1996) Eye and orbit. In: Som PM, Curtin Hd (eds) Head and neck imaging. 3rd ed, Mosby, St Louis
- Maucuso A, Hanafee W (1985) Computed tomography and magnetic resonance of the head and neck. William and Wilkins, Baltimore

- McCord CD (1985) Current trends in orbitral decompression. Ophthalmology 92:21
- Merran S, Cabanis EA, Chevrot A, Cosnard G, Godefroy G, Iba-Zizen MT, Jardin C, Pharaboz C, Pradel J, Pujol J, Tamraz J, Hureau J (1989) Radio-anatomie en coupes TDM-IRM. Pradel, Paris
- Palacios E, Valvassori GE, Shannon M, Reed CF (1990) Magnetic resonance of the temporomandibular joint. Georg Thieme Verlag, Stuttgart, New York
- Reese AB (1976) Tumors of the eye. Harper & Row, New York
- Rhoton Al, Buza R (1975) Microsurgical anatomy of the jugular foramen. J Neurosurg 42:541
- Salamon G, Peretti-Viton P, Faure J, Delannoy L, Fan L, Girard N, Andre P (1992) Imagerie de l'oeil, de l'oreille et du cerveau. Springer-Verlag, Paris
- Salvolini U, Cabanis EA, Rodallec A (1978) Computed tomography of the optic nerve. Part I, Normal results. J Comput Ass Tomogr 2:141-149
- Schenk JF, Hart HR, Foster TH et al (1985) Improved imaging of the orbit at 1.5T with surface coils. A J Neuroradiol 6:193
- Schamlbrock P, Pruski J, Sun L et al (1995) Phased array RF coils for high-resolution imaging of the inner ear and the brain stem. J Comput Assist Tomogr 19:8
- Schmalbrock P, Brogan M, Chakeres DW et al (1993) Optimization of submilimeter resolution MR imaging methods for the inner ear. J Magn Reson Imaging 3:451
- Som PM, Bergeron T (1991) Head and Neck imaging. Boston, New York, St. Louis
- Som PM et al (1984) Parapharyngeal space masses: an updated protocol based upon 104 cases. Radiology 153(1):149
- Stark DD, Bradley WG (1998) Magnetic Resonance Imaging. Mosby, St. Louis, Washington, Toronto
- Steenks MH, Bleys RL, Witkamp TD (1994) Temporomandibular joint structures: a comparison between anatomic and magnetic resonance findings in a sagittal and an angulated plane. J Orofacial Pain 8:120-135
- Swartz JD, Daniels DL, Harnsberger HR et al (1995) The temporal bone. AJNR 16:2081-2084
- Takano K, Utsunomiya H, Ono H et al (1999) Normal Development of the Pituitary Gland: Assessment with Three-dimensional MR Volumetry. AJNR 20:312-315
- Tamraz J, Iba-Zizen MT, Cabanis EA (1984) Atlas d'anatomie céphalique dans le plan Neuro-Oculaire. J Fr Ophtalmol 7(5):371-379
- Tamraz J, Iba-Zizen MT, Atiyeh M, Cabanis EA (1985) Atlas d'anatomie céphalique dans le plan neuro-oculaire. Bull Soc Fr Ophtal 85:853-857
- Tamraz J (1983) Atlas d'anatomie céphalique dans le plan neuro-oculaire. Thèse Médecine. Schering, Paris
- Tamraz J, Iba-Zizen MT, Cabanis EA (1987) Exploration neuroradiologique du chiasma et des bandelettes optiques. Encyclopédie Médico-Chirurgicale. Ophtalmologie 3-C, 1-21008 A20, 1-2
- Tamraz J, Iba-Zizen MT, Cabanis EA (1988) Magnetic resonance imaging of the eyes and the optic pathways. In: Gouaze A, Salamon G (eds) Brain anatomy and magnetic resonance imaging. Springer Verlag, Berlin, pp 71-83
- Tamraz J, Saban R, Reperant J, Cabanis EA (1990) Définition d'un plan de référence céphalique en IRM: le plan chiasmato-commissural. CR Académie des Sciences, Paris, 311, III:115-121
- Tamraz J (1991) Morphométrie de l'encéphale par résonance magnétique: applications à la pathologie chromosomique humaine à l'anatomie comparée et à la tératologie. Université René Descartes, Thèse de Sciences, Paris
- Tamraz J, Saban R, Reperant J, Cabanis EA (1991) A new cephalic reference plane for use with magnetic resonance imaging: the chiasmato-commisural plane. Surg Radiol Anat; 13:197-201
- Tamraz J (1994) Neuroradiologic Investigation of the visual system using MRI. J Clin Neurophysiology 5:500-508
- Tamraz JC, Outin-Tamraz C, Saban R (1999) MR imaging anatomy of the optic pathways. Radiol Clin North Am 1:1-36
- Tamraz J, Comair Y (2000) Atlas of regional anatomy of the brain using MRI. Springer-Verlag, Berlin Heidelberg, New York
- Teresi LM, Kolin E, Lufkin RB et al (1987) MR imaging of the intraparotid facial nerve: normal anatomy and pathology. Am J Neuroradiol 8:253
- Teresi LM, Lufkin RB, Wortham DG et al (1987) MR imaging of the intratemporal facial nerve using surface coils. Am J Neuroradiol 8:49-54
- Thibault F, Halimi P, Bely N et al (1993) Internal architecture of the parotid gland at MR imaging: facial nerve or ductal system? Radiology 188:701
- Tien RD, Hesselink JR, Chu PK et al (1991) Improved detection and delineation of head and neck lesions with fat suppression spin-echo MR imaging. Am J Neuroradiol 12:19
- Tien R, Kucharczyk J, Bassette J et al (1992) MR imaging of the pituitary gland in infants and children: changes in size, shape, and MR signal with growth and development. AJR 153:1151-1154
- Tsunoda A, Okuda O, Sato K (1997) MR height of the pituitary gland as a function of age and sex: especially physiological hypertrophy in adolescence. AJNR 18:551-554
- Tu AS, Geyer CA, Mancall AC, Baker RA (1998) The Buccal Space: A Doorway for Percutaneous CT-Guided Biopsy of the Parapharyngeal Region. AJNR 19:728 -731
- Unsold R, Newton TH, Hoyt (1980) CT examination technique of the optic nerve. J Comput Ass 4:560-563
- Valvassori G, Mafee M, Carter B (1995) Imaging of the head and neck. Thieme, New York
- Van Damme W, Kosman P, Wachenhein C (1977) A standard method for computed tomography of orbits. Neuroradiology 13:139-140
- Van Tassel P, Lee YY (1991) Gd-DTPA enhanced MR for detecting intracranial extension of sinonasal malignancies. J Comput Assist Tomogr 15:387
- Veillon F (1991) Imagerie de l'oreille. Flammarion, Paris

- Vignaud J, Clay C, Lasjaunias P, Fisch C, Salvolini U (1986) Anatomie et imagerie de la loge caverneuse. EMC Paris, Ophtalmologie, 11, 21006 C 10
- Wagner AL, Murtagh FR, Hazlett KS, Arrington JA (1997) Measurement of the normal optic chiasm on coronal MR images. AJNR 18:723-726
- Wolff E (1968) Anatomy of the eye and orbit, WB Saunders, Philadelphia
- Yousem DM (1992) Dashed hopes for MR imaging of the head and neck: the power of the needle. Radiology 184:25
- Zinreich JS, Kennedy Dw, Rosenbaum AE et al (1987) Paranasal sinuses: CT imaging inquirements for endoscopic surgery. Radiology 163:769
- Zinreich JS, Mattox De, Kennedy DW et al (1988) Concha bullosa: CT evaluation. J Comput Assist Tomogr 12:778

Part II

Skull base and face

1 - Anterior skull base and orbito-facial region

a - Orbital region
b - Maxillo-dento-facial region

Orbital region

Case 1: ocular metastasis

Clinical and pathological features

- Adult, mainly females, in fifth and sixth decades.
- Left eye predilection and bilateral in 1/4 cases.
- Common location: choroids at the posterior pole.
- Decreased or loss of vision.
- History of primary cancer.
- Primary sites: breast in women and lung in men.
- Rapid growth and CNS spread.

MR characteristic findings

Morphology

- May be confused with uveal melanoma.
- Involve the posterior pole via short posterior ciliary arteries.
- Usually flat or mottled unlike melanomas.
- May be multiple and bilateral in 1/3 of cases.
- Retina may be involved with extended retinal detachment.
- Intratumoral necrosis possible (adenocarcinoma).

Signal intensity

- SE-T1w: isointense or slightly hyperintense to vitreous (but hyperintense in case of associated subacute hemorrhage or mucinous content in adenocarcinomas).
- SE-PDw: hyperintense to vitreous.
- SE-T2w: variable, hyperintense (usually) or hypointense (mimics melanoma).

Contrast enhancement

- Needed to separate the tumor from the retinal detachment.
- Enhancement common, more frequently and intensely than melanomas.

Cas 1: Intraocular metastasis (breast)

Fig. a: axial cut; SE - T1w + Gd

Fig. b: coronal cut; SE - T1w + Gd

Sudden loss of hemifield of vision. MR shows a choroïdal metastasis from breast carcinoma involving the left eye. The lesion, isointense to vitreous on T1w shows marked enhancement after Gadolinium injection (**a**, ↑,**b**,*). The choroidal mass involves the nasal compartment and more anteriorly, the cilliary body. The metastasis is associated with non enhancing sero-hemorragic detachment (**a**, ↑), shown on the posterior pole of the eyeball. The major differential diagnosis is malignant melanoma which generally shows marked hyperintensity on T1w due to the presence of paramagnetic melanin.

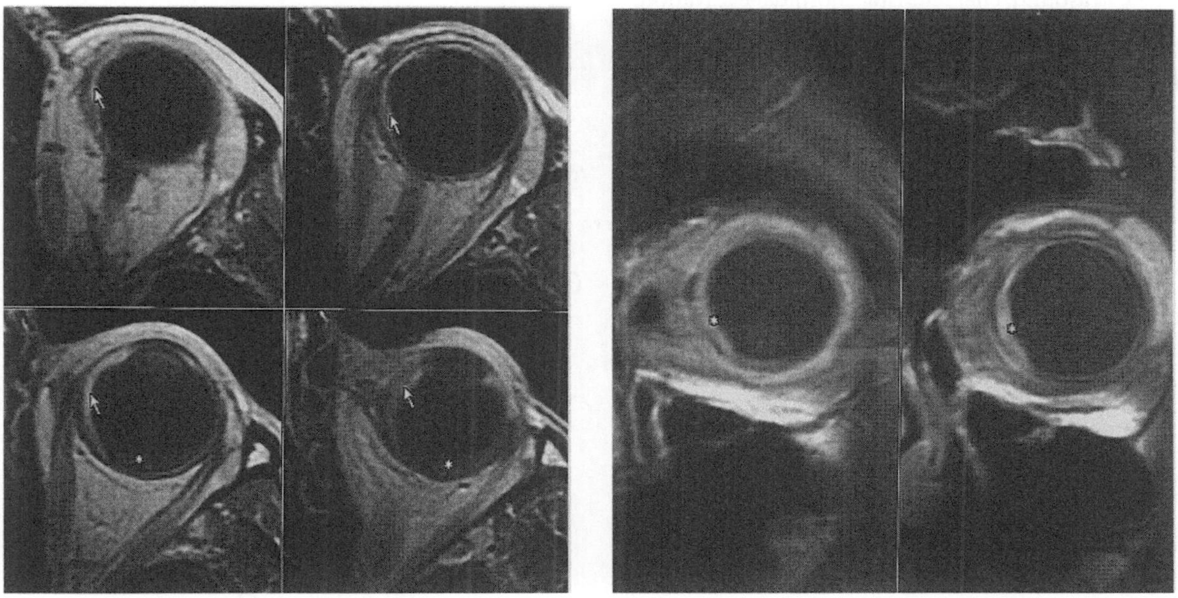

a **b**

Case 2: Uveal melanoma

Clinical and pathological features

- Adults in fourth to seventh decades.
- Most common primary intraocular malignant tumor in adults.
- Location: mainly choroids or ciliary body and iris.
- Decreased vision and visual field defects.
- Slow growth, exophytic extension through sclera.
- Hematogeneous metastasis to liver.
- Ultrasound: characteristic choroidal excavation.

MR characteristic findings

Morphology

- Intravitreous focal mass sharply delineated, frequently biconvexe.
- Biconvexe, polypoid or even crescentic.
- Common associated retinal detachment (separated on lateral gaze acquisitions).
- Note eventual extensions to optic cup or through the sclera in 10-15%.
- May be simulated (MR) by various malignant or benign lesions.

Signal intensity

- Depends on the importance of pigmentation: melanin.
- Characteristic shortening of relaxation times could be due to paramagnetic effects of free radicals present in melanin (SI related to amount of pigmentation).
- Most commonly (melanotic melanomas):
  SE-T1w: hyperintense to vitreous (iso-SI in amelanotic);
  SE-T2w: hypointense to vitreous, or even isointense (hyper-SI in amelanotic).
- SI in melanotic may be similar to early subacute choroidal effusion or hemorrhage.

Contrast enhancement

- Usually mild to moderate enhancement.
- No enhancement of the associated retinal detachment.

Case 2: Uveal melanoma

Fig. c: axial cut; SE - PDw

Fig. d: axial cut; SE - T2w

Middle age patient with visual loss, presenting an intraocular polypoid mass in the temporal aspect of the right globe extending into the vitreous. It shows hypointense signal on PDw (c, *) becoming more hypointense on T2w (d, ↑), with areas of intratumoral necrosis (d,*) and is associated to subretinal fluid collection (d, ↑↑). Note that the tumor is limited to the intraocular space showing no exophytic extension and is well delineated on long TR sequences by the hypointense signal of the sclera. Note that the orbital fat remains slightly hyperintense on FSE sequences. There were no posterior extensions of the tumor to the optic nerve head shown in the neuro-ocular plane, surrounded by the high signal of perioptic subarachnoid space. Melanotic melanomas show shorter T1 and T2 than other malignant tumors due to relaxation enhancement related to the degree of pigmentation. Note also that the distinction between melanoma and subretinal fluid collection is made by MRI without iv contrast as opposed to the experience on CT.

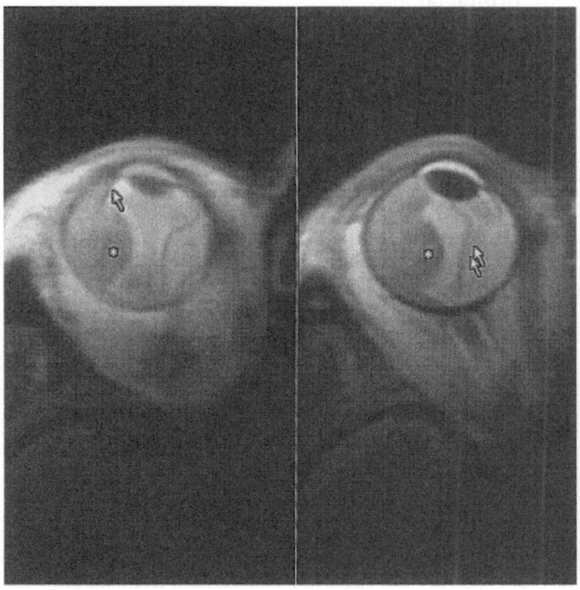

c

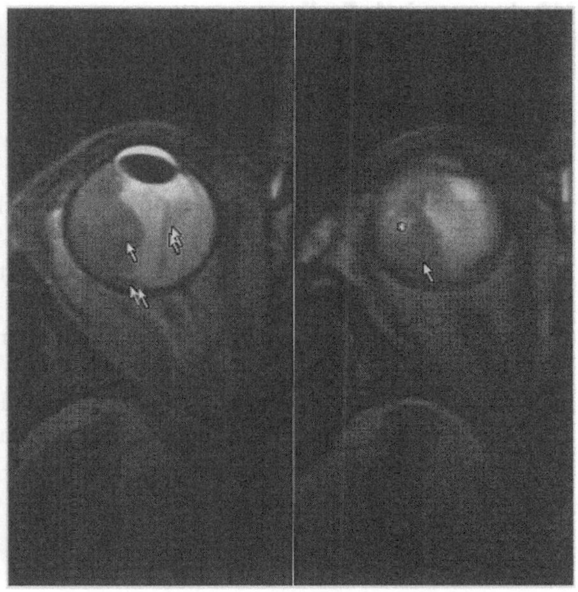

d

Case 3: Perioptic meningioma

Clinical and pathological features

- Arises from optic nerve sheaths.
- Second to sixth decades in middle aged females (80%).
- Corresponds to about 5% of orbital tumors.
- May be bilateral in case of NFR.
- May be aggressive in children, more commonly in NFR.
- *Signs:* transient obscuration, decreased vision, papilledema, proptosis.

MR characteristic findings

Morphology

- Tumor confined to the nerve sheaths.
- Slow growing, perioptic, or crescentric, or flat and tubular.
- Spiculated and irregular margins as they grow out the dura.
- May be calcified partly (MR contrast needed) or completely (CT).
- "Tram-track" sign (axial in NOP), and "Doughnut or Target" sign (coronal), but non specific (perineuritis, inflammatory pseudotumor, granulomatous, metastasis, malignant hemopathy).

MR signal intensity

- "Doughnut or Target" sign (coronal cuts).
- SE-T1w: isointense to orbital muscle.
- SE-PDw: isointense or slightly hyperintense to the ON, may disclose characteristic peripheral spicules transgressing the dura on high resolution images.
- SE-T2w: isointense or slightly hyperintense to orbital fat.

Contrast enhancement

- Most tumors enhance strongly and homogeneously.
- Best evaluated on T1w with fat suppression.
- Perineural enhancement common but non specific.
- "En plaques meningiomas", apical and intracanalicular extensions are best delineated on Gd enhancement and fat suppression.
- Gd infusion is mandatory in patients showing ON enlargement that could be a sheath meningioma.

Case 3: Optic nerve sheath meningioma

Fig. a: axial cut; SE - T1w

Fig. b: axial cut; SE - PDw

Fig. c: coronal cut; SE - PDw

Fig. d: coronal cut; SE - T2w

Middle aged woman presenting loss of vision with proptosis as well as swelling of the optic disc in the ocular fundus. MR shows intraconal mass in the left orbit with isosignal on T1w (**a**, ↑). Note that the outer aspect of the mass is slightly spiculated on T1w (**a**) and PDw (**b, c**) sequences due to dural transgression. The mass does not extend posteriorly to involve the orbital apex and the orbital canal (**a**, ↑↑). On high resolution PDw sequence, the tumor (**b**,*) is well distinguished from the optic nerve (**b**,*). The intraorbital optic nerve is well defined (**b**, ↑; **c**,*). On T2w sequences, the perioptic sheath tumor appears isointense to muscle (**d**,*). After iv contrast the mass enhances markedly and homogeneously (not shown). Note the relative proptosis of the left eyeball, well seen in the neuro-ocular plane. Note that the head coil has been used in this case to better evaluate the intracanalicular optic nerve in order to rule out an eventual intracranial extension on the optic nerve sheath tumor, particularly after iv contrast. The head coil is preferred when evaluating an optic nerve tumor, but using small FOV, thin slices, and when possible high resolution MR images.

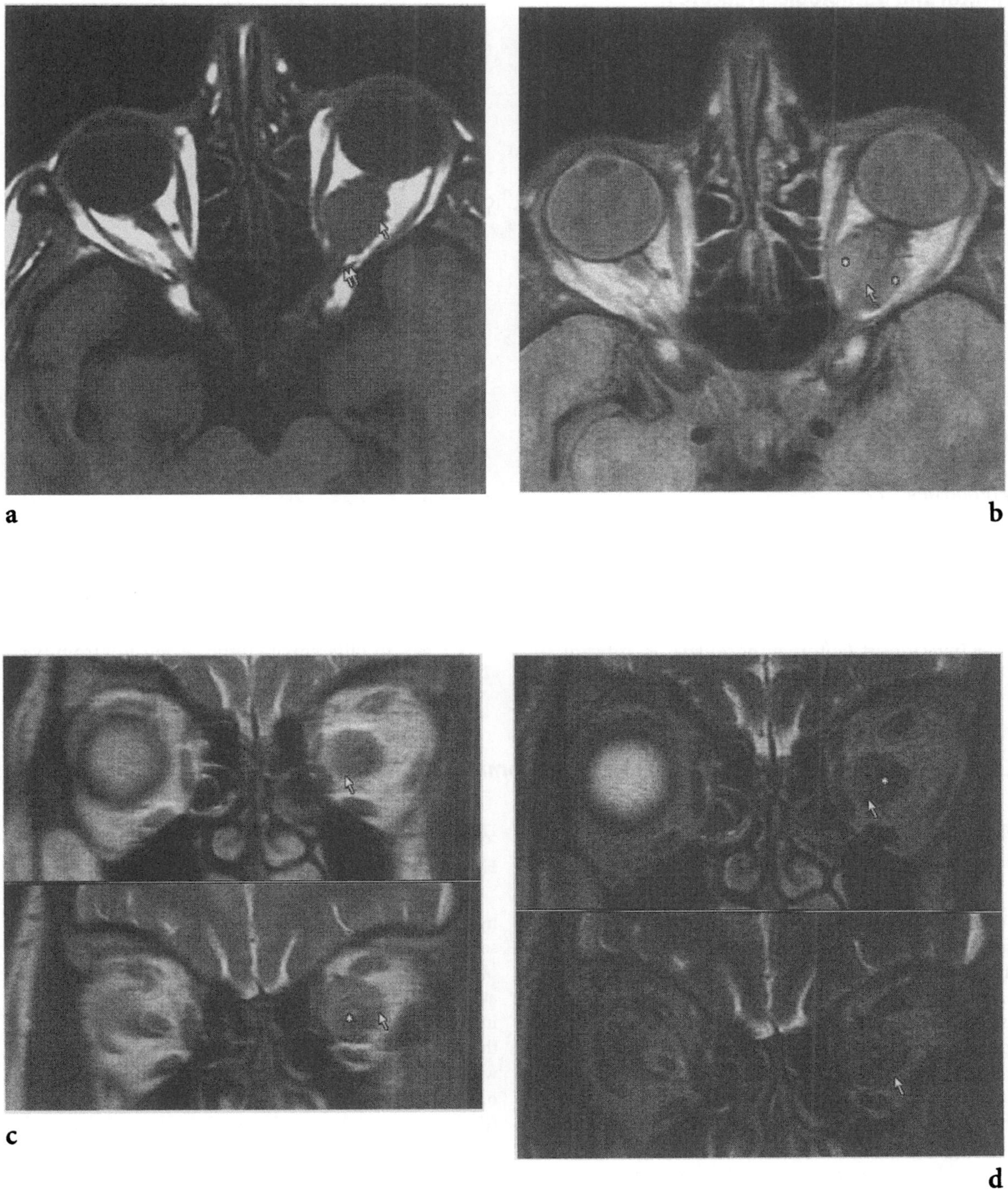
a
b
c
d

Case 4: Bilateral optic gliomas (neurofibromatosis type I, child)

Clinical and pathological features

- Child in first or second decade, peak at 5 years, female predominance.
- Variable behavior: usually benign and slow growing over years.
- Necessitate decades of routine clinical and MR follow up (NFR).
- Frequent association (50-70%) to Von Recklinghausen type I.
- *Signs:* visual loss and/or strabismus, proptosis, optic atrophy.
- Necessitate decades of clinical and MRI follow-up.
- *Histology:* low grade pilocytic astrocytoma.

MR characteristic findings

Morphology

- Tubular enlargement of the optic nerves.
- May be fusiform or even eccentric.
- May show kinking and cystic formation (CSF entrapment).
- Invasion of leptomeningeal spaces, intermixing glial and arachnoid cells.
- Associated arachnoid hyperplasia mimics meningioma.
- Frequent extension to the intracanalicular (even with normal optic canals).
- Cisternal optic nerves and chiasm, optic tracts and radiations, may be involved.

Signal intensity

- SE-T1w: isointense to gray matter.
- SE-PDw: isointense to gray matter.
- SE-T2w: isointense to white matter, homogeneous (mainly tubular forms).
- Heterogeneous on T2w (large fusiform): peripheral high SI due to arachnoid hyperplasia and/or cystic subarachnoid entrapped spaces.

Contrast enhancement

- Unusual in bilateral ON gliomas of the NFR type.
- May be peripheral due to extraneural growth of the glioma.
- Helps to differentiate tumor extension from peripheral arachnoid hyperplasia (no enhancement).

Case 4: Bilateral optic nerves gliomas (neurofibromatosis type I, child)

Fig. a: sagittal cut;	GE - T1w	**Fig. c:** sagittal oblique cut;	SE - PDw
Fig. b: axial cut;	SE - T2w	**Fig. d:** sagittal oblique cut;	SE - T2w

Ten years old child presenting a strabismus and an optic atrophy associated with neurofibromatosis. MR exam showed caracteristic hamartomatous lesions in the globus pallidus, slightly hyperintense on T1w (**a**, right, ↑↑↑) associated to hypointense small masse involving the splenium (**a**, left, ↑) and medulla with mass effect on the fourth ventricle (**a**, left, ↑↑), that could match with low grade astrocytomas necessitating follow-up exams. Note that some hamartomas are also depicted in the brainstem with high signal on T2w (**b**,*). The orbital exploration shows enlargement of both intraorbital optic nerves, well shown in the axial neuro-ocular plane (**b**), as well as on the sagittal oblique cuts, oriented along the optic nerve axis (**c**, **d**). The exam performed using the head coil showed that the enlargement of the optic nerve is readily distinguished from prominent perioptic CSF spaces showing very high signal intensity close to the vitreous on T2w sequences (**b**, left, ↑↑; **d**, up, **). The intraorbital optic nerves showing a hyposignal close brain white matter on T2w, present a somehow fusiform enlargement (**c**, ↑↑). The sagittal oblique cuts passing through the optic canals showed that at the left side, the optic nerve tumor extends into the intracranial cavity (**c**, up, ↑) with a fusiform enlargement of the cisternal portion of the optic nerve at this level but without enlargement of the optic canal itself. The right optic nerve tumor did not involve the cisternal optic nerve (**d**,*). Note the kinking of the intraorbital optic nerve well demonstrated on the sagittal oblique cuts. Note that the apparent enlargement of the perioptic subarachnoid space does not always correspond to CSF trapping but could be due to peritumoral reactive arachnoidal hyperplasia which is rather caracteristic of patients with neurofibromatosis. In NF type I patient, optic gliomas are generally low grade pilocytic astrocytomas, favouring long term MRI and clinical follow-up and conservative treatment.

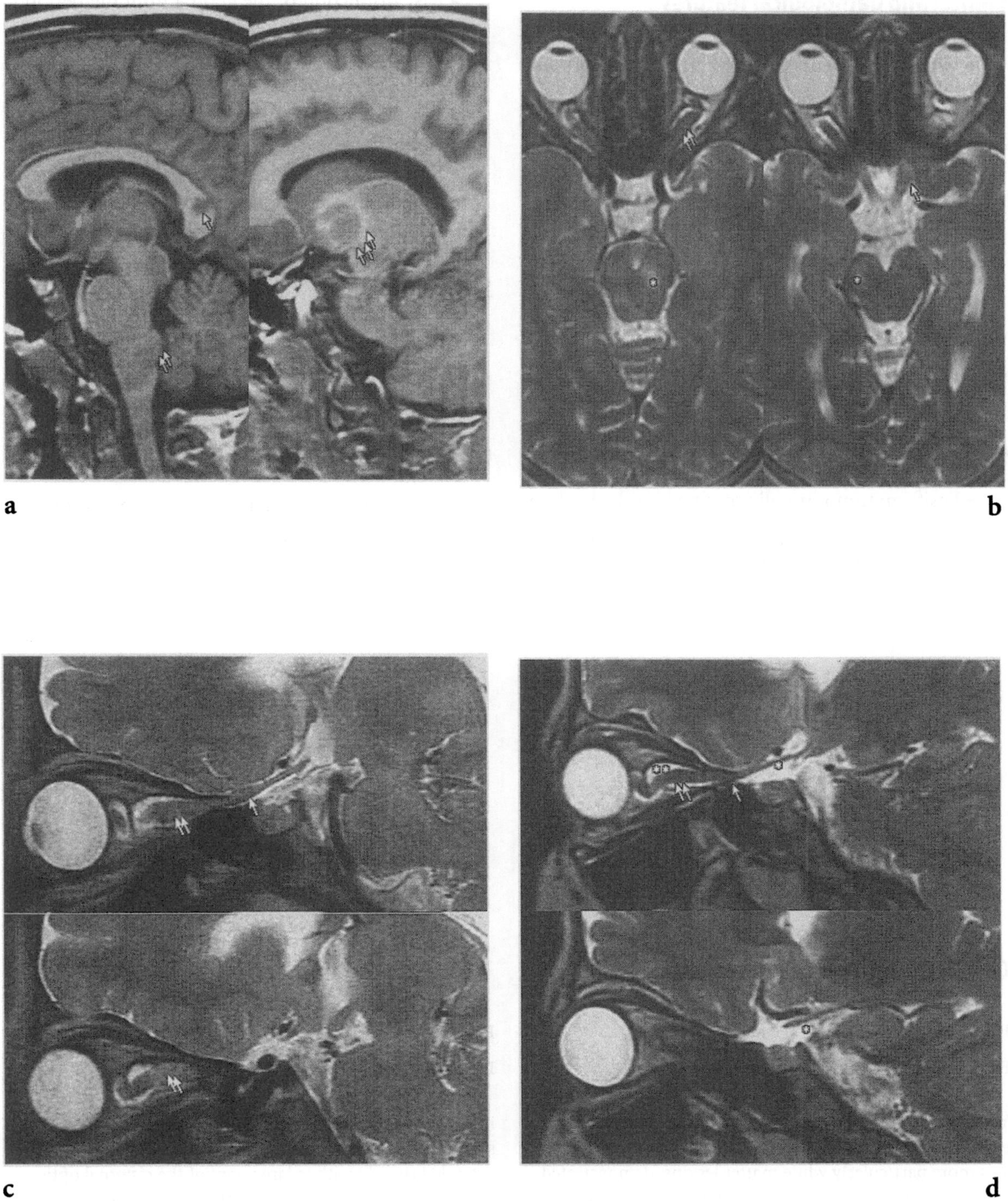

a b

c d

Case 5: Optic nerve glioma (adult)

Clinical and pathological features

- Usually different type and not associated with NFR.
- Aggressive behavior: high grade astrocytomas or glioblastomas.
- Poor long term prognosis, chiasmal and hypothalamic involvement.
- *Signs:* visual loss, variable visual field defects, optic atrophy, non replaceable proptosis, decreased eye movement.

MR characteristic findings

Morphology

- Fusiform tumor usually confined inside the dura.
- Posterior margin tapers; it is virtually impossible to separate from normal nerve.
- Easy distinction of the glioma from the perioptic leptomeningeal spaces.
- Possible peritumoral arachnoidal hyperplasia (more typical of NFR).
- Possible areas of intratumoral necrosis and hemorrhage.
- Intracranial optochiasmatic extension to be evaluated (sagittal oblique cuts parallel to optic nerve axis).

Signal intensity

- SE-T1w: isointense SI to white matter.
- SE-T2w and STIR: hyperintense SI to white matter.

Contrast enhancement

- Variable contrast enhancement of optic glioma.
- Useful for assessment of intracanalicular extension (50%).

Case 5: Optic nerve glioma (adult)

Fig. a: coronal cut (512); SE - T2w

Fig. b: coronal cut (512); SE - T1w + Gd

Fig. c: axial cut; SE - T2w

Fig. d: sagittal oblique cut; STIR - T2w

Young adult presenting visual loss and papilledema. MR exam performed using head coil to rule out eventual brain and more particularly white matter lesions, demonstrated an enlargement of the left intraorbital optic nerve (**c**, **d**) hyperintense on PDw (**c**) as well as on T2w STIR (**d**) sequence. The optic nerve fusiform mass extends into the cranial cavity with enlargement of the optic canal (**d**, down, ≠), involving the intracanalicular and cisternal optic nerve (**a**, ↑↑; ↑↑↑) well shown on coronal high resolution T2w images (**a**). The tumor involving the optic nerve is precisely macroscopically delinated on STIR sequence (**d**, ↑) in the sagittal oblique plane, extending through the optic canal, extended (**b**, ↑↑) with enlargement of the cisternal optic nerve (**d**, ↑↑↑), respecting the optic chiasm of normal thickness and normal hypointense signal isointense to brain white matter (**d**,*). After contrast iv, the tumor enhances in its orbital and cisternal portion (**b**) on T1w. This patient did not show signs of neurofibromatosis and is suspected of high grade glioma.

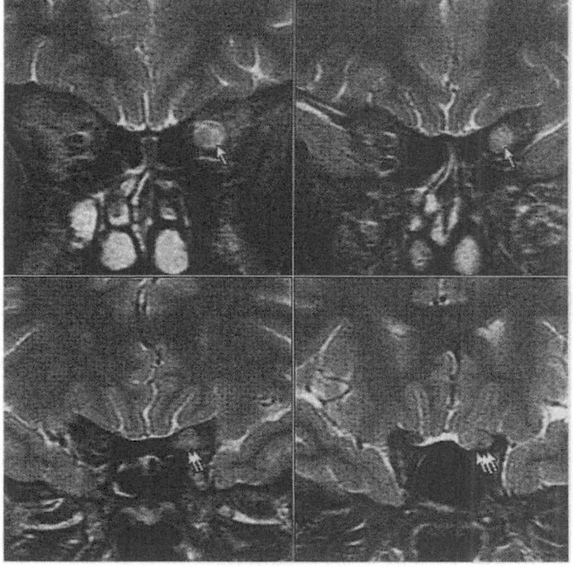
a

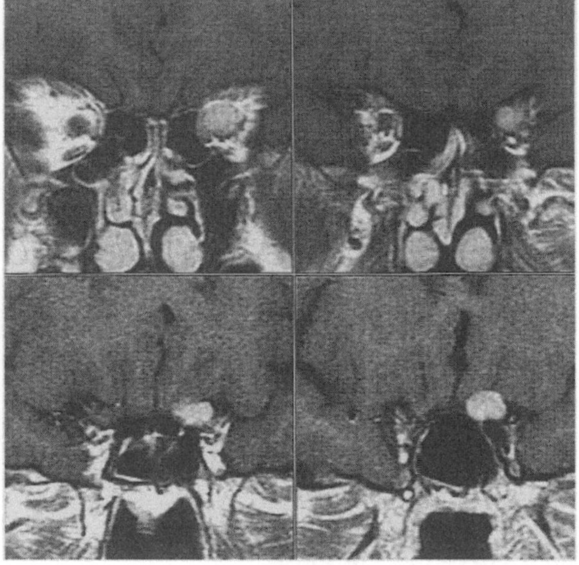
b

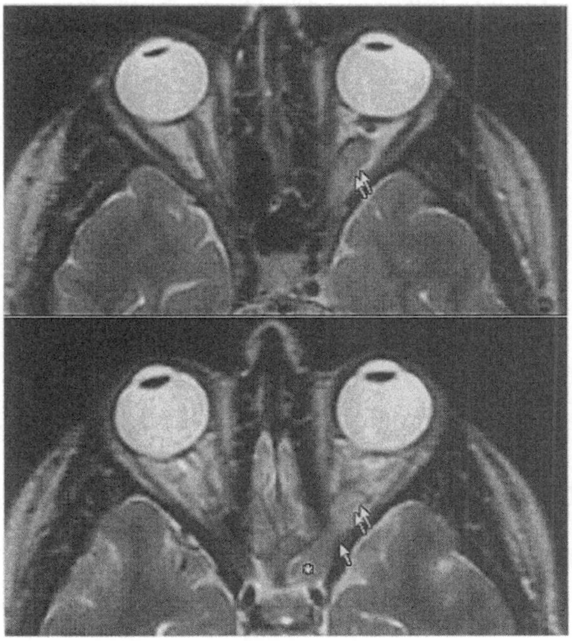
c

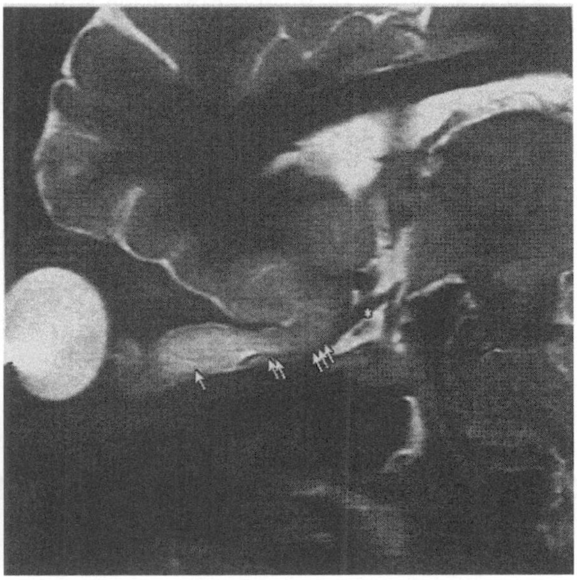
d

Case 6: Inflammatory orbital pseudotumor

Clinical and pathological features

- It is a non specific idiopathic process.
- Most commonly unilateral, in adults.
- Most commonly acute: pain, proptosis, ocular motor disturbance with diplopia.
- May be chronic: orbital mass syndrome.
- One of the most frequent cause of unilateral exophthalmos (apart Basedow's).
- Many clinical forms: scleritis, myositis, dacryoadenitis, orbital fat infiltrate.
- Special entity: Tolosa-Hunt syndrome.
- If bilateral: rule out systemic diseases, such collagen-vascular, Wegener, trichinella, or lymphoid.
- Dramatic response in acute forms to high dose steroid therapy.

MR characteristic findings

Morphology

- Various forms: restricted or diffuse infiltrate.
- Irregular uveal-scleral thickening.
- Orbital muscle enlargement involving typically the tendinous insertion.
- Isolated lacrymal gland enlargement (frequent).
- Ill defined infiltration of the orbital fat "dirty orbit".
- Perioptic involvement similar to lymphoproliferative diseases.
- Sclerosing pseudotumor subtype: in chronic form or as an end stage of the acute form, showing a diffuse mass "freezing the intraorbital structures".
- Tolosa-Hunt subtype: involvement of the orbital apex and the cavernous sinus.

Signal intensity

- Shows some specificity, due to fibro-collagenous composition.
- SE-T1w: isointense to orbital muscles.
- SE-T2w: isointense or hypointense to muscles and orbital fat.
- Lymphoma, myeloma, sarcoidosis or Grave's disease may mimic pseudotumor.
- Usually different SI behavior from metastatic carcinoma (hyperintense on T2).

Contrast enhancement

- Diffuse orbital or focal (muscle, sclera, lacrymal gland, optic nerve sheaths, or cavernous sinus) enhancement.

Case 6: Inflammatory "pseudo-tumor" of the orbit

Fig. a: axial cut; SE - PDw

Fig. b: axial cut; SE - T2w

Patient presenting inflammatory orbital syndrome. Mass infiltrating intraconal and extraconal spaces, presenting a relative hyposignal on PDw sequence (**a**, ↑) and a hypointense signal on heavily T2w (**b**, ↑), surrounding the intraorbital optic nerve and sheaths. Note the bilateral exophthalmos. The lesion showed enhancement after Gadolinium infusion. The hypointensity on T2w probably relates to the presence of desmoplastic fibrosis. Note that some malignant orbital lesions such as lymphoma and myeloma may show similar signal intensity patterns to pseudotumor.

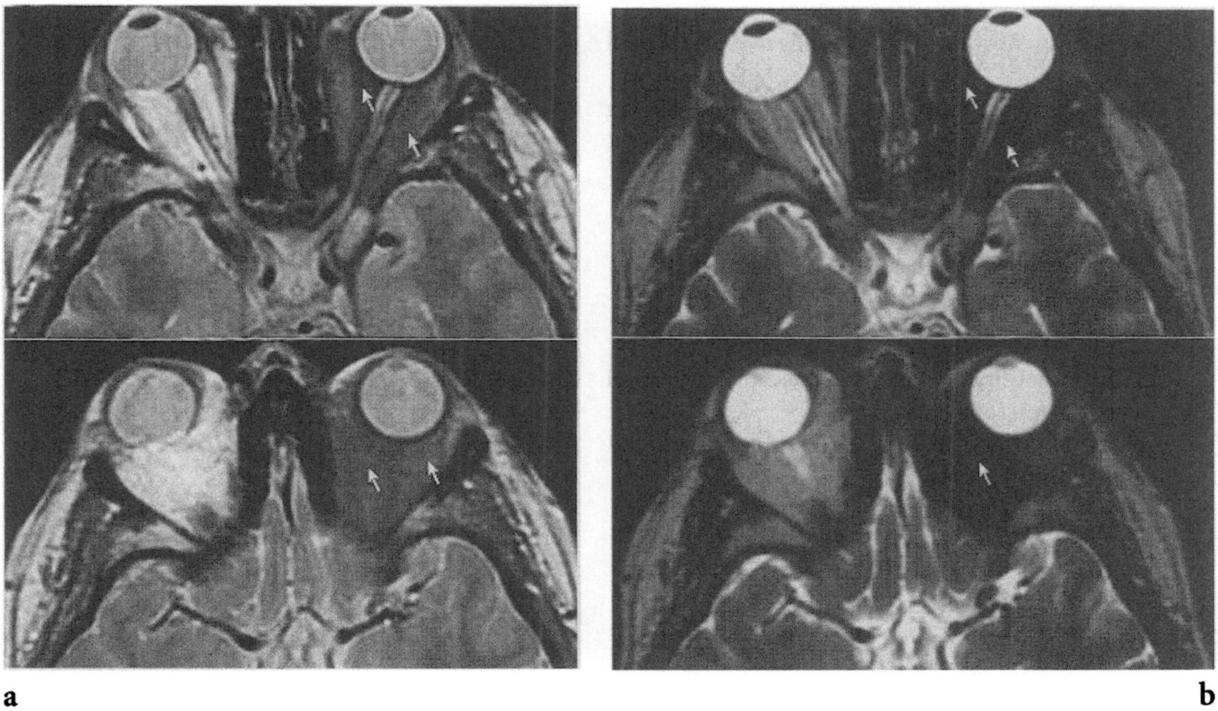
a
b

Case 7: Intraorbital cavernous hemangioma

Clinical and pathological features

- Most frequent orbital tumor in adults.
- Peak incidence in middle age: second to fourth decades.
- Predilection for women (2/3).
- Most frequently unilateral and usually retrobulbar.
- Encapsulated venous malformation (surrounded by a dense fibrous tissue).
- History of slowly progressive proptosis (75%) over years.
- Painless, not pulsating, usually without visual loss.
- No modification in size (proptosis) during Valsalva's maneuver or procubitus CT or MR exams as with orbital varix.

MR characteristic findings

Morphology

- Soft, smooth and well delineated oval or round mass, commonly intraconal.
- Variable size, may displace the optic nerve, and usually do not deform the eyeball.
- Homogeneous and encapsulated, comprising stagnant blood (no flow signs).
- Exceptional evidence of phleboliths.
- Does not involve or extend to the orbital apex.
- May become lobulated or triangular in shape when huge.
- Relationships to optic nerve and globe to be defined (prior to surgical cure).

Signal intensity

- SE-T1w: isointense to orbital muscles, usually homogeneous.
- SE-PDw: slightly hyperintense to orbital muscles.
- SE-T2w: hyperintense to muscles and fat, intermediate to vitreous.
- May show areas of bright SI on T1 due to eventual intravascular thrombosis.
- May show hypointense internal septations on T2 (possible "spongy" appearance).

Contrast enhancement

- Intense early enhancement, could be inhomogeneous (on dynamic imaging).

Case 7: Intraorbital hemangioma

Fig. c: coronal cut; SE - T1w, and axial cut SE - T1w

Fig. d: sagittal oblique cut; S.E. - T2w with fat suppression

Patient presenting orbital pain. MR shows a small well limited rounded mass, developed in the inferior lateral retrobulbar space, showing iso-signal to intraorbital muscles on T1w (c, ↑) and a hyperintense signal on T2w with fat supression technique (**d**, down, ↑). The mass is independent from the optic nerve head and exerces no mass effect on the latter (**d**,*). After contrast injection, the mass showed marked enhancement.

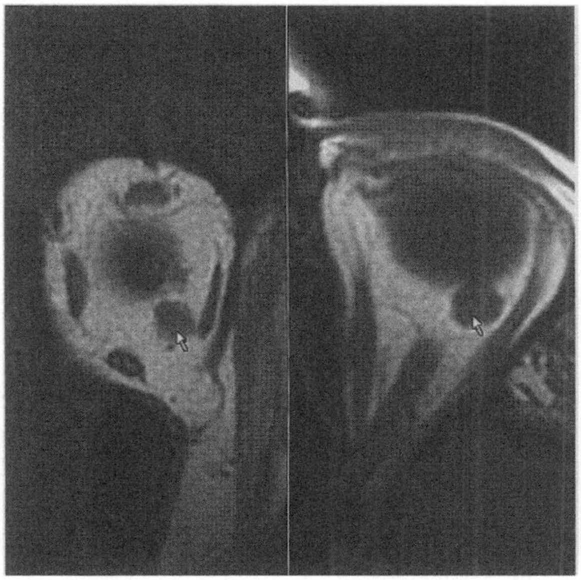

c

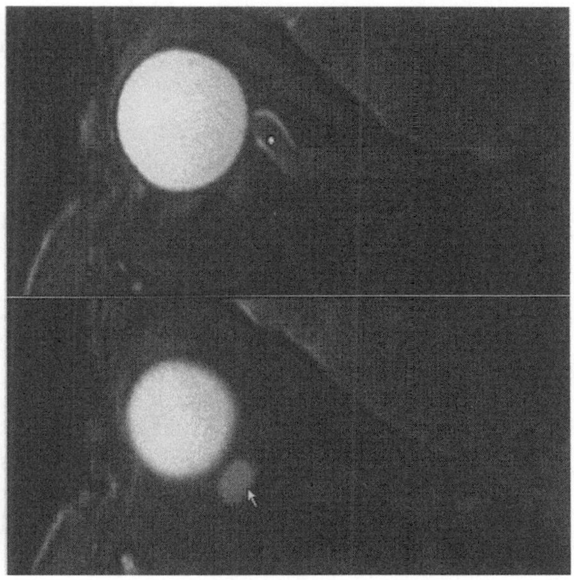

d

Case 8: Retinoblastoma

Clinical and pathological features

- Most common highly malignant intraocular tumor in childhood.
- No sex or race predilection.
- Inherited in about 40% (autosomal dominant, gene defect on long arm of chromosome 13, I-4 band).
- Unilateral in 70%, bilateral in 30% mainly in hereditary forms.
- "Trilateral retinoblastoma" (third tumor in pineal gland): rare form.
- Detection at 15 months; new lesion rare after 24 months.
- *Signs* (over 90%): leukocoria (2/3), strabismus, glaucoma, visual loss, inflammation, systemic manifestations or mental retardation.
- Numerous clinical differential diagnosis (about 30): persistent hyperplastic primary vitreous (PHPV), Coat's disease (persistent of Cloquet canal), retinopathy of prematurity (retrolental fibroplasia).
- Histologically close to primitive neuroectodermal tumors of CNS.

MR characteristic findings

Morphology

- Mass arising from the retina and developing in vitreous.
- Calcifications present in over 90% (US and CT detection).
- May fill the entire eyeball and/or invade the orbit.
- High propensity to spread along perioptic spaces or perineural towards intracranial cavity (brain MR mandatory), rarely hematogeneously.

Signal intensity

- SI may be similar to PNET tumors of the CNS.
- SE-T1w: slightly hyperintense to vitreous.
- SE-T2w: hypointense to vitreous (hypercellularity, iron deposit).
- May show areas of necrosis and hemorrhage.
- Insensitive to calcifications (but include paramagnetic material).
- Differential diagnosis: PHPV and Coat's (hyperintense on T1 and T2).

Contrast enhancement

- Important to evaluate exophytic components and intracranial spread.
- Follow-up: detection of local recurrence and intracranial dissemination.

Case 8: Bilateral retinoblastoma with metastases

Fig. a: coronal cut; SE - PDw

Fig. b: coronal cut; SE - PDw

Very young child, presenting a right leukokoria previously enucleated for the left exophytic retinoblastoma. MR shows a right intraocular tumor with relative hypointense signal on PDw (**a**, ↑), due to partial intratumoral calcification, contrasting with the hyperintense vitreous. On the left side, huge intraorbital mass invading the orbital cavity (**a**, ↑↑), showing an iso-signal to the brain cortex on PDw and associated to an enlargement of the orbital cavity, corresponding to recurrence of retinoblastoma. The mass extended through the optic canal to involve the chiasmal cistern (**b**, ↑↑), probably via the perioptic CSF space. Metastatic lesions are also seen involving the skull voult (**b**, ↑; **b**, ↑↑↑) and the right superior orbital rim (**a**, ↑↑). The lesions enhanced after contrast iv.

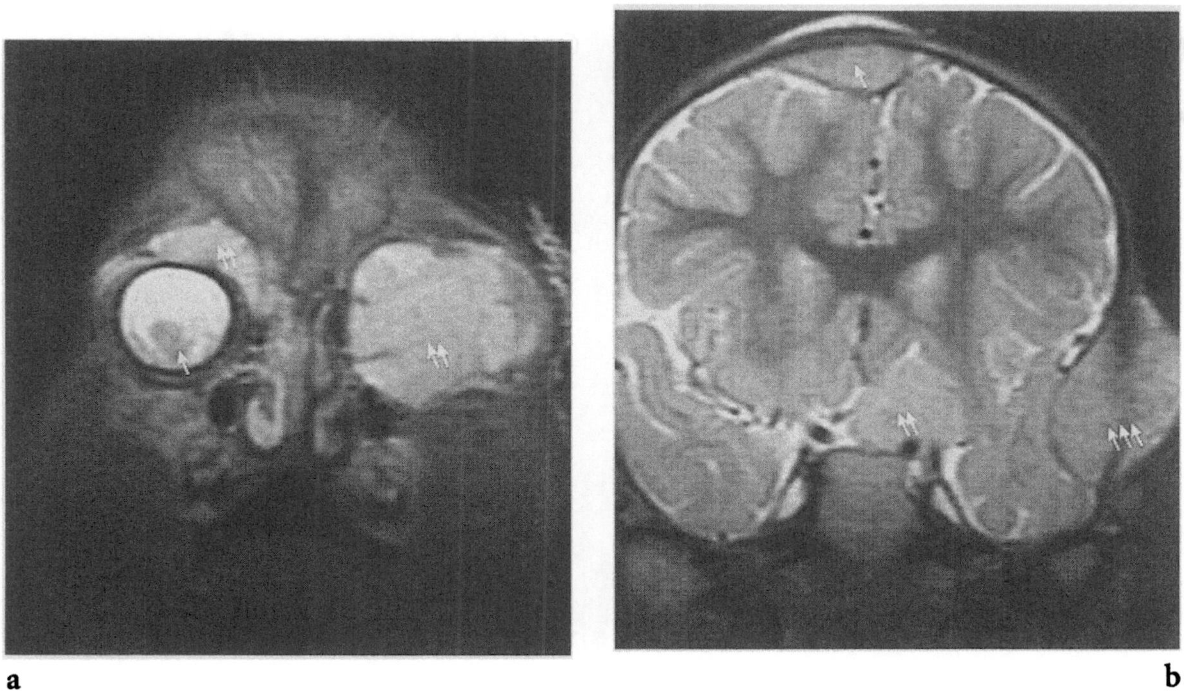

a b

Case 9: Metastatic neuroblastoma

Clinical and pathological features

- Second most common malignancy in infants and children.
- Arises from adrenal medulla or sympathetic chain.
- Frequent metastasis to orbits: 50% (hematogenous).
- Increased urinary catecholamines.
- *Signs:* sudden proptosis, unilateral or bilateral ecchymosis, Horner's syndrome, ocular motor disturbance, ptosis, decreased vision.

MR characteristic findings

Morphology

- Usually bilateral, but commonly unilateral involving orbital walls.
- Bony destruction with extension to the subperiosteal spaces.
- Generally extraconal.

Signal intensity

- SE-T1w: hypointense to fat.
- SE-PDw: slightly hyperintense to fat.
- SE-T2w: hypointense or hyperintense to fat.

Contrast enhancement

- May show marked contrast enhancement.
- Appreciation of intracranial (extradural) and intraorbital extensions.

Case 9: Metastatic neuroblastoma

Fig. c: sagittal cut; SE - T1w

Fig. d: sagittal cut; SE - T2w

Tumors involving the bony orbits and the anterior skull base showing a relatively hyposignal on T1w (**c**, ↑) and a iso-hypo-signal on T2w (**d**, ↑) and associated tc masses involving the skull voult (**c**, ↑↑).

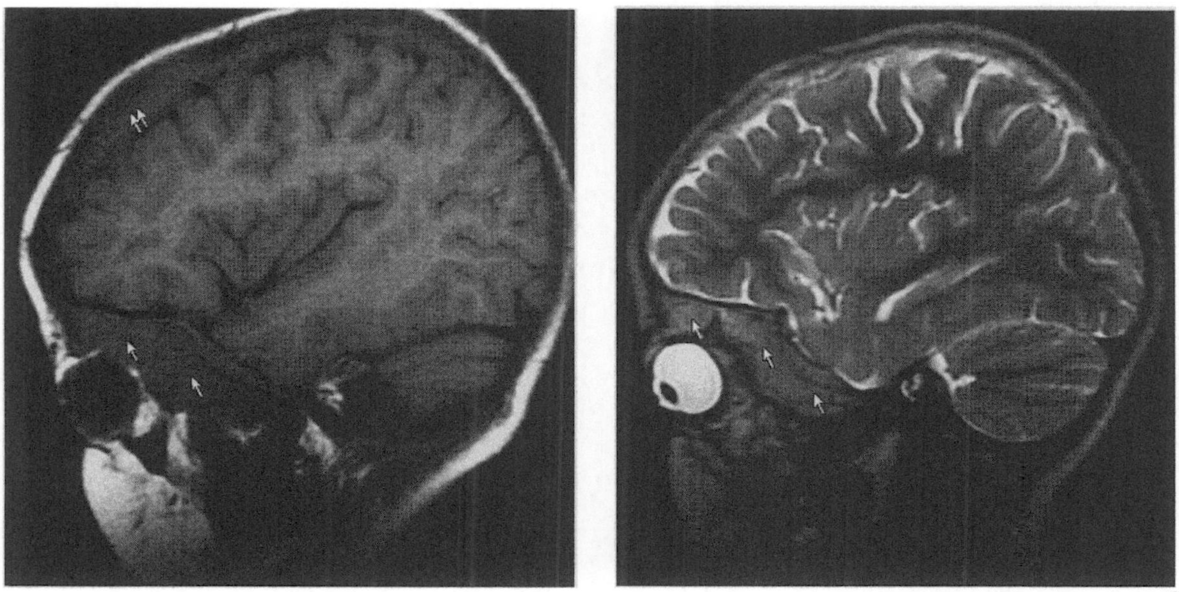

c d

Case 10: Ossifying fibroma of the orbit

Clinical and pathological features

- Affect children and young adults.
- Between least common fibrous tumors of the orbit.
- Highly cellular arising from facial bone.
- Could correspond to a localized form of fibrous dysplasia.
- *Signs:* proptosis, orbitofacial painless swelling.
- Slow growth and smooth margins.

MR characteristic findings

Morphology

- Contains fibroblastic and collagenous stroma, replacing medullary bone.
- Common involvement with expansion of orbital roof.
- May be more aggressive than fibrous dysplasia.

Signal intensity

- SI depends on dense fibrous matrix.
- SE-T1w: hypointense to fat.
- SE-PDw: hypointense to fat and cortex.
- SE-T2w: markedly hypointense to muscle.

Contrast enhancement

- Variable contrast enhancement.

Case 10: Ossifying fibroma of the orbit

Fig. a: sagittal cut;

Fig. b: sagittal oblique cut; SE - T2w

Patient presenting with unilateral proptosis with slight downward displacement of the eye. MR shows a very well demarkated mass involving orbital roof, showing a very low signal on PDw (**a**) as well on T2w (**b**) sequences. The mass exerts mass effect on the supraorbital frontal lobe, raised (**a**, **b**, ↑↑) by the extradural mass. The mass involved the roof of the optic canal as well as the anterior clinoid process and slightly compresses the intracanalicular optic nerve (**a**, ↑↑↑) very well shown on this high resolution proton density sequence performed in the sagittal oblique plane (**a**). Note the fluid retention in the orbito-frontal sinus, displaying a very high signal on PDw and T2w.

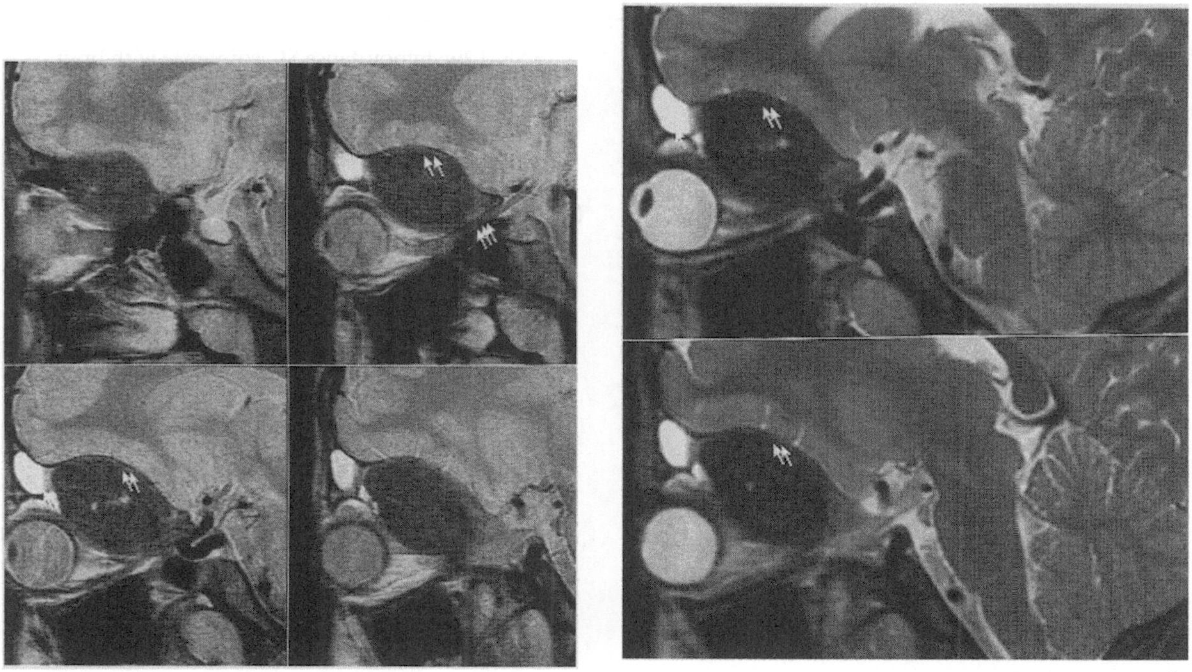

a b

Cases 11: Esthesioneuroblastoma

Clinical and pathological features

- Rare neurogenous tumor originating from neurosensory receptor cells of the olfactory nerves and mucosa (olfactory neuroblastomas).
- Affects both sexes, predominant in young males.
- Develops in the upper paranasal cavities and grows in the ethmoidal sinus.
- *Signs:* nasal obstruction, epistaxis, anosmia, rhinorrhea.
- Slow growing and locally invasive.

MR characteristic findings

Morphology

- Simulate primary nasal tumors.
- Localization in upper nasal cavity at the cribriform plate.
- Invariable local bony destruction.
- Early intracranial and intraorbital extension.

Signal intensity

- Usually non specific: similar to other neoplastic lesions.
- Se-T1W: intermediate signal.
- SE-T2w: hypointense SI, isointense with brain.
- Note that almost most inflammatory lesions are hyperintense on T2.

Contrast enhancement

- Moderate to marked enhancement.
- Useful to evaluate intracranial extent and/or intraorbital (fat suppression).

Case 11: Esthesioneuroblastoma

Fig. c: coronal cut; SE - T2w

Fig. d: sagittal cut; SE - T1w + Gd

Patient previously operated for an upper nasal esthesioneuroblastoma. MR exam shows (**c**) a mass involving the upper nasal cavity and occluding the left ethmoidal cells as well as the left orbitofrontal sinus showing a hyperintense signal on T2w (**c**). After contrast iv (**d**), the T1w sequence showed a heterogeneous contrast enhancement involving the upper part of the nasal cavity, the middle turbinate and the level of the cribriform plate associated to leptomeningeal enhancement of the anterior skull base. Note the metastatic lesion along the frontal squama due to metastatic seeding intracranially.

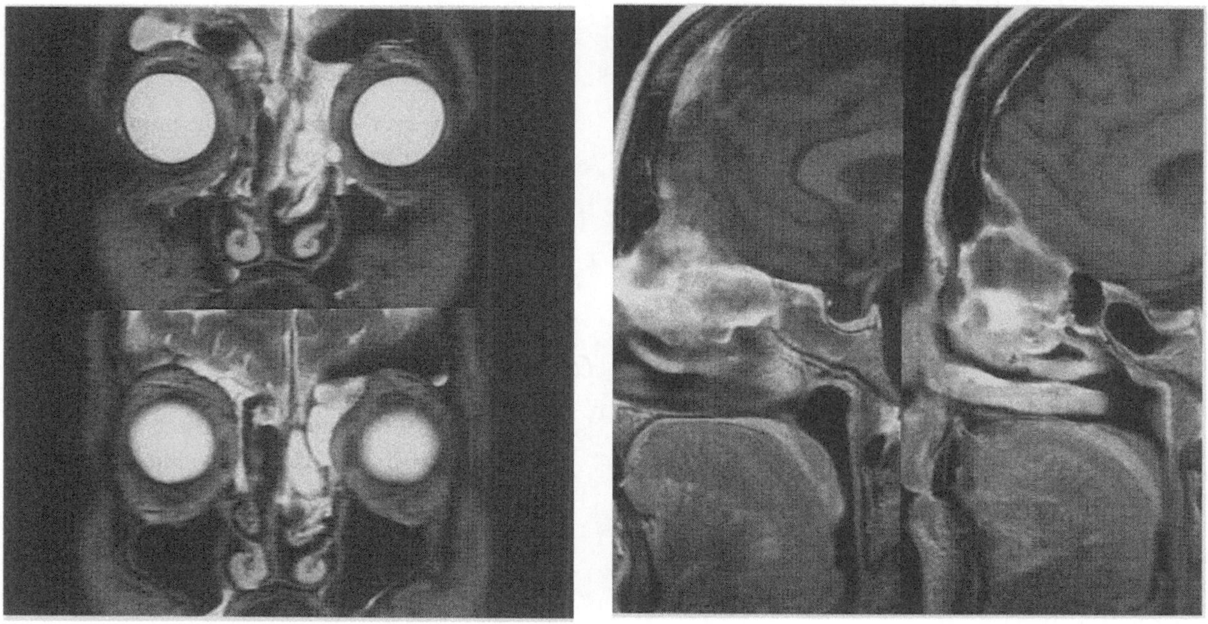

c d

Case 12: Mucocele

Clinical and pathological features

- Fairly infrequent lesion.
- Due to sinus ostium blockage.
- History of infection (50%), or trauma or allergy.
- Expansile paranasal sinus processes fluid filled.
- Lined by a respiratory epithelium.
- Most frequent in the frontal (60%) and ethmoidal (30%) sinuses.
- Relatively rare in the maxillary or sphenoidal sinuses.
- *Signs:* proptosis, ocular motor palsy, field defects, nasal obstruction.
- Variable size, may become huge.
- Regional invasion of the orbits and optic canals or cavernous sinus and sella.
- Differentiate from subperiosteal abscess secondary to sinusitis.

MR characteristic findings

Morphology

- Well defined mass expanding the sinusal cavity.
- Thinning of the bowed sinus walls and even bony erosion and dehiscence.

Signal intensity

- Variable according to the composition of the fluid content.
- SI depends on the amount of the protein content and associated hemorrhage.
- SE-T1w: hyperintense to brain or isointense.
- SE-PDw: hyperintense to CSF.
- SE-T2w: uniformly hyperintense to brain and CSF.

Contrast enhancement

- Usually do not enhance unless infected.
- May help to disclose mucosal modifications.

Case 12: Fronto-ethmoidal mucocele

Fig. a: sagittal cut (512); SE - T1w

Fig. b: sagittal cut (512); SE - T2w

Patient with facial pain. High resolution T1w images obtained using head coil, showing slightly hyperintense mass filling and expanding the frontal sinus (**a**, left, ↑↑; **a**, right, ↑). The mass shows a hyperintense signal on T2w (**b**, ↑), slightly more hyperintense than intracranial CSF. The posterior aspect of the fluid-filled mass erodes the anterior skull base, with intracranial extension and mass effect on the para-median frontal lobe (**a**, **b**,*).

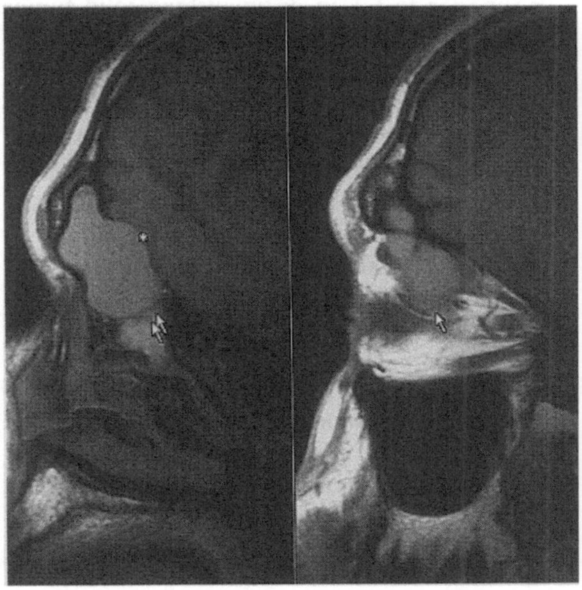
a

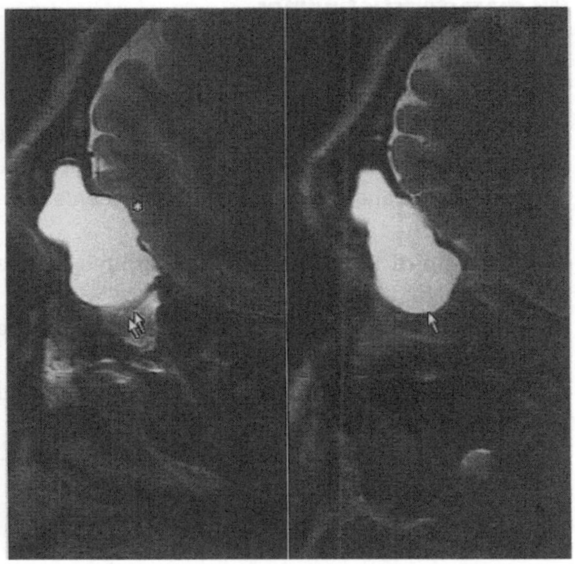
b

Case 13, 63: Ethmoido-nasal carcinoma

Clinical and pathological features

- Squamous cell carcinoma of the sinuses, most frequent type (about 90%).
- Epithelial tumor arising from the maxillary (80%) or ethmoid (15%) sinuses.
- Rare origin: sphenoidal or frontal sinuses.
- *Signs:* nasal obstruction, epistaxis, facial pain, diplopia, facial hypesthesia.

MR characteristic findings

Morphology

- Intrasinusal mass expanding and destroying the sinus walls, with interruption of signal void of the bony walls and retained osseous fragments peripherally.
- Invasion of the orbital cavities through lamina papyracea.
- Spread into the nasal fossae and maxillary antra or anterior cranial fossa.
- Solid infiltrating tumor, may show fluid-filled areas of necrosis.
- May be associated to retentional cysts in obstructed adjacents sinus cells.
- Search for cervical lymphadenopathy.
- *Differential diagnosis:* extensive bony destruction may be found in mucormycosis or Wegener's granulomatosis and remodeling in lymphomas or mucoepidermoids.
- Need to perform SE sequences in the coronal, axial and sagittal planes.

Signal intensity

- Superior to CT in distinguishing solid tumor from retentional fluid.
- SE-T1w: hypointense to brain, mass difficult to separate from fluid filled air cells.
- SE-T2w: hypointense or isointense or slightly hyperintense to brain, well differentiated from retentional fluid in obstructed sinus cells markedly hyperintense.
- SI may be heterogeneous in case of associated secondary inflammation.

Contrast enhancement

- Variable and moderate enhancement of the tumor.
- Associated inflammatory process may show dense enhancement.

Case 13: Ethmoido-nasal carcinoma

Fig. c: axial cut; SE - T2w with fat suppression

Fig. d: coronal cut; SE - T1w

Patient referred after CT examination for abnormal tissue involving the paranasal sinuses with orbital extension and proptosis. MR shows on T1w (**d**) a huge mass filling the left ethmoidal, nasal and maxillary cavities, homogeneously hypointense, invading the adjacent orbital cavity (**b**, ↑) with mass effect on the eyeball, the intraorbital optic nerve and the medial orbital muscle (**d**, right, *). The mass displaces laterally and invades the nasal septum (**d**, left, ↑↑↑) and destroys superiorly the left cribriform plate (**d**, right, ↑↑). On T2w sequence, with fat supression technique, the mass remains hypointense and shows the destruction and displacement of the lamina papyracea (**c**, up, ↑) extending into the extraconal nasal compartment with mass effect on the eyeball displaced anteriorly and laterally (**c**, down, ↑↑). The ethmoidal cells and the ipsilateral maxillary antrum showed hyperintense signal on T2w due to the presence of retentional fluid well differentiated from tumoral extension. In fact, with MRI, inflammatory/obstructive lesions are differentiated from neoplastic changes. After iv contrast, the mucosal lining of the sinus enhances contrarily to the central mucus-filled cavity. Anyway, using multiple pulse sequences, the signal of the tumor will differ enough from the involved sinus helping to accurate differentiation.

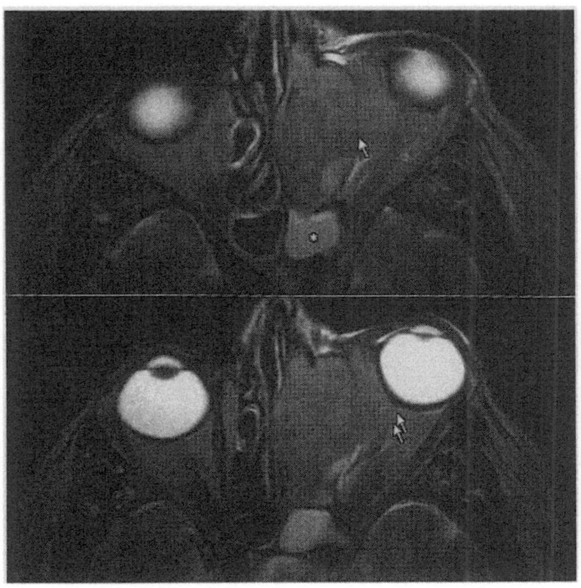

c

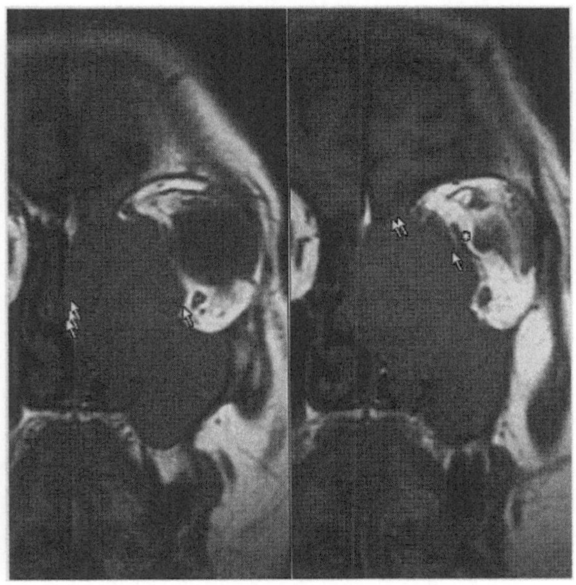

d

Maxillo-dento-facial region

Cases 14-17: temporomandibular joints dysfunctions

Clinical and pathological features

- Very common complaint, young adult population, female prevalence.
- Most frequently non specific: pain in the jaw or hemiface or ear or cephalgia.
- Craniomandibular dysfunctions: dysocclusion, reduce mouth opening, clicking.
- Less common causes: arthritis and degenerative arthropathies, avascular necrosis, infections, trauma (fracture of condylar neck), synovial chondromatosis or sarcoma, gout, craniofacial malformation.

MR characteristic findings

Morphology

- MR is the imaging modality of choice: disk and anterior and posterior bands well depicted, as well as the soft tissue in and around the TMJ and the glenoid cavity (see normal anatomy, case 14).
- Use of dedicated coupled surface coils (circular, 3 inches) is mandatory.
- Acquisitions obtained with open and closed mouth, eventually completed with cinematic imaging, performed in the sagittal (or sagittal oblique) plane, using small FOV (10-12 cm) and thin slices (2-3 mm).
- Coronal acquisitions to determine medial or lateral disk displacement.
- Image analysis (evaluation of shape and position of the meniscus): two forms of internal derangement, most frequently due to disk subluxation and dislocation (with respect to the head of the condyle):
  - reductible (recapture) anterior displacement (case 15);
  - non reductible anterior displacement (case 16).

Signal intensity

- SE-PDw sequence more informative (and signal to noise) than SE-T1w on high fields (morphology).
- Very low SI of dense cortical bone, meniscus (disk formed of fibrous tissue with proteoglycans) and ligaments in normal state, on all sequences.
- SE-T2w useful to disclose hyperintense SI from associated joint effusions.
- Confrontation of MR results with standard X-ray (lateral transcranial radiographs) is very helpful.
- Best bony details are obtained, if needed with CT.

Contrast enhancement

- May be needed in some special conditions: tumoral and infectious or inflammatory.

Case 14: Temporomandibular joint (normal anatomy)

Fig. a: sagittal oblique cut; SE - T1w

Fig. b: sagittal oblique cut; SE - T1w

T1w images in the sagittal or mildly sagittal oblique orientation showing the low signal appearance of the fibrocartilaginous disc (5, 6) in the glenoid cavity. The slices are obtained using small 3-inch-diameter coupled surface coils allowing very small fields of view. Normal anatomical structures are observed: (1) articular eminence; (2) mandibular condyle; (3) superior head of the lateral pterygoid muscle; (4) inferior head of the lateral pterygoid muscle; (5) anterior band of the meniscus or disk; 6-posterior band of the disk; CAE: external auditory canal. A contiguous multislice acquisition is obtained on each side (**b**), showing the disk (**b**, ↑).

Case 15: Subluxation with reduction (TMJ)

Fig. c: sagittal oblique cut; SE - PDw

Fig. d: sagittal oblique cut; SE - PDw

The exploration is performed while the mouth is closed (**c**) and open (**d**). In closed mouth position (**c**), the disk is located anterior to the articular surface of the condyle (**c**, ↑). As the condyle translates anteriorly (**d**, open mouth position) there is normal excursion of the condyle opposing the eminence and a reduction of the disk. This is obtained because the condyle passes underneath the posterior band of the disk during mouth opening in this early form of internal derangement.

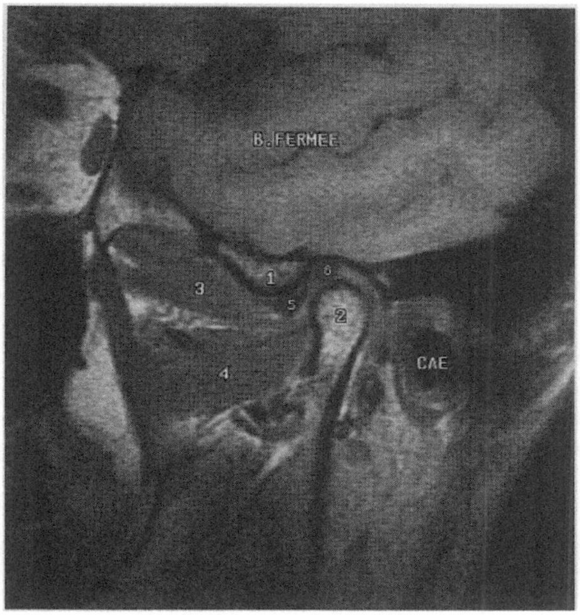

a

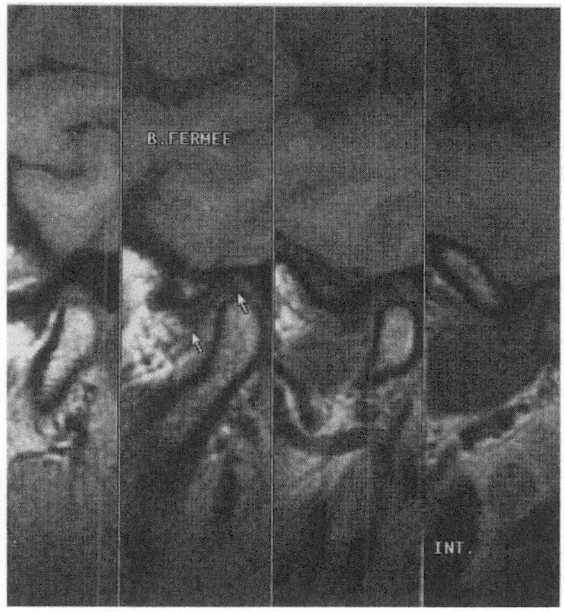

b

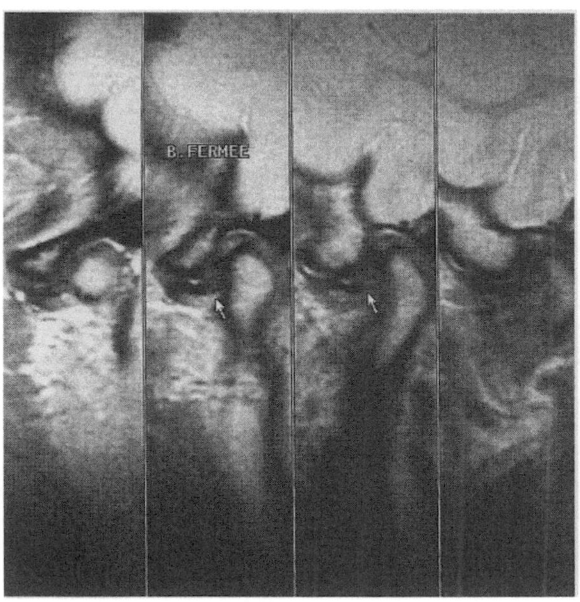

c

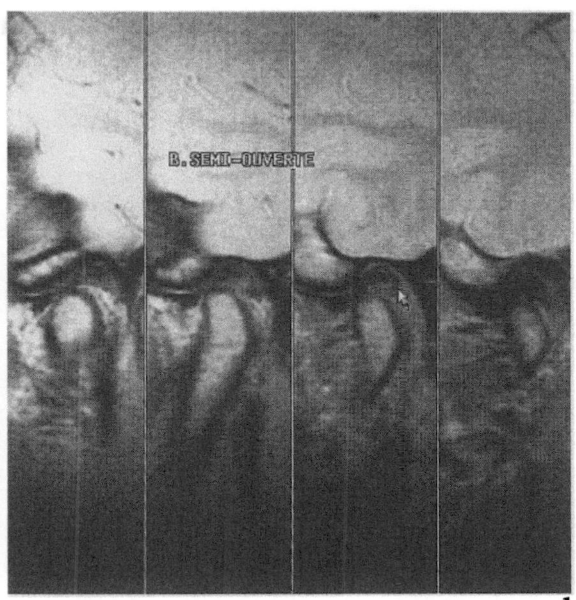

d

Case 16: Subluxation without reduction (TMJ)

Fig. a: sagittal oblique cut; SE - PDw

Fig. b: sagittal oblique cut; SE - PDw

In closed mouth position, the disk is deformed and anterior to the mandibular condyle (**a**, ↑). During mouth opening (**b**), the condyle forces the disk anteriorly which shows greater deformation (**b**, ↑). Note that the condyle is mechanically prevented from moving through its full forward excursion. Note that the mouth is not fully opened in this case (**b**). In open mouth position, the normal location of the condyle is anterior to the eminence.

Case 17: Subluxation with reduction and TMJ arthritis

Fig. c: sagittal oblique cut; SE - PDw

Fig. d: sagittal oblique cut; SE - PDw

In closed mouth position performed on PDw sequence (**c**), the MR exam showed degenerative changes of the condyle (**c**,*) with a posterior localization within the fossa and a disk located anterior to the articular surface (**c**, ↑↑) associated to an adjacent inflammatory exsudate (**c**, ↑, **d**, ↑). In open mouth position, the disk is pushed anteriorly and deformed, losing its normal biconcave configuration (**d**, right, ↑↑). The hyperintense signal of the associated inflammatory suffusion is also displaced anteriorly and upward (**d**, left, ↑). Inflammatory reactions are well shown on these long TR-PDw sequences and could be missed on T1w.

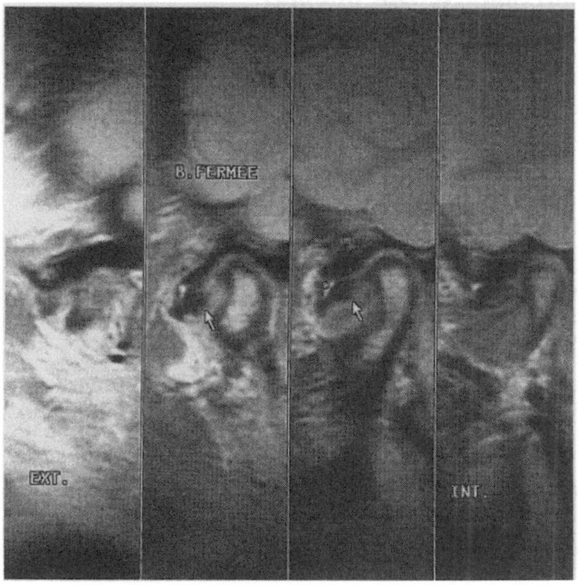

a

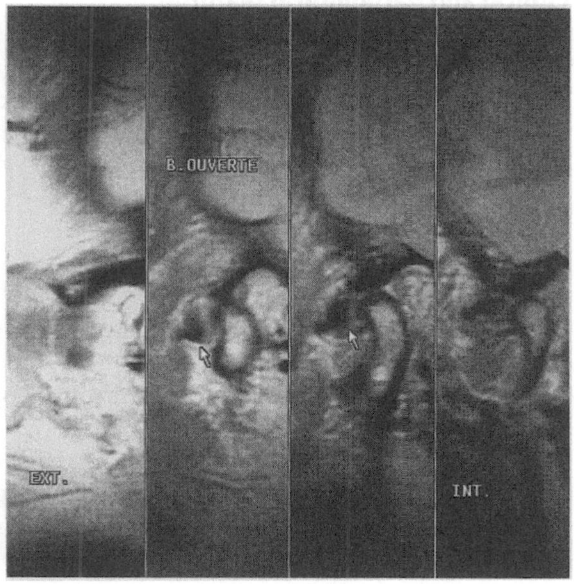

b

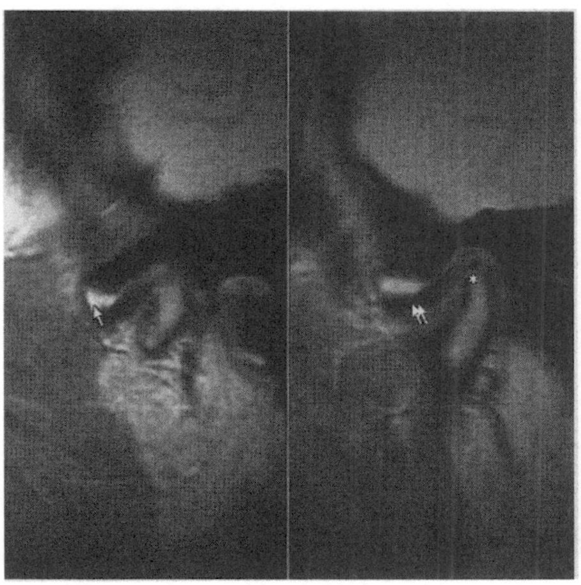

c

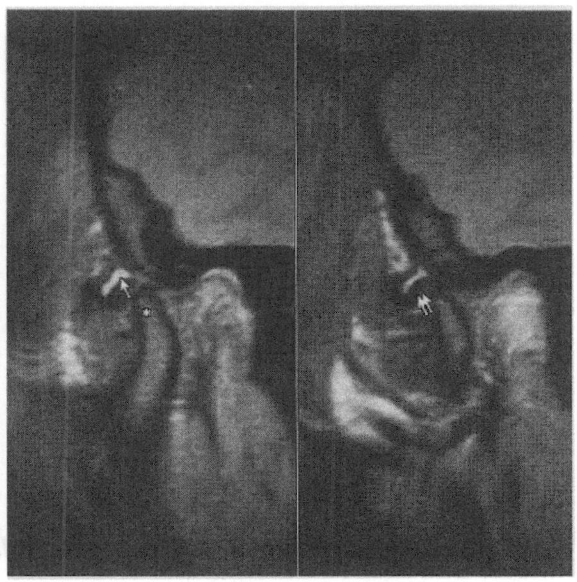

d

Case 18 : Parotid papillary cystadenoma lymphomatosum

Clinical and pathological features

- Also called Warthin's tumor.
- Second most common benign exclusively parotid tumor (after parotid pleiomorphic adenoma: 65%).
- Account for 5-10% of parotid neoplasms.
- Occurs in adults in 4th to 7th decades.
- More frequent in whites and males (5:1).
- Bilateral in 10% of cases.
- Histologically: proliferation of salivary ducts inside parotid lymph nodes.
- *Signs:* diagnosis usually delayed because of absence of symptoms.
- Uptake of Tc 99m (technetium 99m pertechnetate), as oncocytoma (rare 1%), but excluding pleomorphic adenoma and squamous cell carcinoma.

MR characteristic findings

Morphology

- Well defined mass (3-4 cm), possibly lobulated, inhomogeneous.
- Usually located superficially to facial nerve.

Signal intensity

- SE-T1w: hypointense to normal gland.
- SE-T2w: heterogeneous with hypointense and hyperintense (proteinaceous cyst).

Contrast enhancement

- Absence or slight contrast enhancement.

Case 18: Cystadenolymphoma of the parotid (Warthin)

Fig. a: axial cut; SE - T1w

Fig. b: axial cut; SE - T2w

Small cystic mass located in the left parotid gland, as shown clinically and on MRI, at the posterior lateral side, with a very hypointense signal on T1w (**a**, ↑) and a hyperintense signal on T2w (**b**). The cyst wall is smooth except for a very small nodule at its posterior medial side shown on T2w (**b**). The ovoid tumor is encapsulated. Its posterior irregularity shown on T2w suggests that the lesion is a cystic tumor rather than a simple cyst.

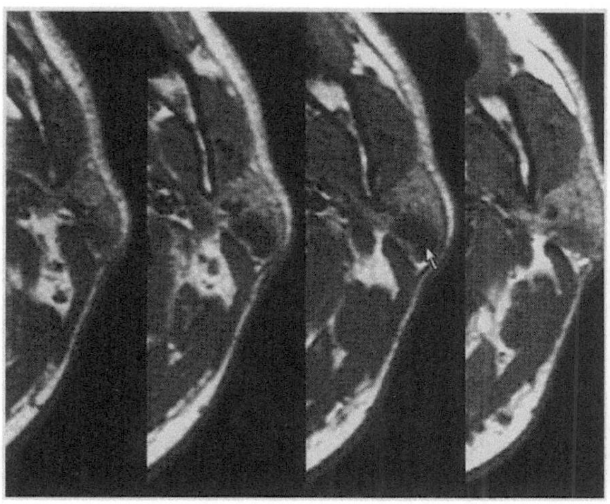

a

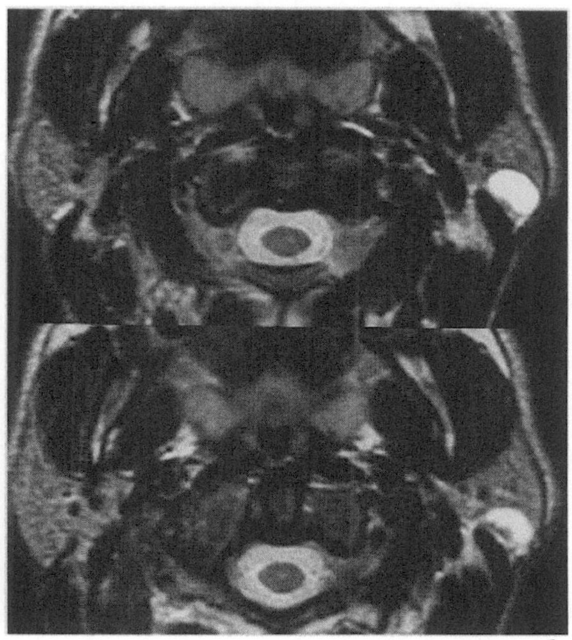

b

Cases 19-20: odontogenic cysts

Clinical and pathological features

- Include radicular and dentigerous cysts.
- *Radicular cyst:* most frequent odontogenic cyst
  - periapical location (non vital tooth);
  - well circumscribed;
  - rapid extension to maxillary sinus.
- *Dentigerous cyst:* second most common entity
  - results from degenerative enamel;
  - cyst associated to a partially developed tooth;
  - located at 3rd mandibular molar or maxillary cuspid.
- *Differential diagnosis:* ameloblastoma, most frequent odontogenic tumor
  - derives from ectodermal remnants;
  - 80% in the mandibule and 20% in the maxilla;
  - 15% grossly cystic (mandibular molar).

MR characteristic findings

Morphology

- CT evaluates efficiently bony changes of a radiolucent process (X-rays).
- MR allows analysis of intralesional soft tissue and helps to differentiate between solid and cystic material (to rule out ameloblastomas, usually mixed solid-cystic tumor).
- MR provides information about the presence of blood, proteinaceous material, and shows eventual extension to the adjacent soft tissue.
- MR defines cortical expansion, the relation of the cyst with the apex of the tooth (radicular) or the partially developed crown (dentigerous).
- Absence of limitating susceptibility artifacts from bone or teeth.

Signal intensity

- SE-T1w: intermediate SI (depending on proteinaceous content or subacute hemorrhage with blood breakdown material in the cyst).
- SE-T2w: hyperintense, bright SI (cystic fluid).

Contrast enhancement

- Smooth enhancement of the cystic wall of a dentigerous cyst.
- MR shows enhancement of the mural tumor of a mixed solid-cystic ameloblastoma (possible differentiation from a dentigerous cyst).

Case 19: Dental apical granuloma

Fig. c: axial cut; SE - T1w

Fig. d: coronal cut; SE - PDw

Left anterior maxillary cystic formation with smooth very well demarkated margins presenting a hyperintense signal on T1w (**c**) and a hyperintense signal on PDw (**d**) sequences due to the presence of intracystic cholesterine. The lesion exerts mass effect on the floor of the nasal cavity raised (**d**, ↑) and on the anteromedial margin of the left maxillary antrum displaced posteriorly (**c**, ↑↑), involving the anterior parasagittal portion of the hard palate. Note on T1w the hypointense signal of the inflammatory polypoid formations within the maxillary antrum (**c**,*) contrasting with the hyperintense signal of the cystic formation.

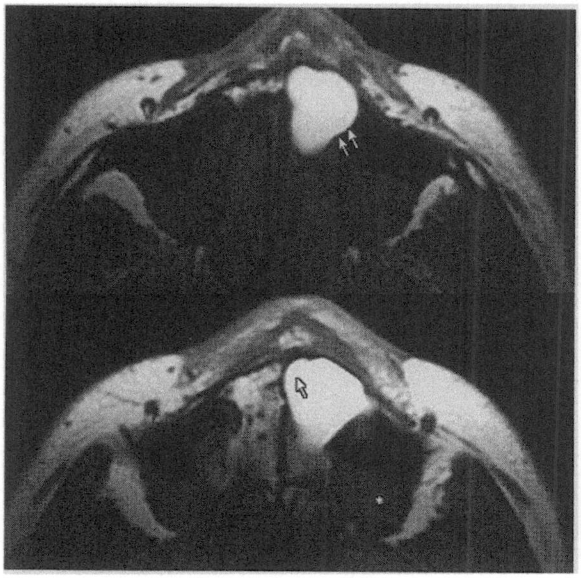

c

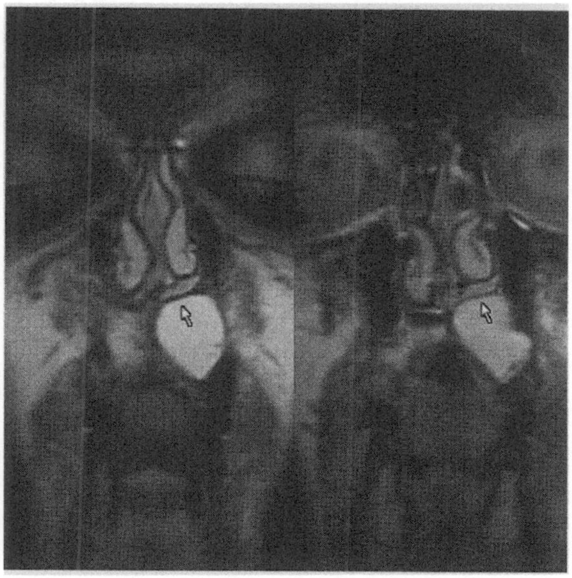

d

Case 20: Dental apical xanthogranuloma

Fig. a: sagittal cut (512); SE - T1w

Fig. b: sagittal cut (512); SE - T2w

Fig. c: axial cut; SE - T2w

Fig. d: coronal cut; SE - T2w

Patient presenting a mass of the right maxillary alveolar ridge associated to gengivitis and swelling of the right superior lip. MR shows a heterogeneous mass developed in the right maxilla, partly cystic with a hypointense signal on T1w (**a**, ↑), and a very hyperintense signal on T2w (**b**, ↑; **d**, ↑↑) correponding to the cystic portion. The lesion exerts a mass effect on the floor of the ipsilateral maxillary sinus raised (**d**, right, *) and on the floor of the nasal fossa (**d**, left, **). The mass involves the right portion of the alveolar ridge (**c**, ↑), showing hypointense signal replacing the hyperintense signal of the normal bony fat and eroding the maxillary bone with extension to the adjacent soft tissues (**c**, *, ↑↑). Note the very close relationship of the lesion with the adjacent dental root (**a**, **d**), showing signal void on all sequences due to its cristalline nature.

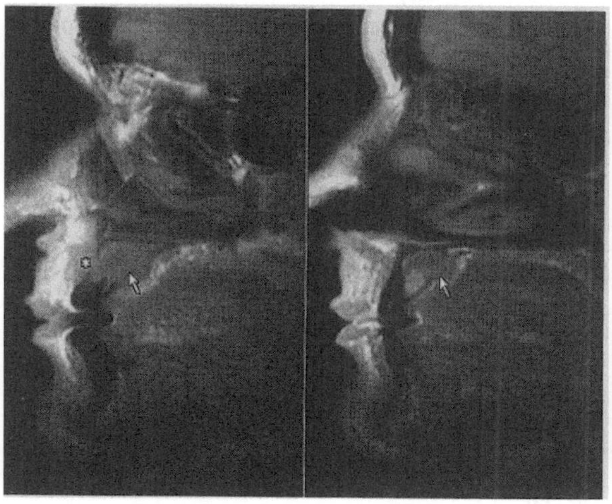

a

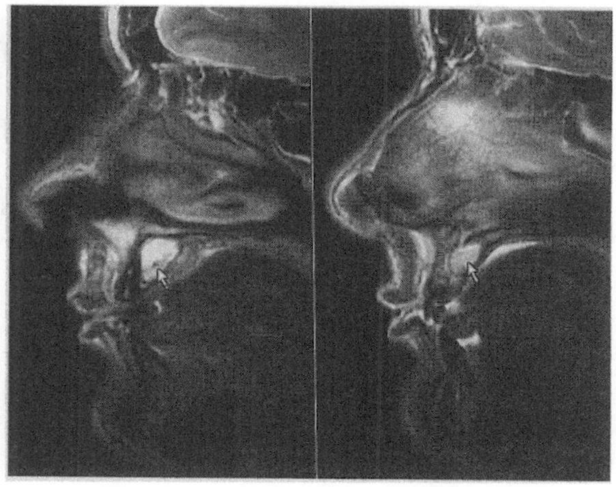

b

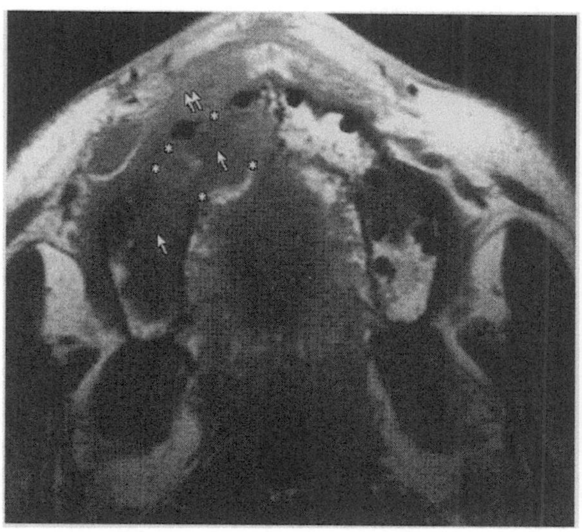

c

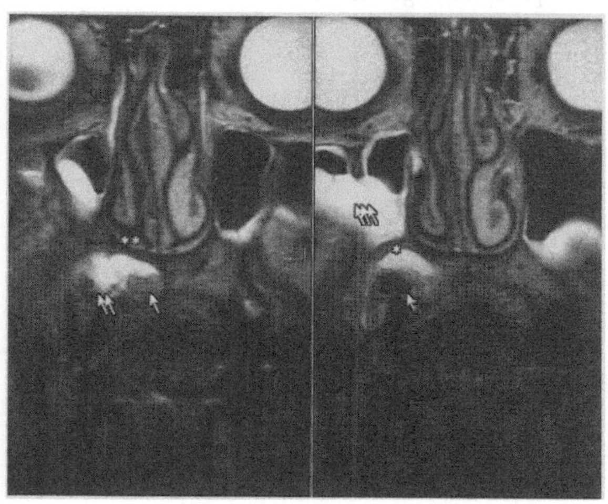

d

Case 21: Branchial cleft cyst

Clinical and pathological features

- Congenital cystic lesions resulting from failure of regression of a branchial cleft.
- Found usually in the anterior or lateral aspect of the neck.
- First BCC uncommon (8%): misdiagnosed as lower parotid tumor and second BCC (over 90%); found at the mandibular angle region.
- *Signs:* young adult presenting with a parapharyngeal neck mass (usually second BCC) or middle-aged woman showing a periparotid mass (first BCC).

MR characteristic findings

Morphology

- Largely cystic, well delineated, oval or rounded parapharyngeal mass.
- Posterior displacement of the sternocleidomastoid muscle (second BCC) and medial displacement of carotid space structures.
- Absence of infiltration of fat planes.
- *Differential diagnosis* of cystic masses: ranulas (mucous retention) and cystic hygromas (fibrous septa).

Signal intensity

- SE-T1w: isointense or slightly hyperintense to muscles.
- SE-PDw: hyperintense to muscles and homogeneous.
- SE-T2w: markedly hyperintense to muscle and fat.

Contrast enhancement

- No enhancement of the cystic mass.
- Possible peripheral enhancement of the cyst wall in case of inflammatory reaction.

Cas 21: Branchial cleft cyst

Fig. a: coronal cut; SE - pT1

Fig. b : axial cut; SE - pT2

Young adult presenting with lateral neck mass. MR shows a huge bilobated smooth walled cystic formation located in the retro and inframandibular region, presenting an iso-signal to the neck muscles on T1w (**a**, ↑) and a hyperintense signal on T2w (**b**).
The cystic formation lies just lateral to the pharynx involving the parapharyngeal spaces and the carotid sheaths with mass effect on the internal jugular vein and the common carotid artery shown at the medial border of the cyst with flow void (**a**). More laterally and superiorly, the mass comes close to the inferior border of the parotid gland, normally slightly hyperintense on T1w (**a**).

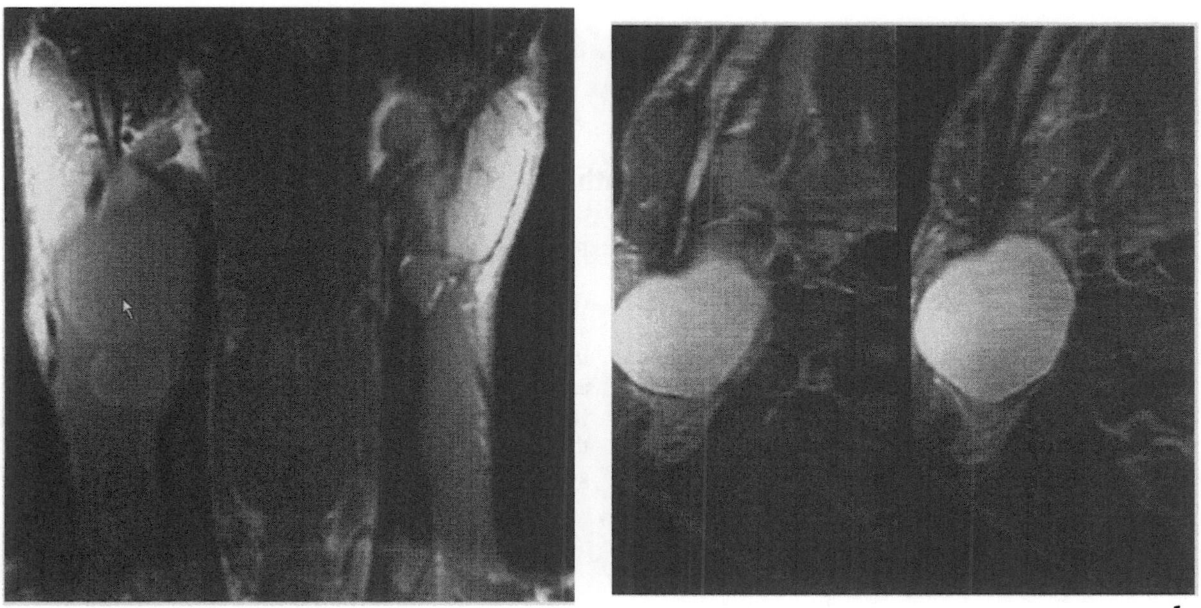

a b

Case 22 : Venous angioma of the floor of the mouth

Fig. c : axial cut; SE - pDP

Fig. d : coronal cut; SE - pDP

Floor of the mouth: polycyclic mass relatively well delineated showing a hyperintense signal on T2w sequences (**c**, **d**) with multiple signal voids rounded and curvilinear images due to vessels. The mass invades the floor of the mouth (**c**, ↑), extending posteriorly and inferiorly to involve and compress the submaxillary gland on the right side (**d**, ↑↑). The medial aspect of the mass involves the ipsilateral pharyngeal wall (**d**, ↑). Note the anterior and right extension of the mass in the floor of the mouth with mass effect on the adjacent geniohyoid muscle (**c**, ↑).

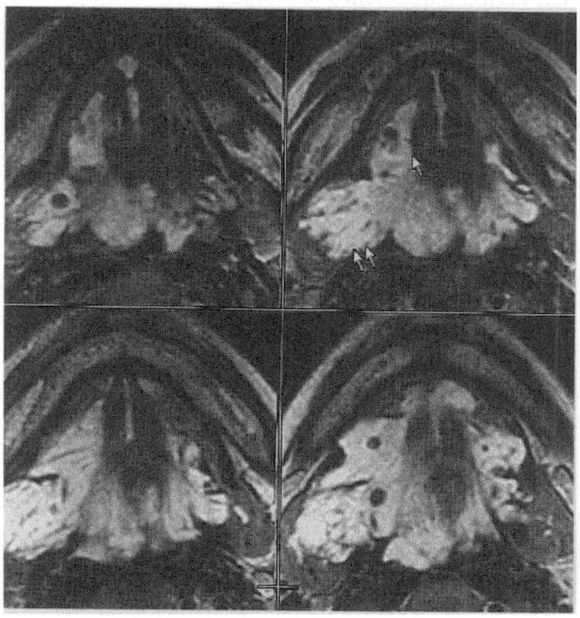

c

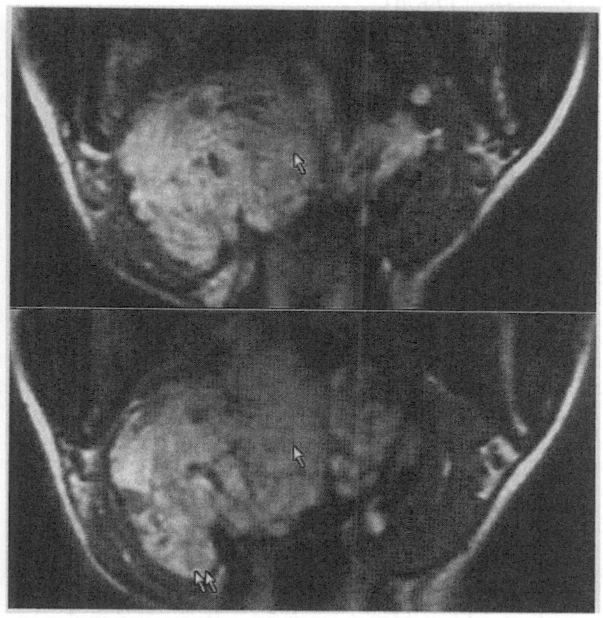

d

Selected References

- Albernaz VS, Castillo M, Hudgins PA et al (1997) Imaging Findings in Patients with Clinical Anophthalmos. AJNR 18:555-561
- Albert DM, Rubenstein RA, Scheie HG (1967) Tumor metastases to the eye. I. Incidence in 213 adult patients with generalized malignancy. Am J Ophthalmol 63:723
- Alvord EC, Lofton S (1988) Gliomas of the optic nerve or chiasm. J Neurosurg 68:85
- Appelboom T, Durso F (1967) Tumor metastases to the eye in patients with generalized malignancy. Am J Ophthalmol 63:7
- Beardsley T, Brown S, Sydnor C et al (1984) Eleven cases of sarcoidosis of the optic nerve. Am J Ophthalmol 97:62-67
- Bernardino ME, Zunn RD, Citrin CM et al (1977) Scleral thickening: a sign of orbital pseudotumor. Am J Roentgenol 129:703
- Bilaniuk LT (1999) Orbital vascular lesions. Radiol Clin North Am 37(1):169-183
- Bilaniuk LT, Schenck JF, Zimmerman RA et al (1985) Ocular and orbital lesions: surface coil MR imaging. Radiology 156:669
- Brooks SL, Westesson PL (1993) Temporomandibular joint: value of coronal MR images. Radiology 188:317-321
- Brown EW, Riccardi VM, Mawad M et al (1987) MR imaging of optic pathways in patients with neurofibromatosis. Am J Neurodadiol 8:1031
- Carmody RF, Mafee MF, Goodwin JA et al (1994) Orbital and optic pathway sarcoidosis: MR findings. AJNR 15:775-783
- Daniels DL, Yu S, Perch P et al (1987) Computed tomography and magnetic resonance imaging of the orbital apex. Radiol Clin North Am 25:803
- Dillon WP, Som PM, Fullerton GD (1990) Hypointense MR signal in chronically inspissated secretions. Radiology 174:73
- Fascerrelli FW (1969) Maxillary sinus abnormalities: radiographic evidence in an asymptomatic population. Arch Otol Laryngol 90:190
- Font RL, Gamel JW (1978) Epithelial tumours of the lacrimal gland: an analysis of 265 cases. In: Jakobiec FA (ed) Ocular and adnexal tumours. Aesculapius, Birmingham
- Fox AJ, Debrun G, Vinuela F et al (1979) Intrathecal metrizamide enhancement of the optic nerve sheath. J Comput Assist Tomogr 3:653
- Franks AS (1969) Temporomandibular joint in adult rheumatoid arthritis. A comparative evaluation of 100 cases. Am Rheum Dis 28:139-145
- Friedman DP, Rao V, Flanders AE (1993) Lesions causing a mass in the medial canthus of the orbit: CT and MR features. Am J Roentgenol 160:1095
- Glaser JS, Hoyt WF, Corbett J (1971) Visual morbidity with chiasmal glioma. Arch Ophthalmol 85:3
- Glasier CM, Ascher DP, Williams KD (1986) Incidental paranasal sinus abnormalities on CT of children: clinical correlation. Am J Neuroradiol 7:861
- Gomori JM, Grossman RI, Shields JA et al (1986) Choroidal melanomas: correlation of NMR spectroscopy and MR imaging. Radiology 158:443
- Graeb D, Rootman J, Robertson W (1990) Orbital lymphangiomas: clinical, radiologic and pathologic characteristics. Radiology 175:417-421
- Green B, Kraft S, Carter K (1990) Intraorbital wood: detection by magnetic resonance imaging. Ophthalmology 97:608-611
- Guyon J, Brant-Zawadski M, Steiff S (1984) CT demonstration of optic canal fractures. AJR 143:1031-1034
- Haik BG, St Louis L, Smith ME et al (1985) Magnetic resonance imaging in the evaluation of leukocoria. Ophthalmology 91:1143
- Hammerschlag S, Hughes S, O'Reilly G (1982) Another look at blow-out fractures of the orbit. AJR 139:133-137
- Helms CA, Kaplan P (1990) Diagnostic imaging of the temporomandibular joint: recommendation for use of the various techniques. AJR 154:319-322
- Hesselink JR, Weber AL, New PFJ et al (1979) Evaluation of mucocele of the paranasal sinuses with computed tomography. Radiology 133:397
- Ho V, McGuckin J, Smergel E et al (1996) Intraorbital wood foreign body: CT and MR appearance. AJNR 17:134-136
- Hoffmann KT, Hosten N, Lemke AJ et al (1998) Septum Orbitale: High-Resolution MR in Orbital Anatomy. AJNR 19:91-94
- Hosten N, Sander B, Cordes M et al (1989) Graves Ophthalmopathy: MR imaging of the orbits. Radiology 172:759-762
- Hoyt WF, Baghdassarian SA (1969) Optic glioma of childhood: natural history and rationale for conservative management. Br J Ophthalmol 53:793
- Hudgins PA, Gussack GS (1992) MR imaging in the management of extracranial malignant tumors of the head and neck. Am J Roentgenol 159:161
- Jakobiec FA, McLean I, Font R (1979) Clinicopathologic characteristics of orbital lymphoid hyperplasia. Ophthalmology 86:948
- Johnson DL, Chandra R, Fisher WS et al (1985) Trilateral retinoblastoma: ocular and pineal retinoblastomas. J Neurosurg 63:367
- Johnson G, Miller DM, MacManus D et al (1987) STIR sequences in NMR imaging of the optic nerve. Neuroradiology 29:238
- Jones I, Jakobiec F (1979) Diseases of the orbit. Harper and Row
- Kelly WM, Paglen PG, Pearson JA et al (1986) Ferromagnetism of intraocular foreign body causes unilateral blindness after MR study. Am J Neuroradiol 7:243
- Knowless DM, Jakobiec FA (1980) Orbital lymphoid neoplasms: a clinicopathologic study of 60 patients. Cancer 46:576

- Kondo M, Horiuch M, Shiga H et al (1982) Computed tomography of malignant tumors of the nasal cavity and paranasal sinuses. Cancer 50:226
- Koorneef L, Zonneveld F (1987) The role of direct multiplanar high resolution CT in the assessment and management of orbital trauma. Radiol Clin North Am 25:753-766
- Liang R, Loke SL, Chiu E (1991) A clinicopathological analysis of seventeen cases of non Hodgkin's lymphoma involving the orbit. Acta Oncologica 30:335-338
- Linder B, Campos M, Schafer M (1987) CT and MRI of orbital abnormalities in neurofibromatosis and selected craniofacial anomalies. Radiol Clin North Am 25:787
- Lufkin R et al (1986) Tongue and oropharynx: findings on MR imaging. Radiology 161:69
- Lustrin ES, Brown JH, Novelline R, Weber AL (1996) Radiologic assessment of trauma and foreign bodies of the eye and orbit. Neuroimaging Clin North Am 6:219-237
- Mafee MF, Goodwin J, Dorodi S (1999) Optic nerve sheath meningiomas. Role of MR imaging. Radiol Clin North Am 37(1):37-58
- Mafee M, Goldberg M (1987) CT and MR imaging for diagnosis of persistent hyperplastic primary vitreous. Radiol Clin North Am 25:683
- Mafee MF, Linder B, Peyman GA et al (1998) Choroidal hematoma and effusion: evaluation with MR imaging. Radiology 168:781
- Mafee MF, Peyman GA, McKusick MA (1987) Malignant resonance imaging in the evaluation and differentiation of uveal melanoma. Ophthalmology 94:341
- Mafee MF, Peyman GA, Grisolano JE et al (1986) Malignant uveal melanomas and simulating lesions: MR imaging evaluation. Radiology 160:773
- Mafee MF, Putterman A, Valvassori GE et al (1987) Orbital space-occupying lesions: role of computed tomography and magnetic resonance imaging: analysis of 145 cases. Radiol Clin North AM 25:529
- Mafee M, Ainbinder D, Afshani E, Mafee R (1996) The eye Neuroimaging. Clin North Am 6:29-59
- Marx H, Coletti P, Raval (1990) Magnetic resonance imaging features in melanoma. Magn Reson Imag 8:223-229
- Nemzek WR, Katzberg RW, Van Slyke MA et al (1995) A reappraisal of the radiologic findings of acute inflammation of the epiglottis and supraglottic structures in adults. Am J Neuroradiol 16:495
- Ni C, Cheng SC, Dryja TP et al (1982) Lacrimal gland tumors: a clinicopathological analysis of 160 cases. Int Ophtalmol Clin 22(1):99
- Nugent RA, Rootman J, Robertson WD et al (1981) Acute orbital pseudotumors: classification and CT features. Am J Roentgenol 138:957
- Onofri M, Tartaro A, Thomas A et al (1996) Long echo time STIR sequence MRI of optic nerves in optic neuritis. Neuroradiol 38:66-69
- Palacios E, Valvassori GE, Shannon M, Reed CF (1990) Magnetic Resonance of the Temporomandibular Joint. Clinical considerations, Radiography, Management. Georg Thieme Verlag, Stuttgart, New York
- Perugini S, Pasquin V, Menichelli F et al (1982) Mucucele in the paranasal sinuses involving the orbit: CT signs in 43 cases. Neuroradiology 23:133
- Peyman G, Juarez C, Diamond J (1984) Ten years experience with eye wall resection for uveal melanoma. Ophthalmology 91:1720-1724
- Peyster RG, Augsburger JJ, Shields JA et al (1988) Intraocular tumors evaluation with MR imaging. Radiology 168:773
- Peyster RG, Ginsburg F, Silber J et al (1986) Exophthalmos caused by excessive fat: CT volumetric analysis and differantial diagnosis. Am J Neuroradiol 7:35
- Pomeranz SJ, Shelton JJ, Tobias, J et al (1987) MR of visual pathways in patients with neurofibromatosis. Am J Neuroradiol 8:831
- Preling NJ, Molenaar WM, Vermey A et al (1992) Malignant parotid tumors: clinical use of MR imaging and histological correlation. Radiology 185:691
- Rejowski JE, Caldarelli DD, Campanella RS et al (1982) Nasal polyposis causing bone destruction and blindness. Otolaryngol Head Neck Surg 90:505
- Romett JL, Newman RK (1982) Aspergillosis of the nose and paranasal sinuses. Laryngoscope 92:764
- Rootman J, Damji KF, Dimmick JE (1989) Malignant rhabdoid tumor of the orbit. Ophthalmology 96:1650
- Rothstein J, Maisel RH, Berlinger NT et al (1984) Relationship of optic neuritis to disease in the paranasal sinus. Laryngoscope 94:1501
- Sandberg-Wolheim M, Bynke H, Cronqvist S et al (1990) A long term prospective study of optic neuritis: evaluation of risk factors. Ann Neurol 27:386
- Seidenberg KB, Leib ML (1990) Orbital myositis with Lyme disease. Am J Ophthalmol 109:13
- Shaha AR (1991) Preoperative evaluation of the mandible in patients with carcinoma of the floor of the mouth. Head Neck 13:398
- Shields J, Bakewell B, Augsburger J (1984) Classification and incidence of space occupying lesions of the orbit: a survey of 645 biopsies. Arch Ophthalmol 102:1606-1611
- Shih L, Hawkins DB, Stanley RB (1988) Acute epiglottitis in adults: a review of 48 cases. Ann Otol Rhinol Laryngol 97:527
- Siegel MJ, Glazer HS, St Amour TE et al (1989) Lymphangiomas in children: MR imaging. Radiology 170:467
- Sigal R, Zagdanski AM, Schwaab G et al (1996) Imaging of squamous cell carcinomas of the tongue and floor of the mouth. Radiographics 16:787
- Som PM, Biller HF (1989) High-grade malignancies of the parotid gland: identification with MR imaging, Radiology 173:823
- Som PM, Shapiro MD, Biller HF et al (1988) Sinonasal tumors and inflammatory tissues: differentiation with MR imaging. Radiology 17:803
- Som PM, Shugar JM, Sacher M et al (1991) Benign and malignant parotid gland in patients with Sjögren's syndrome. J Rheumatol 18:1372

- Som PM, Shugar JMA, Cohen BA et al (1981) The nonspecificity of the antral bowing sign in maxillary sinus pathology. J Comput Assist Tomogr 5:350
- Spiro RH, Koss LG, Hajdu SI et al (1973) Tumors of minor salivary gland origin: a clinico-pathological study of 492 cases. Cancer 31:117
- Som PM, Silvers AR, Catalano PJ et al (1997) Adenosquamous Carcinoma of the Facial Bones, Skull Base, and Calvaria: CT and MR Manifestations. AJNR 18:173-175
- Stone JA, Castillo M, Neelon B, Mukherji S (1999) Evaluation of CSF Leaks: High-Resolution CT Compared with Contrast-Enhanced CT and Radionuclide Cisternography. AJNR 20:706-712
- Takashima S, Takeuchi N, Morimoto S et al (1991) MR imaging of Sjogren syndrome: correlation with sialography and pathology. J Comput Assist Tomogr 15(30):393
- Tien RD, Chu PK, Hesselink JR et al (1990) Intra and periorbital lesion: value of fat-suppression MRI with paramagnetic contrast enhancement. Am J Roentgenol 156:1059
- Unger J (1984) Orbital apex fractures: the contribution of computed tomography. Radiology 150:713-717
- Vogl TJ, Steger W, Ihrler S et al (1993) Cystic masses in the floor of the mouth: value of MR imaging in planning surgery. Am J Roentgenol 161:183
- Warner MA, Weber AL, Jakobiec FA (1996) Benign and malignant tumors of the orbital cavity including the lacrimal gland. Neuroimaging Clin North Am 6:123-142
- Weber AL, Jakobiec FA, Sabates NR (1996) Lymphoproliferative disease of the orbit. Neuroimag Clin North Am 6:93
- Weber AL, Jakobiec FA, Sabates NR (1996) Pseudotumor of the orbit. Neuroimag Clin North Am 6:73-92
- Weber A, Dallow R, Sabates N (1996) Grave's disease of the orbit. Neuroimag Clin North Am 6:61-72
- Wright JE, Rose GE, Garner A (1992) Primary malignant tumours of the lacrimal gland. Br J Ophthalmol 76:401
- Yasumoto M, Shibuya H, Takeda M et al (1995) Squamous cell carcinoma of the oral cavity: MR findings and value of T1-versus T2-weighted fast spin-echo images. Am J Roentgenol 164:981
- Yousem DM (1993) Imaging of sinonasal inflammatory disease. Radiology 188:303
- Zilkha A (1982) Computed tomography in facial trauma. Radiology 144:545
- Zimmerman RA, Bilaniuk LT (1980) CT of orbital infection and its cerebral complications. Am J Roentgenol 134:45
- Zimmermann C, Schatz N, Glaser J (1990) Magnetic resonance imaging of optic nerve meningiomas. Ophthalmology 97:585-591
- Zinreich SJ, Kennedy DW, Malat J et al (1988) Fungal sinusitis: diagnosis with ct and MR imaging. Radiology 169:439
- Zingg M, Chowdhury K, Ladrach K et al (1991) Treatment of 813 zygoma-lateral orbital complex fractures. Arch Otolaryngol Head Neck Surg 117:611-622

Part II

Skull base and face

2- Central skull base

a - Sellar region
b - Parasellar and cavernous regions
c - Nasopharyngeal region

Sellar region

Cases 23, 24 and 33: Pituitary microadenoma

Clinical and pathological features

- Defined as adenomas which do not exceed 10 mm.
- More frequent than macroadenomas.
- Different hypersecreting types giving endocrine effects.
- *Signs:* variable according to endocrine activity:
  - prolactin hormone (35%): amenorrhea, galactorrhea, Forbes-Albright;
  - growth hormone (25%): acromegaly;
  - adrenocorticotropin hormone (5%): Cushing's or Nelson's syndromes;
  - thyrotropin hormone (less than 1%): hyper- or hypothyroidism;
  - gonadotropin hormone (less than 1%): menstrual disorders;
  - multiple hormones (PRL, GH, ACTH, TSH): (10%).
- Signs are caused by hormone hypersecretion or hyposecretion.
- Subclinical microadenomas may not evolve into autonomous adenoma.
- Incidental pituitary microadenomas (2-3 mm) may be found in about 7-10%.

MR characteristic findings

Morphology

- Located in the adenohypophysis in preferential sites:
  - prolactin hormone: basal lateral part;
  - growth hormone: anterior lateral part;
  - adrenocorticotropin hormone: central part or posterior lobe;
  - thyrotropin hormone: basal central part;
  - gonadotropin hormones: central part.
- Boundary between microadenoma and gland may be sharp.
- If laterally placed, microadenomas may cause bulging of the superior or inferior aspect of the pituitary with eventual stalk deviation ipsilaterally.
- ACTH secreting microadenomas are the smallest (3 mm): difficult to visualize.
- About 95% of microadenomas are detected on high resolution MRI (coronal cuts).
- Clinical and biological findings are essential in order to interpret abnormal (particularly incidental) morphological and SI findings.
- Follow-up MR exams in the same technical conditions (coronal) is a common attitude in incidental lesions as well as during medical treatment (bromocriptine).

Signal intensity

- SE-T1w: usually hypointense to normal gland, may be isointense (false negative).
- SE-T1w high resolution images may be conclusive in detecting the microadenoma.
- SE-T2w: may be slightly hyperintense to gland.
- No distinction of the various secreting types on the basis of MR signal intensity.

Contrast enhancement

- Very useful if plain SE-T1w images are non conclusive.
- Immediate SE-T1w acquisition after contrast infusion (half dose), will depict the microadenoma as a hypointense lesion within the enhancing normal gland.
- Delayed MR images (30 mn) may show contrast enhancement of the microadenoma after contrast washout from the normal pituitary.
- Dynamic MR acquisitions with fast imaging may be of help to depict microadenomas as hypointense lesions (due to delayed contrast uptake with respect to the normal glandular tissue rapidly enhancing).
- Note that focal susceptibility artifacts may mimic microadenomas, frequently observed close to the sellar floor at the level of a sphenoid sinus sagittal septum.

Case 23: Pituitary microadenoma (Cushing)

Fig. a: coronal cut; SE - T1w

Fig. b: coronal cut; SE - T1w + Gd

Patient presenting Cushing's syndrome. MR shows a very small, about 3 mm, hypointense lesion on the left side of the pituitary gland (**a**, ↑) seen on T1w. Following iv contrast, the hypointense lesion is still seen on the left side of the gland (**b**, ↑), while the remaining normal pituitary gland enhances homogeneously. Note the absence of displacement of the pituitary stalk.

Case 24: Pituitary microadenoma (prolactinoma)

Fig. c: coronal cut; SE - T2w

Fig. d: coronal cut; SE - T1w + Gd

Young patient explored for galactorrhea. MR shows a microadenoma presenting on T2w sequence as a 5 mm hyperintense lesion on the left side of the pituitary gland (**c**, ↑) well demarcated from the normal hypointense pituitary (**c**, *). The small mass is associated with contour deformities in the superior surface of the gland and right displacement of the pituitary stalk (**d**, *). These latter signs aid in the confirmation of the intrapituitary tumor. Immediately after contrast administration (about 35 seconds after), the dynamic coronal acquisition shows a homogeneous enhancement of the normal gland, the microadenoma remaining hypointense (**d**, ↑).

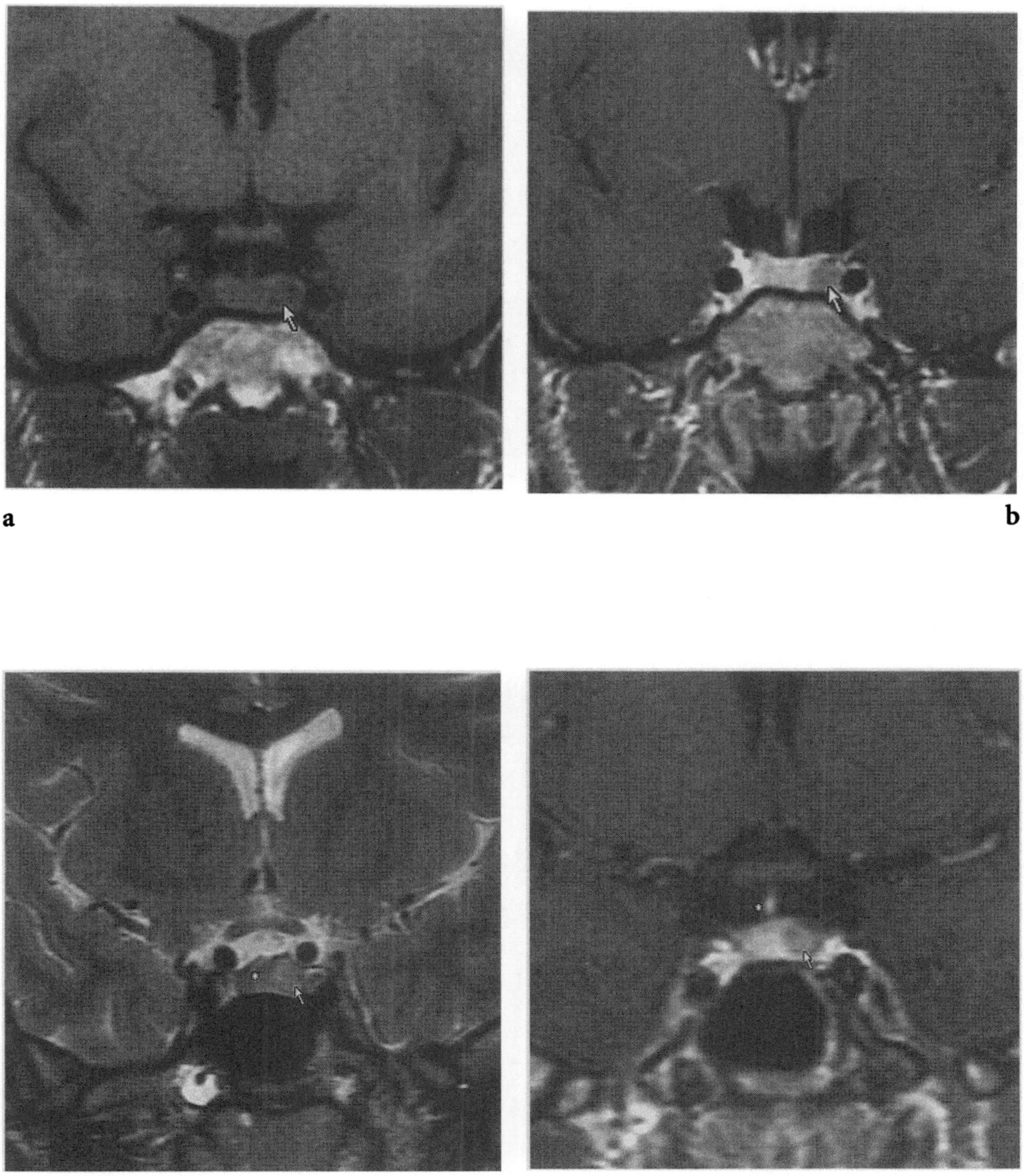

a b c d

Cases 25-29: pituitary macroadenomas

Clinical and pathological features

- Pituitary adenomas are the most frequent tumors of the sellar region.
- They correspond to about 15% of all intracranial tumoral processes.
- Macroadenomas correspond to the tumors exceeding 10 mm in diameter.
- Most of macroadenomas are non functional (delayed diagnosis).
- Clinical signs more commonly observed from 4th to 7th decades.
- Signs due to suprasellar or laterosellar expansion of the macroadenoma:

- various visual field defects, secondary to compression of the cisternal optic pathways: central scotoma and visual loss (compression of one optic nerve), or junctional hemianopia with blindness of one eye and temporal hemianopic defect in the other eye (compression of the anterior chiasmal notch); bitemporal hemianopia (compression of the chiasm); incongruous hemianopia (optic tract compression);
- cranial nerve palsies (10%), III nerve most frequently followed by VI and IV, secondary to invasion and/or compression of the cavernous sinuses (in huge or laterally invasive macroadenomas);
- endocrine signs in functional adenomas.

- After incomplete surgery or radiation therapy: annual visual field and MR follow-up is important in order to detect tumoral regrowth or arachnoidal adhesion with consecutive ptosis of the cisternal optic paths (visual deterioration).

MR characteristic findings

Morphology

- Macroadenomas may be intrasellar and may show variable suprasellar extent.
- Variable shape and suprasellar extent: ovoid or dumb-bell shaped or lobulated, with subfrontal, retrosellar or lateral expansions.
- Variable compression of the cisternal optic pathways (nerves, chiasm, tracts).
- Invasion of the cavernous sinuses more frequently encountered in large tumors and difficult to demonstrate (encasement of intracavernous internal carotid artery).
- Frequent intratumoral necrosis, hemorrhage (10-15%) or cystic degeneration which may show at times a fluid-fluid level.
- Superior tumoral extents are usually surrounded by slit-like CSF at brain interface.

Signal intensity

- SE-T1w: isointense or slightly hypointense to brain.
- T1w sequences are morphologically highly informative and diagnostic.
- SE-T2w: isointense or moderately hyperintense to brain.
- T2w sequence do not contribute to further evaluation of tumoral extensions, but may help to define relationships with the supraclinoid arteries and their bifurcations, and better evaluate intratumoral necrosis or hemorrhage.
- T2w sequence may provide informations regarding tumoral consistency: the macroadenoma showing iso-intense SI to brain white matter due presumably to the presence of intratumoral fibrosis, may consist of a very firm mass difficult or almost impossible to remove surgically using a trans-sphenoidal approach.

Contrast enhancement

- Most adenomas show homogeneous or heterogeneous enhancement, usually to a lesser degree than the normal pituitary tissue (coronal and sagittal cuts).
- Earlier contrast uptake of the adenomas than normal gland on dynamic MR acquisition, suggesting direct blood supply to the tumor.

Case 25: Intrasellar pituitary adenoma

Fig. a: coronal cut; SE - T1w

Fig. b: coronal cut (512); SE - T2w

Large intrasellar adenoma involving most of the intrasellar space lateralized to the right, showing a hypointense signal on T1w (**a**, ↑) and a hypointense signal, iso-intense to the cortex on T2w (**b**, ↑). The mass displaces the normal gland to the left side, well seen on the T1w sequence as a relatively hyperintense mass and displaces the pituitary stalk to the left (**a**, *). There is a very moderate superior extension of the adenoma into the suprasellar cistern and a lateral extension with slight compression of the medial right cavernous sinus wall with subtle displacement of the intracavernous artery (**a**).

Case 26: Intrasellar pituitary adenoma

Fig. c: sagittal cut; SE - T2w

Fig. d: coronal cut; SE - T2w

Similar case showing an intrasellar mass with hyperintense signal on T2w sequence (**c**, ↑↑; **d**, ↑↑) displacing posteriorly the normal pituitary gland remaining hypointense (**c**, ↑). The mass extends slightly into the suprasellar cistern but remains at distance of the inferior aspect of the optic chiasm well shown on the sagittal cuts (**c**, *) and on the coronal cuts (**d**) with interposed hyperintense CSF. The mass shows no lateral invasion of the cavernous sinuses. Surgically, this was proved to be an ACTH-secreting adenoma.

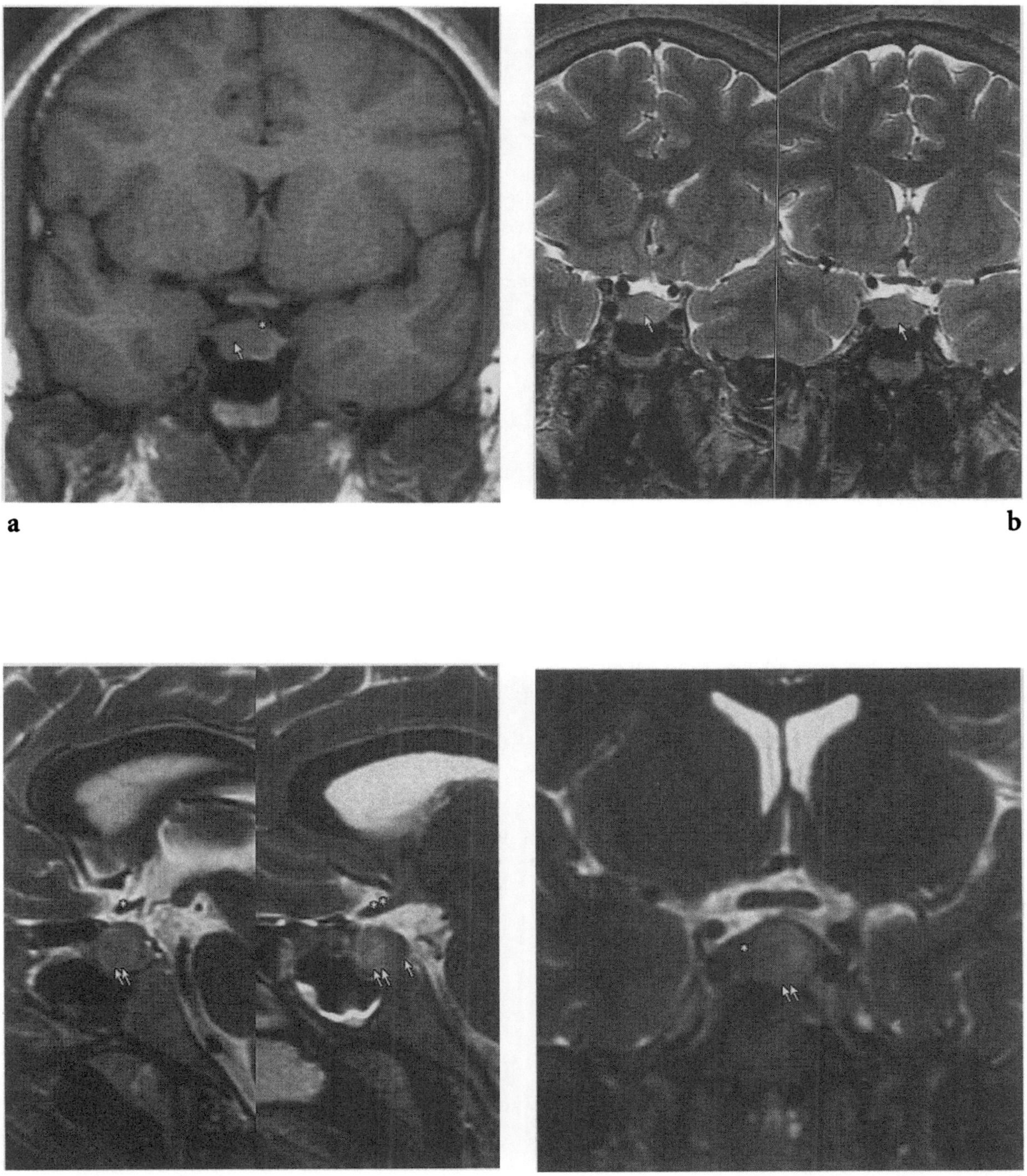

a b

c d

Case 27: Pituitary adenoma (ACTH, Cushing)

Fig. a: sagittal cut; SE - T2w

Fig. b: coronal cut; SE - T1w + Gd

Patient presenting a cushingoid facies. T2w-mid-sagittal MR cut shows a macroadenoma slightly heterogeneous extending to the suprasellar cistern (**a**, ↑; **b**, ↑) and reaching the chiasm (**a**, *). After contrast iv, the mass enhances very slightly and shows a lateral right extension (**b**). Note the lateral inflexion of the floor of the sella (**b**, ↑↑). There is no invasion of the cavernous sinuses well delimitated after contrast iv, due to normal enhancement of the medial venous plexuses (**b**, *).

Case 28: Pituitary adenoma (TSH)

Fig. c: coronal cut; SE - T1w

Fig. d: coronal cut (512); SE - T2w

Pituitary macroadenoma enlarging the sella turcica and expanding in the suprasellar cistern. The mass is iso-intense on T1w (**c**, ↑↑) becoming slightly hyperintense and heterogeneous on T2w sequence (**d**, ↑↑). The suprasellar extension of the adenoma reaches the inferior aspect of the chiasm slightly raised (**c**, **; **d**, *) but does not affect the cisternal optic nerve (**c**, left, *). The sphenoid sinus is normal (**c**, ↑↑↑).

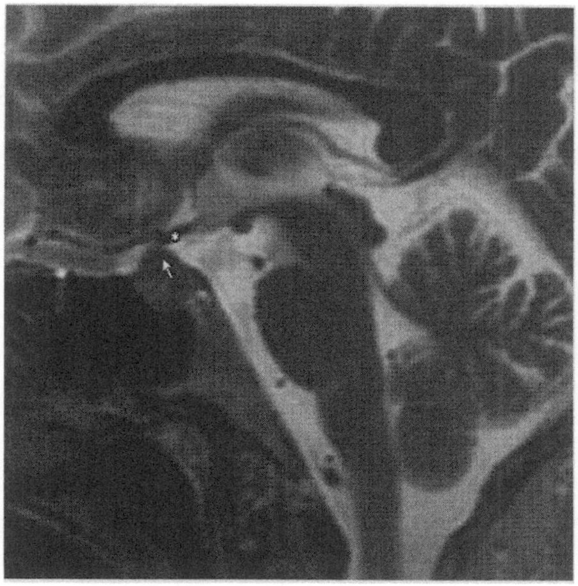

a

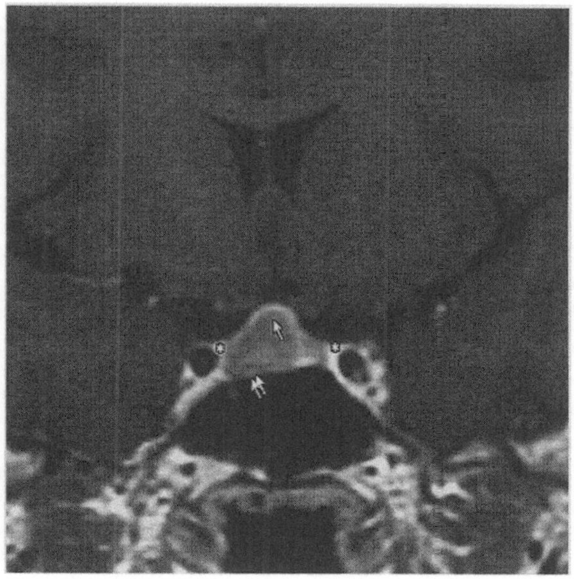

b

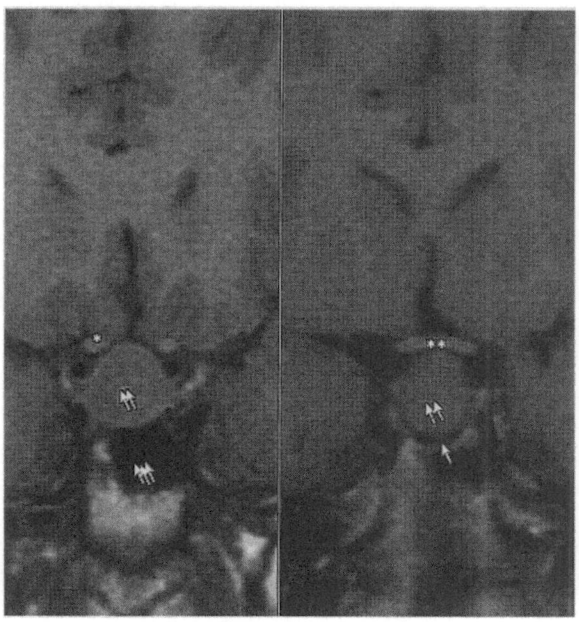

c

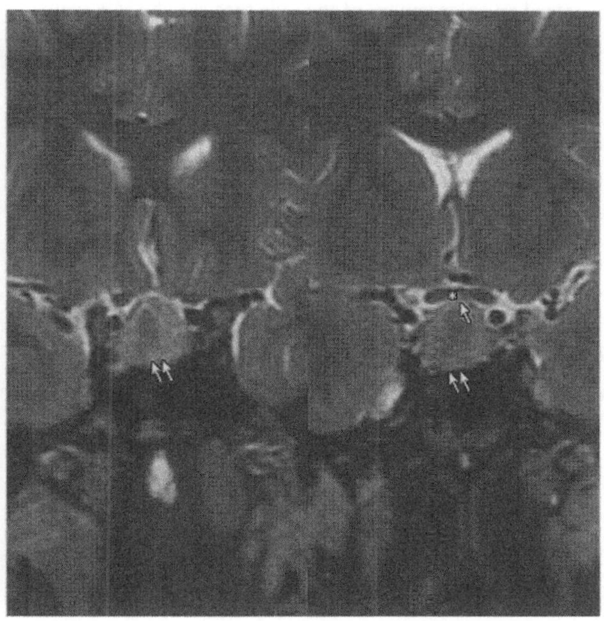

d

Case 29: Invasive suprasellar adenoma (GH)

Fig. a: coronal cut (512); SE - T2w

Fig. b: coronal cut; SE - T1w + Gd

Patient presenting with acromegaly. Standard X-rays showed an enlargement of the sella turcica. The MR exam showed an important soft tissue mass in the sella turcica and the suprasellar cistern with an important expansion of the sella itself which may be considered as a good sign that the tumor originates from within the sella rather than above it. The mass is iso-intense to the cortex on T2w (**a**) and enhances homogeneously after contrast iv (**b**). The chiasm and the anterior portion of the 3rd ventricle are elevated and displaced posteriorly (not shown). The intrasellar portion of the mass displaces to the right the normal portion of the gland (**b**), hyperintense after iv contrast. The proximal portions of the anterior cerebral and pericallosal arteries are raised by the suprasellar portion of the macroadenoma (**a**, *; **b**, **). Note that it is difficult to assess the eventual subtle degree of left cavernous sinus invasion (**b**). Note also the absence of infiltration of the adjacent brain parenchyma and of reactive edema.

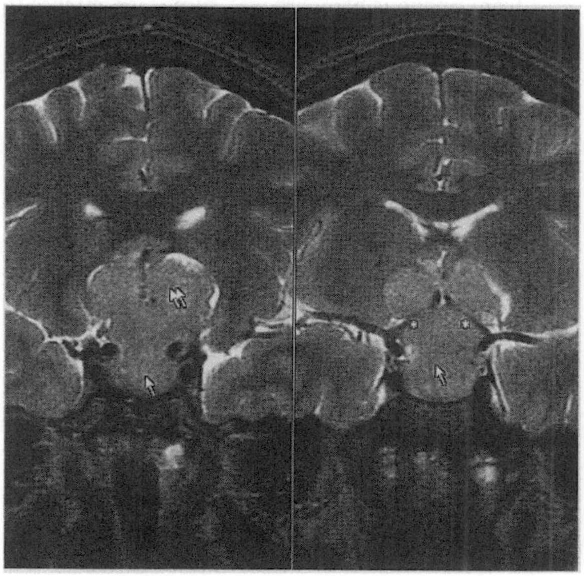

a

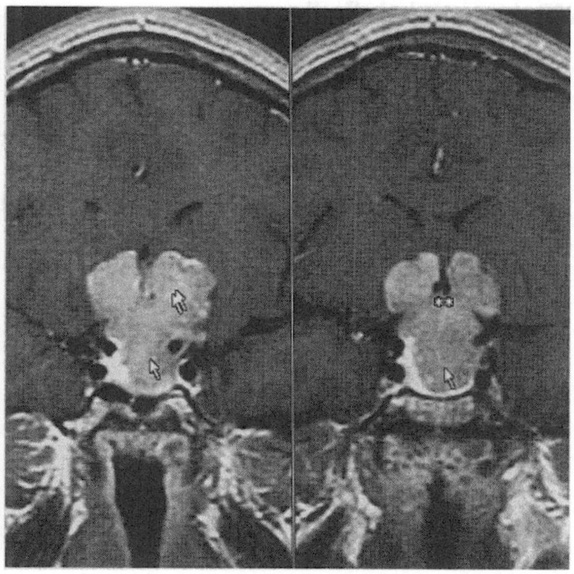

b

Case 30: Pituitary apoplexy

Clinical and pathological features

- Due to spontaneous necrosis and hemorrhage into a macroadenoma.
- May give rise to secondary ballooned empty sella.
- *Signs:* acute syndrome including severe headache, ophthalmoplegia, visual loss, signs of subarachnoid hemorrhages; endocrine signs such hypopituitarism; rapid obtundation or stupor.
- Sheehan's syndrome corresponds to postpartum ischemic necrosis and hemorrhage, and involving a normal sized pituitary gland in a normal sella turcica.

MR characteristic findings

Morphology

- Commonly macroadenoma with important suprasellar extension.
- Intratumoral necrosis or cystic degeneration and hemorrhage usually contained within the tumor.
- Hemorrhage occurs in about 25% of macroadenomas (commonly subclinical).
- Note that intratumoral bleeding is frequent during treatment by bromocriptine.

Signal intensity

- SI is tributary to the presence of blood breakdown products and cystic formations.
- SE-T1w: intratumoral subacute (2-3 days from bleeding) hemorrhage shows hyperintense SI (methemoglobin deposits), remaining tumoral process iso-intense.
- SE-T2w: intratumoral subacute bleeding shows hypointense SI (presence of deoxyhemoglobin, intracellular methemoglobin, or hemosiderin from previous hemorrhage), tumor iso-intense to brain.

Contrast enhancement

- Adenoma shows marked enhancement (T1).
- May not be necessary in case of subacute hemorrhage (hyperintense on T1), unless invasive adenoma, to evaluate the lateral and superior extents (coronal).

Case 30: Pituitary adenoma, hemorrhagic necrosis

Fig. c: sagittal cut; SE - T2w

Fig. d: coronal cut; SE - T1w

Young woman presenting a syndrome of amenorrhea-galactorrhea accompanied by hyperprolactinemia. MR demonstrated an intra- and suprasellar macroadenoma showing heterogeneous signal intensity on T1w (**b**) and on T2 w (**c**). MR shows a hyperintense focus occupying the major central portion of the mass lateralized to the right side on T1w (**d**, ↑↑) and T2w (**c**, ↑↑), encircling a very small area of iso-hypointense signal (**d**, ↑) more hypointense on T2w (**c**, ↑) corresponding to fresh blood, the remaining major portion of hyperintense signal being due to subacute hemorrhage with methemoglobin deposition inside the mass. The non hemorrhagic portion of the macroadenoma, at the periphery of the hemorrhagic focus as well as at the left intrasellar region (**d**) is iso-intense to brain on T1w (**d**) and T2w (**c**). The suprasellar extent of the macroadenoma elevates the chiasm (**d**, up, *).

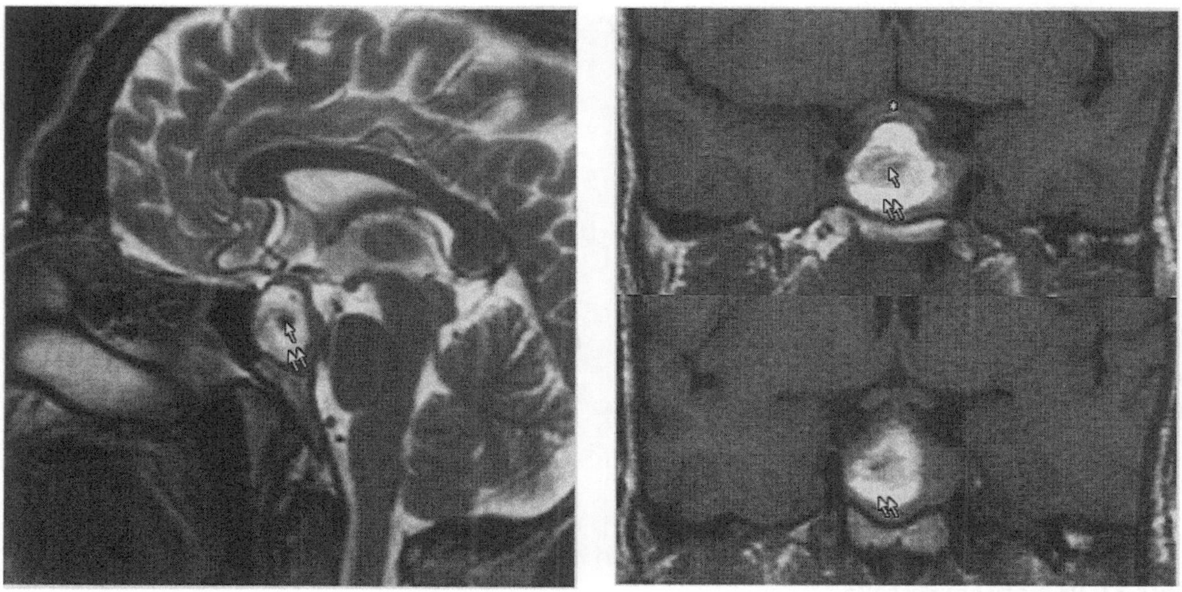

c d

Cases 31-32: Empty sella

Clinical and pathological features

- Commonly misdiagnosed cause of non tumoral enlargement of the sella.
- Corresponds to an intrasellar arachnoidocele through a deficient diaphragm.
- Dehiscent diaphragm greater than 5 mm is found in 40% of normal autopsies.
- The result is a remodeling and enlargement of the sella due to CSF pressure.
- 2 varieties: primary and secondary empty sellas.
- *Signs:* headache, visual field defects, hypopituitarism, rarely rhinorrhea.
  - *Primary empty sella:* obese female, aged 50, multiparous, often hypertensive, presenting headaches, blurred vision, diplopia, bilateral papilledema, commonly asymptomatic.
  - *Secondary empty sella:* follow surgery or radiotherapy or infarction of a normal pituitary or an intrasellar tumor, visual field defects secondary to adhesion retracting downward the chiasm and optic nerves and posterior gyri recti, into the enlarged sella.

MR characteristic findings

Morphology: primary empty sella

- Enlarged sella with smoothly curved walls and flattened pituitary gland against the sellar floor which is smoothly curved, the pituitary stalk shown extending down at the bottom of the sella (coronal).
- Remodeling of the sellar floor is usually symmetrical, flat or concave (coronal cuts), but may be asymmetrical due to an eventual presence of a parasagittal septum in the sphenoid sinus.
- The enlarged sella maintains a normal "closed" shape (sagittal) without straigthening and erosion of the dorsum sellae or erosion of the sellar floor.
- CSF filled intrasellar space communicating with the suprasellar cistern.

Signal intensity

- SI follows similar to CSF on all sequences (primary empty sella).
- SE-T1w: hypointense to brain.
- SE-T2w: hyperintense to brain.

Contrast enhancement

- No contrast enhancement (primary empty sella).
- Enhancement of eventual residual necrotic adenoma (secondary empty sella).

Case 31: Primary empty sella

Fig. a: sagittal cut; SE - T2w

Middle age obese woman. MR showed an apparent enlarged empty sella turcica, filled with cerebrospinal fluid (**a**, ↑), with hyperintense signal on T2w, showing CSF-like signal on all sequences. The pituitary gland is severely flattened in the sellar fundus.

Case 32: Secondary empty sella

Fig. b: coronal cut; SE - T2w

T2w coronal slices showing an enlarged slightly deformed sella turcica filled with cerebrospinal fluid and largely communicating with the chiasmatic cistern. The patient was treated medically for a prolactin macroadenoma.

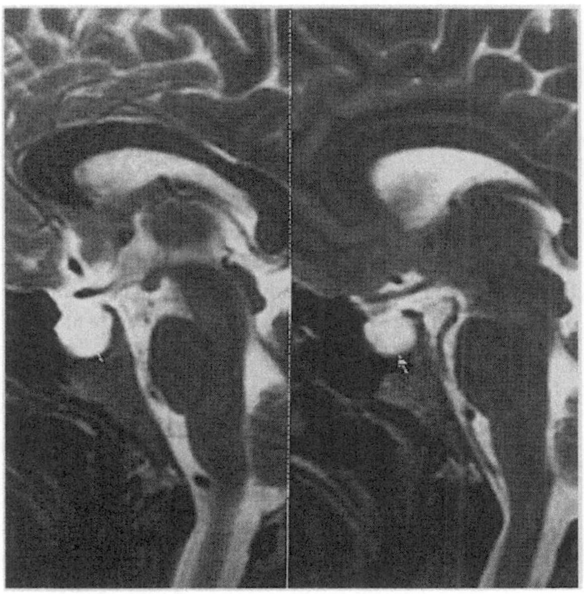

a

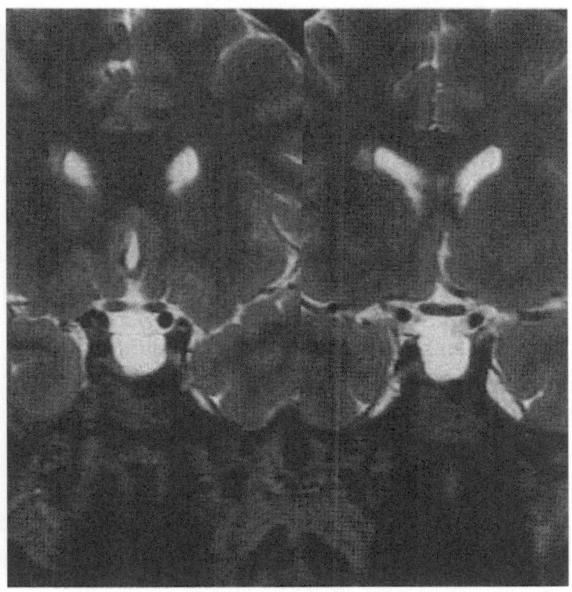

b

Case 33: Microadenoma and meningiomatosis

Fig. c: sagittal cut; SE - T1w + Gd

Fig. d: coronal cut; SE - T1w + Gd

Incidental finding associating a pituitary microadenoma located at the right infero-lateral border of the pituitary gland adjacent to the medial wall of the right cavernous sinus (**d**, ↑), presenting as a hypointense non enhancing mass contrasting with the hyperintense normal gland. It is difficult in this case to predict the right cavernous sinus invasion reliably. Its medial wall seems merely displaced laterally. MR exam also evidences two extra-axial frontal convexity masses showing marked enhancement with hyperintense signal on T1w (**c**, ↑↑). These tumors are dural-based and show for the larger one, small adjacent tails of dural thickening extending on both sides (**c**, down, ↑).

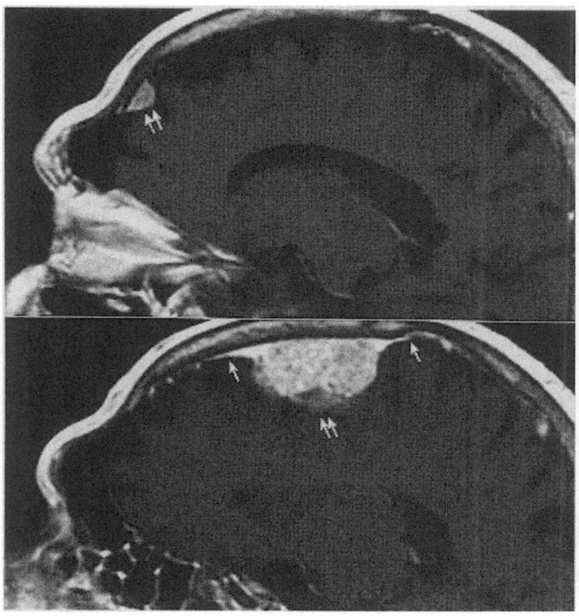

c

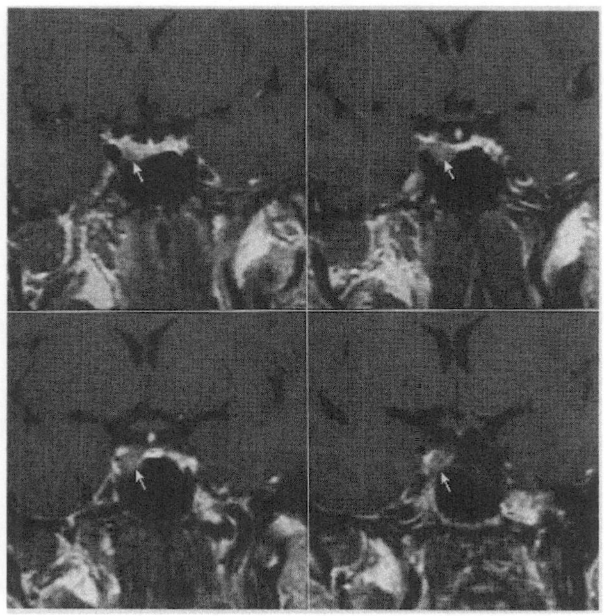

d

Case 34: Ectopic neurohypophysis

Clinical and pathological features

- Observed in pituitary dwarfism.
- Prevalence: 2 males/1 female.
- Found in over 40% patients presenting a growth hormone deficiency.
- Associated multiple hormone deficiencies.
- *Signs:* short stature (dwarf), delayed skeletal and dental development.

MR characteristic findings

Morphology

- Small sella turcica and small pituitary gland.
- Absence (or hypoplasia) of the pituitary stalk and transection of infundibulum.
- Ectopic neurohypophysis located in tuber cinereum or the infundibulum presenting as a "bright spot" on T1.
- Ectopia of post-hypophysis, associated to an absence of the olfactory sulcus in the orbito-frontal lobe, may be observed in Kallmann's syndrome: genetic disease of neuronal migration. Clinical signs: anosmia and hypogonadotropic hypogonadism.

Signal intensity

- SE-T1w: hyperintense ectopic posthypophysis (bright spot in median eminence).
- SE-T2w: isointense to brain.

Contrast enhancement

- No need to perform contrast infusion (obvious bright signal on T1).

Case 34: Ectopic neurohypophysis (dwarf)

Fig. a: sagittal cut; GE - T1w

Fig. b: coronal cut; SE - T1w

Sagittal and coronal T1w sequences showing a small sella turcica containing a small pituitary gland. The pituitary stalk is not visible and there is a focal area of high intensity (**a, b**, ↑) at the proximal infundibulum (**a**, ↑) corresponding to the ectopic location of the posterior pituitary lobe. The high intensity nodule is well shown at the level of the infundibular recess between the optic tracts (**b**) diplayed laterally. These findings are specific for the disorder.

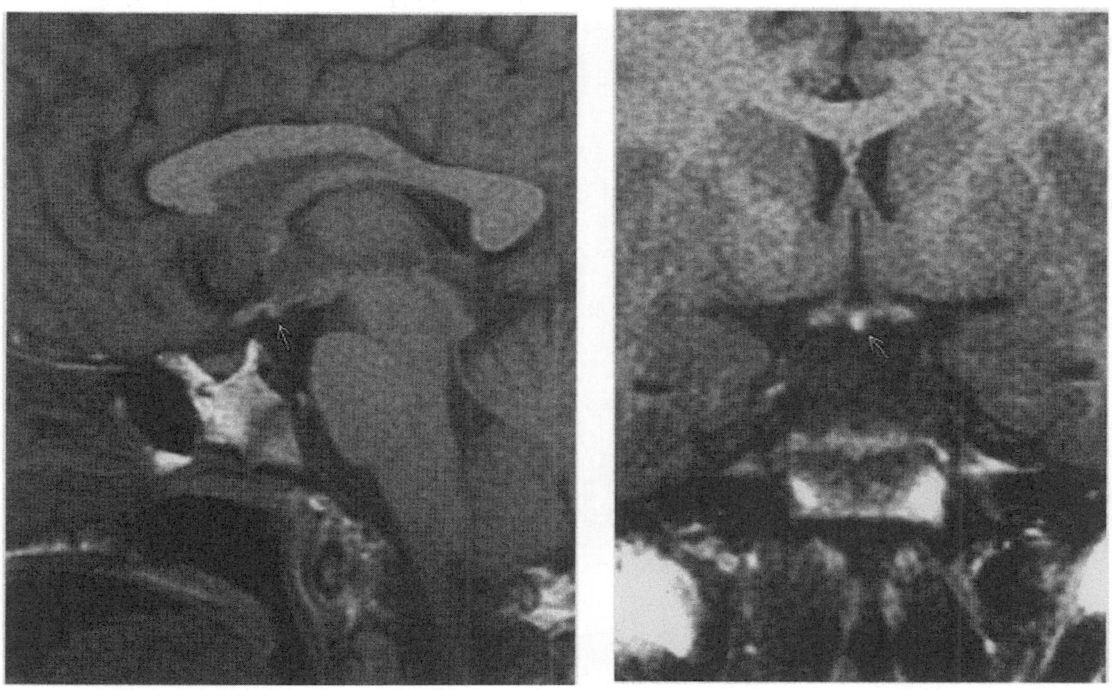

a b

Case 35: Rathke's cleft cyst

Clinical and pathological features

- May be observed in all age groups.
- Derive from remnants of Rathke's pouch, related to craniopharyngiomas.
- Benign epithelium-lined mass containing mucoid or serous fluid and cellular debris.
- Located in the anterior sella and suprasellar cistern.
- Found in about 10% of pituitary glands, more often in its median part.
- Small cysts are usually found incidentally, patients being asymptomatic.
- *Signs* (large cysts): endocrine disorders, visual field defects, headaches.

MR characteristic findings

Morphology

- Sharply delineated mass, located in the pars intermedia of the hypophysis (found between the anterior and posterior pituitary) or anterior to the pituitary stalk.
- Variable size ranging from 2 mm to 3 cm (intra- and suprasellar).
- Content: serous or mucoid fluid (high protein), cellular debris (mimic solid mural nodule, difficult to separate from cystic craniopharyngioma).

Signal intensity

- Variable signal intensity depending on fluid content.
- SE-T1w: hyperintense to brain (may be iso- to hyperintense to CSF).
- SE-T2w: hyperintense to brain (may be iso- to hypointense to CSF).

Contrast enhancement

- No contrast enhancement of the cystic mass.
- Exceptional rim enhancement of cystic wall.

Case 35: Rathke's pouch cleft cyst

Fig. c: sagittal cut; GE - T1w

Fig. d: axial cut; SE - PDw

Intra- and suprasellar cystic mass presenting a homogeneous hyperintense signal on T1w (**c**, ↑) and a hyperintense signal to CSF on PDw sequence (**d**, ↑). The mass elevates the chiasm as well as the floor of the third ventricle displaced superiorly and posteriorly (**c**, *). The content of the cyst was a mucoid fluid, explaining the hyperintense signal on T1 and T2w sequences. Note that this pattern may also be observed in case of cystic craniopharyngioma.

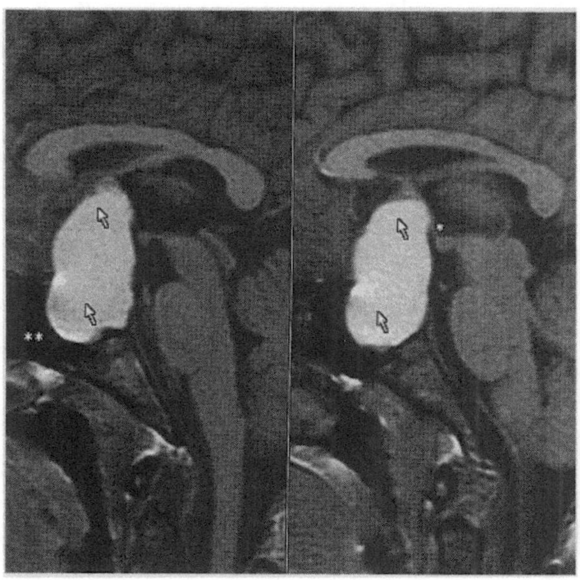

c

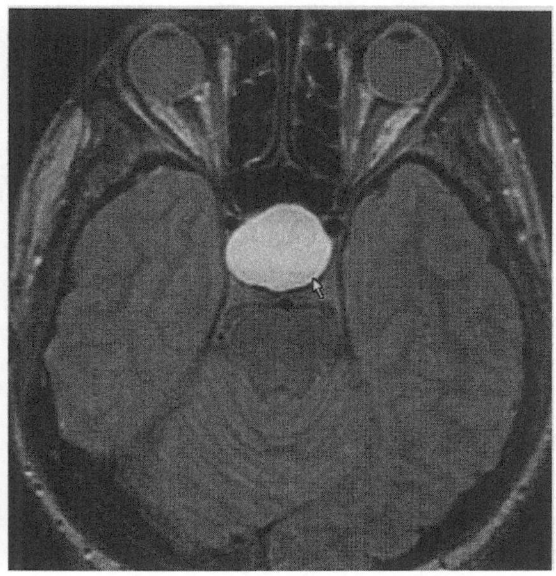

d

Parasellar and cavernous regions

Cases 36, 37: Chordomas of skull base

Clinical and pathological features

- Arise from notochordal remnants, in the clivus.
- Less than 1% of all intracranial tumors, in adults aged 25-50.
- Affect the sacrococcygeal region (50%) and the skull base (40%).
- Clival chordomas originate from the spheno-occipital synchondrosis.
- Slow growing locally aggressive tumors.
- Chondroid chordoma: less aggressive behavior and better prognosis.
- *Signs:* headaches, cranial nerves palsies, endocrine signs.
- High recurrence rate after surgery.

MR characteristic findings

Morphology

- Histologically benign but locally invasive.
- Though usually located in the midline are paramedian in 1/3 of cases.
- Frequent extension to the basisphenoid, petrous apex and greater sphenoid wing.
- Contain coarse calcifications and bony sequestra from bone destruction (50%).
- Intratumoral hemorrhages (highly vascular) or cysts may be found.
- Heterogeneous aspect: tumor vascularity (flow voids), calcifications.
- Differential diagnosis: metastasis, chondroma, chondrosarcoma.
- Chondrosarcoma (malignant cartilaginous tumor) may be difficult to differentiate from chordomas: usually slow growing paramedian neoplasm, arising at the spheno-occipital synchondrosis, sharply marginated, with bony erosion, showing low T1 and T2 and marked contrast enhancement.

Signal intensity

- SE-T1w: hypointense to brain (unless presence of methemoglobin hyperintense).
- SE-PDw: hyperintense to brain, may show curvilinear vascular flow voids.
- SE-T2w: hyperintense to brain, highly heterogeneous, rarely homogeneous.

Contrast enhancement

- Most frequently show marked or moderate contrast enhancement.

Case 36: Sphenoid chordoma

Fig. a: sagittal cut; GE - T1w

Fig. b: sagittal cut; SE - T1w + Gd

Fig. c: axial cut; SE - T2w

Huge mass developed in the body of the sphenoid bone destroying most of the sphenoid body and the sellar region with posterior extension in the prepontine cistern (**a**, ↑) and anterior and inferior extension toward the nasopharynx (**a**, ↑↑). The mass is hypointense on T1w (**a**) and is surrounded anteriorly by a hyperintense material corresponding to subacute hemorrhage confirmed at surgery (**a**). On T2w sequence (**c**) the mass shows a hyperintense signal (**c**, left) different from the hypointense signal of the pituitary gland raised into the chiasmal cistern (**c**, right, ↑). After iv contrast, the mass enhances markedly showing some intratumoral curvilinear hypointensities most likely representing tumor vascularity (**b**). The pituitary gland is raised into the suprasellar cistern reaching the chiasm (**b**; **c**, right, ↑). The tumor was developed extradurally.

Case 37: Clival chordoma

Fig. d: sagittal cut; SE - T1w + Gd

Heterogeneous huge mass destroying the upper portion of the clivus and extending anteriorly (**d**, ↑) into the sellar region and backwards into the posterior fossa, occupying the prepontine and interpeduncular cisterns, with mass effects on the brainstem displaced posteriorly (**d**, ↑). The mass is heterogeneously hyperintense after contrast injection showing multiple small rounded and corkscrew-like hypointensities representing tumor vascularity. The upper portion of the lesion reaches the chiasm and the floor of the 3rd ventricle (**d**, *). The anterior inferior aspect of the clivus shows no abnormality (**d**, **).

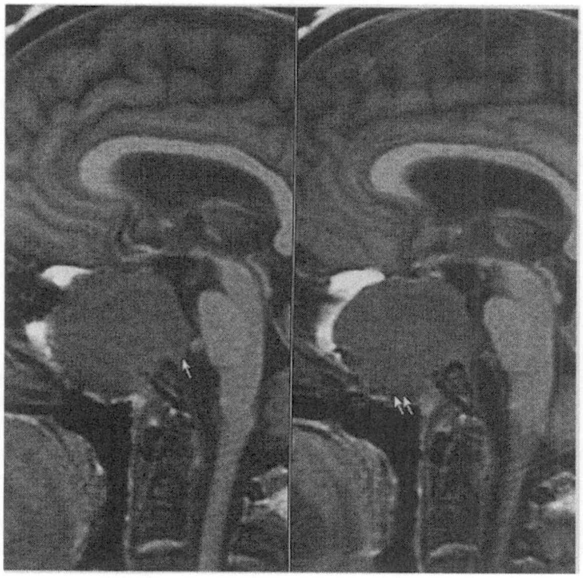

a

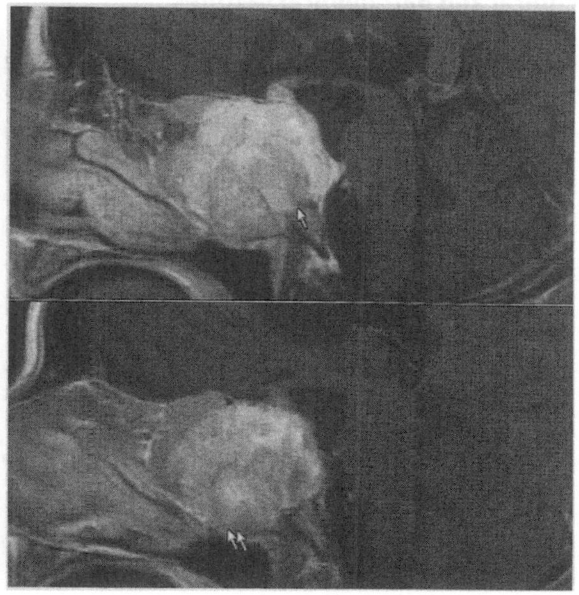

b

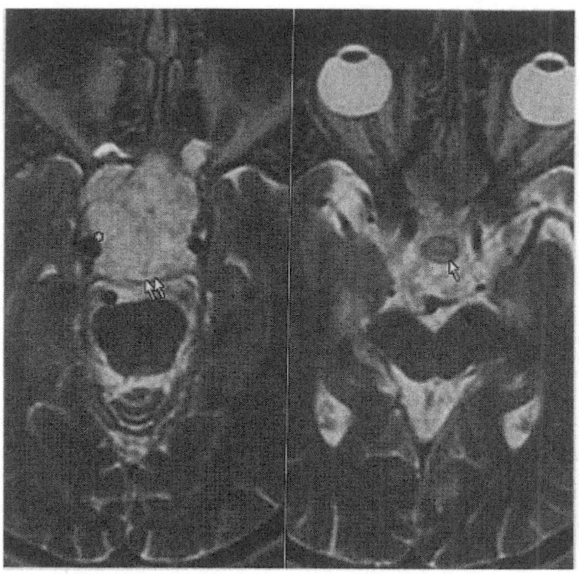

c

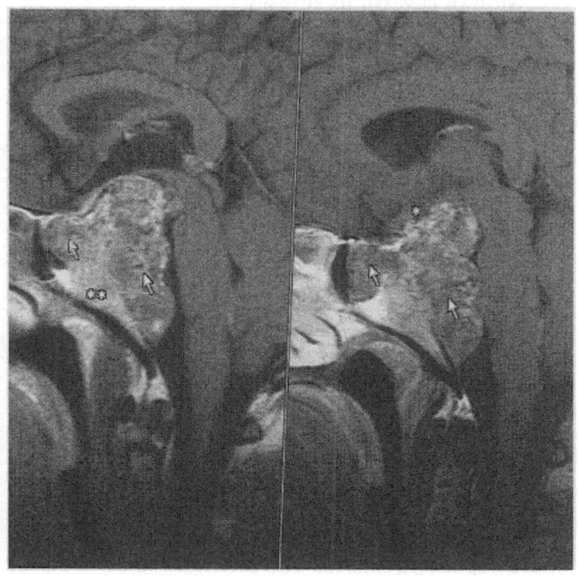

d

Case 38 : Sphenoiditis and sinus mucocele

Clinical and pathological features

- Isolated sinusitis and mucocele of the sphenoid sinus are frequent.
- Rule out focal causes: nasal mass, deviated nasal septum which may block the sinus ostium and compromise drainage, osteitis, or consider systemic origin: altered immunity, allergy.
- Expansion of the sinus due to mucus accumulation leading to mucocele or mucopyocele formation.
- *Signs:* vertex, periorbital or frontal cephalgias, may include ocular motor palsies and visual loss (regional expansion of the mucocele with compression of the cavernous sinuses or extent to orbital apex).

MR characteristic findings

Morphology

- *Large mucocele:* expansile lesion bowing sinus walls, with erosion of sellar floor and compression of cavernous sinuses.
- *Mucopyocele:* extension of the infectious process to the other sinus cavities and to the leptomeninges, with eventual meningitis, subdural (or epidural) empyema, cavernous sinus thrombosis, cerebritis.

Signal intensity

- Variable signal intensity on SE-T1 and T2.
- SI depends on secretions content (proteinaceous material, associated hemorrhage).

Contrast enhancement

- No contrast enhancement unless infected: rim enhancement of the mucopyocele; enhancement of the intracranial leptomeninges.

Case 38: Sphenoiditis

Fig. a: sagittal cut;	SE - T2w	**Fig. c:** coronal cut;	SE - T1w
Fig. b: sagittal cut;	SE - T1w + Gd	**Fig. d:** coronal cut;	SE - T1w + Gd

Unusual sphenoid sinusitis with cavernous sinus extension and intracranial invasion. On the sagittal T2w sequence, MR shows thickening and hyperintense signal of the sphenoidal sinus mucosa associated with a fluid-filled cavity (**a**, right, ↑↑) and a subfrontal high intensity collection (**a**, ↑). After Gadolinium injection, the sphenoid sinus mucosa shows enhancement delineating the intrasinusal fluid collection (**b**, ↑↑). The contrast enhancement is also observed along the anterior cranial base due to extension of the infection through the walls of the sinus with meningeal contrast enhancement (**d**, up, ↑↑), associated with fluid collections remaining hypointense on T1w (**c**) and after contrast injection. Note that the process extends also to invade the intrasellar region raising the pituitary gland (**d**, **) hyperintense after contrast iv as well as into the right cavernous sinus (**d**, *). The abnormalities and fluid formations are well differentiated after contrast (**d**) in comparison to pre-contrast T1w sequences (**c**).

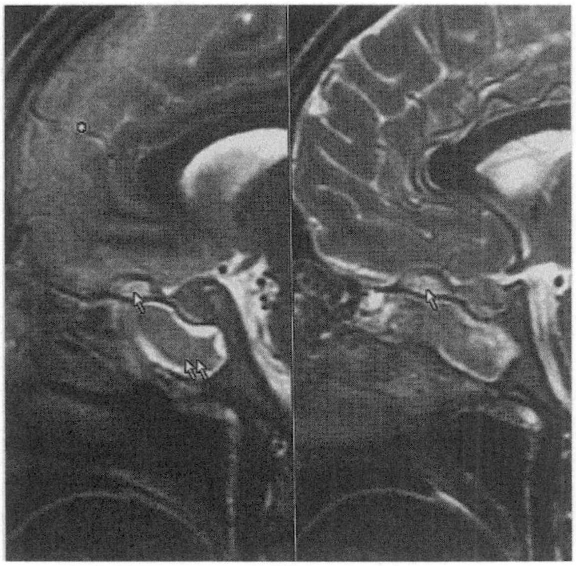

a

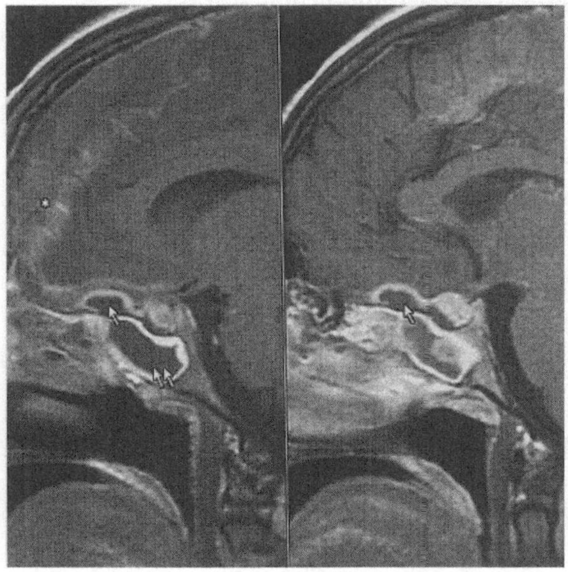

b

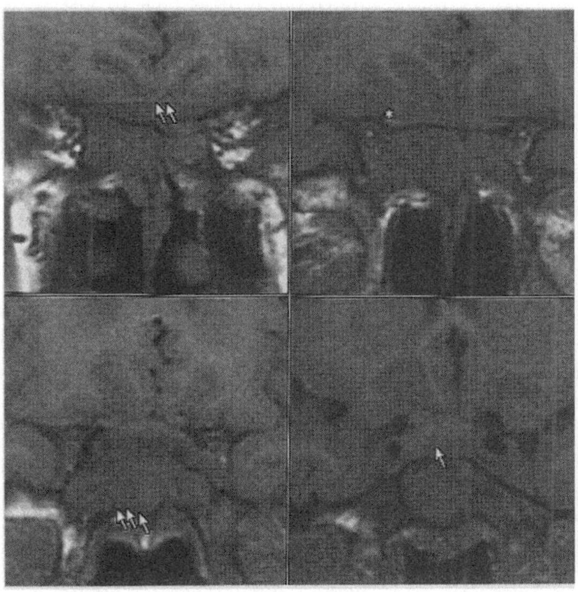

c

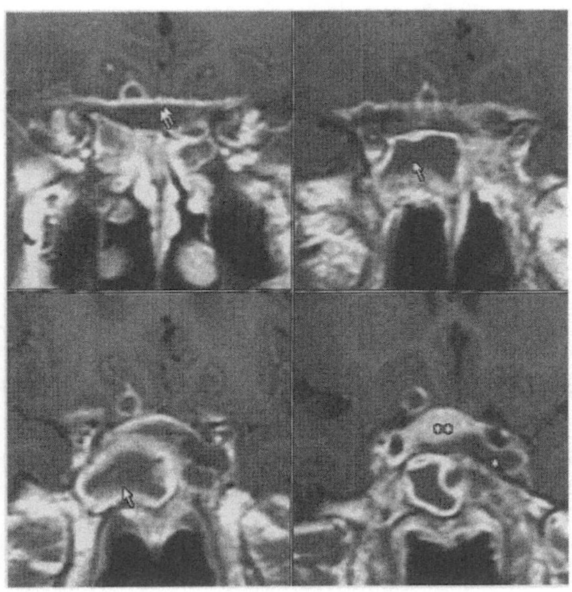

d

Case 39: Giant parasellar aneurysms of ICA

Clinical and pathological features

- Intracavernous aneurysms arise from the internal carotid as it traverses the cavernous sinus; "giant" when exceeding 2.5 cm.
- Uncommon, show slowly progressive enlargement with compression effect.
- May cause carotid-cavernous sinus fistula.
- *Signs:* ipsilateral pain in or about the eye, single or multiple ocular motor palsies: VI, III or IV trigeminal nerve involvement, mid-dilated pupil.
- Important to identify on MR: life threatening.

MR characteristic findings

Morphology

- Found in the parasellar or even developing in the sellar area, may become huge.
- Must not be confused with a solid sellar mass (may simulate a macroadenoma).
- Variable aspect related to laminated clot formation with partially or completely filled lumen (flow void of flowing blood in patent ectatic sac associated with ghost artifacts along the phase-encoding axis proving vascular origin and/or heterogeneous high SI of turbulent flow: intraluminal high SI may be due to refocusing of slow flow by gradient moment nulling on SE-T2w).
- Cause erosion of the anterior clinoid, optic foramen, and superior orbital fissure.
- May extend posteriorly to erode petrous apex.
- Saccular intracavernous aneurysms may be found bilaterally.

Signal intensity

- Characteristic MR aspect (SE-T1 and SE-T2), variable (thrombosis).
- Flow void of flowing blood in patent lumen (ghost pulsation artifacts, diagnostic); gradient echo sequence depicts inflow enhancement inside the lumen.
- Intrasaccular laminated clot: high SI on T1 (stripes of hyperintensity).
- Low SI on T2 due to hemosiderin deposits (peripheral rim).
- MR angiography: visualizes the patent lumen and discloses the neck of the aneurysm and its relation with the parent vessel.

Contrast enhancement

- 3D TOF MR angiography may lack to visualize slowly filling giant aneurysm (slow intraluminal flow saturation effect), showing only its peripheral aspect.
- Contrast infusion (3D TOF) helps to demonstrate the entire saccular aneurysm.
- Conventional SE sequences may suffice to the positive diagnosis.

Case 39: Intrasellar thrombosed giant aneurysm of the ICA

Fig. a: sagittal cut;	SE - T1w	**Fig. c:** coronal cut;	SE - T2w
Fig. b: sagittal cut;	SE - T2w	**Fig. d:** angio, coronal view;	2D-TOF

Large mass developed in the sella turcica and extending into the chiasmal cistern highly heterogeneous with mixed high and low signal within the tumor-like formation. The hyperintense signal on T1w (**a**, ↑↑) is due to the presence of methemoglobin in the thrombus. The hypointense areas on T2w (**b**, **c**) correspond to hemosiderin deposit. Note that at the anterior portion of the intrasellar mass, sagittal MR shows a small signal void mass developed into the upper portion of the sphenoid sinus with no associated flow artifacts along the phase encoding direction. This mass is demonstrated on MR angio (**d**, ↑) as corresponding to a small remaining patent lumen at the upper and anterior portion of the thrombosed intrasellar aneurysm (**d**, ↑↑↑). Flow into the aneurysm is well shown coming from the cavernous portion of the right internal carotid artery (**d**, ↑↑). This is an unusual, rather exceptional case due to median intrasellar location of a largely thrombosed giant aneurysm mimicking a macroadenoma, with a small patent lumen contained in air cells.

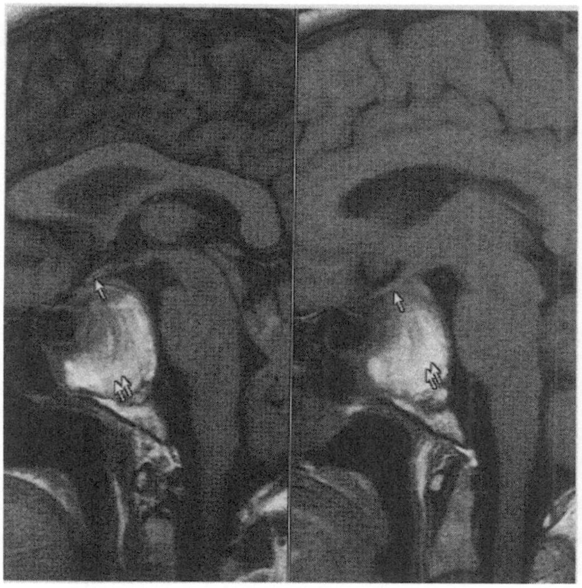

a

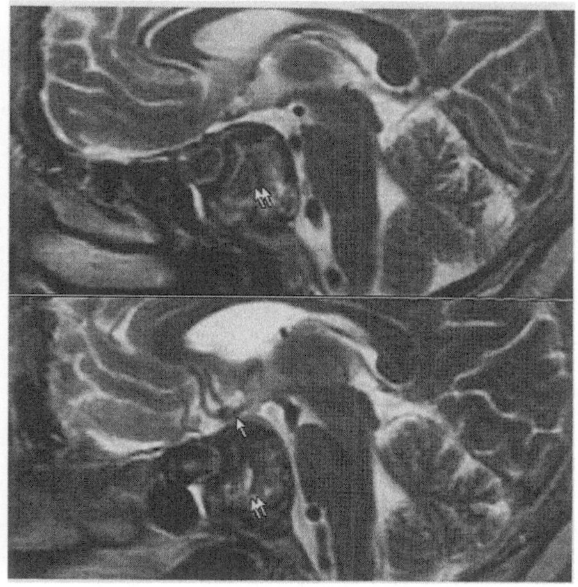

b

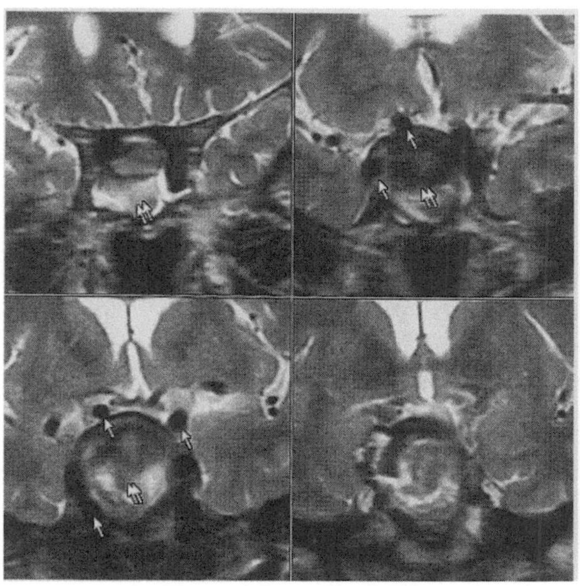

c

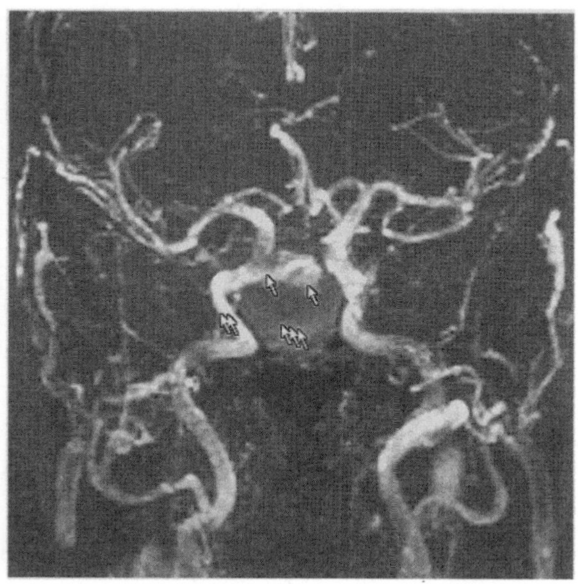

d

Cases 40, 43 and 46: Jugo-optic meningiomas

Clinical and pathological features

- Meningiomas arising from the tuberculum sellae and adjacent region of the planum sphenoidale, the optic canal, the anterior clinoid process or the diaphragma sellae.
- Meningiomas of perisellar region account for 10% of intracranial meningiomas.
- Most common site is the tuberculum sellae.
- Most clino-optic meningiomas are very small (less than 2 cm) at the time of detection, producing early visual deficits.
- Predilection to occur in middle age women (around 45-50 years).
- *Signs:* slowly progressive with fluctuations over months of a monocular visual loss, secondary to ipsilateral optic nerve compression, and/or bilateral asymmetric visual field defects secondary to compression of the optic nerve and adjacent anterior chiasm (junctional defects); headaches, periorbital pain (mimicking retrobulbar neuritis), optic pallor or atrophy (late sign), rarely a Foster-Kennedy syndrome (optic atrophy and contralateral papilledema).
- Surgery through frontal craniotomy approach is indicated in most circumstances.

MR characteristic findings

Morphology

- Bony hyperostosis and/or "blistering" of the planum sphenoidale or the tuberculum sellae at the site of dural attachment, are characteristic radiological signs (frequent finding on lateral X-ray or midsagittal MR cut) in jugo-optic meningiomas (excluding those arising from diaphragma sellae).
- MR shows a benign suprasellar extra-axial mass, lying over the jugum or the tuberculum sellae, or encircling the anterior clinoid process, usually separated from adjacent brain parenchyma by a cleft of CSF.
- Relationships with the optic canal, cisternal optic pathways, the pituitary gland and the supraclinoid internal carotid artery and its branches, must be evaluated (best achieved using SE-T2 and enhanced SE-T1 sequences in the sagittal and coronal planes and the sagittal oblique plane parallel to the optic canal axis).
- Small jugo-optic meningiomas may frequently be overlooked ("en plaques").

Signal intensity

- SE-T1w: usually iso-intense to gray matter or moderately hypointense to white matter and pituitary gland.
- SE-PDw: slightly hyperintense to brain cortex.
- SE-T2w: variable, iso-intense to gray matter or slightly hyperintense (about half the cases) mainly on FLAIR.

Contrast enhancement

- Jugo-optic meningiomas are best evaluated on enhanced MR images: usually show intense and homogeneous contrast enhancement, including "dural tail".
- Best evaluation of eventual extension to the optic canal, ipsilaterally or even contralaterally (late diagnostic stage), using sagittal oblique cuts.
- Helps to evaluate relationships with the optic nerves and chiasm and eventual encasement ICA branches (coronal and axial in the CH-PC plane).
- Allows differentiation from the pituitary gland (sagittal and coronal).
- Delineate the areas of hyperostosis and/or associated blistering.

Case 40: Meningioma of the jugum and tuberculum sellae

Fig. a: sagittal cut (512); SE - T1w + Gd

Fig. b: sagittal cut (512); SE - T1w + Gd

Middle age adult presenting for a Foster-Kennedy syndrome. MR shows small "plaque" meningioma inserted on the tuberculum sellae with marked enhancement (**a, b**, ↑) on high resolution T1w images. The small meningioma extends downward along the anterior wall of the sella (**a**) and involves anteriorly the chiasmatic groove, the limbus sphenoidalis and the adjacent planum sphenoidale with associated blistering of the sphenoid sinus (**a**, ↑) laterally. The meningioma involves also the anterior clinoid process (**b**, up, ↑) extending to the cranial aperture of the optic canal (**b**, down, *). On the mid-line, the small meningioma elevates the optic chiasm (**a**, *).

Case 41: Osteomeningioma of the planum sphenoidale

Fig. c: sagittal cut; GE- T1w + Gd

Fig. d: sagittal cut; GE- T1w + Gd

Post contrast T1w sequence (**c, d**) showing a hyperintense mass well demarcated, involving the planum sphenoidale (**c, d**, ↑) associated with a subjacent osteoma (**b**, ↑↑) and a blistering of the sphenoidal sinus (**c**, ↑; **d**, ↑↑↑). The osteomeningioma does not reach posteriorly the chiasmatic groove nor the tuberculum sellae. The patient was free from visual defect.

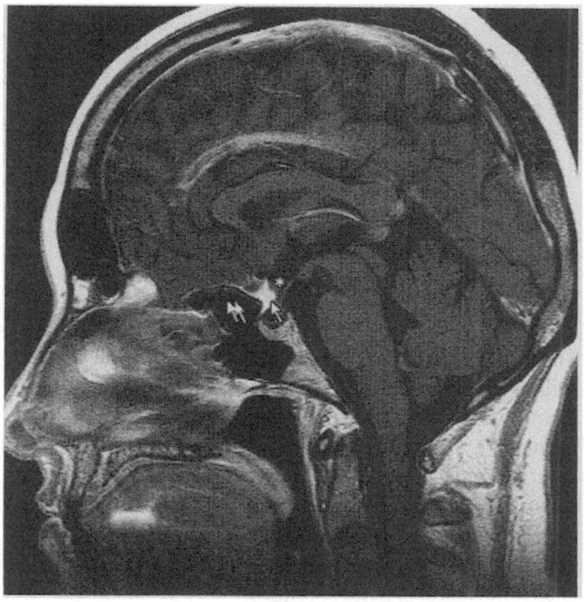

a

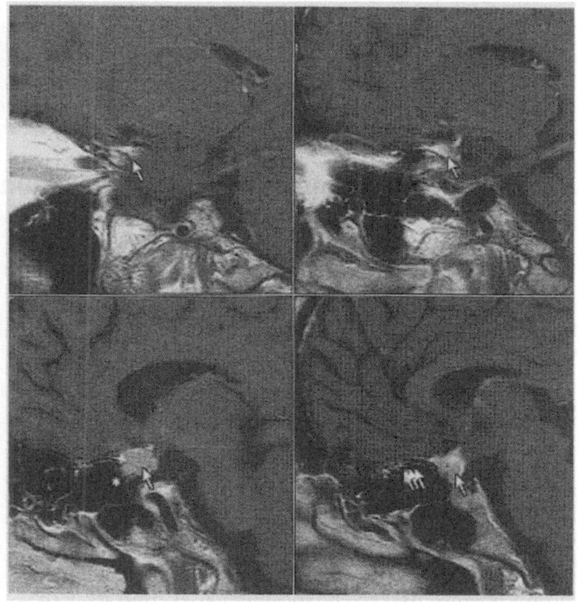

b

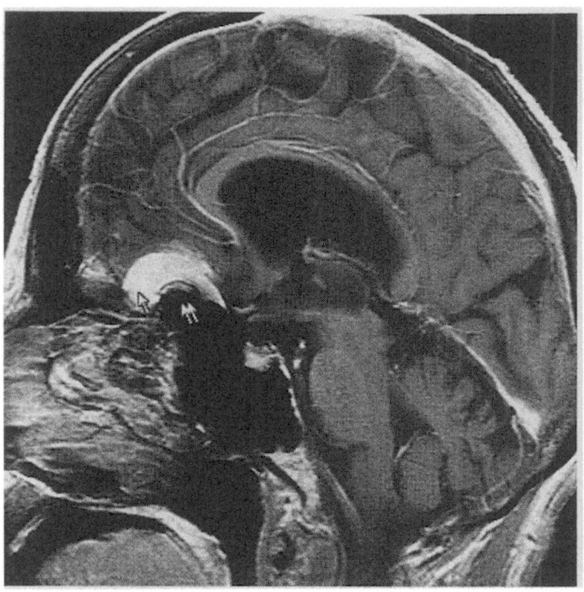

c

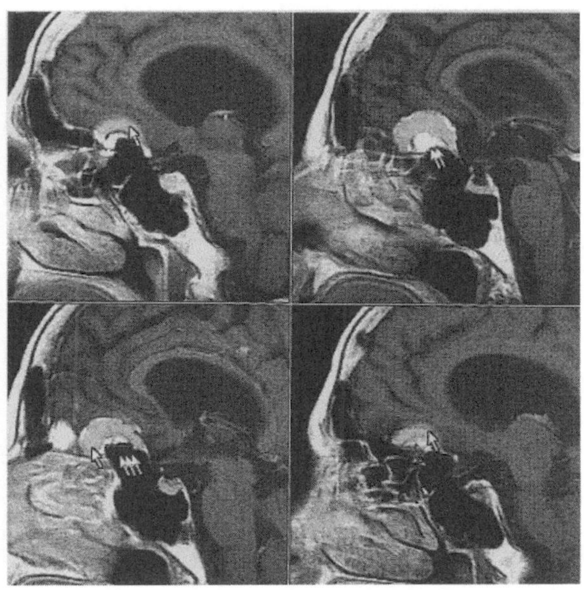

d

Case 42: Jugo-optic meningothelial meningioma

Fig. a: sagittal cut; SE - T2w

Fig. b: sagittal cut (512); SE - T1w + Gd

Fig. c: coronal cut (512); SE - T1w

Fig. d: coronal cut (512); SE - T2w

Middle age women presenting a chiasmal syndrome with optic atrophy at the ocular fundus. The MR exam shows a well demarcated mass readily visible on the T2w sagittal cuts (**a**, ↑↑) displacing posteriorly the optic chiasm (**a**, ↑). The mass involves the tuberculum sellae, the chiasmatic groove and extends on to the jugum sphenoidale with associated blistering (**a**, ↑↑↑). The mass is iso-intense on T1w (**c**, ↑↑) and hypointense on T2w (**d**, ↑↑), showing homogeneous marked enhancement after contrast iv (**b**).

The posterior inferior extension of the mass is well shown, spilling over into the sella turcica (**b**), well separated from the pituitary gland, showing a brighter signal on T1w (**c**, right). The meningiomatous mass reaches laterally the cisternal optic nerves (**c**, *) slightly compressed, and surrounds the supraclinoid portions of the internal carotid arteries with compression of the right one (**c**, right, ↑). Superiorly, the mass raises the proximal segments of the anterior cerebral arteries (**d**, right).

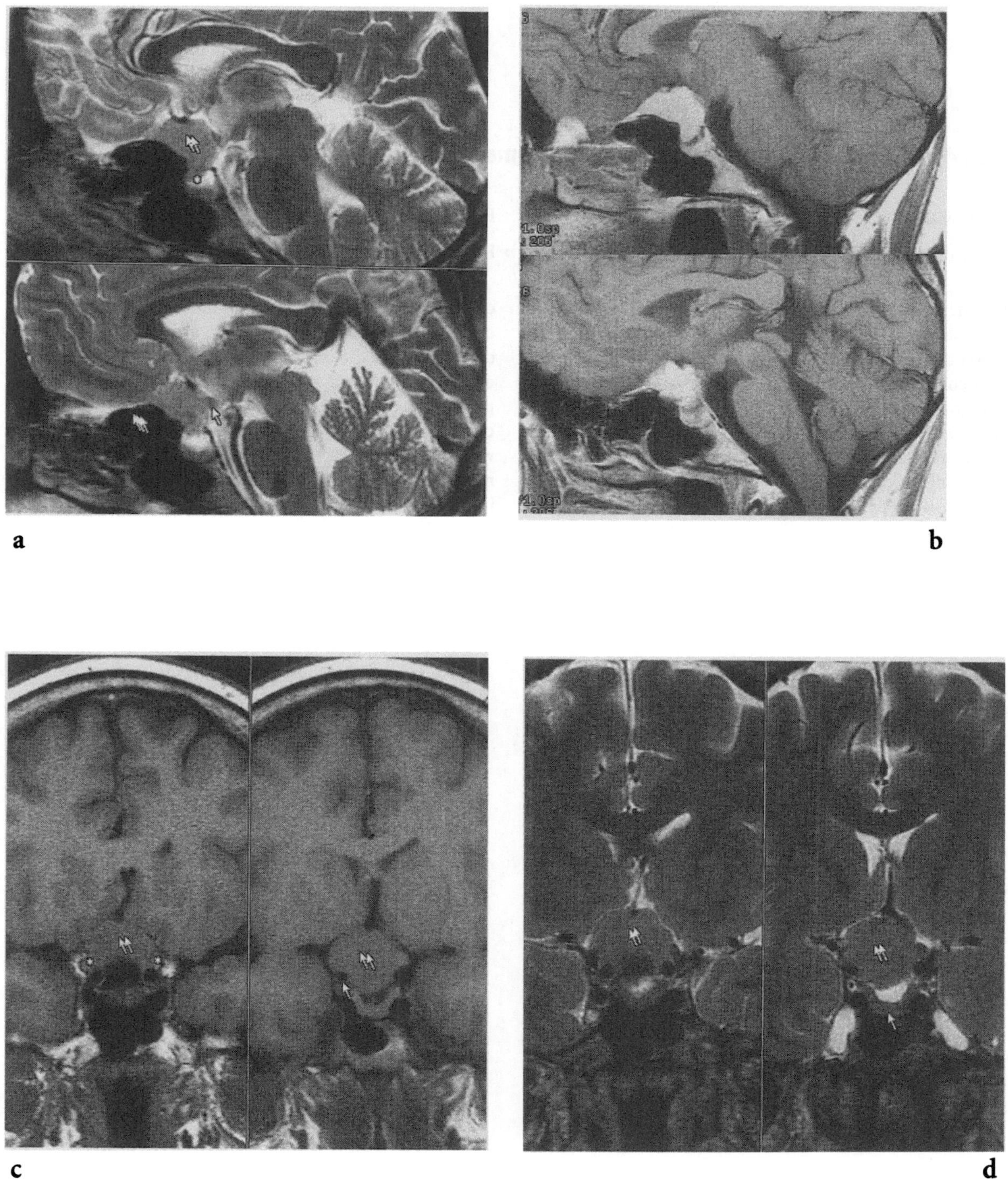
a
b
c
d

Case 43: Clinoid meningothelial osteo-meningioma

Fig. a: axial cut;	SE - T1w	**Fig. c:** sagittal cut;	SE - T1w + Gd
Fig. b: axial cut;	SE - T1w + Gd	**Fig. d:** coronal cut;	SE - T1w + Gd

Patient presenting with left optic neuropathy. MR exam discloses a mass involving the left anterior clinoid process presenting an iso-signal on T1w (**a**, ↑) enhancing intensly after contrast infusion (**b**, **c**, **d**). The mass surrounds the anterior clinoid process, extending laterally along the left sphenoid wing and medially along the chiasmatic groove well demonstrated mainly after contrast (**b**). The bony texture of the clinoid process is modified with osteoma formation presenting a hypointense signal on all sequences (**a**, *; **b**). Moreover, contrast enhancement helped to demonstrate a subtle extension of the mass within the left optic canal, expanding along its lateral wall with compression of the intracanalicular optic nerve, which obviously explains the optic neuropathy (**b**, up; **c**, ↑). More posteriorly, the mass exerts mass effect on the ipsilateral cisternal optic nerve, displaced medially and downward, with modification of its shape as compared to the opposite nerve (**d**, ↑). There was no obvious extension to the ipsilateral cavernous sinus, its superior border being involved exclusively (**d**).

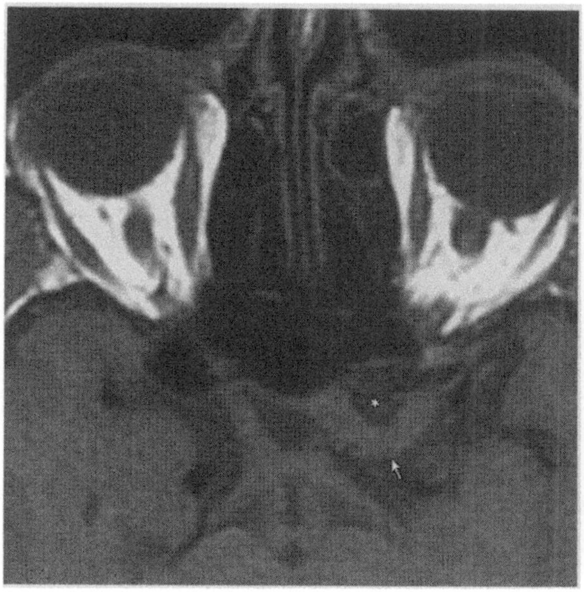

a

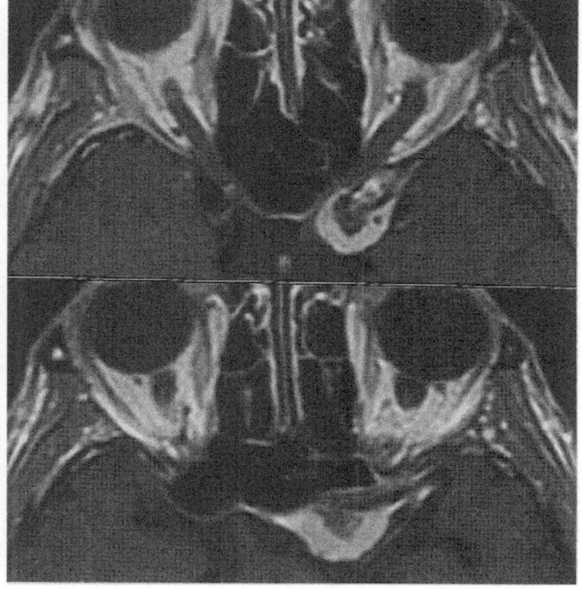

b

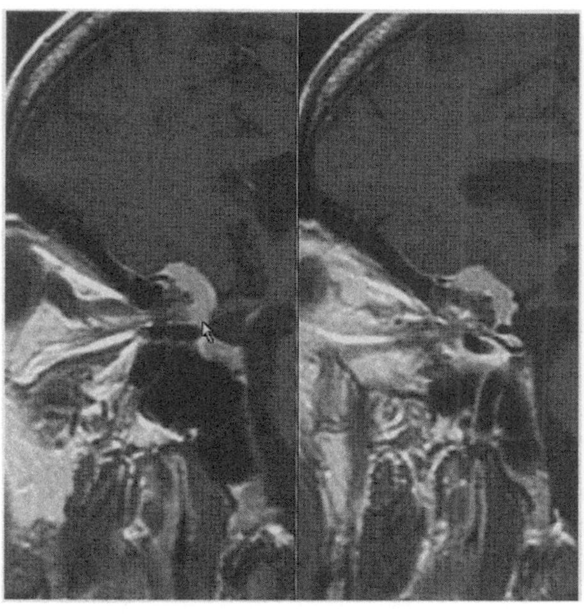

c

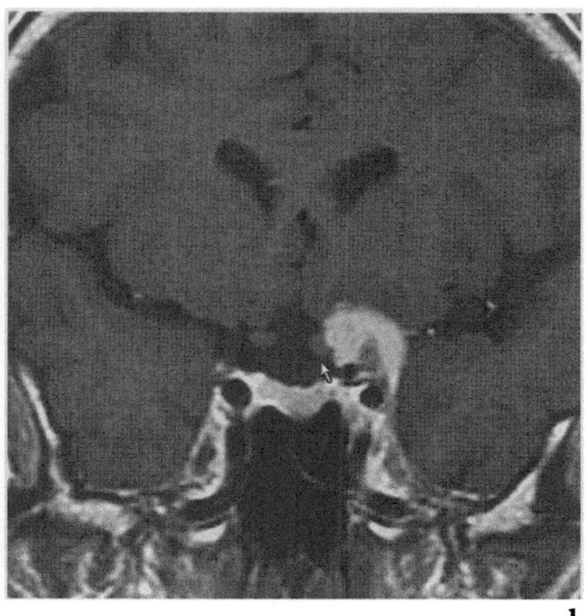

d

Cases 44, 45, 47, 48: cavernous and sphenocavernous meningiomas

Clinical and pathological features

- Meningiomas of the parasellar region account for 10-15% of all intracranial meningiomas.
- Benign slow growing tumors, twice as common in adult females; increasing incidence with increasing age.
- Cavernous meningiomas varieties are relatively rare.
- Sphenoid meningiomas with intracavernous extension more common.
- *Signs:* partial or complete cavernous sinus syndrome, including: ophthalmoplegia due to involvement of one or more of the ocular motor nerves (diplopia being the most common initial symptom); facial hypoesthesia (V1 and/or V2 trigeminal roots) and pain (less common than with cavernous carotid aneurysm); proptosis with conjunctival or lid edema.
- Particular features of intracavernous meningioma: primary aberrant regeneration of the oculomotor (III) nerve. Miosis due to compression of the sympathetic plexus surrounding the intracavernous carotid (but frequently masqued by IIIrd nerve palsy).

MR characteristic findings

Morphology

- Mass originating from the cavernous sinus or from the sphenoid wings.
- Well circumscribed tumors, though may be locally aggressive.
- Possible extensions: ventrally to the orbital cavity (with orbital venous occlusion) through the superior orbital fissure and/or optic foramen, and/or may develop paramedially in the paracavernous region and the adjacent temporal fossa, and/or course posteriorly along the tentorium cerebelli ("posterior tail sign" typical of intracavernous meningiomas).
- Intracavernous meningiomas encase and straighten the intracavernous siphon segment of the ICA, displaced anteriorly and laterally, and constrict or even occlude its arterial lumen.
- Meningiomatous process of the sphenocavernous region may be associated with thickened hyperostotic greater or lesser wings of the sphenoidal bone and may show calcifications and thickening of the lateral cavernous sinus dura (about 20% of cases). Superior orbital fissure and optic foramen may be enlarged or eroded.

Signal intensity

- SE-T1w: isointense to brain gray matter.
- SE-T2w: isointense or slightly hyperintense to white matter.
- Thickened dura of the cavernous sinus showing a hypointense signal on T2.

Contrast enhancement

- Use of contrast infusion showing an intense and homogeneous enhancement of the tumor and the thickened adjacent dura, is mandatory.
- It shows tumoral extents: to mid and inferior orbital cavity through superior orbital fissure and optic foramen (SE-T1 with fat suppression), to middle cranial fossa along the sphenoid wings (detect "en plaques extensions"), to the tentorium cerebelli ("posterior tail" in intracavernous meningiomas).
- Whole intracranial study to rule out other localization of the meningiomatous disease (case 46).

Case 44: Sphenocavernous meningioma

Fig. a: coronal oblique cut; SE - T1w + Gd

Fig. b: sagittal cut; SE - T1w + Gd

Fig. c: coronal cut (512); SE - T2w

Fig. d: axial cut; SE - T1w + Gd

Patient presenting with ocular motor disturbance of the left eye associated with hyperesthesia of the upper left hemiface. MR shows an asymetric, enlarged left cavernous sinus (**c**, ↑↑) with apparent thickening of its lateral wall (**c**, ↑) associated to a deformation of the upper lateral wall of the sphenoid sinus (**c**) readily shown on the T2w sequence. After contrast injection (**a**, **b**, **c**) T1w sequences show a homogeneous enhancement of the mass involving the anterior portion of the left cavernous sinus (**a**, ↑; **d**, ↑↑) extending more laterally to involve the lateral wall of the cavernous sinus (**a**, ↑↑). The tumor, filling the region of the cavernous sinus, extends into the orbital apex through the superior orbital fissure, enlarged (**d**, *; **b**, ↑) with mass effect on the lateral rectus muscle displaced medially.

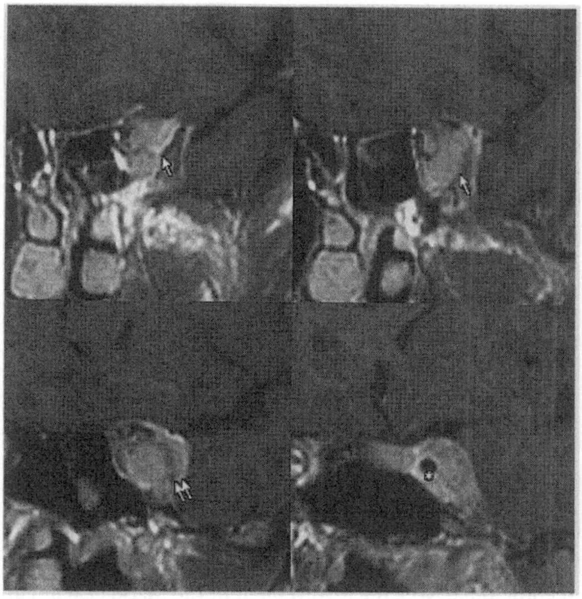

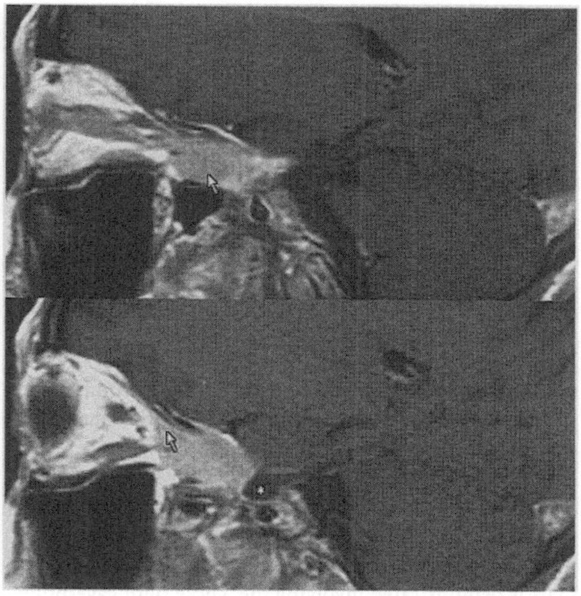

a

b

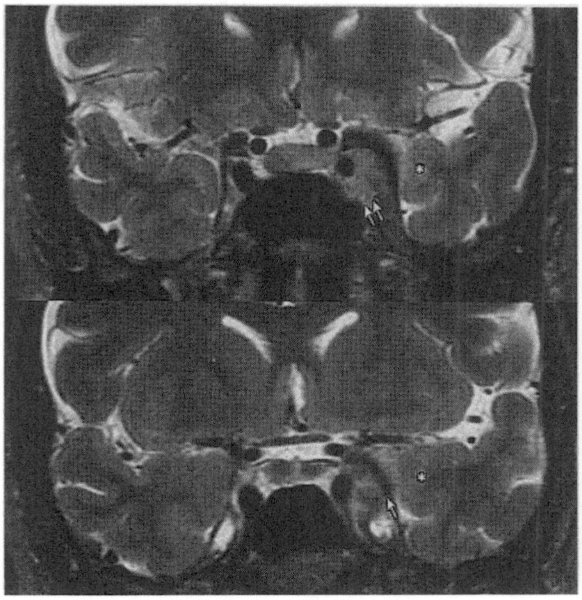

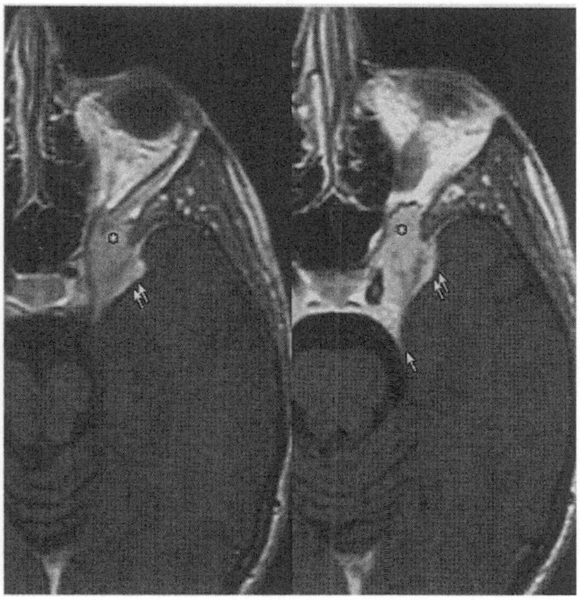

c

d

Case 45: Orbito-sphenocavernous meningioma

Fig. a: axial cut; SE - T1w

Fig. b: axial cut (512); SE - T1w + Gd

Fig. c: coronal cut (512); SE - T1w + Gd

Fig. d: sagittal cut; SE - T1w + Gd

Patient presenting with right proptosis and a cavernous syndrome associated with an orbital apex syndrome. MR showed a mass involving the intraconal and the lateral extraconal orbital compartments (**a**, ↑) showing a hypointense signal on T1w well seen in the hyperintense orbital fat, enhancing after contrast infusion (**b**, ↑). Note the right exorbitism of the eye (**a**, **b**). The mass extends from the sphenocavernous region in which an enhanced mass is demonstrated involving the whole right cavernous sinus and the sellar and suprasellar regions (**c**, down; **b**, ↑↑), to invade laterally the middle cranial fossa involving the right medial sphenoid wing with associated bony thinckening of the lateral orbital wall (**a**, *; **d**, *) enlarged with bone marrow replacement (**b**, **c**). The cavernous tumor extends anteriorly as shown, into the orbital cavity through the superior sphenoidal fissure (**d**, ↑↑) and posteriorly, well shown after enhancement, along the tentorium cerebelli (**d**, ↑↑↑; **b**, ↑↑↑) with a typical “tail” appearance. This is highly caracteristic of intracavernous meningioma. Note the involvement of the Meckel cavum at the level of the cisternal trigeminal nerve (**d**, ↑).

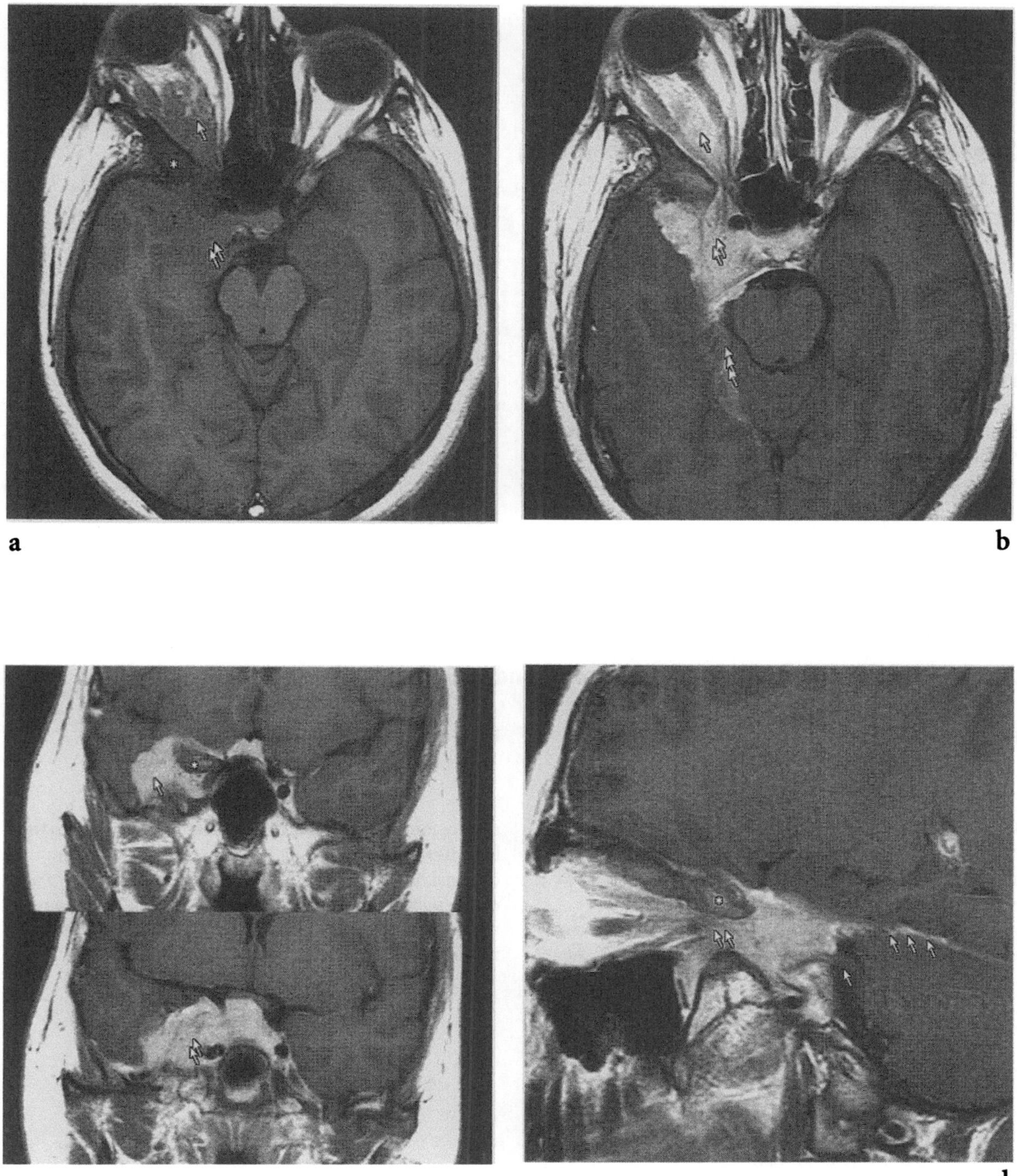
a
b
c
d

Case 46: Meningiomatosis (clino-optic and voult)

Fig. a: sagittal cut; SE - T1w + Gd

Fig. b: coronal oblique cut; SE - T1w + Gd

Patient presenting with optic atrophy. Sagittal oblique images oriented along the optic canal axis and coronal oblique images perpendicular to the previous shown on T1w after contrast infusion, a small enhancing hyperintense lesion involving the optic canal (**a**, ↑, **b**, ↑) extending laterally to involve the adjacent chiasmatic groove (**b**, down, *) and the anterior clinoid process (**b**, up, *). This small optico-clinoid meningioma is associated to another frontal convexity meningioma partly shown after contrast enhancement (**a**, ↑↑) surrounded by hypointense perilesional edema (**a**) on T1w sequence. The patient did not present features of neurofibromatosis.

Case 47: Meningiomatosis (bilateral cavernous and voult)

Fig. c: axial cut; SE - T1w + Gd

Fig. d: coronal cut; SE - T1w + Gd

Patient presenting a diplopia due to a left abducens palsy. The physical inspection disclosed large "café-au-lait" spots suggesting neurofibromatosis. MR exam showed on T1w enhanced slices, bilateral almost symmetric masses involving the cavernous sinuses (**c**, ↑; **d**, ↑). The contrast enhancement also depicts a very thin extension of the meningiomas to the tuberculum sellae (**c**, right) and to the left sphenoid wing (**d**, ↑↑). Note the relationship of the cisternal optic nerve (**d**, *) and the incidental finding of an empty sella (**d**, ↑↑). Associated to the bilateral cavernous meningiomas, MR also disclosed a left temporal meningioma (**c**, ↑↑).

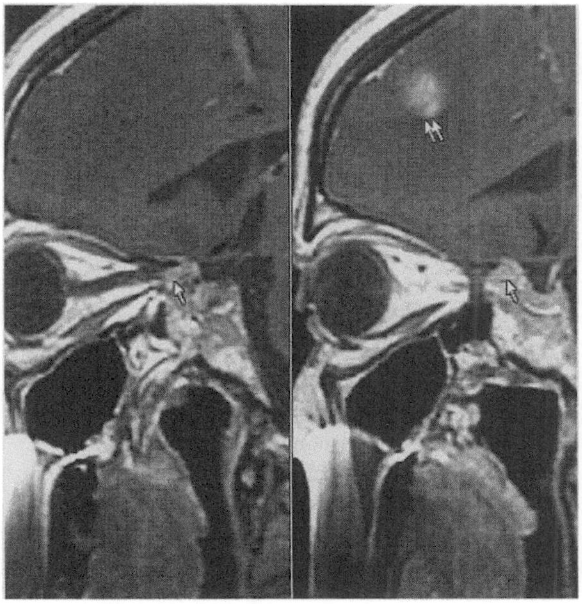

a

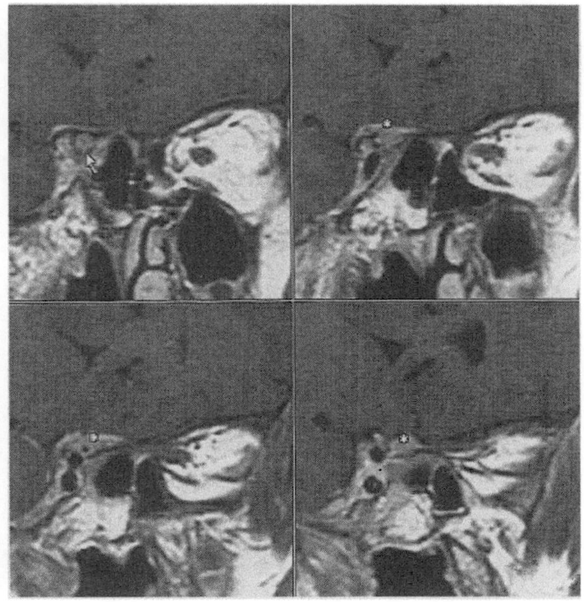

b

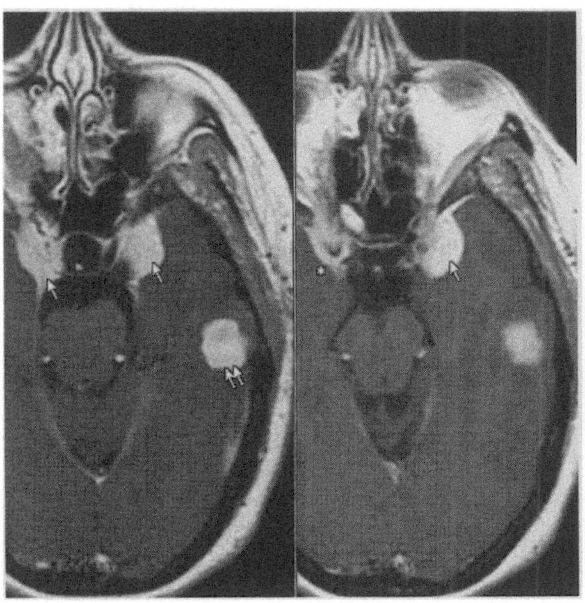

c

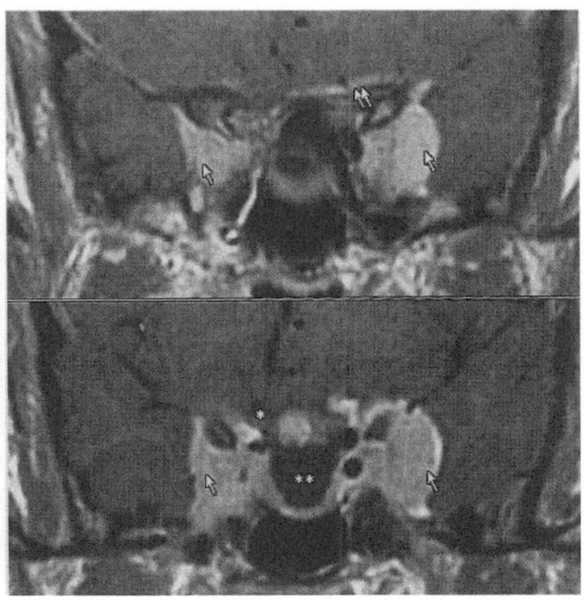

d

Case 48: Posterior cavernous meningioma

Fig. a: coronal cut; SE - T2w

Fig. b: coronal cut; SE - T1w + Gd

Middle age woman presenting for many years an extrinsinc oculomotor nerve palsy. MR showed a small mass presenting a hypointense signal on T2w involving the posterior portion of the right cavernous sinus (**a**, ↑) which shows marked contrast enhancement after Gadolinium iv on T1w (**b**, up, ↑) extending more posteriorly along the tentorium cerebelli (**b**, down, ↑) partly thickened. The mass did not involve inferiorly the cavum of Meckel showed on T2w (**a**, ↑↑) due to the hyperintense signal of the CSF and the corresponding hyposignal on T1w (**b**). The mass did not extend to the sellar region well shown on T2w. The marked hypointensity on T2w corresponds partly to intrinsic microcalcifications.

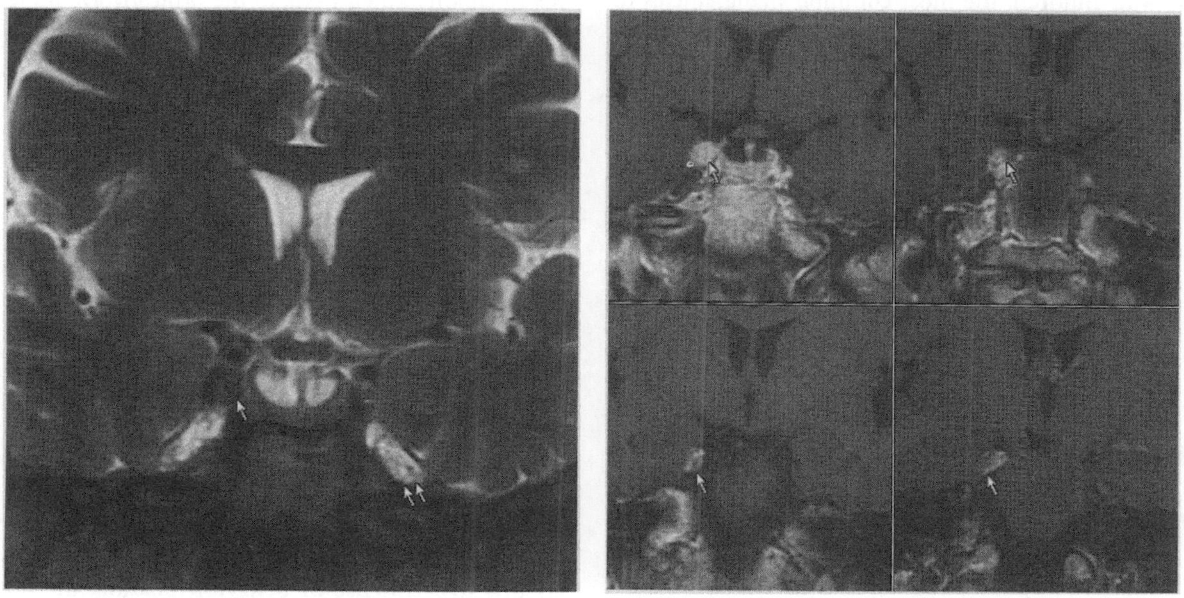

a b

Case 49: orbito-sphenoidal metastases

Clinical and pathological features

- Metastatic lesions to the orbital region are uncommon.
- In adults: most frequent origin are carcinomas of the breast and lung (about half the cases), the remaining occurring in older patients arise from carcinomas of the prostate, kidneys, gastrointestinal tract and thyroid gland.
- Metastasis from breast carcinoma occurs in a middle age woman known to have a primary tumor or leads to the discovery of the primary process, or may be observed many years after treatment.
- In children: the most common malignancies to metastasize to the orbit are the neuroblastoma (bilateral in 40%), leukemias, and Ewing's sarcoma.
- *Signs:* rapid onset of proptosis (except the scirrhous carcinoma of the breast associated with an enophthalmos), pain, visual loss and variable ocular motor palsies (orbital muscles being a frequent site of involvement).
- Considering orbital lymphomas (about 10% of orbital tumors), they are most frequently located in the anterior orbit (palpable mass), and are usually part of a systemic disease. Retrobulbar localization may simulate an inflammatory pseudotumor or a metastatic carcinoma.

MR characteristic findings

Morphology

- Different localizations in adults and children, commonly bilateral.
- In adults: metastases may be found in the eyeballs, the orbital muscles (mimicking myositis), the lacrymal gland, the extraconal spaces more frequently infiltrated than the retrobulbar fat and perioptic area (mimicking a pseudotumor) or may involve the lateral orbital walls, commonly the lateral and/or the anterior clinoid; actually, any intraorbital structure may be involved.
- Sphenoidal greater wing is a localization of choice of metastatic prostatic carcinoma, showing associated hyperostotic changes to the soft tissue mass, developing in the lateral orbital cavity as well as intracranially in the anterior temporal fossa (do not misdiagnose with a meningioma); renal and gastrointestinal metastatic carcinomas cause osteoblastic changes as well; most of the other metastatic processes in this site show lytic bony modifications.
- Metastatic tumors to central skull base (spheno-cavernous region), usually heterogeneous, are more frequent than primary tumors, arising from lung, prostate, breast or renal carcinomas or hematogenous.
- Metastases to the pituitary gland or stalk are not uncommon, originating from lung or breast carcinomas and observed in about 3% of the cases.
- In children: metastases to the orbit involve more frequently the orbital cavity than the eyeball, with predilection to the orbital walls which show bony destruction with subperiosteal tumoral expansion.

Signal intensity

- SE-T1w: usually iso-intense to orbital muscles, unless hemorrhagic (melanoma, renal cell carcinoma or neuroblastoma) or paramagnetic melanin containing tumor such as melanotic melanomas. T1w sequences are highly informative in the evaluation of skull base metastases, demonstrating bone marrow replacement (absence of bright signal from fat associated to bony destruction).
- SE-T2w: usually shows characteristic hyperintense SI to orbital fat, which may help to differentiate most metastases from benign conditions like the orbital pseudotumor (usually hypointense). Multiple myeloma and lymphomas may not show this hyperintense SI behavior, possibly due to dense hypercellularity (remaining iso-intense to fat) as well as blastic metastasis from prostatic carcinoma and metastatic neuroblastoma, remaining hypointense.

Contrast enhancement

- Usually moderate enhancement is observed.
- Fat suppression techniques are mandatory to better evaluate extensions to the orbital spaces or localizations to the central skull base, in addition to the precontrast SE-T1w cuts.

Case 49: Jugo-optico-clinoid metastasis (breast)

Fig. c: axial cut; SE - T1w + Gd

Fig. d: coronal cut; SE - T1w + Gd

Sudden visual right loss in an adult patient. Post contrast T1w sequence disclosed a small enhancing mass involving the chiasmatic groove and extending to the right optic canal (c). The posterior extension of the mass in the suprasellar cistern, elevates the right cisternal optic nerve (**d**, *) partly enlarged due to reactive edema and/or infiltration. The anterior extension is well shown beyond the chiasmatic groove and the limbus toward the jugum (**d**, up, ↑). Moreover, brain slices show after contrast enhancement a small rounded subcortical mass in the left hemisphere (**d**, ↑↑) corresponding to another metastatic focus.

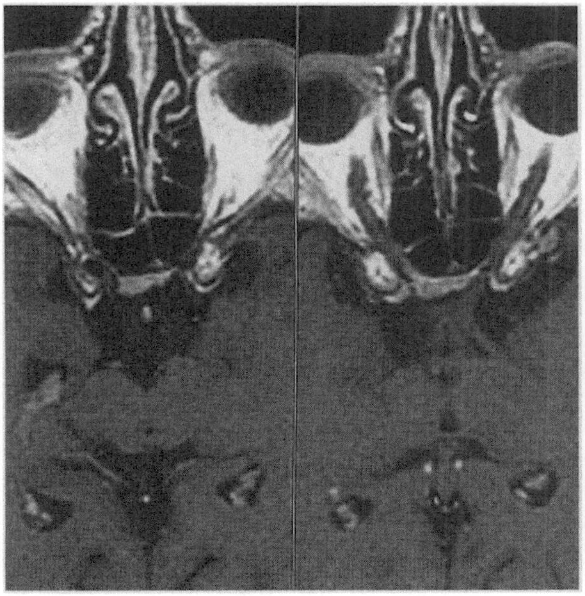

c

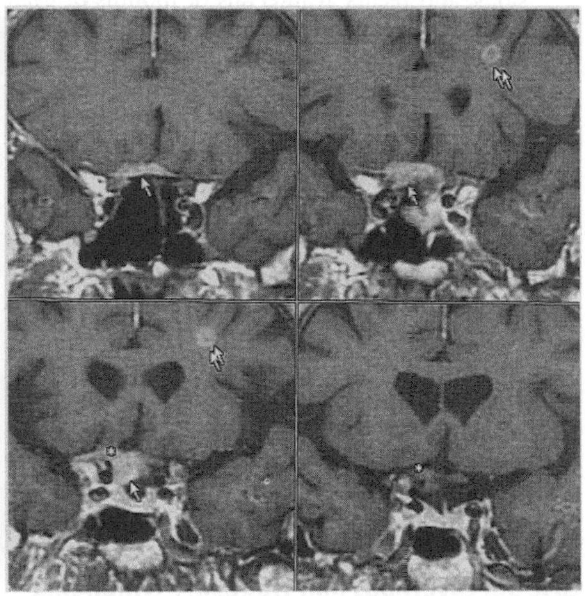

d

Cases 50, 51: tolosa-hunt syndrome

Clinical and pathological features

- It is a recurrent "painful ophthalmoplegia syndrome" (case 51).
- Caused by a non specific granulomatous inflammatory process, related and similar to idiopathic orbital pseudotumor.
- *Signs:* acute installation of pain in or about the eye, partial or total ophthalmoplegia, variable pupil size (small or partially dilated), ophthalmic (V1) hypesthesia.
- The syndrome may show spontaneous remissions, and recurs at intervals of months or even years. It responds dramatically to high doses of steroids.
- Other tumoral conditions located in or about the cavernous sinus or the Meckel's cave, with therapeutic response to steroids, may present as a Tolosa-Hunt-like syndrome (i.e. lymphoma, case 50, or sarcoidosis).

MR characteristic findings

Morphology

- Tumoral process located in the region of the cavernous sinus and/or the superior orbital fissure (in idiopathic Tolosa-Hunt syndrome), most frequently extending to the orbital apex.
- Tumors in the cavernous sinus, or the Meckel's cave or the orbital apex may be found (mimicking the clinical syndrome).
- The cavernous sinus is usually enlarged, with bulging of its lateral dural wall (coronal and axial). Return to normal size and shape after steroid therapy.

Signal intensity

- SE-T1w: iso-intense to muscle.
- SE-T2w: hypointense to fat.

Contrast enhancement

- Relatively marked contrast enhancement of an enlarged cavernous sinus.
- Fat suppression needed to assess the eventual orbital extent.
- Obliteration of low SI from CSF in tumoral conditions developed in Meckel's cave (lymphoma, meningioma, trigeminal neuroma).

Case 50: Cavernous non-Hodgkin's lymphoma

Fig. a: axial cut; SE - T2w

Fig. b: axial cut; SE - T1w + Gd

Hypesthesia of the right hemiface orienting to a trigeminal nerve deficit. On T2w sequence, MR shows a lack of hyperintense signal of CSF filled right Meckel's cavum (**a**, ↑). The normal left side shows normal hyperintense signal (**a**, ↑↑). After contrast enhancement, the T1w sequence discloses the presence of a mass partly enhancing and involving the Meckel's cavum (**b**, ↑) with a convex deformation of its lateral wall as compared to the left side. The MR aspect is non specific and the diagnosis was consistent with a non-Hodgkin's lymphoma.

Case 51: Cavernous granuloma

Fig. c: axial cut (512); SE - T1w + Gd

Fig. d: axial cut (512); SE - T1w + Gd

Patient presenting with an isolated left abducens nerve palsy. Post Gadolinium T1w shows in the axial plane a very small mass involving the most anterior portion of the left cavernous sinus at the superior orbital fissure (**c**, **d**, ↑) without adjacent bony modification, nor intracavernous carotid artery narrowing. A non specific granuloma was retained as the underlying disease process at surgery.

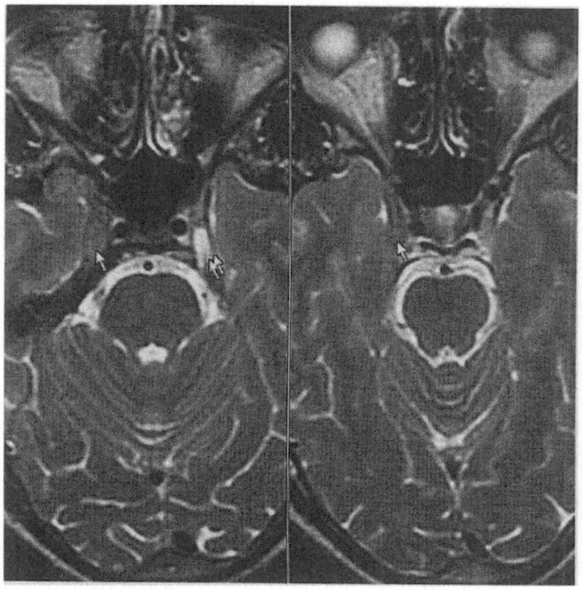

a

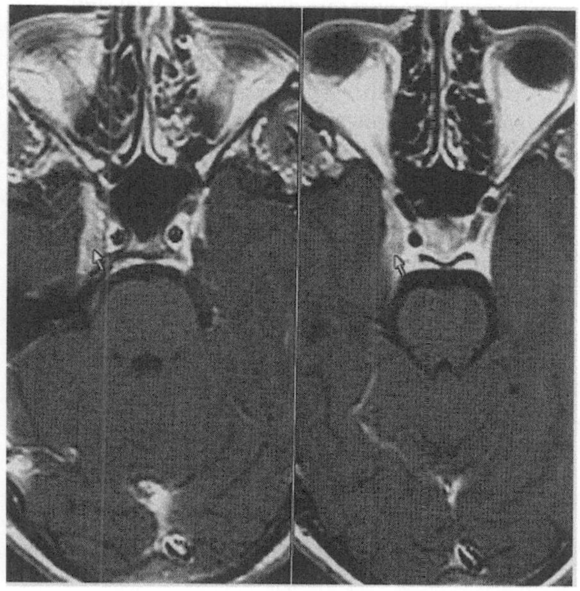

b

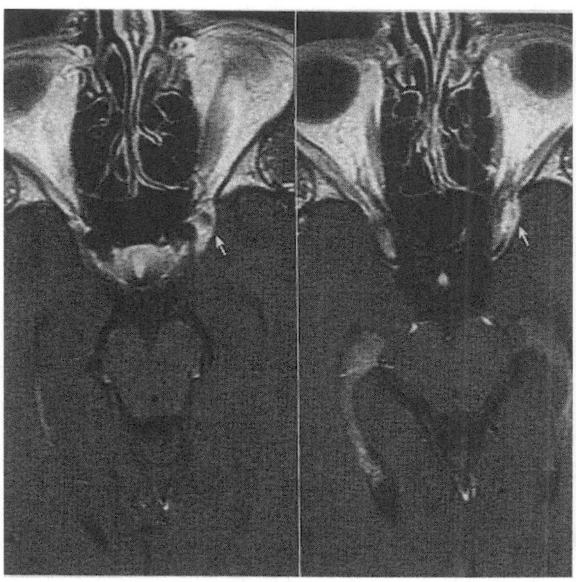

c

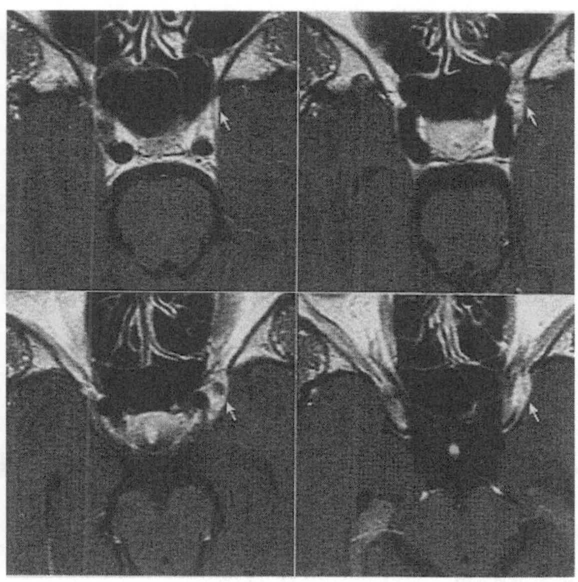

d

Cases 52-54: sphenocavernous tumors

Clinical and pathological features

- Tumoral processes involving the central skull base may be classified as intrinsic lesions, or extrinsic invading secondarily the basisphenoidal region, arising from the sellar region endocranially or from the nasopharyngeal, parapharyngeal or prevertebral spaces exocranially.
- *Intrinsic lesions:* schwannoma or neurofibroma, glomus jugulare tumor, chordoma (cases 36, 37), chondroma and chondrosarcoma, giant cell tumor, metastasis, lymphoma; osteoma, fibrous dysplasia and Paget's disease.
- *Extrinsic neoplasms:* these may arise endocranially (from the parasellar region such as meningiomas, or from the sella such as invasive pituitary adenomas), or exocranially originating from deep facial soft tissues (squamous cell carcinomas in adults, rhabdomyosarcomas in children, juvenile angiofibroma, and mucoceles.
- *Symptomatology:* the initial signs are commonly cranial nerve palsies, affecting the Vth and VIth nerves most frequently. Pain due to invasion of the trigeminal ganglion is another prominent clinical feature, presenting most often as a continuous and intractable facial neuralgia associated with paresthesia, sensory loss, and abolition of the corneal reflex.

MR characteristic findings

Morphology

- MRI is crucial, complementing CT (bone analysis), to determine the site of origin and the tumoral extents as well as imaging characteristics of central skull base lesions, in order to contribute to the surgical planning.
- MR must be performed in the sagittal, coronal and axial planes using SE-T1 and T2 weighted sequences.
- In the evaluation process of intrinsic neoplasms, T1w images without contrast enhancement are mandatory in order to appreciate the invasion of bone marrow within the skull base and investigation of the osseous foramina (to rule out perineural and/or perivascular spread of the tumoral processes).
- Endocranial tumors are mainly represented by parasellar meningiomas (cases 44, 45, 47 and 48), or laterally invasive pituitary adenomas extending to the cavernous sinus and compressing or encasing the ICA siphon (case 52), or less frequently, schwannomas of the trigeminal roots and ganglion developing in the Meckel's cave and/or the lateral prepontine cistern. Best explored on coronal cuts.
- Exocranial neoplasms (squamous cell carcinomas case 53, 62 and 63, cylindroma case 54; lymphoepithelioma case 61; juvenile angiofibroma case 60; rhabdomyosarcoma case 59; non-Hodgkin's lymphomas), or infectious processes (sphenoid mucocele or sphenoiditis case 38), invade the central skull base through direct extensions or perineural spread (2/3 of cases) along the maxillary or the mandibular divisions of the trigeminal nerve or the facial nerve. Coronal cuts with contrast enhancement and fat suppression are highly informative.

Signal intensity

- SE-T1w: most commonly hypointense to fat marrow and brain.
- SE-PDw: iso-intense or slightly hyperintense to brain.
- SE-T2w: most frequently hypointense but may be slightly hyperintense to brain, more or less heterogeneous.

Contrast enhancement

- Commonly show intense enhancement.
- Need to obtain non enhanced T1w coronal cuts.
- Need to use fat suppression technique with contrast infusion.

Case 52: Invasive sphenocavernous pituitary adenoma

Fig. a: sagittal cut;	SE - T2w	**Fig. c:** coronal cut;	SE - T2w
Fig. b: axial cut.;	SE - T2w	**Fig. d:** coronal cut;	SE - T2w

Young adult presenting a bitemporal hemianopia with diplopia. Skull X-rays show marked enlargement of the sella with a double floor on the lateral projection corresponding to a left inferior lateral inflexion of the floor. MR shows on T2w sequence a large soft tissue mass in the sella turcica (**c**, **b**, ↑↑) and in the suprasellar cistern (**c**, **d**, ↑) elevating the optic chiasm (**d**, ↑). The mass extends laterally through the medial wall into the right cavernous sinus encircling the carotid artery (**c**; **d**, *) with bulging of the lateral wall (**c**, ↑↑↑). Note the clear visualization of the lateral cavernous sinus dura displaced laterally with mass effect on the medial aspect of the adjacent temporal lobe. Note the left inflexion of the floor of the sella turcica (**d**, ↑↑) and the development of the mass into the sphenoid sinus reduced to its anterior portion (**a**, *).

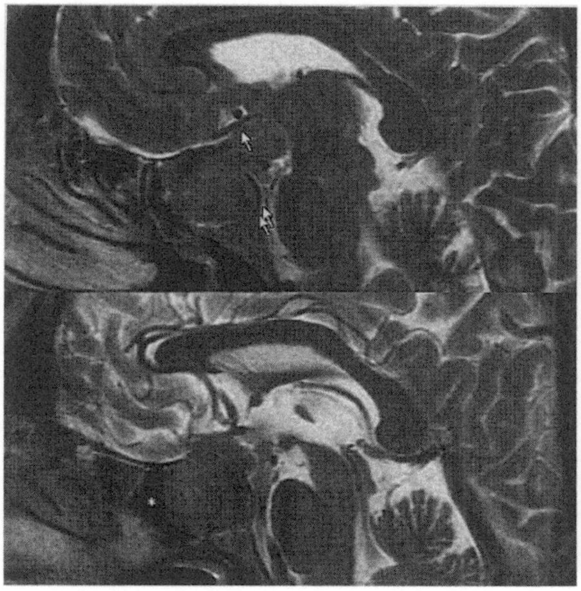

a

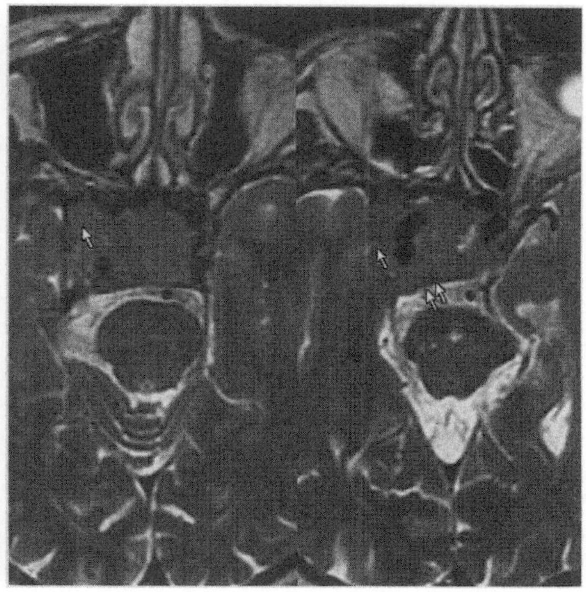

b

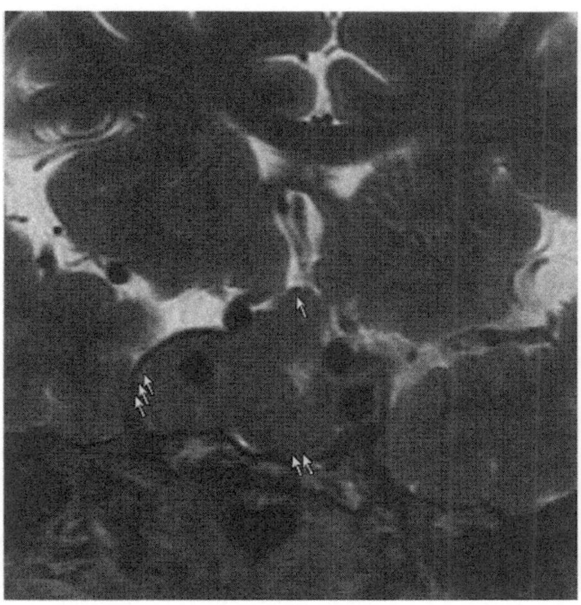

c

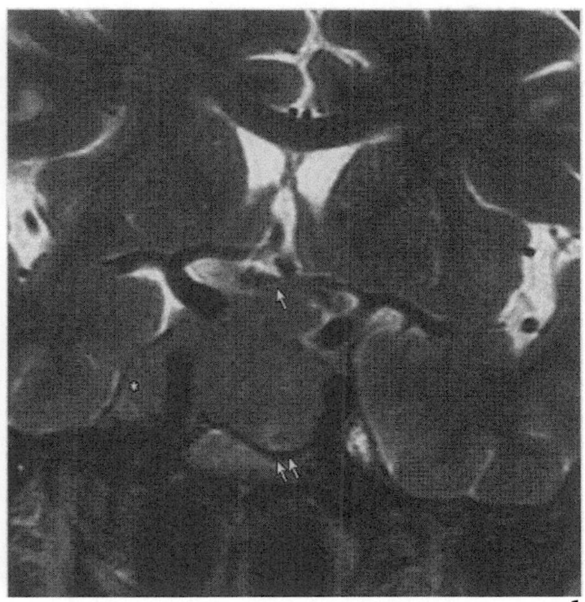

d

Case 53: Nasopharyngeal carcinoma

Fig. a: axial cut; SE - PDw

Fig. b: axial cut; SE - T2w

Heterogeneous destructive tumor mass filling the sphenoid body and the posterior ethmoid sinus and invading laterally the left cavernous sinus as well as the Meckel's cavum. The lesion is heterogeneous on PDw sequence with areas of isosignal to cortex associated with areas of hyposignal intensity (**a**) largely hypointense on T2w (**b**). The lesion encircles the left intracavernous carotid artery (**a**, *) and bulges in the temporal fossa compressing the temporal pole (**a**, right, ↑) and in the prepontine cistern with slight compression of the pons (**a**, left, ↑; **b**, left, ↑). Anteriorly, the tumor extends into the posterior ethmoidal cells (**b**, left, ↑↑) surrounded by hyperintense material corresponding to associated retentional fluid (**b**, right). Note the normal cavum of Meckel (**b**, **) with hyperintense CSF-like signal on T2w.

Case 54: Cylindroma

Fig. c: axial cut; SE - PDw

Fig. d: coronal cut; SE - T1w

Large tumor mass filling the sphenoid body and sinus and involving the left infratemporal fossa extending to the skull base. The mass shows a homogeneous hypointense signal on T1w (**d**) and a slightly hyperintense signal on PDw (**c**). Superiorly, the lesion invades the sellar region reaching the left cisternal optic nerve (**d**, left) and extending laterally toward the left middle cranial fossa with mass effect on the left temporal pole (**d**, left; **c**).

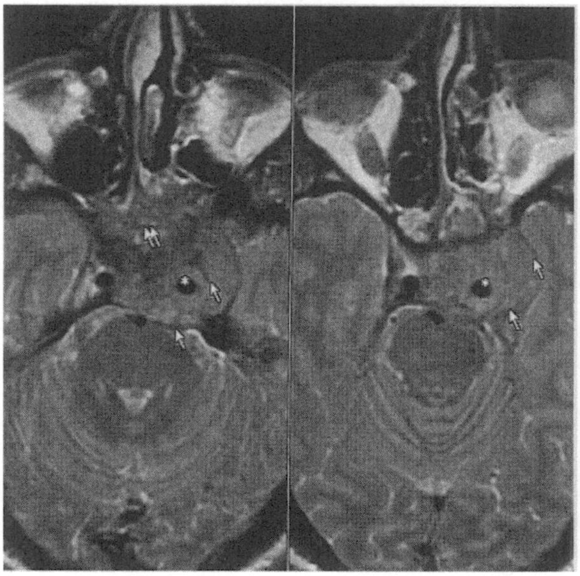

a

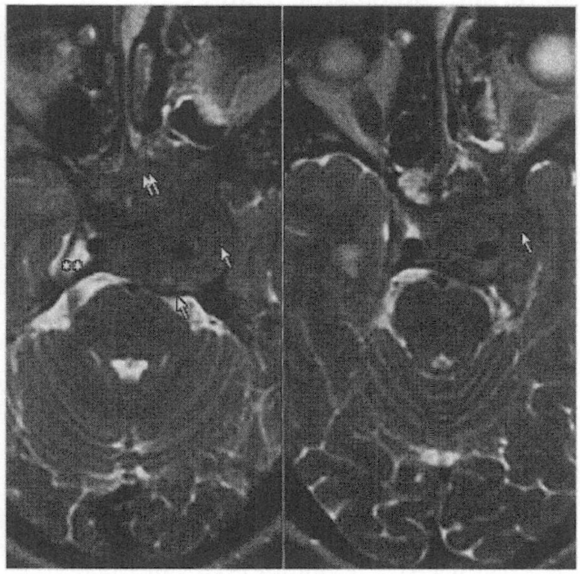

b

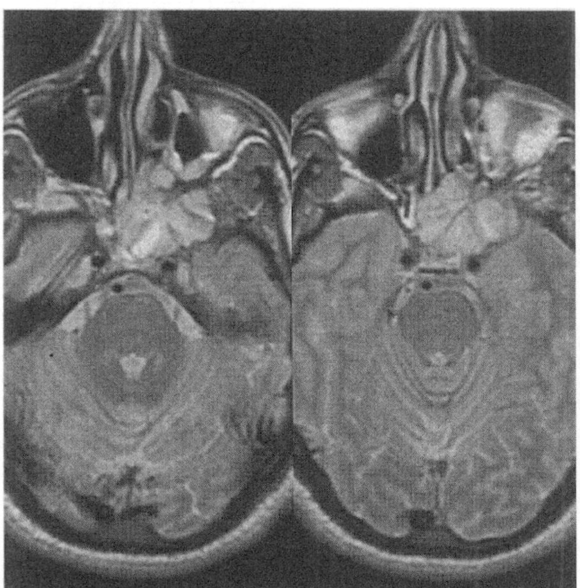

c

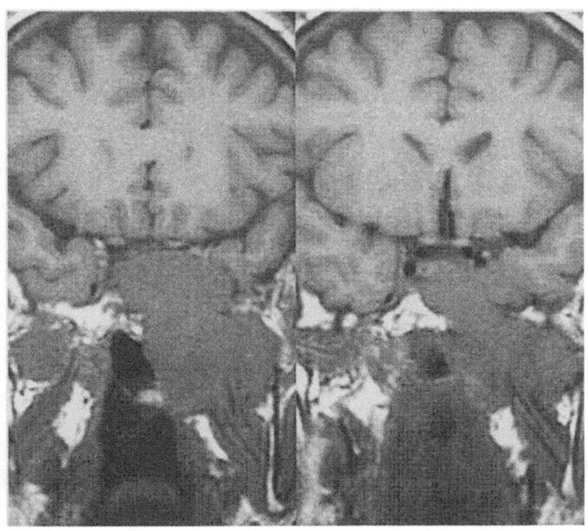

d

Case 55: chiasmatic gliomas

Clinical and pathological features

- Primary astrocytic tumors of the opto-chiasmatic pathways.
- Two major clinical and pathological patterns according to age.
- Childhood (3-10 years, 75% of chiasm gliomas): 5% of primary brain tumors
  - low grade pilocytic astrocytomas;
  - usually non invasive, infiltrating optic nerves and tracts.
  - *Signs:* decreased vison, strabismus, amblyopia, disc pallor, nystagmus (bilateral and mixed horizonto-rotatory or spasmus nutans-like with head dodding), insidious proptosis (optic nerve tumor), hypothalamic syndrome (hypersomnia, diabetes insipidus, precocious puberty), visual fields defects (commonly central scotoma);
  - course: non progressive or very slow growing;
  - neurofibromatosis type I frequently associated (25-50%): hamartomas.
- Adulthood (40-50 years, commonly males): rare (2-5% of primary brain tumors)
  - malignant astrocytomas (glioblastomas);
  - invasive hypothalamo-chiasmatic.
  - *Signs:* rapid loss of vision with orbital pain, papilledema, chiasmal syndrome (junctional scotomas, rarely bitemporal hemianopia), bilateral blindness, increased intracranial pressure (obstructive hydrocephalus);
  - course: rapid development and regional invasion with spread to hypothalamus and midbrain, fatal outcome (less than 2 years).

MR characteristic findings

Morphology

- Chiasmatic or hypothalamic origin (depends on epicenter).
- More invasive than tumors arising from cisternal optic nerves or tracts.
- When small may be difficult to separate from chiasmal neuritis (follow-up).
- Variable shape, depending on size and regional infiltration (hypothalamus, tracts).
- Superior and lateral extensions best shown on sagittal and coronal cuts.
- Cystic components in larger tumors (some may regress on follow-up MRs).

Signal intensity

- SE-T1w: hypointense or even iso-intense (small lesion) to brain.
- SE-T2w: usually hyperintense to brain and homogeneous, may become heterogeneous with intratumoral cystic formations in larger tumors.

Contrast enhancement

- Variable and marked contrast enhancement (solid components of larger lesions).
- Homogeneous or heterogeneous enhancement.

Case 55: Opto-chiasmatic glioma

Fig. a: sagittal cut; SE - T2w

Fig. b: coronal cut; SE - T1w + Gd

Young patient presenting a chiasmal syndrome with a rapid decrease of vision. T2w sagittal MR readily shows on T2w sequence the important enlargement of the optic chiasm with a slightly hypointense signal (**a**, ↑) occupying the suprasellar cistern and apparently well delineated. After contrast iv, T1w coronal slices show heterogeneous enhancement of the enlarged optic chiasm (**b**, ↑↑) extending anteriorly to involve the cisternal optic nerves (**b**, ↑) and posteriorly into the right optic tract (**b**, **). Note the small area of intratumoral necrosis in the right portion of the chiasm (**b**, *). Note that the gliomatous process showed high signal intensity on proton density images (not shown).

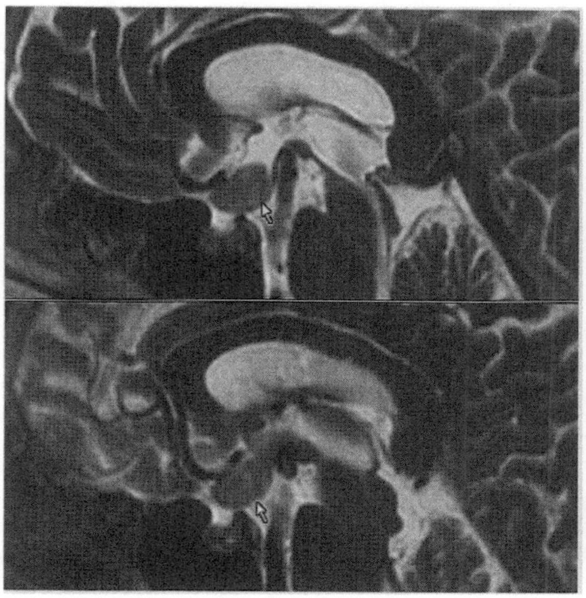

a

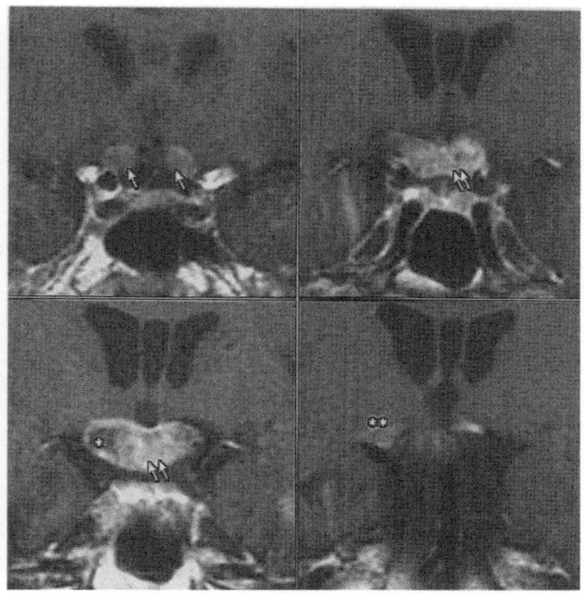

b

Case 56: hypothalamic hamartoma

Clinical and pathological features

- Benign rare congenital slow growing process.
- Composed of heterotopic neuroepithelial cells.
- Found in the hypothalamus, involving the tuberal region, mamillary bodies.
- More frequent in males.
- *Signs:* most frequently precocious puberty, gelastic seizures, hyperactivity.

MR characteristic findings

Morphology

- Solid well demarcated mass of neuronal tissue.
- Attached to the floor of the 3rd ventricle, posterior to the pituitary stalk.
- Usually pedunculated (coronal) or sessile (sagittal).
- Very rarely calcified, and no hemorrhage.
- It does not infiltrate the adjacent neural tissue.

Signal intensity

- SE-T1w: isointense to gray matter.
- SE-PDw: slightly hyperintense to brain.
- SE-T2w: slightly hyperintense or iso-intense (1/3 cases) to brain.

Contrast enhancement

- Show no contrast enhancement (rule out other tumoral conditions if enhancing).

Case 56: Hypothalamic hamartoma

Fig. c: axial cut; SE - T1w

Fig. d: axial cut; SE - T2w

Huge mass occupying the opto-chiasmatic cistern, presenting a slightly heterogeneous hypointense signal on T1w (**c**, ↑), becoming slightly hyperintense to cerebellar white matter on heavily T2w with some necrotic areas showing hyperintense signal on T2w (**d**, ↑). The mass compresses the mesencephalon (**c**, **d**, **) and mainly the cerebral peduncle with compression of the cerebral aqueduct (**c**, *). Note also the compression of the 3rd ventricle (**d**, *). The tumor shows no enhancement after contrast iv and was associated with hydrocephalus with marked enlargement of the lateral ventricles as shown here with periventricular CSF resorption.

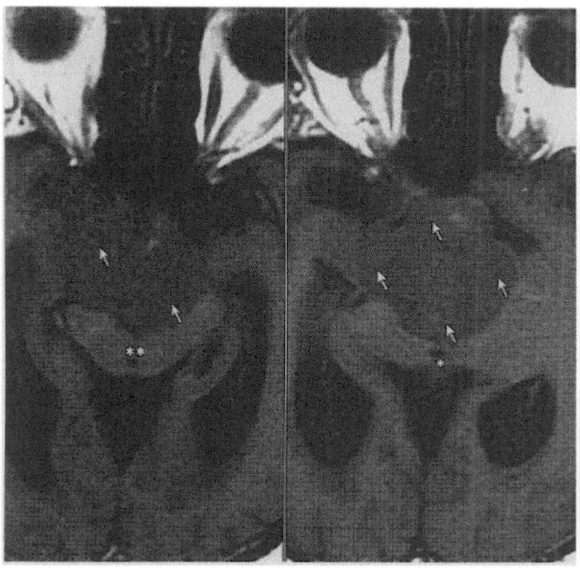

c

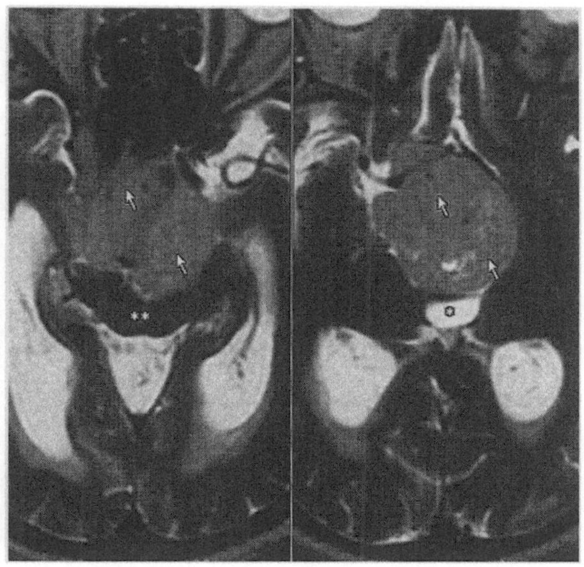

d

Cases 57, 58: craniopharyngiomas

Clinical and pathological features

- Account for about 3% of intracranial tumors.
- Account for 50% of sellar region tumors in children and 20% in adults.
- No sex predilection, they are found in patients less than 20 years (50-70%).
- Peak incidence: 5-10 years old (adamantinomatous type, most frequent, case 57) and later at 50-60 years old (papillary type, case 58).
- Maldevelopmental tumor arising from the Rathke's pouch, found in the suprasellar region, but may be found in the sphenoid or even the nasopharynx.
- *Histology:* solid and cystic components formed of epithelial tissue, surrounded by fibrous collagenous tissue (which may cause adjacent glial reaction).
- Histological similarity with adamantinoma (rudimentary tooth may be found in craniopharyngiomas): adamantinomatous craniopharyngiomas.
- *Signs:* variable, including: visual deficit and field defects (75%), headache (50%), vomiting, diabetes insipidus, seizures (15%), endocrine dysfunction (70%), cranial nerve palsies, papilledema or optic atrophy (40-60%), psychiatric syndrome, recurrent chemical meningitis.
- In children, prominent signs: headache (80%) and vomiting due to increased intracranial pressure, endocrine dysfunction (75%), impairement of growth (25%), optic atrophy, transitory visual loss mimicking retrobulbar optic neuritis.

MR characteristic findings

Morphology

- Located most frequently in the suprasellar cistern.
- About 20% are intrasellar (5%) with suprasellar component.
- May be divided into 3 varieties based on location: intrasellar, prechiasmatic and retrochiasmatic, and 3 varieties according to morphological texture: solid, cystic (containing cholesterol) and mixed solid-cystic, multilobular and multicystic.
- Usual size 2 to 10 cm; huge tumors may extend to the subfrontal or temporal regions or even in the posterior fossa.
- Displace the chiasm posteriorly (or anteriorly in prefixed chiasm) or raise up the floor of the 3rd ventricle.
- Papillary type in adults, often solid intraventricular mass (3rd ventricule), no cholesterol crystals content or calcifications, but squamous differentiation (case 58).
- Enlargement of the sella (60%) with presence of intra- and/or suprasellar calcifications mottled and/or curvilinear (75%), shown on standard X-ray (mainly children).

Signal intensity

- SE-T1w: hyperintense to CSF and brain (due to very high protein content).
- SE-PDw: hyperintense to brain or CSF.
- SE-T2w: hyperintense to brain (cystic portion showing higher SI than solid).
- Heterogeneous solid component including calcifications; it shows low signal intensities on T2 and T1.
- Differential diagnosis of adamantinomatous type from Rathke's cyst (both hyperintense on T1w).
- Differentiation of papillary type from intraventricular or suprasellar glioma (both show non specific SI pattern, hypointense on T1 and hyperintense on T2).

Contrast enhancement

- Show contrast enhancement in over 90% of the cases.
- Marked heterogeneous enhancement of the solid component and rim enhancement of the cystic components (variable thickness of the cystic wall, usually thin).

Case 57: Craniopharyngioma (adamantinomatous type)

Fig. a: sagittal cut;	GE- T1w	**Fig. c:** coronal cut;	SE - T2w
Fig. b: coronal cut;	SE - T1w	**Fig. d:** coronal cut;	SE - T1w with fat suppression

Very young adult with growth failure. MR exam shows a huge suprasellar mass presenting a hyperintense signal on T1w (**a, b**, ↑↑), mainly in its posterior portion associated to a slightly less hyperintense signal of the anterior portion. The posterior hyperintense portion remains hyperintense after fat supression (**d**, ↑↑). The anterior portion shows a hypointense signal on T2w (**c**, ≠) as the inferior intrasellar portion (**c**, ↑↑). The posterior portion shows a hyperintense signal (**c**, ↑↑↑). This high signal intensity on T1w and T2w corresponds to the "machine-oil" material, frequently observed in such cystic tumoral components, containing suspended cholesterol crystals. The portions slightly hyperintense on T1w and markedly hyperintense on T2w presented fresh hemorragic foci as well as microcalcifications as disclosed on histology. The mass is well separated from the pituitary gland showing a hypointense signal on T1w (**a**, *). The tumor appears attached to the region of the pars tuberalis in the floor of the 3rd ventricle (**d**, ↑), developing into the suprasellar cistern, elevating the chiasm (**a**, ↑; **b**; right, ↑) and displacing more laterally the cisternal optic nerves (**b**, left, ↑).

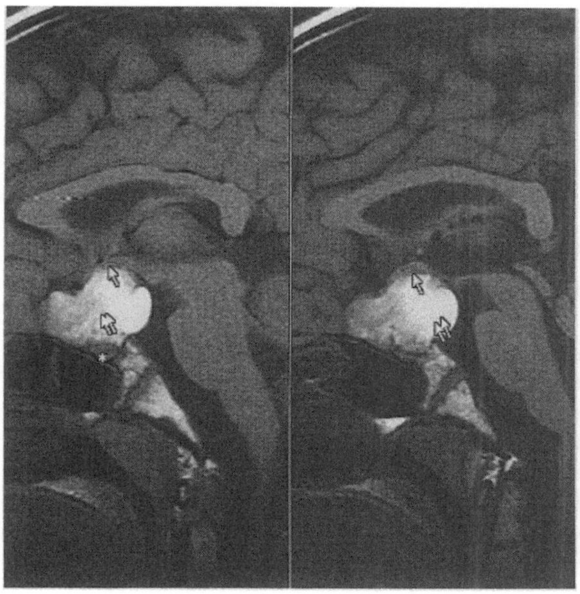

a

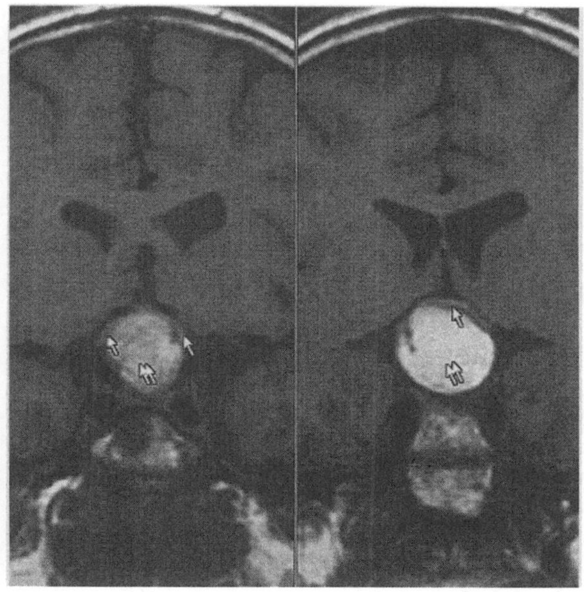

b

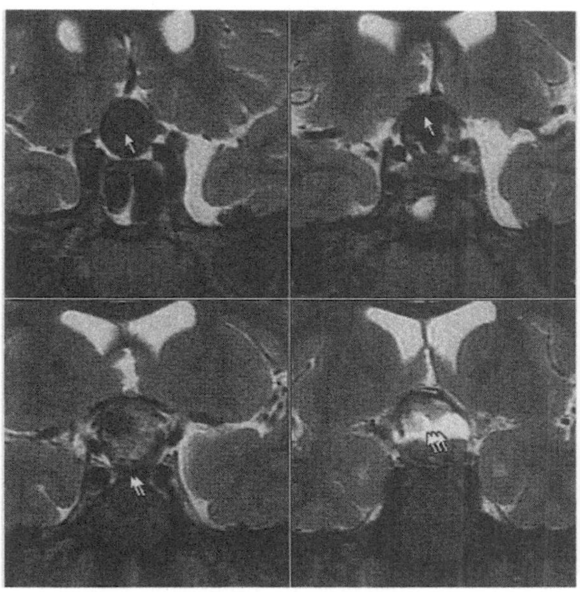

c

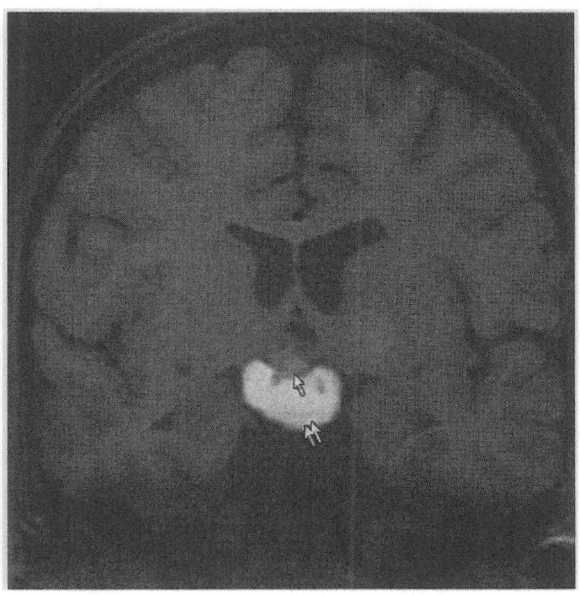

d

Case 58: Craniopharyngioma (papillary type)

Fig. a: sagittal cut; GE- T1w

Fig. b: sagittal cut; SE - T2w

Fig. c: coronal cut; GE- T1w

Fig. d: coronal cut; SE - T2w

Adult patient in the fifth decade presenting a diabetes insipidus and behavioral disturbances. MR shows highly heterogeneous suprasellar mass presenting a hypointense signal on T1w (**a**, ↑↑) with a right lateral cystic formation markedly hypointense on T1w (**c**, *) and hyperintense CSF-like on T2w (**d**, *). The mass involves the tuberal region of the 3rd ventricular floor occupying the posterior portion of the suprasellar cistern and invading partially the ventricular cavity (**c**, down, ↑↑; **d**, down, ↑). The chiasm is elevated by the anterior portion of the mass (**a**, *; **c**, ↑). Posteriorly, the mass abuts the optic tracts (**c**, ↑↑↑). The pituitary stalk is not visualized, contained into the mass. The pituitary gland is well differentiated from the suprasellar tumor (**a**, ↑; **b**, ↑). Note that the tumor enhanced after administration of contrast iv in its solid portion and tumoral wall.

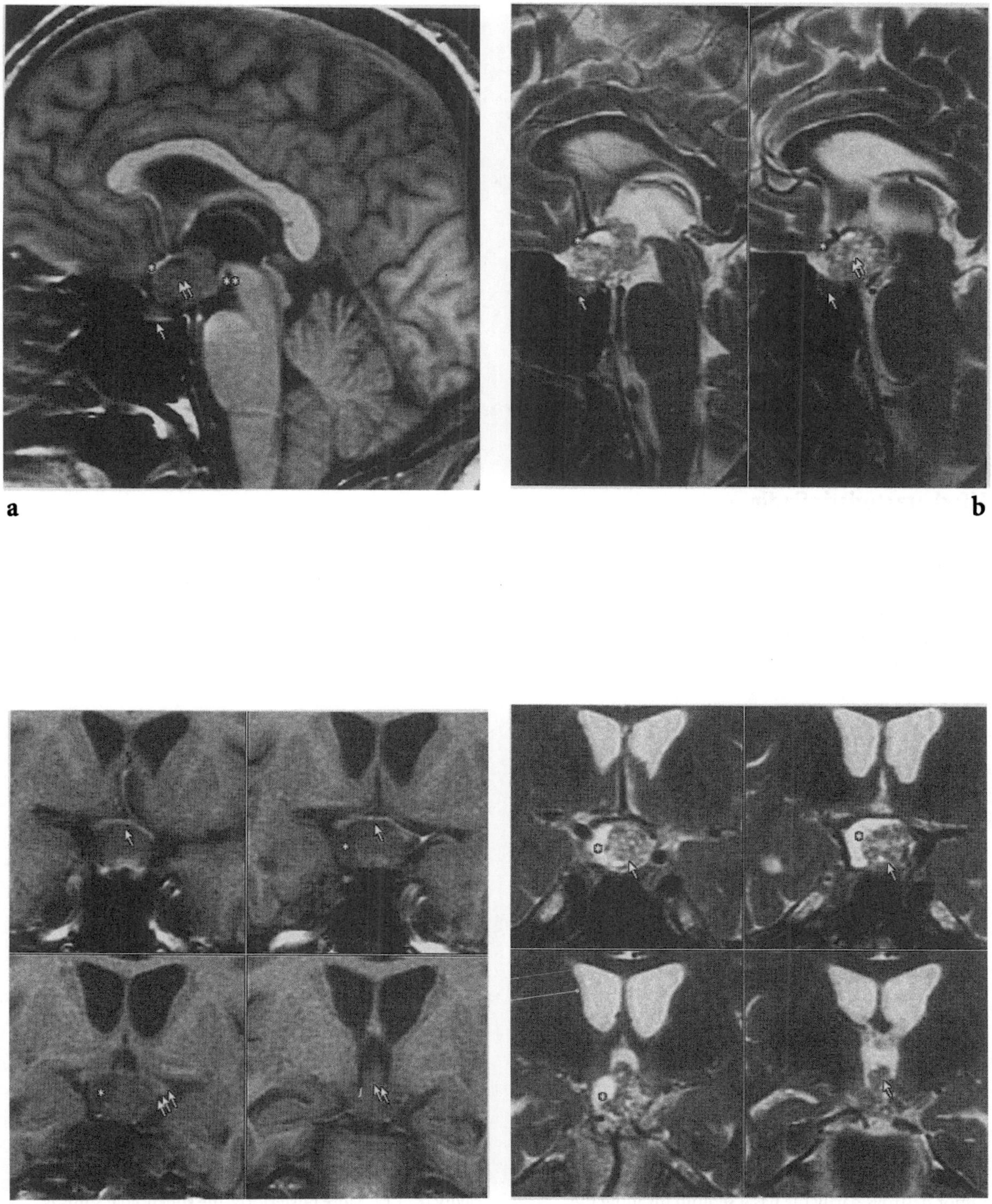

a b

c d

Nasopharyngeal region

Case 59: Rhabdomyosarcoma

Clinical and pathological features

- Most common malignancy in children (younger than 6 years of age).
- Rare mesenchymal malignant tumors (rhabdomyoblasts in nasopharynx muscles).
- Head and neck localization (30%): orbits, nasopharynx and paranasal sinuses.
- Early metastases to the neck (50%).
- Pathological forms: embryonal, alveolar (skeletal muscles), botryoid, pleiomorphic (adults).
- *Signs:* insidious or rapid onset of rhinitis, proptosis, sore throat, cranial neuropathies, otitis media, depending on tumoral spread.

MR characteristic findings

Morphology

- Malignant behavior: loco-regional invasion with aggressive bony destruction.
- Destruction of skull base.
- High water content in loose stroma (high signal on T2).
- Local recurrence frequent and hematogenous metastases (lung, brain, bone).

Signal intensity

- No specificity in SI behavior, heterogeneous.
- SE-T1w: hypo- to iso-intense to brain.
- SE-PDw: slightly hyperintense to brain.
- SE-T2w: heterogeneous, hyperintense and iso-intense to brain.
- Retentional sinonasal fluids: markedly hyperintense on T2.

Contrast enhancement

- Moderate enhancement of the tumor, heterogeneous.
- Marked enhancement of the normal mucosae.

Case 59: Rhabdomyosarcoma

Fig. a: sagittal cut; SE - T2w

Fig. b: axial cut; SE - PDw

Fig. c: coronal cut; SE - T2w

Fig. d: coronal cut; SE - T2w

Huge mass invading the nasal cavity and extending into the nasopharynx, showing heterogeneous hyperintense and mixed signal on T2w sequences (**a**, ↑). The aggressive tumor destroys partly the nasal septum invading the turbinates (**c**, ↑↑; **b**, ↑) and extending to the maxillary sinuses, more prominently on the right side (**c**, *). The heterogeneous tumor showing hyperintense areas of tumoral necrosis remains easily distinguished from the retentional fluids inside the paranasal sinuses markedly hyperintense on T2w and on PDw (**b**, ↑↑; **c**, **). The superior extension of the tumor involves the left ethmoid sinus (**c**, right, ↑) with lateral invasion of the left orbital cavity (**c**, ↑↑). More posteriorly, the sphenoidal sinus is invaded (**b**; **d**, right, ↑). There was no extention to the middle cranial fossa. The frontobasal cortex showed no abnormality. Note the normal aspect of the cisternal optic nerves (**d**, *) and their proximity to the sphenoidal extension of the mass.

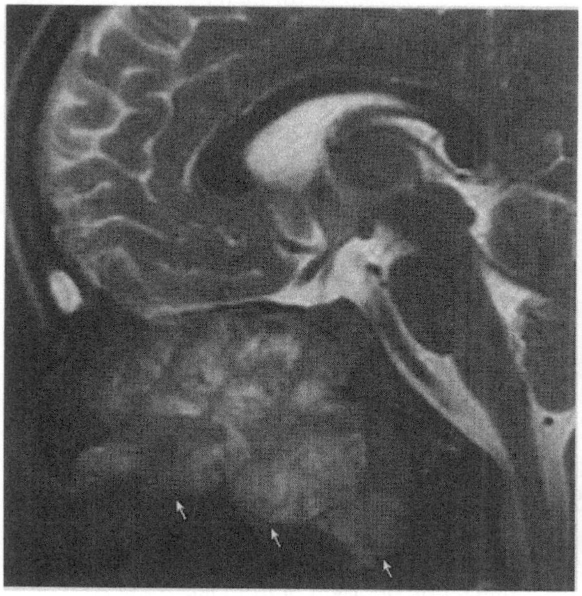

a

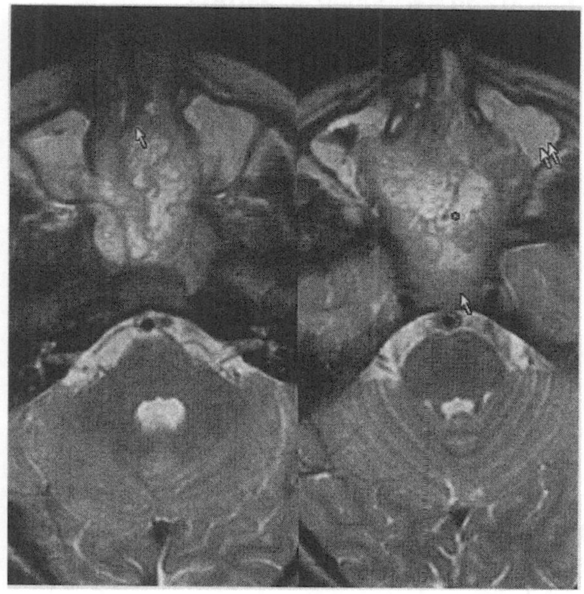

b

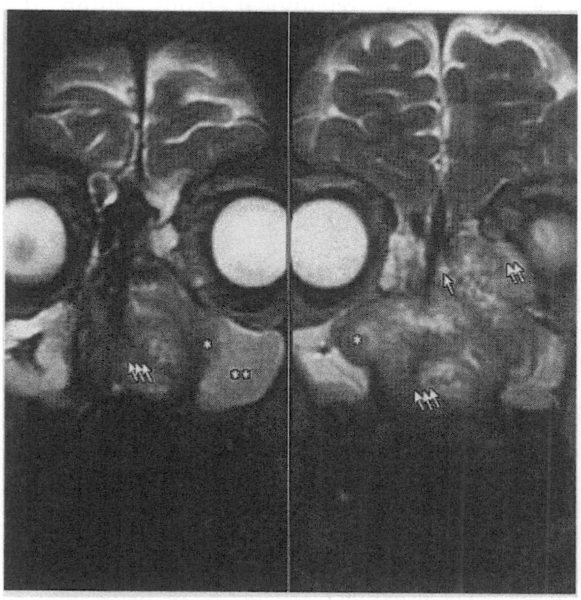

c

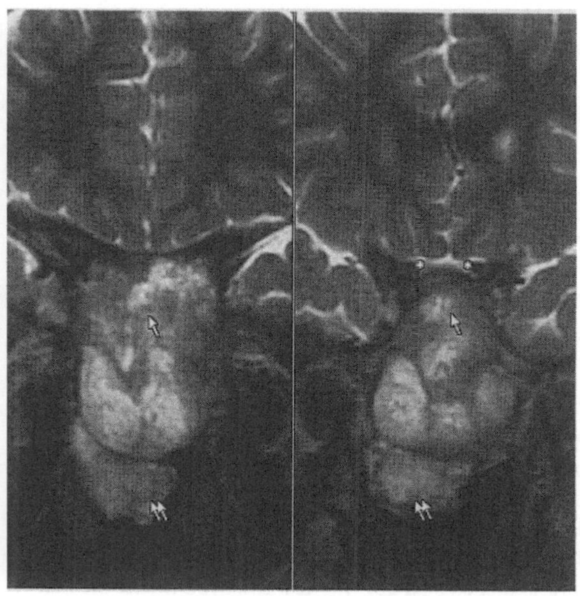

d

Case 60: juvenile angiofibroma

Clinical and pathological features

- Uncommon nasopharyngeal lesion.
- Most frequent benign tumor of the nasopharynx.
- Occurs in adolescent males (10-20 years old).
- Arises at the pterygomaxillary space or sphenopalatine foramen.
- Highly vascular benign tumor, supplied by pedicles of internal maxillary artery.
- Biopsy dangerous, to be strictly avoided (MR features reliably diagnostic).
- Common recurrence if incomplete surgery, performed after selective embolization.
- *Signs:* epistaxis, nasal obstruction, proptosis, sinusitis, anomia.

MR characteristic findings

Morphology

- Histologically benign but locally aggressive, highly vascular tumor (do not confuse with angiomatous polyps: hypervascular nasal benign mass).
- Asymmetrical development and growth from the pterygopalatine fossa (90%).
- Extension: to the sphenoid sinus in 2/3 of the cases (through foramen rotundum), to the orbit (through the inferior orbital fissure), to the anterior cavernous sinus (through the superior orbital fissure), to the ethmoid and maxillary sinuses, and to the infratemporal fossa (through pterygomaxillary fissure).
- Highly suggestive features:
  - scalloping of adjacent bony walls: anterior bowing of the posterior wall of the ipsilateral maxillary sinus;
  - obliteration of fat of the pterygopalatine fossa;
  - enlargement of sphenopalatine foramen;
  - intralesional numerous signal voids corresponding to intralesional vessels;
  - presence of petechial hemorrhages ("salt and pepper" aspect).

Signal intensity

- SE-T1w: hypointense to iso-intense to muscles, obliteration of bright SI from fat.
- SE-PDw: intermediate SI, iso-intense to brain, heterogeneous, with flow voids.
- SE-T2: isointense or slightly hyperintense to brain, with flow voids.
- Hyperintense retentional fluids within the sinuses, or intratumoral, on T2.

Contrast enhancement

- Marked contrast enhancement.

Case 60: Angiofibroma of the nasopharynx

Fig. a: axial cut;	SE - T1w	**Fig. c:** sagittal cut;	SE - T2w
Fig. b: coronal cut;	SE - T2w	**Fig. d:** axial cut;	SE - T2w

Young, teenager male presenting nasal obstruction and epistaxis as the existing symptoms. MR shows a huge extensive heterogeneous and highly vascular mass giving a characteristic appearance on MRI. The mass shows a major development at the junction of the lateral wall and roof of the posterior superior nasal cavity as shown on the parasagittal cuts (**c**, up, *). The mass, hypointense on T1w (**a**) shows a slightly hyperintense signal on PDw (**d**) and a mixed hyposlightly hyperintense signal on T2w (**c**, **b**, *). The tumor extends into the pterygopalatine fossa obliterating its normal fat content (**d**, ↑). The upper extension of the posterior portion of the mass is well shown on coronal cuts, eroding the sphenoid bone, and extending into the sphenoid sinus medially and the middle cranial fossa laterally to the central skull base (**b**, *). The left cavernous sinus seems elevated by the mass. The tumor remains epidural in location elevating the cavernous sinus (**b**, ↑↑) and displacing laterally the medial aspect of the temporal lobe (**b**, ↑) as confirmed at surgery. Note the hypointense rim surrounding the upper border of the mass, corresponding partly to intact dura. Note also on all sequences, but mainly on PDw sequence (**d**) due to the slightly bright signal of the mass, the multiple flowvoids curvilinear images corresponding to high flow vessels more or less large in this highly vascular tumor. Note the excellent differentiation of the tumor from the retentional fluids inside the paranasal cavities on T2w sequence (**d**, *; **b**; **c**). Note the erosion of the posterior wall of the left maxillar antrum (**d**, left).

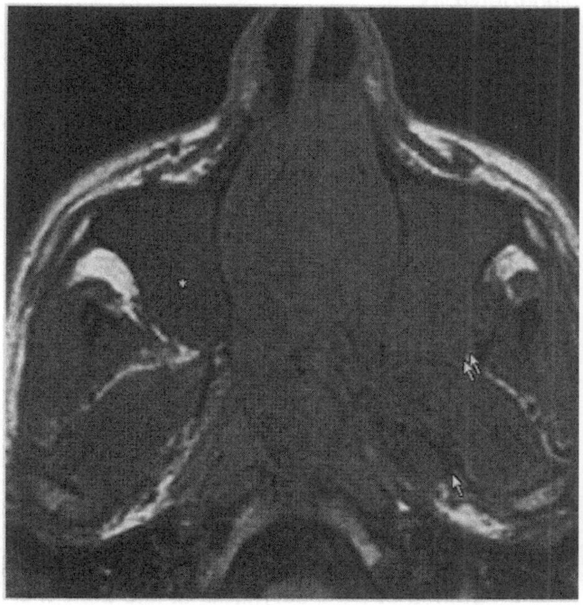

a

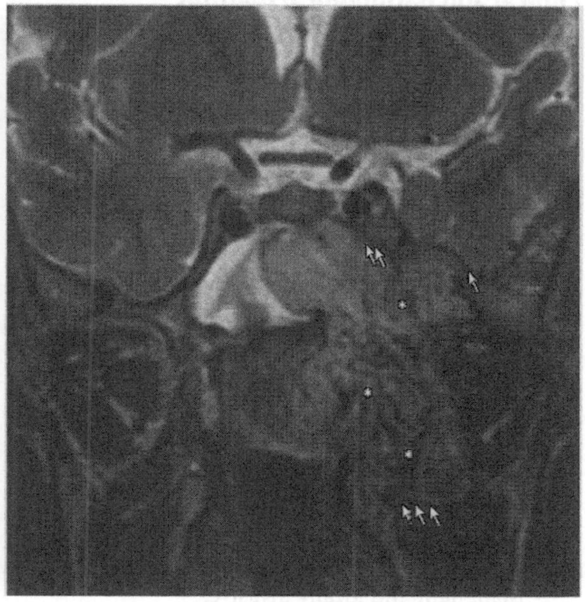

b

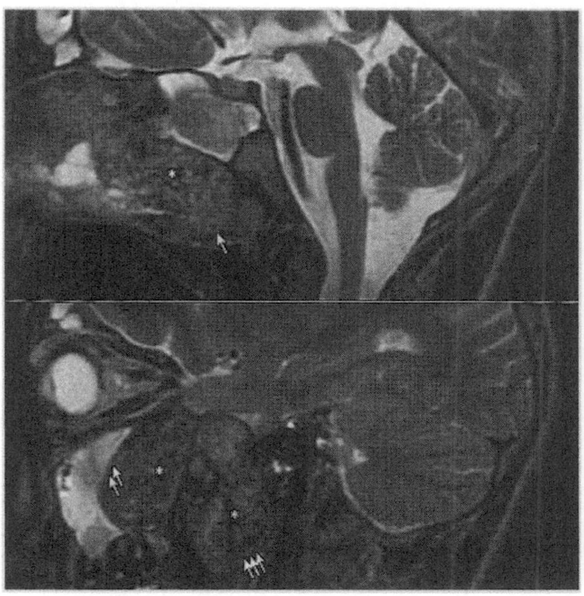

c

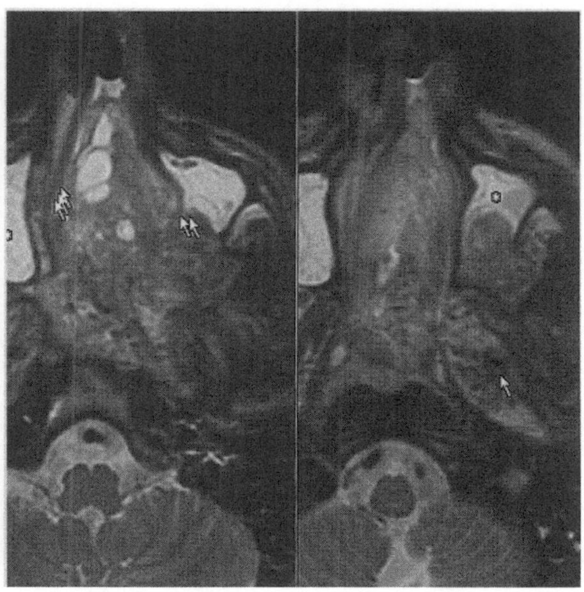

d

Case 61: Lymphoepithelioma

Clinical and pathological features

- Infrequent variety of malignant epithelial carcinoma (less than 1%, including mucoepidermoid carcinoma of the nasal cavity and maxillary sinus).
- Arise in the paranasal sinuses and the nasopharynx.
- Similar presenting signs to squamous cell carcinomas.
- *Signs:* nasal obstruction, epistaxis (paranasal sinus), otitis media (nasopharynx).

MR characteristic findings

Morphology

- Malignant and locally invasive.
- Similar behavior than squamous cell carcinomas.

Signal intensity

- SE-T1w: isointense to muscles (hypointense to fat).
- SE-PDw: isointense or slightly hyperintense to muscles.
- SE-T2w: hypointense or isointense to muscles.

Contrast enhancement

- Marked enhancement (best shown using SE-T1w with fat suppression).
- Show extensions to the parapharyngeal, prevertebral and carotid spaces.
- Identify invasion and lysis of the skull base (petro-clival region).
- Note that T1w sequences without contrast infusion and fat suppression best appreciate tumoral infiltration of parapharyngeal fat and bone marrow.

Case 61: Lymphoepithelioma

Fig. a: axial cut;	SE - PDw and SE - T2w	**Fig. c:** axial cut;	SE - T1w + Gd and fat suppression
Fig. b: sagittal cut;	SE - T1w + Gd	**Fig. d:** axial cut;	SE - T1w + Gd and fat suppression

Extensive infiltrating mass filling the nasopharynx, presenting a slightly hyperintense signal on PDw sequence (a, left) and a hypointense signal on T2w (a, right) but showing marked enhancement after contrast iv (**b**, **c**, **d**). The tumor is mainly developed on the left side, involving the posterior and lateral walls of the nasopharynx and invading the left parapharyngeal space (**c**, ↑; **b**, ↑). Note the extension into the prevertebral muscles (**b**, ↑; **c**, ↑) with invasion of the left occipital condyle and basiocciput on the left side (**c**). The parapharyngeal extension of the tumor is responsible of the lateral displacement of the left carotid artery (**c**, *). Superiorly, the lesion extends posteriorly to the retroclival region (**b**, *) bulging into the prepontine cistern (**d**, ↑↑) and involving laterally the carotid canal (**d**, right, ↑).

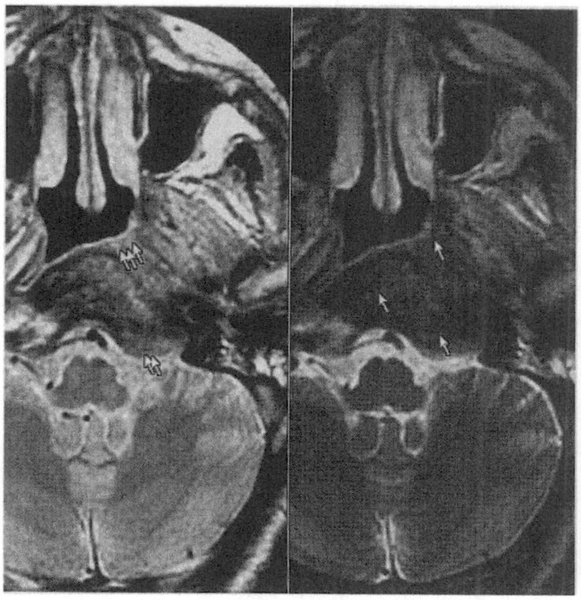

a

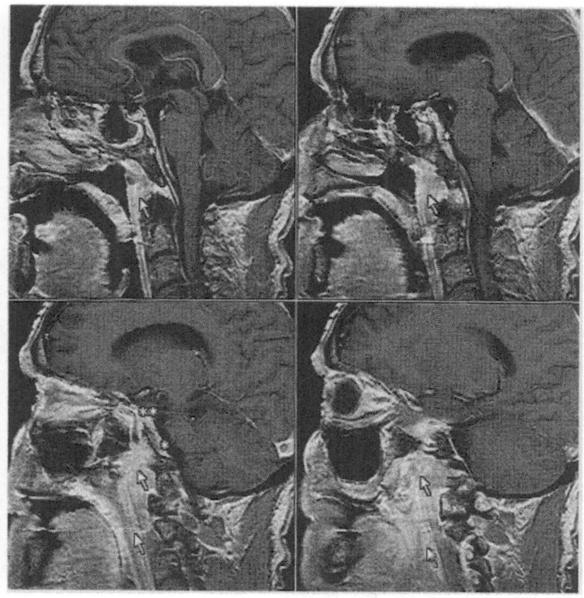

b

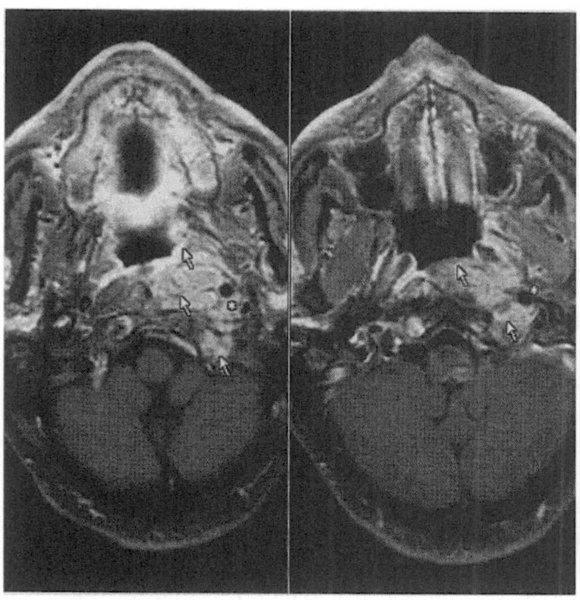

c

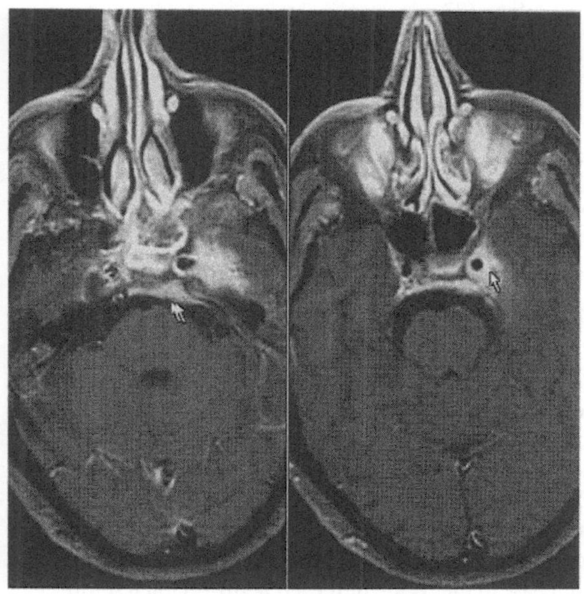

d

Case 62: Nasopharyngeal squamous cell carcinoma

Clinical and pathological features

- Represent more than 85% of primary carcinomas of the nasopharynx.
- Occur in middle age adults over 40, males in 2/3 of cases.
- Prevalence: 20/100 000 in Asians and 1/100 000 in Caucasians.
- Association with Epstein-Barr virus.
- Cervical lymphadenopathy found in 80% at presentation.
- Bilateral nodes found in 50% (suspect over 10 mm).
- *Signs:* serous otitis media due to dysfunction of Eustachian tube in elderly, nasal obstruction, epistaxis, diplopia (abducens palsy), facial hypesthesia or trismus.
- Signs are more frequently delayed with respect to tumoral development.

MR characteristic findings

Morphology

- High resolution SE images mandatory in the axial plane (T1w).
- Differentiation of small tumors from adenoidal tissue (present in young individuals but regressing usually in adulthood) may be difficult in early stages.
- Highly infiltrating tumor, located in or about the fossa of Rosenmüller.
- Regional extension to the intrinsic muscles of the nasopharynx and/or infiltrating the parapharyngeal fatty soft tissue space (T1 without fat suppression).
- Early widespread extension and infiltration of the petro-clival bone posteriorly (sagittal or axial cuts), and/or outside the pharyngeal basilar fascia into the soft tissue of the parapharyngeal space laterally via Morgagni's foramen (coronal cuts).
- Anterior extension along the medial pterygoid muscle towards the pterygo-maxillary and nasal spaces.
- Superior extension to the skull base and cavernous sinus spreading along the mandibular nerve route.

Signal intensity

- SE-T1w: iso-intense or slightly hyperintense to muscle.
- SE-PDw: iso-intense to brain.
- SE-T2w: iso-intense or slightly hyperintense to brain.

Contrast enhancement

- Non specific contrast enhancement of the tumoral process.
- Coronal T1w with fat suppression to evaluate extension to skull base through the foramina (perineural tumoral spread).

Case 62: Nasopharyngeal carcinoma

Fig. a: axial cut;	SE - T2w	**Fig. c:** axial cut (512);	SE - T1w
Fig. b: axial cut (512);	SE - T1w + Gd	**Fig. d:** coronal cut;	SE - T1w + Gd

Patient presenting with epistaxis, nasal obstruction and cranial neuropathy. MR shows a large tumor mass filling the nasopharynx and invading the sphenoid (**a**, **b**) with skull base erosion and left cavernous sinus involvement (**d**). The tumor presents as slightly hyperintense signal on T2w (**a**, ↑), enhancing moderately after contrast iv (**b**, **d**). With contrast, the tumor is well delineated extending from the skull base to the region of the left petro-occipital suture bulging in the lateral pontomedullary cistern (**b**, ↑). This extension, better delineated after contrast iv, abuts the basilar artery as shown on T2w (**a**). The superior extension of the mass is followed into the left cavernous sinus with outward bulging of its lateral wall encircling the intracavernous carotid artery (**d**, ↑). The mass extends to the sellar region displacing the pituitary gland and stalk to the right (**d**, right). Note the soft-tissue masses in the retropharyngeal spaces (**d**, ↑↑), showing little enhancement and suggesting retropharyngeal lymph nodes. Axial T1w sequence shows an enlarged, moderatly enhancing lateral retropharyngeal node of "Rouvière" medial to the internal carotid artery (**c**, ↑↑) with mass effect on the ipsilateral prevertebral muscles, explaining the clinical symptoms. Note the encroachment on the airway by hypointense mass on T2w, well differentiated from the hyperintense benign secretions in the maxillary sinuses as well as the inflammatory turbinates (**a**). This demonstrates the advantage of MR for the evaluation and follow-up of such malignant base of the skull and nasopharyngeal tumors. In fact, on multiple pulse sequences, the signal caracteristics of these lesions differ from the inflammatory reactions in the paranasal sinuses. Multiplanar acquisitions are also needed.

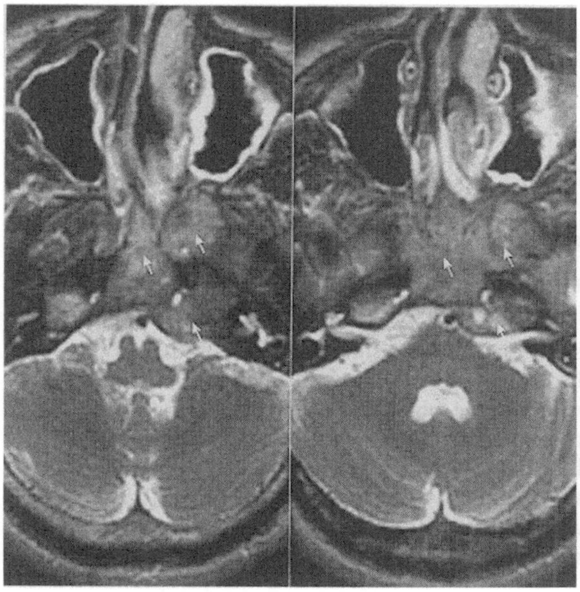

a

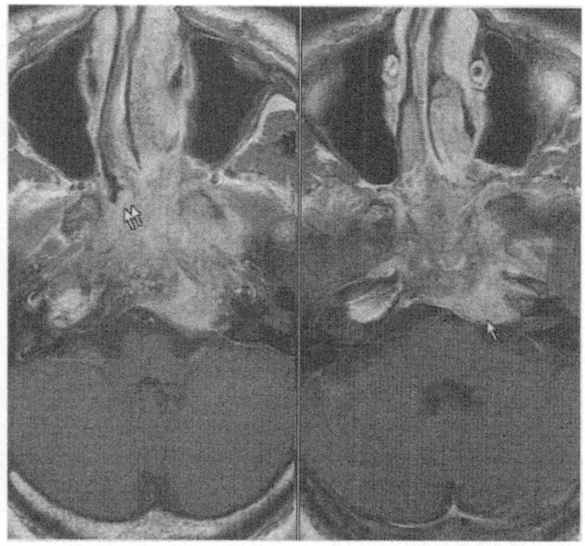

b

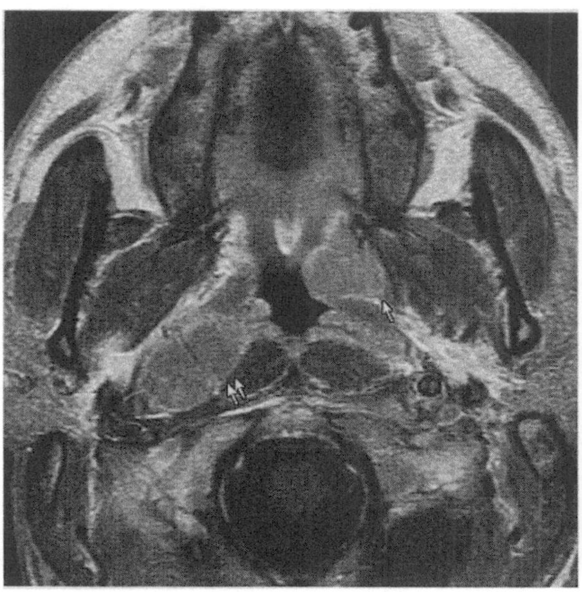

c

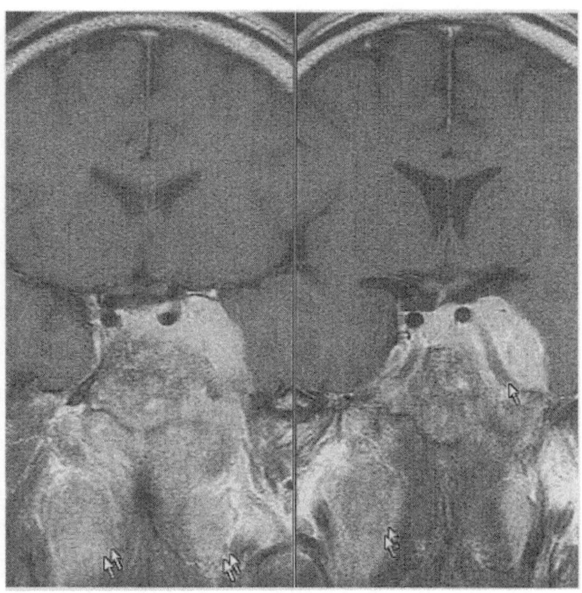

d

Case 63: Adenocarcinoma

Fig. a: axial cut (512);	SE - T1w	**Fig. c:** axial cut;	SE - T2w
Fig. b: coronal cut (512);	SE - T1w	**Fig. d:** coronal cut;	SE - T2w

Abnormal mass involving the ethmoid cells, the nasal fossa and the sphenoid sinus. The abnormal tissue shows a hypointense signal on T1w (**a, b**) remaining moderately hyperintense to white matter on T2w (**c**) and becoming hyperintense on T2w (**d**). The signal of the mass differs enough from the sphenoidal sinus content, more hyperintense on T1w (**a**, *) as well as on PDw (**c**, ↑↑) and is well distinguished from the maxillary sinuses and the non invaded ethmoidal cells (**d**, ↑) showing a hyperintense signal on T2w contrasting with the mass markedly hypointense and iso-intense to brain (**d**, ↑↑). Note that with MRI the tumor can be differentiated from the obstructed sinuses in most cases by comparing T1w (**a**, *; **b**, ↑) and T2w (**c**, ↑↑; **d**, ↑) sequences. The mass did not show extension to the left orbital cavity (**d**). The tumor was a well differentiated adenocarcinoma of the paranasal sinuses.

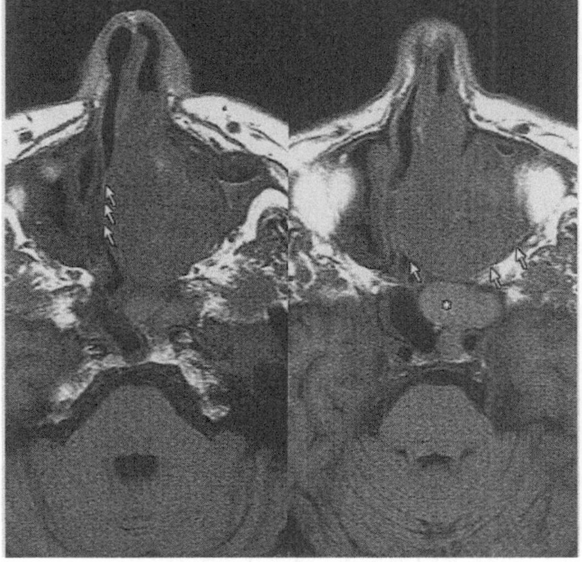

a

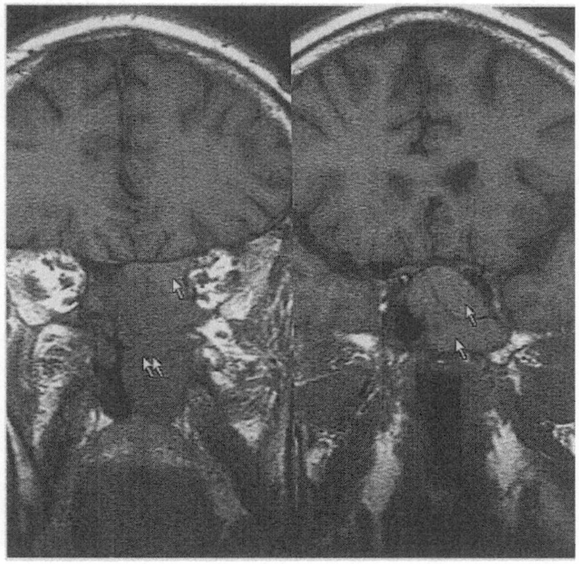

b

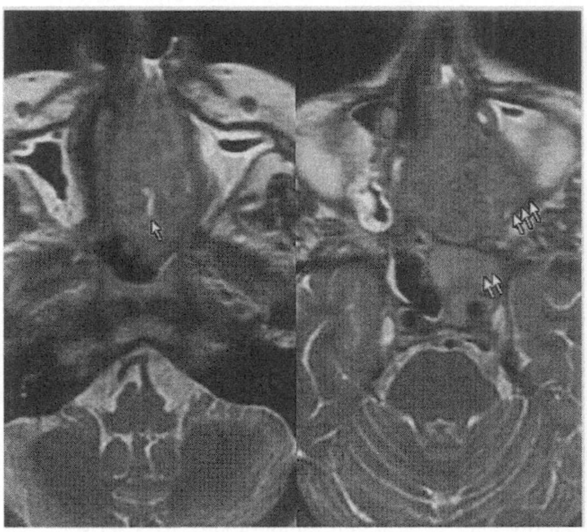

c

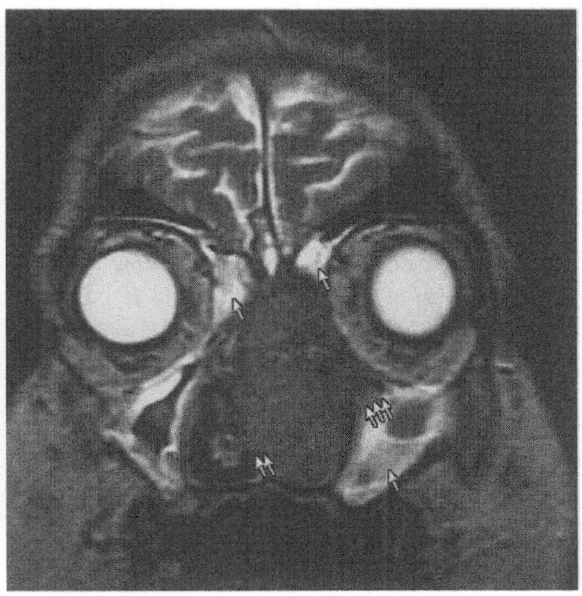

d

Selected References

- Ahmadi J et al (1995) Lymphocytic adenohypophysitis: contrast-enhanced MR imaging in 5 cases. Radiology 195: 30
- Ahmadi JA et al (1992) Cystic fluid in craniopharyngiomas: MR imaging and quantitative analysis. Radiology 182:783
- Bartynski WS, Lin L (1997) Dynamic and Conventional Spin-Echo MR of Pituitary Microlesions. AJNR 18:965-972
- Boyko OB et al (1991) Hamartoma of the tuber cinereum: CT, MR, and pathologic findings. AJNR 12:309
- Brown EW, Megerian CA, McKenna KJ et al (1995) Fibrous dysplasia of the temporal bone: imaging findings. Am J Roentgenol 164(3):679
- Brown RV, Sage MR, Brophy BP (1989) CT and MR findings in patients with chordomas of the petrous apex. Am J Neuroradiol 11:121
- Carmody RF, Rickles DJ, Johnson SF (1983) Giant cell tumor of the sphenoid bone. J Comput Assist Tomogr 7:370
- Chaljub G, Van Fleet R, Guinto FC et al (1992) MR imaging of clival and paraclival lesions. Am J Roentgenol 159:1069
- Chong VF, Fan YF (1996) Skull base erosion in nasopharyngeal carcinoma: detection by CT and MRI. Clin Radiol 51(9):625
- Daniels DL, Schenck JF, Foster T et al (1985) Magnetic resonance imaging of the jugular foramen. Am J Neuroradiol 6:669
- Davis WL et al (1990) Retropharyngeal space: evaluation of normal anatomy and diseases with CT and MR imaging. Radiology 174:59
- Derdeyn CP, Moran CJ, Wippold FJ et al (1994) MRI of esthesioneuroblastoma. J Comput Assist Tomogr 18:16
- Dillon WP, Som PM, Fullerton GD (1990) Hypointense MR signal in chronically inspissated sinonasal secretions. Radiology 174:73
- Fitzpatrick PJ, Briant DR, Berman JM (1980) The nasopharyngeal angiofibroma. Arch Otolaryngol 106:234
- Freeman MP et al (1987) Craniopharyngioma: CT and MR imaging in nine cases. J Comput Assist Tomogr 11:810
- Giletee GR, Phil D, Symon L (1987) Hypothalamic glioma, Surg Neurol 28:291
- Ginsberg LE (1992) Neoplastic diseases affecting the central skull base: CT and MR imaging. Am J Roentgenol 159:581
- Harman SA, Pribram H (1990) Case report 586: giant cell tumor of sphenoid bone. Skeletal Radiol 19:212
- Higinbothan N, Philips R, Farr H et al (1967) Chordoma, 35 year study at Memorial Hospital. Cancer 11(20):1141
- Hudgins PA, Gussack GS (1992) MR imaging in the management of extracranial malignant tumors of the head and neck. Am J Roentgenol 159:161
- Hungerford G, Biggs P, Levine J et al (1982) Lymphoid adenohypophysitis with radiologic and clinical findings resembling a pituitary tumor. AJNR 3:444-449
- Kaplan H, Baker H, Houser O et al (1985) CT of the sella turcica after trans-sphenoidal resection of pituitary adenomas. AJNR 6:723-732
- Kaufman B, Tomsak R et al (1989) Herniation of the suprasellar visual system and third ventricle into empty sellae: morphologic and clinical considerations. AJNR, 10:65-76
- Korneich L et al (1995) Central precocicus puberty: evaluation by neuroimaging. Pediatr Radiol 25:7
- Lacomis D, Johnson L, Mamourian A (1988) Magnetic resonance imaging in pituitary apoplexy. Arch Ophthalmol 106:207-209
- Laine FJ, Braun IF, Jensen ME et al (1990) Perineural extension through the foramen ovale: evaluation with MR imaging. Radiology 174:65
- Laine FJ, Braun IF, Jensen ME et al (1990) Perineural tumor extension through the foramen ovale: evaluation with MR imaging. Radiology 174:65
- Laine FJ, Nadel L, Braun IF (1990) CT and MR imaging of the central skull base. II. Radiographics 10:797
- Lee YY, Van Tassel P (1989) Craniofacial chondrosarcomas: imaging findings in 15 untreated cases. Am J Neuroradiol 10:165
- Levine S, Benzel E, Fowler M et al (1988) Lymphocytic hypophysitis: clinical, radiological, and magnetic resonance imaging characterization. Neurosurgery 22:937-941
- Lewin JS, Curtin HD, Eelkema E, Obuchowski N (1999) Benign Expansile Lesions of the Sphenoid Sinus: Differentiation from Normal Asymmetry of the Lateral Recesses. AJNR 20:461-466
- Lufkin Rb, Hanafee WN, Wortham D et al (1986) Larynx and hypopharynx: MR imaging with surface coils. Radiology 158:747
- Meyers SP, Hirsch WL, Curtin HD (1992) Chondrosarcomas of the skull base: MR imaging features. Radiology 184:103
- Nagahata M, Hosoya T, Kayama T, Yamaguchi K (1998) Edema along the optic tract: a useful MR finding for the diagnosis of craniopharyngiomas. AJNR 19:1753-1757
- Naylor MF, Scheithauer BW, Forbes GS, Tomlinson FH, Young WF (1995) Rathke cleft cyst: CT, MR and pathology of 23 cases. J Comput Assist Tomogr 19:853-859
- Neelon F, Goree J, Lebowitz H et al (1973) The primary empty sella: clinical and radiographic characteristics and endocrine function. Medicine 52:73-92
- Nemzek WR, Hecht S, Gandour-Edwards R et al (1998) Perineural Spread of Head and Neck Tumors: How Accurate Is MR Imaging? AJNR 19:701-706
- Olsen WL, Dillon WP, Kelly WM et al (1987) MR imaging of paragangliomas. Am J Roentgenol 148:201
- Oot RF, Melville GE, New PFJ et al (1988) The role of MR and CT in evaluating clival chordomas and chondrosarcomas. Am J Roentgenol 151:567

- Poussaint TY, Barnes PD, Nichols K et al (1997) Diencephalic Syndrome: Clinical Features and Imaging Findings. AJNR 18:1499-1505
- Ross DA, Norman D, Wilson CB (1992) Radiologic characteristic and results of surgical management of Rathke's cysts in 43 patients. Neurosurgery 30(2):173-178
- Savino PJ, Grossman RI, Schartz NJ et al (1986) High field magnetic resonance imaging in the diagnosis of cavernous sinus thrombosis. Arch Neurol 43:1081
- Schubiger O, Haller D (1992) Metastases to the pituitary-hypothalamic axis: an MR study of 7 symptomatic patients. Neuroradiology 34(2):131-134
- Schuster JJ, Philips CD, Levine PA (1994) MR of esthesioneuroblastoma and appearance after craniofacial resection. Am J Neuroradiol 15:1169
- Scotti G, Yu C, Dillon W et al (1989) MR imaging of cavernous sinus involvement by pituitary adenomas. AJNR 9:657-664
- Shubiger O, Haller D (1992) Metastases to the pituitary hypothalamic axis: an MR study of 7 symptomatic patients. Neuroradiology 34:131-134
- Som PM, Biller HF, Lawson W et al (1984) An approach to parapharyngeal space masses: an updated protocol based upon 104 cases. Radiology 153-149
- Som PM, Braun IF, Shapiro MD et al (1987) Tumors of the parapharyngeal space and upper neck: MR imaging characteristics. Radiology 164:823
- Som PM, Dillon WP, Sze G et al (1969) Benign and malignant sinonasal lesions with intracranial extensions: differentiation by MR imaging. Radiology 172:763
- Som PM et al (1987) Tumors of the parapharyngeal space and upper neck: MR imaging characteristics. Radiology 164:823
- Som PM et al (1988) Common tumors of the parapharyngeal space: refined imaging diagnosis. Radiology 161:81
- Som PM, Lawson W, Lidov MW (1991) Simulated aggressive skull base erosion in response to benign sinonasal disease. Radiology 180:755
- Steiner E, Imhof H, Knosp E (1989) Gd-DTPA enhanced high resolution MR imaging of the pituitary adenomas. Radiographics 9:587-598
- Steiner E, Knosp E, Herold C et al (1992) Pituitary adenomas: findings of postoperative MR imaging. Radiology, 185:521-527
- Sumida M, Uozumi T, Mukada K et al (1994) Rathke cleft cysts: correlation of enhanced MR and surgical findings. AJNR 15:525-532
- Taylor SL, Barakos JA, Harsh GR et al (1992) Magnetic resonance imaging of tuberculum sellae meningiomas: preventing preoperative misdiagnosis as pituitary macroadenoma. Neurosurgery 31:621-627
- Tien RD et al (1990) Thickened pituitary stalk on MR images in patients with diabetes insipidus and Langerhans' cell histiocytosis. AJNR 11:703
- Utz JA, Kransdorf MJ, Jelinek JS et al (1989) MR appearance of fibrous dysplasia. J Comput Assist Tomogr 13(5):845
- Van Tassel P, Lee Y, Jing B et al (1989) Mucoceles of the paranasal sinuses: MR imaging with CT correlation. Am J Roentgenol 153:407
- Voelker J, Campbell R, Muller J (1991) Clinical, radiographic and pathological features of symptomatic Rathke's cleft cysts. J Neursurg 74:535-544
- Wakai S, Fukushima T, Teramoto A et al (1981) Pituitary apoplexy: its incidence and clinical significance. J Neurosurg 55:187-193
- Young SC, Zimmerman RA, Nowell MA et al (1987) Giant cystic craniopharyngiomas. Neuroradiology 29:468-473
- Yuk W, Wright D, Barloon T et al (1988) MR imaging of primary tumors of the trigeminal nerve and Meckel's cave. AJR 151:577-582

Part II

Skull base and face

3 - Posterior fossa

a - Cerebellopontine region
b - Craniocervical region

Cerebellopontine region

Case 64: Cholesterol granuloma (cyst)

Clinical and pathological features

- One of the most frequent benign lesion of the petrous apex.
- May also be found in the middle ear or the mastoid.
- Arises characteristically from obstructed air cells.
- Results from accumulation of inflammatory granulation reacting to the presence of cholesterol crystals and blood breakdown products from recurrent hemorrhages.
- Slow-growing process (over years) lined by a fibrous capsule.
- *Signs:* usually asymptomatic, unless located in the middle ear (earliest symptoms).
- Must not be confused with epidermoid tumors.

MR characteristic findings

Morphology

- Well delineated tumor, in the petrous apex, middle ear or mastoid.
- May be large when arising in the petrous apex.
- Erodes and displaces adjacent structures (carotid, middle ear, mesial temporal).
- Frequent incidental finding (apex lesions).
- To be differentiated from epidermoids, mucoceles, chordomas.
- Similar aspect on CT: cholesterol cyst, epidermoid, mucocele.

Signal intensity

- Characteristic SI due to methemoglobin deposits, cholesterine.
- SE-T1w: hyperintense, close to fat.
- SE-T2w: hyperintense, close to CSF.
- May show intratumoral low signal areas due to hemosiderin deposit.
- Differentiated from epidermoids: intermediate SI on T1, hyperintense on T2.
- May be difficult to differentiate from rare mucoceles: possibly hyperintense on T1 (proteinaceous and hemorrhagic content).

Contrast enhancement

- No need for contrast infusion (bright signal on T1).

Case 64: Cholesterol granuloma

Fig. a: coronal cut;	SE - T1w
Fig. b: axial cut;	SE - T1w
Fig. c: sagittal cut;	SE - T1w
Fig. d: sagittal cut;	SE - T2w

The MR exam was performed using the head-coil (**d**) and a 3 inches surface coil positioned against the right temporal region, centered by the external auditory canal (**a**-CAE, **b**, **c**). The T1w sequences (**b**, **c**) show a high signal mass involving the right petrous bone (**a**, left, ↑; **b**, right, ↑; **c**, ↑), with areas of low signal within the mass (**a**, right, ↑; **b**, left, ↑). On T2w, the mass looks largely hyperintense but shows persistent areas of low signal. The hyperintense signal on both T1 and T2w images is characteristic of cholesterol cysts or cholesterol granulomas, and the dark areas within represent hemosiderin from recurrent hemorrhage. The mass involves the lateral aspect of the petrous bone extending to the apex with its relationship inferiorly with the mastoid (**a**, ↑↑), medially with the petrous bone (**a**, ↑↑↑) and posteriorly with the cochleovestibular apparatus (**d**, ↑↑; **d**, ↑↑↑). The relation of the mass with the middle ear raises the differential diagnosis of cholesteatoma or epidermoid. The latter tends to be hypointense or intermediate on the T1w and hyperintense on T2w. Note that cholesterol granulomas result from a foreign body response to the presence of cholesterol cristals and the bright signal demonstrated on all SE sequences is also due to the presence of extracellular methemoglobin.

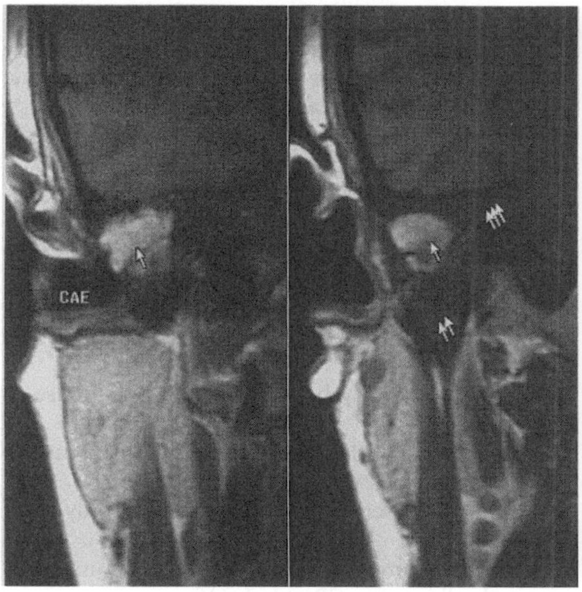

a

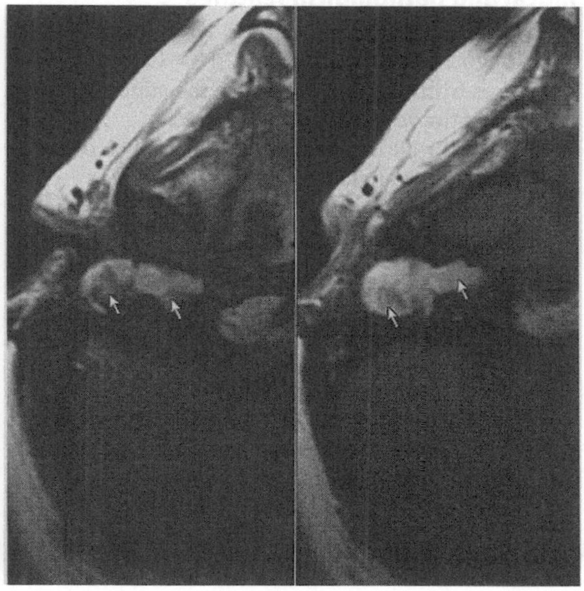
b

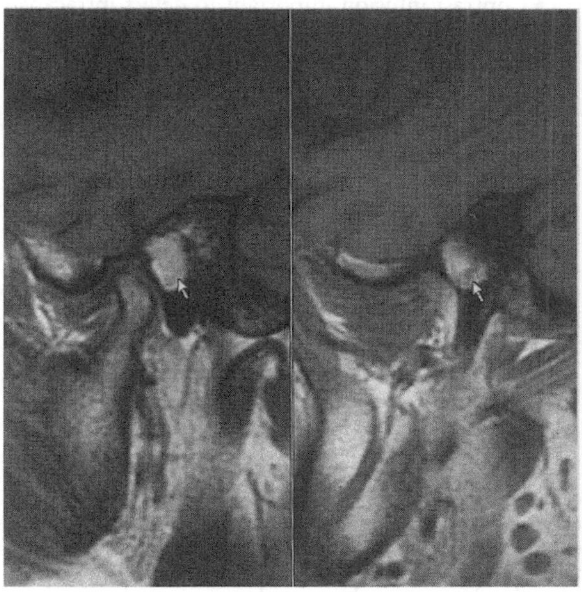
c

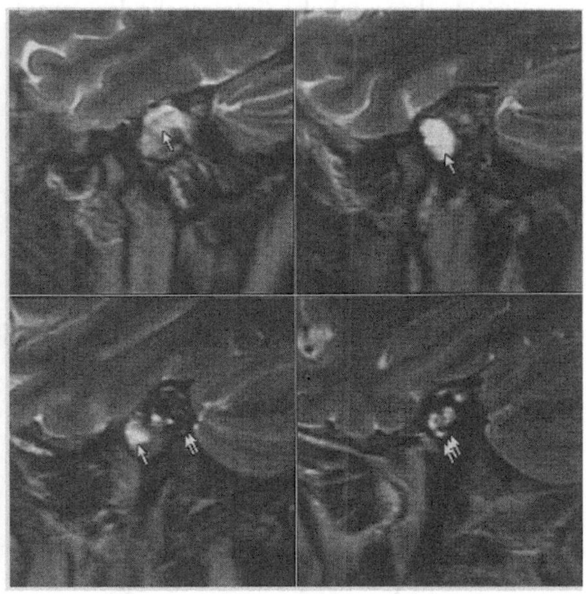
d

Cases 65-70: acoustic neurinomas

Clinical and pathological features

- Benign encapsulated tumors in the CP angle.
- Arises from Schwann's cells of the superior division of the vestibulocochlear nerve.
- 8-10 % of all intracranial and over 80% of cerebellopontine angle tumors.
- Frequent association (95%) with neurofibromatosis type II (usually bilateral).
- Histologically: 2 cellular types (Antoni A and B cells), absence of nerve fibers inside the tumor (contrary to neurofibromas).
- *Signs:* progressive unilateral retrocochlear neurosensory hearing loss and tinnitus.
- Hearing loss involves high frequencies initially.
- Disorders of the vestibular system (disequilibrium or dizziness) are less frequent.

MR characteristic findings

Morphology

- MR is the imaging modality of choice for the diagnosis of CPA masses.
- Circumscribed mass, variable size, encapsulated.
- Variable shape: tubular (intracanalicular), ovoid, mushroom-like, lobulated.
- Usually centered at the intracranial porus acousticus.
- Intracanalicular (auditory canal widening, 20%) and/or cisternal within CPA.
- 2 histological types: compact cellular (spindle cells) texture (Antoni A) and loose texture with microcystic and mucinous components (Antoni B).
- Presence of intratumoral areas of cystic degeneration or hemorrhage (usually more characteristic of schwannomas than CPA meningiomas).
- Eventual presence of intratumoral vessels (curvilinear flow voids).
- CSF entrapment in the vicinity or at the bone-tumor interface.
- Bilateral tumors (95%) in neurofibromatosis type II or type I (5%), see case 70.

Signal intensity

- SE-T1w: slightly hypointense or iso-intense to brain.
- SE-PDw: slightly hyperintense or iso-intense to cerebellar cortex.
- SE-T2w: hyperintense to brain, commonly heterogeneous.
- May be homogeneous, or heterogeneous cystic and solid (larger CPA tumors).

Contrast enhancement

- Usually marked contrast enhancement: most common.
- Contrast infusion important to detect intracanalicular extent or processes (axial).
- Pattern of contrast uptake usually heterogeneous (associated non enhancing cystic formations, which may be large), but sharp boundaries with adjacent tissues.
- High resolution (SE-T2 , 512 matrix) MR may suffice to screen CPA tumors.

Case 65: Intracanalicular acoustic neurinoma (VIII)

Fig. a: axial cut (512);	SE - T1w + Gd	**Fig. c:** axial cut (512);	SE - T2w
Fig. b: axial cut;	GE (3D) - T1w + Gd	**Fig. d:** axial cut (512);	SE - T1w + Gd

Middle age woman presenting a sensoryneural hearing loss with a disequilibrium state. MR shows a small soft tissue mass within the medial aspect of the right internal auditory canal (**c**, ↑) showing an isosignal to brain on high resolution T2w sequence associated to a subtle enlargement of the corresponding portion of the canal. The lesion is readily differentiated from the hyperintense signal of the intracanalicular CSF (**c**) still demonstrating the cochleo vestibular nerves. After contrast iv, high resolution T1w images confirm the oval enhancing mass within the auditory canal (**d, a**, ↑). 3D-GE-T1w sequence showed also enhancement of the lesion (**b**, ↑) but lack of morphological details as observed on SE-T1 and T2w high resolution images, even with millimetric contiguous slices. Gradient echo sequences may be used in case of suspicion of exceptional hemangiomatous lesions. Note also that small enhancing tumoral conditions along the facio-acoustic nerves in the internal auditory canal are not specific of acoustic schwannomas but may also correspond to focal neuritis and/or chronic inflammation. The presence of a dural extension or "tail" may help to distinguish these unusual entities. Of course, metastatic lesions or lymphomas may also involve the internal auditory canal and could mimick small intracanalicular schwannomas. Other foci are generally associated in such cases. Anyhow, follow-up exams, along with or without medical treatement, are recommended.

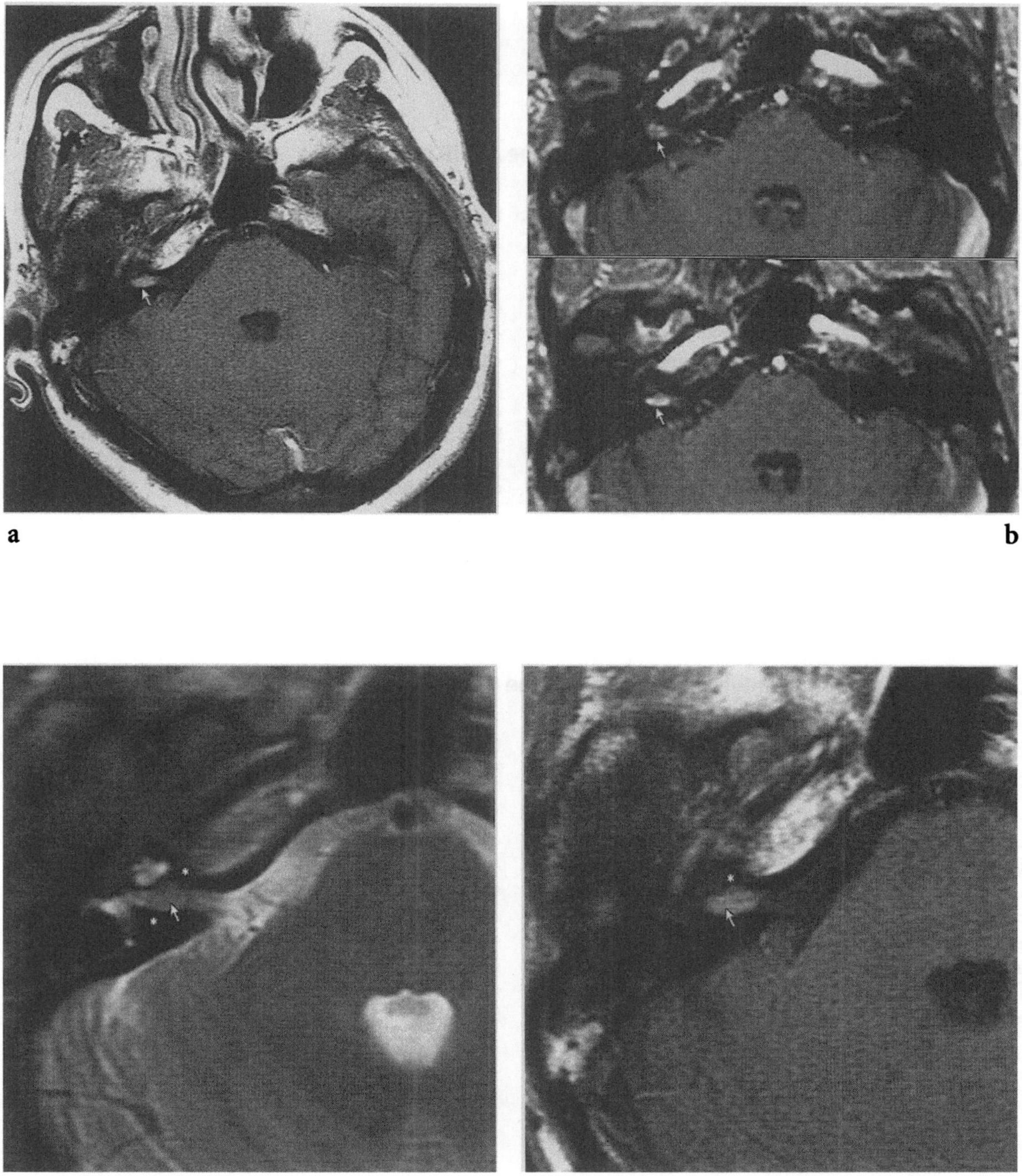

a b

c d

Case 66: Intra and extracanalar schwannoma type A (VIII nerve)

Fig. a: axial cut; SE - PDw

Fig. b: coronal cut; SE - T2w

Right sensoryneural hearing loss and vertigo. MR shows a small well demarcated mass in the right cerebellopontine angle cistern presenting a hyperintense signal on PDw, iso-intense to cerebellar cortex (**a**, ↑↑) and a hypointense signal on T2w (**b**, ↑). The mass is associated to an enlargement of the internal auditory canal (**a**, ↑). This enlargement corresponds to the intracanalicular extension of the mass well shown with hypointense signal as compared to brigh CSF on T2w sequence (**b**). The diagnosis may be difficult on PDw due to slight isointensity to CSF. Note the mass effect on the ipsilateral middle cerebellar peduncle. The left internal auditory canal was normal sized (**b**). The mass enhances homogeneously after contrast iv. It presented histologically a compact texture with numerous fibers within the stroma associated to the bipolar spindle cells corresponding to Anthony type A tissue.

Case 67: Intra and extracanalar schwannoma type B (VIII)

Fig. c: axial cut; SE - PDw

Fig. d: coronal cut; SE - T2w

Left deafness with ipsilateral hypesthesia of the face. MR shows a huge heterogeneous multicystic cerebellopontine angle mass extending into an expanded internal auditory canal (**c**, *; **d**, ↑). The mass is composed of multiple small and very large well defined hyperintense cysts whose intensity is close to CSF on PDw (**c**, ↑) and on T2w (**d**, ↑↑↑) sequences. On FLAIR images (not shown) the cystic foci were hyperintense to CSF indicating that they do not correspond to loculated subarachnoid fluid collections. Note the mass effect on the brainstem displaced to the right side (**c**, **d**) and the close situation of the ipsilateral vertebral artery with flow-void (**d**, *). Note also that the mass is somehow centered by the internal auditory canal axis. At surgery, the tumor presentend a loose texture containing tumor cells as well as mucinous and cystic changes more or less confluent, corresponding to Anthony type B.

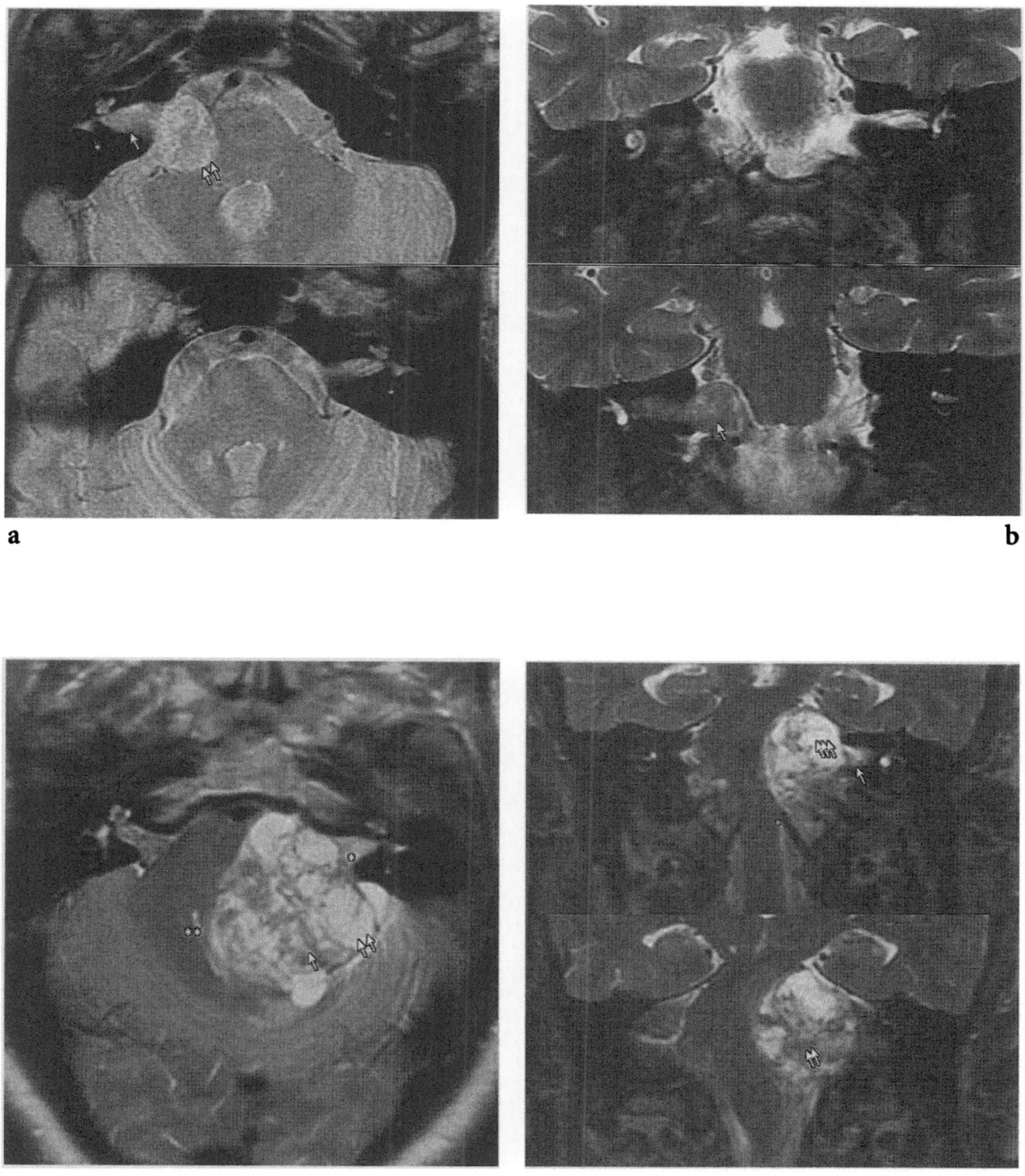

a b

c d

Case 68: Extracanalicular schwannoma (VIII)

Fig. a: axial cut; SE - T1w + Gd

Fig. b: coronal cut; SE - T1w + Gd

Patient presenting with episodes of dizziness. On MR, well demarcated homogeneously enhancing mass (**a, b**, ↑↑), developed in the right cerebellopontine angle region without consistent intracanalicular extension (**a**, ↑) on T1w images. The internal auditory canal is not enlarged. Note that the anterior medial and posterior lateral margins of the mass are well demarcated from the hypointense CSF by curvilinear bands. Note the mass effect on the middle cerebellar peduncle and the enlargement of the cerebellopontine angle.

Case 69: Extracanalicular schwannoma (VIII)

Fig. c: axial cut; SE - PDw (up) and T2w (down)

Fig. d: axial cut; GE (3D) - T1w + Gd

Moderate size left acoustic neurinoma showing slightly heterogeneous intensity pattern mainly on dual echo T2w sequence (**c**) and heterogeneous Gadolinium enhancement (**d**). The mass, well demarcated, shows a slightly hyperintense signal on PDw sequence (**c**, up), becoming more hyperintense on T2w (**c**, down) but less hyperintense than the intracanalicular CSF (**c**, ↑). The left internal auditory canal is mildly expanded as compared to the opposite side (**c**, ↑; **d**, ↑↑) showing mainly on PDw and T2w sequences a small knuckle of tumor extending into its medial aspect (**c**). Note the hypointense non enhancing portions of the tumor (**d**, ↑) and the mass effect on the 4th ventricle displaced posteriorly and to the right side (**d**,*; **c**). Note that about 1/5th of acoustic neurinomas have no intracanalicular components and may not be differentiated from meningiomas in the cerebellopontine angle when becoming as broad based mass against the petrous ridge.

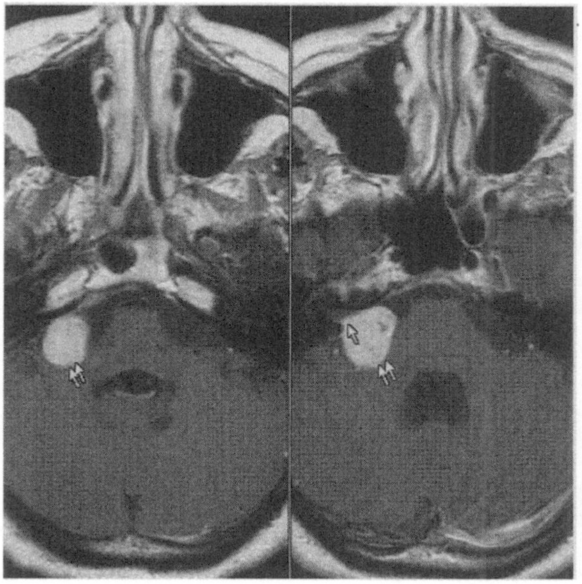

a

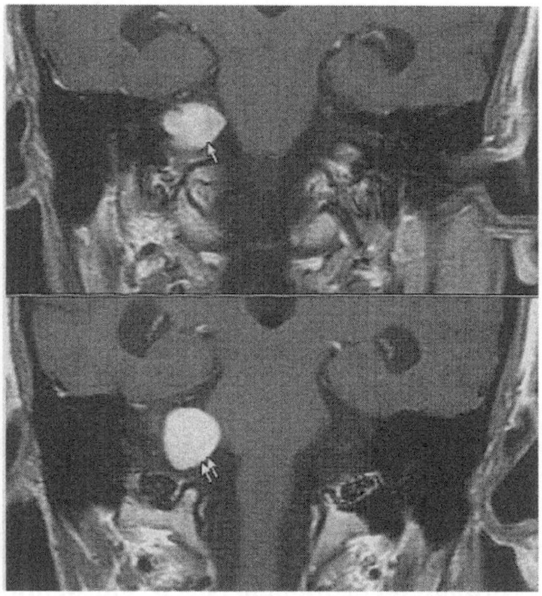

b

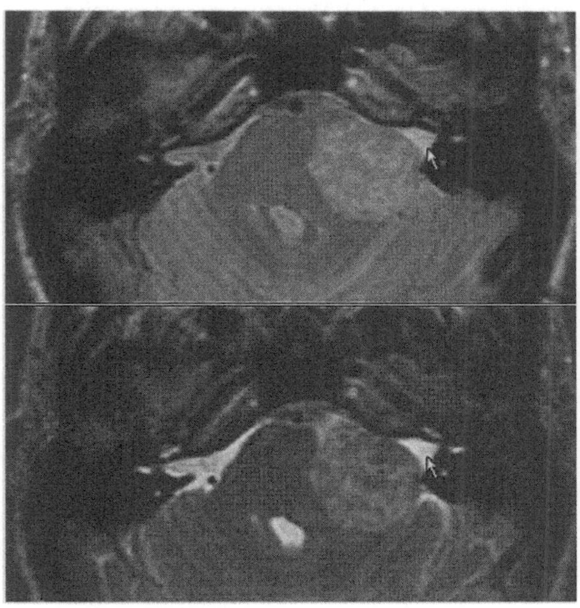

c

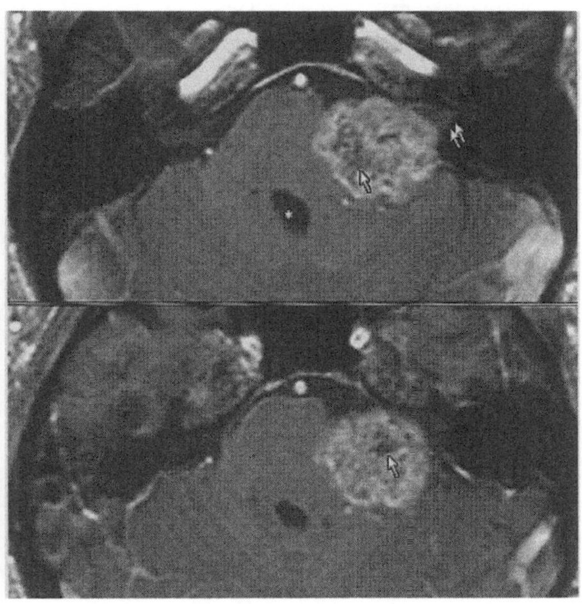

d

Case 70: Bilateral neurinoma type B (VIII; neurofibromatosis)

Fig. a: axial cut; SE - T1w

Fig. b: axial cut; SE - T1w + Gd

Patient with neurofibromatosis and bilateral sensory neural hearing loss associated to a brainstem syndrome. Bilateral huge polylobulated cerebellopontine angle masses presenting a slightly hypointense signal on T1w (**a**, ↑) enhancing heterogeneously after contrast iv (**b**, ↑). Note the absence of pattern enhancement of the cystic portion of the left schwannoma (**b**, ↑↑). These tumors cause a major compression of the brainstem (**a**, ↑↑) accounting for the brainstem syndrome. The intracanalar component of the masses erodes the anterior and medial portion of the petrous bones on both sides, with even extension into the temporal fossa on the right side (**b**, left, ↑).

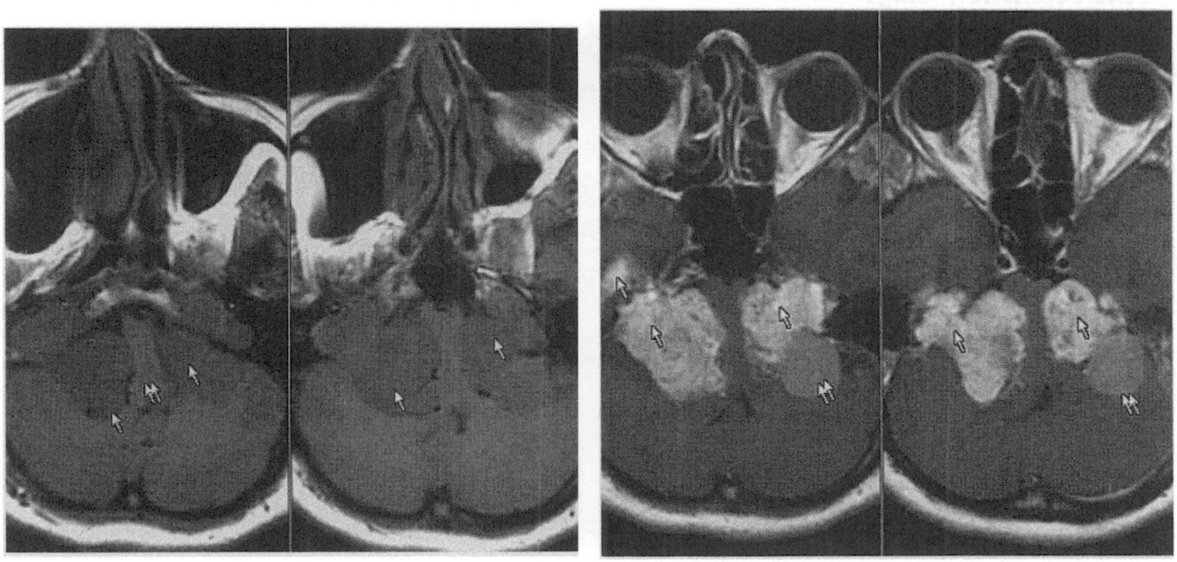

a **b**

Cases 71-73: Petrous and CPA meningiomas

Clinical and pathological features

- Relatively common benign tumor, 10% of intracranial meningiomas.
- Second most frequent tumor of the CPA.
- Middle and later decades, female predilection (2:1).
- Arises usually along the posterior aspect of the petrous bone.
- Develops in the CPA cistern and extends anterioposteriorly.
- Slow growing extra-axial intracranial mass.
- *Signs:* increased intracranial pressure (huge), progressive cerebellar deficit.
- Hearing loss: less constant feature than with acoustic neuromas.

MR characteristic findings

Morphology

- Mass arising from the dura of the petrous bone.
- Frequently close to the porus of the internal auditory canal.
- May arise from the clivus and further extends to the adjacent posterior aspect of the petrous bone.
- Growth in the CPA with compression of the middle cerebellar peduncle.
- May involve or extend to the tentorium cerebelli.
- Suggestive morphological patterns: broad-dural based mass to petrous bone with obtuse angle interface, associated enhanced "dural-tail", eventual hyperostosis.
- 20% show intratumoral calcifications (very rare in schwannomas) and cystic foci.
- May be multiple in case of neurofibromatosis or associated to other cell type tumors (neurinomas).
- Recurrence is not uncommon even after complete removal (8%), intimate attachment to neurovascular structures.

Signal intensity

- SE-T1w: iso-intense or hypointense to cortex, very rarely slightly hyperintense.
- SE-PDw: slightly hyperintense to brain.
- SE-T2w: iso-intense to cortex or hyperintense to white matter and even to cortex.
- Correlation between SI to cortex and histological type on T2w, hyperintense in angioblastic or syncitial types (transitional being iso-intense).
- Vascularity (1/3) and feeding "flow void" vessels may be shown (PD or T2 or T1).

Contrast enhancement

- Marked and usually homogeneous contrast enhancement.
- May be heterogeneous (presence of calcification, fibrous tissue or cystic process).

Case 71: Psammomatous meningioma of the tentorium cerebelli

Fig. c: axial cut; SE - T2w

Fig. d: sagittal cut; SE - T1w + Gd

Young woman presenting with left trigeminal neuralgia. Medium-sized well demarcated mass developed in the left cerebellopontine angle presenting an isosignal on T2w sequence (**c**, ↑) with mass effect on the brainstem and enlargement of the CSF space. The mass enhances markedly (**d**, ↑) and homogeneously and shows close relation with the tentorium cerebelli (**d**, ↑↑). Note the bright CSF cleft between the tumor and the petrous ridge (c, right). Note also the absence of typical "dural-tail sign" corresponding to dural thickening bordering the meningioma on enhanced slices (**d**). Note that "dural-tails" or "meningeal signs" shown on enhanced MRI are found in 1/2 to 1/3 of the cases of meningiomas.

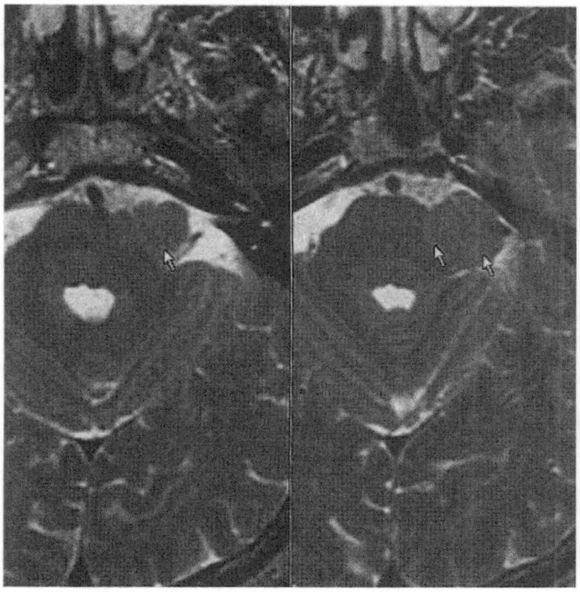

c

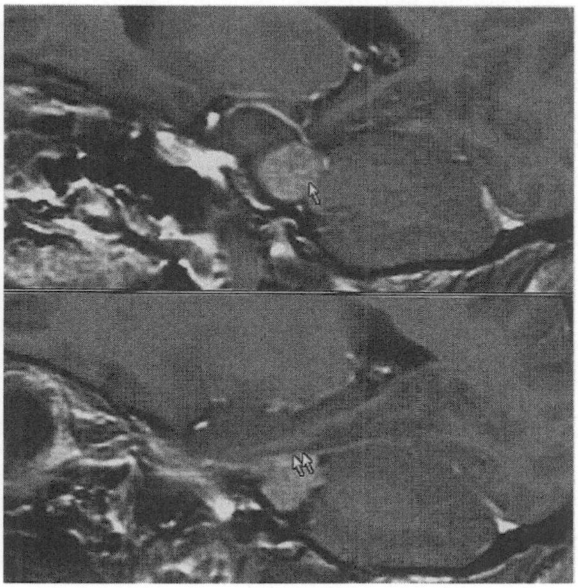

d

Case 72: Fibroblastic petro-occipital meningioma

Fig. a: axial cut; SE - T2w

Fig. b: coronal cut (512); SE - T1w + Gd

Middle age woman presenting with headaches. MR discloses a very large extra-axial mass developed in the posterior fossa with a broad base against the posterior petrous wall and the lateral portion of the occipital squama. The mass is heterogeneous on T2w sequence (**a**) presenting areas of intratumoral hypointense signal on T2w (**a**) consistent with fibrosis and calcifications. The mass is surrounded by hyperintense vasogenic edema (**a**, ↑↑↑) responsible of mass effect on the 4th ventricle and the dentate nucleus (stripe of low signal) displaced to the right side. After Gadolinium iv, the mass enhances heterogeneously. Note non enhancing areas (**b**, *) remaining hypointense CSF-like on T1w. The mass elevates the tentorium cerebelli as shown after enhancement on the coronal plane (**b**). Note the acute bone-tumor angles at the anterior and posterior aspects of the mass and the absence of relation with the ipsilateral internal auditory canal which is normal sized.

Case 73: Petrous (cerebellopontine angle) meningioma

Fig. c: axial cut; SE - PDw

Fig. d: coronal cut (512); SE - T1w + Gd

Patient presenting with hearing loss on the left side. MR shows an extra-axial hemispherical mass with broad base (**c**, ↑↑) against the left petrous wall with a marked posterior extension most excentric to the internal auditory canal. The mass is slightly hyperintense on PDw images, isointense to cerebellar cortex (**c**) and shows homogeneous moderate enhancement (**d**). Note the non enhancing hypointense signal on all sequences with a vermicular shape probably corresponding to a vascular pedicle (**d**, *). Note the CSF cleft at the posterior aspect of the mass due to its extra-axial nature (**d**, ↑). Note the mass effect on the 4th ventricle collapsed and displaced to the right side (**d**, ↑↑). The most reliable signs for CPA meningiomas are the excentric developement to porus acousticus and the obtuse angle against the petrous bone.

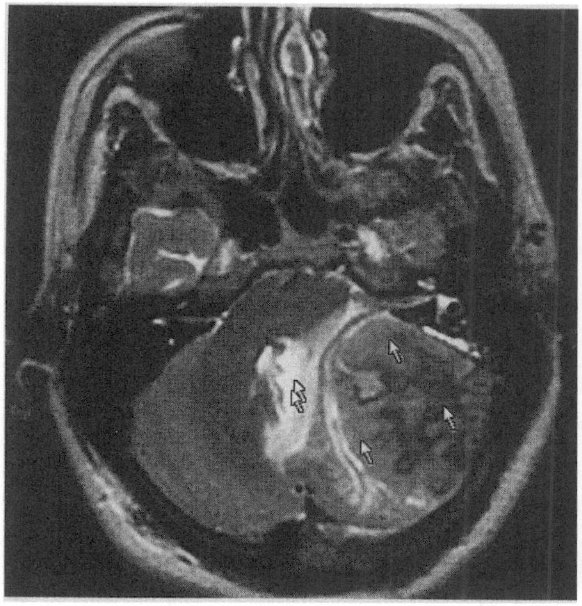

a

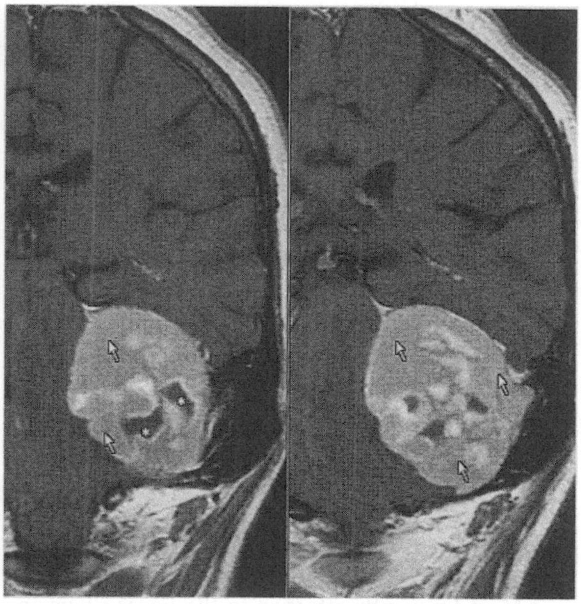

b

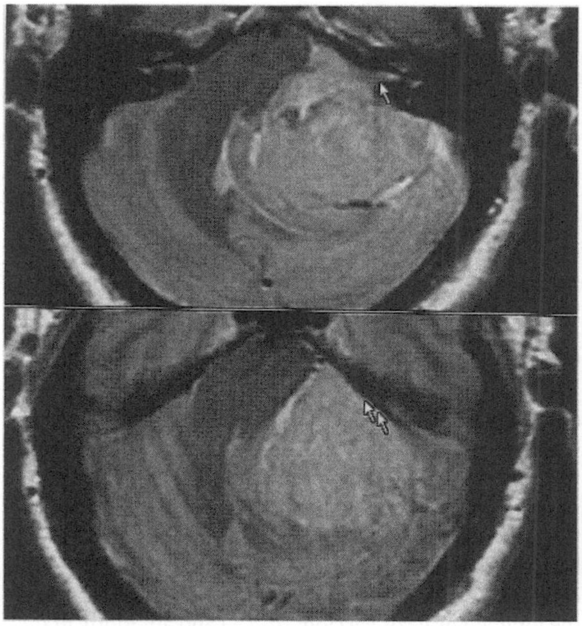

c

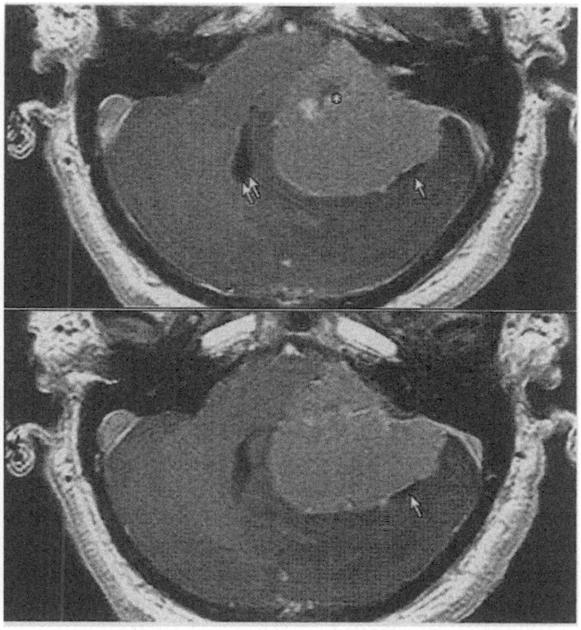

d

Craniocervical region

Cases 74, 75: Foramen magnum tumors

Clinical and pathological features

- Most frequent benign tumor: meningioma (2.5% of intracranial meningiomas).
- Commonly misdiagnosed tumors.
- Mimics cervical spondylosis, myelopathy, multiple sclerosis.
- *Signs:* suboccipital or neck pain, upper limbs paresthesias and sensory loss, clumsiness and gait disturbances, hand atrophy.

MR characteristic findings

Morphology

- Most frequently attached to the lower clivus and anterolateral border of the foramen magnum, close to the vertebral artery.
- Show a flat-base with the dura, with tendency to extend along the dura towards the upper cervical canal or along the posterior aspect of the clivus (+/- "tail").
- Clivus meningiomas may extend to the interpeduncular and chiasmal cisterns, with mass effect on the brainstem displaced posteriorly and/or encasement of the basilar artery (sagittal and axial cuts highly contributive).
- Absence of bony destruction and invasion of the clivus (as in chordomas).
- MR is the modality of choice for the diagnosis and presurgical planning.

Signal intensity

- SE-T1w: iso-intense or hypointense to brain (meningiomas).
- SE-PDw: slightly hyperintense to brain.
- SE-T2w: hypointense or iso-intense to cortex (hyperintense in solid or exceptionally cystic schwannomas).

Contrast enhancement

- Marked and homogeneous contrast enhancement in meningiomas, as in neurinomas or chordomas of cranio-cervical junction.

Case 74: Clivus meningioma

Fig. a: sagittal cut (512);	SE - T2w	**Fig. c:** axial cut (512);	SE - T2w
Fig. b: axial cut (512);	SE - PDw	**Fig. d:** axial cut (512);	SE - T2w

Middle age woman presenting with right trigeminal dysesthesia associated with gait disorder. High resolution MRI showed a huge well demarcated extensive posterior fossa mass developed in the prepontine cistern extending to the posterior suprasellar and interpeduncular cisterns. The mass presents a broad base (**a**, *) against the clivus and the right petro-clival wall extending to the dorsum sellae and reaching the optic tracts (**a**, ↑↑; **b**, ↑). The mass is hyperintense to the cortex on PD w sequence (**b**) and slightly hypointense on T2w (**a**, **c**, **d**). The mass at its mid-portion invades the right cavum of Meckel (**c**, ↑), lacking of the bright signal of normal filling CSF as shown on the opposite side (**c**). The right trigeminal nerve is embedded within the tumor shown only on the left side (**c**) crossing the cerebellopontine angle cistern (**c**). The mass displaces posteriorly and compresses the brainstem (a, b, c) as well as the cerebellar vermis (**b**, left, ↑↑; **d**, ↑↑↑) and encircles the basilar artery (**a**, ↑; **c**, ↑↑). The 4th ventricle is partly collapsed (**c**). Note the upper extension of the lesion into the foramen ovale of Paccioni (**d**, ↑↑) and anteriorly in the suprasellar cistern (**d**, ↑). Note the slight edematous reaction at the mesencephalic level (**b**, right, ↑↑), and note that high resolution multiplanar images may suffise for the surgical planning avoiding the need of contrast use.

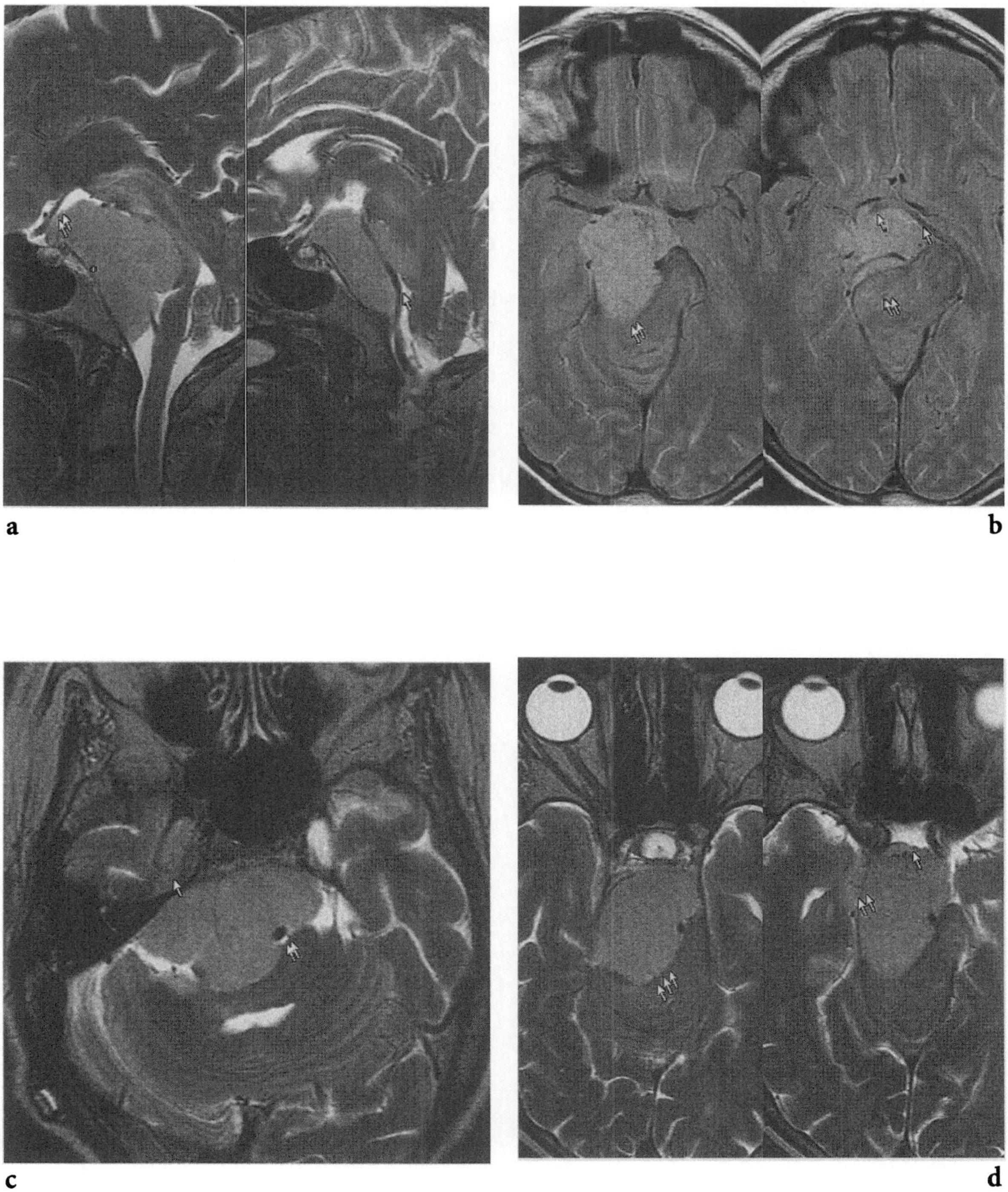

a b c d

Case 75: Cystic neurinoma

Fig. a: sagittal cut;	SE - T1w + Gd	**Fig. c:** axial cut;	SE - PDw
Fig. b: axial cut;	SE - T1w + Gd	**Fig. d:** axial cut;	SE - T2w

Huge craniocervical very well demarcated cystic mass developed in the prebulbar cistern and extending through the foramen magnum into the upper cervical canal (**a**, ↑↑; **b**, **c**, **d**, ↑). The mass is slightly less hypointense than CSF on T1w sequence (**a**, **b**), showing a linear rim of contrast enhancement (**a**, ↑) mainly marked at its left postero-lateral aspect contiguous to the medulla (**b**, ↑↑). On PDw sequence the cystic formation is hyperintense to the cortex and CSF (**c**, ↑) remaining hyperintense on T2w (**d**, ↑). The bulbo-medullary junction is compressed and displaced posteriorly (**c**, ↑↑) and the vertebral arteries displaced laterally (**c**, *). Note the compression of the 4th ventricle, the posterior displacement of the bulbo-medullary junction and the compression of the vermis (**a**). The patient was free of lower cranial nerves deficit. On surgery, the mass appears entirely cystic, confirmed on histopathology as being an unusual cystic schwannoma.

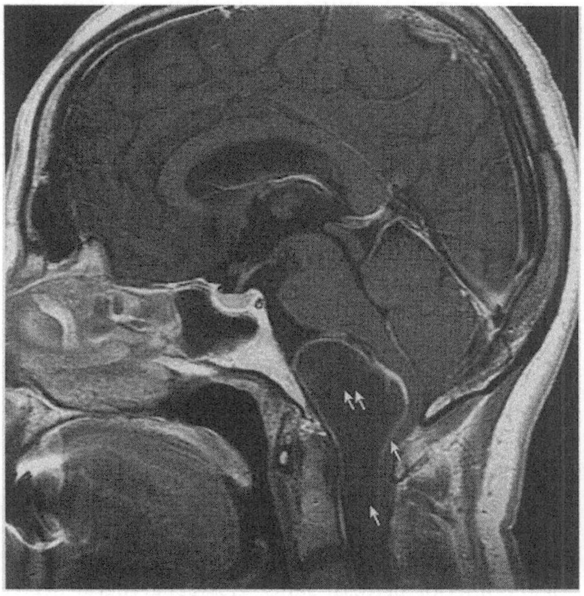

a

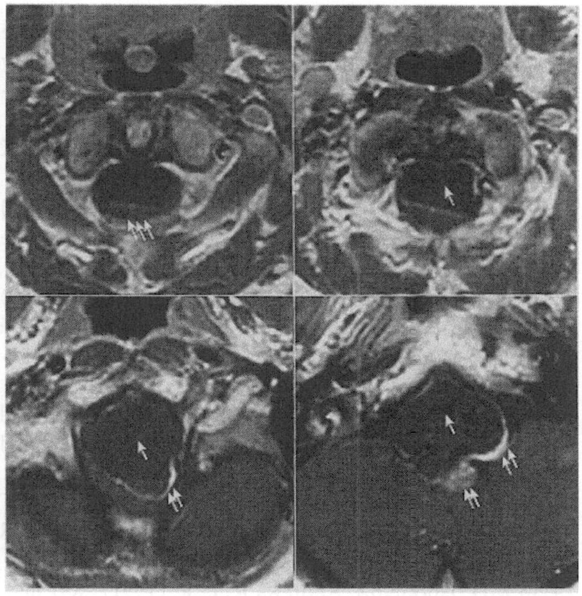

b

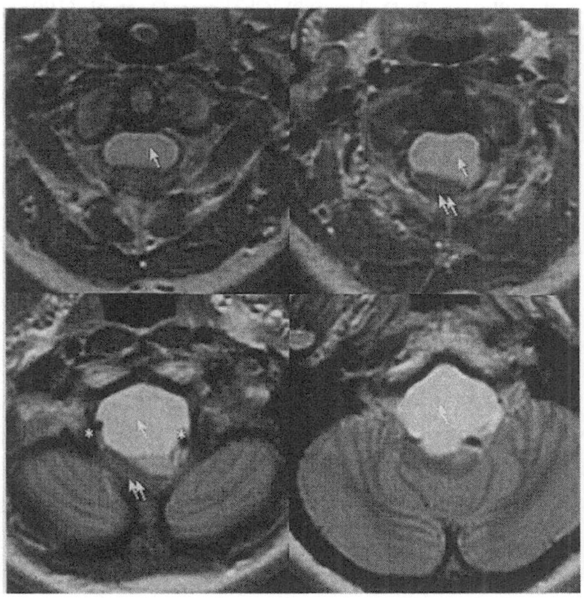

c

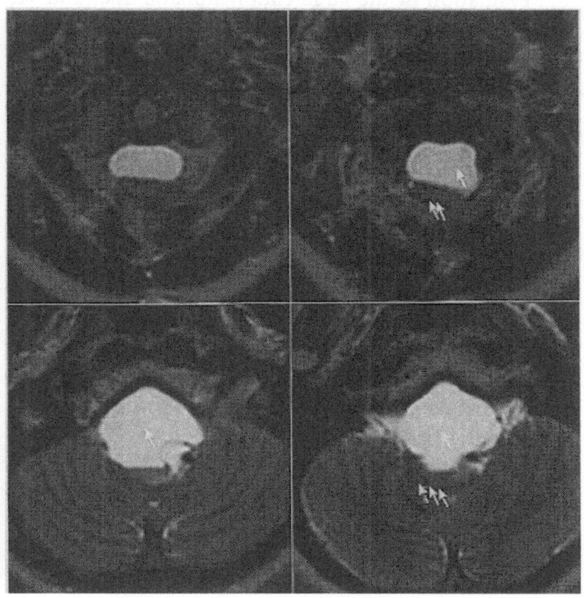

d

Selected References

- Arriage MA, Carrier D, Houston GD (1995) False positive magnetic resonance imaging of small internal auditory canal tumors: a clinical radiologic and pathologic correlation study. Otolaryngol Head Neck Surg 113:61-70
- Bonafe A, Joomye H, Jaeger P et al (1994) Histiocytosis X of the petrous bone in the MRI. Neuroradiology 36:330
- Dahlen RT, Harnsberger HR, Gray SD et al (1997) Overlapping thin section fast spin echo MR of the large vestibular aqueduct syndrome. AJNR 18:67-75
- Davidson C, Harnsberger R, Lemmerling M et al (1999) MR Evaluation of Vestibulocochlear Anomalies Associated with Large Endolymphatic Duct and Sac. AJNR 20:1435-1441
- Greenberg JJ, Oot RF, Wismer GL et al (1988) Cholesterol granuloma of the petrol MR and CT evaluation. Am J Neuroradiol 9:105
- Griffin C, DeLapaz R, Enzmann D (1987) MR and CT correlation of cholesterol cysts of the petrous bone. AJNR 8:825-929
- Han MH, Jabour BA, Andrews JC et al (1991) Non-neoplastic enhancing lesions mimicking intracanalicular acoustic neuroma on gadolinium-enhanced MR images. Radiology 179:795-796
- Harnsberger H, Dahlen R, Shelton C et al (1995) Advanced techniques in magnetic resonance imaging in the evaluation of the large endolymphatic duct and sac syndrome. Laryngoscope 105:1037-1042
- Iwayama E, Naganawa S, Ito T et al (1999) High-Resolution MR Cisternography of the Cerebellopontine Angle: 2D versus 3D Fast Spin-Echo Sequences. AJNR 20:889-895
- Larson TL, Wong ML (1992) Primary mucocele of the petrous apex. Am J Neuroradiol 13:203
- Mafee MF, Valvassori GE, Deitch RL (1985) Use of CT in the evaluation of cochlear otosclerosis. Radiology 156:703-708
- Mark AS, Fitzgerald D (1993) Segmental enhancement of the cochlea on contrast-enhanced MR: correlation with the frequency of hearing loss and possible sign of perilymphatic fistula and autoimmune labyrinthitis. Am J Neuroradiol 14:4491
- Martin N, Sterkers O, Nahum H (1990) Chronic inflammatory disease of the middle ear cavities: Gd-DPTA-enhanced MR imaging. Radiology 176:339
- Martin N, Sterkers O, Mompoint D et al (1992) Facial nerve neuromas: MR imaging. Neuroradiology 34:62
- Mayer TE, Brueckmann H, Siegert R (1997) High resolution CT of the temporal bone in dysplasia of the auricle and external auditory canal. AJNR 18:53-65
- McGhee R, Chakeres DW, Schmalbrock P et al (1993) The extra-cranial facial nerve: high resolution three-dimensional Fourier transform MR imaging. Am J Neuroradiol 14:464
- Mulkens TH, Parizel PM, Martin J-J et al (1993) Acoustic schwannoma: MR findings in 84 tumors. Am J Roentgenol 160:395-398
- Okamoto K, Ito J, Furusawa T et al (1997) Large vestibular aqueduct syndrome with high CT density and high MR signal intensity. AJNR 18:482-484
- Olsen WL et al (1986) MR imaging of Paragangliomas. Am J Neuroradiol 7:1039
- Press G, Hesselink J (1988) MR imaging of cerebellopontine angle and internal auditory canal lesions at 1.5 T. AJR 150:1271-1281
- Robert Y, Cracasset S, Rocourt N et al (1995) Congenital cholesteatoma of the temporal bone: MR findings and comparison with CT. Am J Neuroradiol 16:755
- Rubinstein D, Burton BS, Walker AL (1995) The anatomy of the inferior petrosal sinus, glossopharyngeal nerve, vagus nerve, and accessory nerve in the jugular foramen. Am J Neuroradiol 16:185
- Saeed SR, Birzgalis AR, Ramsden RT (1994) Intralabyrinthine schwannoma shown by magnetic resonance imaging. Neuroradiology 36:63
- Schmalbrock P, Dailiana T, Chakeres D et al (1996) Submilimeter-resolution MR of the endolymphatic sac in healthy subjects and patients with Meniere's disease. Am J Neuroradiol 17:1707
- Seltzer S, Mark AS (1991) Contrast enhancement of the labyrinth on MR scans in patients with sudden hearing loss and vertigo: evidence of labyrinthine disease. Am J Neuroradiol 12:13
- Swartz JD (1989) Otosclerosis, diagnosis and differential diagnosis. Semin Ultrasound CT MRI 10:251-261
- Tali ET, Yuh WTC, Nguyen HD et al (1992) Cystic acoustic schwannomas: MR characteristics. Am J Neuroradiol 14:1241
- Tien R, Dillon WP, Jackler RK (1990) Contrast-enhanced MR imaging of the facial nerve in 11 patients with Bell's palsy. Am J Neuroradiol 11:735
- Weissman JL (1997) A Pain in the Ear: The Radiology of Otalgia. Special Report AJNR 18:1641-1651

Part III

Spine and cord

Imaging technique and normal anatomy

Magnetic Resonance Imaging is the modality of choice for studying the spine and more particularly the spinal cord, which benefits a direct evaluation without any need of administration of contrast media intrathecally. Major developments have been acquired with respect to surface coils as well as new pulse sequences, improving images and contrast resolution. Basic principles and strategies in MRI technique need to be preserved and used in spinal imaging.

It is beyond the scope of this monograph to develop all pulse sequences which may be used in MR exploration of the spine and cord. We feel that it is important to settle on standard protocols in the clinical practice even if it could vary considerably regarding the anatomical region studied or the disease process. Signal-to-noise ratios depend of several physical factors including field strength (images shown are acquired on a high field 1.5 T system), MR imaging parameters such as TE, TR, Flip angle, slice thickness, pixel size, matrix, and bandwidth, and the surface coil used (linear, quadrature or phased-array). An important determinant factor is related to the patient's cooperative state, motion during image acquisition degrading image quality and generating motion artifacts. The result obtained will certainly depend upon the imaging protocol. But one must keep always in mind that the clinical data will most often orient the procedure and point out to the region of interest.

In most circumstances, an accurate evaluation of the spine and cord will be obtained using a standard protocol comprising four series performed in the sagittal and in the axial planes with at times an additional coronal acquisition. The use of contrast agent (Gadolinium chelate iv) may be of great help in a number of clinical states, particularly in spinal or cord tumors and in the postoperative spine, in order to demonstrate a leptomeningeal invasion, for example.

From a practical, technical point of view we will focus in the following on the contrast appearance of the images as obtained using T1w, T2w (including STIR or FLAIR) or PDw imaging sequences. In most cases, spin echo images (commonly fast SE) will be the major pulse sequences used routinely, to which other pulse sequences are compared. The advantages of the spin echo sequences are their high signal-to-noise and contrast-to-noise ratios. Moreover, SE sequences lack of sensitivity to magnetic field inhomogeneities contrarily to gradient echo sequences which are sensitive to chemical shift artifacts. These latter pulse sequences will be therefore restricted to the evaluation of postoperative spines. Nevertheless, in gradient-echo sequences, the signal-to-noise ratio is maintained at an accurate level due to the use of limited flip angles. But it is important to note in this case that the choice of the flip angle will influence the contrast between bone, cord, CSF and disc.

T1 weighted images

Obtained using a short TE (less than 20 msec) and a short TR (less than 600 msec), will best demonstrate morphology. Bone marrow is nicely evaluated on T1w images and most of its disorders (hypointense signal intensities) are assessed. Such sequence is used whenever a contrast enhancement is expected, usually in association with fat suppression techniques.

T2 weighted images

Obtained with fast spin echo T2 weighted pulse sequences, using a long TE (around 100 msec) and a long TR (generally greater than 3000 msec and usually around 4000 to 5000 msec). This pulse sequence is very sensitive to detect intrinsic lesions of the spinal cord. It is also efficient in the evaluation of the hydration state of the intervertebral disc or in the depiction of various spinal and paraspinal extradural disease processes. FSE-T2 weighted sequences are also useful in the evaluation of marrow disorders, rendered even more sensitive with the adjunction of fat suppression techniques. STIR sequences with a short TI inversion recovery and long TR may be used as an alternative suitable for the detection of bone marrow diseases as well.

Proton Density (PD) weighted images

Obtained using a short TE (around 20 msec or less) and a long TR (around 3000 msec). These PDw sequences are very helpful for the detection of cord abnormalities such as those found in myelitides or myelopathies. They are for instance very sensitive for the detection of demyelinating or edematous areas, associated with multiple sclerosis or other demyelinating, inflammatory or ischemic myelopathies.

The majority of spinal examinations are performed and ought to be performed using dedicated surface coils (quadrature or phased-array coils). In obese patients, an exploration using the body coil is the alternative because the patient may not fit inside the magnet bore when a surface coil is used. Use of a cervico-thoracolumbar coil is useful if the physician is unable to precise with enough confidence the spinal level involved by the questionable disease process, or if the patient presents a multifocal such as a metastatic disease. The MR strategy adopted in our routine practice, comprises two sagittal acquisitions (using FSE-T2w sequence followed by an SE-T1w sequence, eventually complemented by a PDw sequence in case of spinal cord disease), followed by two axial series performed usually parallel to the intervertebral discs with small gap between the slices (using SE-T1w and SE-T2w pulse sequences, and in case of degenerative disc disease of the cervical spine a 3D or 2D gradient echo T2*w or T1w sequences). Fat suppression techniques are used in case of marrow disease and in metastatic diseases of the spine after intravenous contrast administration. STIR sequence (short TI inversion recovery) may be used in the same clinical circumstances. Fast-FLAIR sequence is improving the detection of intramedullary as well as intradural diseases due to the excellent contrast shown between the disease process usually of hyperintense signal intensity (long T2) with respect to CSF remaining hypointense. Saturation pulses are mandatory to use in order to decrease artifacts from flow (CSF), movements (abdominal gastrointestinal tract) and pulsations (blood vessels). Cardiac or peripheral gating are exceptionally used in our practice found to be useless and time-consuming.

Despite the large array of pulse sequences available on MR systems, new software developments and new technical improvements, it is mandatory to try to focus on the information needed from each exam and determine the most accurate and realistic strategy to adopt, in order to obtain maximum informations in the most time-efficient non invasive way. Ultrafast imaging sequences and high field systems allow us to explore most of claustrophobic patients and pediatric population, without the need of any sedation, in most circumstances.

Normal anatomy

Plate I: Anatomy of the cervical spine

Fig. a: mid sagittal and lateral sagittal cuts; SE - T1w

Fig. b: mid and paramedian sagittal cuts; SE - T2w

Fig. c: lateral and foraminal sagittal cut; SE - T2w

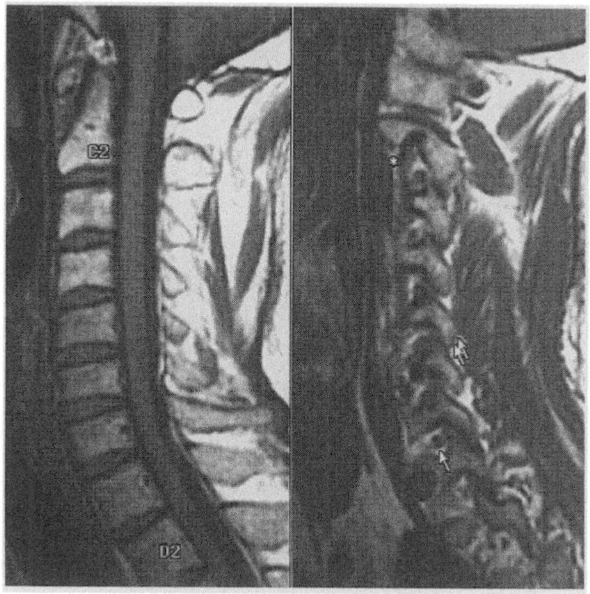

a

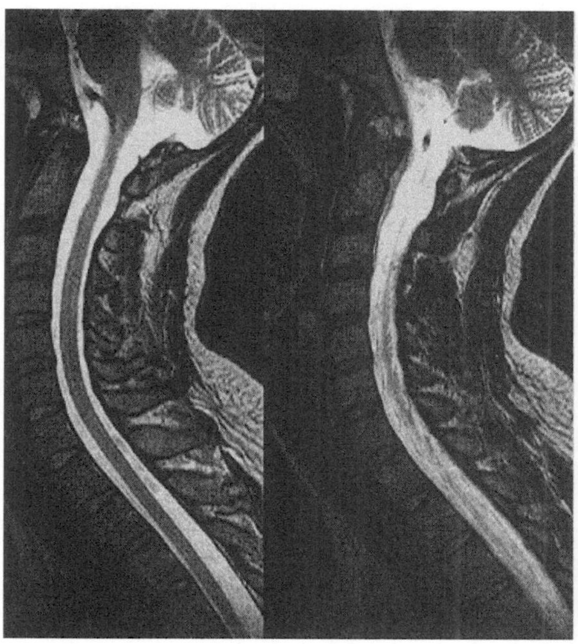

b

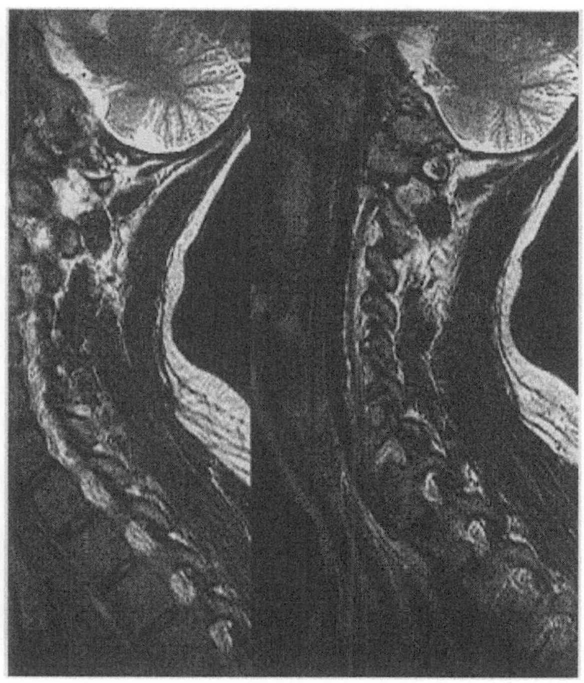

c

Normal anatomy

Plate II: Anatomy of the thoracolumbar spine

Fig. a: lateral and mid-sagittal cuts; SE - T1w

1 *Pedicle*
2 *Inferior articular process*
3 *Superior articular process*
4 *Neural foramen*
5 *Spinous process*
(*) *nerve root*

Fig. b: coronal cut (cauda equina); SE - T2w

(↑) *Conus medullaris*
(↑↑) *Cauda equina*

Fig. c: axial cuts (cauda equina); SE - T2w

1 *Conus medullaris*
2 *Filum terminale*
3 *Cauda equina*
4 *Thecal sac*
5 *Basivertebral vein*

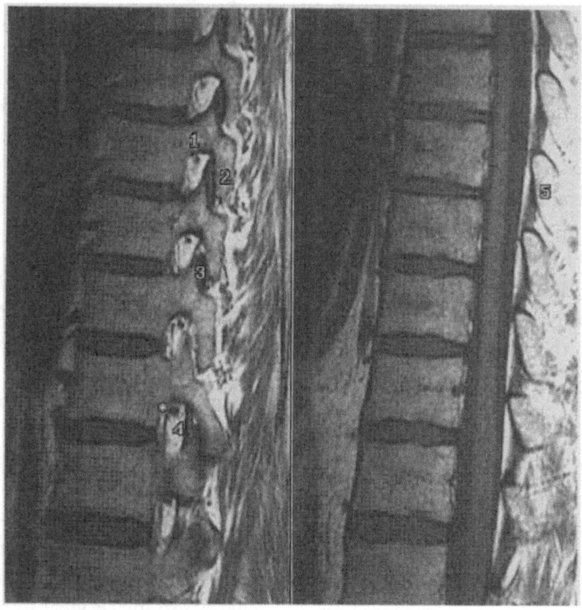

a

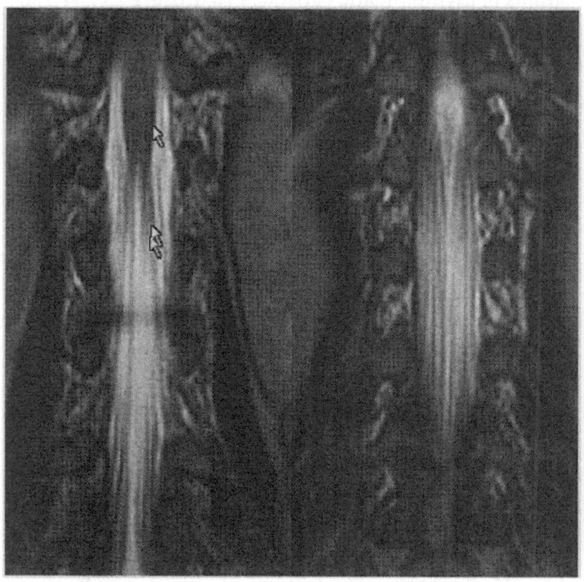

b

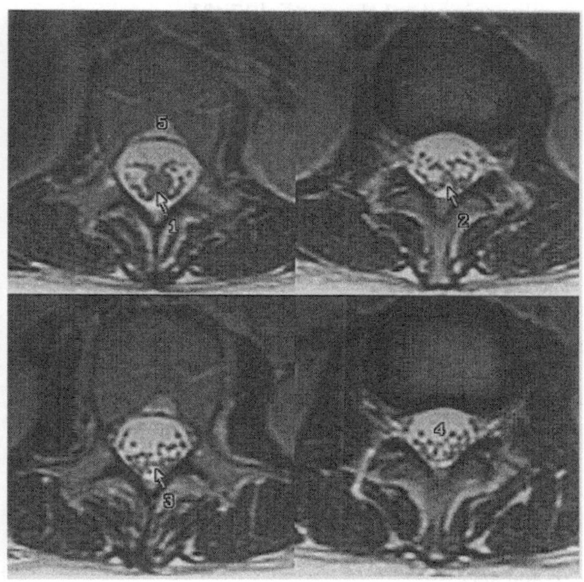

c

Selected References

- Aguila LA, Piraino DW, Modic MT (1985) Magnetic resonance imaging of the intranuclear cleft. Radiology 155: 155-158
- American Spinal Injury Association (1989) Standards for neurologic classification of spinal injury patients. Chicago
- Batson OV (1957) The vertebral vein system. AJR 78:192-212
- Brazis PW, Masdeu JC, Biller J (1990) Localization in clinical neurology. Little Brown, Boston, pp 76-78
- Bland JH, van Bushirk FW, Tampas JP et al (1965) A study of roentgenologic criteria for rheumatoid arthritis of the cervical spine. AJR 95:949
- Boden SD, Davis DO, Dina TS et al (1990) Abnormal magnetic resonance scans of the lumbar spine in asymptomatic subjects. J Bone Joint Surg 72A:403-408
- Bozzao A, Galluci M, Masciocchi C et al (1995) Interobserver and intraobserver variability in interpretation of lumbar disc abnormalities: a comparison of two nomenclatures. Spines 20:1257
- Brown CW, Deffer PA Jr, Akmakjian J et al (1992) The natural history of thoracic disk herniation. Spine 17:97-102
- Bowen BC, DePrima S, Pattany PM et al (1996) MR angiography of normal intradural vessels of the thoracolumbar spine. Am J Neuroradiol 17:483
- Burton CV, Krikadly-Willis WH, Yong-Hing K, Heithoff KB (1981) Causes of failure of surgery on the lumbar spine. Clin Orthop 157:191-199
- Bush K, Cowan N, Katz DE, Gishen P (1992) The natural history of sciatica associated with disc pathology: a prospective study with clinical and independent radiologic follow-up. Spine 17 (10):1205-1212
- Coussement A, Faure C, Coussement-Beylard N (1990) Repères et mesures en imagerie médicale. 4th ed, L'expansion Scientifique Française
- Coventry MB (1969) Anatomy of the intervertebral disk. Clin Orthop 67:9
- Coventry MB, Ghormley RK, Kernohan JW (1945) The intervertebral disc its microscopic anatomy and pathology. I. Anatomy, development and physiology. J Bone Joint Surg 27:105-112
- Cowen NC, Bush K, Katz DE et al (1992) The natural history of sciatica: a prospective radiological study. Clin Radiol 46:7
- Daniels DL, Hyde JS, Keelqnd JB et al (1986) The cervical nerves and foramina: local-coil MR imaging AJNR 7:129-133
- Doyon D, Laval-Jeantet M, Halimi Ph, Cabanis EA, Frija (1988) Tomodensitométrie. Masson, Paris
- Doyon D, Cabanis EA, Laval-Jeanter M, Frija J, Pariente D, Idy-Peretti I (1994) Imagerie par Résonance Magnétique. Masson, Paris
- Dvorak J, Jassen B, Grob D (1990) The neurologic work up in patients with cervical spine disorders. Spine 15:1017-1022
- Enzmann D, Rubin JB, Short TR (1989) Variable flip angle, gradient echo scans of the cervical spine, comparison of IDFT and 3DFT techniques. Neuroradiology 31:213-216
- Enzmann DR, Rubin JB (1988) Cervical spine: MR imaging with a partial flip angle, gradient-refocused pulse sequence. Part II. Spinal cord disease. Radiology 166:473-478
- Fisher WS III (1993) Selection of patients for surgery. Neurosurg Clin North Am 4:35-44
- Gehweiler JA, Osborne RL, Becker RF (1980) The radiology of vertebral trauma. WB Saunders, Philadelphia, p 102
- Greenberg JO, Schnell RG (1991) Magnetic resonance imaging of the lumbar spine in asymptomatic adults. J Neuroimag 1:2-7
- Halimi P (1992) Imagerie de la moelle et du rachis. Flammarion-Médecine-Sciences, Paris
- Hassler O (1970) The human intervertebral disc. A micro angiographical study of its vascular supply at various ages. Acta Orthop Scand 40:765-772
- Hayashi H, Okada K, Hamadu M et al (1987) Etiologic factors of myelopathy. A radiologic evaluation of the aging changes in the cervical spine. Clin Orthop 214:200-209
- Helms CA, Dorwart RH, Gray M (1982) The CT appearance of conjoined nerve roots and differentiation from a herniated nucleus pulposus. Radiology 144:803-807
- Hitselberger WE, Witten RM (1968) Abnormal myelograms in asymptomatic patients. J Neurosurg 28:204-206
- Kambin P, Schaffer JL (1989) Percutaneous lumbar discetomy: review of 100 patients and current practice. Clin Orthop 238:24-34
- Kelsey JL, Githens PB, Walter SD et al (1984) An epidemiologic study of acute prolapsed cervical intervertebral discs. J Bone Joint Surg 66A:907-914
- Kostelic JK, Haughton VM, Sether LA (1991) Lumbar spinal nerves in the neural foramen: MR appearance. Radiology 178:837-839
- Le Bihan D (1985) Imagerie par Résonance Magnétique. Bases physiques. Masson, Paris
- Lipson SJ (1989) Cervical disc disease: pathogenesis and natural history. Semin Spine Surg 1:190-193
- Long DM, (1991) Failed back surgery syndrome. Neurosurg Clin North Am 2:899-919
- Maigne J, Rime B, Delignet B (1992) Computed tomographic follow-up study of forty-eight cases of nonoperatively treated lumbar intervertebral disc herniation. Spine 27(9):1071
- Manelfe C (1989) Imagerie du rachis et de la moelle.Vigot, Paris
- Manelfe C (1992) Imaging of the spine and spinal cord. New York, Raven Press
- Maravilla KR, Cohen WA, Wessbecher FW (1991) Imaging studies in the assessment of low back pain Neurosurg Clin North Am 2:817-837
- Maroon JC, Kopitinik TA, Schulhof LA, Abla A, Wilberger JE (1990) Diagnosis and microsurgical approach to far-

lateral disc herniation in the lumbar spine. J Neurosurg 72:378-382
- Marsault CL, Gaston CL, Lebras A (1991) Imagerie du système nerveux central. Flammarion, Paris
- Mayoux-Benhamou MA, Revel M, Aaron C, Chomette G, Amor B (1989) A morphometric study of the lumbar foramen. Sug Radiol Anat 11:97-102
- McGregor M (1948) The significance of certain measurements of the skull in the diagnosis of basilar impression. Br J Radiol 21:171
- Milette PC (1997) The proper terminology for reporting lumbar intervertebral disk disorders. AJNR 18:1859-1866
- Mirvis SE, Diaconis JN, Chirico PA et al (1989) Protocol driven radiologic evaluation of suspected cervical spine injury: efficacy study. Radiology 170:831-834
- Modic M, Masaryk TJ, Ross JS (1996) Magnetic resonance of the spine. Year Book Medical Publishers, Chicago, London
- Nabors MW, Pait TG, Byrd EB et al (1988) Updated assessment and current classification of spinal meningeal cysts. J neurosurg 68:366
- Nashold BS, Hrubec Z (1971) Lumbar disc disease: a twenty year clinical follow-up study. CV Mosby, St Louis
- National Institutes of Health Consenus Development Conference (1988) Neurofibromatosis conference statement. Arch Neurol 45:575-578
- Oda J, Tanaka H, Tsuzuki N (1988) Intervertebral disc changes with aging of human cervical vertebral from the neonate to the eighties. Spine 13:1205-1211
- Postachini F (1989) Lumbar spinal stenosis. Springer-Verlag, New York
- Rabischong P, Manelfe C, Halimi P, Francke J-P (1992) Normal anatomy of the spine, spinal cord, and nerve roots. In: Manelfe C (ed) Imaging of the spine and spinal cord. Raven Press, New York, pp 1-91
- Ross JS, Delamarter R, Hueftle MG et al (1989) Gadolinium-DTP-enhanced MR imaging of the postoperative lumbar spine: time course and mechanism of enhancement. AJR 152:825-834
- Saal JA, Saal JS, Herzog RJ (1990) The natural history of lumbar intervertebral disc extrusions treated nonoperatively. Spine 15(7):683-686
- Shermann JL, Nassaux PY, Citrin CM (1990) Measurements of the normal cervical spinal cord on MR imaging. AJNR 11:369-372
- Shapiro R (1984) Myelography. 4th ed, Year Book Medical Publishers, Chicago, pp 226-230
- Schiebler ML, Grenier N, Fallon M et al (1991) Normal and degenerated intervertebral disk: in vivo MR imaging with histopathologic correlation. AJR 157:93-97
- Sigal R, Doyon D, Halimi PH, Atlan H (1988) Imagerie par résonance magnétique. Bases d'interprétation. Springer-Verlag, Berlin Heidelberg New York
- Solsberg MD, Lemaire C, Resch L, Potts DG (1990) High-resolution MR imaging of the cadaveric human spinal cord: normal anatomy. AJNR 11:3-7
- St-Amour T, Hodges S, Laakman R, Tamas D (1994) MRI of the spine. Raven Press, New York
- Stark DD, Bradley WG (1998) Magnetic Resonance Imaging. C.V. Mosby, St. Louis, Washington, Toronto
- Teplick JG, Haskin ME (1985) Spontaneous regression of herniated nucleus pulposus. AJR 145:371-375
- Tersti LM, Lufkin RB, Reicher MA et al (1987) Asymptomatic degenerative disk disease and spondylosis of the cervical spine: MR imaging. Radiology 164:83-88
- Thron AK (1988) Vascular anatomy of the spinal cord. Springer-Verlag, New York
- Tkach JA, Haacke EM (1988) A comparison of fast spin echo and gradient echo sequences. Magn Reson Imaging 6:373-389
- Tsuruda J, Norman D, Dillon W et al (1989) Three-dimensional gradient-recalled MR imaging as a screening tool for the diagnosis of cervical radiculopathy. AJNR 10:1263-1271
- Weinstein JN, Wiesel SW (1990) The lumbar spine. WB Saunders, Philadelphia
- Wiesel SW, Tsourmas N, Feffer HL et al (1984) A study of computer assisted tomography. I. The incidence of positive CAT scans in an asymptomatic group of patients. Spine 9:549-551
- Yasuma T, Makino E, Saito S, Inui M (1986) Histological development of intervertebral disc herniation. J Bone Joint Surg 68 A:1066-1072

Part III

Spine and cord

1- Tumoral conditions

a - Extradural tumors
b - Intradural extramedullary tumors
c - Spinal cord tumors

Extradural tumors

Case 1: Aneurysmal bone cyst

Clinical and pathological features

- Benign lesion of bone described by Jaffe and Lichtenstein (1942).
- Represents up to 6% of all primary bone tumors.
- Occurs in patients under age 20 in almost 80% of the cases. A female predominance has been at times reported.
- The etiology remains unclear: may arise from a pre-existing process or could arise *de novo*. Association with other bone tumoral processes is observed in almost 1/3 of the cases, such as: giant cell tumor, chondroblastoma, osteoblastoma, chondromyxoid fibroma, non ossifying fibroma and fibrous dysplasia.
- *Pathology:* multicystic, thin walled bony mass formed from subperiosteal new bone containing blood breakdown products. A solid component may be found formed of osteoid material. Giant cells may be found on histology. The aneurismal cyst may mimic malignant tumors such as telangiectatic osteosarcoma.
- *Signs:* the most frequently reported symptom is back pain which may be also of the radicular type (2/3 of the patients). A palpable mass may be found in almost 25% of the cases. Neurological signs are found in half of the spinal tumors mostly due to cord impingement or compression.

MR characteristic findings

Morphology

- Occurs most frequently in the cervical or thoracic spine. The lumbosacral region may be involved except the coccyx.
- Spinal aneurysmal bone cysts involve the neural arch in 3/5 of the cases and the vertebral bodies in 2/5 of the cases.
- Aneurysmal bone cysts appear as well delineated and lobulated cystic masses, filled with unclotted blood of different ages and showing fluid-fluid levels. Some of the lesions are mostly solid.
- Intratumoral fluid-fluid levels may also be found in osteosarcomas, giant cell tumors, chondroblastomas and in fibrous dysplasia.
- About 1/5 of these lesions extend to the perivertebral soft tissues.

Signal intensity

- Heterogeneous SI due to the presence of fine septations (cortical bone) delineating cystic areas, solid portions (osteoid and fibrous material) and fluid-fluid levels (containing hemorrhagic material of various stages).
- SE-T1w: heterogeneous aspect with low SI of bony septations and trabeculae and eventually high SI from blood breakdown products.
- SE-T2w: heterogeneous to homogeneous hyperintense SI with fluid-fluid levels of variable SI.

Contrast enhancement

- Possible enhancement of the septations within the tumor.

Case 1: Aneurysmal bone cyst

Fig. a: sagittal cut (512); SE - T2w

Fig. b: axial cut (512); SE - T2w

Young woman presenting with back pain and leg pain. Multiloculated spinal mass demonstrating multiple well defined cystic cavities surrounded by a rim of low signal intensity (**a**, *) most likely representing the cortical rim. The cystic portions show on T2w heterogeneous signal intensities with a fluid-fluid level (**a**, **b**, ↑) of hyperintense signal superiorly, and intermediate signal inferiorly due to the presence of intracystic hemorrhage of different ages. The mass involves the posterior elements of the spine at L2 level and here the right pedicle. Note the expansile lytic characteristic of the mass with a scalloping of the posterior wall of the vertebral body (**a**, *). The dural sac containing the cauda equina is compressed and displaced to the left (**b**, down).

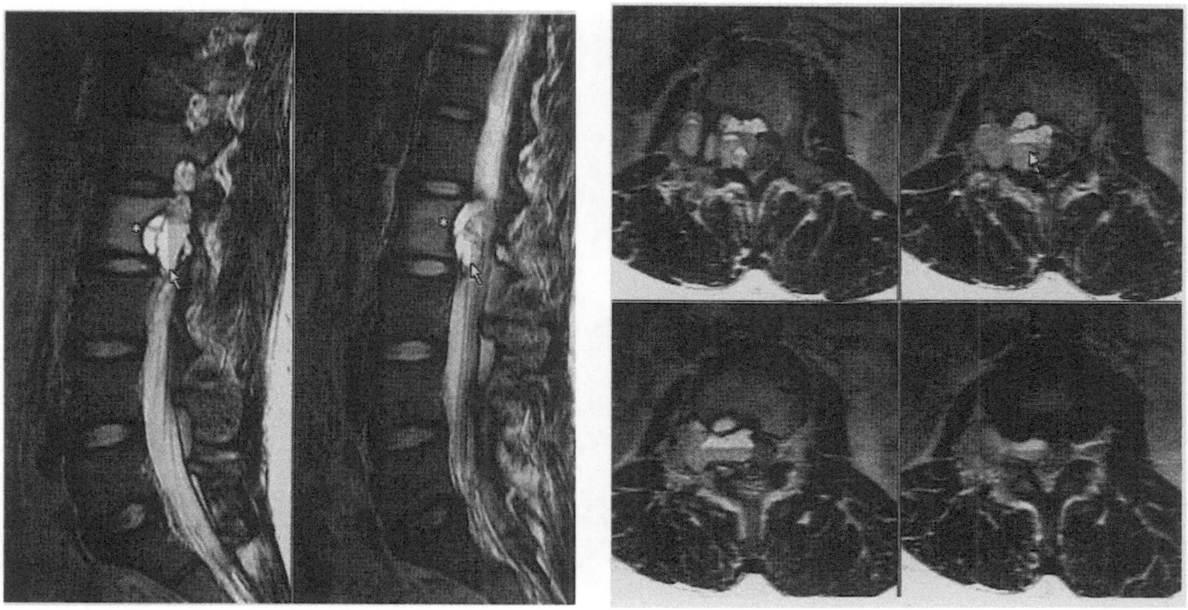

a b

Cases 2, 3: vertebral hemangiomas

Clinical and pathological features

- Common benign spinal vascular tumors, incidentally found on at least 10% of MR exams and with the same frequency on autopsy series. They usually do not require further investigation or follow-up.
- Hemangiomas are most frequently found in the thoracic region and most are solitary. A slight female predominance is reported.
- An increase in incidence with age has been reported and an enlargement of the lesion may be observed during pregnancy.
- *Histopathology:* spinal hemangiomas are formed of dilated vascular structures lined by endothelium associated with fatty marrow. Most are of the cavernous type but may be of the capillary or the venous types. Bony trabeculae are partially destroyed with thickening of the remaining due to a compensatory effect.
- *Signs:* usually asymptomatic, the vertebral hemangiomas may be associated with pain and muscle spasms or may cause neurological signs secondary to entrapment radiculopathies or cord compression, in large expansile tumors. Most symptomatic compressive hemangiomas are encountered between T3 and T9 thoracic levels. The occurrence of a compression fracture or an epidural hematoma may give rise to a sudden onset of neurological signs.

MR characteristic findings

Morphology

- The spine is the most frequent site for hemangiomas, 60% of which are thoracic and 30% lumbar in location.
- Vertebral hemangiomas present as solitary tumors in almost 2/3 of the cases, being multiple up to five in number, in roughly 1/3 of the cases.
- Hemangiomas are located preferentially within the vertebral body but may extend to the posterior elements, particularly those associated with compression of the neural structures.
- Hemangiomas are usually small rounded or oval lesions that may even involve the entire vertebral body.
- Thickened linear vertebral trabeculae giving rise to a "jail bar" aspect, are highly characteristic of vertebral hemangiomas (best seen on CT or MR using PDw or T2*w sequences; plain X-rays often suffice to confirm the diagnosis).
- Asymptomatic inactive hemangiomas have been shown to contain large amounts of fat (hyperintense SI on T1w) and conversely, active or aggressive tumors appear to include a much larger vascular component (decrease SI on T1w).
- Extravertebral development of a hemangioma to the epidural and/or paravertebral spaces may be observed with cord compression (showing an intermediate SI on T1w due to presence of intratumoral fibrous tissue and angiomatous vessels and a paucity of fat, and a hyperintense SI on T2w). In such cases, vertebral hemangiomas may appear similar to angiolipomas, which are most often thoracic (80%) and posterior in location, arising and developing mainly epidurally, and showing usually a relatively high SI on T1w due to the greater amount of intratumoral fat tissue.

Signal intensity

- SE-T1w: hyperintense to adjacent marrow in benign forms (due to fat content); the extraosseous component does not usually show a hyperintense SI (as would appear an angiolipoma) and may be hypointense or isointense (more aggressive forms with less amount of fat and a greater vascular component).
- SE-PDw: hyperintense to normal marrow.
- SE-T2w: hyperintense to normal marrow; usually helps to differentiate hemangiomas from focal areas of fat within bone marrow (more hyperintense SI and remaining of bright SI with fat suppression technique).
- GE-T2*w: thickened trabeculae are more obviously demonstrated (axial plane).

Contrast enhancement

- Usually mild enhancement may be observed, but aggressive forms may show marked contrast enhancement.

Case 2: Aggressive vertebral hemangioma

Fig. c: coronal cut (512); SE - T1w

Fig. d: sagittal cut (512); SE - T1w

Patient presenting with rapidly evolving thoracic cord compression syndrome and a history of thoracic pain with muscle spasm. T1w high resolution MR sequences disclose a hyperintense mass involving the vertebral body at the mid-lower thoracic spine associated to thickened vertical trabeculae causing parallel linear hypointensities with a "jail bar" appearance (**d**, ↑). The high signal on T1w reflects the adipose tissue contained in the mass and associated to area of iso-hypointense signal (**d**, ↑↑; **c**, ↑). The low signal intensity associated areas shown on T1w are encountered in symptomatic hemangiomas as in this case and correspond to areas of significant hypervascularity. Note the posterior epidural extension of the mass with compression of the spinal cord on both sides well shown on the coronal slices (**c**, right; **d**, left). Note the craniocaudal compression of the vertebral body (**c**, left; **d**).

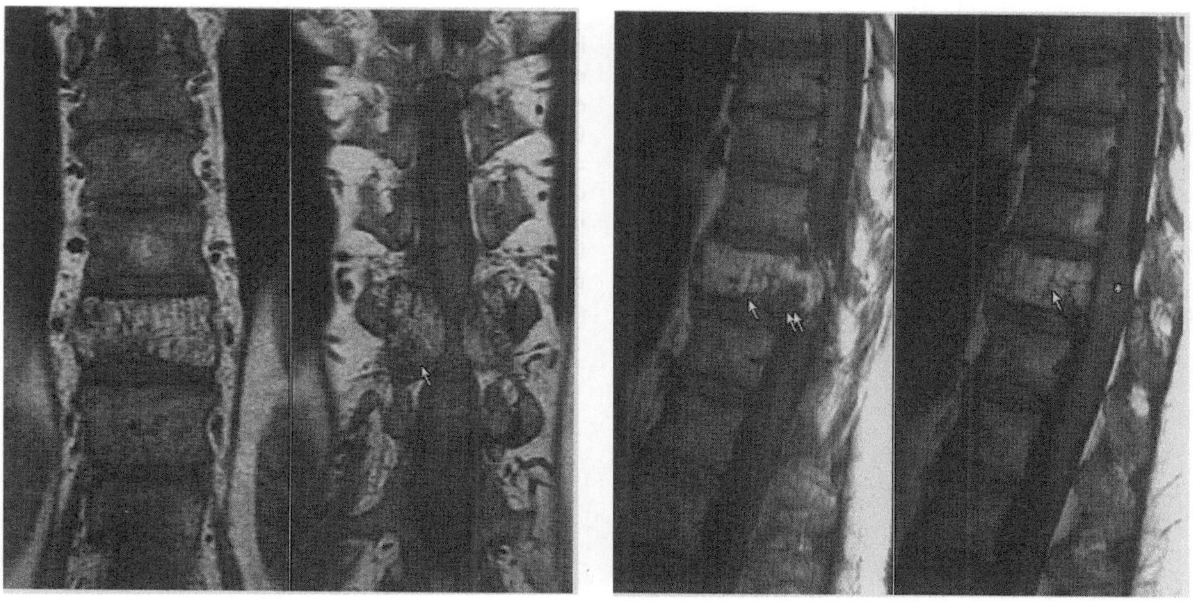

c d

Case 3: Vertebral angiolipoma

Fig. a: sagittal cut (512); SE - T1w

Fig. b: sagittal cut; SE - T2w

Fig. c: axial cut; SE - T1w

Fig. d: axial cut; SE - T2w

Young woman presenting with rapidly evolving spinal cord compression syndrome associated with focal thoracic back pain. MRI showed a well demarcated posterior epidural mass in the thoracic region involving the posterior neural arch and the posterior spinal process (**a**, **c**, ↑↑). The epidural development of the mass (**a**, **b**, ↑) presented a heterogeneous isosignal on T1w (**a**, ↑) as the corresponding vertebral body (**a**, ↑) and a hyperintense signal on T2w (**b**, ↑) with compression of the corresponding spinal cord segment well shown on the sagittal cuts (**a**, **b**) but without intramedullary edema or myelomalacia (**b**) on T2w. The epidural mass is associated with an increase of the adjacent epidural fat well shown with hyperintense signal on T1w (**a**) adjacent to the cranial and the caudal extremities of the mass. On T2w the mass looks hyperintense as well as the corresponding vertebral body and the posterior spinous process (**b**, ↑ and ↑↑). Note in the axial plane that the intravertebral extension of the infiltrating angiolipoma produces a trabeculated bony pattern similar to the typical spotted appearance of the vertebral hemangiomas (**c**, **d**, ↑).

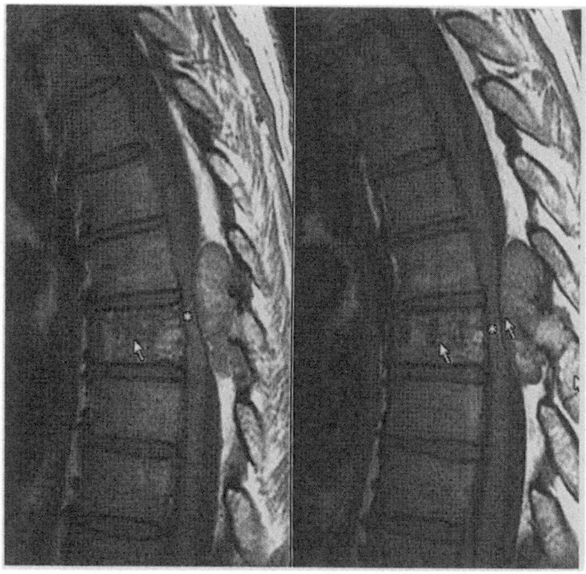

a

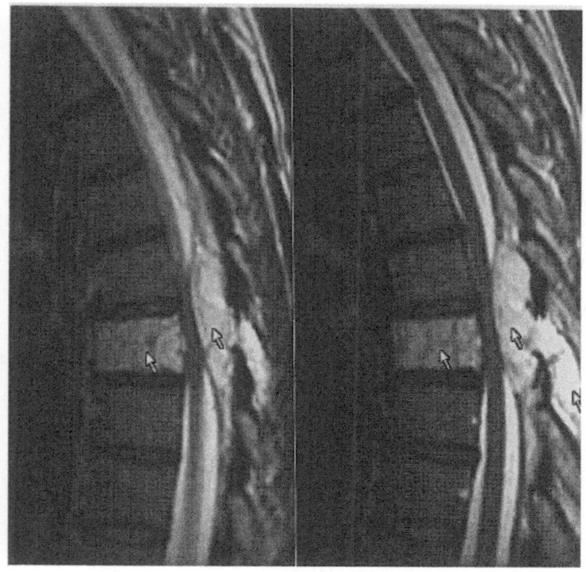

b

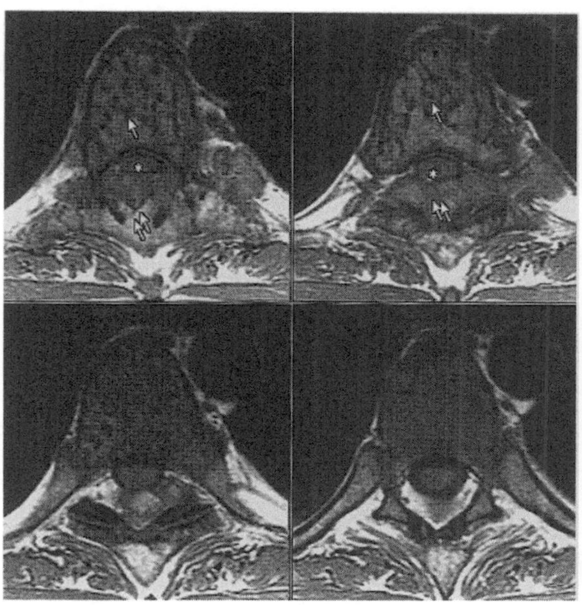

c

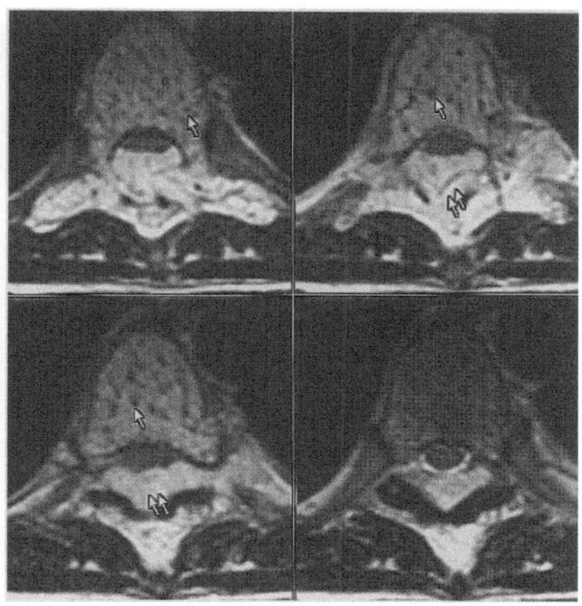

d

Cases 4, 5: spinal sarcomas

Clinical and pathological features

- Primary malignant tumors of the spine are very uncommon.
- Primary malignant bone tumors involving the spine are the following: osteosarcoma (3%), chondrosarcoma (7%), Ewing's sarcoma (7%), malignant fibrous histiocytoma (4%).
- Spinal osteosarcomas occur in the second decade with a male predilection, and are mainly localized in the vertebral bodies. Their incidence is higher in irradiated bones, in pagetic bones and may develop from osteochondromas. Histopathologically they are formed of sarcomatous connective tissue with associated bony and osteoid formation.
- Spinal chondrosarcomas occur preferentially during the fifth decade, most patients ranging from 30 to 60 years of age, with a male predominance, involving most commonly the posterior elements in the thoracic spine. These tumors arising from the cartilage may develop from enchondromas or osteochondromas, or secondary to previous irradiation.
- Ewing's sarcomas are highly cellular tumors formed of round cells separated by fibrous tissue, occuring most commonly in adolescents (10-20 years old) with a male predilection, involving mainly the sacrum but also the vertebral bodies.
- Malignant fibrous histiocytomas are tumors of mesenchymal origin occurring most commonly in adults aged between 40-70 years. They may arise in the spine after radiation therapy or in pagetic patients.
- *Signs:* the most frequent presenting symptom is pain which may increase at night, eventually associated with a local tenderness and/or a palpable mass.

MR characteristic findings

Morphology

- MR shows a great advantage over CT in demonstrating the extension of the tumoral process within the bone marrow or toward the perispinal soft tissues or in the epidural space with or without associated cord compression.
- CT is more efficient than MR to depict intratumoral ossification or calcifications and to demonstrate bone destruction.

Signal intensity

- MRI usually shows non specific, more or less heterogeneous changes in SI which do not help to differentiate between malignant bone tumors.
- SE-T1w: usually hypointense to marrow.
- SE-T2w: usually hyperintense to marrow and paravertebral soft tissues. SI may be hypointense in osteoblastic osteosarcomas.
- Intratumoral heterogeneity is due to the presence of areas of necrosis, hemorrhage, fibrous and osteoid formation, and mottled calcifications.

Contrast enhancement

- Bony extent of the tumors is usually well evaluated using T1w and T2w sequences. Fat suppression techniques must be used if a contrast enhanced scan is needed.
- Contrast infusion is not mandatory but may be of help in order to best appreciate the presence of necrotic areas or mottled calcifications. It is informative to demonstrate the more vascular tumoral process with respect to adjacent normal soft tissue or muscle, and orient the biopsy.

Case 4: Chondrosarcoma grade I

Fig. a: sagittal cut (512);	SE - T1w	**Fig. c:** sagittal cut;	SE - T2w with fat suppression
Fig. b: sagittal cut (512);	SE - T2w	**Fig. d:** axial cut (512);	SE - T2w

Adult male operated ten years ago for a huge tumor of the spine, presently referred for a recurrence of the tumor. MR shows a non specific low signal intensity on T1w (**a**) and a high signal intensity on T2w images (**b**, **d**), even more prominent on fat suppression techniques (**c**). The tumor involves the right lateral portion of the posterior neural arch at the L3 level. It extends posteriorly into the vertebral canal, remaining epidural with anterior displacement of the cauda equina well seen on sagittal T2w high resolution images (**b**). The major development of the mass is in the right lateral and posterior soft tissues invaded and infiltrated as shown on the axial cuts (**d**, ↑). Note on T2w sequences mainly after fat suppression technique (**c**) the hyperintense signal intensity of the mass showing numerous signal septa as could be observed in the majority of chordomas. At surgery, a well differentiated grade I chondrosarcoma with myxoid cartilage matrix is confirmed, explaining the long term survival.

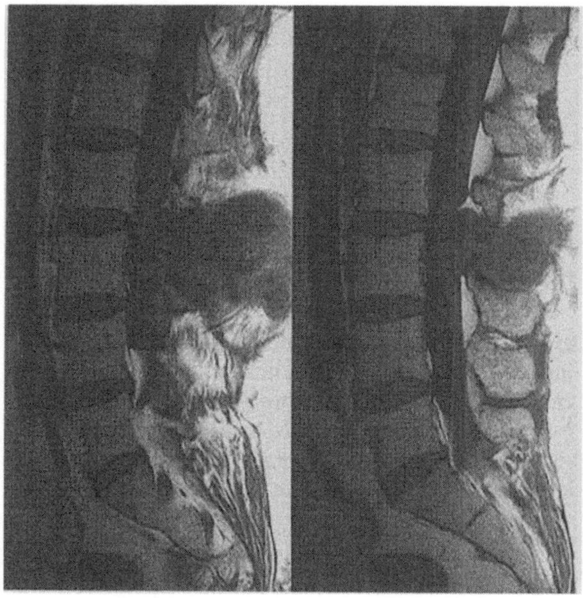

a

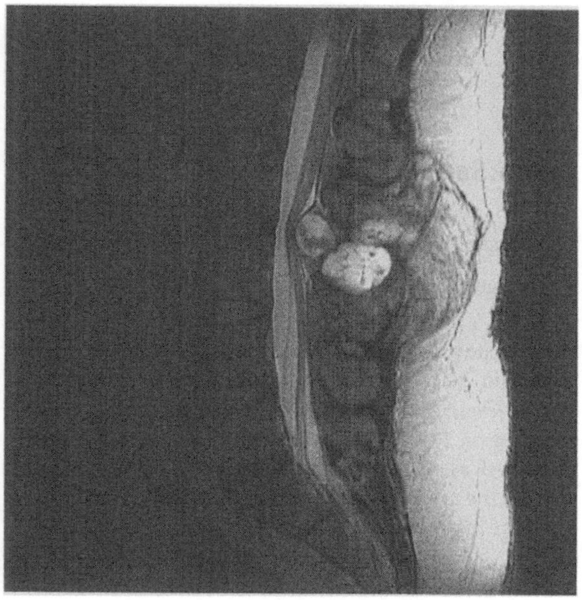

b

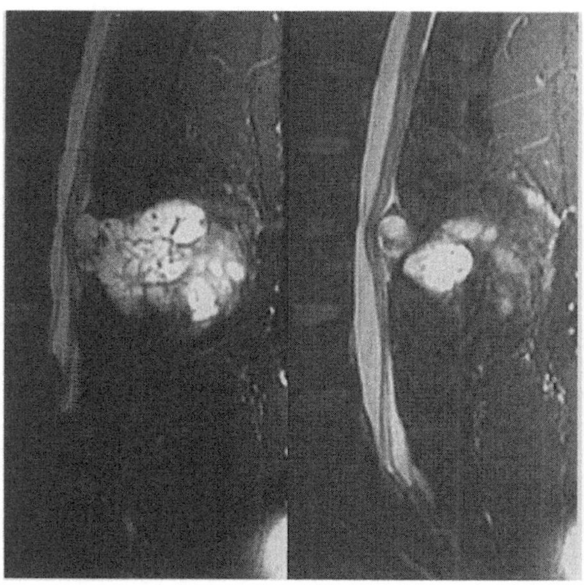

c

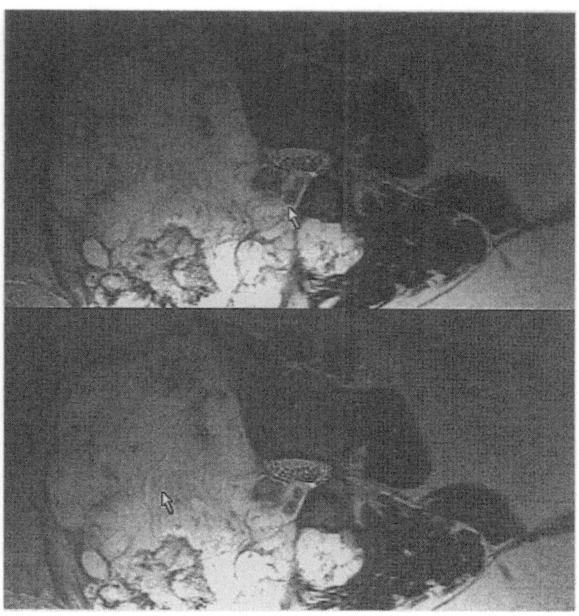

d

Case 5: Sacral osteosarcoma

Fig. a: sagittal cut (512);	SE - T1w	**Fig. c:** coronal cut;	SE - T2w with fat suppression
Fig. b: sagittal cut;	SE - T2w	**Fig. d:** coronal and axial cut;	SE - T1w + Gd

Adult male presenting a cauda equina syndrome with lower back pain. MR shows a mass involving the sacrum at the S3 and S4 levels. The mass shows non specific decreased signal intensity on T1w (**a**) and increased signal on T2w images (**b**) as well as on T2w with fat suppression (**c**). The mass invades the presacral soft tissues well separated on T1w from the hyperintense signal of the fat (**a**). The posterior extension of the mass is well identified on sagittal T2w images (**b**) and on the oblique images after contrast enhancement (**d**). The mass shows a mixed pattern with osteolytic and sclerotic areas. Note the involvement of the right iliac bone (**c, d**, up). Note also the compression of the lower sacral foramina with root compression (**a**, right).

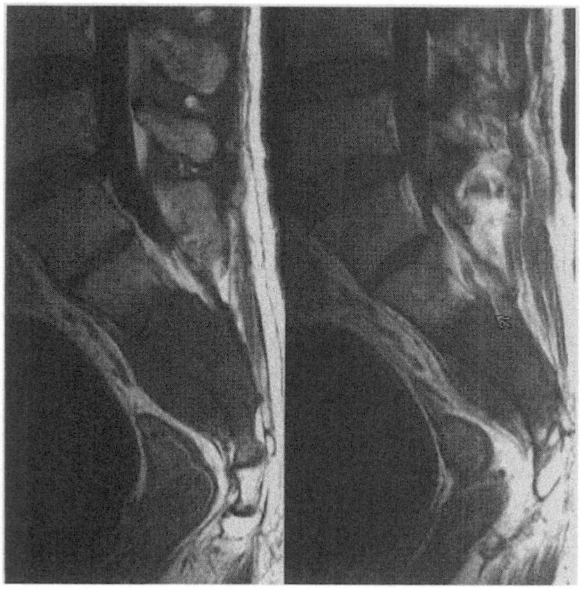

a

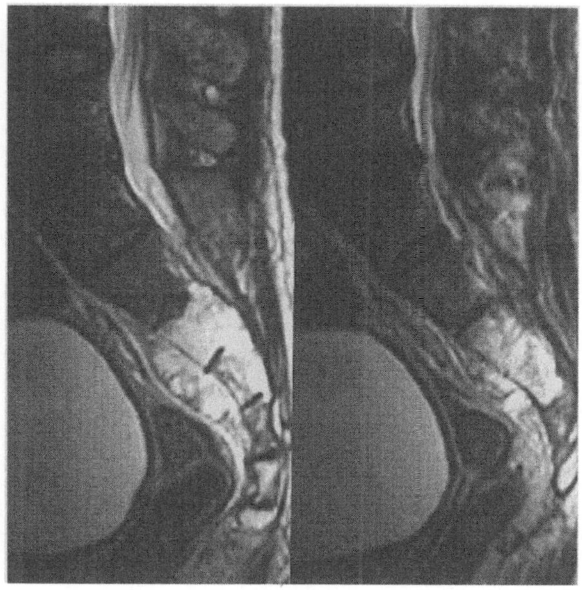

b

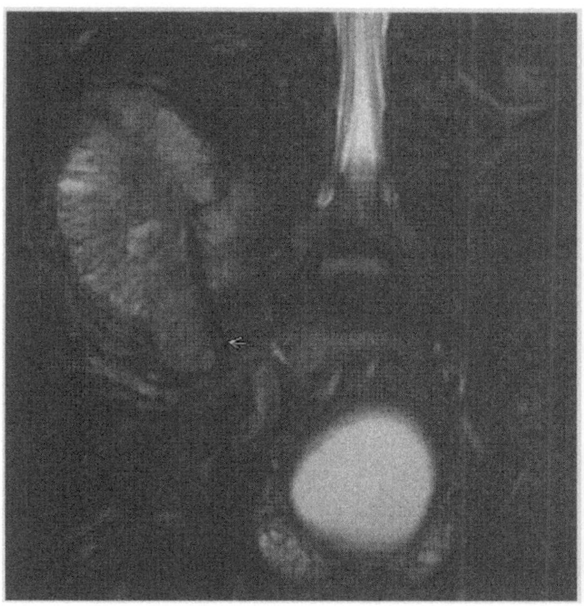

c

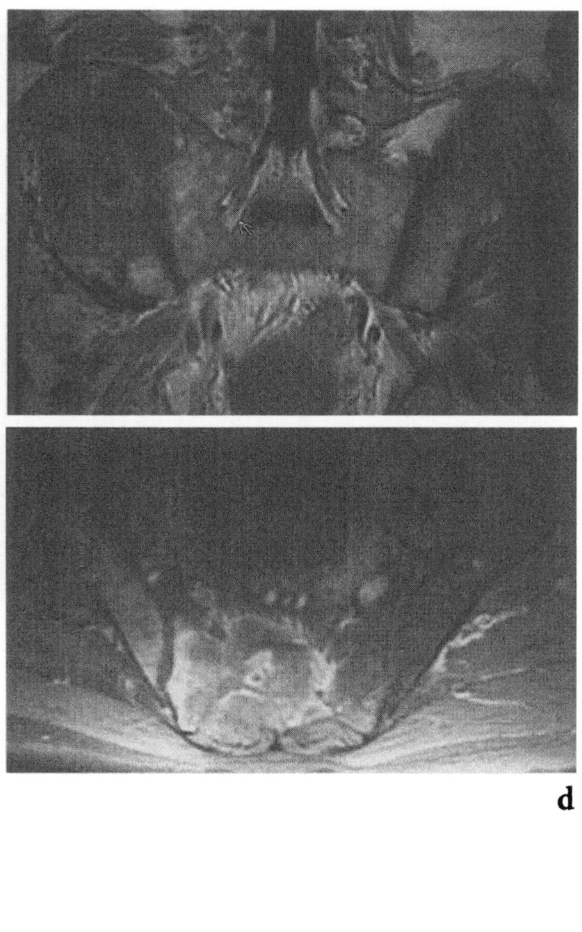

d

Cases 6-8: spinal lymphomas and leukemias

Clinical and pathological features

- Lymphomas are represented by the Hodgkin's disease accounting for almost 1/3 of the cases and non Hodgkin's in about 2/3 of the cases. Bone involvement is more commonly encountered in non Hodgkin lymphomas.
- Non Hodgkin's lymphomas occur most frequently in adults and are more common in males (2:1 male-female ratio).
- Involvement of the spine by lymphomas is usually due either to direct infiltration of vertebral bone marrow or regional invasion from paraspinal tumor.
- In early childhood, leukemia is the most frequent malignancy, mainly represented by the acute lymphoblastic leukemia accounting for more than 2/3 of the patients, the myelogenic type being found in 10% of the cases between other rare forms.
- *Signs:* usually non specific and variable according to the site of involvement.

MR characteristic findings

Morphology

- In spinal involvement, there is an infiltration with replacement of the bone marrow by the malignant cells, accounting for the SI changes on MRI. Note that a diffuse osteoporotic appearance of the spine may be evidenced in more than half the patients with leukemia and a "moth-eaten" lytic aspect may be shown in non Hodgkin's lymphomas, on plain X-rays.
- Focal paravertebral mass invading the spine and originating from retroperitoneal adenopathies may be found in lymphomas. SI patterns shown are similar to other solid tumors (long T1 and T2). An "ivory vertebra" may be seen on plain films. Compression deformity of the infiltrated vertebral body may occur.
- Lymphomas and leukemias show a great propensity to diffusely infiltrate the CNS. Leptomeningeal involvement appears more common in non Hodgkin's lymphomas in adults and in acute lymphoblastic leukemia of childhood. Tumoral deposits in nerve root sleeves may be also shown.
- Spinal cord involvement may be observed resulting from hematogeneous spread.

Signal intensity

- SE-T1w: hypointense to bone marrow (focal or diffuse areas in lymphomas and usually diffuse decreased SI in leukemias). In childhood leukemia the appreciation of the decrease in SI may be difficult due to the small amount of normal fat (relative bright SI on T1).
- SE-T2w: mild to marked non specific hyperintensity to normal bone marrow.
- Exceptions to this non specific patterns of SI changes are not rare: hyperintensity on T1w and decrease in SI on T2w sequences have been shown in lymphomas.

Contrast enhancement

- Contrast infusion is helpful in leukemia, showing the diffuse infiltration of bone marrow. Multiple focal foci of enhancement are shown in spinal lymphomas (using fat suppression techniques).
- Leptomeningeal infiltration is best evidenced after contrast enhancement showing a diffuse augmentation of the subarachnoid spaces.
- The extension of focal paravertebral masses to the vertebral bodies, and of the neural foramina the epidural space and the "plaque-like" infiltrating lesions, usually in non Hodgkin's lymphomas are more easily evidenced after contrast infusion.

Case 6: Hodgkin's lymphoma

Fig. a: sagittal cut (512); SE - T2w

Fig. b: sagittal cut (512); SE - T2w

Fig. c: axial cut; SE - PDw and T1w + Gd (left)

Fig. d: sagittal cut; SE - T2w

Slightly hyperintense mass involving the lower cervical spine on t2w (**a, b**), involving C7, D1, and D2 vertebral levels and extending to the paraspinal soft tissues well shown on the high resolution T2w parasagittal cuts (**b**, ↑↑) and on the axial cuts after contrast iv (**c**, ↑). The axial T1w enhanced slices demonstrate the epidural extension of the masses (**c**, ↑↑) with compression of the adjacent spinal cord (**c**, *). The epidural and foraminal extension of the masses is also shown on the sagittal T2w sequences obturating the hyperintense signal of the perimedullary CSF. Note also the dissemination of the disease involving at different levels the thoracic spine and showing a slightly hyperintense signal on T2w (**d**, ↑) with erosion of the posterior vertebral cortical wall (↑↑). These findings as well as the MRI signal caracteristics of the lymphomatous tissue are somehow similar to other solid tumors showing long T1 and T2 signal intensity.

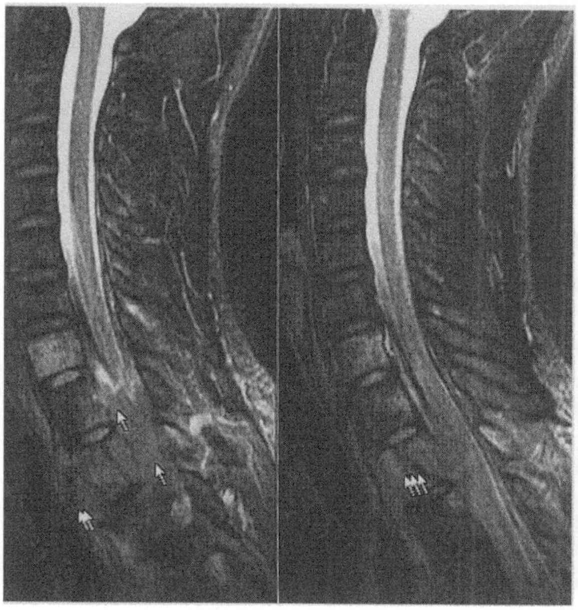

a

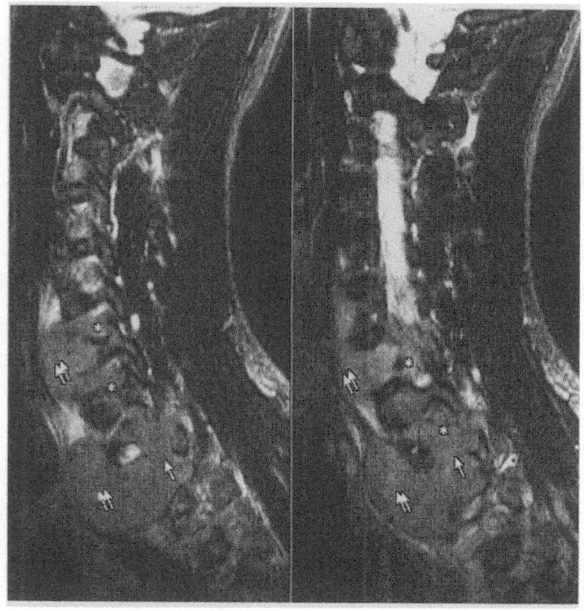

b

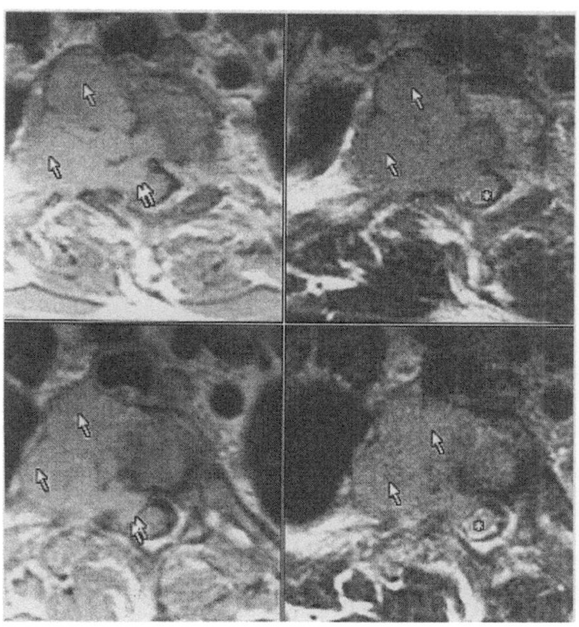

c

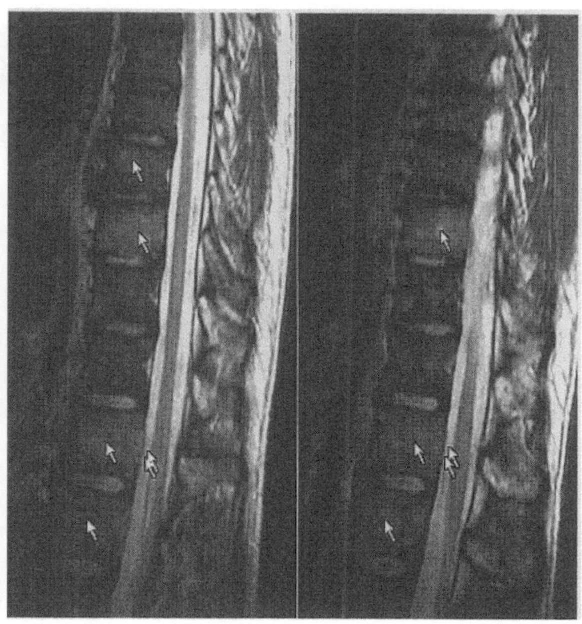

d

Case 7: Malignant non Hodgkin's lymphoma

Fig. a: sagittal cut;	SE - T1w	**Fig. c:** sagittal cut;	SE - T2w
Fig. b: sagittal cut;	SE - T2w	**Fig. d:** axial cut;	SE - T2w

Patient presenting progressive symptoms of thoracic spinal cord compression. MR shows a relatively hypointense signal of the thoracic vertebral bodies on T1w (**a**) due to diffuse infiltration of the normal hyperintense fatty marrow of the vertebral bodies. The bright area involving the two vertebral bodies observed on T1w (**a**, *) and T2w (**b**, *) corresponds to incidental benign hemangiomas. The disease causes at the lower dorsal level a vertebral fracture (**b**, ↑) associated to an expansion of the vertebral body and retropulsion of the posterior wall into the spinal canal (**a**, **b**, ↑↑) with moderate cord compression (**a**,**; **b**). The mass extends to the paravertebral soft tissues (**d**, ↑↑) and invades the anterior and the left lateral epidural space (**d**, ↑) as well as the intervertebral foramina displacing the spinal cord posteriorly and to the right side (**d**, *). Note the absence of bright signal of CSF in the intervertebral foramina obstructed by the disease process (**c**, **). Note also at a lower level a compression-facture of the vertebral body (**b**, ↑↑↑) with slight hyperintense signal on T2w and moderate narrowing of the spinal canal at the level of the cauda equina.

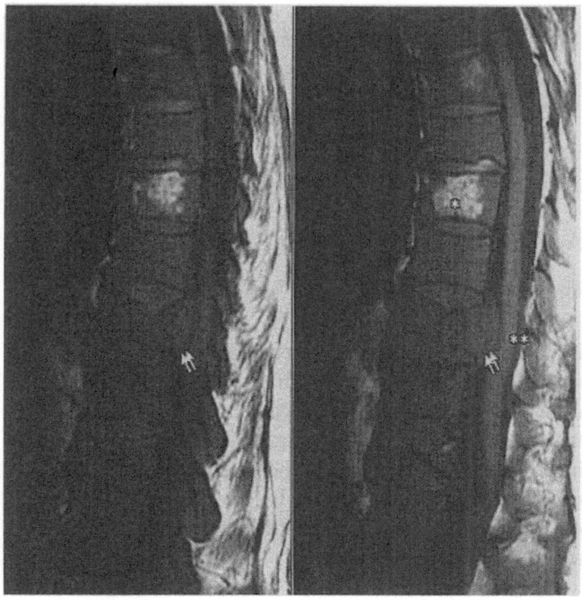

a

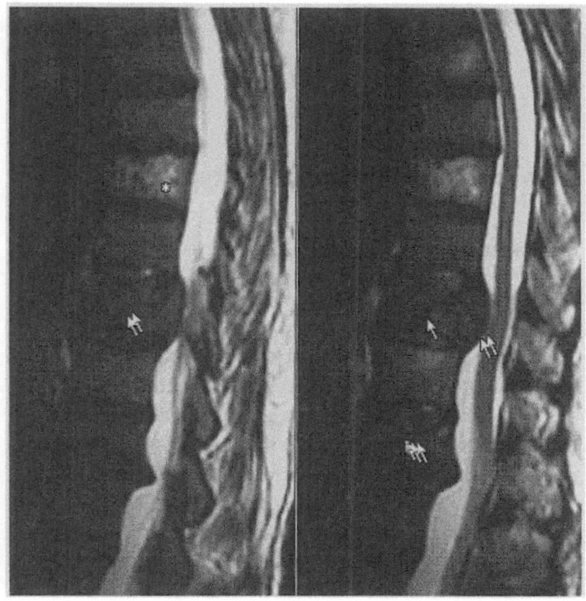

b

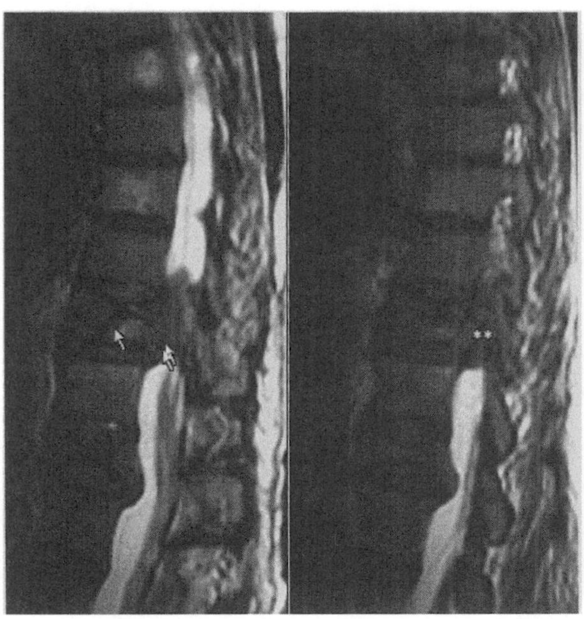

c

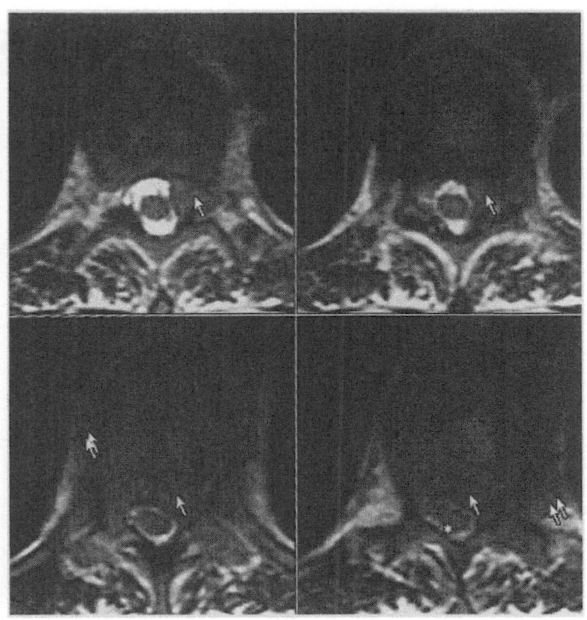

d

Case 8: Malignant lymphoma type B

Fig. a: sagittal cut; SE - T1w + Gd

Fig. b: sagittal cut; SE - T2w

Fig. c: coronal cut; SE - T2w

Fig. d: axial cut; SE - T2w

Adult male with a prior history of non Hodgkin's malignant lymphoma presenting with localized pain and weakness in the right leg. MR shows an aggressive lytic tumoral lesion hyperintense on T2w (**b, c, d**) involving L2 and L3 levels as well as L5 and S1, showing a heterogeneous hyperintense signal on T2w (**b, c, d,** ↑) and extending through the vertebral bodies to invade the right lateral paravertebral soft tissues (**c,** ↑; **d,** ↑↑). The mass also extends posteriorly to involve the lateral and the posterior epidural spaces (d,*) as well as the posterior neural arch (**b**). After contrast iv, the mass presents moderate enhancement (**a,** ↑) as shown at the sacral level with compression of the L5 nerve root in the intervertebral foramen (**a,***). The signal and the enhancement patterns observed are non specific. In some cases, the lesions may show mottled contrast enhancement after Gadolinium injection.

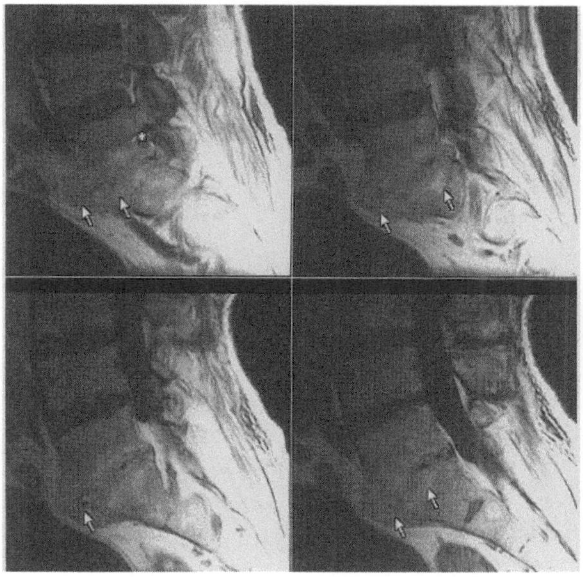

a

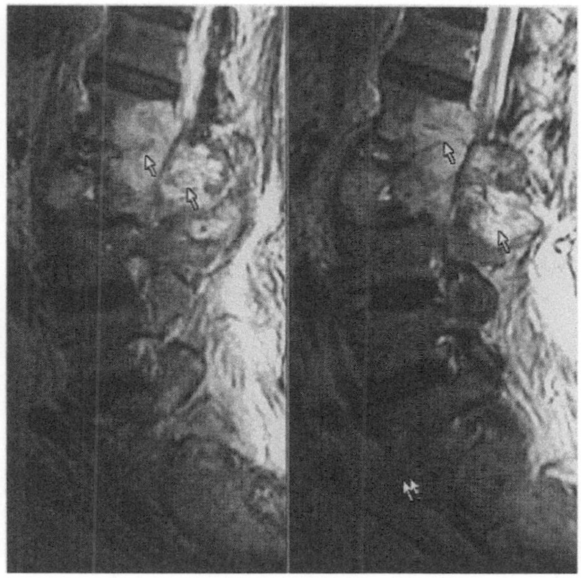

b

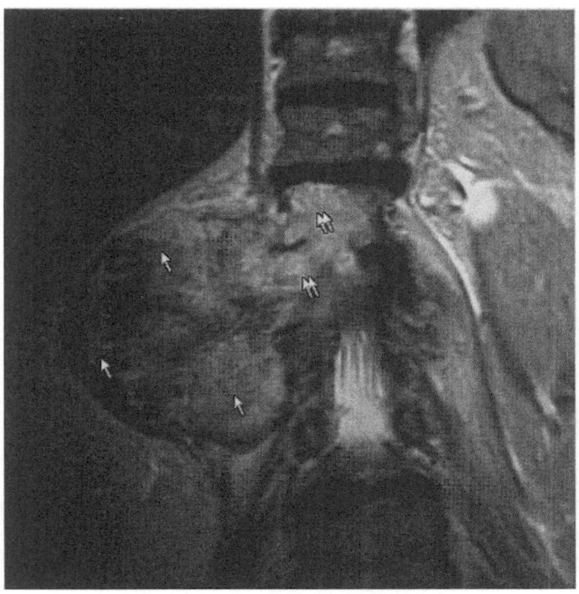

c

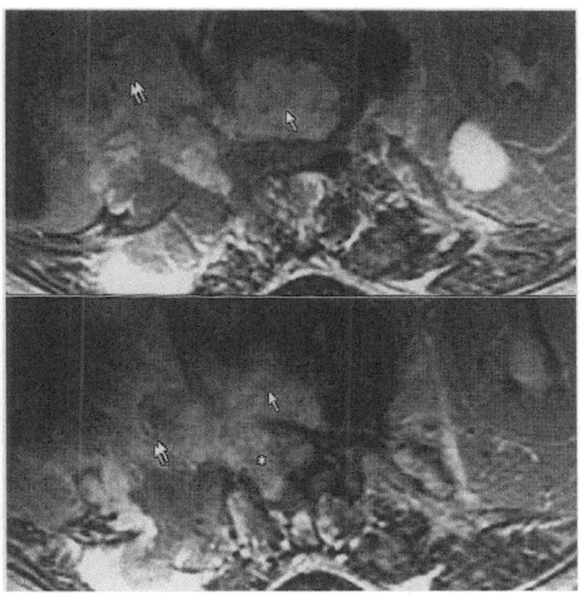

d

Cases 9, 11: multiple myeloma

Clinical and pathological features

- Multiple myeloma is a malignant tumor of plasma cells of unknown etiology.
- It is the most frequent primary tumor of bone and represents almost 1/6 of malignant hemopathies.
- It occurs in adulthood and predominates during the seventh decade of life.
- There is a marked male predominance (3:1).
- The diagnosis is made on bone marrow study showing an increase of the amount of plasma cells (20%).
- *Histopathology:* bone marrow is replaced by clusters of myeloma cells which consist in abnormally large and pleiomorphic plasma cells with basophilic cytoplasm.
- Biologically, a paraproteinemia with Bence-Jones' proteinuria is evidenced. It is sometimes found in the solitary plasmacytoma as well. The sedimentation rate is usually increased.
- *Signs:* the most common complaint is pain found in 2/3 of the cases. With disease progression, weight loss and cachexia, anemia and fatigue become prominent. Signs of peripheral neuropathy with motor deficits may concern almost half of the patients. Pathological fractures are frequently reported involving the spine or the ribs. Later in the course of the disease, bacterial infections and terminal renal failure secondary to Bence-Jones' nephrosis develop in more than half of the patients.

MR characteristic findings

Morphology

- Multiple myeloma most commonly involves the spine with a predilection to the thoracolumbar region, the ribs, mandibles and maxillas, long bones, pelvis and skull.
- Spinal involvement predominates in the vertebral bodies but may later, in the course of the disease, affect the posterior elements and the pedicles as well. A solitary plasmacytoma, usually a lytic process, involves the spine in more than half the cases, and may precede for several years the development of the systemic disease.
- The disease process may extend to the paravertebral soft tissues and at times even show disc involvement.
- Plain films are poorly contributive to the diagnosis showing most often signs of osteoporosis or bone destruction, or appear normal (almost 15% of the cases).
- Bone scans lack of sensitivity for a preliminary diagnosis, and the abnormalities eventually shown are non specific, particularly in elderly patients which may present spinal degenerative disease.
- Replacement of the spinal red marrow may be focal in one third of the patients with lesions ranging in size from 0.5 to 3 centimeters, or show a diffuse pattern in about 1/4 of the cases. MR diagnosis is sometimes difficult to reach in the elderly (paucity of fat within marrow and consequently lack of relative contrast) or to differentiate from other diffuse or metastatic diseases.
- Compression fracture with vertebral collapse are frequently encountered in multiple myeloma, occurring in almost 2/3 of the patients and involving commonly the thoracic region, explaining the frequently associated paraplegia due to spinal cord compression. Nerve roots compressions are also common. The morphological appearance of vertebral invasion and epidural extent of the disease process is almost impossible to differentiate from a metastasis, although the latter shows usually a more patchy distribution pattern.

Signal intensity

- The MR exam may appear as normal in almost 1/3 of the cases early in the course of the disease and at time of diagnosis.
- SE-T1w: hypointense SI (focal or diffuse) to normal marrow.
- SE-T2w: hyperintense SI to normal marrow (GE less sensitive).
- Following radiation therapy, the lesions may show a decrease in SI, mainly on T1.

Contrast enhancement

- Focal or diffuse contrast enhancement is usually demonstrated using T1w sequences with fat suppression techniques.
- Enhancement may be relatively diminished in treated patients, as compared with the pretherapeutic MR.

Case 9: Multiple myeloma

Fig. a: sagittal cut; SE - T2w with fat suppression

Fig. b: sagittal cut; SE - T1w + Gd with fat suppression

Adult patient presenting with cord compressive syndrome. Plain X-rays and bone scan were negative. On MR, T2w sequence with fat suppression readily shows a posterior epidural mass presenting a hypointense signal on T2w slightly more hyperintense than spine (**a**, ↑) enhancing markedly after contrast iv (**b**, ↑). The mass involves the vertebral body as well as the posterior spinous process (**a**, **b**, ↑↑). The spinal cord is compressed by the mass with some reactive intramedullary edema. Note the small bulging of the posterior wall of the vertebral body and the anterior epidural component of the lesion. Other levels were also involved.

Case 10: Metastatic lung hemangiopericytoma

Fig. c: sagittal cut; SE - T1w

Fig. d: sagittal cut; SE - T2w with fat suppression

Patient presenting back pain and rapidly progressive gait disorder. T1w MR images show abnormal bone marrow signal involving multiple vertebral levels extending to the pedicles as well as to the spinous process. The intravertebral masses present a hypointense signal on T1w (**c**, *) and a hyperintense signal on T2w with supressed fat (**d**, *). Note for the lower spinal mass the extension to the pedicle and the posterior elements which is a common manifestation of spinal metastasis. Note for all intracorporeal masses the bulging of the posterior walls with compression for the lowest localization (**c**, **d**, ↑). The spinal cord presents probably a reactive edema shown on T2w (**d**, right,↑↑) due to the compression with slight enlargement of the involved cord segment. The lowest mass presents with intratumoral spotted and curvilinear images shown on all sequences which may correspond to thickened trabeculae as well as to flow-void intratumoral vessels.

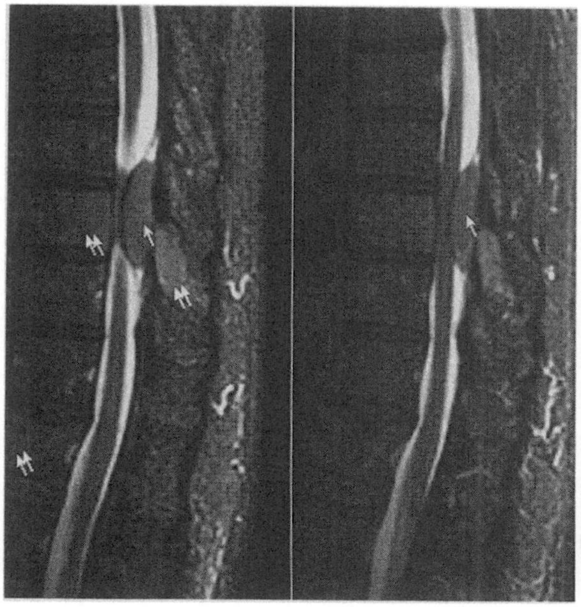

a

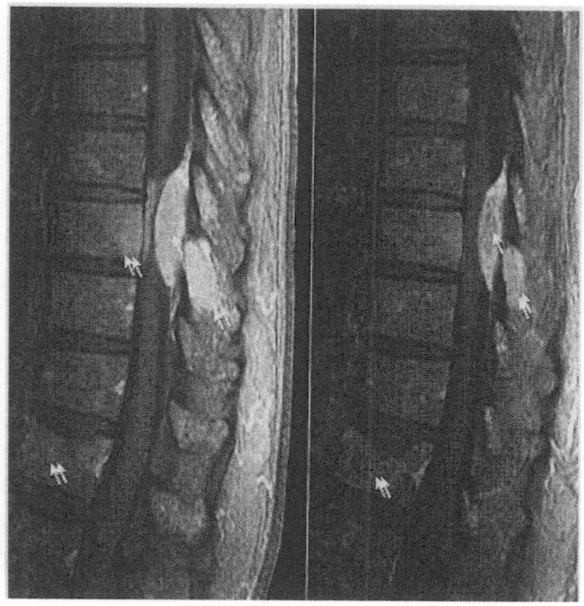

b

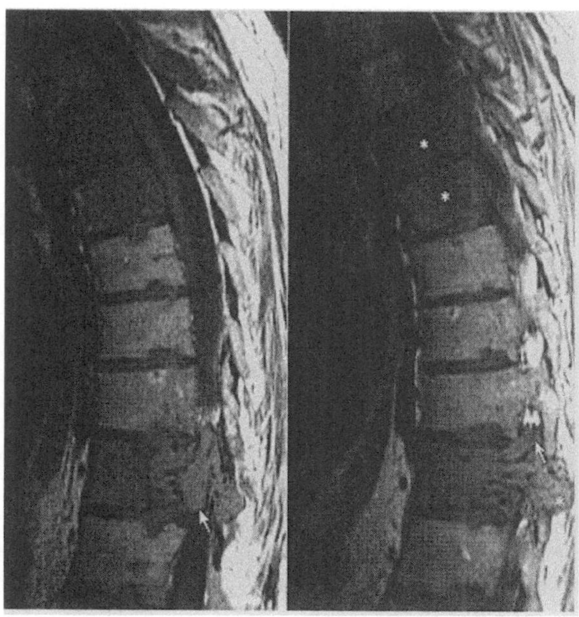

c

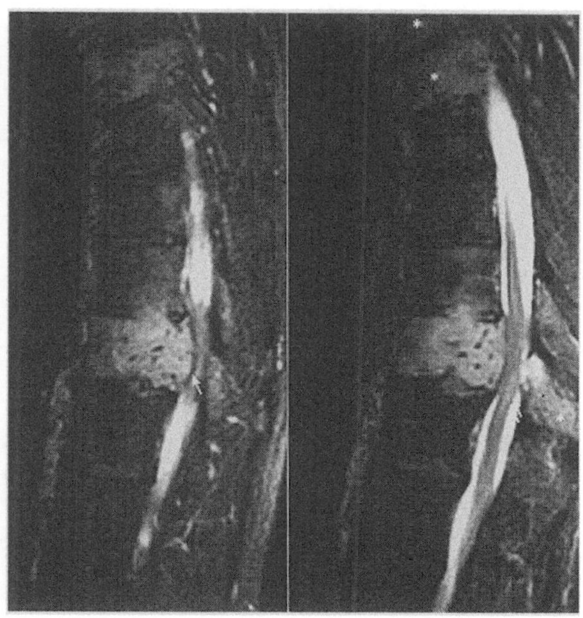

d

Case 11: Multiple myeloma

Fig. a: sagittal cut; SE - T1w

Fig. b: sagittal cut; SE - T2w with fat suppression

Unenhanced T1w images revealing multiple low signal lesions disseminated in nearly every vertebral body (**a**, ↑). These lesions become hyperintense on T2w sequence with suppressed fat (**b**, ↑). Enhanced T1w images (not shown) demonstrated variable enhancement, even better shown on fat suppressed images. In fact, the lesions may become iso-intense to the adjacent marrow after contrast iv. A bone scan failed to show any abnormal uptake in the lumbar spine.

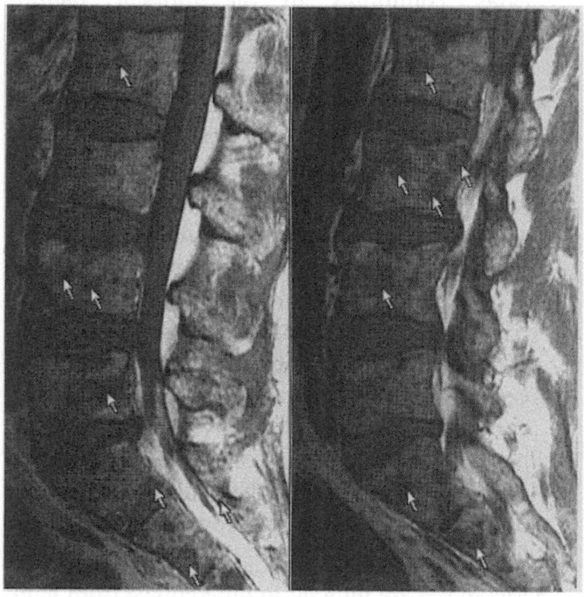

a

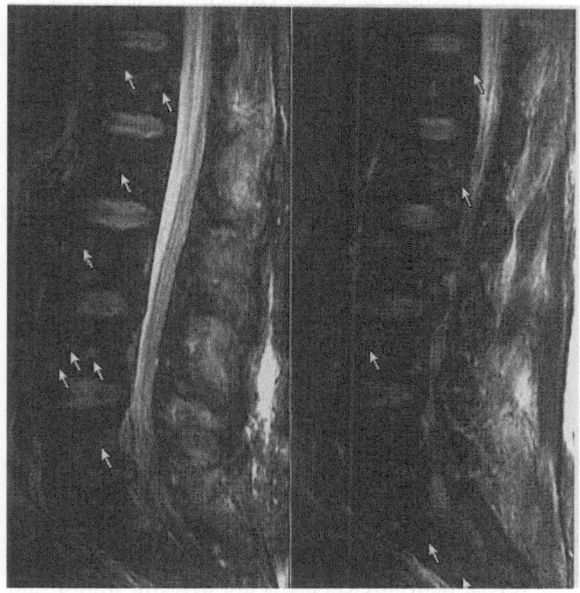

b

Cases 10, 12-14: spinal metastases

Clinical and pathological features

- Metastases to the spine are the most frequent extradural malignancy.
- MRI stands as the modality of choice in the evaluation of metastatic disease involving the spine, being more sensitive and more specific than bone scan. This latter technique still remains the screening method for a cost-efficient evaluation of the entire skeleton.
- The most frequent malignancies which metastasize in the spine are: the breast, lung and prostate carcinomas, followed by the gastrointestinal, renal and bladder, melanomas, and pharynx.
- Intraspinal red marrow is the site of predilection of hematogenous metastases. Venous route of spreading is also facilitated due to lack of valves within the vertebral venous plexuses. Locoregional invasion from adjacent paraspinal tumors may be observed.
- *Signs:* the most common complaint is back pain which may increase during night and may be associated with local tenderness under pressure over the spinous processes. In case of associated cord compression due to extension of the mass to the epidural space, usually at the thoracolumbar levels, neurological signs are present: motor and sensory deficits, radicular signs and pain, Brown-Sequard syndrome, bladder or bowel dysfunctions, or cauda equina syndrome. Serum alkaline phosphatase is usually increased in spinal metastases and proteinorachy may also be increased in case of spinal block.

MR characteristic findings

Morphology

- Metastases are commonly multiple but may be solitary. The morphological and signal intensity patterns lack of specificity for any type of lesion.
- Spinal metastases are usually located in the lumbar region, but may involve, in decreased frequency, the thoracic and the cervical region, and the sacrum.
- Metastases are preferentially located in the vertebral body but may involve or extend to the pedicles and posterior elements, due to the usual hematogenous spread of the metastases. They usually do not invade the intervertebral discs and tend to be limited by the dura.
- Metastases are usually of the lytic type, but may be osteoblastic.
- Extraosseous extension of a metastasis to the spinal canal with infiltration of the epidural spaces and cord compression occur in about 15% of the cases, commonly in the thoracic region. Metastases from lung, breast and prostate carcinomas are most frequently found.
- Vertebral collapse may occur expanding the anterior and/or the posterior vertebral walls with secondary spinal canal stenosis and eventual cord compression. The vertebral walls commonly show erosion and rupture. In benign osteoporotic compression fractures, the vertebral body appears as compressed usually without expansions or destructions of the anterior or posterior walls. Areas of residual normal bone marrow are still evidenced within the collapsed vertebra. Note that contrast enhancement may even be seen in benign burst fractures secondary to degenerative diseases, sparing usually the pedicles. In litigious cases follow-up MR exams may be of great help (1-3 months).
- Metastases are difficult to differentiate from lymphomas or multiple myeloma.

Signal intensity

- Metastases show no specific signal intensity changes as compared to most primary spinal tumors.
- SE-T1w: usually hypointense to bone marrow, but may be at times iso-intense and exceptionally hyperintense (breast carcinoma). Focal hyperintense lesions are most frequently hemangiomas or correspond to focal fatty marrow.
- SE-T2w: usually hyperintense to bone marrow, unless blastic (prostatic or breast carcinomas).
- STIR: usually hyperintense due to fat signal suppression in normal marrow.

Contrast enhancement

- Metastases usually show contrast enhancement (T1w sequences with fat suppression technique).

- Osteoblastic metastases may show moderate enhancement. Contrast uptake may decrease following treatment.
- Contrast administration is needed to demonstrate or rule out a leptomeningeal spread or an intramedullary localization.

Case 12: Lytic metastases

Fig. c: sagittal cut; SE - T1w

Fig. d: sagittal cut; SE - T2w with fat suppression

Unenhanced T1w sequence, without fat suppression technique, showing on high resolution images multiple low signal lesions involving the vertebral bodies, shown here in the lumbar spine (**c**, ↑) becoming slightly hyperintense on T2w sequence best shown after fat supression technique (**d**, ↑), the latter sequence also showing the involvement of the spinous process with hyperintense signal of the adjacent soft tissue at the L3 level (**d**). Note also the hyperintense rims surrounding the mass at L2 and L4 levels (**d**) giving a target appearance. Note normal spinal canal and cauda equina. Note the absence of involvement of the intervertebral discs.

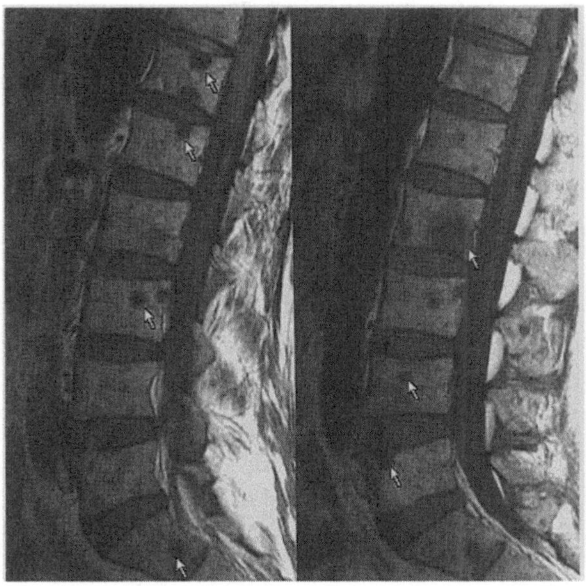

c

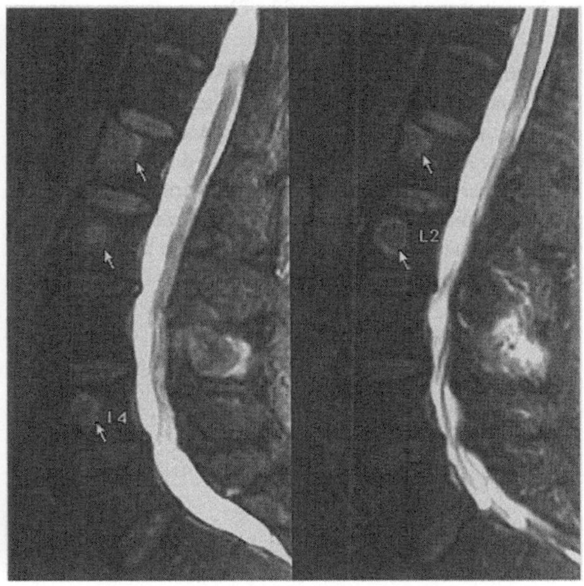

d

Case 13: Sclerotic metastases (prostate adenocarcinoma)

Fig. a: sagittal cut;	SE - T1w	**Fig. c:** sagittal cut;	SE - T1w
Fig. b: sagittal cut;	SE - T2w	**Fig. d:** sagittal cut;	SE - T2w

Old man with prostate carcinoma presenting back pain. Unenhanced T1w images of the dorsolumbar spine revealed multiple areas of marked decreased signal involving most of the vertebral bodies (**a**, **c**, ↑) well depicted in the hyperintense vertebral marrow and showing only mild patchy enhancement after contrast iv (images not shown). On T2w sequences, the small lesions remain hypointense, some of which being more difficult to depict (**b**, **d**, ↑). Note the hyperintense area at the L5 level due presumably to degenerative changes (**d**, ↑↑). Such blastic metastases generally demonstrate a decrease signal on both T1w and T2w images. Note that it is not uncommon to detect spinal metastasis on MRI although bone scan is negative. MRI is more sensitive and specific than a bone scan for spinal metastasis.

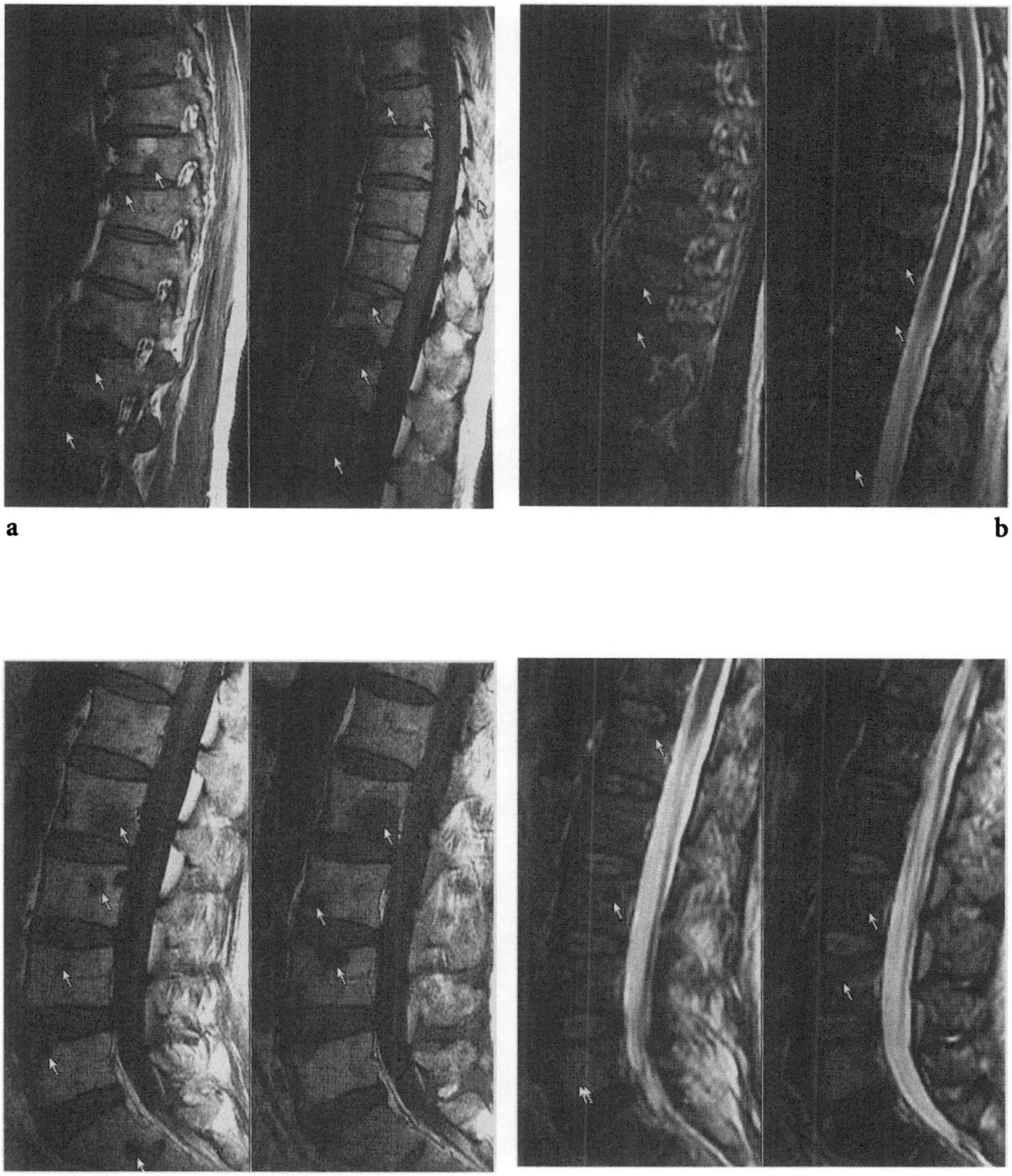

a b c d

Case 14: Meningeal metastases (breast carcinoma)

Fig. a: sagittal cut;	SE - T2w	**Fig. c:** coronal cut;	SE - T1w + Gd
Fig. b: sagittal cut;	SE - T1w + Gd	**Fig. d:** axial cut;	SE - T1w + Gd

Middle age woman with known breast carcinoma presenting with a sudden onset of cervical neuralgia with neck pain. T2w sagittal cuts show a slightly hyperintense signal involving the cervical cord (**a**, ↑) associated to a disappearance of the perimedullary hyperintense subarachnoid spaces mainly from C6 to Th1 levels. After contrast iv, T1w sequence shows the presence of an enhanced perimedullary lesion (**b**, ↑, **c**, ↑, **d**, ↑) corresponding to a metastatic seeding in the leptomeningeal perimedullary spaces as well as in the epidural space.

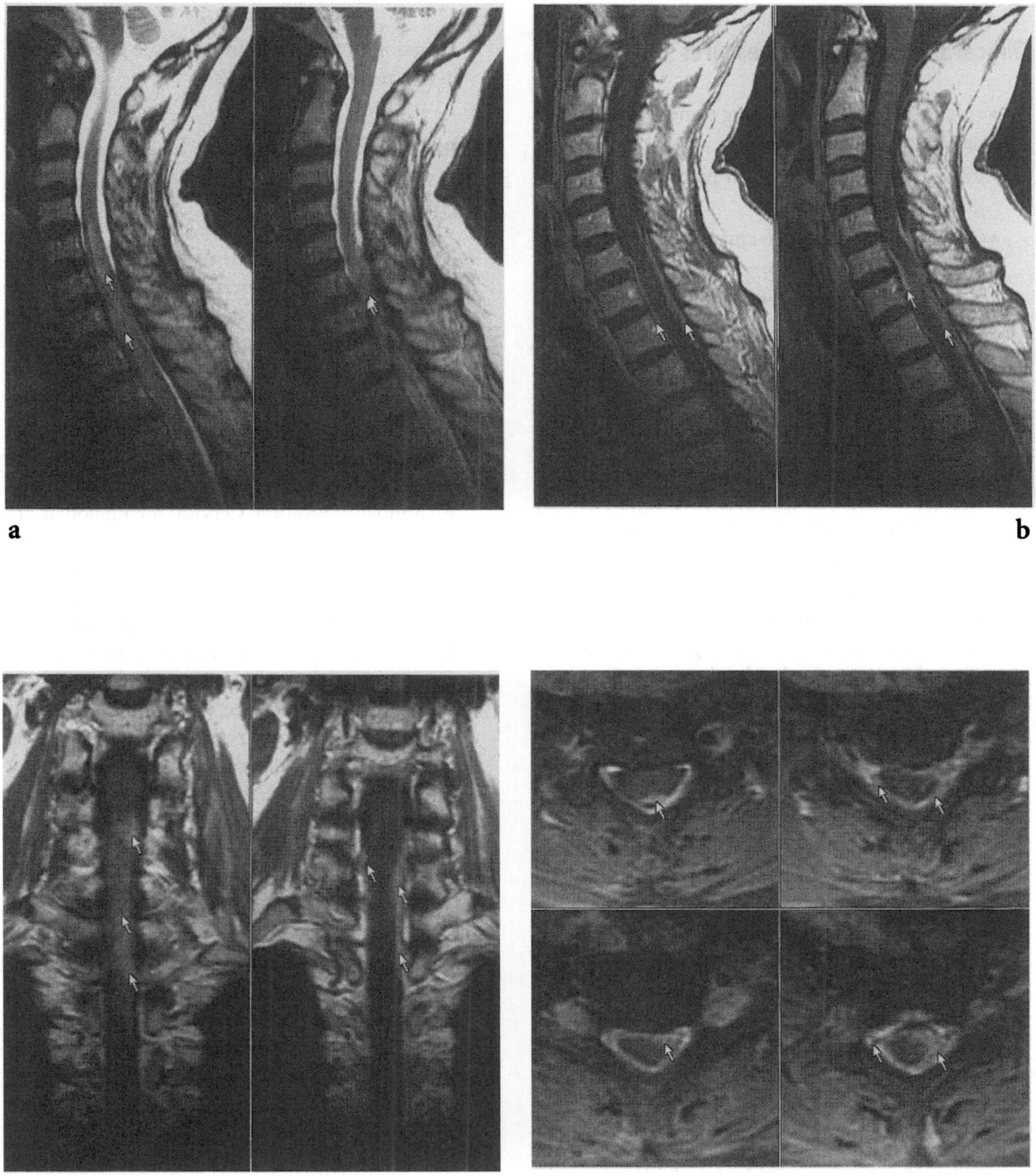

a b c d

Intradural extramedullary tumors

Cases 15-18: spinal meningiomas

Clinical and pathological features

- Usually benign slow growing tumors arising from neoplastic arachnoidal cells.
- Represent almost 1/4 of all primary intraspinal processes.
- Occur in middle-aged and elderly during the sixth to seventh decades.
- A marked female predominance is found (female:male ratio of 3:2). In children, meningiomas are rare (about 5%) often anaplastic or atypical with a male predominance.
- In neurofibromatosis type 2 (locus on chromosome 22q), meningiomas may be multiple, occurring in younger patients and equally found in women and males. Multiple meningiomas may also be found in other hereditary diseases.
- *Histopathology:* various histological subtypes are recognized, the most common of which are the meningothelial, fibroblastic and transitional or mixed (see intracranial meningiomas). The role of sex hormones in the development of meningiomas is unsettled, although 2/3 of the tumors express progesterone receptors, mainly in females, and estrogen receptors are usually absent. Multiple meningiomas may be of different subtype varieties.
- *Signs:* as any intradural-extramedullary tumor, the symptomatology commonly associates a medullary syndrome and radicular signs (radicular pain is more common in nerve sheath tumors). More than 2/3 of the patients present motor corticospinal tract signs associated with increased fatigability and stiffness and hyperreflexia. Posterior column signs and symptoms with paresthesias and dysesthesias are found early, in more than half the cases. A Brown-Sequard syndrome is common, with ipsilateral motor deficit and deep and tactile sensory impairement, and contralateral spino-thalamic thermoalgesic impairement of sensation (due to the frequent lateral location of the meningiomas mainly at the mid-thoracic levels). Gait disturbance and genitourinary dysfunctions may be present. Pain is a very frequent complaint, worse at night. Duration of symptoms before diagnosis averages 2 years. Elevation of protein amount in the CSF is found in the majority of the cases.

MR characteristic findings

Morphology

- Almost 90% of spinal meningiomas are intradural-extramedullary in location.
- Meningiomas may rarely involve the intra- and extradural spaces or even develop extradurally with a malignant behavior.
- Spinal meningiomas appear most common in the thoracic region (80% in females and 50% in males). A 7:1 ratio of thoracic to cervical meningiomas is shown in females. Only 3% of the cases are lumbar in location. Cervical meningiomas are found mainly at the foramen magnum (2:1 female to male ratio) and at levels C3 and C4.
- The majority of the thoracic meningiomas are postero-lateral in location, close to the nerve root sleeves, whereas the cervical tend to be found anterior to the spinal cord.
- Spinal meningiomas are very well demarcated rounded or oval tumors, with smooth margins, showing a broad dural attachment, displacing with compression of the spinal cord. A marked compression of the cord with eventual associated syrinx are found at time of diagnosis due to their very slow grow.
- Spinal meningiomas appear to be less vascular than intracranial. Unlike intracranial meningiomas, spinal tumors calcify very rarely (rare heterogeneous appearance). Areas of hyperostosis or bone invasion are also very rarely found (except in very unfrequent extradural meningiomas), but bone erosion (pedicles) may occur. "En plaque" meningiomas are rarely uncountered in the spinal canal.
- Spinal meningiomas have to be differentiated from nerve sheaths tumors. The latter tumors are usually larger, commonly dumb-bell shaped, and usually hyperintense to cord on T2w sequences. Areas of cystic degeneration and hemorrhagic changes are more common in nerve sheaths tumors, but calcifications are more frequent in meningiomas.

Signal intensity

- SE-T1w: usually iso-intense to spinal cord.
- SE-T2w: usually iso-intense or slightly hyperintense to cord. May be hypointense in case of presence of dense calcifications.

Contrast enhancement

- Meningiomas show intense and homogeneous contrast enhancement. Enhancement is immediate and faster than in nerve sheaths tumors. Intratumoral areas of very low signal may correspond to calcifications rarely found in intraspinal meningiomas as compared to intracranial (heterogeneous appearance, commonly found in psammomatous variety, to be at times differentiated from ossification of the ligamentum flavum).
- A suggestive but not specific dural "tail sign" may be demonstrated adjacent to the dural base due to reactive changes and/or tumor extension.

Case 15: Intradural cervical meningioma

Fig. a: sagittal cut; SE - T1w

Fig. b: axial cut; SE - T1w

Middle age woman presenting a progressive gait imbalance with weakness. MR shows an intradural upper cervical mass iso-intense to the cord on T1w (**a**, ↑↑) developed in the spinal canal at C2 level, displacing posteriorly and compressing the adjacent spinal cord (**a**, ↑↑↑). The cord slightly hyperintense to the mass on T1w is shown in the axial plane laminated at the left posterolateral aspect of the dural sac (**b**, ↑↑↑). Note the anterior broad base of the mass (**a**, left). The mass showed marked homogeneous enhancement (not shown).

Case 16: Calcified dorsal meningioma

Fig. c: sagittal cut; SE - T2w

Fig. d: coronal cut; SE - T2w

Patient presenting pain in the leg consistent with a radiculopathy. T2w MR sequences in the sagittal (c) and in the coronal (**d**) planes did not disclose any lower lumbar disc herniations but a small partly calcified intradural mass (**c, d**, ↑) partly enhancing (not shown) durally based at the level of the conus medullaris (**d**, *). The small mass displaces the nerve roots of the cauda equina to the left side (**d**, left, ↑↑) and compresses moderatly the conus (**d**, right, *).

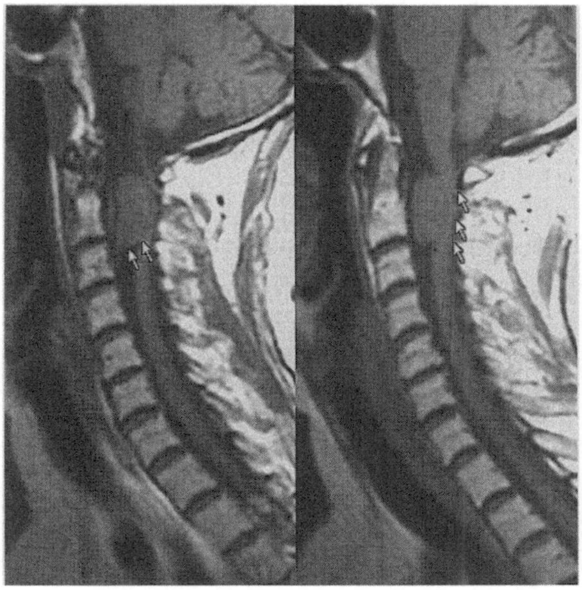

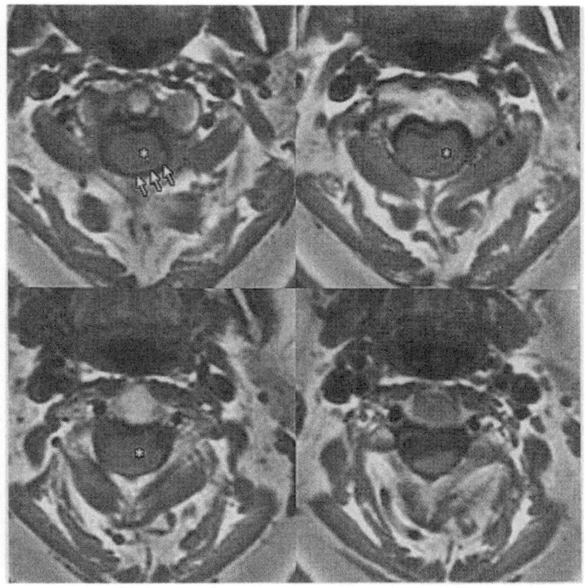

a b

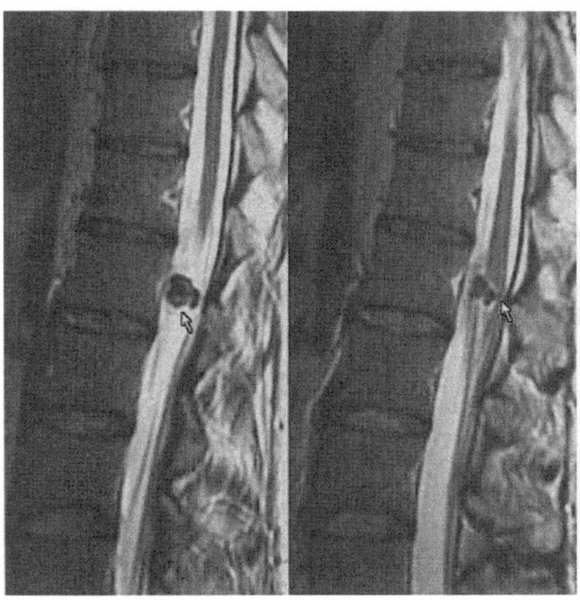

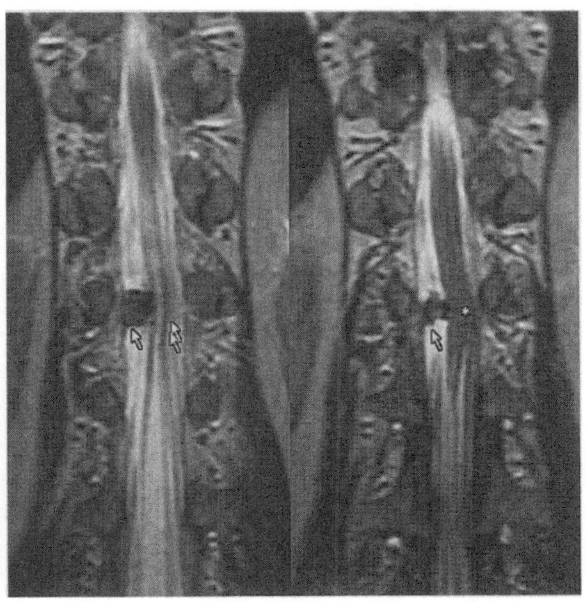

c d

Case 17: Dural sac meningioma

Fig. a: sagittal cut; SE - T2w

Fig. b: sagittal cut; SE - T1w + Gd with fat suppression

Patient presenting with a cauda equina syndrome. T2w MR sequence shows an intradural extramedullary mass iso-intense to the spinal cord (**a**, ↑) with subsequent marked homogeneous enhancement (**b**, ↑↑). The mass displaces anteriorly and posteriorly the nerve roots (**b**, ↑) without compression of the conus medullaris (**a**). Note the absence of "dural tails" on enhanced scans (**b**) cephalad or caudad from the dural base of the tumor.

Case 18: Dorsal meningioma

Fig. c: sagittal cut; SE - T1w

Fig. d: coronal cut; SE - T1w + Gd with fat suppression

Middle age woman presenting with a conus medullaris syndrome. MR images through the lower thoracic spine show a soft tissue mass anterolateral to the spinal cord, iso-intense in signal intensity with the spinal cord itself on T1w (**c**,↑), inhomogeneously enhancing after contrast iv (**d**, ↑). The conus is displaced to the left side and compressed (**d**, ↑↑). The lesion demonstrated a very high signal intensity on T2w images that could suggest a nerve sheath tumor. Enhanced MR did not reveal the dural "tails" but a broad right dural base.

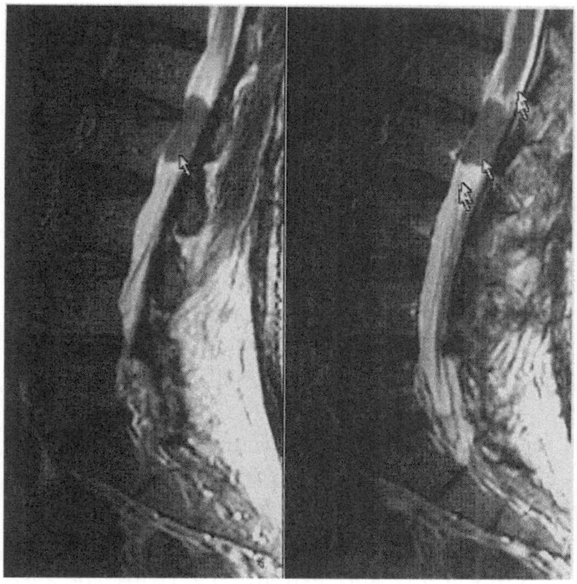

a

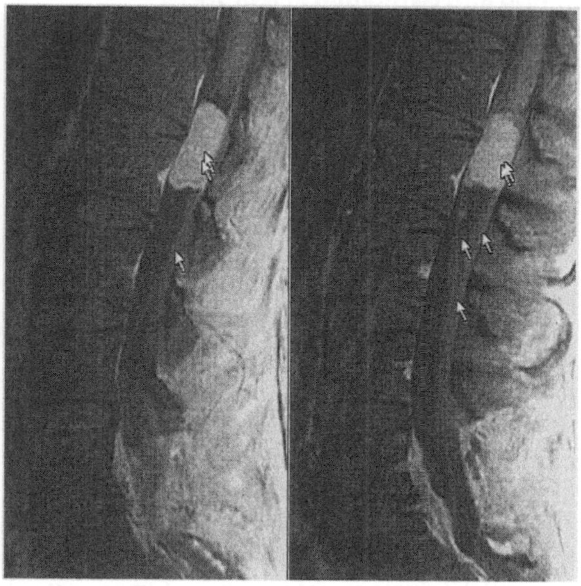

b

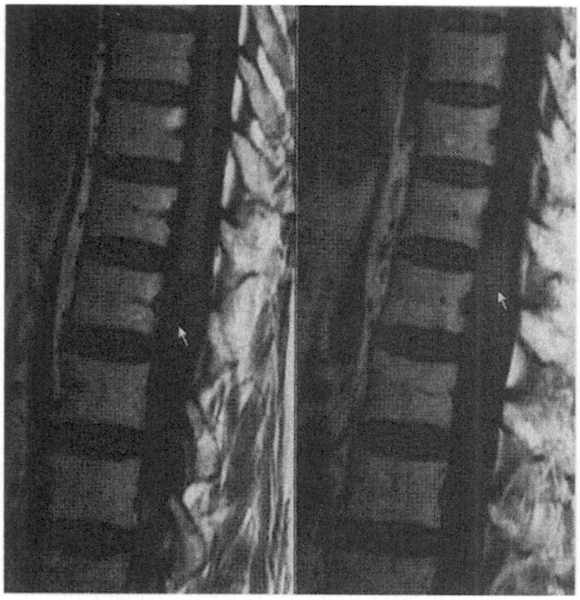

c

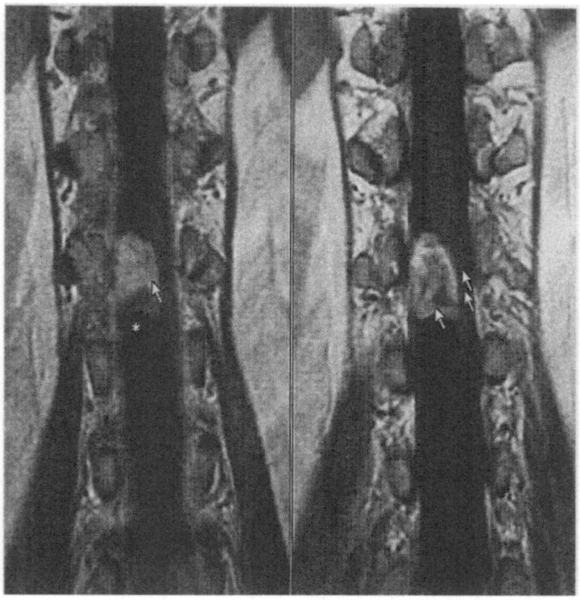

d

Cases 19-22: nerve sheaths tumors

Clinical and pathological features

- Nerve sheaths tumors account for 16 to 30% of all primary intraspinal neoplasms.
- They include schwannomas or neurinomás and neurofibromas.
- Almost 2/3 of theses nerve sheaths tumors are intradural-extramedullary in location. They occur with the same frequency as meningiomas.
- Nerve sheaths tumors are found at all ages, being most common in the 4th to 6th decades, with no sex predilection. In childhood they are much less common (less than 10% of all intraspinal tumors).
- *Histopathology:* nerve sheaths tumors are divided into 3 clinicopathological entities: the schwannomas (WHO grade I), the neurofibromas (WHO grade I) and the malignant form (WHO grade III or IV). **Schwannomas** are benign tumors formed of neoplastic Schwann's cells, giving rise to 2 different patterns: a compact densely cellular Antoni A type, and a loosely arranged cellular tissue, the Antoni B type. Lipid-laden cells may be found in either type. **Neurofibromas** are non encapsulated fusiform tumors comprising a mixture of Schwann's cells and fibroblasts, associated to collagen fibers, and showing an encasement of the nerve fibers. May be solitary but are multiple especially in association with neurofibromatosis type 1. Nerve sheaths tumors usually arise from the dorsal sensory roots. **Malignant degeneration** is reported in 1-2% of the cases and almost 2/3 arise from a neurofibroma and particularly from a plexiform neurofibroma in NF1 patients in the 3rd or 4th decades (younger age group than in sporadic cases). Using immunohistochemistry, the tumor cells express the S-100 protein, more extensively in schwannomas.
- *Signs:* the most common complaints are radicular pain, typically worse at night, which may last more than 2 years before diagnosis. In case of lumbar location, the symptoms may mimic disc herniation. A cauda equina syndrome may be observed. Signs of cord compression may accompany the radicular syndrome. Patients presenting neurofibromas must be investigated for neurofibromatosis type 1 (search for: café-au-lait spots, axillary freckling, iris Lish nodules, sphenoid wing dysplasia, other neurofibromas). Screening of the spine and brain with contrast administration is highly recommended. An increase in CSF protein is usually demonstrated.

MR characteristic findings

Morphology

- Abnormalities on plain films may be found in almost half the cases, comprising: enlargement of a neural foramen or the spinal canal, scalloping of the posterior vertebral wall, or erosion of a pedicle.
- Considering MRI, schwannomas show similar morphological characteristics to neurofibromas.
- Almost 2/3 of nerve sheath tumors are intradural, the remaining develop equally extradurally or both intra- and extradurally (with so-called "dumb-bell" apperance).
- Nerve sheath tumors may be found all along the whole spine, from the cervical region to the sacrum, with a predilection to the posterior and lateral intraspinal compartment, in relation to the dorsal nerve roots. Spinal meningiomas occur most frequently in the thoracic region and mainly in women.
- The majority of these tumors are solitary schwannomas. Multiple tumors (schwannomas and neurofibromas) are commonly found in patients with neurofibromatosis.
- Schwannomas may show more frequently than neurofibromas a heterogeneous aspect due to cystic and xanthomatous changes.
- A "dumb-bell" shape is characteristic of nerve sheath tumors developing intra- and extradurally, most commonly in the thoracic region, with enlargement of the related neural foramen. It is not pathognomonic of such neoplasms.
- Calcifications are rare in nerve sheath tumors as compared to spinal meningiomas.
- Plexiform schwannomas (variant) usually involve the skin or the subcutaneous soft tissue of the head and neck or trunk, and spinal nerves are usually uninvolved.
- Rare melanotic schwannomas of the non psammomatous type, containing melanosomes in the neoplastic Schwann's cells occur in younger

patients, and may show a malignant course in almost 10% of the cases.

- Rare malignant nerve sheath tumors are usually difficult to differentiate from benign lesions but have the propensity to enlarge (more than 5 cm) and become more irregular and inhomogeneous with areas of necrosis. The MR findings are almost similar to sarcomas of the soft tissues.

Signal intensity

- SE-T1w: usually iso-intense or slightly hypointense to spinal cord.
- SE-T2w: hyperintense to spinal cord.
- Schwannomas vs neurofibromas: schwannomas are almost impossible to differentiate from neurofibromas on the basis of SI behavior, though they may show intratumoral areas of cystic formations which evidence higher hyperintense signal on T2 and at time areas of hemorrhages which may show a decrease of the SI. A relative decrease in SI on T2 may be associated in areas of dense cellularity and collagen deposit. Neurofibromas may show an iso-intense SI on T2 mainly in their central portion due to the presence of fibrous material.

Contrast enhancement

- Nerve sheath tumors usually show an intense and almost homogeneous contrast enhancement.
- In some cases, schwannomas may show a heterogeneous or even a peripheral enhancement sparing the central region, mainly in large cystic tumors or in those which show xanthomatous degenerative changes.
- Neurofibromas may also show ring enhancement patterns in case of myxomatous degeneration or intratumoral hemorrhage.

Case 19: Dural sac neurofibroma

Fig. a: sagittal cut (512); SE - T1w + Gd

Fig. b: coronal cut; SE - T2w and coronal cut (512); SE - T1w + Gd

Patient presenting with pain in the hip and groin with nocturnal exacerbations. MR shows an intradural extramedullary round mass at the right lateral aspect of the lumbar canal at the level of L2, slightly heterogeneous on T2w (**b**, left, ↑) enhancing heterogeneously after contrast iv (**a**, **b**, ↑), with central area of hypointensity (**a**, *) non enhancing, that may correspond to a fibrous component. The small well demarcated mass displaces the cauda equina to the left side (**b**, ↑↑) with nerve roots compression. Note the relation of the tumor with the ipsilateral nerve root (**b**, right, *).

Case 20: Dural sac neurofibroma

Fig. c: sagittal cut; SE - T2w

Fig. d: coronal cut; SE - T2w

Patient presenting a radiculalgia referred to rule out a disc herniation. MR disclosed a very small round soft tissue mass developed in the dural space at the level of L2-L3 (**c**, ↑↑) occupying the dural sac and displacing the intradural nerve roots anteriorly as shown on the sagittal cuts (**c**, ↑). The conus (**d**, ↑↑) is slightly displaced to the right. The coronal cuts also showed the emerging root caudally from the mass (**b**, ↑). A neurofibroma was found at surgery, confirmed histologically.

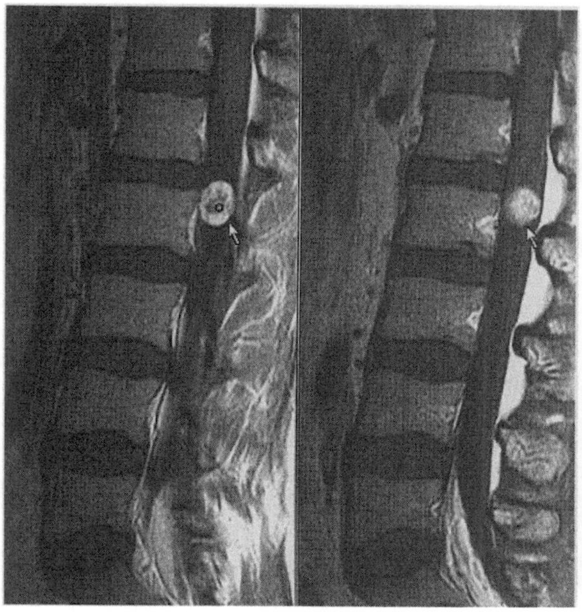

a

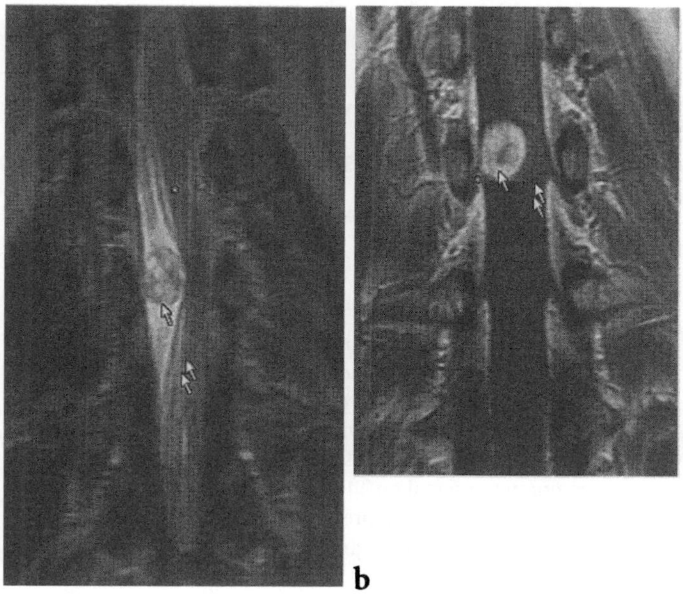

b

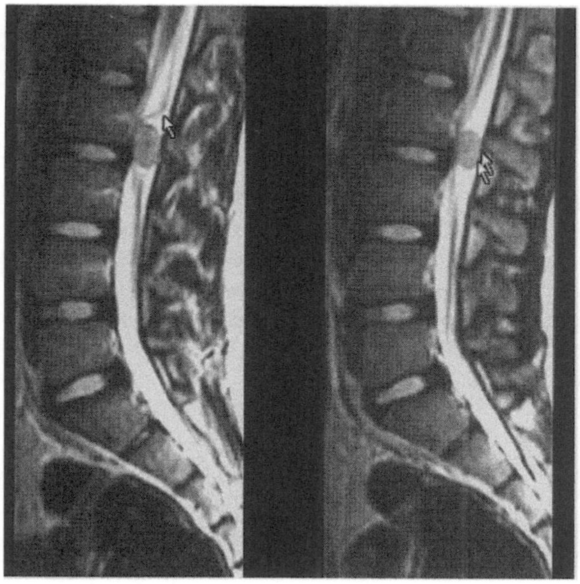

c

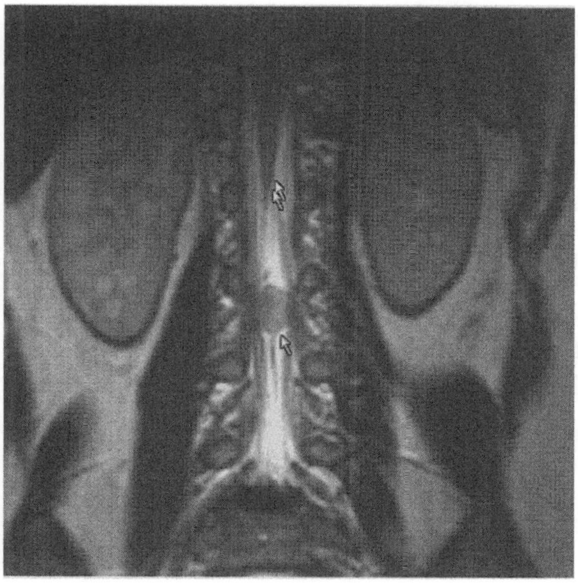

d

Case 21: Dumbbell-shaped cystic dorsal schwannoma

Fig. a: sagittal cut (512); SE - pT1

Fig. b: coronal cut; SE - pT1 + Gd with fat suppression

Patient presenting with intractable left thoracic neuralgia. MR exam shows a well delineated oval mass developed in the thoracic spinal canal posteriorly, showing a large extradural component with increased fat amount at its cranial extremity (**a**, *). The mass shows a homogeneous hypointense signal on T1w (**a**, ↑) approximately iso-intense to CSF. The mass compresses the spinal cord displaced anteriorly and to the right side (**a**; **b**, *). After contrast iv the mass shows peripheral contrast enhancement demonstrating its dumb-bell shape with extension through the left intervertebral foramen (**b**,↑↑) into the paraspinal region. The major portion of the mass did not enhance and remains hypointense due to its cystic nature. Note the enlargement of the corresponding intervertebral foramen (**b**, **).

Case 22: Dumbbell-shaped dorsal neurofibroma

Fig. c: sagittal cut (512); SE - pT1

Fig. d: coronal cut; SE - pT1 + Gd

Patient presenting with a pain in the leg associated with weakness. MR shows a hypointense soft tissue mass obturating the right L3-L4 intervertebral foramen (**c**, ↑) on T1w associated with an enlargement of the neural foramen in question (**c**). The foraminal fat hyperintense on T1w is still recognized at the lower extremity of the interventricular foramen at this level. On T2w sequences, the anterior cuts showed the dumb-bell-shaped mass (**d**, ↑) extending to the right paravertebral region, presenting a slightly hyperintense signal on T2w and displacing more laterally the psoas muscle. There were no areas of intratumoral cystic degeneration and Gadolinium enhancement was noted (not shown).

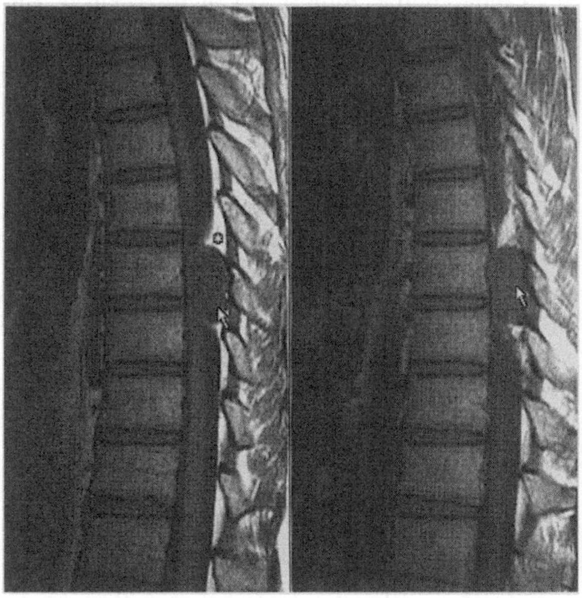

a

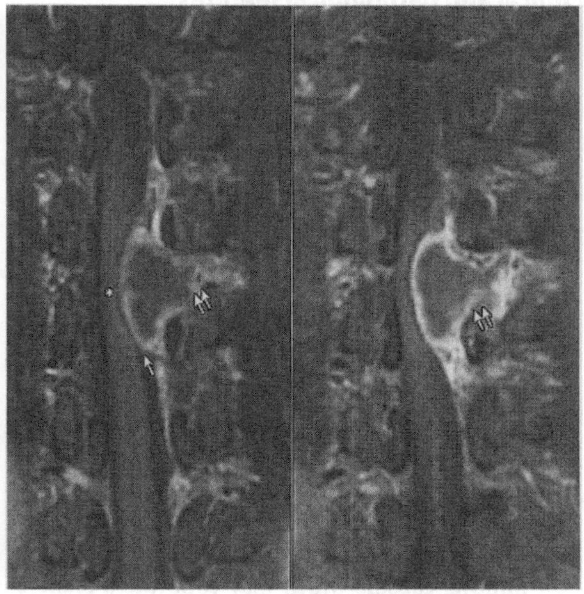

b

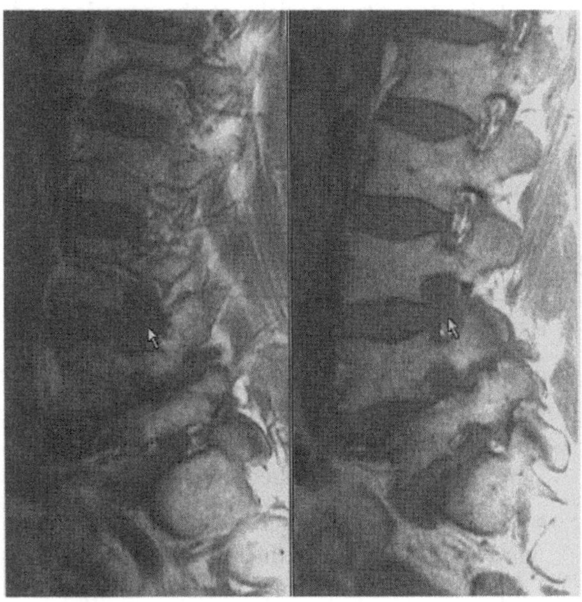

c

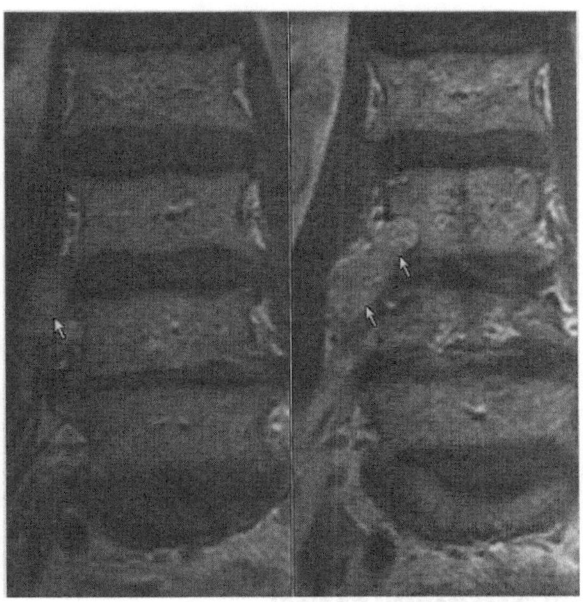

d

Cases 23, 24: spinal ependymomas

Clinical and pathological features

- Ependymomas are neoplasms arising from the ependymal cells lining the central canal of the spinal cord.
- Spinal ependymomas account for almost 1/3 of CNS ependymomas.
- They occur in all age groups with a peak in young adults in the 4th and 5th decades, thus older than with intracranial tumors.
- Spinal ependymomas show a slight male predominance (3:2 male:female ratio).
- *Histopathology:* ependymomas are moderately cellular neoplasms arranged in perivascular pseudorosettes, and which may show areas of hemorrhages, necrosis and myxoid degeneration. Macroscopically ependymomas may show a thin limiting capsule which may contitute a plane of cleavage with the normal cord. Spinal ependymomas may involve several segments with a preferential posterior growth pattern. A characteristic ependymoma variant occurring almost often in the conus and the filum terminale, is the myxopapillary type (WHO grade I).
- *Signs:* symptomatology is commonly non specific. Persistent back pain of ill defined type is the most frequent complaint. Bilateral but usually asymmetric sensory motor deficits, involving the lower limbs particularly, associated with sphincter disturbances are the most common presenting signs at time of diagnosis. In the myxopapillary ependymomas the most frequent clinical signs are back pain and sciatica, mimicking a degenerative disc herniation. Such ependymomas of the filum terminale may present with an associated subarachnoid hemorrhage of acute onset. Spinal ependymomas are the most frequent tumor associated with hydrocephalus due to increase in CSF protein content.

MR characteristic findings

Morphology

- Spinal ependymomas are the most frequent cord tumors arising from the lumbar cord and the conus as well as from the filum terminale and the cauda equina. Almost 1/3 of these are of the myxopapillary histological type.
- Ependymomas arising from remnants of the central canal of the spinal cord tend to develop centrally showing a fusiform expansion of the cord with a segmental extension averaging five vertebral levels. Widening of the spinal canal and the interpedicular distance and scalloping of the vertebral bodies, may be present.
- The tumor is usually well delineated with a thin "pseudocapsule" (surgical resection enabled) and tend to grow posteriorly involving the posterior columns.
- Associated cystic formations are common occurring in almost half the cases. Such benign reactive cysts are found mainly rostrally or caudally to the solid tumor.
- Intratumoral cystic degeneration with necrosis, showing peripheral contrast enhancement are less common than in spinal astrocytomas.
- The propensity to bleed explains the frequent identification of areas of hemorrhage and hemosiderin deposits especially at the superior and inferior borders of the tumor. Ependymomas are more hemorrhagic than spinal cord infiltrating low grade astrocytomas.
- Calcifications are rare in spinal ependymomas in comparison with intracranial.

Signal intensity

- SE-T1w: usually hypointense or iso-intense to spinal cord.
- SE-T2w: usually slightly hyperintense to spinal cord but may be iso- or hypointense in case of intratumoral hemorrhages.
- GE-T2*w: better shows the hemorrhagic changes and more specifically the areas of hemosiderin deposits at the rostral and caudal extremity of the tumor.
- Ependymomas of the filum terminale may show a relatively higher SI to CSF.
- Reactive benign cysts which may be due to focal enlargement of the central canal may show variable SI usually similar to CSF.

Contrast enhancement

- Spinal ependymoma shows intense homogeneous enhancement (virtually all spinal cord gliomas enhance, even low grade tumors). Contrast enhancement delineates most of the solid portion of the tumor distinguished from the eventually associated cyst and the edematous changes, guiding the biopsy or the surgery. Note that the depiction of a densely enhancing mural nodule within a non enhancing cystic formation is more suggestive of a hemangioblastoma (3-10% of intramedullary tumors associated with Von-Hippel Landau in 1/3 of the cases) than ependymoma.
- Intratumoral cystic formations or necrosis show peripheral enhancement.
- Benign reactive cysts do not show peripheral enhancement.
- Contrast infusion helps to identify recurrent ependymomas after surgery and/or radiotherapy and rule out eventual myelomalacia. Enhancing tumor must be distinguished from postoperative changes (follow-up exams delayed after 3 months, showing increase in contrast uptake and/or cord enlargement associated with progressive neurological deficits).

Case 23: Myxopapillary ependymoma of the cauda equina

Fig. a: sagittal cut; SE - T2w and SE - T1w + Gd

Fig. b: axial cut; SE - T2w

Young patient presenting with a history of long standing low back pain extending into the buttocks as well as the legs. MR shows a homogeneous mass filling the inferior portion of the dural sac with the upper extremity reaching the L3-L4 level, and presenting a slightly hypointense signal on T2w as compared to CSF (**a**, left, ↑) with marked enhancement after Gadolinium iv as shown on T1w (**a**, right, ↑). The mass displaces the intradural nerve roots anteriorly as shown on the sagittal cuts (**a**, ↑↑) and on the axial cuts (**b**, up, right, ↑) at its upper extremity. At the level of the mass (**b**, ↑↑) the nerve roots are displaced laterally on both sides (**b**, down, ↑). At surgery the tumor was an intradural ependymoma of the characteristic myxopapillary type.

Case 24: Ependymoma of the cauda equina

Fig. c: sagittal cut (512); SE - T1w

Fig. d: sagittal cut; SE - T2w

Patient presenting an intermittent, long standing low back pain and more recently a cauda equina syndrome. MR exam shows an intradural mass developed in the lower thoracic upper lumbar spine presenting an isosignal to the spinal cord on T1w (**c**, ↑↑) and a heterogeneous hypointense signal on T2w (**d**, ↑↑) with some intratumoral hyperintense areas. Note that the spinal canal is slightly enlarged with scalloping of the posterior walls of the adjacent vertebral bodies (**c**, **d**, *). Surgery disclosed a filum ependymoma which was resected.

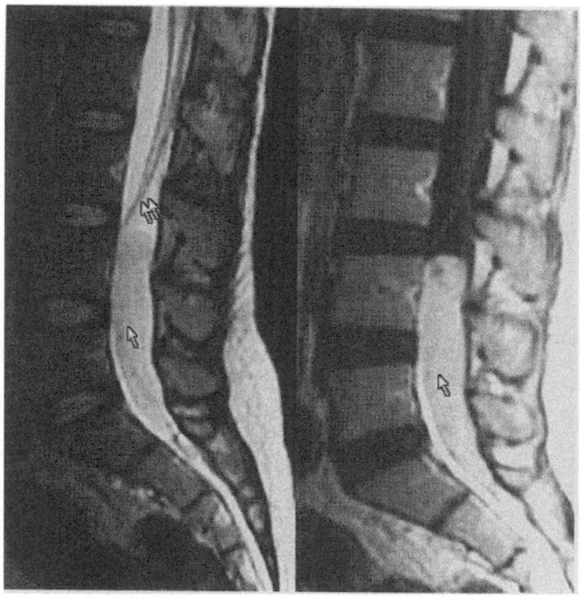

a

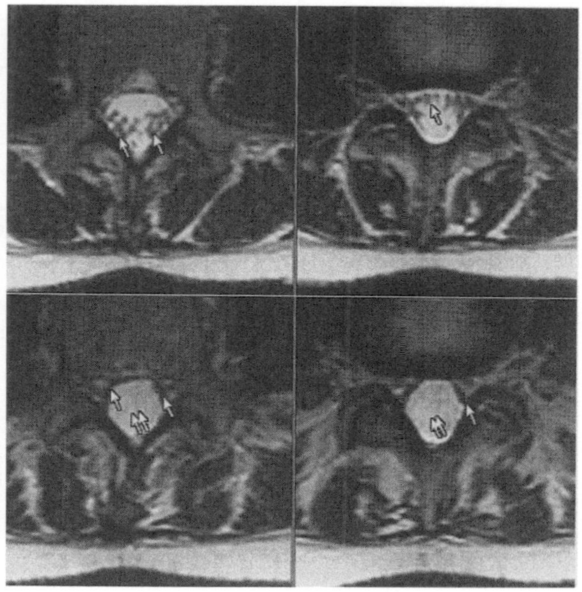

b

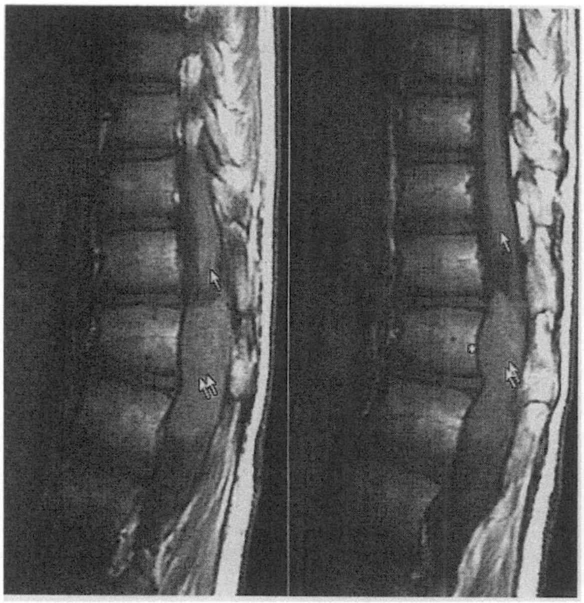

c

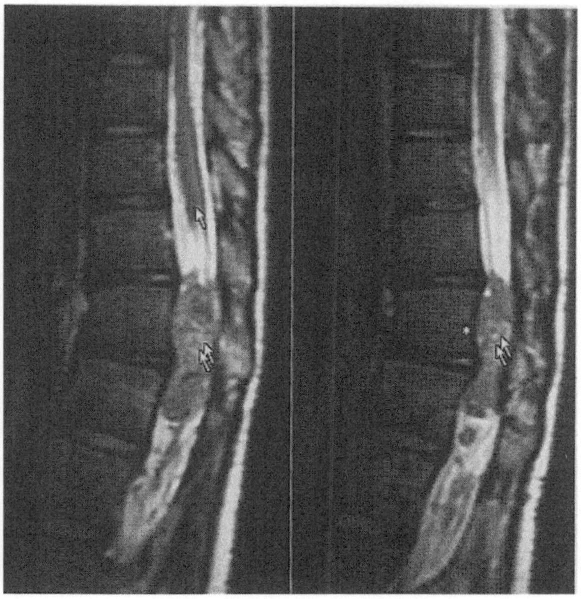

d

Cases 25, 26: leptomeningeal metastases

Clinical and pathological features

- Primary malignant visceral, hematological or CNS tumors may give rise leptomeningeal metastastatic dissemination. Its incidence is increasing with improvement of survival of cancer patients.
- The most common visceral malignant tumors to seed the subarachnoid spaces are: lung, breast, gastric and genitourinary carcinomas and melanomas.
- The malignant hemopathies incriminated are non Hodgkin's lymphomas in adults and leukemias particularly in children.
- Primary CNS tumors such as medulloblastoma, glioblastoma multiformis and high grade astrocytomas, ependymoma, germinoma, pineoblastoma and choroids plexus papilloma, show a propensity to seed the leptomeninges.
- *Signs:* spinal leptomeningeal metastatic dissemination are commonly associated with back pain (about 25% of the cases) which may be of the radicular type and lower limbs paraparesis, sensory disturbances and loss, bladder or bowel dysfunctions. A cauda equina syndrome is elicited in about 30% of the patients. Signs of intracranial meningeal carcinomatosis may be associated as well, with headaches (about 30%), confusional states and cognitive disorders and cranial nerve palsies. CSF study may show abnormal cytological findings in less than 2/3 of the cases.

MR characteristic findings

Morphology

- Most leptomeningeal metastases are found in the thecal sac in the lumbosacral region secondary to gravity effect. In the cervical or even the thoracic regions, the lesions predominate in the posterior leptomeningeal compartment following normal CSF circulation.
- The intradural metastatic lesions may appear as nodular usually multiple, or more diffuse covering more or less smoothly the outer aspect of the cord and/or the cauda equina nerve roots, or may be ill defined filling the almost entire subarachnoid spaces (more common in malignant hemopathies).
- There is no specific pattern of meningeal invasion.

Signal intensity

- SE-T1w: may show mild inhomogeneous hyperintense SI to normal hypointense CSF or a more diffuse iso-intense SI of CSF. The SI appearance is similar to that occurring in case of high protein content. Bright nodular processes may correspond to persistent entrapped lipiodol (previous myelography).
- SE-PDw: may be hyperintense to CSF in the nodular periradicular pattern.
- SE-T2w: nodular lesions are commonly relatively hypointense to CSF.

Contrast enhancement

- Contrast infusion is mandatory to demonstrate leptomeningeal metastases.
- Enhanced MR may show multiple intradural periradicular enhanced nodules, or a perimedullary linear enhancement or a more diffuse subarachnoid enhancement almost filling the entire thecal sac or the perimedullary space.
- Leptomeningeal enhancement may be shown also in other intradural pathological conditions such as: arachnoiditis, pachymeningitis, previous subarachnoid hemorrhage, infectious, parasitic and granulomatous diseases (sarcoidosis and tuberculosis). Rarely, pachymeningitis hypertrophica with dural thickening in the cervico-thoracic region and neurocutaneous melanosis with melanosis of the leptomeninges, may mimic metastatic meningeal seeding. Meningeal thickening may be observed secondary to shunting procedures.

Case 25: Subarachnoid malignant lymphoma

Fig. a: sagittal cut (512); SE - T1w and sagittal cut; SE - T2w

Fig. b: sagittal cut; SE - T1w + Gd

Extensive leptomeningeal disease identified on sagittal T1w sequence (**a**, left) showing an iso-intense signal involving the spinal canal and content and no hyposignal intensity from subarachnoid CSF. On T2w sequences, the perimedullary CSF appears bright and MR discloses a hyperintense intramedullary lesion (**a**, right, ↑) with no enlargement of the cord. After contrast infusion, the T1w sequence shows an extensive homogeneous enhancement of the subarachnoid spaces associated to areas of intramedullary enhancement (**b**, ↑). This corresponds to extensive lymphoma almost filling the entire spinal canal and dural sac with intradural foci.

Case 26: Metastatic seeding of lung carcinoma

Fig. c: sagittal cut; SE - T1w + Gd

Fig. d: axial cut; SE - T1w + Gd

Patient treated for lung carcinoma and developing a cervical painful neuralgia. Contrast enhanced MR images disclosed multiple enhancing lesions involving the cervical spinal canal at the C4-C5 level (**c**, ↑), extending to the left intervertebral foramen as demonstrated on the axial cuts (**d**, ↑↑) following the nerve roots. This focal lesion was associated to linear enhancing lesions along the cervical spinal cord, well shown on its posterior aspect on sagittal cuts (c, ↑↑) but involving also the entire outer aspect of the cord as shown on the axial cuts (**d**, ↑), corresponding to metastatic seeding along the leptomeninges.

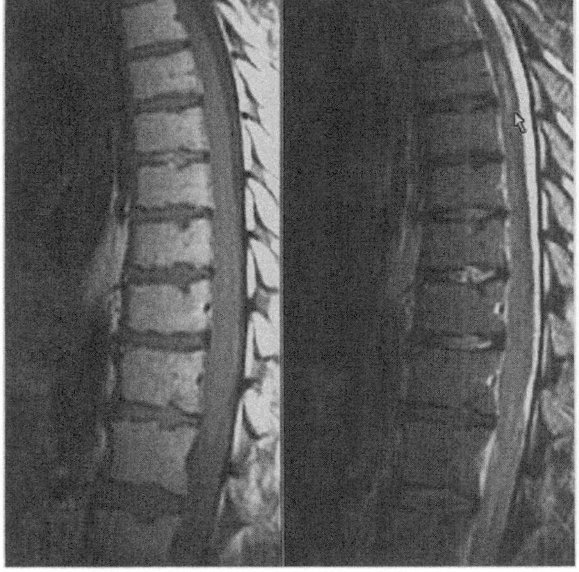

a

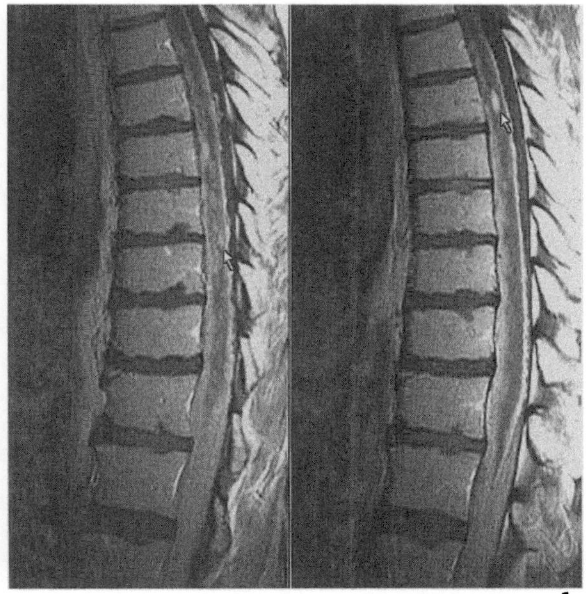

b

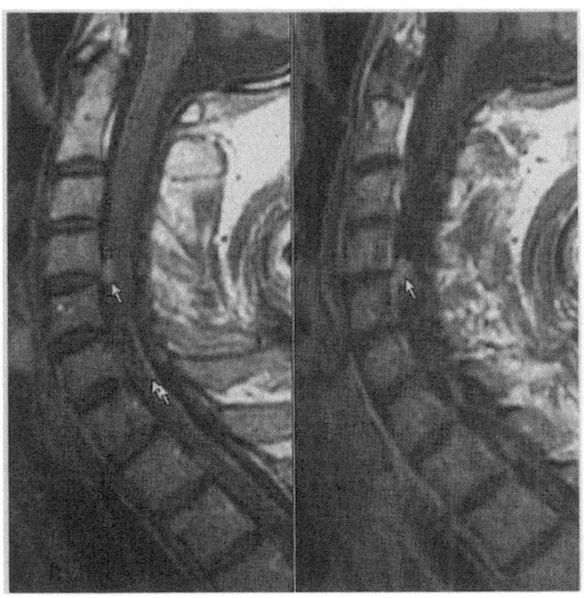

c

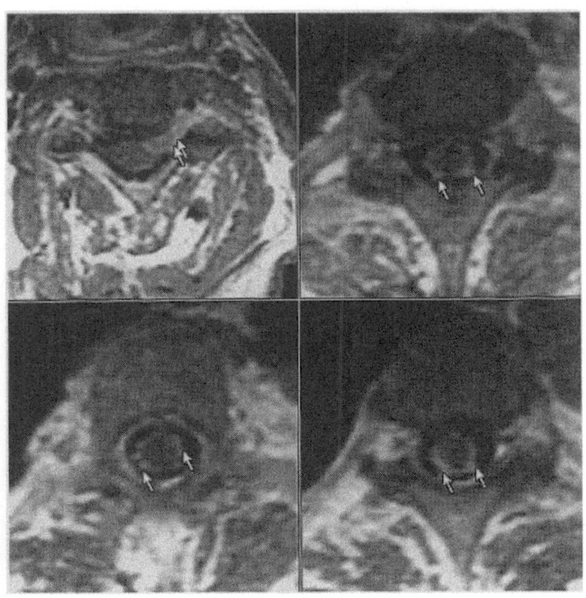

d

Spinal cord tumors

Cases 27, 28: spinal astrocytomas

Clinical and pathological features

- Spinal astrocytomas are the other most common primary tumor of the spinal cord.
- Astrocytomas are less frequent than ependymomas in adults, but are relatively more common in children.
- They occur in the third and fourth decades in adulthood with no sex predilection.
- *Histopathology:* the majority of spinal astrocytomas are low grade (WHO grade II, Kernohan grade I or II). Mixed types are common. High grade tumors may be encountered in about less than 10% of the cases. On histological diagnosis is difficult on biopsy material, the periphery of the tumor being possibly of lower grade. On immunohistochemistry these tumors express the glial fibrillary acidic protein (GFAP).
- *Signs:* the signs and symptoms are similar to the other intramedullary tumors (see ependymoma), including pain as one of the chief complaints associated with sensory and motors disturbances and deficits, and sphincter dysfunctions. Hydrocephalus may be associated with these spinal tumors in almost 1/5 of the cases (see ependymomas). The duration of clinical history prior to diagnosis is commonly long, averaging 3 years in low grade tumors, and as short as a few months in high grades (WHO grades IV or III).

MR characteristic findings

Morphology

- Spinal astrocytomas are frequently difficult to differentiate from spinal ependymomas on MRI.
- On plain films, in almost half the cases, astrocytomas may cause an anterior-posterior enlargement of the spinal canal and/or an increase of the interpedicular distance, with possible bony erosion.
- Astrocytomas are usually infiltrating tumors enlarging the cord usually asymmetrically, with no clear-cut demarcation, but may also appear relatively well delineated.
- They may be eccentric in location and usually posteriorly within the cord.
- They are more frequently found in the thoracic and cervical regions and much less common in the lumbar region or in the conus unlike ependymomas.
- Astrocytomas may be associated with cystic components in almost 1/3 of the cases. Such glial-lined cysts are benign of the reactive type located rostrally or caudally in the central part of the cord. Malignant intratumoral cystic formations may be found as well.
- Astrocytomas with or without extensive associated cysts, may involve almost the entire spinal cord ("holocord astrocytomas").

Signal intensity

- SE-T1w: hypointense to iso-intense to the cord.
- SE-PDw: may help to separate tumor from associated cystic formations; fluid within the cyst may show a lower SI than the solid tumoral component.
- SE-T2w: hyperintense to the cord.

Contrast enhancement

- Unlike intracranial low grade astrocytomas, virtually all spinal astrocytomas, although all of low grade, show contrast enhancement, regardless of grade. Extent of the enhancing zone may extend with delayed acquisitions.
- Intratumoral cysts unlike benign reactive cysts, do show peripheral contrast uptake and helps to their identification (in case of high protein content and/or hemorrhagic material).

Case 27: Low grade fibrillary cervical astrocytoma

Fig. a: sagittal cut (512); SE - T1w

Fig. b: sagittal cut; SE - T2w

Young patient presenting a history of abnormal gait with long tracts signs on the neurological exam. MR shows an important enlargement of the upper cervical cord extending to C4-C5 (**a, b**, ↑) level and involving at its upper extremity the bulbo-medullary junction (**a, b**, ↑↑). The mass is slightly hypointense on T1w (**a**, ↑↑↑) becoming hyperintense on T2w (**b**). Note the slight enlargement of the upper cervical bony canal. Enhanced T1w sequence (**a**) did not show significant tumoral enhancement. Delayed scans have not been performed. Note that even low grade spinal astrocytomas, as in this case, enhance in the vast majority of cases unlike brain gliomas.

Case 28: Cervical exophytic astrocytoma grade IV

Fig. c: sagittal cut; SE - T2w

Fig. d: coronal cut; SE - T2w

Patient previously operated for a cervical tumor. The symptoms presented by the patient were motor and sensory disturbances in the legs associated to bladder and bowel dysfunction. MR showed an enlargement of the cervical cord with decreased signal on T1w and focal enhancement after Gadolinium iv (not shown). A follow-up exam after surgery shows the pseudomeningocele (**c**, *) at the posterior aspect of the spine, developed in the soft tissues and showing a hyperintense signal, CSF-like on T2w. The cervical cord is irregularly enlarged with a bright signal on T2w (**c, d**, ↑↑), extending from the level of C2-C3 to the level of Th2-Th3 (**c**). Note the eccentrically located portion of the tumor (**d**, ↑↑) with an exophytic component to the right side extending from the posterior aspect of the cord.

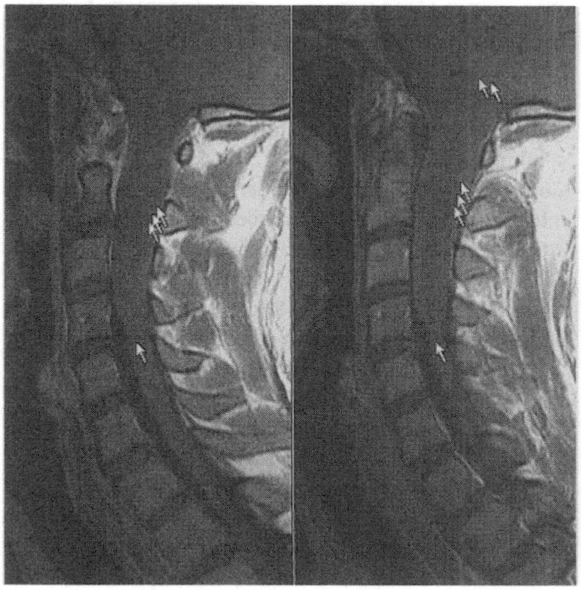

a

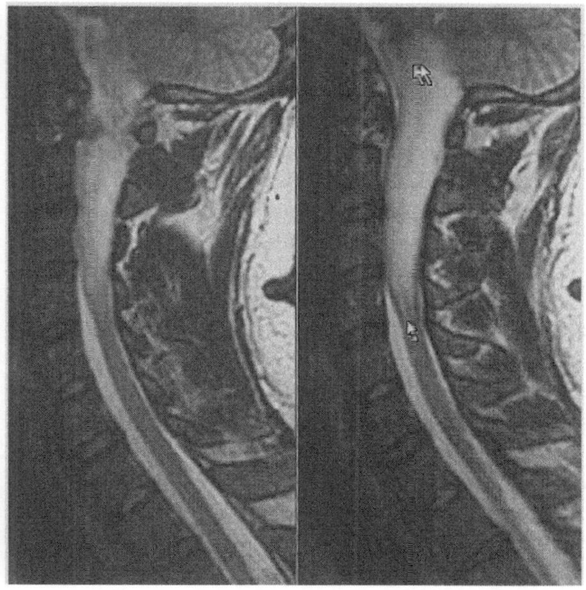

b

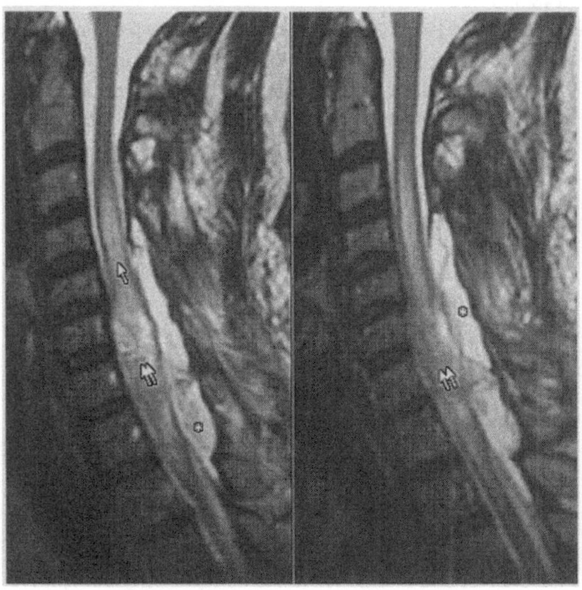

c

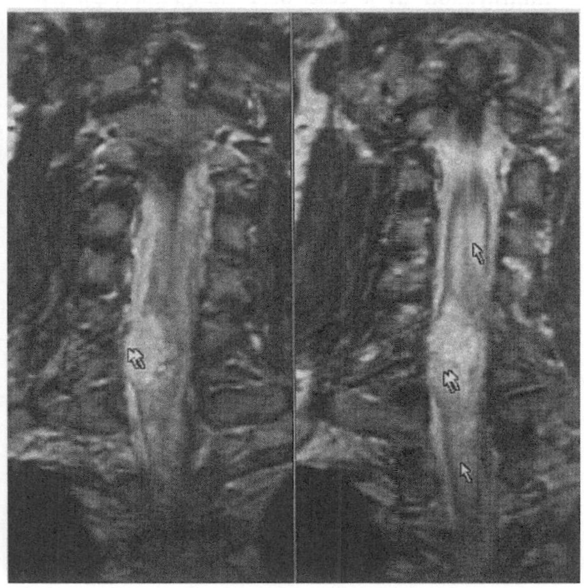

d

Selected References

- Algra PR, Bloem JL, Tissing H et al (1991) Detection of vertebral metastases: comparison between MR imaging and bone scintigraphy. Radiographics 11:219-232
- Ashkenazi E, Costantin S, Pappo O et al (1991) Hypertrophic spinal pachymeningitis: report of two cases and review of the literature. Neurosurgery 28:730-732
- Baker LL, Goodman SB, Perkash I et al (1990) Benign versus pathologic compression fractures of vertebral bodies: assessment with conventional spin-echo, chemical shift, and STIR MR imaging. Radiology 174:495-502
- Barba D, Marshall LF (1990) Intradural tumors. Semin Spine Surg 2:228-242
- Baston OV (1940) The function of the vertebral veins and their role in the spread of metastases. Ann Surg 112:138-148
- Beltran J, Simon DC, Levry ML et al (1986) Aneurysmal bone cysts: MR imaging at 1.5 tesla. Radiology 158:689-690
- Bensaid AH, Dietemann JL, Kastler B et al (1992) Neurofibromatosis with dural ectasia and bilateral symmetrical pedicular clefts; report of two cases. Neuroradiology 34:107-109
- Berger B, Oritz O, Gold A et al (1992) Total cerebrospinal fluid enhancement following intravenous Gd-DTPA adminstration in a case of meningiomatosis. Am J Neuroradiol 13:15
- Blaser SI, Ross JS et al (1988) MR imaging with gadolinium DTPA in leptomeningeal spread of lymphoma. J Comput Assist Tomogr 12(3):499
- Borne J, Daniels DL (1991) Guidelines for differentiating vertebral marrow abnomalities on MRI. MRI Decisions 5(6):2-18
- Bourgouin PM, Tampieri D, Johnston W et al (1992) Multiple occult vascular malformations of the brain and spinal cord. Neuroradiology 34:110-111
- Bradway JK, Pritchard DJ (1990) Ewing's sarcoma of the spine. In Sundaresan N, Schimedek HH, Schiller AL, Rosenthal DI (eds) Tumors of the spine diagnosis and clinical management. WB Saunders, Philadelphia, pp 235-239
- Branwell SL, Dowd CF, Davis RL et al (1990) Cryptic vascular malformations of the spinal cord: diagnosis by magnetic resonance imaging and outcome of surgery. J Neurosurg 72:403-407
- Brotchi J, Dewite O, Levivier M et al (1991) A survey of 65 tumors within the spinal cord: surgical results and the importance of preoperative magnetic resonance imaging. Neurosurgery 129:651-657
- Browne TR, Adams RD, Roberson GH (1976) Hemangioblastoma of the spinal cord, review and report of five cases. Arch Neurol 33:435-441
- Camins MB, Duncan AW, Smith J, Marcove RC (1978) Chondrosarcoma of the spine. Spine 3:202-209
- Cassidy JR, Ducker TB (1991) Intradural neural tumors. In: Frymoyer JW (ed) The adult spine: principles and practice. Raven Press, New York, pp 889-903
- Cheung Y, Fung C, Chan F, Leong L (1991) MRI features of spinal ganglioglioma. Clin Imaging 15:109-112
- Ciapetta P, Salvati M, Capoccia G et al (1991) Spinal glioblastomas: report of seven cases and review of the literature. Neurosurgery 28:302-306
- Coman DR, DeLong RP (1951) The role of the vertebral venous system in the metastasis of cancer to the spinal column. Cancer 4:610-618
- Cooper JR (1989) Outcome after operative treatment of intramedullary spinal cord tumors in adults: intermediate and long-term results in 51 patients. Neurosurgery 25: 885-889
- Cooper PR, Epstein F (1985) Radical resection of intramedullary spinal cord tumors in adults. Recent experience in 29 patients. J Neurosurg 63:492-499
- Cosgrove GR, Bertrand G, Fontaine S et al (1988) Cavernous angiomas of the spinal cord. J Neurosurg 68: 31-36
- Costigan DA, Winkelman MD (1985) Intramedullary spinal cord metastasis. A Clinicopathological study of 13 cases. J Neurosurg 62:227-233
- Cuenold CA, Laredo JD, Chicheportiche V et al (1990) Vertebral collapses: distinction between porotic and malignant causes on MR images before and after Gd-DPTA enhacement. Radiology 177(P):240
- Daffiner RH, Lupetin AR, Dash N et al (1986) MRI in the detection of malignant infiltration of bone marrow. AJR 146:353-356
- Dahin DC (1978) Bone tumors: general aspects and data on 6221 cases. 3rd ed, Charles C Thomas, Springfield (IL)
- Dahlin DC (1977) Giant cell tumor of vertebrae above the sacrum: a review of 31 cases. Cancer 39:1350-1356
- Dahlin DC (1985) Giant cell tumor of bone: highlights of 407 cases. AJR 144:955-960
- Dahlin DC, McLeod RA (1982) Aneurysmal bone cyst and other non-neoplastic conditions. Skeletal Radiol 8:243
- De Brune FT, Kroom HM (1988) Spinal chordoma: radiologic features in 14 cases. AJR 150:861-863
- Devita VT, Hellman S, Jaffes ES, Longo DL (1993) Hodgkin's disease and lymphomas In: devita VT, Hellman S, Rosénberg SA (eds) Cancer: principles in practice of oncology, 4th ed, JB Lippincott, Philadelphia, pp 1819-1927
- Dillon WP, Norman D, Newton TH et al (1989) Intradural spinal cord lesions; Gd-DPTA-enhanced. Radiology 170:229-237
- Dillon WP, Norman D, Newton TM, Bolla K, Mark A (1989) Intradural spinal cord lesions gadolinium-DTPA-enhanced MR imaging. Radiology 170(1):229-239
- Djindjian M, Nguyen JP, Gaston A et al (1992) Multiple vertebral hemangiomas with neurological signs: case report. J Neurosurg 76:1025-1028
- Domingues RC, Mikulis D, Swearinging B et al (1991) Subcutaneous sacrococcygeal myxopapillary ependymona: CT and MR findings. AJNR 12:171-172

- Dunn RC, Kelly WA, Wohns RNW, Howe JF (1980) Spinal epidural neoplasia: a 15 year review of the results of surgical therapy. J Neurosurg 52:47-51
- Egelhoff JC, Bates DJ, Ross JS et al (1992) Spinal MR findings in neurofibromatosis types 1 and 2. AJNR 13:1071-1077
- Elster AD (1992) Radiologic screening in the neurocutaneous syndromes: strategies and controversies. AJNR 13: 1078-1082
- Enomoto H, Goto H (1991) Spinal epidural cavernous angioma: MRI finding. Neuroradiology 33:462
- Fessler RG, Johnson DK, Brown FD et al (1992) Epidural lipomatosis in steroid-treated patients. Spine 17:183-188
- Fincher EF (1951) Spontaneous subarachnoid hemorrhage in intradural tumors of the lumbar sac. A clinical syndrome. J Neurosurg 8:576-584
- Firoozina H, Pinto RS, Lin JP et al (1976) Chordoma: radiologic evaluation of 20 cases. AJR 127:797-805
- Fontaine S, Melanson D, Cosgrove R, Bertrand G (1988) Cavernous hemangiomas of the spinal cord: MR imaging. Radiology 166:839-841
- Fornasier VL, Czitrom AA (1978) Collapsed vertebrae: a review of 659 autopsies. Clin Orthop 131:261-265
- Fortuna A, Celli P, Palma L (1980) Oligodendrogliomas of the spinal cord. Acta Neurochir 52:305-329
- Frank JA, Ling A, Patronas NJ et al (1990) Detection of malignant bone tumors: MR imaging vs scintigraphy. AJR 155:1043-1048
- Gilbert RW, Kim J-H, Posner JB (1978) Epidural spinal compression from metastatic tumor. Diagnosis and treatment. Ann Neurol 3:40-51
- Gindre-Barrucand T, Charleux F, Turjman F et al (1991) Magnetic resonance imaging contribution to the diagnosis of spinal cord compression by a subdural arachnoid cyst. Neuroradiology 33:87
- Giuffre R (1966) Intradural spinal lipomas: review of the literature (99 cases) and report of an additional case. Acta Neurochir 14:69-95
- Gottieb A, Severi P, Ruelle A, Lasio G (1986) Exostosis as a cause of spinal cord compression. Surg Neurol 26:581-584
- Gouliamos AD, Kontogiannis DS, Androulidakis J et al (1993) Spinal neurilemmomas and neurofibromas: central dot sign in postgadolinium MRI. J Comput Assist Tomogr 17:446
- Grant JW, Kaech D, Jones DB (1986) Spinal cord compression as the first presentation of lymphoma- a review of 15 cases. Histopathology 10(11):1191-1202
- Grem JL, Burgess J, Trump DL (1985) Clinical features and natural history of intramedullary spinal cord metastases. Cancer 56:2305-2314
- Guideti B, Mercuri, Vagnozzi R (1981) Long-term results of the surgical treatment of 129 intramedullary spinal gliomas. J Neurosurg 54:323-330
- Guidetti B, Fortuna A (1967) Surgical treatment of intramedullary hemangioblastomas of the spinal cord. Report of six cases. J Neurosurg 27:530-540
- Guthrie BL, Ebersold MJ, Scheithauer BW et al (1989) Meningeal hemangiopericytoma: histopathological features, treatment, and long-term follow-up of 44 cases. Neurosurgery 25:514
- Haddad FS, Abla A, Allam CK (1986) Extradural spinal angiolipoma. Surg Neurol; 26:473-486
- Haddad SF, Hitchon PW, Godersky JC (1991) Idiopathic and glucocorticoid-induced spinal epidural lipomatosis. J Neurosurg 74:38-42
- Haimes AB, Krol G (1991) Dumb-bell-shaped spinal cavernous hemangioma: a case report. AJNR 12:1021-1022
- Halliday AL, Sobel RA, Martuza RL (1991) Benign spinal nerve sheath tumors: their occurrence sporadically and in neurofibromatosis types 1 and 2. J Neurosurg 74:248-253
- Harkin JC (1986) Pathology of nerve sheath tumors. Ann NY Acad Sci 486:147-154
- Hautzer NW, Aiyesimoju A, Robitaille Y (1983) "Primary" spinal intramedullary lymphomas: a review. Ann Neurol 14:6-66
- Heiser S, Schwartzman JJ (1952) Variations in the Roentgen appearance of the skeletal system in myeloma. Radiology 58:178-191
- Hekster REM, Endtz LJ (1987) Spinal-cord compression caused by vertebral haemangioma relieved by percutaneous catheter embolization: 15 year later. Neuroradiology 29:101
- Higinbotham NL, Philips RF, Farr HW, Hustu HO (1967) Chordoma: thirty-five year study at Memorial Hospital. Cancer 20:1841-1850
- Hirsch LF, Thanki A, Spector HB (1984) Primary spinal chondrosarcoma with eighteen-year follow-case report and literature review. Neurosurgery 14:747-749
- Horner NB, Pinto RS (1989) The fat-cap sign: an aid to MR evaluation of extradural spinal tumors. AJNR 10:S93
- Horowitz S, Fine M, Azar-Kia B (1990) Differentiation of benign versus metastatic vertebral body compression fractures. Radiology 177(P):370
- Hu HP, Huang QL (1992) Signal intensity correlation of MRI with pathological findings in spinal neurinomas. Neuroradiology 34:98
- Huvos AG (1991) Bone tumors: diagnosis, treatment, and prognosis. 2nd ed, WB Saunders, Philadelphia, pp 653-676
- Issu T, Iwasaki Y, Akino M et al (1989) Mobile schwannoma of the cauda equina diagnosed by magnetic resonance imaging. Neurosurgery 25:968-971
- Jaffe HL, Lichtenstein L (1942) Solitary unicameral bone cyst with emphasis on the Roentgen picture, the pathologic appareance, and the pathogenesis. Arch Surg 44:1004-1025
- Kalyanarman UP, Heuderson JP (1982) Intramedullary ganglioneuroma of signal cord: a clinicopathologic study. Hum Pathol 13:952
- Kamholtz R, Sze G (1990) MRI of spinal metastases. MRI Decisions 4(6):2-15
- Kaplan RD, Coons S, Drayer BP et al (1992) MR characteristics of meningioma subtypes at 1.5 tesla. J Comput Assist Tomogr 16:366-371
- Katoh S, Ikata T, Inoue A et al (1995) Intradural extramedullary ependymoma: a case report. Spine 20:2036

- Klein SL, Sanford RA, Muhlbauer MS (1991) Pediatric spinal epidural metastases. J Neurosurg 74:70-75
- Kochi M, Takada A, Mihara Y et al (1991) MRI of subarachnoid dissemination of medulloblastoma. Neuroradiology 33:264
- Krol G, Sze G, Malkin M, Walker R (1988) MR of cranial and spinal meningeal carcinomatosis: comparaison with CT and myelography. AJNR 9:709-714
- Kudo H, Tamaki N, Kim S et al (1987) Intraspinal tumors associated with hydrocephalus. Neurosurgery 21:726-731
- Kuroda S, Abe H, Akino M et al (1990) Infiltrating spinal angiolipoma causing myelopathy: case report. Neurosurgery 27:315-318
- Lantos G, Epstein F, Kory LA (1987) Magnetic resonance imaging of intradural spinal lipoma. Neurosurgery 20:469-472
- Lanzieri CF, Sacher M, Solodnik P et al (1987) Unusual patterns of solitary sacral plasmacytoma. AJNR 8:556-567
- Lapointe JS, Graeb DA, Nugent RA, Robertson WD (1985) Value of intravenous contrast enhancement in the CT evaluation of intraspinal tumors. AJNR 6:939-943
- Laredo JD, Assouline E, Gelbert F et al (1990) Vertebral hemangiomas: fat content as a sign of aggressiveness. Radiology, 177:467-472
- Laredo JD, Reizine D, Bard M et al (1986) Vertebral hemangiomas: radiologic evaluation. Radiology 161:183
- Leaney BJ, Rowe PW, Klug GL (1985) Neurocutaneous melanosis with hydrocephalus and syringomyelia. J Neurosurg 62:148-152
- Lee M, Rezai AR, Abbot R et al (1995) Intramedullary spinal cord lipomas. J Neuroradiol 82:394
- Levine AM (1990) Operative techniques for treatment of metastatic disease of the spine. Semin Spine Surg 2:210-227
- Levy WJ, Bay J, Dohn D (1982) Spinal cord meningioma. J Neurosurg 57:804-812
- Levy WJ, Latchaw J, Hahn JF et al (1986) Spinal neurofibromas: a report of 66 cases and a comparison with meningiomas. Neurosurgery 18:331-334
- Li KC, Poon PY (1988) Sensitivity and specificity of MRI detecting malignant spinal cord compression and in distinguishing malignant from benign compression fractures of vertebrae. Magn Reson Imaging 6:547-556
- Li MH, Holtas S (1991) MR imaging of spinal intramedullary tumors. Acta Radiol 32:505-513
- Li MH, Holtas S (1991) MR imaging of spinal neurofibromatosis. Acta Radiol 32:279-285
- Lipson SJ, Neheedy MH, Kaplan MM, Bienfang DC (1980) Spinal stenosis caused by epidural lipomatosis in Cushing's syndrome. N Engl J Med 302:36
- Lossos A, Siegal T (1992) Spinal subarachnoid hemorrhage associated with leptomeningeal metastases. J Neurooncol 12:167-171
- Ludwing H, Fruhwald F, Tscholakoff D et al (1987) Magnetic resonance of the spine in multiple myeloma. Lancet 15:364-366
- Mascalchi M, Arnetoli G, Pozzo GD et al (1991) MR of spinal epidural angiolipoma. AJNR 1:744-745
- Mascalchi M, Torselli P, Falaschi F et al (1995) MRI of spinal epidural lymphoma. Neuroradiology 37:303
- Matsuhima K, Shinohara Y, Yamamoto M et al (1987) Spinal extradural angiolipoma: MR and CT diagnosis. J Comput Assist Tomogr 11:1104-1106
- McCormick PC, Michelsen WJ, Post KD et al (1988) Cavernous malformations of the spinal cord. Neurosurgery 23:459-463
- McCormick PC, Por KD, Stein BM (1990) Intradural extramedullary tumors in adults. Neurosurg Clin North Am 1:591-608
- McCormick PC, Stein BM (1990) Intramedullary tumors in adults. Neurosurg Clin North Am 1:609-630
- McCormick PC, Torres R, Post KD, Stein BM (1990) Intramedullary ependymoma of the spinal cord. J Neurosurg 72:523-532
- McLone DG, Naidich TP (1986) Laser resection of fifty spinal lipomas. Neurosurgery 18:611-615
- Meyer FB, Ebersold MJ, Reese DF (1984) Benign tumors of the foramen magnum. J Neurosurg 61:136-142
- Modic MT, Steinberg PM, Ross JS et al (1988) Degenerative disk disease: assessment of changes in vertebral body marrow with MR imaging. Radiology 166:193-199
- Mohan V, Gupta SK, Tuli SM et al (1980) Symptomatic vertebral hemangiomas. Clin Radiol 31:575
- Mork SJ, Loken AC (1977) Ependymoma. A follow-up study of 101 cases. Cancer 40:907-915
- Moulopoulos LA, Dimopoulos MA, Alexanian R et al (1994) Multiple myeloma: MR pattern of response to treatment. Radiology 193:441
- Moulopoulos LA, Varma DGK, Dimopoulos MA et al (1992) Multiple myeloma: spinal MR imaging in patients with untreated newly diagnosed disease. Radiology 185:833-840
- Munk PL, Helms CA, Holt RG et al (1989) MR imaging of aneurysmal bone cysts. AJR 153:99-101
- Murota T, Symon L (1989) Surgical management of hemangioblastoma of the spinal cord: a report of 18 cases. Neurosurgery 25:699-708
- Nabors MW, Pait TG, Byrd EB et al (1988) Updated assessment and current classification of spinal meningeal cysts. J Neurosurg 68:366
- Nanda A, Hirsh LF, Antoiniades K (1991) Malignant fibrous histiocytoma in a recurrent thoracic chordoma: case report and literature review. Neurosurgery 28:588-592
- Nemoto Y, Inoue Y, Tashiro T et al (1992) Intramedullary spinal cord tumors: significance of associated hemorrhage at MR imaging. Radiology 182:793-796
- Nobler MP, Higinbotham NL, Philips RF (1968) The cure of aneurysmal bone cyst: irradiation superior to surgery in an analysis of 33 cases. Radiology 90:1185-1192
- Palkovic S, Wassmann H, Bone R, Kashab M (1988) Angiolipoma of the spinal cord: magnetic resonance imaging and microsurgical management. Surg Neurol 29:243-245
- Parizel PM, Balériaux D, Rodesch F et al (1989) Gd-DPTA-enhanced MR imaging of spinal tumors. AJNR 10:249-258

- Patel U, Pinto RS, Miller DC, Handler MS, Rorke LB, Epstein FJ, Kricheff II (1998) MR of spinal cord ganglioglioma. AJNR 19:879-887
- Raffel C, Neave VCD, Lavine S, McComb JG (1991) Treatment of spinal cord compression by epidural malignancy in childhood. Neurosurgery 28:349-352
- Reimer R, Onofrio BM (1985) Astrocytomas of the spinal cord in children and adolescents. J Neurosurg 63:669-675
- Resenfeld JV, Kaye AH, Davis S, Gonzales M (1987) Pachymeningitis cervicalis hypertrophica. Case report. J Neurosurg 66:137-139
- Resnick D, Niwayama G, Guerra J et al (1981) Spinal Vacuum phenomena: anatomical study and review. Radiology 139:341-348
- Rhodes RE, Friedman HS, Hatten HP Jr et al (1991) Contrast-enhanced MR imaging of neurocutaneous melanosis. AJNR 12:380-382
- Rich TA, Schiller A, Suit HD, Mankin HJ (1985) Clinical and pathologic review of 48 cases of chordoma. Cancer 56:182-187
- Roberts MC, Kressel HY, Fallon MD et al (1989) Paget's disease: MR imaging findings. Radiology 173:341-345
- Roda JM, Gutierrez-Molina M (1995) Multiple intraspinal low-grade astrocytomas mixed with lipoma (astrolipoma): case report. J Neurosurg 82:891
- Rodesch G, Van Bogaert P, Mavroudakis N et al (1990) Neuroradiologic findings in leptomeningeal carcinomatosis: the value interest of gadolinium-enhanced MRI. Neuroradiology 32:26-32
- Roscoe MWA, Barrington TW (1984) Acute spinal subdural hematoma. A case report and review of the literature. Spine 9:672-675
- Ross DA, Edward MSB, Wilson CB (1986) Intramedullary neurilemomas of the spinal cord: report of two cases and review of the literature. Neurosurgery 19:458-464
- Ross JJ, Masaryk TJ, Modic MT et al (1987) Vertebral hemangiomas: MR imaging. Radiology 65:165-169
- Roux A, Mercier C, Larbrisseau A et al (1992) Intramedullary epidermoid cysts of the spinal cord. J Neurosurg 76:528-533
- Roy-Camille R, Mazel CH, Husson JL, Saillant G (1991) Symptomatic spinal epidural lipomatosis induced by long term steroid treatment. Review of the literature and report of two additional cases. Spine 16:1365-1371
- Russo CP, Katz DS, Corona RJ et al (1995) Gangliocytoma of the cervicothoracic spinal cord. Am J Neuroradiol 16:889
- Salvati M, Artico M, Lumardi P, Gagliardi FM (1992) Intramedullary meningioma: case report and review of the literature. Surg Neurol 37:42-45
- Sato Y, Waziri M, Smith W et al.(1988) Hippel-Lindau disease MR Imaging 166:241-246
- Schellinger D, LeBihan D, Rajan SS et al (1992) MR of slow CSF flow in the spine. Am J Neuroradiol 13:1393
- Schroth G, Thron A, Guhl L et al (1987) Magnetic resonance imaging of spinal meningiomas and neurinomas. Improvement of imaging by paramagnetic contrast enhancement. J Neurosurg 66:695-700
- Scotti G, Sciafia G, Colombo N, Landoni L (1987) Magnetic resonance diagnosis of intramedullary tumors of the spinal cord. Neuroradiology 29:130-135
- Sibila L, Martelli A, Farina L et al (1995) Ganglioneuroblastoma of the spinal cord. Am J Neuroradiol 16:875
- Siegal T (1989) Current considerations in the management of neoplasic spinal cord compression. Spine 14:223-228
- Sim FH, Frassica FJ, Wold Le, McLeod RA (1990) Chondrosarcoma of the spine: Mayo clinic experience. In: Sundaresan N, Schmidek HH, Schiller AL, Rosenthal DI (eds) Tumors of the spine: diagnosis and clinical management. WB Saunders, Philadelphia, pp 155-162
- Simoene FA (1990) Spinal cord tumors in adults. In : Youmans JR (ed) Neurological surgery 3rd ed, WB Saunders, Philadelphia, pp 3531-3547
- Slasky BS, Bydder GM, Niendorf HP, Young IR (1987) MR imaging with gadolinium-DTPA in the differentiation of tumor, syrinx, and cyst of the spinal cord. J Comput Assist Tomogr 11:845-850
- Smith, Ludwing RL, Marcove RC (1987) Sacrococcygeal chordoma: a clinicoradiological study of 60 patients. Skeletal Radiol 16:37-44
- Solero CL, Fornari M, Giombini S et al (1989) Spinal meningiomas review of 174 operated cases. Neurosurgery 25:153-160
- Sonneland PRL, Scheithauer BW, Onofrio BM (1985) Myxopapillary ependydoma. A clinicopathologic and immunocytochemical study of 77 cases. Cancer 56:883-893
- Sze G (1988) Gadolinium-DPTA in spinal disease. Radiol Clin North Am 26(5):1009-1024
- Sze G, Abramson A et al (1988) Gadolinium-DTPA in the evaluation of intradural extramedullary spinal disease. Am J Radiol 150(4):911-921
- Sze G, Krol G, Zimmerman RD, Deck MDF (1988) Malignant extradural spinal tumors: MR imaging with Gd-DTPA. Radiology 167:217-223
- Sze G, Krol G, Zimmerman RD, Deck MDF (1988) Intramedullary disease of the spine: diagnosis using gadolinium-DTPA-enhanced MR imaging. Am J Radiol 151(6): 1193-1204
- Sze G, Krol G, Zimmerman RD, Deck MDF (1988) Intramedullary disease of the spine: diagnosis using gadolinium-DTPA-enhanced MR imaging. AJNR 9:847-858
- Sze G, Krol G, Zimmerman RD, Deck MDF (1988) Malignant extradural spinal tumors: MR imaging with Gd-DPTA. Radiology 167:217-223
- Takemoto K, Matsumura Y, Hashimoto H et al (1988) MR imaging of intraspinal tumors-capability in histological differentiation and compartmentalization of extramedullary tumors. Neuroradiology 30:303-309
- Tan SB, Kozak JA, Mawad ME (1991) The limitations of magnetic resonance imaging in the diagnosis of pathologic vertebral fractures. Spine 16:919-923
- Tognetti F, Lanzino G, Calbucci F (1988) Metastases of the spinal cord from remote neoplasms. Study of five cases. Surg Neurol 30:220-227

- Valk J (1988) Gd-DTPA in MR of spinal lesions. AJNR 9: 345-350
- Vaquero J, Martinez P (1988) Cavernous of the spinal cord: report of two cases. Neurosurgery 22:143-144
- Volger JB III, Murphy WA (1988) Bone marrow imaging. Radiology 168:679-693
- Von Hanwehr R, Apuzzo MLJ, Ahmadi J, Chandrasoma P (1985) Thoracic spinal angiomyolipoma: case report and literature review. Neurosurgery 16:406-411
- Wagle VG, Villemure JG, Melanson D et al (1987) Diagnostic potential of magnetic resonance in cases of foramen magnum meningiomas. Neurosurgery 21:622-626
- Weiss SW, Enzinger FM (1978) Malignant fibrous histiocytoma: an analysis of 200 cases. Cancer 41:2250-2266
- Williams AL, Haughton VM, Pojunas KW et al (1987) Differentiation of intramedullary neoplasms and cysts by MR. AJR 149:159-164
- Yasargil MG, Antic J, Laciga R et al (1976) The microsurgical removal of intramedullary spinal hemangioblastomas. Report of twelve cases and a review of the literature. Surg Neurol 6:141-148
- Yousem DM, Patrone PM, Grossman RI (1990) Leptomeningeal metastases: MR evaluation J Comput Assist Tomogr 14(2):255
- Yuh WTC, Flickinger FW, Barloon TJ, Montgomery WJ (1988) MR imaging of unusual chordomas. J Comput Assist Tomogr 12:30-35
- Yuh WTC, Zachar CK, Barloon TJ et al (1989) Vertebral compression fractures: distinction between benign and malignant causes with MR imaging. Radiology 172:215-218
- Zentner J, Hassler W, Gawehn J, Schroth G (1989) Intramedullary cavernous angiomas. Surg Neurol 31:64-68
- Zimmerman RA, Bilaniuk LT (1988) Imaging of tumors of the spinal canal and cord. Radiol Clin North Am 26(5): 965-1007

Part III

Spine and cord

2- Degenerative diseases of the spine

a- Cervical spine
b- Thoracic spine
c- Lumbar spine

Cervical spine

Cases 29-34: cervical disc herniation

Clinical and pathological features

- Cervical degenerative disease is a common indication for MRI.

- Cervical spine spondylosis is commonly found in more than half to two thirds of the adults over 50 years old, involving more specifically the most mobile C5-C6 and C6-C7 vertebral levels. Only a small number of these patients will come to clinical attention.

- Intervertebral disc herniations occur most commonly in early adulthood with a male predominance, found in the 4th decade, and involving usually the same most mobile levels.

- *Signs:* neck pain with or without brachial neuralgia due to a cervical radiculopathy are the most common long standing complaints indicating the MR exam. Herniated disc may be found in asymptomatic patients (20% of the population). More than 1/3 of soft disc herniations are consecutive to an automobile trauma. Associated soft to hard disc herniations are more commonly found in elderly patients in their sixth decade. Herniated disc involving two levels are uncommonly found. Cervical root compression gives rise to segmental neurological sensory-motor deficits with hyporeflexia, involving the related dermatome. An isolated radiculopathy is usally secondary to a lateral paracentral or foraminal disc herniation. Compression of C6 root (1/4 of the cases) induces pain and sensory paresthesias in the external surface of the forearm and arm with depression of the biceps reflex. Compression of C7 (3/5 of the cases) involves the forearm and middle finger with depression of the triceps reflex. In compression of C8 root the neurologic signs are located in the medial aspect of the hand (5th digit) extending to the forearm and arm. In C5 root compression, the distribution of the sensory changes involve the shoulder and the lateral aspect of the arm with possible depression of the biceps and the brachioradialis reflexes.

MR characteristic findings

Morphology

- Spondylosis is a common degenerative process affecting the spine and including narrowing of the disc space with formation of osteophytic spurs and ridges affecting preferentially the mobile vertebral levels.
- Soft disc herniation with extrusion of the nucleus pulposus through the annulus fibrosus occurs much less frequently in the cervical than in the lumbar spine, although midline disc herniations are more frequent in the cervical region as compared with the lumbar (case 34).
- Lateral (case 31) and foraminal (case 32) herniated discs and foraminal are relatively less common. Lateral herniated discs may show an upward or even downward migration (parasagittal cuts).
- Most of the cervical herniated discs extrude through the annulus, and displace posteriorly the posterior longitudinal ligament. Hernia impingement of the dural sac (axial cuts) may cause a variable compression of the cord, ranging from mild deformation of its anterior surface without associated edema, to severe compression and flattening of the involved cord segment with edematous changes. A mild cord compression with a reduction cord area of less than 10-15% may be found in asymptomatic patients.
- A large herniated disc may extrude through the annulus and further through a defect in the posterior longitudinal ligament showing a "double fragment sign" (case 33). The dark line bounding posteriorly the extruded disc material corresponds to the displaced posterior longitudinal ligament and dura (best seen using GE sequence in the axial plane). The interrupted inner dark line separating the disc material into two superimposed fragments corresponds to the torn annulus.
- Even more frequent than cervical disc herniations are the degenerative hypertrophy of the uncinate processes, causing a narrowing of the neural foramina (obliteration of the epidural fat

within the neural foramen as shown on MR in the axial or oblique parasagittal planes) with nerve root entrapment and radiculopathy (case 30). Narrowing of the neural foramen may even be secondary to degenerative facet joint hypertrophy or osteophytic bone production. Neural foraminal narrowing is more frequent at the entrance of the neural foramen, and the osteophytes are found more commonly in the upper part of the foramen (note that nerve roots are located inferiorly).

Contrast enhancement

- Enhancement of the foraminal epidural veins within the cervical neural foramina may be of great help for the depiction of protruded disc material.
- Using T1w sequences and very thin slices, the sensitivity of MRI has been reported to be equally sensitive to contrast-CT and CT-myelography for the detection of lateral and foraminal disc herniations.

Case 29: Disco-osteophytic protrusion

Fig. a: sagittal cut; SE - T2w

Fig. b: axial cut; GE (3D) - T2*w

Middle age patient presenting with cervicobrachial neuralgia. T2w sagittal cuts show a small disco-osteophytic protrusion at the C5-C6 level lateralized to the left side and indenting the thecal sac (**a**, ↑). The spinal cord does not show abnormal signal. The 3DFT-GE-T2*w axial contiguous slices (**b**) identify the bulging disc material with peripheral low signal intensity compressing moderately the left cervical hemicord. The lateral cervical X-rays showed bone density material at the same level, consistent with posterior osteophytic formation. Note that if using 3DFT-GE sequences, magnetic susceptibility artifacts may accentuate foraminal stenosis which may even appear progressively more stenotic as the TE values are increased.

Case 30: Foraminal stenosis

Fig. c: sagittal cut; SE - PDw

Fig. d: axial cut; SE - T2w

Elderly man presenting with neck pain. Sagittal PDw images (**c**) demonstrate segmental narrowing of the spinal canal at C3-C4 level due to posterior disco-osteophytic formation (**c**, ↑) associated to hypertrophic posterior ligaments (**c**, *). T2w axial cuts showed severe narrowing of the right neural foramina (**d**, *), due to uncinate and facet hypertrophy. Note the slight relative hyperintense signal of the spinal cord (**d**, ↑↑) on PDw sequences (**c**) that could correspond to slight edema. Note that SE-T2w sequence may be preferred to axial 3DFT-GE sequences due to lack of magnetic susceptibility artifacts that might accentuate the foraminal stenosis.

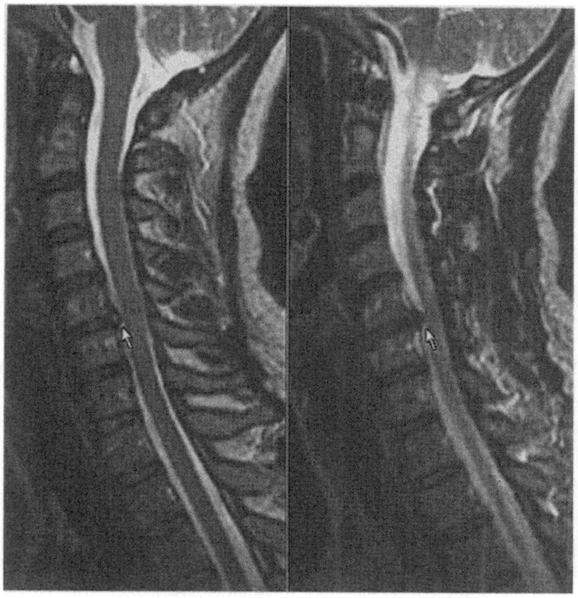

a

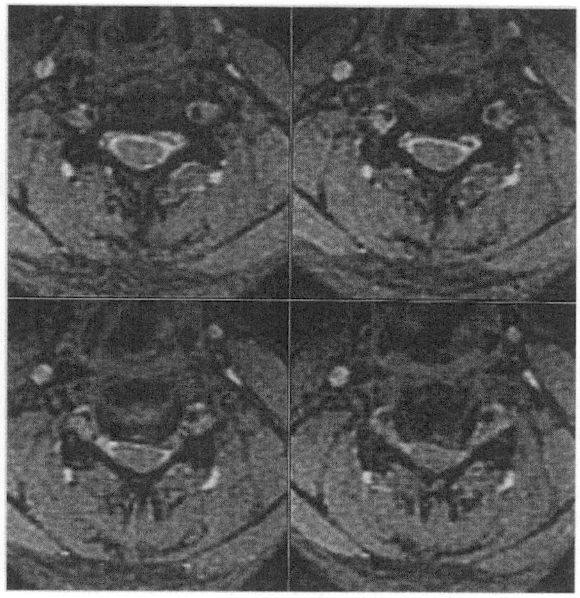

b

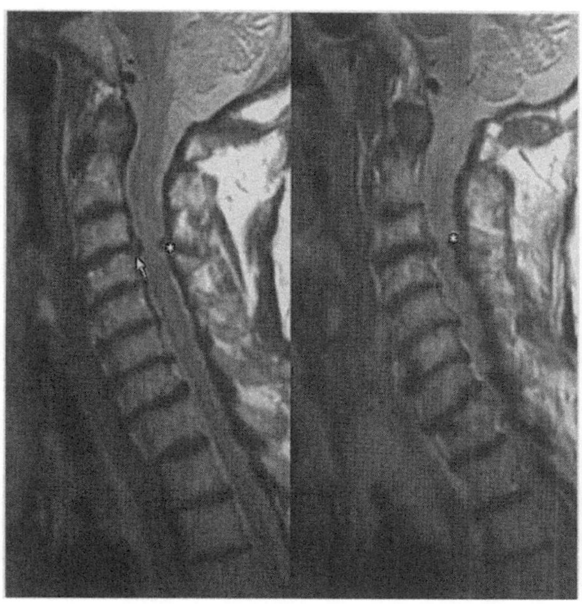

c

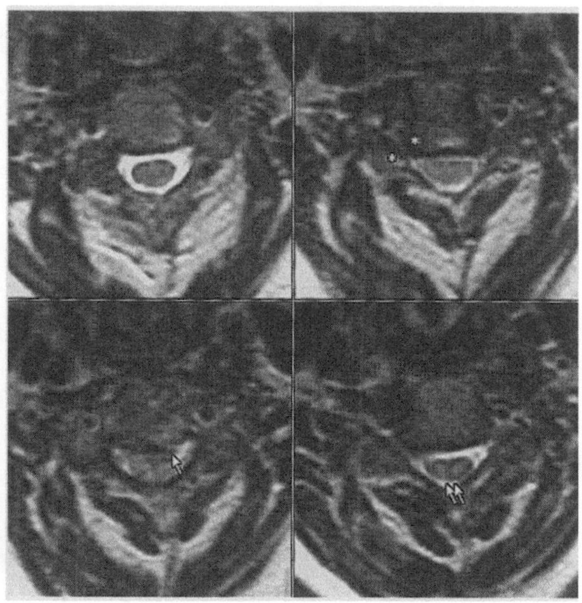

d

Case 31: Paracentral juxtaforaminal disc herniation

Fig. a: sagittal cut;	SE - T2w
Fig. b: axial cut;	SE - T2w
Fig. c: sagittal cut;	EG - T2*w
Fig. d: axial cut;	SE - T1w

Large left paracentral disc herniation (**a**, ↑; **b**, ↑) well visualized on the T2w sequences and filling the left paracentral spinal canal and effacing the ipsilateral subarachnoid space (**b**, ↑). The slightly hyperintense signal of the herniated disc must not be confused with the flow-void artifacts as could be observed on FSE-T2w sequences in the axial plane (**b**, *). Note the normality of the right paracentral canal on T1w sequence (**d**). Note on the parasagittal cuts the linear low signal of the posterior longitudinal ligament. On T1w axial slices (**d**), the paracentral herniated disc is well shown as an iso-intense mass (**d**, ↑↑) occluding the hypointense signal of CSF in the left lateral recess. The disc is well shown compressing the anterior nerve root (**d**, ↑) displaced posteriorly. Note the normal nerve roots on the opposite side (**b**, up, left). Note also the similar hyperintense signal of the disc material protruted, on GE images with the associated hyperintense adjacent turgescent venous plexuses (**c**, ↑), as shown in another patient, that may help to identify the protruding disc material with its relation with the parent disc, and the eventual existence of a defect in the PLL.

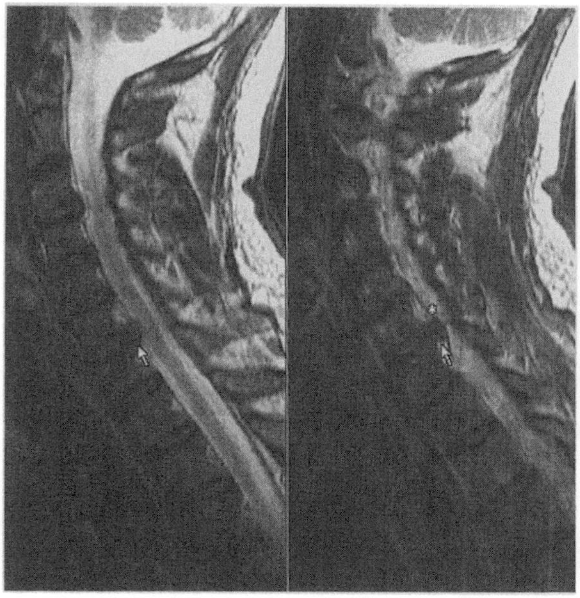

a

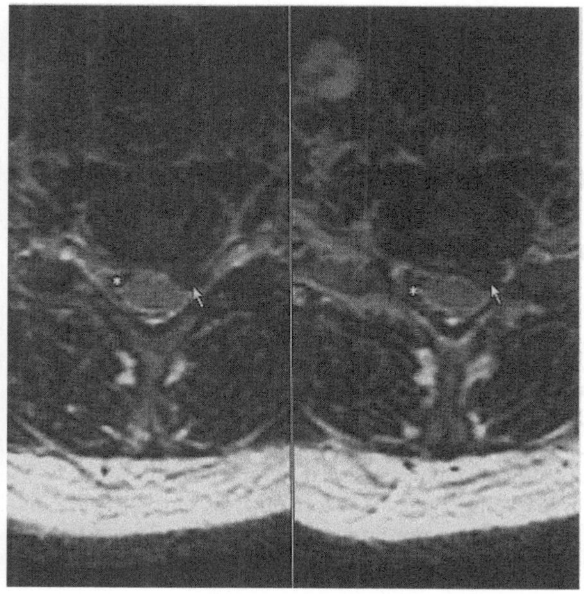

b

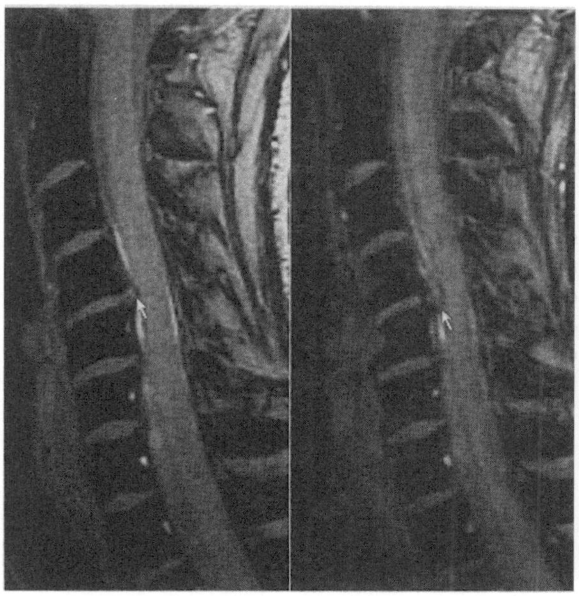

c

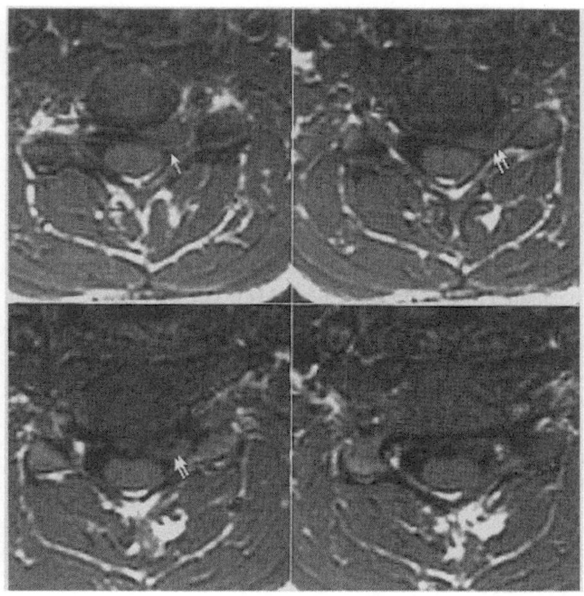

d

Case 32: Foraminal disc herniation

Fig. a: sagittal cut; SE - T1w

Fig. b: axial cut; GE (3D) - T2*w

Patient presenting with an acute right cervicobrachial neuralgia following a whiplash trauma. Parasagittal T1w MR slices showed at the C6-C7 level the presence of a small soft tissue mass involving the right foramen, related to the adjacent neighbouring disc (**a**, ↑) and obtruding the hyperintense signal of the foraminal fat. On 3DFT-GE-T2*w sequence, the 2 mm contiguous slices demonstrated a slightly hyperintense signal, herniated disc extending into the medial aspect of the right neural foramen (**b**, ↑) and causing no cord deformity at the same level. The dark line located between the disc fragment and the cord (**b**, ↑) represents displaced posterior longitudinal ligament and dura.

Case 33: Extruded central disc herniation

Fig. c: sagittal cut; SE - T1w and sagittal cut; SE - T2w

Fig. d: axial cut; SE - T1w and axial cut; SE - T2w

Young man presenting a post-traumatic neck pain with numbness in the right hand. MR sagittal cuts show a large central and slightly right paracentral mushroom-shaped herniated disc at the C5-C6 level (**c**). The herniated material is iso-intense on T1w (**c**, left, ↑), slightly hypointense on T2w with spinal cord compression (**c**, ↑↑). The axial cuts (**d**) show on T1w (**d**, up, left) a central and paracentral disc herniation (**d**, up, ↑) compressing the spinal cord laminated posteriorly and failing to show the subarachnoid spaces on T2w (**d**, right and down). Note on the sagittal cuts, mainly on the T1w (**c**, left), the caracteristic two apparently superimposed disc fragments, corresponding to the caracteristic "double fragment sign" due to a defect in the annulus and the posterior longitudinal ligament.

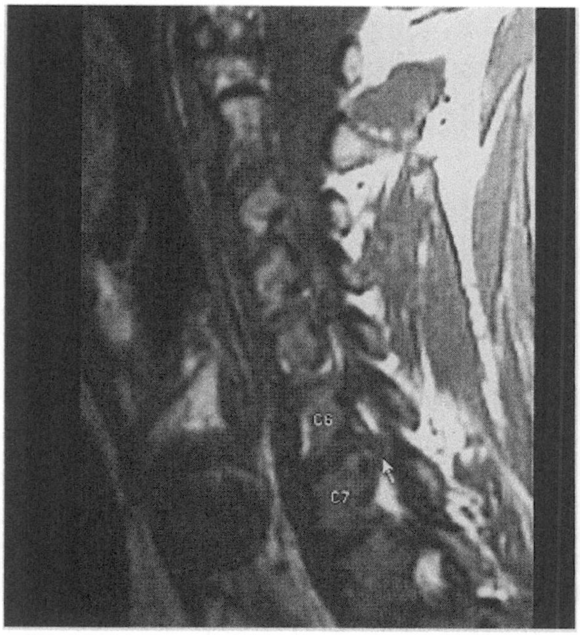

a

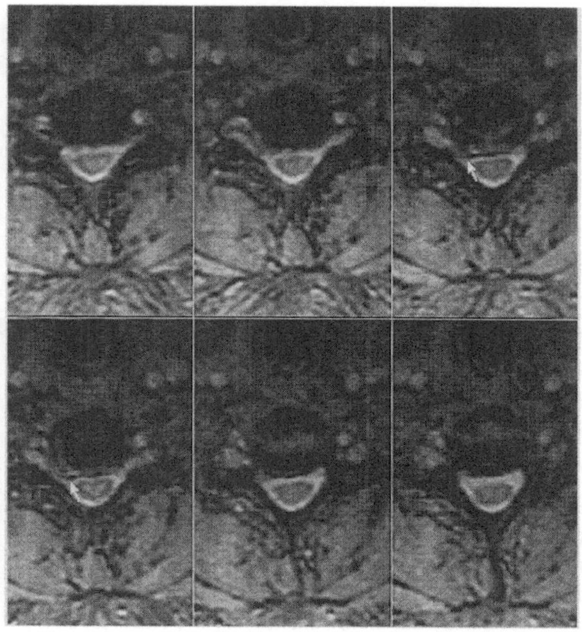

b

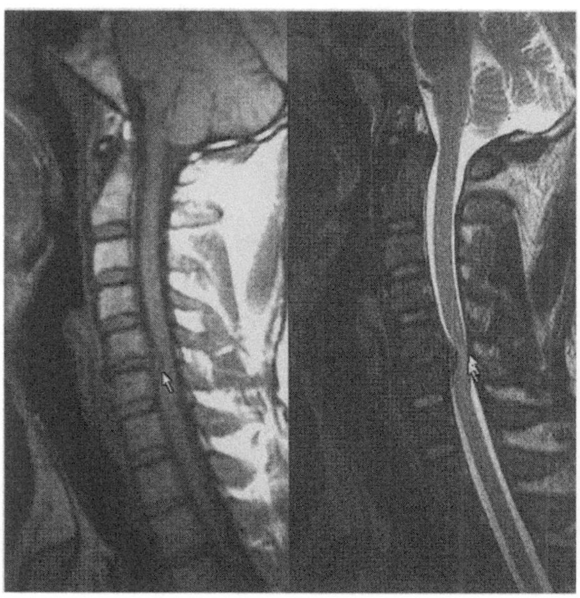

c

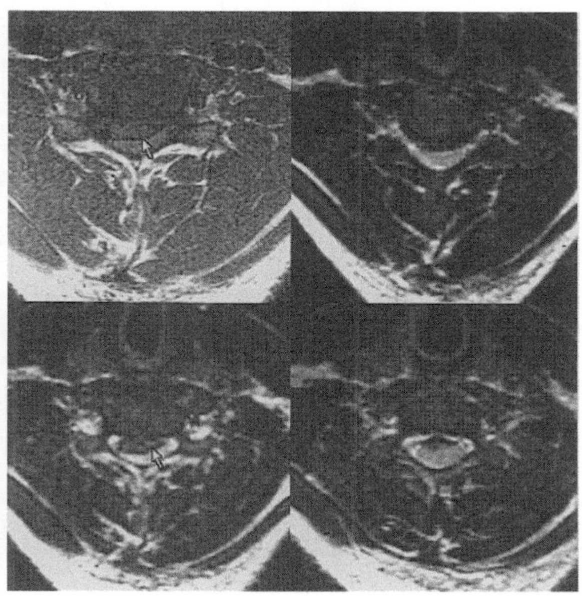

d

Case 34: Central disc herniation with cord compression

Fig. a: sagittal cut; SE - PDw

Fig. b: sagittal cut; SE - T2w

Patient presenting with a mild cervical myelopathic syndrome. PDw sequence (**a**) and T2w sequence (**b**) showed an approximately 50% narrowing of the anterior-posterior dimension of the cervical spinal canal at C3-C4 level associated with an intramedullary hyperintense signal on both sequences, corresponding to cervical myelopathy, due to segmental cord compression (**a**, ↑↑; **b**, right, ↑). Note also the associated indentation of the anterior subarachnoid spaces at C5-C6 level (**b**). Note on PDw images the important thickening of the posterior longitudinal ligament and dural shown as a dark band between the herniated disc and the iso-intense CSF-cord complex on PDw, and corresponding presumably to ossification of the PLL (**a**, left, ↑) and better shown on GE images due to susceptibility effects of the heterotopic bone formation from C3 to C6 levels.

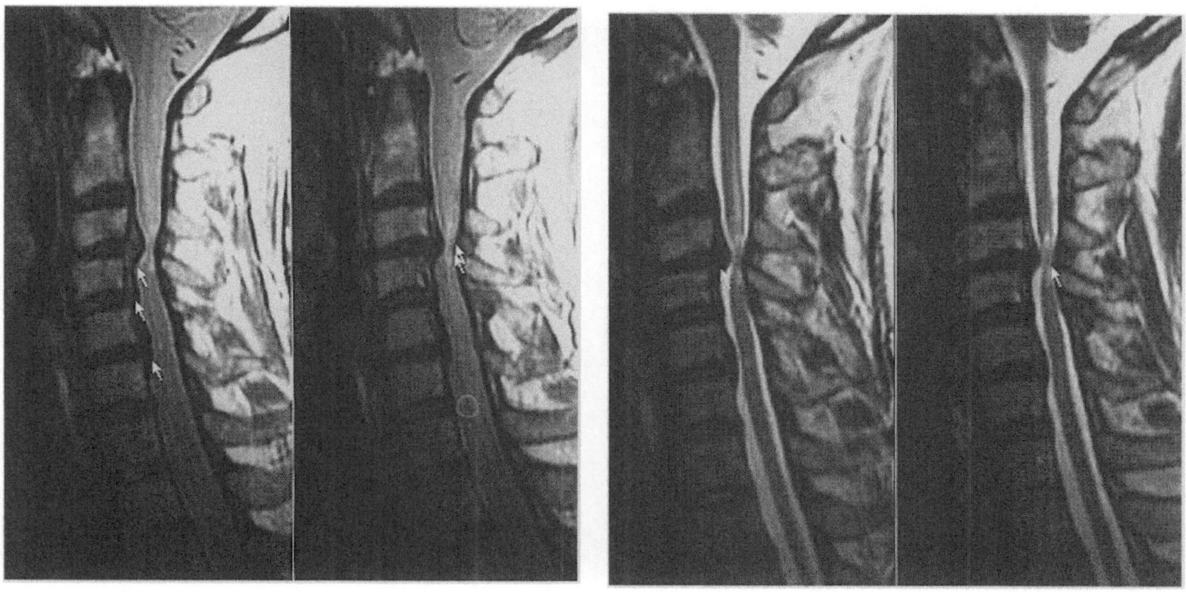

a b

Cases 35-37: cervical myelopathy

Clinical and pathological features

- Cervical spondylotic myelopathy is estimated to be the most common myelopathy causing a progressive injury to the cord and its roots.
- It is found in association with cervical disc herniations in almost half the cases.
- Spinal cord injury results from long standing cord compression in a degenerative stenotic spinal canal showing a reduction of its anterior-posterior diameter to less than 10 mm (normal average dimension is 12 mm in the lower spine).
- *Signs:* the usual syndrome observed include: pain in the neck often radiating to the scapular region, brachialgia with paresthesias along the forearm and hand and gait disturbance with spastic inferior limbs weakness of variable importance. Neurological exam discloses signs of nerve root and cord involvement, with depression in biceps and supinator reflexes in the superior limbs and increase in tendon reflexes and changes in vibratory sense in the lower limbs, Babinski's sign, hand and forearm atrophy. A partial Brown-Sequard syndrome may be present. A Lhermitte's sign may be elicited on neck flexion. With progression of the spondylotic myelopathy, gait ataxia becomes more severe with increase of spasticity and sphincteric dysfunctions.

MR characteristic findings

Morphology

- The abnormalities shown within the cord are commonly non specific.
- These intramedullary changes secondary to degenerative diseases of the spine may correspond to either edema, ischemia ("snake's eyes"), demyelination, gliosis, necrosis, myelomalacia.
- Signal changes (bright lesions on T2w) has been shown in more than half the patients with mild or severe cord compression, occurring usually at or below the level of maximal compression.
- Spinal stenosis is well demonstrated using sagittal T2w images with a myelographic effect. Multiple contiguous bulging and/or herniated discs, bony ridges and osteophytic formations, hypertrophy and/or ossification of the posterior longitudinal ligament, thickening and/or ossification of the ligamentum flavum, contribute to narrow the spinal canal and perimedullary subarachnoid spaces with secondary cord compression.
- The presence of these cord signal changes seems not to be correlated with the severity of the clinical status.
- Considering prognosis and post surgical outcome, reported series tend to show a negative correlation with presence of bright signal abnormalities and their persistence after decompressive surgery. A poor clinical recovery seems to be also associated with cord enlargement following surgery.
- Segmental cord atrophy may be present in chronic cases. MRI is helpful in diagnosing irreversible segmental gliosis and atrophy of the cord.

Signal intensity

- SE-T1w: iso-intense to hypointense (myelomalacia) to cord.
- SE-PDw: hyperintense (demyelination) to cord and CSF.
- SE-T2w: hyperintense to cord.
- It is important to keep in mind that the signal abnormalities shown are not highly specific and in doubtful cases other etiologies must be ruled out.
- The bright signal abnormalities must not be confused with truncation artifacts or flow artifacts due to CSF pulsations or a cord syrinx.

Contrast enhancement

- There is usually no need to perform contrast exams for degenerative myelopathy, unless there is strong evidence of other disease process such as: myelitis, infarct, cord tumor.

Case 35: Cervical myelopathy

Fig. c: sagittal cut; SE - T2w

Fig. d: sagittal cut; SE - PDw

Patient presenting with long standing symptoms associating neck pain irradiating in both arms, and more recently weakness in the legs and bladder dysfunction. Sagittal MR cuts showed on T2w (c) a cervical spinal stenosis due to multilevel central herniated discs indenting the CSF subarachnoid spaces from C3 to C6 levels. MR shows a higher degree of central canal stenosis at C3-C4 level with associated marked cord compression and hyperintense signal intramedullary (c, *). The spinal cord was also moderately hyperintense on T2w sequence (d) between C3 and C5 levels.

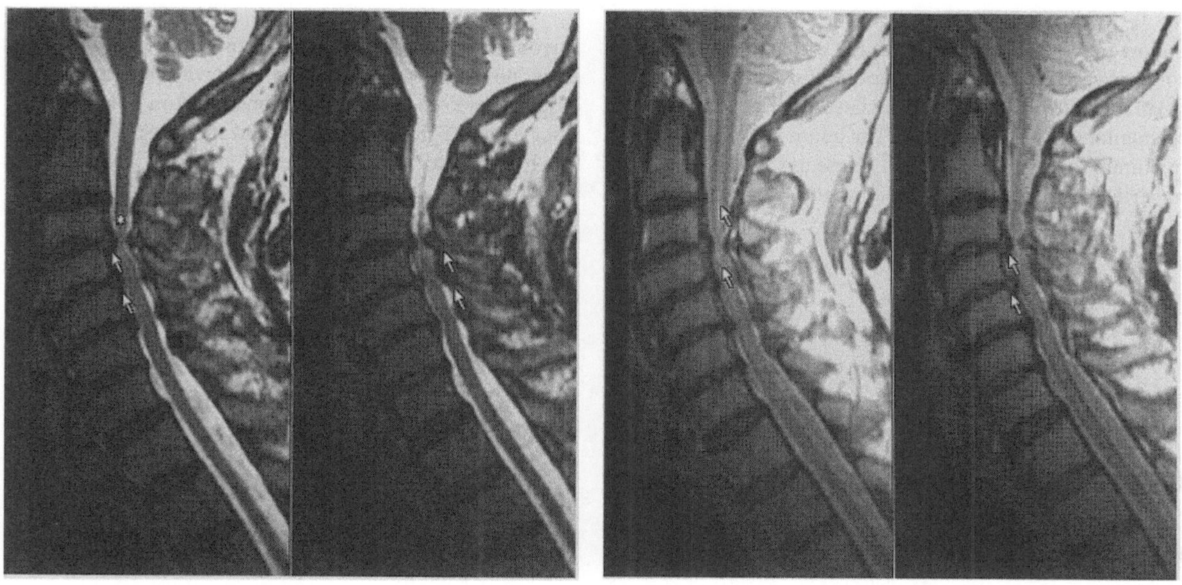

c d

Case 36: Cervical myelomalacia (post-surgery)

Fig. a: sagittal cut; SE - T1w

Fig. b: sagittal cut; SE - T2w

Follow-up cervical exam of a patient who underwent surgery with cervical vertebral bodies fusion. MR shows a small intramedullary lesion associated with a moderate segmental atrophy of the cord. The lesion shows a hypointense signal on T1w (**a**, ↑) close to CSF and a hyperintense signal on T2w (**b**, ↑) slightly larger and surrounded by a small area of intermediate signal shown on PDw (not shown). The small central cavity corresponds to a sequelar myelomalacia well demonstrated on T1w, associated to a rim of demyelination shown on T2w and much better on PDw.

Case 37: Ischemic myelomalacia ("snake's eyes")

Fig. c: sagittal cut; SE - T2w

Fig. d: axial cut; SE - T2w

Cervical follow-up exam showing a segmental atrophy of the spinal cord on T2w (**c**) at the C5 and C6 levels, associated to a longitudinal area of hyperintense signal on T2w shown on the sagittal cuts (**c**, ↑). The axial SE-T2w images demonstrated central hyperintense lesions involving roughly symmetrically both hemicords (**d**, ↑) corresponding to areas of junctional ischemic myelomalacia and/or demyelination, giving the caracteristic "snake's eyes" appearance. Note the vertebral fusion and remodeling at the C5-C6 level (**c**).

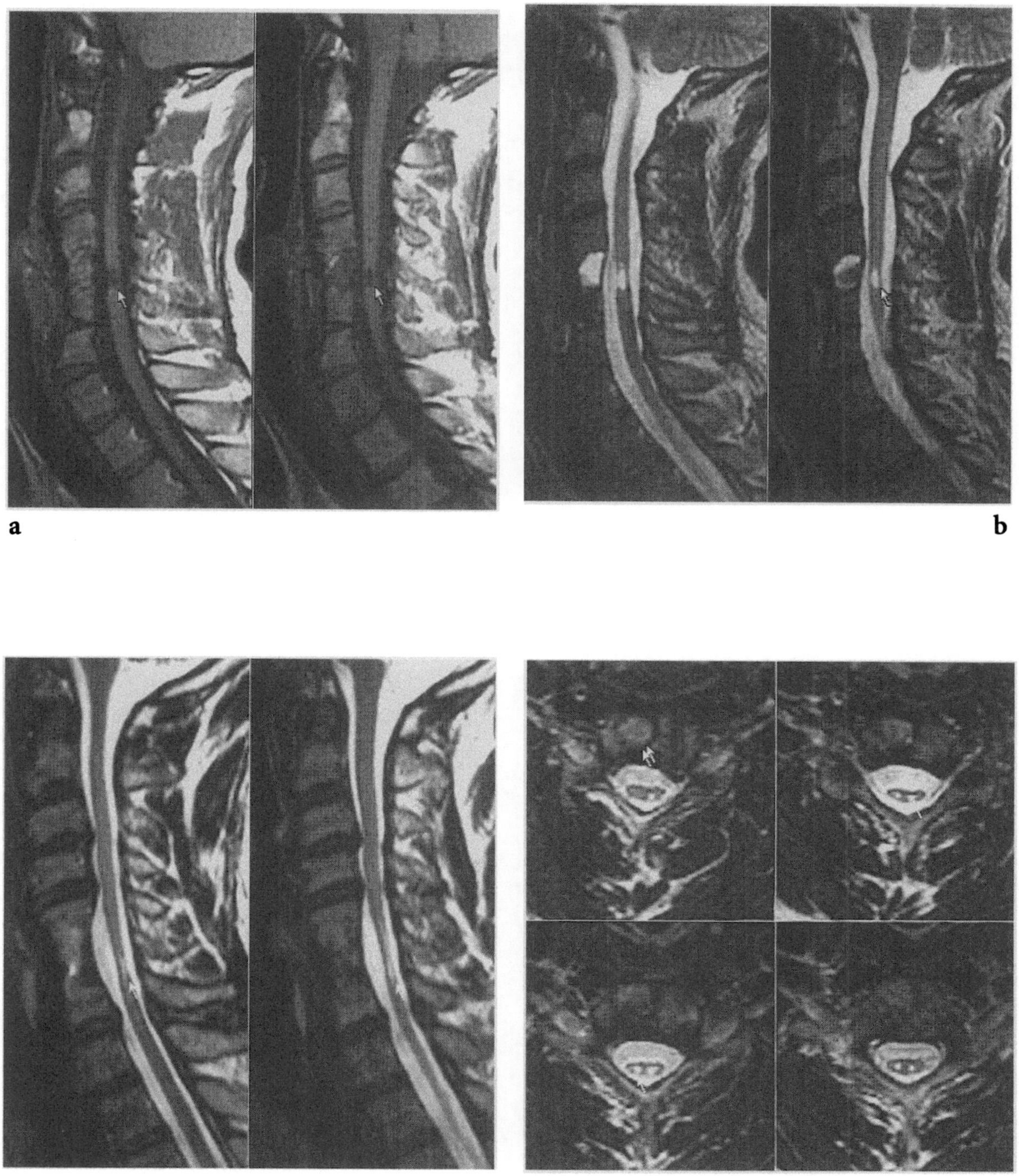

a b c d

Thoracic spine

Cases 38, 39: thoracic disc herniation

Clinical and pathological features

- Thoracic disc herniations have a very low incidence as compared to lumbar and cervical, estimated to average 1 per million population.
- This relatively low incidence is due to the relative stability of the thoracic spine owing to rigidity of the chest cage and to the smaller volume with thicker annulus of the intervertebral discs.
- Herniations occur during the fourth decade, slightly more frequently in males.
- *Signs:* the most common complaint is back pain of insidious onset and frequently poorly defined, associated with radiculalgia. Numbness or pain in the lower limbs are commonly reported. Pain radiation into the flank, abdomen or pelvis may cause a misdiagnosis. Signs of spinal cord compression may be present including important sensory signs, genitourinary dysfunctions, a Brown-Sequard syndrome.

MR characteristic findings

Morphology

- MRI is the most sensitive noninvasive technique for the depiction of thoracic disc herniations, being superior to myelography and CT and at least as sensitive as CT myelography.
- The relative immobility of the upper thoracic spine explains that most (2/3) of the herniated discs are found in the lower region between T8 and L1 levels.
- Disc herniations are most commonly central or paracentral.
- Thoracic disc herniation may commonly calcify (more than half of the cases).
- Degenerative spondylotic changes are less common than at the cervical or lumbar levels. Ossification of the ligamentum flavum mainly in the lower thoracic spine between T10-T12 and osteophyte formations and/or ossification of the posterior longitudinal ligament in the upper thoracic spine, may contribute to cause a segmental stenosis of the thoracic canal.
- Motion artifacts from chest wall and cardiovascular pulsations or CSF flow may render difficult the investigation of the mid-thoracic region.
- Scheuerman's disease commonly involve the mid-thoracic region in young people.

Contrast enhancement

- Contrast infusion is seldom useful for the detection of disc herniation.
- It may help sometimes to depict the herniated disc in enhancing the adjacent congested epidural venous plexuses.
- Differentiation of calcified discs and small meningiomas may benefit from contrast administration.

Case 38: Multiple dorsal disc herniations

Fig. a: sagittal cut; SE - T2w

Small thoracic disc herniations involving two adjacent levels in the lower thoracic spine associated to another disc herniation at the mid-thoracic spine (**a**, right, ↑) well shown here on T2w sagittal slices (**a**). The upper thoracic disc herniation abuts the spinal cord which is compressed at the lower levels with segmental swelling and slightly hyperintense signal (**a**,*)

Case 39: Calcified herniated disc with myelomalacia

Fig. b: sagittal cut; SE - T1w

Fig. c: sagittal cut; SE - PDw

Fig. d: sagittal cut; SE - T2w

Very large calcified thoracic disc herniation compromising largely the thoracic canal with marked compression of the spinal cord. The sagittal T2w sequence (**d**) demonstrates a mass developed in the spinal canal in the mid-thoracic region presenting a heterogeneous signal with peripheral hypointense signal (**d**, left, *) and central slightly hyperintense signal iso-intense to vertebral bone marrow (**d**, right, *), corresponding to peripheral calcification surrounding the brighter inner zone. Note slight amount of bright fat in the epidural space at the upper and lower margins of the mass (**b**, left). On PD images (**c**) and T2w images (**d**), the mass remains heterogeneous, the calcified portion becoming more hypointense (**c**, **d**, left, *). The spinal cord shows a segmental enlargement with a hyperintense signal to normal cord and to CSF on PDw (**c**, ↑) becoming more hyperintense on T2w (**d**, right, ↑). This portion of compressed cord adheres to the posterior wall of the spinal canal (site of a previous posterior approach surgery 20 years ago). The hyperintense signal with apparent enlargement of the spinal cord are probably due to compressive and/or ischemic demyelination and myelomalacia. T1w sequences show area of hypointense signal due to a central syrinx.

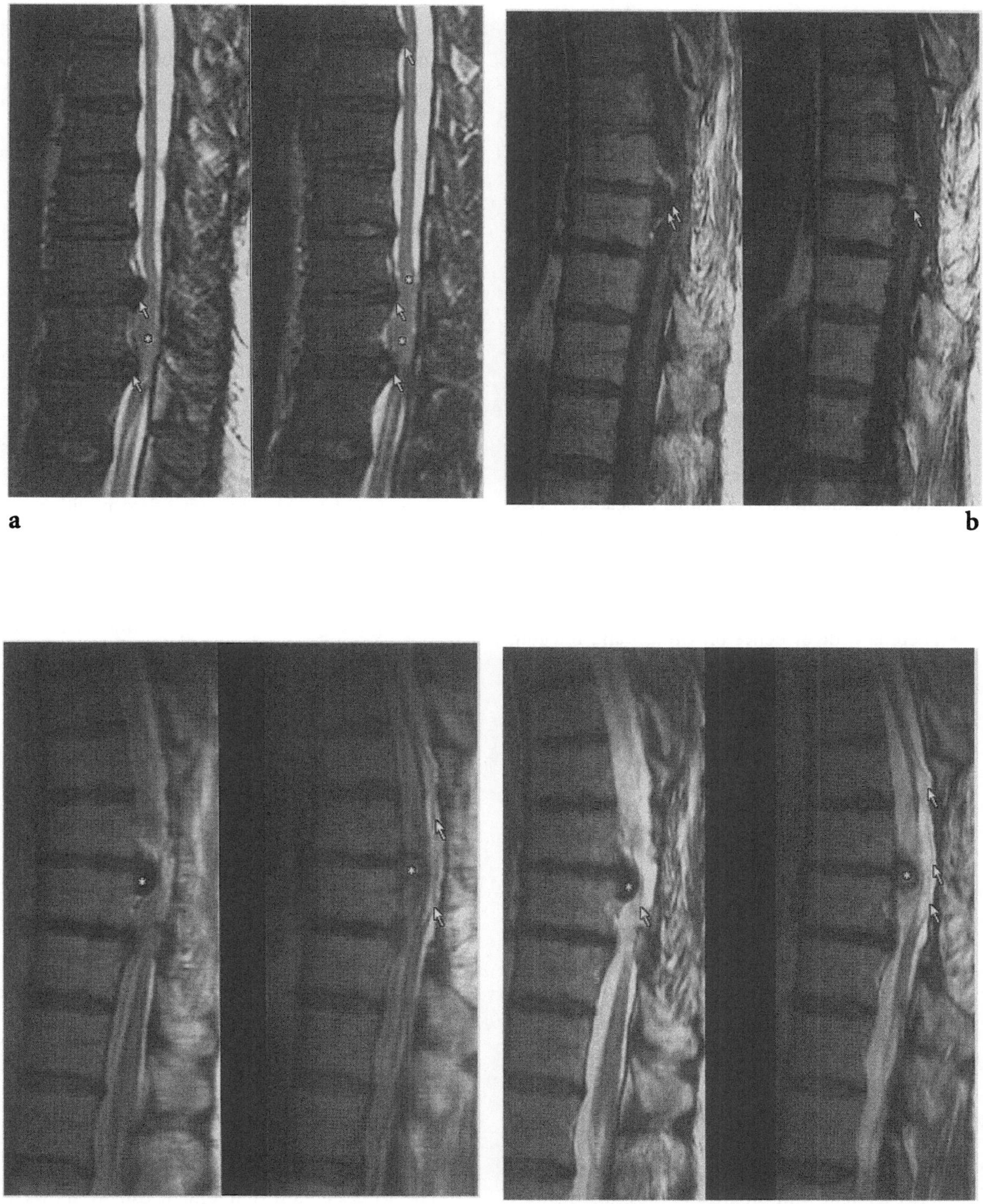
a
b
c
d

Lumbar spine

Cases 40-52: degenerative lumbar disc herniation

Clinical and pathological features

- It constitutes a major cause of low back pain in the adult population.
- Its incidence varies according to occupation of the individual. A prevalence of about 20% is reported in adulthood. Herniated discs are rare in children.
- The natural history of sciatica due to herniated lumbar discs is that of spontaneous improvement over weeks or months.
- Surgery is indicated in less than 10% of the patients. Retrospective studies have shown little long-term differences in the clinical outcome of patients treated medically or surgically.
- It has been shown that almost half the patients with disc herniation causing a sciatica, have improved with a conservative treatment. Complete or partial regression of herniated disc (2/3 of the cases) or bulging discs (1/3 of the cases) has been demonstrated as well.
- There is wide evidence on clinical and imaging data for spontaneous regression of herniated discs in patients presenting a sciatica. The largest herniated discs and the free fragment herniations appear to be the more likely to regress.
- The majority of degenerated discs do not show associated herniation.
- *Signs:* disc herniation may be asymptomatic (10-30%) or may present with long standing mild to severe back pain and sciatica. Increase in pain commonly occur in walking or standing. The neurological exam may demonstrate a positive Lassegue's sign and a loss of a deep tendon reflex in the lower limb.

MR characteristic findings

Morphology and signal intensity

- Disc degeneration is associated with decrease in water content with progressive dessication (from 90 to 70%) and tearing in the annulus. Disc disruption accounts for almost 40% of low back pain, reproduced with discography. Calcifications may be present most commonly in the annulus, and vacuum phenomena may be found as well in degenerated lumbar discs, occurring in less than 5% of the cases.
- Degenerative discs may be associated with subchondral osteochondrosis of the adjacent vertebral bodies. These degenerative changes have been classified by Modic into 3 types (cases 40-42). Vertebral marrow changes are observed around the vertebral end plates. Type I corresponds to the presence of an active inflammatory fibrovascular process, showing a hypointense SI on T1w and a hyperintense SI on T2w, and may enhance. Type II corresponds to a fatty yellow marrow replacement, showing a hyperintense SI on T1w and an iso-intense to hyperintense SI on T2w. Contrast enhancement may be observed. Type III corresponds to bony sclerosis with relative absence of marrow, showing a hypointense SI on both T1w and T2w.
- Disc herniations may be classified as follows: 1- bulging disc (case 43): corresponding to a diffuse and symmetrical extension of an intact annulus fibrosis beyond the posterior vertebral end plate margins. It has been reported that a bulge greater than 2,5 mm is associated with radial tearing; 2- protruding disc: consists in a partial annular fibers disruption, with or without secondary focal contour deformity of the posterior disc margin; 3- extruded disc (case 44-46): corresponds to a total disruption of the annulus fibrosis with extrusion of nuclear and even par of annular disc material into the epidural fat. The herniated material may lie anteriorly (subligamentous) or posteriorly to the posterior longitudinal ligament; 4- free fragment or sequestered herniated disc (cases 47-50): corresponds to fragments of extruded disc material separated from the parent disc (may be reattached to some extent by a tiny pedicle), which may also migrate (elongated shape) inferiorly or less often superiorly. Multiple fragments may even be found. The sequestered disc fragment usually shows a hyperintense SI on T2w (probably due to increase in water content) and a peripheral contrast enhancement. The identification of free fragments is important for the surgeon since

they do not respond well to conservative therapy. Percutaneous discectomy is contraindicated in such patients; 5- intradural disc herniation (cases 51, 52): it is a rare entity (less than 0.5%) most frequently found in the lumbar spine affecting mainly men at the L4-L5 level in general. The free disc material extends almost totally into the thecal sac, showing a relative hyperintense SI on T1 and T2w; 6- Schmorl's node herniation: occurring through the cartilaginous end plates (may be secondary to microfractures), with associated edema within bone marrow and showing eventually contrast enhancement.

- Topographically, disc herniations may be central or paracentral, lateral, and foraminal or extraforaminal, in location.
- Almost 90% of lumbar disc herniations are found at L4-L5 and L5-S1 levels, followed by L3-L4, L1-L2 or L2-L3. They are well demonstrated on SE-T1w sequence in the axial plane, presenting as small masses of soft tissue (low SI on T1 and T2) displacing and obliterating the epidural fat (bright SI on T1). Nerve root and thecal sac compression are well demonstrated also on T1w axial cuts.
- Foraminal and extraforaminal disc herniations (more than 10%, most overlooked) are more frequently found at L4-L5 level followed by L3-L4, and are uncommon at L5-S1. Therefore, the most common complaint involve L3 or L4 nerve roots, or is typically biradicular. They concern older adults and mainly males. Over half of the herniated discs are free fragments. Foraminal herniations are best seen on parasagittal cuts, found superiorly within the foramen (almost 2/3 of the cases) and extraforaminal are shown on axial cuts. These must be differentiated from nerve sheath tumors which typically enhance. Peripheral contrast enhancement due to peridiscal granulation tissue may be observed.

Contrast enhancement

- Spinal nerve roots may enhance in association with disc protrusion, due to BBB breakdown or venous stasis.
- Contrast administration is mandatory to differentiate a sequestered disc (no enhancement but possibly peripheral, unless delayed scan after 5 minutes) from scar tissue, and disc material from other enhancing soft tissue processes in the epidural spaces.

Case 40: Degenerative disc diseases (4 types)

Fig. a: coronal cut; SE - T2w

Fig. b: sagittal cut; SE - T1w and sagittal cut; SE - T2w

Fig. c: sagittal cut (512); SE - T1w

Fig. d: sagittal cut (512); SE - T1w

(**a**): coronal T2w images showing: normal discs in an adult with a relatively hyperintense signal of hydrated nucleus and inner annulus (**a**, ↑), surrounded by the peripheral annulus containing Sharpey's fibers showing lower signal and the intranuclear cleft seen as a band in the disc (**a**, ↑↑). The degenerated disc has a low signal intensity associated with a more or less narrowing of the disc space related to a disc degeneration. The loss of signal intensity of the degenerative disc is more apparent on SE-T2w than on GE-T2*w images.
(**b**): degenerated L5-S1 disc with narrowing of the disc space showing a high linear increased signal intensity on T2w (**b**, right, ↑) presumably secondary to fluid within a fissured disc, less evident on T1w (**b**, left, ↑).
(**c**): narrowing of the posterior portion of L5-S1 disc showing a linear hyperintense band on T1w (**c**, ↑), probably due to the presence of calcifications, associated to end plates degenerative abnormalities (**c**, *).
(**d**): T1w MR showing degenerative L5-S1 disc with marked narrowing of the disc height associated to a linear intra-discal hypointense band corresponding to the presence of gas (probably nitrogen) and to the vacuum phenomenon (↑). Note that gas may be better seen on GE-T2*w images due to signal loss related to magnetic susceptibility effects.

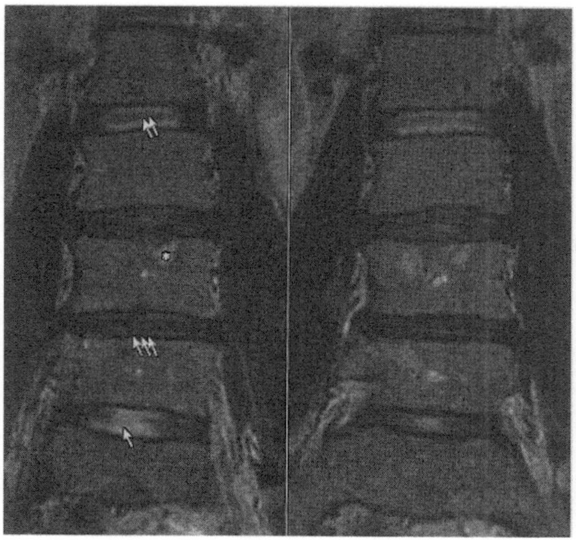
a

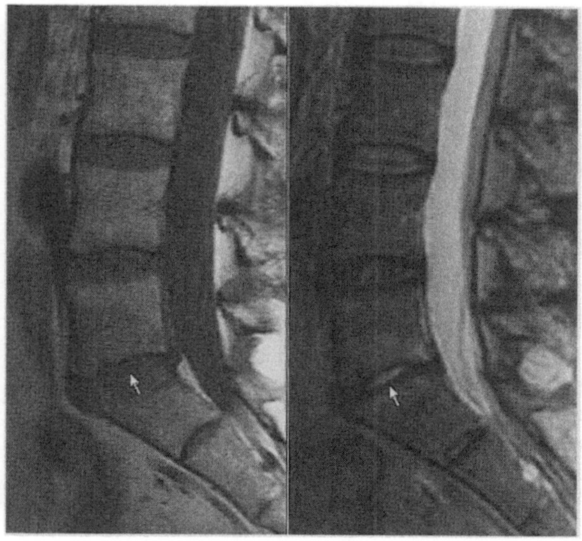
b

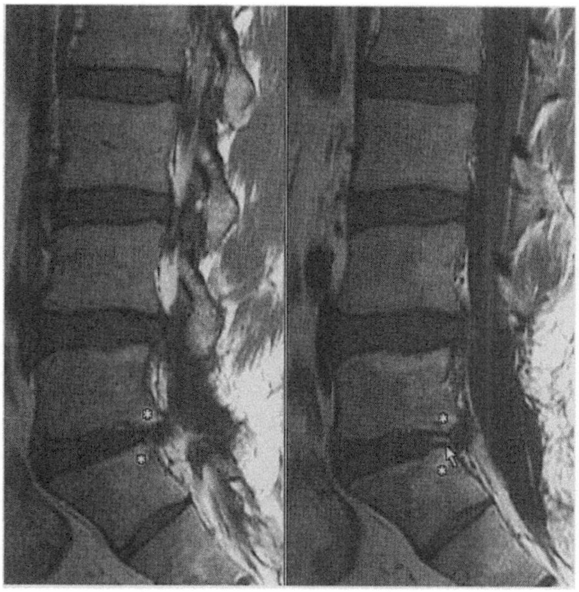
c

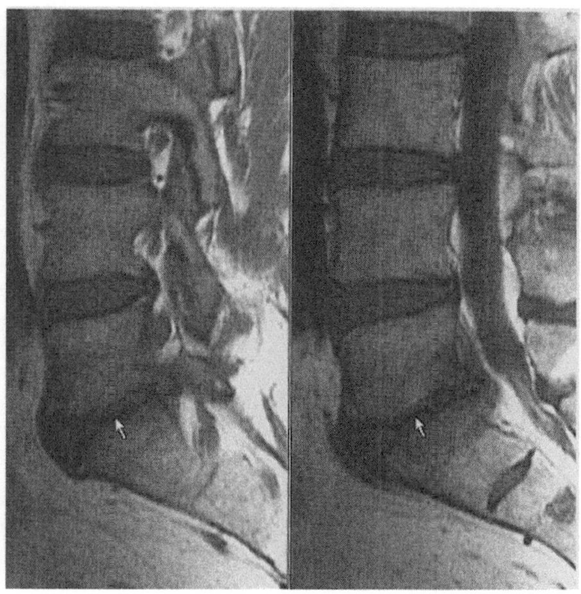
d

Case 41: Degenerative disc type I

Fig. a: sagittal cut (512); SE - T1w

Fig. b: sagittal cut; SE - T2w

T1w (**a**) and T2w (**b**) sagittal SE images showing type I of Modic, marrow changes at the L3-L4 level with a mild narrowing of the interspace. T1w sequence shows areas of decreased signal intensity of the adjacent portions of the L3 and L4 vertebral bodies (**a**, ↑↑), becoming hyperintense on T2w (**b**, ↑↑). These end plate changes represent vascularized marrow with associated edema. Note the posterior bulging disc (**a**, **b**, ↑). These modifications are observed in about 5% of the cases.

Case 42: Degenerative disc type II

Fig. c: sagittal cut (512); SE - T1w

Fig. d: sagittal cut; SE - T2w

Sagittal T1w (**c**) and T2w (**d**) SE images showing degenerative vertebral body changes type II of Modic represented by an increased signal intensity on T1w (**c**, ↑) and a hyperintense signal on T2w (**d**, ↑) involving end plates at the L4-L5 and L5-S1 levels. These changes are due to yellow marrow replacement in the adjacent portions of the vertebral bodies of the disc spaces in question. These modifications are observed in about 15% of the cases.

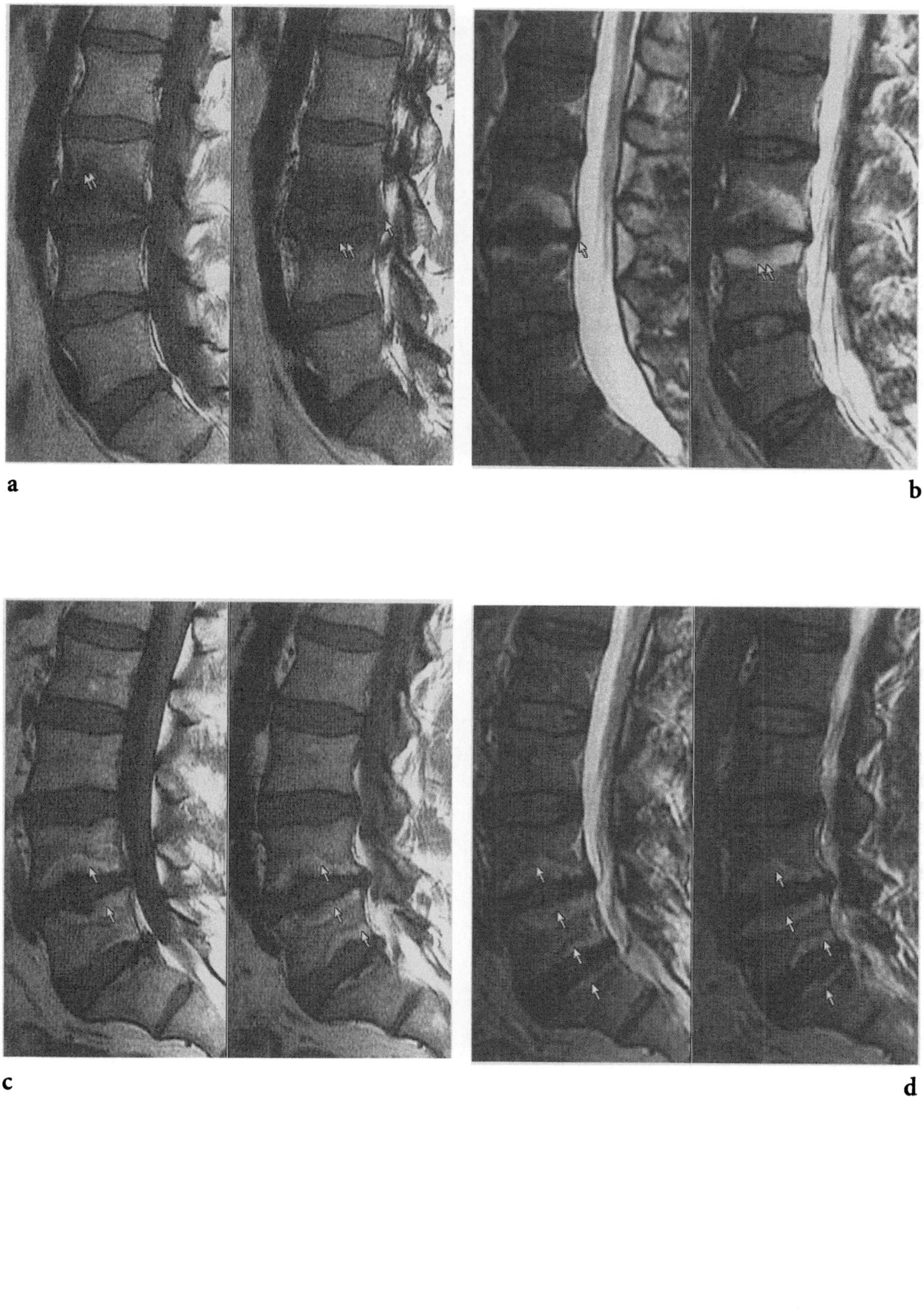

a b c d

Case 43: Disc protrusion

Fig. a: sagittal cut; SE - T2w

Fig. b: axial cut; SE - T1w and SE - T2w

SE-T2w images showing a degenerated L4-L5 disc associated with a moderately diffuse bulging disc (**b**, ↑) associated with a central protrusion recognized as a focal extension of the disc margin due to herniation of nuclear material through a defect in the annulus and beyond the margin of the adjacent vertebral end plates (**a**, ↑). Note that the herniated material is contiguous to the parent nucleus having a slightly hyperintense signal on T2w.

Case 44: Central extruded disc

Fig. c: sagittal cut; SE - T1w and sagittal cut; SE - T2w

Fig. d: axial cut; SE - T1w

Large L3-L4 central extruded disc showing a sub-ligamentous mass of nuclear material (**c**, ↑; **d**, ↑) which remains contiguous to the parent disc and showing a slightly increased signal on T2w (**c**, right, ↑; **d**, down, ↑). Note that the disc presents an iso-intense signal on T1w (**c**, left, ↑; **d**, up, ↑) with narrowing of the spinal canal. Note that this extruded disc to the rupture annulus fibrosus is still contained by the posterior longitudinal ligament, visualized as the peripheral dark rim at the posterior border of the mass. Note the stretching of this PLL (**c**, ↑↑) and that the decreased signal intensity outlining the herniated material may also represent portions of the carried ruptured annulus fibrosus.

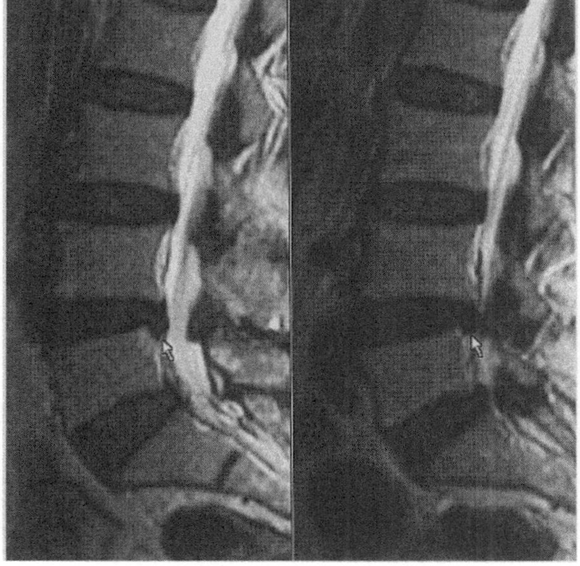

a

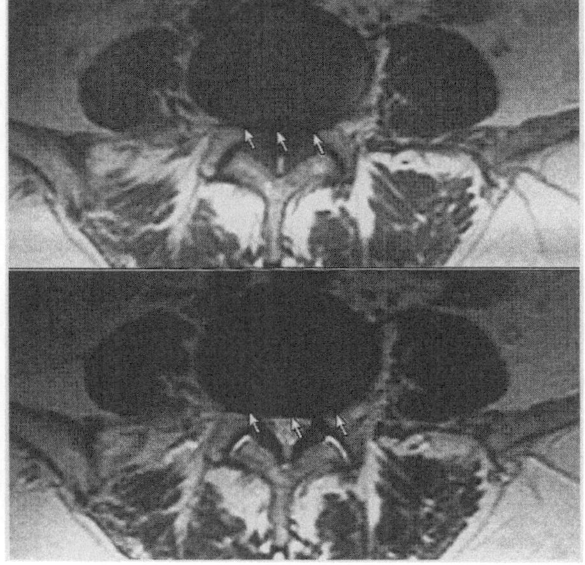

b

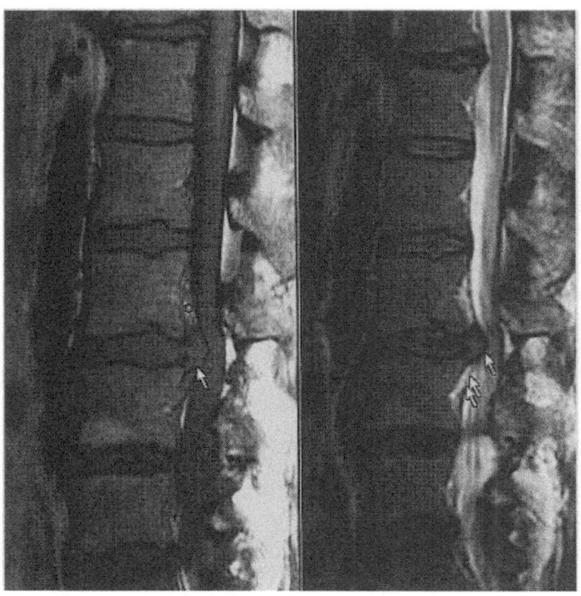

c

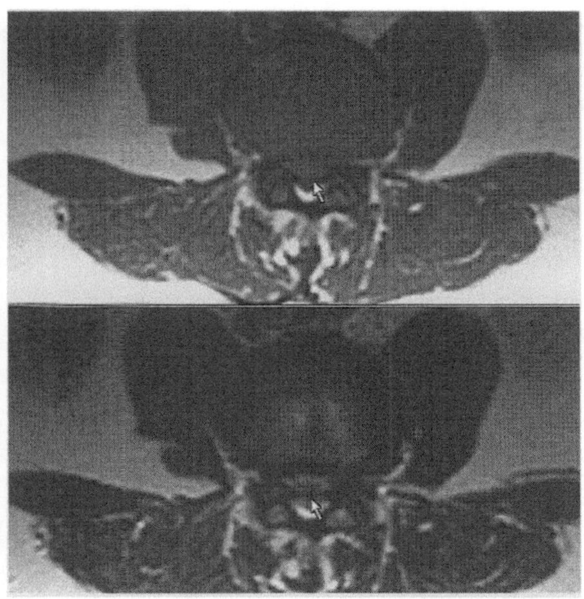

d

Case 45: Paracentral extruded disc

Fig. a: sagittal cut; SE - T1w

Fig. b: axial cut; SE - T1w and axial cut; SE - T2w

Right paracentral extruded L5-S1 disc showing an isosignal on T1w (**a**, *; **b**, up, ↑) and a slight hyperintense signal on T2w (**b**, down, ↑) effacing the ipsilateral epidural fat and compressing the adjacent nerve root (**b**, ↑↑). Note the curvilinear rim of decreased signal intensity outlining the herniated disc segment mainly shown on T2w representing stretched posterior longitudinal ligament as well as probable portions of the annulus fibrosus. Note the partial downward migration of the extruded disc (**a**, ↑).

Case 46: Lateral and extraforaminal extruded disc

Fig. c: sagittal cut (512); SE - T1w

Fig. d: axial cut; SE - T1w

T1w-SE sequence showing a large lateral disc herniation on the left side, evidencing an iso-intense signal on T1w (**c**, ↑; **d**, ↑). The lateral extruded disc is well shown on the parasagittal cuts, obtruding the inferior portion of the intervertebral foramen (**c**, ↑) compressing and elevating the corresponding nerve root at its upper border (**c**, left). Note in the axial plane (**d**, down, ↑) the obliteration of the normal fat within the neural foramen and difficulty to separate the nerve root ganglion from the soft tissue mass.

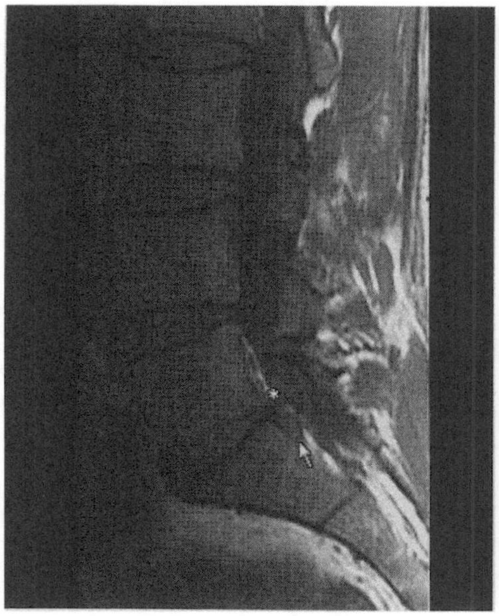
a

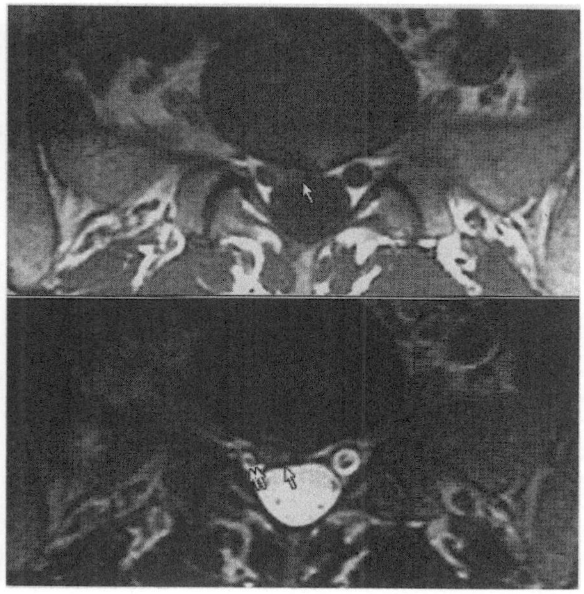
b

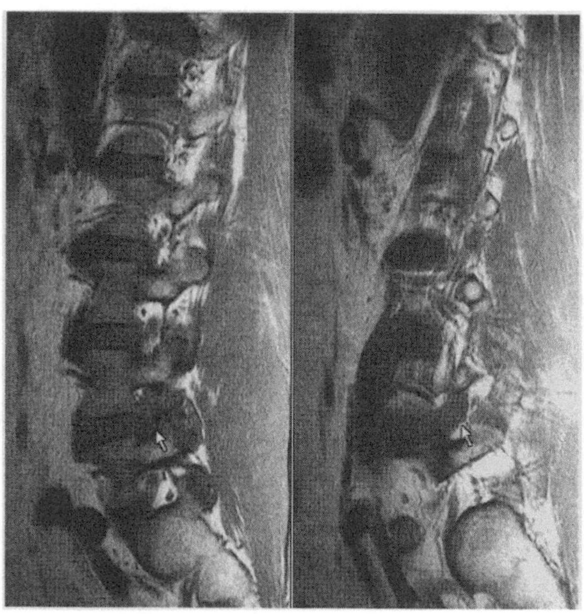
c

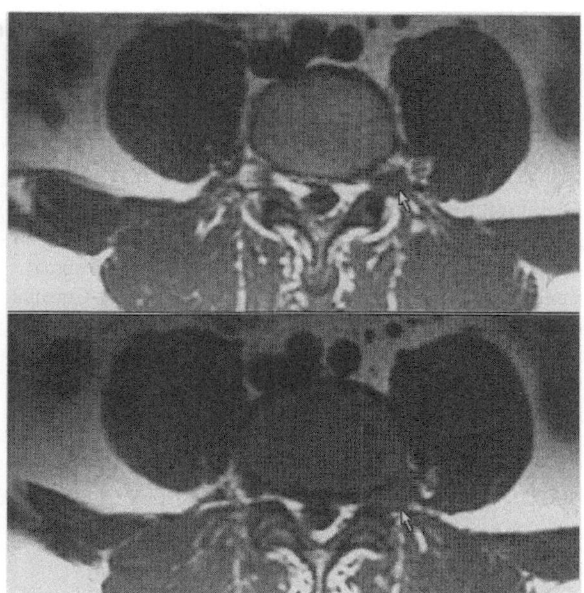
d

Case 47: Upward migrated herniated disc

Fig. a: sagittal cut (512); SE - T1w

Fig. b: axial cut; SE - T1w

Paracentral L4-L5 herniated disc (**a**, ↑; **b**, ↑) associated to a soft tissue mass identified posterior to the inferior aspect of the L4 vertebral body (**a**, *; **b**, *), corresponding to the partial upward migration. Note the posterior surrounding hyperintense signal of fat in the paracentral region (**a**, right). There is no compression of the dural sac as shown on the axial cuts (**b**).

Case 48: Upward migrated herniated free fragment

Fig. c: sagittal cut (512); SE - T1w

Fig. d: coronal cut; SE - T1w and coronal cut; SE - T2w

Ovoid soft tissue mass identified as posterior to the inferior aspect of the L4 vertebral body (**c**, ↑) on T1w presenting an isosignal to the parent disc and surrounded by high signal intensity representing epidural fat (**c**, right; **d**, left). Coronal cuts on T1 and T2w showed mass effect on the thecal sac displaced to the right, and compressed and demonstrated nicely the compression of the left L5 root (**d**, *) located at its inner border, displaced in its route along the lateral recess, medial to the pedicle.

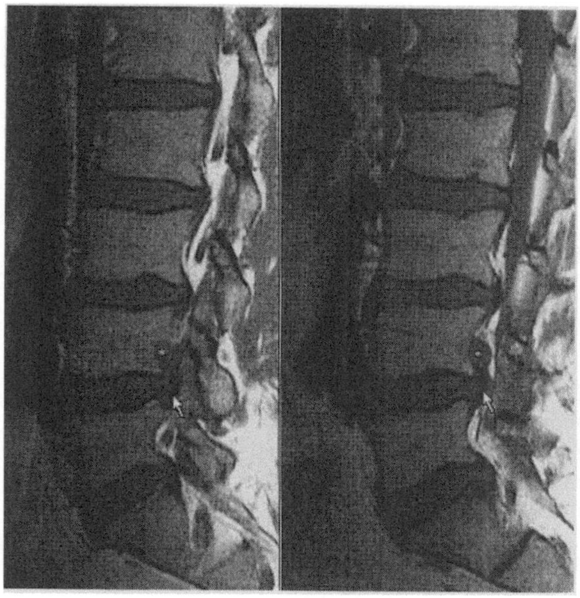

a

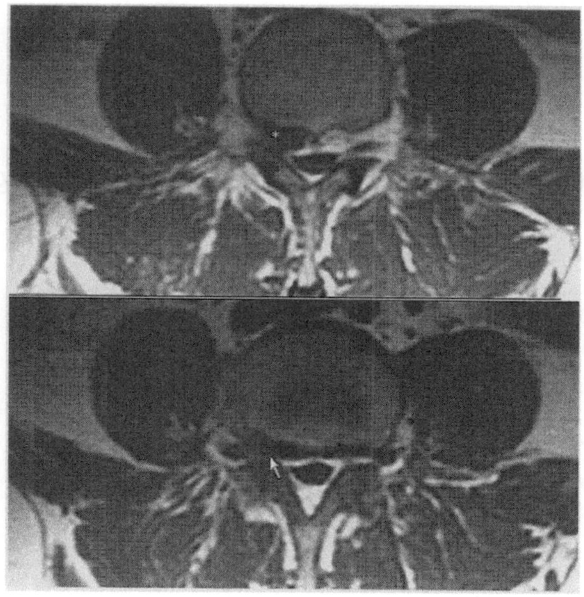

b

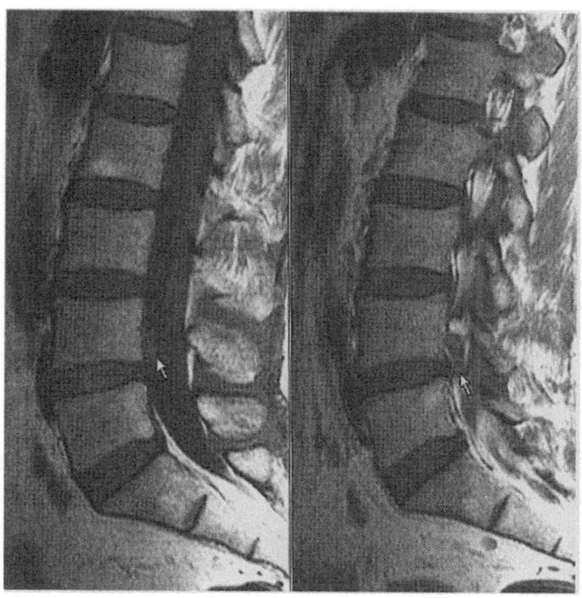

c

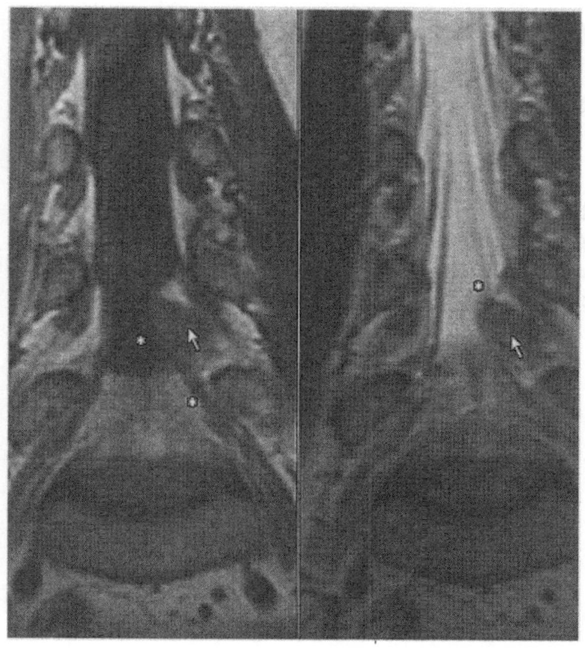

d

Case 49: Downward migrated herniated free fragment

Fig. a: sagittal cut (512); SE - T1w and sagittal cut; SE - T2w

Fig. b: axial cut; SE - T2w

Huge left paracentral lumbar disc free fragment migrating downward and seen as an iso-intense mass on T1w (**a**, left, ↑) and a heterogeneous hypo- and hyperintensity mass on T2w (**a**, right, ↑) extending posteriorly to the superior aspect of the L4 vertebral body. The migrated fragment is surrounded inferiorly by a hyperintense rim. Note the marked narrowing of the L3-L4 disc and even more of the L4-L5 disc with anterior disc herniation. The axial T2w cut shows the mass effect on the dural sac and the intradural roots (**b**, ↑↑). Note also the paracentral location of the migrated disc material (**b**, ↑) and its relation with the midline septum (**b**, *).

Case 50: Migrated herniated free fragment

Fig. c: sagittal cut (512); SE - T1w

Fig. d: axial cut; SE - T2w

Large mushroom-like soft tissue mass iso-intense to the parent disc (**c**, ↑) and slightly hyperintense on T2w (**d**, ↑) extruding at L4-L5 level and migrating into the left paracentral area. A thin hyperintense rim of surrounding epidural fat is still identified on the T1w lateral sagittal cuts (**c**). The mass compresses the ipsilateral nerve root displaced medially in the left lateral recess (**d**, ↑↑) and does not show mass effect on the thecal sac.

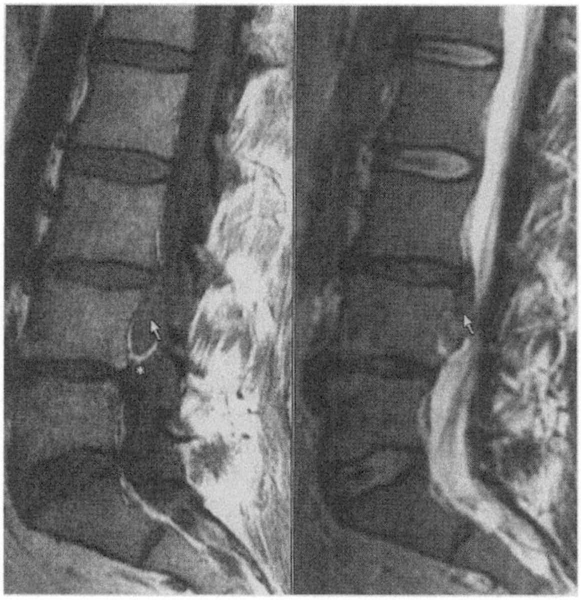

a

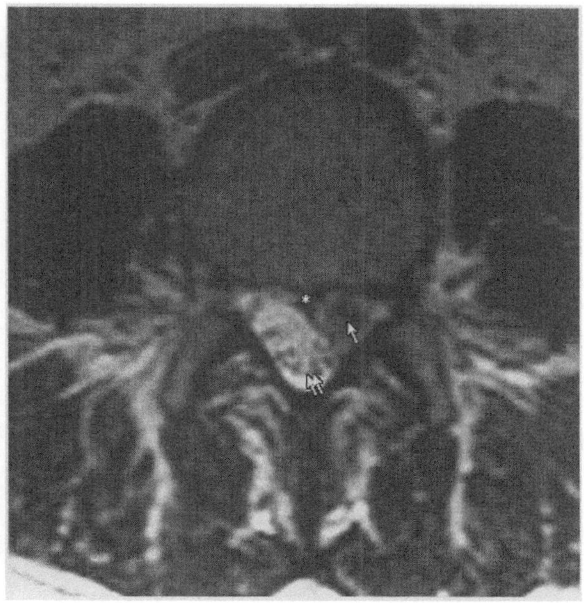

b

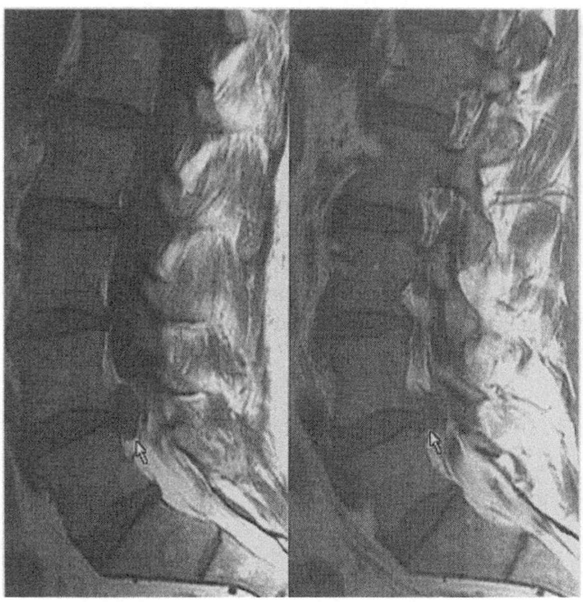

c

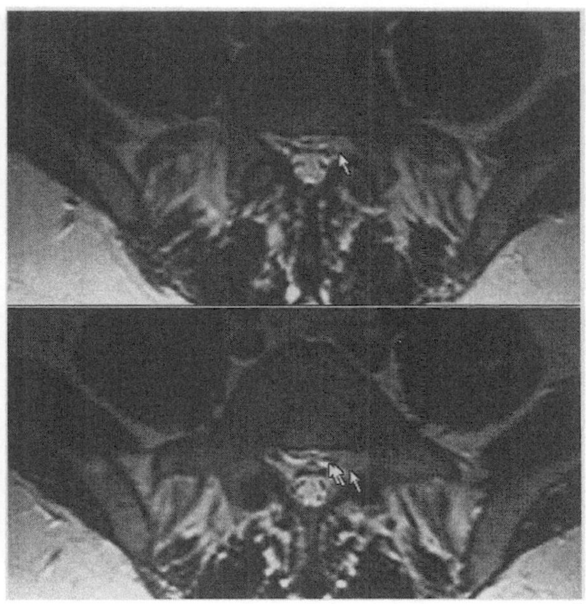

d

Case 51: Transligamentous disc herniation

Fig. a: sagittal cut; SE - T2w and sagittal cut; SE - T1w

Fig. b: axial cut; SE - T1w

MR shows a free fragment that has ruptured through the posterior longitudinal ligament at the L3-L4 level showing an iso-signal on T1w (**a**, right, ↑) and a slightly hyperintense signal similar to the parent disc on T2w (**a**, left, ↑) with stretching of the posterior longitudinal ligament shown on the axial plane (**b**, up). The PLL is seen between two apparent disc fragments as a dark line (**b**, up, ↑) separating the parent disc and the free fragment, and giving a "double fragment sign" due to a tear in the outer annulus and the posterior longitudinal ligament. Note that the free fragment paracentrally located at the right side compresses the dural sac (**b**, ↑).

Case 52: Transligamentous intradural disc herniation

Fig. c: sagittal cut (512); SE - T1w

Fig. d: sagittal cut (512); SE - T2w

Sagittal high resolution T1w (**c**) and T2w (**d**) SE sequences showing a large soft tissue mass presenting a slightly hyperintense signal as compared to the parent disc at L3-L4 level (**c**, left, ↑) and an intermediate signal on T2w (**d**, ↑) showing a transgression of the posterior longitudinal ligament with indentation of the dural sac as evidenced mainly on T2w slices (**d**, *) reaching the cauda equina. The posterior longitudinal ligament is stretched superiorly and inferiorly, displaced posteriorly as well as the subjacent epidural fat, hyperintense on T1w (**c**, right, *). The disc material did not migrate completely within the dural sac but only perforated and entered the subarachnoid space through a tear in the dura.

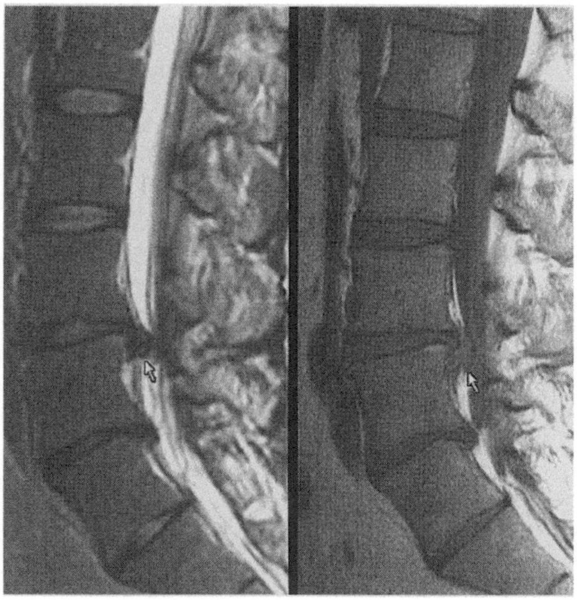

a

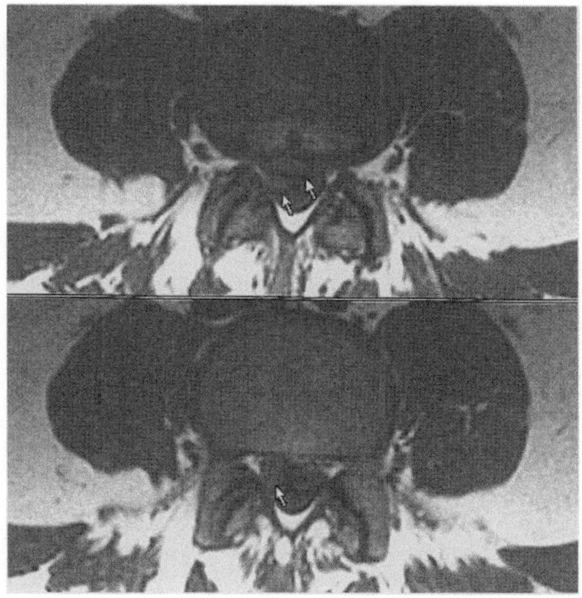

b

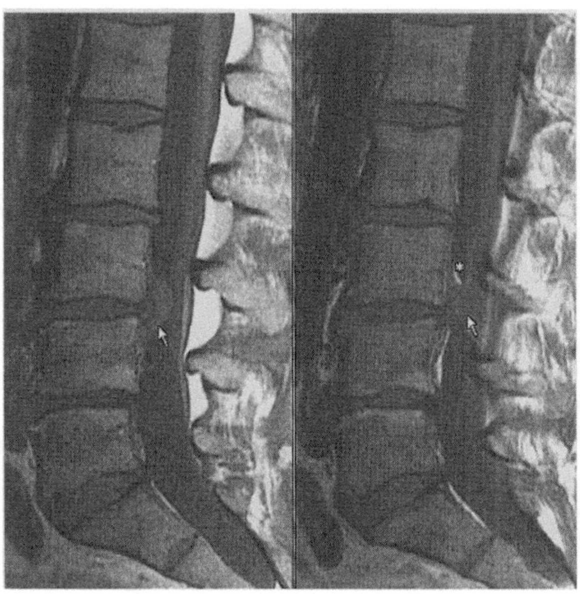

c

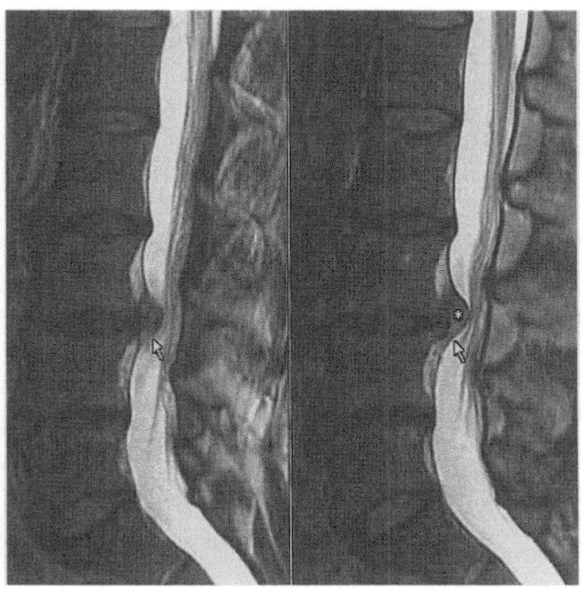

d

Cases 53, 54: postoperative spine

Clinical and pathological features

- Postoperative complications responsible of the so-called failed back surgery syndrome, are variable, occurring in as much as 10% of the cases.
- Almost one third of the operated patients may continue to experience back pain or sciatica following surgery.
- The most frequent causes found are: recurrent disc herniation or failure to complete removal of the hernia, epidural fibrosis formation, lateral or central canal stenosis and arachnoiditis. Other less common causes may occur such as: discitis and/or abscess formation, nerve injury, pseudomeningocele due to tear of the dura, and at times a surgery at a wrong level.
- Imaging of the postoperative spine is critical because the success of a repeated surgery is not obvious.

MR characteristic findings

Morphology

- MRI performed immediately after surgery, is often confusing, rendering almost impossible the diagnosis of a residual disc herniation. The eventual presence of a soft tissue with a similar appearance of the preoperative disc herniation is misdiagnosing. Postoperative hemorrhage may be present at the acute postoperative stage as well, showing no enhancement (less than 1 month after surgery). Vertebral body changes may be found secondary to curettage of the end plates, and may show enhancement, mimicking a discitis.
- Recurrent disc herniation or sequesterd free residual fragment appear as small epidural soft tissue, fairly well delineated, eventually linked with the parent disc, and causing a mass effect on the adjacent nerve root or thecal sac. The disc fragment usually shows an iso-intense or hypointense SI to the disc on all sequences.
- Scar formation evidenced at the discectomy site, above or below the disc level, appears as an epidural soft tissue mass, replacing the fat, usually without mass effect but with eventual retraction of the dural sac toward it. The epidural scar is formed of fibroblasts and collagen and contains capillaries (contrast enhancement observed after 4 to 6 weeks following surgery). Being iso-intense or hypointense on T1w, the epidural fibrosis is commonly hyperintense to the annulus on T2w sequence, in the postoperative period. Increase in fibrous content with time will lead to a decrease in SI on T2w images.
- In most cases, contrast enhanced MR with immediate slice acquisition, permits to differentiate scar from recurrent disc. The overall accuracy of Gd-MR in demonstrating scar tissue is superior to enhanced CT.
- A pseudomeningocele filled with CSF (similar SI) may develop in the posterior soft tissues secondary to tearing of the dura, to be differentiated from postoperative hematoma.
- Long term postsurgical damages may consist in paraspinal denervation atrophy with fatty degeneration of the denervated muscles.

Contrast enhancement

- In the acute postsurgical stage, the disc space shows contrast enhancement in more than 4/5 of the cases with nerve root enhancement in 1/5.
- Contrast administration (with scanning using fat suppressed SE-T1w sequence, performed less than 5 minutes after iv injection) is needed in order to differentiate between recurrent disc and scar formation which shows contrast enhancement. Note that in case of delayed scanning after contrast administration, small discs may enhance as well.
- Peripheral contrast enhancement may be seen bordering the herniated disc.
- In arachnoiditis, clumped nerve roots show variable mild to intense enhancement.

Case 53: Recurrent disc herniation and postoperative scar

Fig. a: sagittal cut; SE - T1w and SE - T2w

Fig. b: axial cut; SE - T1w + Gd with fat suppression

Postoperative sagittal T1w (**a**, left) and T2w (**a**, right) slices showing L5-S1 laminectomy and huge soft tissue mass, hypointense on T1w (**a**, ↑↑) and an intermediate signal on T2w (**a**, *) different from the corresponding L5-S1 disc. After contrast injection, T1w axial slices with fat suppression showed an enhancement of the scar tissue (**b**, ↑), extending to the right posterior soft tissue, surrounding a non enhancing recurrent disc fragment which remained hypointense (**b**, *). This small disc material is also seen on the parasagittal T2w image extruded into the scar tissue (**a**, right, ↑) and differentiated in this case with a slightly hyperintense surrounding scar (**a**, right, *).

Case 54: Sequestered disc and fibrosis

Fig. c: sagittal cut; SE - T2w

Fig. d: sagittal cut; SE - T1w + Gd with fat suppression

Postoperative L5-S1 herniated disc shown as a round soft tissue presenting a slightly higher intensity than the parent disc (**c**, ↑) that seems extruded through the PLL (**c**, *) compressing the ipsilateral nerve root enlarged (**c**, ↑↑). After contrast enhancement, the T1w SE images with fat suppression showed the round hypointense soft tissue corresponding to a sequestered free fragment (**d**, ↑) surrounded by the enhanced scar tissue (**d**, left). Note also the enhancement of the compressed nerve root (**d**, ↑↑).

Contrast enhanced MR is still mandatory for the evaluation of the postoperative lumbar spine in order to differentiate scar vs recurrent of residual sequestered herniated disc as in this case. Fat suppression technique associated to Gadolinium injection appears even more sensitive. Note that it is not always possible as in this case or the previous one to distinguish the scar from the disc material even if the epidural scar may be slightly hyperintense relatively to the disc on T2w images.

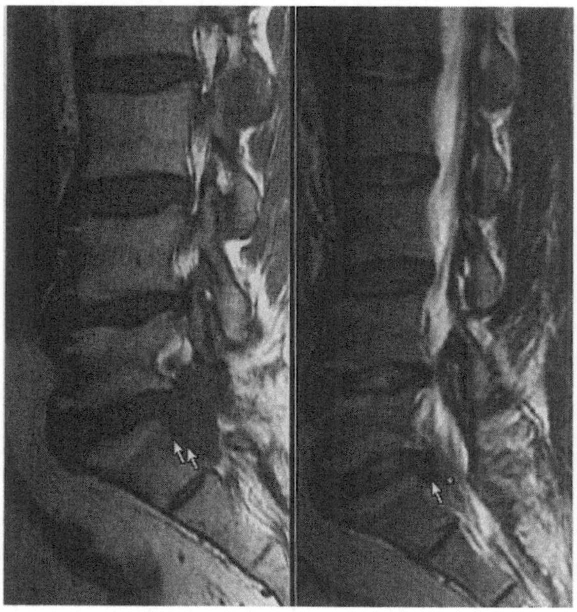

a

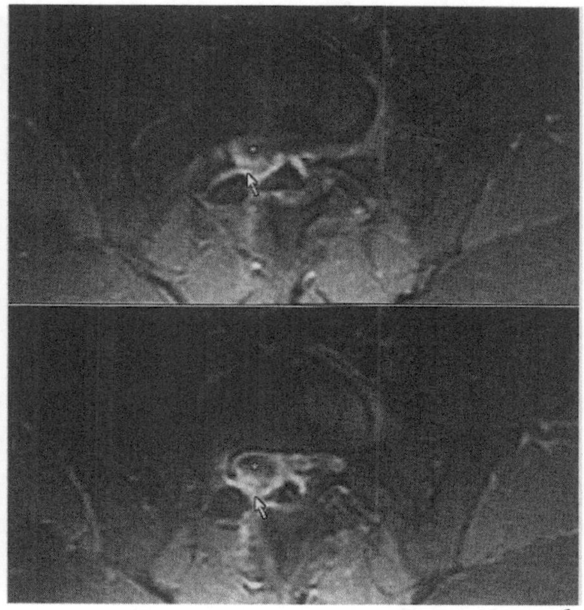

b

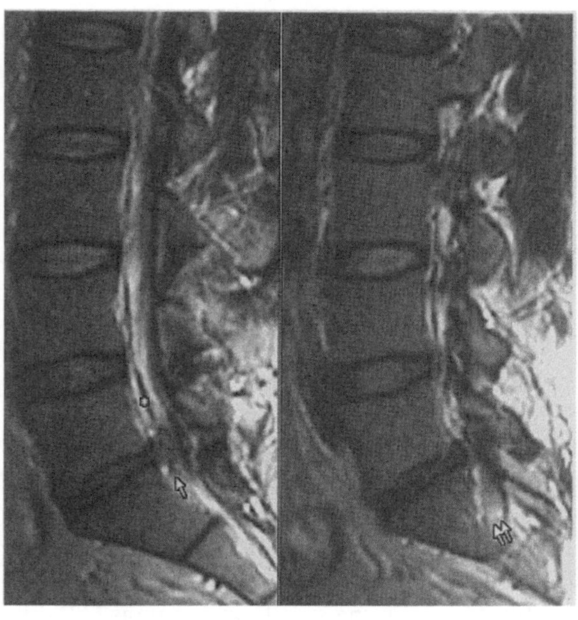

c

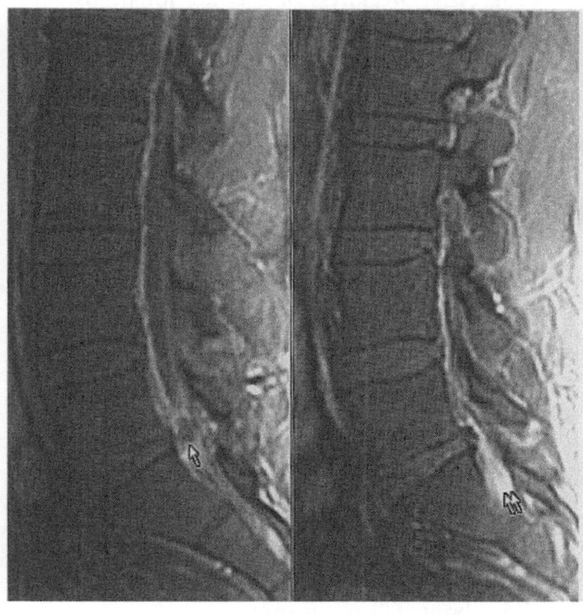

d

Cases 55, 56: spondylolysis and spondylolisthesis

Clinical and pathological features

- Spondylolysis is defined by the presence of a cleft in the pars interarticularis.
- Spondylolisthesis referring to a slippage of a vertebral body with respect to the adjacent one, occurs secondarily to a spondylolysis.
- Different categories of spondylolisthesis may be found: isthmic, degenerative, traumatic (caused by a fracture of the neural arch or following surgical facetectomy or laminectomy), tumoral (due to a metastasis or Paget's disease), and congenital (rare dysplastic type).
- Degenerative spondylolisthesis usually occurs after the sixth decade, with a marked predominance in females presenting more particularly a hyperlordosis and an overweight.
- *Signs:* patients may be asymptomatic or may experience back pain or a neural claudication syndrome. The pain may radiate into the buttocks or to the posterior thighs, bilaterally with worsening of the symptoms with activity or after long standing or walking. The pain is relieved at rest. Progressive weakness of the inferior limbs with paresthesias occur secondarily to associated lumbar spinal stenosis.

MR characteristic findings

Morphology

- Spondylolysis is found in roughly 5% of the general population, developing during childhood.
- The defect of the pars interarticularis involves L5 in almost 2/3 of the cases and L4 in about 1/3 of the cases.
- Spondylolytic defects rarely involve more than one level.
- MR may show variable changes in signal intensity of the pedicles adjacent to the spondylosis.
- Spondylolysis is associated with a spondylolisthesis in about 60% of the cases.
- Spondylolisthesis corresponds to a slipping of one vertebra with respect to the adjacent lower one. It is bilateral in about 2/3 of the cases. In the degenerative type, resulting of a severe degeneration of the facet joints and discs, the anterior vertebral slippage usually does not exceed 10 mm.
- In case of unilateral defect (less than 15%), the contralateral side usually shows a hypertrophy with sclerosis of the pars interarticularis.
- The commonly associated morphological abnormalities include a diffuse bulging of disc material (more common in case of isthmic spondylolisthesis than degenerative), and a deformity of the neural foramina with reduced height, the latter showing a horizontal orientation instead of the normal vertical, paralleling the importance of anterior vertebral slippage. Lateral bulging of the disc contribute to reduce the size of the intervertebral neural foramina.
- In degenerative retrospondylolisthesis, the vertebra is displaced posteriorly in relation to the lower vertebra.
- Plain films commonly show a radiolucent defect eventually associated with sclerosis, best evaluated on lateral oblique incidences. It will demonstrate the degree of the subluxation if present, as well. Dynamic views performed in flexion and extension are mandatory in order to appreciate the abnormal spinal motion. The apparent severity of the disease may vary significantly between the erect and supine positioning.
- With the advent of MR, there is no more indication to perform a myelography to evaluate a spondylolysis with spondylolisthesis.
- CT remains an important imaging tool to evaluate spondylolytic defects and demonstrate the severity of the associated morphological abnormalities of the spinal canal and neural foramina. Synovial cyst formations may be depicted also.
- MR will show the degree of the spondylolisthesis easily displayed on the sagittal cuts. Bulging of the intervertebral disc is also evidenced. Axial cuts will add informations concerning associated changes such as: hypertrophy of the articular processes, thickening of the ligamenta flava, and subluxation of the vertebral bodies, all leading to a stenosis of the spinal canal and the intervertebral foramina. Note that MR and CT are equally effective for measuring spinal canal stenosis (AP diameter of less than 11.5 mm).
- Many causes may be responsible of spinal canal stenosis other than spondylolisthesis: some correspond to an idiopathic narrow spinal canal or are congenital (Klippel-Feil syndrome); the others are of metabolic or bony and cartilaginous etiolo-

gies (acromegaly, fluorosis, Paget's disease, Morquio's syndrome, achondroplasia), but the majority are of the acquired degenerative form. In spinal stenosis, CT and MR demonstrate a reduction of both the midsagittal diameter of the spinal canal and of the interarticular distance, as measured in the axial plane.

- Spondylolytic spondylolisthesis account for about 10% of cases of the lumbar spinal stenosis.

Contrast enhancement

- Contrast administration is needed in case of pathological spondylolisthesis, as in case of metastatic diseases.

Case 55: Spondylolysis with spondylolisthesis

Fig. a: sagittal cut (512); SE - T1w

Fig. b: sagittal cut (512); SE - T1w

Parasagittal T1w sequence through the level of the L5-S1 neural foramina (a, left) and the midsagittal plane (**a**, right) showing a very mild anterior displacement of L5 on S1. Note the interruption of the pars interarticularis (**a**, left, ↑) with separation of the superior and inferior articular processes of L5 vertebra. Note the apparent diffuse disc bulging (**a**, ↑↑) encroaching on the L5 neural foramen narrowed, consistent with foraminal stenosis compressing the nerve root (**a**, left, *). Note the vacuum phenomenon in the degenerated L5-S1 disc (**a**, ↑) showed as a linear hypointense signal.
A similar case is also shown on the parasagittal T1w cuts (**b**) with a spondylolysis L5-S1 with nerve root compression in the L5-S1 neural foramen.

Case 56: Spondylolysis

Fig. c: sagittal cut (512); SE - T1w

Fig. d: axial cut; SE - T1w

Marked degenerative lesions involving the intervertebral disc and plates of the ajdacent vertebral bodies at L4-L5 level showing a hypointense signal on T1w (**c**). The parasagittal slices show an interruption of the pars interarticularis separating the superior and the inferior articular processes of L4 with cephalocaudal narrowing of the L4-L5 neural foramen. Axial slices performed using SE-T1w sequence demonstrated the associated presence of a rounded hypointense lesion developed in the right lateral recess, extending to the ipsilateral neural foramen (**d**, *), compressing the adjacent nerve root with loss of the normal epidural fat surrounding it in the neural foramen. This soft tissue mass corresponds to Gilles'nodule.

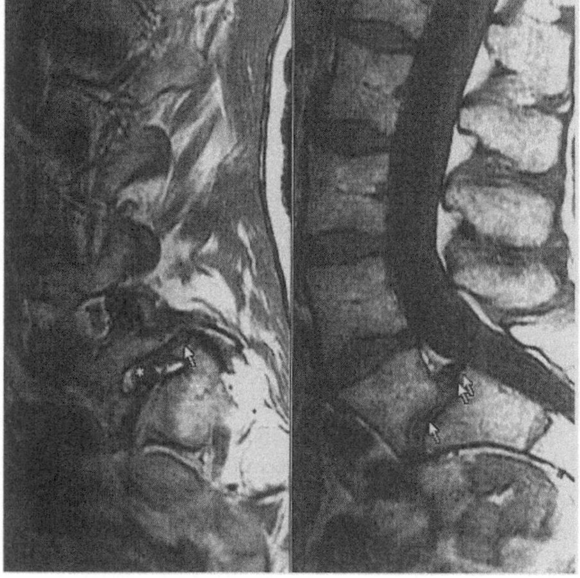
a

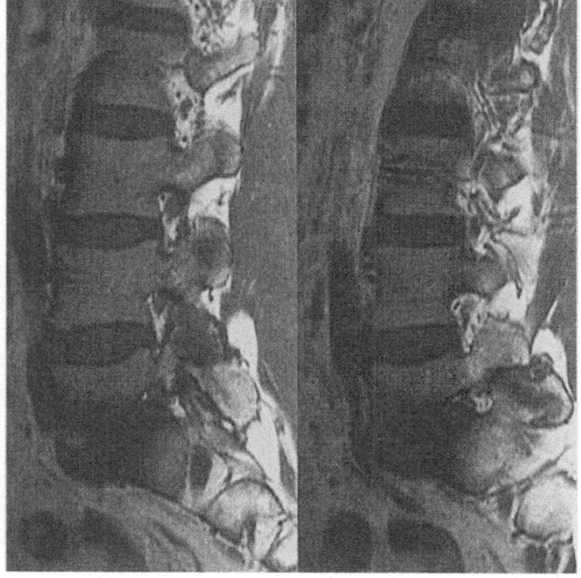
b

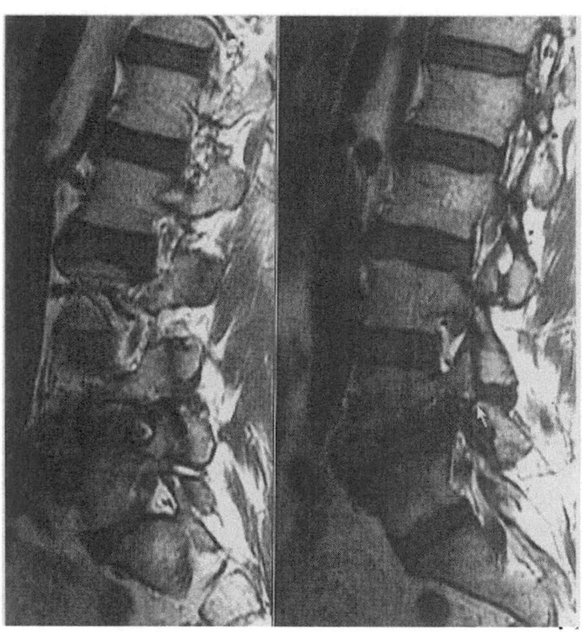
c

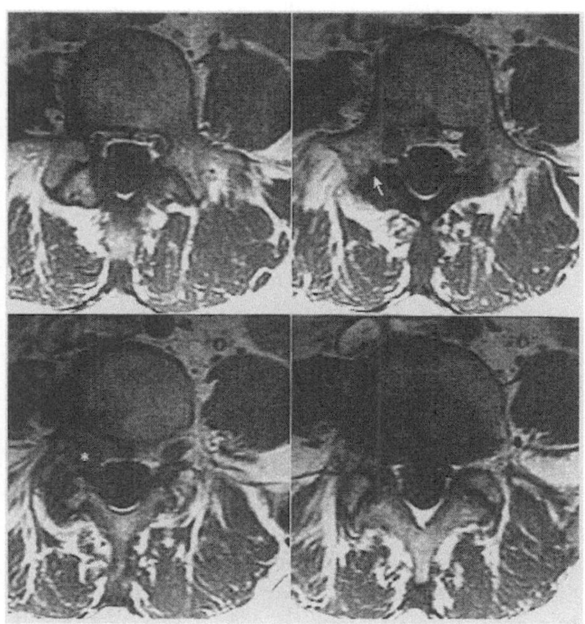
d

Selected References

- Abdullah AF, Wolbert PGH, Warfield JK, Krist Gunadi I (1988) Surgical management of extreme lateral lumbar disc herniations; review of 138 cases. Neurosurgery 22: 648-653
- Aoki J, Yamamoto I, Kitamura N et al (1987) End plate of the discovertebral joint: degenerative change in the elderly adult. Radiology 164:411-414
- Araki Y, Ootani M, Furukawa T et al (1992) Pseudo-enhancement of intervertebral disc herniation. Neuroradiology 34:271-272
- Arasaki K, Kanazawa I, Anno I (1992) Hypertrophy of the posterior longitudinal ligament: relationship with ossification of the posterior longitudinal ligament. J Neurosurg 2: 158-161
- Arce CA, Dohrmann GJ (1985) Thoracic disc herniation. Surg Neurol 23:356-361
- Arce CA, Dohrmann GJ (1985) Herniated thoracic disks. Neurol Clin 3:383-392
- Awwad EE, Martin DS, Smith KR (1992) The nuclear trail sign in thoracic herniated disks. AJNR 13:137-143
- Batzdorf U, Flannigan BD (1991) Surgical decompressive procedures for cervical spondylotic myelopathy: a study using magnetic resonance imaging. Spine 16:123-127
- Bernick S, Walder JM, Paule WJ (1991) Age changes to the anulus fibrosus in human intervertebral discs. Spine 16:520-524
- Berns DH, Ross JS, Kormos D, Modic MT (1991) The spinal vacuum phenomenon: evaluation by gradient echo MR imaging. J Comput Assist Tomogr 15:233-236
- Blumenkopf B (1988) Thoracic intervertebral disc herniations. Diagnostic value of magnetic resonance imaging. Neurosurgery 23:36-40
- Bundschuh CV, Modic MT, Ross JS et al (1988) Epidural fibrosis and recurrent disk herniation in the lumbar spine: assessment with MR. AJNR 9:169-178
- Ciapetta P, Delfini R, Cantore GP (1981) Intradural lumbar disc hernia: description of three cases. Neurosurg 8: 104-107
- Crisi G, Carpeggiani P, Trevisan C (1993) Gadolinium-enhanced nerve roots in lumbar disk herniation. Am J Neuroradiol 14:1379
- Delauche Cavallier MC, Budet C, Laredo JD (1992) Lumbar disc herniation. Spine 17(8):927
- Djukic S, Genant HK, Helms CA et al (1990) Magnetic resonance imaging of the postoperative lumbar spine. Radiol Clin North Am 28:341-360
- Eisenberg RA, Bremer AM, Northup HM (1986) Intradural herniated cervical disk: a case report and review of the literature. AJNR 7:492-494
- Epstein NE, Epstein JA, Carras R, Hyman RA (1990) Far lateral lumbar disc herniations and associated structural abnormalities. Spine 15:534-539
- Epstein NE, Syrquin MS, Epstein JA, Decker RE (1990) Intradural disc herniations in the cervical, thoracic and lumbar spine: report of three cases and review of the literature. J Spinal Disord 3:396-403
- Firoozina H, Rafii M, Golimbuc et al (1984) Computed tomography of the calcification and ossification of the posterior longitudinal ligament of the spine. J Comput Tomogr 8:3117-3124
- Fridenberg ZB, Miller WT (1963) Degenerative disc disease of the cervical spine: a comparative study of asymptomatic and symptomatic patients. J Bone Joint Surg 45A:1171-1178
- Grenier N, Greselle J-F, Douws C, Vital JM et al (1990) MR imaging of foraminal and extraforaminal lumbar disk herniations. J Comput Assist Tomogr 14:243-249
- Hamanishi C, Kawabata T, Yosii T et al (1994) Schmorl's modes on magnetic resonance imaging: their incidence and clinical relevance. Spine 19:450
- Harada A, Mimatsu K (1992) Prospective changes in the spinal cord in cervical myelopathy demonstrated by magnetic resonance imaging. Spine 17:1275-1280
- Harsh GR, Sypert GW, Weinstein PR et al (1987) Cervical spine stenosis secondary to ossification of the posterior longitudinal ligament. Neurosurg 67:349-357
- Hassine D, Perret C, Paraire F et al (1992) Gadolinium-enhanced MR imaging of inflammation associated with preoperative herniated disks: new approach to sequestered disks? Radiology 185:403
- Hsu K, Zucherman J, Shea W et al (1990) High lumbar disc degenerative: incidence and etiology. Spine 679-682
- Hueftle M, Modic MT, Ross JS et al (1988) Lumbar spine: post-operative MR imaging with Gd-DTPA. Radiology 167:817-824
- Jackson RP, Glah JJ (1987) Foraminal and extraforaminal lumbar disc herniation: diagnosis and treatment. Spine 12:577-585
- Jensen M, Brant-Zawadzki M, Obuchowski N et al (1994) MRI of lumbar spine in the asymptomatic subject, N Engl J Med 331:69
- Jinkins JR (1993) MR of enhancing nerve roots in the unoperated lumbosacral spine. AJNR 14:193-202
- Jinkins JR, Bashi R, Al-Mefdy O et al (1986) Cystic necrosis of the spinal cord in compressive cervical myelopathy: demonstration by iopamidol CT-myelography. AJR 147:767-775
- Jinkins RJ (1993) MR of enhancing nerve roots in the unoperated lumbosacral spine. Am J Neuroradiol 14:193
- Kataoka O, Nishibayashi Y, Sho T (1989) Intradural disc herniation: report of three cases with a review of the literature. Spine 14:529-532
- Krause D, Drape JL, Maitrot D et al (1991) CT and MRI of disk herniations. Springer-Verlag, Berlin, pp 3-38
- Logue V (1952) Thoracic intervertebral disc prolapse with spinal cord compression. Neurol Neurosurg Psychiatry 15: 227-241
- Manelfe C, Chevrot A, Arrue P et al (1992) Disc herniations. In: Manelfe (ed) Imaging of the spine and spinal cord. Raven Press, New York, pp 269-332

- Masaryk TJ, Ross JS, Modic MT et al (1998) High resolution of sequestered lumbar intervertebral disks. Am J Roentgenol 150:1155
- Massaryk TJ, Ross JS, Modic MT et al (1988) High resolution MR imaging of sequestered lumbar intervertebral disks. AJNR 9:351-358
- Mehalic TF, Pezzuti RT, Applebaum BI (1990) Magnetic resonance imaging and clinical spondylotic myelopathy. Neurosurgery 26:217-227
- Milier JAA, Schmartz C, Schultz AB (1988) Lumbar disk degeneration: correlation with age, sex, and spine level in 600 autopsy specimens. Spine 13:173
- Miller P, Elder F (1969) Meningeal pseudocysts (meningocele spurius) following laminectomy. Report of 10 cases. J Bone Joint Surg 50A:268-276
- Modic MT, Masaryk TJ, Mulopulos GP et al (1986) Cervical radiculopathy: prospective evaluation with surface coil MR imaging CT with metrizamide and metrizamide myelography. Radiology 151:573-579
- Modic MT, Masaryk TJ, Ross JS, Carter JR (1988) Imaging of degenerative disk disease. Radiology 168:177-186
- Modic MT, Ross JS, Obuchowski N et al (1995) Contrast enhanced MR and acute lumbar radiculopathy: a pilot study of the natural history. Radiology 195:429
- Modic MT, Steinberg PM, Ross JS et al (1988) Degenerative disk disease: assessment of changes in vertebral body marrow with MR imaging. Radiology 166:193-199
- Modic MT (1989) Degenerative disorders of the spine. In: Modic MT, Masaryk TJ, Ross JS (eds) Magnetic resonance imaging of the spine. Year Book Medical Publishers, Chicago, pp 75-119
- Montaldi S, Fankhauser H, Schynyeler P, de Triboley N (1989) Computed tomography of the postoperative intervertebral disc and lumbar spinal canal: investigation of twenty-five patients after successful operation for lumbar disc herniation. Neurosurg 22:1014-1022
- Mortensen WW, Thorne RP, Donaldson WF III (1991) Symptomatic gas-containing disc herniation: report of four cases. Spine 16:190-192
- Murakami N, Muropa T, Sobue I (1978) Cervical myelopathy due to ossification of the posterior longitudinal. Ann Neurol 35:33-36
- Murphey F, Simmons JCH, Brunson B (1973) Ruptured cervical disks, 1939-1972. Clin Neurosurg 20:9-17
- Nabors MW, Pait TG, Byrd EB et al (1988) Updated assessment and current classification of spinal meningeal cysts. J Neurosurg 68:366
- Nagata K, Kiyonaga K, Ohasti T et al (1990) Clinical value of magnetic resonance imaging for cervical myelopathy. Spine 15:1088-1096
- Okamoto A, Shinomya BD, Furuya K, Yamaura I (1991) Postoperative magnetic resonance imaging in patients with cervical myelopathy. Spine 16:S30-S35
- Osborne AG, Hood RS, Sherry RG, Smoker WRK, Harnsberger HR (1988) CT/MR spectrum of far lateral and anterior lumbosacral disc herniations. AJNR 9:775-778
- Osti OL, Vernon-Roberts B, Moore R, Fraser RD (1992) Annular tears and disc degeneration in the lumbar spine: a post-mortem study of 135 discs. J Bone Joint Surg 74B: 678-682
- Otake S, Matsuo M, Nishizawa S et al (1992) Ossification of the posterior longitudinal ligament MR evaluation. AJNR 13:1059-1067
- Parizel PM, Rodesch G, Baleriaux D, Zegers de Beyl D, D'Haens J, Noeterman J, Segebarth C (1989) Gd-DTPA-enhanced MR in thoracic disc herniations. Neuroradiology 31:75-79
- Pavlov H, Torg JS, Robie B, Jahre C (1987) Cervical spinal stenosis determination with vertebral bod radio method. Radiology 164:771-775
- Ramos Remus C, Gomez-Vargas A, LeClercq S et al (1993) Radiologic features of DISH may mimic ankylosing spondylitis, Clin Exp Rheumatol 11:603
- Resnick D, Guerra J, Robinson C, Vint V (1978) Association of diffuse idiopathic skeletal hypertrophy (DISH) and calcification and ossification of the posterior longitudinal ligament. AJR 131:1049-1053
- Ridenour T, Haddad S, Hichon P, VanGilder JC, Menezes A, Tayneils (1991) Herniated thoracic disks: operative procedures and clinical results. J Spinal Discord 4:397-398
- Roosen N, Dietrich U, Nicole N, Irlich G, Gahlen D, Stork W (1987) MR imaging of the calcified herniated thoracic disk. J Comput Assist Tomogr 11:733-735
- Ross JS, Masaryk TJ, Modic MT et al (1987) Lumbar spine: postoperative assessment with surface-coil MR imaging. Radiology 164:851-860
- Ross JS, Masaryk TJ, Schrader M et al (1990) MR imaging of the postoperative lumbar spine: assessment with gadopentetate dimeglumine. AJNR 11:771-776
- Ross JS, Modic MT, Masaryk TH et al (1989) Assessment of extradural degenerative disease with Gd-DTPA-enhanced MR imaging: correlation with surgical and pathologic findings. AJNR 10:1242-1249
- Ross JS, Modic MT, Masaryk TJ (1990) Tears of the anulus fibrosus: assessment with Gd-DTPA-enhanced MR imaging. AJR 10:1251-1254
- Ross JS, Modic MT, Masaryk TJ et al (1989) Assessment of extradural degenerative disease with Gd-DTPA-enhanced MR imaging correlation with surgical and pathologic findings. AJNR 10:1243-1249
- Ross JS, Modic MT, Massaryk TJ (1989) Tears of the anulus fibrosus assessment with Gd-DTPA-enhanced MR imaging. AJNR 10:1251-1254
- Ross JS, Modie MT, Masaryk TJ et al (1989) Assessment of extradural degenerative disease with Gd-DTPA-enhanced MR imaging: correlation with surgical and pathologic findings. Am J Neuroradiol 10:1243
- Ross JS, Perez-Reyes N, Massaryk TJ, Bohlman H, Modic MT (1987) Thoracic disk herniation: MR imaging. Radiology 165:511-515
- Sakamoto R, Ikata T, Murase M et al (1991) Comparative study between magnetic resonance imaging and histopathologic findings in ossification or calcification of ligaments. Spine 16:1253-1261
- Sakow T, Taketomi E, Matsunaga S et al (1991) Genetic study of ossification of the posterior longitudinal ligament in the cervical spine. Spine 16:1249-1252

- Schneider SJ, Grossman RG, Bryan RN (1988) Magnetic resonance imaging of transdural herniation of a cervical disk. Surg Neurol 30:216-219
- Silverman CS, Lenchik L, Shimkin PM et al (1995) The value of MR in differentiating subligamentous form supraligamentous lumbar disk herniations. Am J Neuroradiol 16:571
- Spangfort EV (1972) The lumbar disc herniation. A computer-aided analysis of 2504 operations. Acta Orthop Scand (suppl) 142:40-44
- Takahasi M, Yamashita Y, Sakamoto Y et al (1989) Chronic cervical cord compression; clinical significance of increased signal intensity on MR images. Radiology 173: 219-224
- Wasserstrom R, Momourian AC, Black JF, Lehman RAW (1993) Intradural lumbar disk fragment with ring enhancement. AJNR 14:401-404
- White JG, III, Strait TA, Binkley JR, Hunter SE (1982) Surgical treatment of 63 cases of conjointed nerve roots. J Neurosurg 56:114-117
- Wilmink JT (1989) Clinical relevance of cervical disk herniation diagnosed on the basis of MR imaging. AJNR 10: 1278-1279
- Yamashita Y, Takahashi M, Matsuro Y et al (1990) Spinal cord compression due to ossification of ligaments: MR imaging. Radiology 175:843-848
- Yone K, Sakou, Yanane M, Ijiri K (1992) Preoperative and postoperative magnetic resonance image evaluations of the spinal cord in cervical myelopathy. Spine 17 S388-S392

Part III

Spine and cord

3- Infectious and inflammatory diseases

a- Infectious diseases
b- Inflammatory spinal diseases
c- Inflammatory spinal cord diseases

Infectious diseases

Cases 57-63: spondylitis and discitis

Clinical and pathological features

- Spinal infectious disease occurs in about 3% of skeletal pyogenic osteomyelitis. Spondylitis may be due to infection caused by pyogenic or non pyogenic (tuberculosis, brucellosis) organisms.
- Spinal infections are relatively uncommon, affections arising secondary to hematogeneous spread or direct involvement postsurgery.
- Pyogenic infections are caused by gram positive aerobic cocci as *Staphylococcus aureus* (more than 50%) and *Streptococcus pneumoniae,* and gram negative germs (almost 30%) as *Escherichia coli, Pseudomonas aeruginosa* and *Salmonella.* Polymicrobial infections and anaerobic infections are uncommon.
- Several predisposing factors have been reported; among these, diabetes mellitus, intravenous drug abuse and immunosuppressed state appear as the commonest.
- MRI is the imaging modality of choice for the diagnosis and follow-up of spondylodiscitis. Plain radiographs are initially normal showing abnormalities (irregular erosion of vertebral end plates) only late detecting osteomyelitis one or two months after the onset of the symptomatology. MRI is at least as sensitive as combined technetium and gallium bone scans and is more specific.
- *Signs:* spinal osteomyelitis causes a severe back pain of acute or progressive onset, usually associated with focal tenderness on the neurological exam. Genitourinary tract infections (half the cases) followed by skin infections (one fifth) and pneumonia (less than one fifth) are the most frequent primary sources. A propensity for adults affecting men in the six and seventh decades is reported. In childhood girls younger than 2 years old are affected. Postsurgical infections, causing a recrudescence of back pain, are uncommon occurring with an incidence ranging between 2-9%, usually 2 weeks to 3 months after surgery. A marked increase in the erythrocyte sedimentation rate is usually found, showing an accurate correlation with the course of the disease, decreasing with improvement. White blood cell count may be normal in more than half the cases. Needle biopsy may be negative in half to almost two thirds of the cases.

MR characteristic findings

Morphology

- Pyogenic spondylodiscitis shows a predilection for the lumbar spine involved in almost half the cases, followed by the dorsal and the cervical. In nonpyogenic infections, tuberculous spondylitis involves most commonly the lower thoracic spine and the thoracolumbar junction, and brucellosis tends to involve the lower lumbar spine (L4 particularly).
- The imaging features of infectious spondylodiscitis include: edematous marrow changes involving end plates subchondral bone of both adjacent vertebrae, with a more or less symmetrical distribution (hypointense on T1w and hyperintense on T2w, similar to that seen in degenerative end plate disease type I), associated later with disc involvement showing loss in height, absence of visualization of the intranuclear cleft and signal changes (hyperintense on T2w).
- The intervertebral discs are involved early in the majority of patients with pyogenic spondylitis, but tend to be spared in brucellosis and only later in the course of the disease are involved and markedly destroyed in tuberculosis. Tuberculosis may involve multiple vertebral levels (3 or more) secondary to subligamentous spread, with sparing of the discs mimicking metastatic disease which usually does not cross the disc space.
- The posterior vertebral elements are uncommonly involved in pyogenic infections except in advanced forms, whereas they are affected in about 50% of the cases and even solely in tuberculosis, though tend to be spared in brucellosis. In case of involvement of the posterior elements with associated soft tissue mass, one must rule out a metastatic process.
- Associated changes may be present such as: paravertebral and/or epidural extension of the granulation tissue, paraspinal abscess, or leptomeningitis. Paraspinal soft tissue infiltration and/or psoas abscesses are commonly found and huge in tuberculosis, being relatively more prominent than with pyogenic (20% of the cases) infections and rarely encountered in brucellosis. Epidural extension (contrast administration with fat sup-

pression is mandatory) of the disease process is variably present in pyogenic as well as in tuberculous spondylodiscitis. Intraosseous abscesses and kyphotic spinal deformities due to vertebral body and disc collapse are common (60%) in tuberculosis.

Contrast enhancement

- In pyogenic spondylodiscitis, contrast infusion leads to marked enhancement of vertebral endplates bone marrow (commonly as horizontal bands on either side of the disc), and of the disc space. The enhancement is usually much more intense in infectious than in degenerative vertebral and disk disease (Modic type I). Disc enhancement in discitis is variable, patchy or diffuse, densely marginal or central linear in earliest stages.
- Contrast administration may help to depict and evaluate associated extensions of the disease process to the epidural and paraspinal soft tissue spaces, or demonstrate leptomeningeal involvement.
- Epidural abscesses usually found in the anterior epidural space show 2 different patterns of contrast enhancement (T1w with fat suppressed): a homogeneous contrast uptake corresponding to a granulation tissue mass (phlegmon) and a peripheral enhancement surrounding a non enhancing central fluid (drainable pus). A combined pattern may be shown. Signs of spinal osteomyelitis are found in almost one third of the cases.

Case 57: Spondylodiscitis (staphylococcus)

Fig. a: sagittal cut (512); SE - T1w

Fig. b: sagittal cut; SE - T2w with fat suppression

Fig. c: sagittal cut (512); SE - T1 + Gd with fat suppression

Fig. d: axial cut; SE - T1w + Gd with fat suppression

MR shows a decreased signal intensity from the L4-L5 intervertebral disc and the adjacent vertebral body and plates (**a**, ↑↑). T2w sequence with fat suppression demonstrated a correlated hyperintense signal involving the end plates as well as the disc. These signal modifications most likely result from increased water content of the inflammatory exudate replacing the normal marrow signal. Note that the disc margins are slightly indistinct, and on T2w with fat suppression note the hyperintense signal of the posterior infiltrated soft tissues (**b**, ↑↑↑). After contrast iv, T1w sequences with fat suppression technique showed on the sagittal (**c**) and axial cuts (**d**) a marked contrast enhacement of the discovertebral junction at the inferior and superior end plates (**c**, ↑↑) as well as a slight enhancement of the intervertebral disc with a ventral epidural collection (**c**,**d**, ↑) associated to a herniated material (**a**, **c**, *). Note also the epidural extension of the infectious process (**c**, ↑) behind L5 vertebral body, involving the epidural lateral recess (**d**, ↑↑) on both sides and mainly on the right side. Note that in discitis the intranuclear cleft is not shown and the disc involved shows a hyperintense signal on T2w unlike metastatic disease which usually lacks of involvement of the disc space.

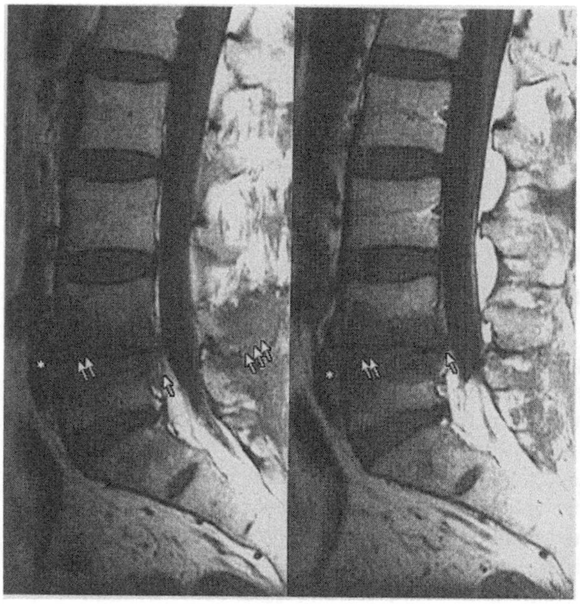

a

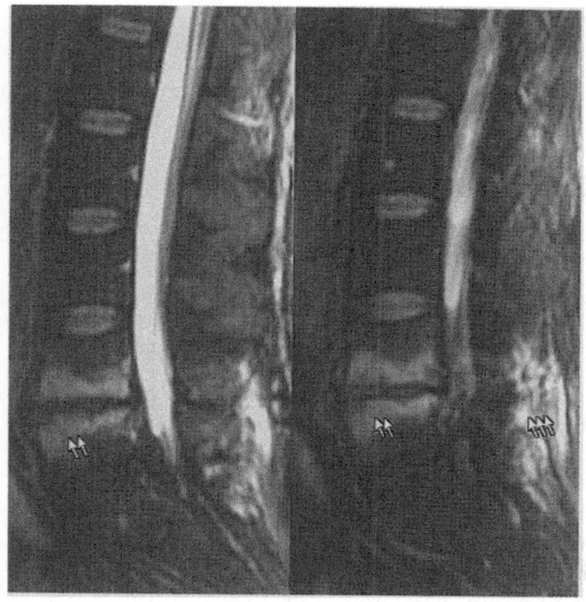

b

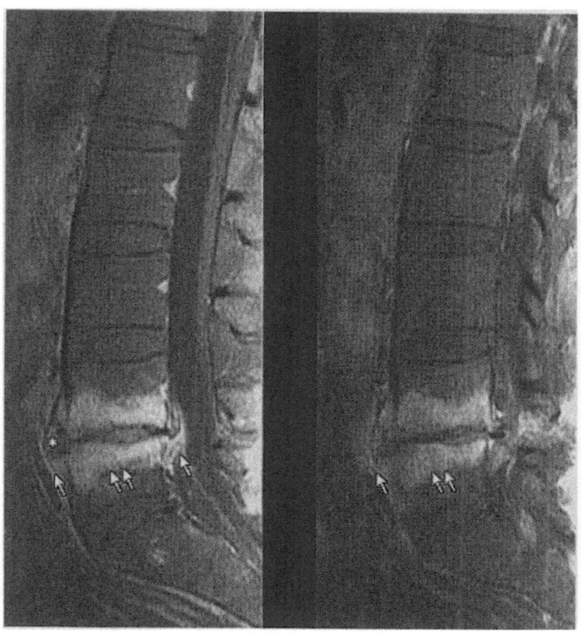

c

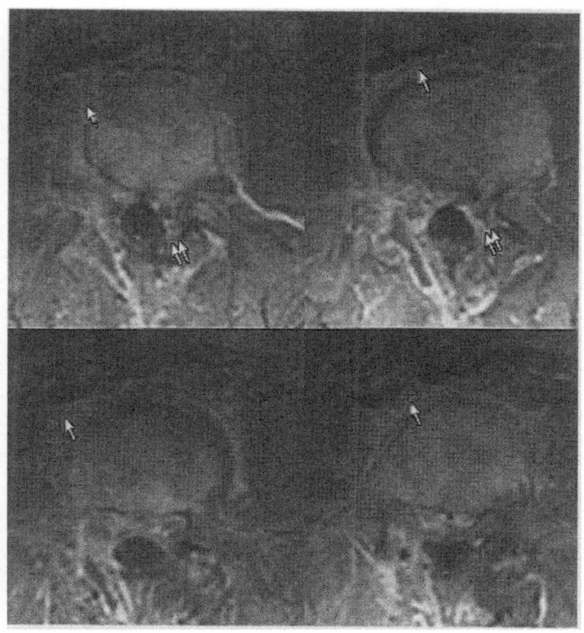

d

Case 58: Spondylodiscitis (brucella)

Fig. a: sagittal cut (512); SE - T1w

Fig. b: sagittal cut; SE - T2w with fat suppression

Fig. c: axial cut; SE - T2w

Fig. d: axial cut; SE - T1w + Gd with fat suppression

MR shows decreased intensity of the L4 and L5 vertebral body marrow on T1w (**a**, ↑↑). Note the areas of indistinct disc margin (**a**, *, **b**, ↑) and the hypointense signal of the epidural spaces behind L4 and L5 bodies (**a**, ↑) with erasement of the bright epidural fat. On T2w sequence with fat suppression, the disc appears slightly heterogeneous with areas of hyperintense signal (**b**, right) associated to areas of indistinct margins (**b**, ↑) and extensive hyperintense signal involving the L4 body (**b**, left, ↑↑) as well as the superior end-plate of the L5 vertebra (**b**, right, ↑↑). Note the hyperintense signal of the epidural spaces and the prevertebral soft tissue. On the axial cuts, T2w sequence shows the hyperintense soft tissue developed in the left lateral recess (**c**, ↑) with mass effect on the thecal sac associated to a facet joint effusion (**c**, *). After Gadolinium injection, there is marked contrast enhancement involving the epidural spaces and the lateral recesses (**d**, ↑) as well as the adjacent interspace disc (**d**, *) with extension of the inflammatory process to the pre- and latero-vertebral soft tissues (**d**, ↑↑). Note that on sagittal T2w, the intranuclear cleft is still observed.

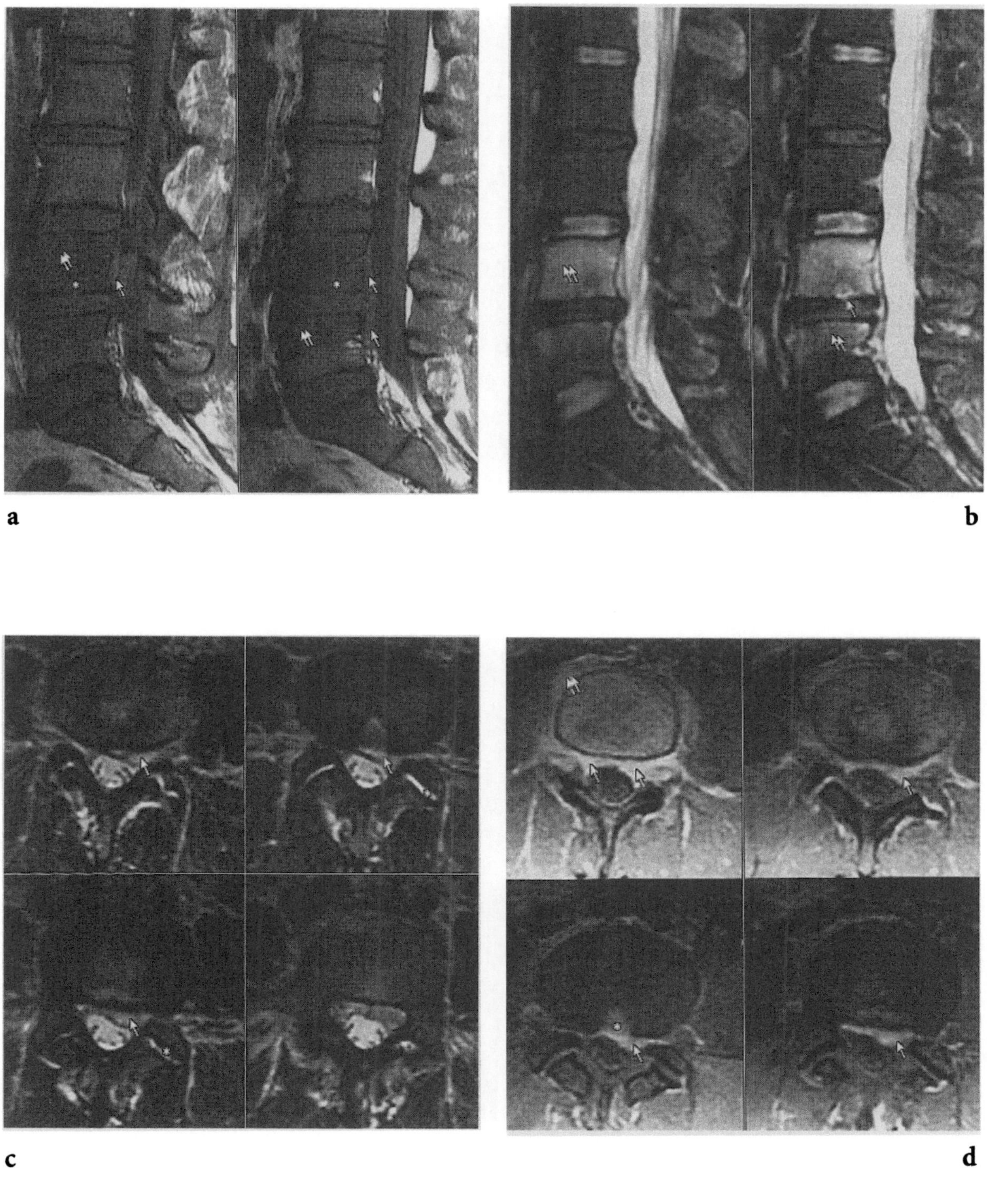

a b c d

Case 59: Spondylodiscitis (brucella)

Fig. a: sagittal cut; SE - T2w with fat suppression

Fig. b: sagittal cut (512); SE - T1w + Gd with fat suppression

Abnormal slightly heterogeneous hyperintense signal involving the adjacent end plates anteriorly and posteriorly (**a**, ↑↑) as well as the intervertebral disc markedly laminated at the L5-S1 level (**a**,↑) as shown on T2w with fat suppression. Note also the hyperintense signal coming from the L5-S1 disc (**a**, ↑). On T1w with fat suppression after Gadolinium iv, sagittal images show enhancement of the anterior and posterior end plates at L4-L5 level (**b**, ↑) as well as restricted areas of disc margin (**b**, *). A diffuse contrast enhancement is also observed involving the anterior paravertebral soft tissues (**b**, ↑↑) as well as the posterior epidural space (**b**, ↑). Note the marked degenerative appearance of L4-L5 disc space and the degenerative changes involving also the L5-S1 disc.

Case 60: Spondylodiscitis with epidural abscess

Fig. c: sagittal cut; SE - T2w with fat suppression

Fig. d: sagittal cut (512); SE - T1w and sagittal cut; SE - T1w + Gd with fat suppression

Sagittal MR T2w with fat suppression (**c**) and T1w (**d**, left) showing the usual T1 and T2 pattern of the vertebral end plates at L4-L5 level seen in infectious spondylitis which is similar to that seen in type I degenerative end-plate changes. A hypointense signal on T1w (**d**) and a hyperintense signal on T2w with fat suppression (**c**) is shown involving the L5 vertebral body and the anterior portion of the L4 end-plate associated to a hyperintense signal of the intervertebral disc (**c**, *) with absence of visualization of the intranuclear cleft and indistinct identification of the L5 superior vertebral margin (**c**,↑). Note the enhancement of this area of the intervertebral disc and the superior L5 end-plate after contrast IV (**d**, right, *) associated to an enhancing soft tissue in the posterior epidural space (**d**, ↑↑) with enhancement of the prevertebral soft tissue (**d**, ↑). Note the normal hyperintense signal of the L3-L4 disc and the presence of the intranuclear cleft.

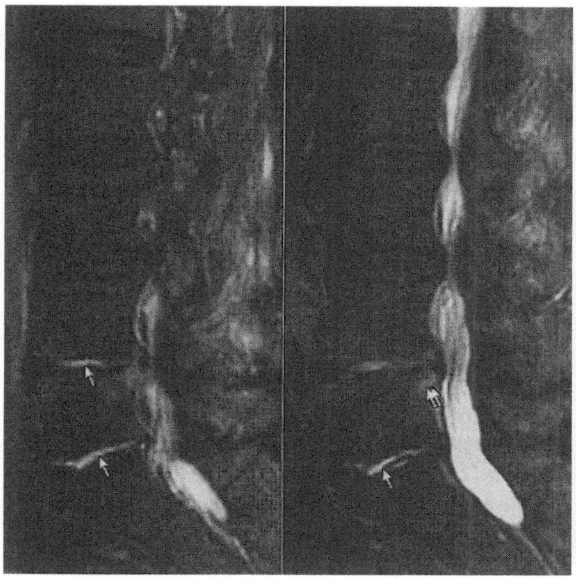
a

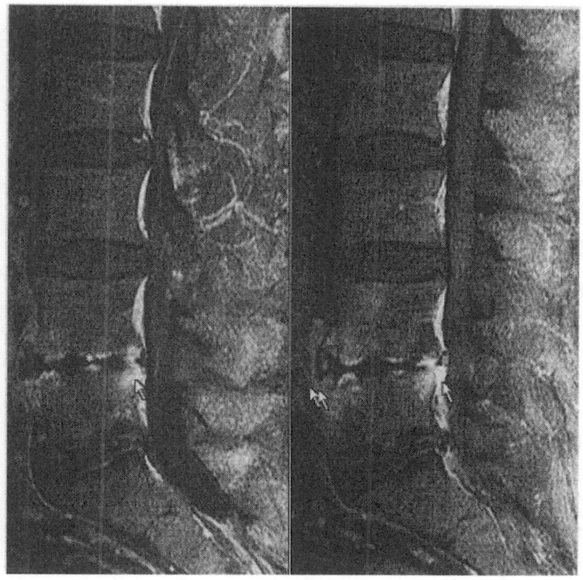
b

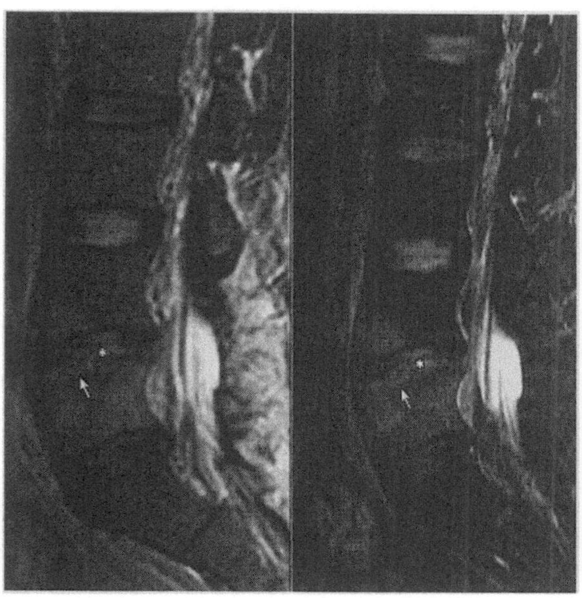
c

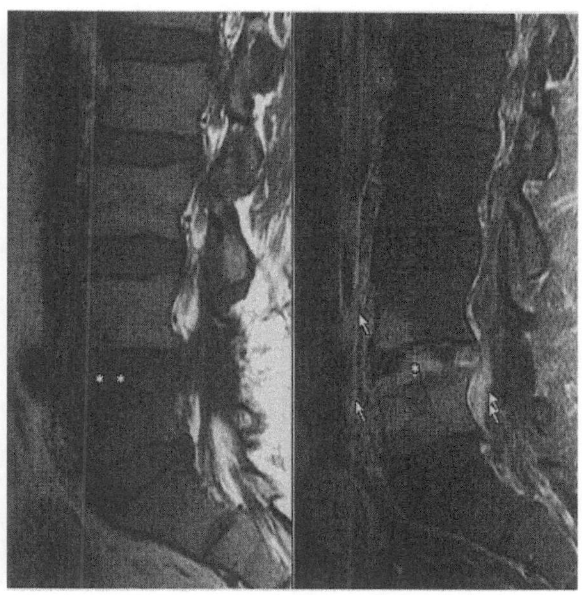
d

Case 61: Epidural abscess (brucella)

Fig. a: sagittal cut; SE - T1w

Fig. b: sagittal cut; SE - T1w + Gd with fat suppression

Fig. c: sagittal cut; SE - T2w

Fig. d: axial cut; SE - T2w

Important soft tissue mass developed in the epidural space presenting a heterogeneous signal on T1w (**a**,↑), apparently multicystic, obturating the hyperintense epidural fat (**a**, ↑). The mass slightly hypointense on T2w (**c**, ↑) shows hyperintense cystic formations (**c**, ↑↑) developed mainly in the left lateral recess, as shown on the axial cuts with mass effect on the dural sac (**d**, ↑). The L4-L5 degenerated disc shows hypointense signal on T1 and mainly on T2w without any sign of discitis or contrast enhancement after Gadolinium injection (**b**). After contrast iv, the epidural mass enhances (**b**, ↑) inhomogeneously with persistent areas of hypointense signal corresponding to abscess formation containing fluid material consistent with pus. Despite anterior artifacts, a slight enhancement of the anterior L5 and S1 end plates may be retained.

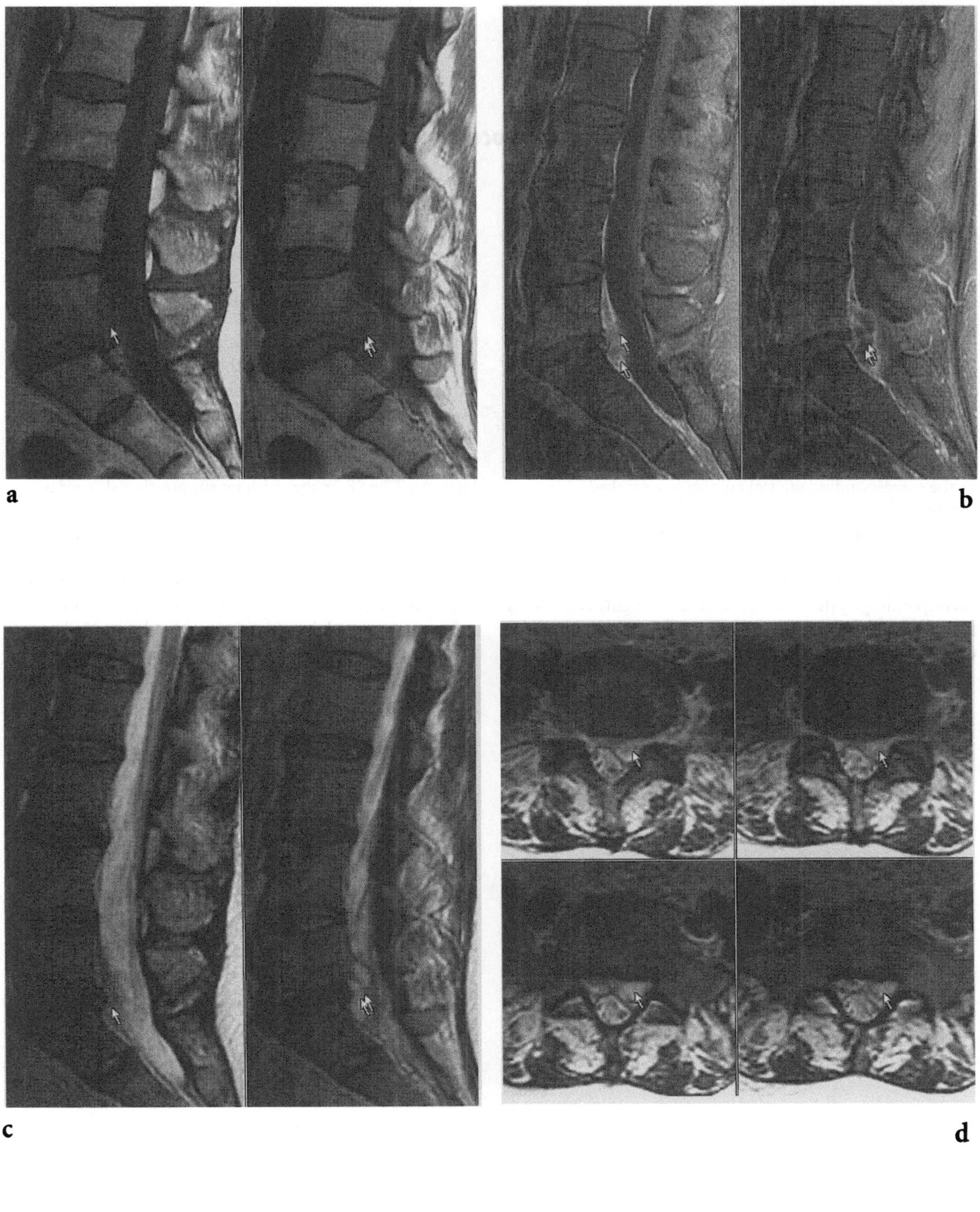
a
b
c
d

Case 62: Evolution of an epidural abscess (enterococcus)

Fig. a: sagittal cut (512); SE - T2w

Fig. b: sagittal cut (512); SE - T1w + Gd

Fig. c: sagittal cut (512); SE - T2w

Fig. d: sagittal cut; SE - T2w with fat suppression and sagittal cut (512); SE - T1w + Gd with fat suppression

T2w sequence with fat suppression demonstrates a slightly hypointense L4-L5 disc with present intranuclear cleft and relatively hyperintense protruded material (**a**, ↑) extending to the epidural space (**a**, ↑↑), narrowing the dural sac. The L4 and L5 vertebral bodies do not show any marked abnormal signal intensity. On T1w sequence with contrast iv, the corresponding sagittal cuts show a marked enhancement of the epidural soft tissue (**b**, ↑↑) encircling the nerve root (**b**, ↑) and extending superiorly to involve the anterior border of L4-L5 intervertebral disc with apparent interruption of the limitating dark rim corresponding to the annulus fibrosus (**b**, right) confirming the T2w aspect (**a**, right). Note the absence of enhancement of the intervertebral disc and adjacent vertebral end plates.

Ten days later, T2w high resolution images showed a hyperintense signal involving the L4-L5 disc space surrounding the intranuclear cleft still visible (**c**, right) associated to a diffuse slight hyperintense signal involving the adjacent verterbral end plates. The patient was already treated for his epidural abscess and improving. One week later, the T2w SE images with fat suppression showed a marked decrease of volume of the epidural abscess still enhancing (**d**, right) but disclosed marked extension of the hyperintense signal of the disc (**d**, *) associated to symmetric large indentation of the disc margins. Note the hyperintense signal of marrow showing diffuse contrast enhancement involving the adjacent vertebral end plates. The latter aspect is characteristic of spondylodiscitis.

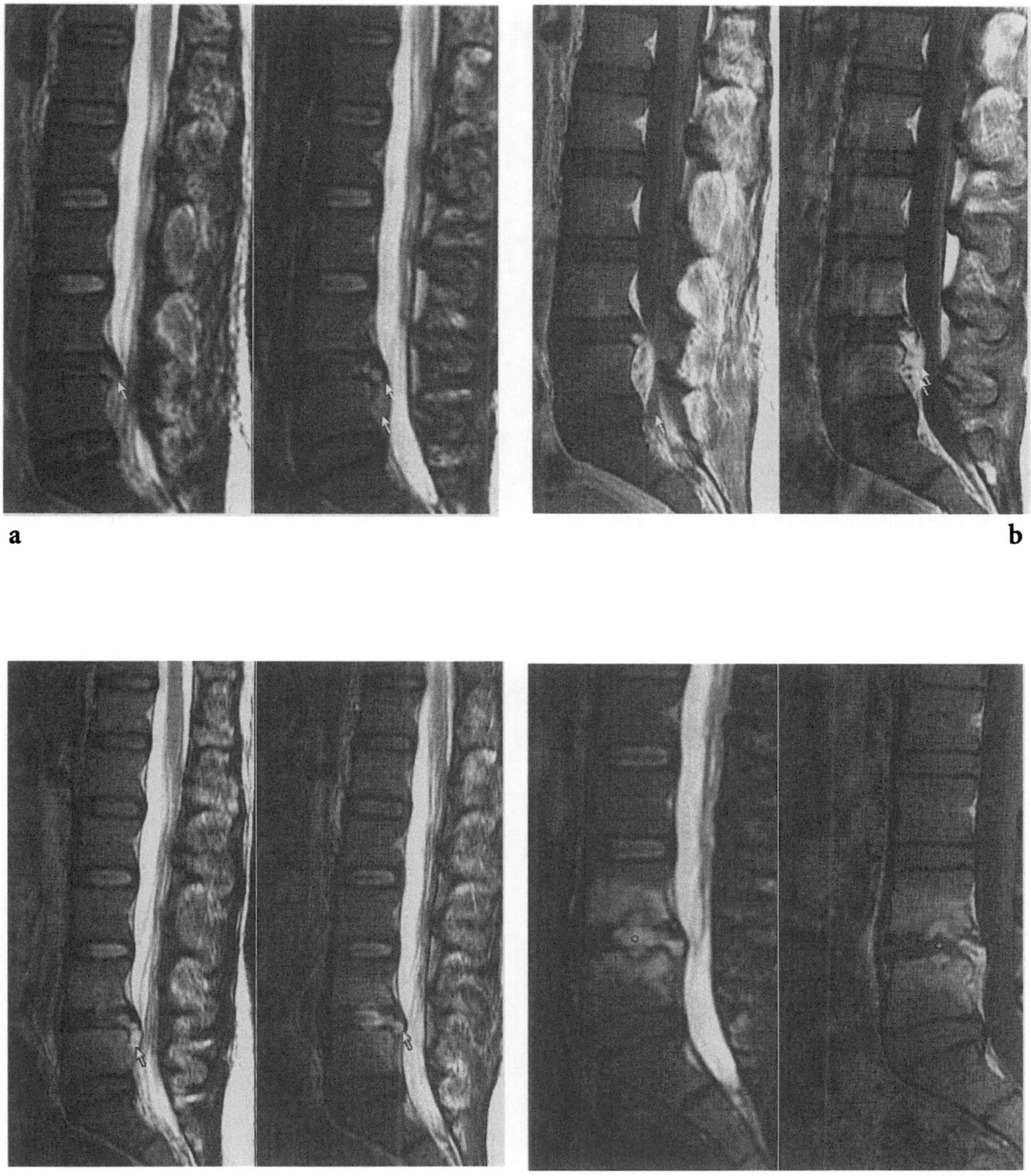

a b c d

Case 63: Spondylodiscitis (candida)

Fig. a: sagittal cut; SE - T1w

Fig. b: sagittal cut; SE - T2w

Patient treated for *Candida* spondylodiscitis. MR Control exam few months later shows on T1w images a heterogeneous hypointense space occupying lesion involving two adjacent dorsal vertebral bodies and mainly their adjacent end plates (**a**, *) remaining slightly hypointense on T2w (**b**) surrounded by hyperintense signal from the remaining portion of the vertebral bodies (**b**, *). The intervertebral space is hardly recognized being largely destroyed and dessicated (**b**, ↑). The marked hypointense signal on T1w and T2w sequences may correlate with bony sclerosis.

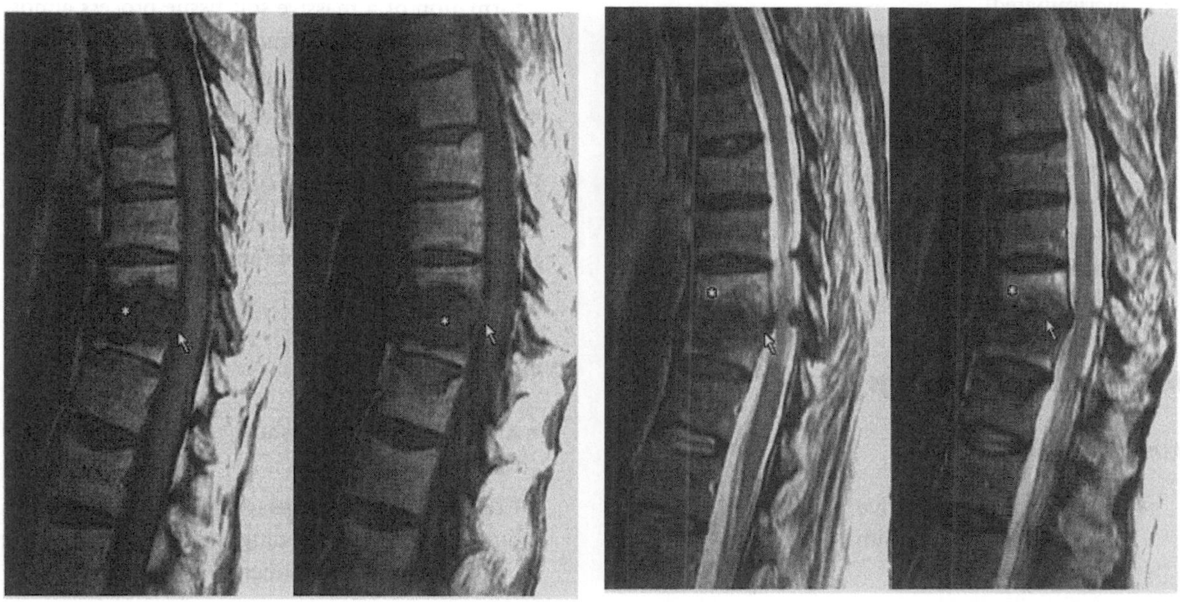

a b

Case 64: Arachnoiditis

Clinical and pathological features

- Arachnoiditis is a rare cause of back pain and disability secondary to spinal surgery (less than 15% of the cases). In the past, myelography using Pantopaque was a common cause of arachnoiditis. Infectious and neoplastic processes are still incriminated.
- *Histopathology:* arachnoiditis is secondary to inflammation of the leptomeninges with swelling of the nerve roots, followed by proliferation of fibroblasts and collagen deposition leading to the formation of an adhesive arachnoiditis.
- *Signs:* arachnoiditis causes a long standing invalidating back and legs pain associated with sudden spasms in the legs, cramps, usually found in the days following spinal surgery. A fever is present.

MR characteristic findings

Morphology

- MRI is the non-invasive modality of choice to demonstrate arachnoiditis, being specific in almost all the cases.
- The intrathecal nerve roots show different patterns. They usually appear conglomerated and found centrally or posteriorly within the thecal sac, best shown using high resolution heavily weighted T2w SE sequences performed in the axial plane.
- Extensive adhesive arachnoiditis may lead to the formation of a massive soft tissue process almost filling the dural sac and mimicking neoplastic conditions seeding the subarachnoid spaces.
- Arachnoidal thickening and adhesion may lead to the formation of intrathecal cystic formations which may enlarge and compress the nerve roots.
- Spinal stenosis may mimic arachnoiditis secondary to compression of the nerve roots contained in a narrow canal.

Contrast enhancement

- Arachnoiditis may show mild to moderate contrast enhancement.
- Contrast injection helps to differentiate arachnoiditis from tumors seeding the CSF spaces, filling the thecal sac and/or adhering to the nerve roots.

Case 64: Arachnoiditis

Fig. c: sagittal cut; SE - T2w

Fig. d: axial cut; SE - T2w

Heavily weighted T2 FSE images showing in the lumbar spinal canal a markedly hyperintense cystic formation developed anterolaterally (**c, d,** ↑) associated to conglomerated nerve roots displaced posteriorly in the thecal sac (**d,** ↑↑). Note the global decrease of the normal hyperintense signal of the CSF on these T2w sequences. The patient was followed up for an unidentified meningitis following surgery at a lower lumbar level. Note at this level a heterogeneous hypointense soft tissue due to thickened meninges and/or posterior epidural fibrosis. Note the post-surgical scar in the area of laminectomy (**c,** right).

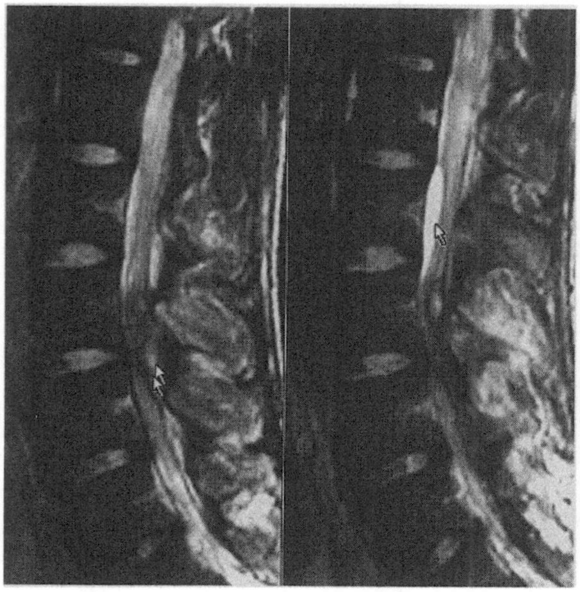

c

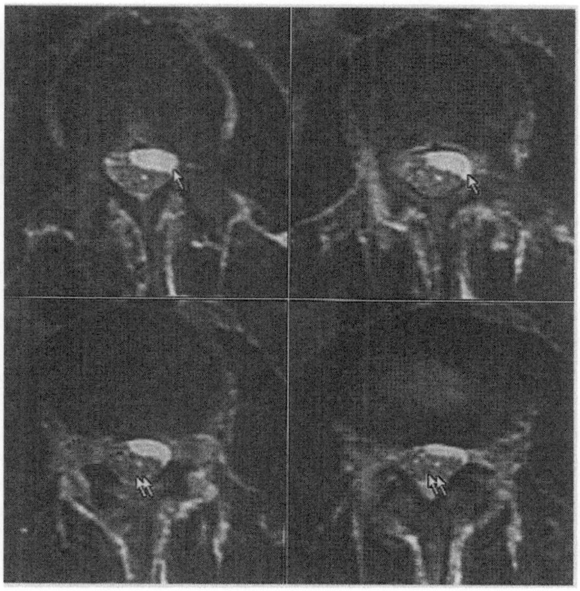

d

Inflammatory spinal diseases

Case 65: Interapophyseal joint effusion

Fig. a: sagittal cut (512); SE - T1w and sagittal cut; SE - T2w

Fig. b: axial cut (512); SE - T1w and axial cut (512); SE - T2w

Fig. c: sagittal cut; SE - T2w

Fig. d: axial cut; SE - T2w

Facet arthrosis is one of the most common causes of low back pain. With advancing disease, joint effusions may be observed presenting a hyperintense signal on T2w as shown in these cases: the first one presenting also a paracentral herniated disc (**a**, **b**, ↑) shown on T1 (**b**, up) as well on T2w (**b**, down) sequences. The T2w sequences show joint effusions with hyperintense signal on T2w (**b**, ↑↑). Note that on T1w, the effusion is difficult to differentiate from the adjacent cortical bones. In another patient complaining of long standing low back pain, the only finding disclosed is an important facet joint effusion well shown on the parasagittal cuts using T2w sequence (**c**, ↑) as well as on the axial T2w cuts (**d**, ↑). Note also the presence of subchondral bony sclerosis involving the facets (**d**, ↑↑), and that there is no hypertrophy of the facet joints that could lead to lumbar stenosis considering that the main degenerative changes of the facet joint generally involve the superior articular facet with hypertrophy and osteophyte formation. Note that GE sequences may demonstrate with better conspicuity the articular cartilage well differentiated from the adjacent bony structures and help on high resolution images to visualize eventual thinning and/or irregularities. Magnetization transfer (MTR) may add some advantages.

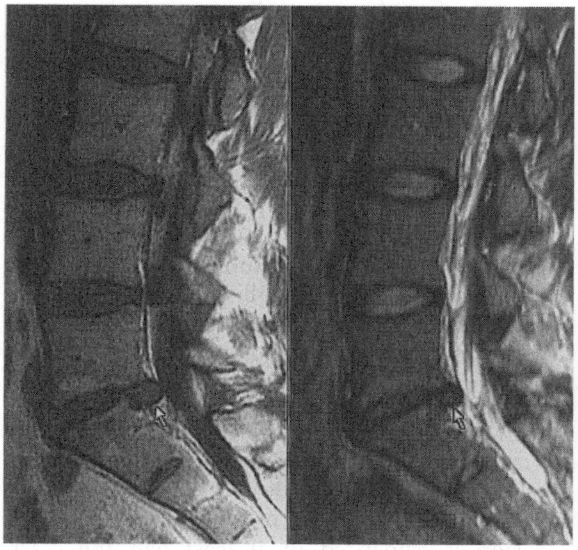

a

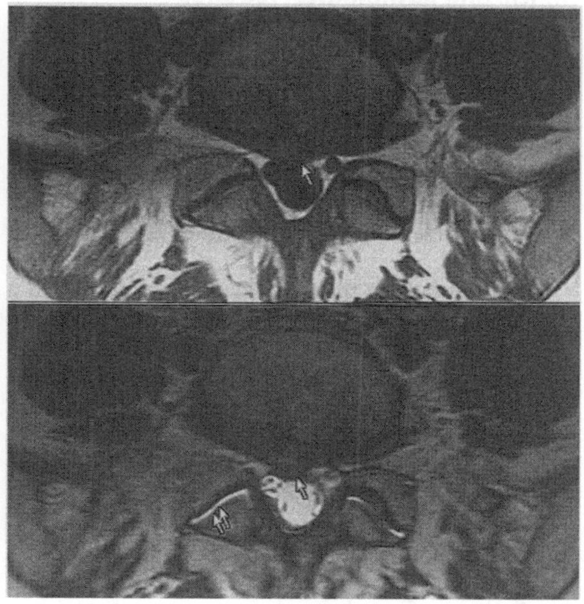

b

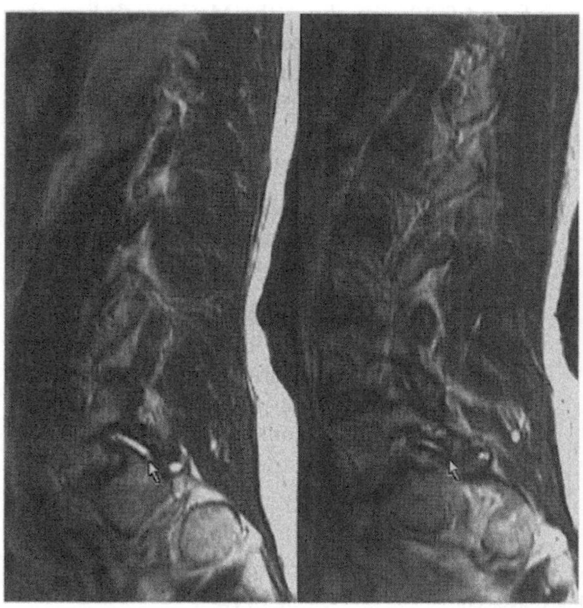

c

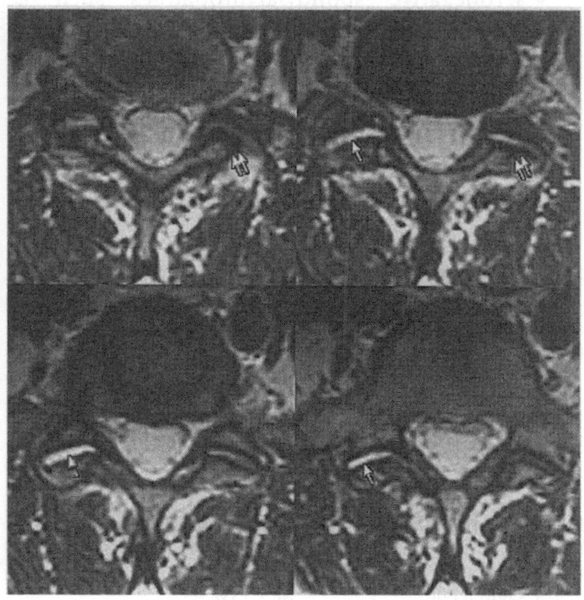

d

Case 66: rhumatoid arthritis

Clinical and pathological features

- Rheumatoid arthritis is an immune mediated systemic disease.
- Its incidence averages 0.25%, being more frequent in females.
- The disease affects the cervical spine, most commonly with involvement of hand and feet joints. The thoracolumbar spine is much less commonly involved.
- *Histopathology:* the primary changes consist in a proliferative synovitis named "pannus". Later in the course of the disease, extension of the pannus to the articular surface causes a destruction of the articular cartilage and bone, resulting in a variety of subluxations involving predominantly the cranio-cervical junction.
- *Signs:* the most frequent complaints are related to the metacarpophalangeal and the metatarsophalangeal joints. The atlanto-axial joint is the most common site of cervical spine involvement. Patients affected may be asymptomatic or may complaint of suboccipital headaches. In case of associated myelopathy secondary to cord compression, tetraparesis and limbs paresthesias, or a Brown-Sequard syndrome are reported.
- Seronegative spondylarthropathies such as Reiter's syndrome, ankylosing spondylitis and psoriatic arthritis, differ from rheumatoid arthritis and most frequently involve the sacroiliac joints and the subchondral bone. Spinal focal erosions at the disco-vertebral junction and calcifications of the anterior longitudinal ligament with syndesmophytes are also present.

MR characteristic findings

Morphology

- Rheumatoid arthritis involve preferentially the upper cervical spine, and more particularly the atlanto-axial joint leading to a subluxation of C1 on C2, best evaluated with MR in the sagittal plane (flexion and extension).
- The pannus formation usually shows a hypointense SI on T1w and a hypointense to mild hyperintense (due presumably to presence of inflammation with increase in water content) SI on T2w.
- Three types of C1-C2 subluxation may occur resulting in spinal instability. The most frequent is the anterior atlanto-axial subluxation followed uncommonly by the posterior or the vertical subluxations.
- In the anterior type, the atlanto-axial distance shows a diastasis superior to 2.5-3 mm as evaluated on plain films with neck flexion. When the diastasis equals 9 mm or more, a cord compression is evidenced in at least half the cases. Associated lateral subluxations with asymmetry of the synovial reaction may coexist.
- The posterior type usually occurs secondary to a fracture or a resorption of the odontoid process.
- In vertical subluxation, the tip of the dens is found 8-10 mm above the McGregor's line, the line joining the hard palate to the inferior aspect of the basioccipital bone. This type of subluxation may cause a compression of the cervicomedullary junction and eventually the vertebral artery.
- Subaxial subluxations of the cervical vertebrae below C1-C2 level give rise to multiple vertebral retrolisthesis ("stair-like appearance").
- Later in the disease course, involvement of the lower cervical spine may lead to spinal canal stenosis secondary to multiple vertebral subluxations. The cervical discs are progressively resorbed with vertebral fusions in end stages.

Case 66: Rheumatoid arthritis pannus (C1 - C2)

Fig. a: sagittal cut;	SE - T1w	**Fig. c:** axial cut;	SE - T1w
Fig. b: coronal cut;	SE - T1w	**Fig. d:** axial cut;	SE - T2w

The sagittal T1w sequence shows a separation of the dens and the anterior arch of C1 by a pannus (**a**, ↑) observed also on the coronal T1w sequences (**b**, *). Note the narrowed spinal canal due to dens displacement, with spinal cord compromised. Axial slices on T1w (**c**) and on T2w (**d**) do not show any compression of the bulbomedullary junction. Note that the pannus is best shown on the axial T1w slices (**c**) and the relation between the dens and the spinal cord is best shown on T2w with bright CSF in the subarachnoid spaces (**d**).

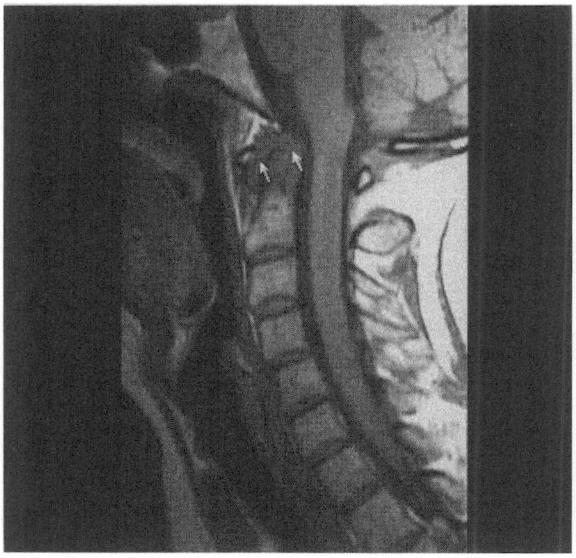

a

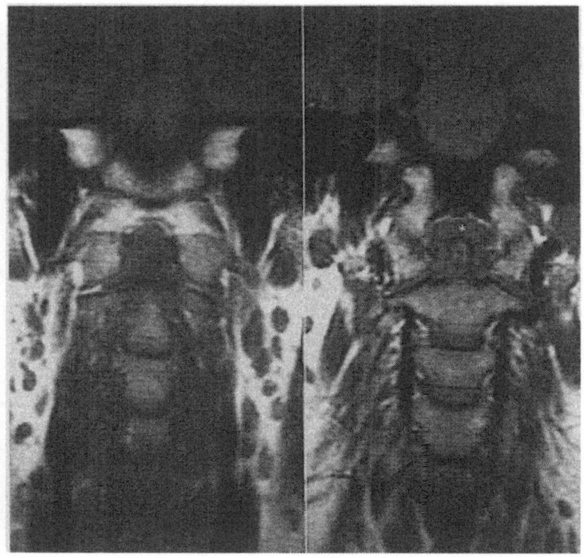

b

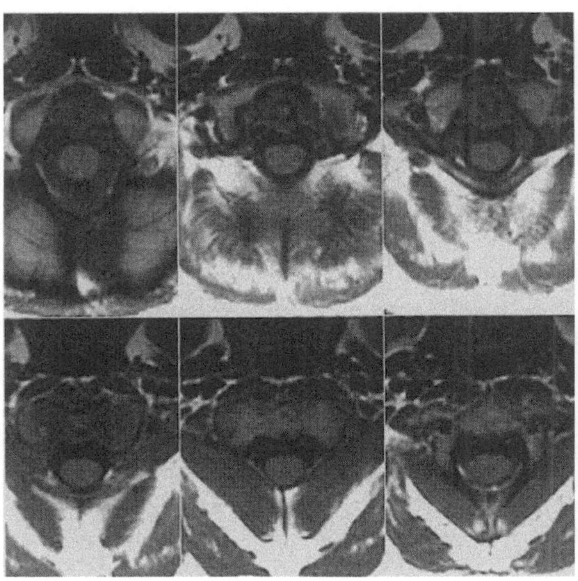

c

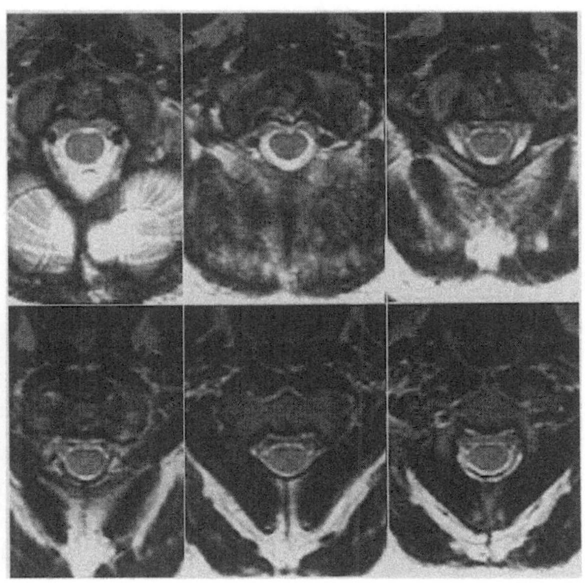

d

Cases 65, 67, 68: facet joint arthritis and synovial cysts

Clinical and pathological features

- Osteoarthritis is a degenerative affection eventually associated with synovial cyst, involving the facet joints.
- Bony changes and hypertrophy may result contributing to narrow the spinal canal or the neural foramen.
- *Histopathology:* a synovial cyst is lined by a synovial membrane, and usually shows a communication with the apophyseal joint. Cyst content is formed of myxoid proteinaceous material. Gas or hemorrhage may be found.
- *Signs:* facet joint disease may cause chronic low back pain and even sciatica. No specific clinical history may be elicited. The disease may remain asymptomatic in almost half the cases.

MR characteristic findings

Morphology and signal intensity

- Degenerative facet joint disease include: facet joint hypertrophy, osteophytic formations, narrowing of the joint space. Secondary rotary subluxation of the joints may occur. Signal void vacuum phenomenon within the fact may be shown.
- Facet joint synovial cysts are well delineated rounded formations, found extradurally within the spinal canal and medial to the facet (case 67). They may uncommonly arise also from the posterior aspect of the apophyseal joint and develop in the paraspinal soft tissue (case 68). The cyst content show a hypointense SI on T1w and a hyperintense SI on SE-T2w, similar or superior to CSF. The cyst wall may show a hypointense SI on T2 due to hemosiderin deposit and/or calcification. Cystic component may be absent in case of chronic inflammation leading to fibrosis.
- Almost 2/3 of the synovial cysts occur at the L4-L5 level, the remaining being found at L5-S1 or L2-L3. They are rare in the cervical or thoracic spine. Compression of the thecal sac and/or of the adjacent nerve root in the lateral recess may be demonstrated.
- The differential diagnosis include: cysts of the ligamentum flavum, perineural cysts, or eventual schwannoma.

Contrast enhancement

- Fat suppressed SE-T1w sequences are very sensitive to depict inflammatory osteoarthritis.
- Synovial cysts show a peripheral enhancement, eventually associated with enhancement of the facet joint.

Case 67: Synovial cyst in the lateral recess

Fig. a: sagittal cut; SE - T2w and sagittal cut; SE - T1w

Fig. b: axial cut; SE - T1w

Sagittal MR shows a rounded lesion intimately related to the medial aspect of the right apophyseal joint presenting a hypointense signal on T1w as shown on the parasagittal cuts (**a**, right,↑) and on the axial cuts (**b**, down, ↑) and a hyperintense signal on T2w (**a**, left, ↑; **b**, up, ↑). Note the slight peripheral enhancement shown as a thin rim of hyperintense signal on T1w (**b**, ↑, left). The mass is responsible of a right foraminal and central canal stenosis. Note the associated inflammatory effusion in the ipsilateral facet joint (**b**, *).

Case 68: Synovial cyst of the posterior interapophyseal joint

Fig. c: sagittal cut; SE - T2w

Fig. d: sagittal cut (512); SE - T2w with fat suppression

Parasagittal slices performed using T2w sequence (**c**) showing hyperintense well circumscribed cystic mass related to the apophyseal joint (**c**, ↑↑). The small cystic mass is better shown on T2w with fat suppression technique (**d**, ↑). Note the associated joint effusion (**c**, ↑).

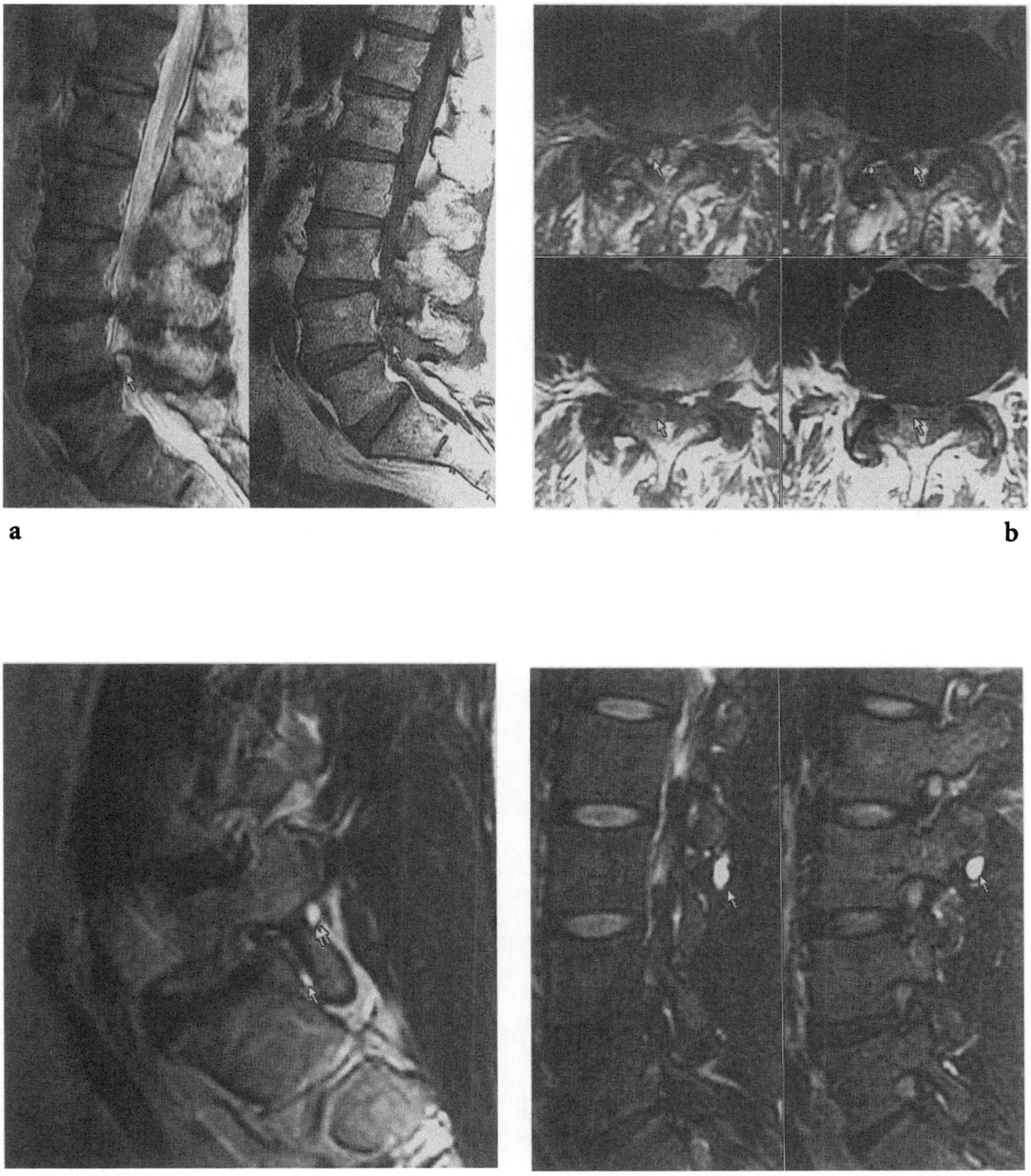

a b c d

Inflammatory spinal cord diseases

Cases 69-72: multiple sclerosis

Clinical and pathological features

- MS is a demyelinating disease which affects the entire central nervous system.
- MS develops in about 2/3 of the cases between 20-40 years of age, but in a small number of cases it occurs late in the fifth decade.
- MS shows a higher incidence in women than in men (1.7:1, female:male ratio).
- *Histopathology* (see brain MS): the demyelinating lesions or "plaques" are scattered throughout the CNS affecting mainly the white matter and varying in size from 1mm to several centimeters particularly in the cord. The plaques are sharply delineated showing a predilection to the myelin where it abuts the pial veins. This explains the propensity to occur in the white tracts of the spinal cord, optic nerves and chiasm, and brainstem. The plaques demonstrate 3 types of microscopic changes: a demyelination with a myelino-axonal dissociation; the presence of an inflammatory perivenular reaction, and a proliferation of fibrillary astrocytic cells with rarefaction of oligodendrocytes. Two main etiologies are retained to explain pathogenesis of MS: an autoimmune disorder secondary to an environmental cause in patients showing a genetic predisposition, and an immune reaction to a low-grade virus infecting the CNS.
- In the neuromyelitis optica of Devic, the disease shows a predilection to the optic nerves (bilateral involvement in about 90%) and spinal cord in which the pathologic process is more intense being of the necrotizing type that may lead to microcystic cavitations (with consequently more likely permanent clinical effects). The development of transverse myelitis usually occurs less than 8 weeks after the visual impairment. Symptoms of viral illness preceding the development of the disease are found in almost 2/3 of the cases. Note that an acute disseminated encephalomyelitis or an acute necrotizing hemorrhagic leucoencephalitis may present as a Devic's disease.
- *Signs* (see brain MS): difficulty in MS diagnosis arises in pure spinal forms of the disease, in which the patient presents as a progressive spastic paraparesis with variable sensory signs. MS patients presenting with a progressive course of the disease account for almost 10% of the cases. The clinical syndrome may simulate spinal cord compression secondary to: cervical spondylosis, platybasia, basilar impression, Chiari I malformation, foramen magnum tumor, which would be ruled out using cranio-cervical MR. Other medical differential diagnoses have to be sometimes considered such as subacute combined degeneration of the cord or even amyotrophic lateral sclerosis or a heredodegenerative disorder. Careful clinical correlations with MR results are mandatory in order to achieve an accurate diagnosis, especially when associated etiologies are present, as may be commonly found in late adulthood presenting a spinal form of MS.

MR characteristic findings

Morphology

- Pure spinal cord involvement with no associated brain lesions may be found at autopsy as well as on MRI. Brain MRI may be normal in as much as 10-50% of patients demonstrating cord lesions at time of diagnosis. The diagnosis should not be excluded if brain MR is normal.
- Most plaques involving the spinal cord are found in the cervical region which appears to be twice as frequent as the thoracolumbar region. The most common reported levels are C3, C4 and C7.
- The "demyelinating plaque" within the cord are usually elongated and almost sharply delineated, sometimes associated in the acute stage with segmental smooth enlargement of the cord. Cord swelling due to edema is common in acute transverse myelitis associated with neuromyelitis optica, and may be followed by segmental atrophy on follow-up exams.
- Topographically the lesions show a preponderance to the white matter tracts of the lateral columns and the posterior columns, with a roughly symmetrical proneness to distribution. The demyelinating lesions tend to develop preferentially at the outer aspect of the lateral hemicords about the insertion of the denticulates ligaments (areas of tethering of the cord in the

regions of more mobility, where presumably secondary perivenous microvascular mechanical damage may occur).

- The cervical cord is much more frequently involved than the remaining part of the spinal cord as observed on MR or autopsy series.
- Detection sensitivity of medullary plaques is limited to some extent by anatomical and physical constraints which reduce the signal-to-noise ratio, such as motion and flow artifacts necessitating to adapt compensatory techniques (flow and respiratory compensation, presaturation, cardiac or peripheral gating).

Signal intensity

- SE-T1w: iso-intense to slightly hypointense to cord.
- SE-PDw: hyperintense to cord and CSF.
- SE-T2w: hyperintense to cord.
- FLAIR: hyperintense to cord and CSF.

Contrast enhancement

- Contrast enhancement may be shown in acute plaques. Enhancement is more obviously demonstrated with delayed imaging (about 1 hour) or with T1w sequence associated with magnetization transfer suppression sequence.
- Differential diagnosis of enhancing plaques may necessitate, particularly when brain MR is unremarkable, to perform follow-up exams (2 months) in order to rule out other inflammatory (case 72) or vascular diseases and small primary spinal tumor (which virtually all of show enhancement).
- Brain MRI may be of great help for the diagnosis of MS cord lesions.

Case 69: Isolated cervical "plaque" (MS)

Fig. a: sagittal cut; SE - T1w and SE - T2w

Fig. b: axial cut; SE - T2w

Cervical spinal cord exam showing an ill-defined small area of hypointense signal on T1w (**a**, left, ↑) becoming very hyperintense on T2w (**a**, right, ↑) surrounded by linear areas of hyperintense signal involving the upper cord posterior to C2 level. Note the absence of enlargement of the corresponding cord segment. On the axial T2w slices, the small area of bright signal involves mainly the right lateral hemicord without any modification of cord morphology (**b**, ↑). The lesion was also hyperintense on PDw sequence (not shown). There were no other localizations on the cord, neither on brain white matter.

Case 70: Isolated pseudotumoral cervical "plaque" (MS)

Fig. c: sagittal cut; SE - T2w and sagittal cut; SE - PDw

Fig. d: sagittal cut; SE - T2w

Small oval lesion involving the cervical spinal cord at C4-C5 level showing a hyperintense signal on PDw sequence (**c**, right, ↑) and a hyperintense signal on T2w (**c**, left, ↑). The lesion is much better delineated on high resolution T2w sequence (**d**, ↑) showing a slight segmental enlargement of the cord. The MR exam of the brain depicted also several hyperintense lesions disseminated in the white matter, consistent with multiple sclerosis.

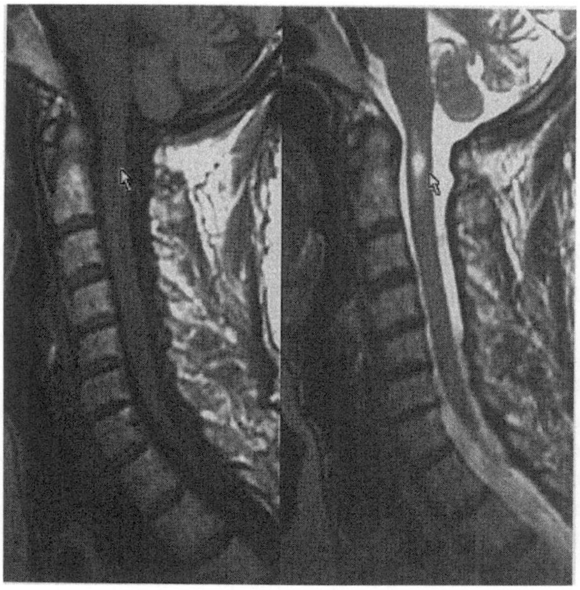

a

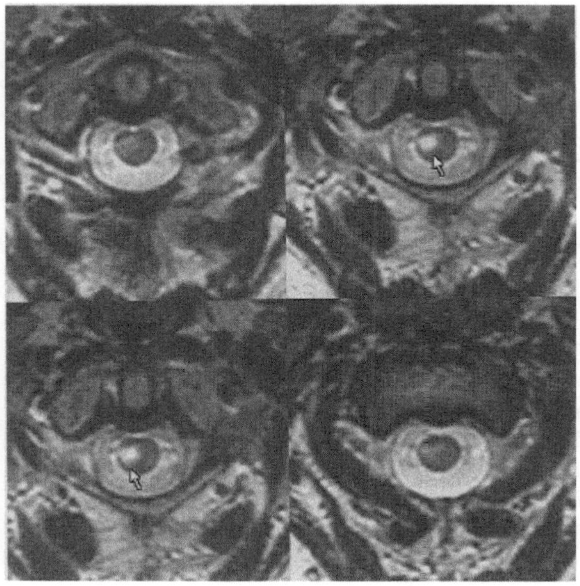

b

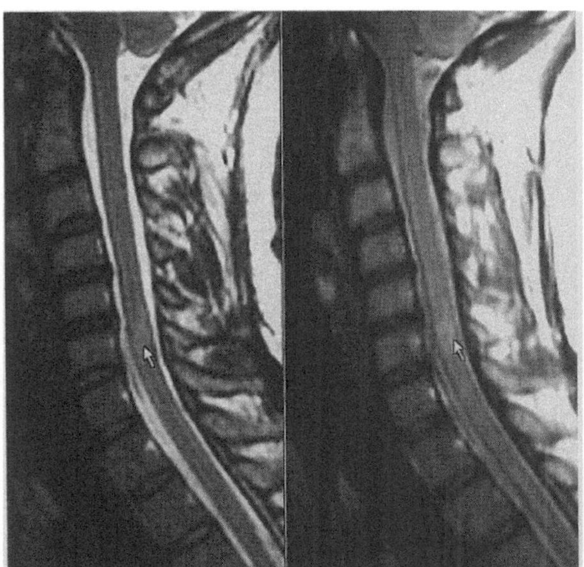

c

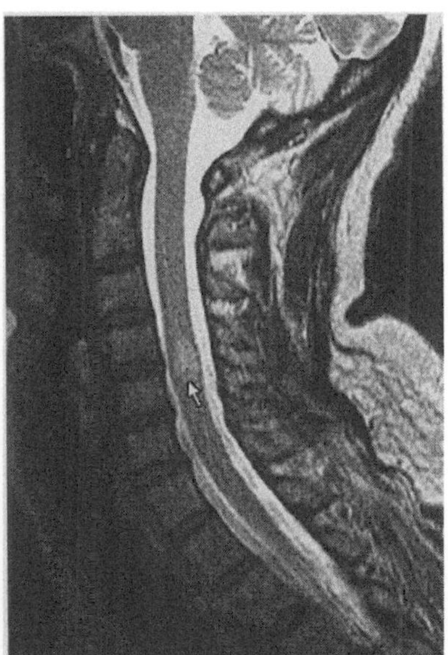

d

Case 71: Cervical, dorsal and conus "plaques" (MS)

Fig. a: sagittal cut; SE - T2w

Fig. b: sagittal cut; SE - PDw

Fig. c: sagittal cut; SE - PDw

Fig. d: sagittal cut; SE - T2w

Patient with MS showing multiple disseminated lesions in brain white matter. The spinal cord MR exploration shows multiple nodular and band hyperintense foci disseminated throughout the spinal cord from the cervical level (**a**, **b**) to the conus medullaris (**c**, **d**). The lesions are depicted on T2w sequences (**a**, **d**) but are slightly better shown on PDw images (**b**, **c**). Some extensive lesions are associated to a slight segmental cord atrophy (**a**, ↑↑) (presumably an old lesion with residual demyelination and gliosis). Other more acute lesions are associated with a slight enlargement of the corresponding cord segment as shown at the level of the conus (**d**, ↑). FLAIR images, most efficient as demonstrated routinely on MS brain exams, would be most valuable for the detection of small plaques in the spinal cord.

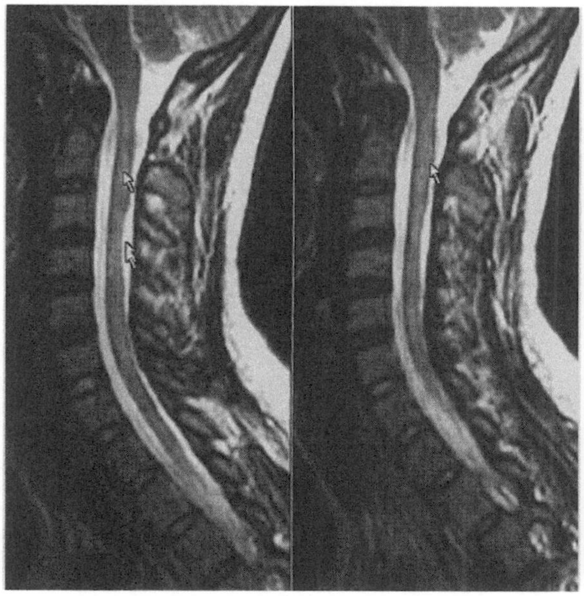

a

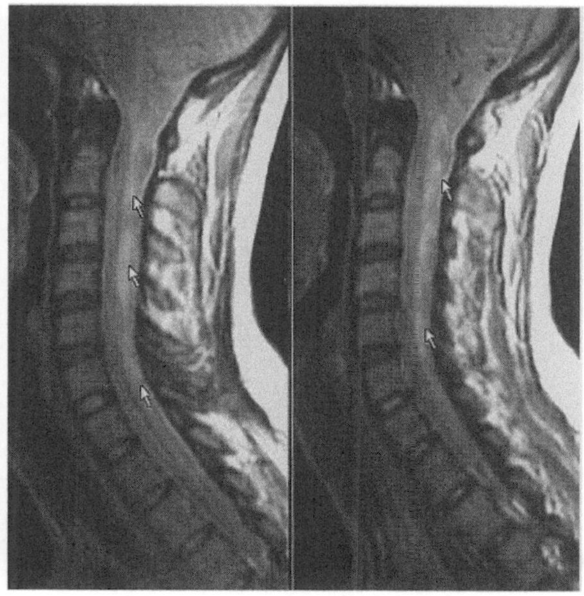

b

c

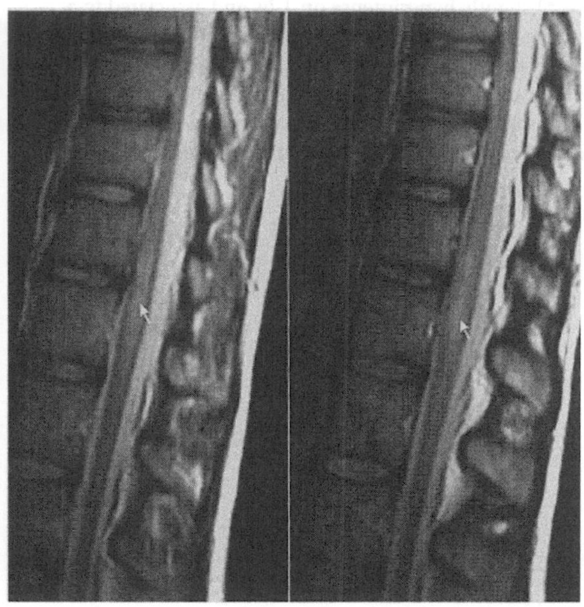

d

Case 72: Cervical granuloma

Fig. a: sagittal cut; SE - T2w

Fig. b: axial cut; SE - T2w

Focal hyperintense lesion involving the cervical spinal cord (**a**, *) slightly hyperintense on T2w and associated to a segmental enlargement of the corresponding portion of the cord (**a**, ↑). The lesion involves the right hemicord as shown on the axial slices (**b**, ↑) and mainly the posterior column with slight mass effect on the posterior median sulcus erased at the level of the lesion and only shown on the upper extremity (**b**, up, left) due to the bright CSF in the groove. Note laterally in the subarachnoid spaces, black areas of flow-void corresponding to flow artifacts in this FSE-T2w sequence. The lesion, biopsied, showed a non specific granulomatous inflammatory process. It was isolated and the brain MR exam was normal at this time. The etiology remains unsettled.

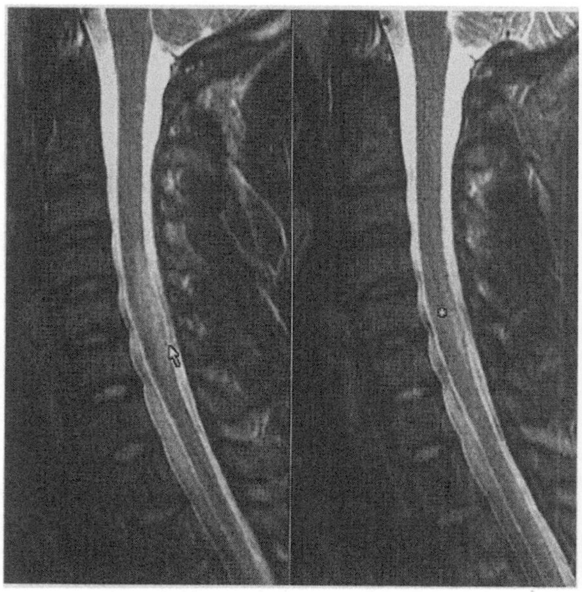

a

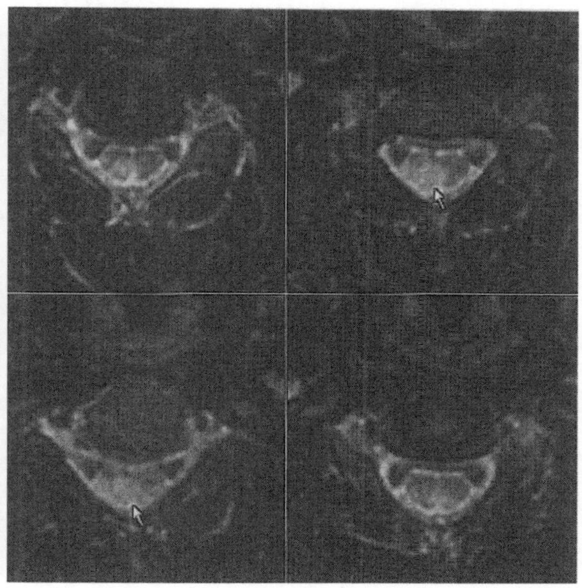

b

Cases 73, 74: transverse myelitis

Clinical and pathological features

- Transverse myelitis is a rare disease involving the spinal cord, related to several different etiologies.
- It includes: acute disseminated encephalomyelitis (ADEM), viral and infectious myelitis, multiple sclerosis (about 5% of the cases, usually at C2-C3 levels), and myelitis of indeterminate etiology (diagnosis of exclusion).
- ADEM usually develops few days to 2 weeks following a prior viral infection (herpes, chickenpox, measles, rubella, mumps, polio, influenza) or vaccination (polio, tetanus, smallpox). The perivenular inflammatory lesions affect the white matter of the brain and spinal cord. The clinical syndrome usually regresses in less than 2 months. In acute hemorrhagic leucoencephalopathies, showing aggressive lesions of hemorrhagic necrosis, the clinical course is more severe and the outcome usually fatal.
- Transverse myelitis affects both sides of the spinal cord, more or less totally at one or more segmental levels, resulting in an almost complete transection of the cord.
- *Signs:* the clinical syndrome of transverse myelitis consists in impairement of all motor and sensory functions below the cord lesion level (loss of sensation commonly found at the thoracic level). The onset of the neurological deficits may be acute or subacute, sometimes preceded by a backache. Sphincteric dysfunctions or paralysis are most commonly associated. A band of hyperesthesia may be found at the upper level of the lesion. The clinical course is non progressive. Variable pathophysiological processes have been incriminated and are considered in the differential diagnosis of transverse myelitis: autoimmune and allergic, demyelinating, viral and infectious, vasculitis, paraneoplastic, and idiopathic. The diagnosis remains a diagnosis of exclusion. Treatment is usually based on administration of high doses of steroids.

MR characteristic findings

Morphology and signal intensity

- MR discloses spinal cord lesions in almost half the patients presenting a syndrome of acute transverse myelitis. In more than half the cases the etiology remains unknown (cases 73 and 74), MS representing less than 1/4 of such patients (dissemination of the lesions in the CNS demonstrated with brain screening).
- Spinal cord lesion involves predominantly the cervical cord (almost 2/3 of the cases) and the thoracic cord.
- The MR abnormalities found are not specific, including an increased SI on T2w sequences in virtually all the patients with a mild cord enlargement found in almost half the cases at the acute stage. The high SI may extend over several vertebral segments due to associated edema.
- Long term MR follow-up may show a residual segmental cord atrophy.

Contrast enhancement

- Enhancement is found in about 60% of the patients with transverse myelitis.
- The pattern of enhancement may vary according to the subjacent etiology, being usually focal and nodular in granulomatous diseases (sarcoidosis), and more diffuse in demyelinating or viral diseases.
- On follow-up exams (3-6 months and 12 months) contrast uptake tends to resolve in the following weeks. In case of clinical progression and increase in lesion size and extension of area of enhancement, a spinal cord tumor must be ruled out.

Case 73: Myelitis

Fig. c: sagittal cut; SE - T1w and sagittal cut; SE - T2w

Sagittal T1w (c, left) and T2w (c, right) cuts showing an oblong cord lesion involving the mid thoracic region and presenting a hyperintense signal on T2w (c, right, ↑) and an isosignal on T1w (c, ↑), associated to a slight segmental enlargement of the cord. The lesion is isolated in the spinal cord and brain MR exam was normal. The lesion shows a slight contrast enhancement after Gadolinium injection. The disease process involves the entire sectional area of the spinal cord at this level as shown on the axial slices. The patient presented an acute loss of sensory and motor function with a transverse segmental level and a non progressive clinical course. The most likely clinical diagnosis was transverse myelitis. Neurologic explorations showed absence of any known neurologic disease including syphilis, encephalitis, malignancy or others.

Case 74: Myelopathy

Fig. d: sagittal cut; SE - T2w

T2w sagittal MR, 3 mm slice thickness, showing a hyperintense oblong focus involving the mid thoracic cord (d, ↑). The lesion is nicely delineated and is associated with a slight enlargement of the corresponding cord segment. There were no other foci on the remaining cord and MR brain exam was normal. The patient presented a progressive development of paraparesis associated to some sphincter disturbances. The lesion is still unidentified and clinical and MR exams are completed.

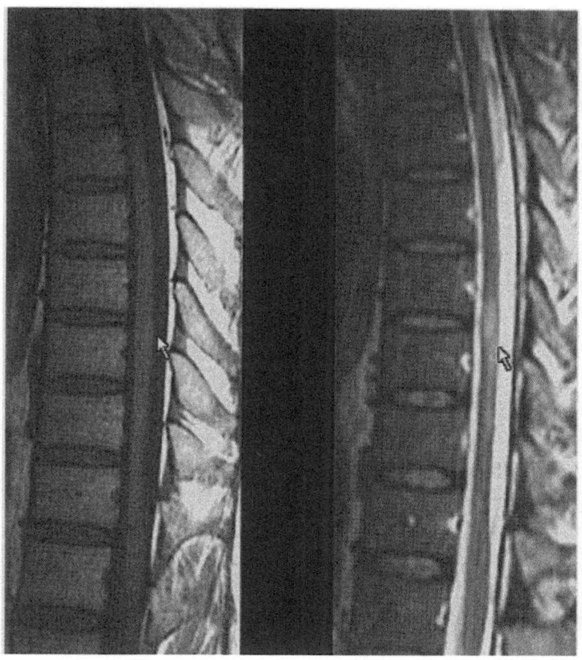

c

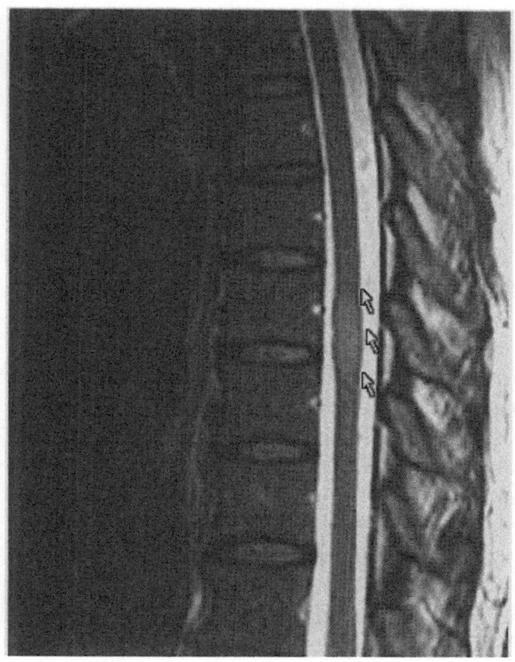

d

Selected References

- Abdullah AF, Chambers RW, Daut D (1984) Lumbar nerve root compression by synovial cysts of the ligamentum flavum. Report of four cases. J Neurosurg 60:617-620
- Abramsky O, Teitelbaum D (1977) The autoimmune features of acute transverse myelopathy. Ann Neurol 2:36-40
- Ahlstrom H, Feltelius N, Y man R et al (1990) Magnetic resonance imaging of sacroiliac joint inflammation. Arthritis Rheum 33:1763-1769
- An HS, Vaccaro AR, dolinsdas CA et al (1991) Differentiation between spinal tumors and infections with magnetic resonance imaging spine 6:S334-S338
- Andrianakos AA, Duffy J, Suzuki M, Sharp JT (1971) Transverse myelopathy in systemic lupus erythematosus: report of three cases and review of the literature. Ann Intern Med 83:616-624
- Apel DM, Lorenz MA, Zindrick MR (1989) Symptomatic spondylolisthesis in adults: four decades later. Spine 14:345-348
- Auls AA (1978) Chronic spinal arachnoiditis. Spine 3:88-91
- Awwad EE, Martin DS, Bucholz RD, Sundaram M (1991) Synovial cyst of the mid thoracic spine. AJNR 12:562-563
- Baker AS, Ojemann RG, Swartz MN et al (1975) Spinal epidural abscess. N Engl Med 293:463-468
- Baker AS, Ojemann RG, Swartz MN, Richardson EP Jr (1975) Spinal epidural abscess. N Engl J Med 293:463-468
- Barakos JA, Mark AS, Dillon WP, Norman D (1990) MR imaging of acute transverse myelitis and AIDS myelopathy. J Comput Assist Tomogr 14:45-50
- Barakos JA, Mark AS, Dilon WP et al (1990) MR imaging of acute transverse myelitis and AIDS myelopathy. J Comput Assist Tomogr 14:45
- Bates DJ (1991) Inflammatory diseases of the spine. Neuroimaging Clin North Am 1:231-250
- Berman M, Feldman S, Alter M et al (1981) Acute transverse myelitis: incidence and etilogic considerations. Neurology 31:966-971
- Bertino RE, Porter BA, Stimac GK, Tepper SJ (1988) Imaging spinal osteomyelitis and epidural abscess with short T1 inversion recovery (STIR). AJNR 9:563-564
- Beyer GA, Lipman BT, Haughton VM, Ho KC (1987) Effect of arachnoiditis on pain threshold. Invest Radiol 22:781-785
- Binkert CA, Kollias SS, Valavanis A (1999) Spinal Cord Vascular Disease: Characterization with fast three-dimensional contrast-enhanced MR angiography. AJNR 20:1785–1793
- Bizan M (1987) Rubella myelitis and encephalomyelitis in childhood: a report of two cases with magnetic resonance imaging. Neuropediatrics 18:84-87
- Boden SD, Davis DO, Dina TS et al (1992) Postoperative diskitis: distinguishing early MR imaging findings from normal postoperative dick space changes. Radiology 184(3):765
- Boden SD, Davis DO, Dina TS et al (1990) Abnormal magnetic resonance scans of the lumbar spine in asymptomatic subjects. J Bone Joint Surg 72A(3):403
- Bozzali M, Rocca MA, Iannucci G, Pereira P, Comi G, Filippi M (1999) Magnetization-transfer histogram analysis of the cervical cord in patients with multiple sclerosis. AJNR 20:1803–1808
- Bricher MD, Tasker T, Crawshaw C, Mulholland RC (1988) Discitis following lumbar surgery. Spine 13:98-102
- Burton CV (1978) Lumbosacral arachnoiditis. Spine 3:24-30
- Campi A, Filippi M, Comi G et al (1995) Acute transverse myelopathy: spinal and cranial MR study with clinical follow-up. Am J Neuroradiol 16:115-123
- Campi A, Filippi M, Comi G, Scotti G (1998) Recurrent acute transverse myelopathy associated with anticardiolipin antibodies. AJNR 19:781-786
- Casselman JW, Jolie E, Dehaene I, Meeus L, St-Jan AZ (1991) Gadolinium-enhanced MR imaging of infarction of the anterior spinal cord. AJNR 12:562
- Chan KM, Leung PC, Lee SY et al (1988) Pyogenic osteomyelitis of the spine: a review of 16 consecutive cases. J spinal Disord 1:224-231
- Chiang KS, Lee YY, Mawad ME (1991) MR of intraspinal synovial cyst: rim enhancement with gadopentetate dimeglumine. AJR 157:416
- Choi KH, Lee KS, Chung So et al (1996) Idiopathic Transverse myelitis: MR characteristics. AM J Neuroradiol 17:1151
- Chong J, Di Rocco A, Tagliati M, Danisi F, Simpson DM, Atlas SW (1999) MR findings in AIDS-associated myelopathy. AJNR 20:1412-1416
- Ciric I, Mikhael MA, Tarkington JA et al (1980) The lateral recess syndrome: A variant of spinal stenosis. J Neurosurg 53:433-443
- Colombo N, Berry I, Norman D (1992) Infections of the spine. In: Manelfe C (ed) Imaging of the spine and spinal cord. Raven Press, New York, pp 489-512
- Crisi G, Carpeggiani P, Trevisan C (1993) Gadolinium-enhanced nerve roots in lumbar disk herniation. Am J Neuroradiol 14:1379
- Curling OD, Gower DJ, McWhorter JM (1990) Changing concepts in spinal epidural abscess: a report of 29 cases. Neurosurgery 27:185-192
- Danner RL, Hartman BJ (1987) Update of spinal epidural abscess: 35 cases and review of the literature. Rev Infect Dis 9:265-274
- Darouiche Ro, Hamill RJ, Greenberg SB et al (1992) Bacterial spinal epidural abscess. Review of 43 cases and literature survey. Medicine 71:369-385
- Del Curing O, Gower DJ Jr, McWhorten JM (1990) Changing concepts in spinal epidural abscess: a report of 29 cases. Neurosurgery 27:185-192
- Dillon WP, Norman D, Newton TH, Bolla K, Alexander M (1989) Intradural spinal cord lesions: Gd-enhanced MR imaging. Radiology 170:229-237

- Floris R, Bianco F, Rossi F et al (1987) The role of spinal cord MRI in the diagnosis of multiple sclerosis. Rev Neurol 57:120
- Fouquet B, Goupille P, Jattiot F et al (1992) Discitis after lumbar disc surgery features of aseptic forms. Spine 17:356-358
- Friedman DP (1992) Herpes zoster myelitis: MR appearance. AJNR 13:1404-1406
- Gieron MA, Martinez CR, Dipasquale KJ, Korthals JK (1992) Transverse myelitis mimicking intramedullary spinal cord tumor. J Neuroimaging 2:39-41
- Grenier N, Kressel HY, Schiebler ML, Grossman RI (1989) Isthmic spondylosis of the lumbar spine MR imaging at 1.5 tesla. Radiology 170:489-493
- Guyer DW, Wiltse LL, Eskay ML, Guyer BH (1989) The long-range prognosis of arachnoiditis. Spine 12:1332-1341
- Hashemi RH, Bradley WG et al (1995) Suspected multiple sclerosis: MR imaging with a thin-section fast FLAIR pulse sequence. Radiology 196:505
- Haughton VM, Cheng HK, Boedecker RA (1979) The contracting cord sign of multiple sclrosis. Neuroradiology 17:207-209
- Hlavin ML, Kaminski HJ, Ross JS, Ganz E (1990) Spinal epidural abscess ten year perspective. Neurosurgery 27:177-184
- Honig LS, Sheremata WA, Shapiro L et al (1986) Spinal cord abnormalities without cerebral lesions: a magnetic resonance imaging study of demyelinating disease (abstract). Ann Neurol 20:152
- Jeffery DR, Mandler RN, Davis LE (1993) Transverse myelitis: retrospective analysis of 33 cases, with differentiation of cases associated with multiple sclerosis and parainfectious events. Arch Neurol 50:532-535
- Jinkins JR, Matthes JC Sener RN, Venkatappan S et al (1992) Spondylolysis, spondylolisthesis, and associated nerve root entrapment in the lumbosacral spine: MR evaluation. AJR 159:799-803
- Jinkins RJ (1993) MR of enhancing nerve roots in the unoperated lumbosacral spine. Am J Neuroradiol 14:193
- Johnson CE, Sze G (1990) Benign lumbar arachnoiditis MR imaging with gadopentetate dimeglumine. AJNR 11:763-770
- Johnson RT, Richardson EP (1996) The neurological manifestations of systemic lupus erythematosus: a clinical-pathological study of 24 cases and review of the literature, Medicine 47:337
- Kricum R, Shoemaker EI, Chovanes GI, Stephens HW (1992) Epidural abscess of the cervical spine: MR findings in the five cases. AJR 158:1145-1149
- Kulowski J (1936) Pyogenic osteomyelitis of the spine: an analysis and discussion of 102 cases. J Bone Joint Surg 18:343-364
- Kumar J, Kimm J (1994) MR in Toxocara canis myelopathy. Am J Neuroradiol 15:1918
- Kuntz D, Bertrend N, Bardin T et al (1984) Destructive spondyloarthropathy and the hemodialized patient. A new syndrome. Arthritis Rheum 27:369
- Kuroiwa Y (1970) Neuromyelitis (Devic's disease and Devic's syndrome) In: Vinken and Bruyn GW (eds) Handbook of clinical neurology. Vol 9, North-Holland, Amsterdam, pp 426-436
- Larson E-M, Holtas S, Nilsson O (1989) Gd-DTPA-enhanced MR of suspected spinal multiple sclerosis. AJNR 10:1071-1076
- Lee H-J, Bansil S, Cook SD et al (1992) MRI of the spine in multiple sclerosis. J Neuroimaging 2:61
- Lipson SJ (1985) Cervical myelopathy and posterior atlantoaxial subluxation in patients with rheumatoid arthritis. J Bone Joint Surg 67A:593
- Liu LX (1993) Spinal and cerebral schistosomiasis, Semin Neurol 13:189
- Liu SS, Williams KD, Drayer BP, Spetzler RF, Sonntag VKH (1990) Synovial cysts of the lumbosacral spine: diagnosis by MR imaging. AJNR 10:1239-1242
- Martel W (1977) Pathogenesis of cervical discovertebral destruction in rheumatoid arthritis. Arthritis Rheum 20:1217-1225
- Martin D, Delacollete M, Collignon J et al (1994) Radiation-induced myelopathy and vertebral necrosis. Neuroradiology 36:405
- Mawad ME, Rivera V, Crawford S, Ramirez A, Breitbach W (1990) Spinal cord ischemia after resection of thoracoabdominal aortic aneurysms: MR findings in 24 patients. AJNR 11:987-991
- McGee-Collert M, Johnston IH (1991) Spinal epidural abscess: presentation and treatment. A report of 21 cases. Med J Aust 155:14-17
- Melhem ER, Wang H (1982) Intramedullary spinal cord tuberculoma in a patient with AIDS. Am J Neuroradiol 13:986
- Miller DH, Mcdonald WI, Blumhardt LD et al (1987) Magnetic resonance imaging in isolated noncompressive spinal cord syndromes. Ann Neurol 22:714-723
- Mirich DR, Ducharczyk W, Keller MA, Deck J (1992) Subacute necrotizing myelopathy: MR imaging in four pathologically proved cases. AJNR 158:163-169
- Modic MT, Feiglin DH, Piraino DW et al (1985) Vertebral osteomyelitis: assessment using MR. Radiology 157:157-166
- Modic MT, Ross JS, Obuchowski N et al (1995) Contrast enhanced MR and acute lumbar radiculopathy: a pilot study of the natural history. Radiology 195:429
- Modic MT, Steinberg PM, Ross JS et al (1988) Degenerative disk disease: assessment of changes in vertebral body marrow with MR imaging. Radiology 166:193-199
- Murphy KJ, Brunberg JA, Quint DJ, Kazanjian PH (1998) Spinal cord infection: myelitis and abscess formation. AJNR 19:341-348
- Nabatame H, Nakamura K, Matuda M et al (1992) MRI of syphilitic myelitis. Neuroradiology 34:105
- Naidich JB, Mossey RT, McHeffey-Atkinson B et al (1988) Spondyloarthropathy from long-term hemodialysis. Radiology 167:761-764
- Nijeholt GJL, Barkhof F, Scheltens P, Castelijns JA, Ader H, van Waesberghe JH, Polman C, Jongen SH, Valk J (1997)

MR of the spinal cord in multiple sclerosis: Relation to clinical subtype and disability. AJNR 18:1041-1048
- Nijeholt GJL, Castelijns JA, Weerts J, Ader H, van Waesberghe JH, Polman C, Barkhof F (1998) Sagittal MR of multiple sclerosis in the spinal cord: fast versus conventional spin-echo imaging. AJNR 19:355-360
- Nussbaum ES, Rigamonti D, Standiford H et al (1992) Spinal epidural abscess: a report of 40 cases and review. Surg Neurol 38: 225-231
- Patel SC, Sanders WP (1988) Synovial cyst of the cervical spine: case report and review of the literature. AJNR 9: 602-603
- Paty DW Jr, Oger JJF, Kastrukoff LF et al (1988) MRI in the diagnosis of MS: a prospective study with comparison of clinical evaluation evoked potentials, oligoclonl banding, and CT. Neurology 38:180-185
- Pendleton B, Carl B, Pollay M (1983) Spinal extradural benign synovial or ganglion cyst: cases report and review of the litterature. Neurosurg 13:322-325
- Poon TP, Tchertkoff V, Pares GF et al (1992) Spinal cord Toxoplasma Lesion in AIDS: MR findings, J comput Assist Tomogr 16:817
- Post MJD, Quencer RM, Montalvo BM et al (1988) Spinal infection: evaluation with MR imaging and intraoperative US. Radiology 169:765-771
- Post MJD, Sze G, Quencer RM et al (1990) Gadolinium-enhanced MR in spinal infection. J Comp Assist Tomogr 14:721-729
- Provenzale JM, Barboriak DP, Gaensler EH et al (1994) Lupus-related myelitis: serial MR findings. Am J Neuroradiol 15:1911
- Provenzale JM, Bouldin TW (1992) Lupus-related myelopathy: report of three cases and review of the literature. J Neurol Neurosurg, Psychiatr 55:830
- Provenzale JM, Barboriak DP, Gaensler EH et al(1994) Lupus-related myelitis: serial MR findings. Am J Neuroradiol 15:1911
- Quiles M, Marchisello PJ, Tsairis P (1978) Lumbar adhesive arachnoiditis. Spine 3:45-50
- Ramos-Remus C, Gomez-Vargas A, LeClercq S et al (1993) Radiologic features of DISH may mimic ankylosing spondylitis. Clin Exp Rheumatol 11:603
- Resnik D (1989) Inflammatory disorders of the vertebral column: seronegative spondyloathropathies, adult onset rheumatoid arthritis and juvenile chronic arthritis. Clin Imaging 13:253-268
- Reynolds H, Carter SW, Merthefr et al (1987) Cervical rheumatoid arthritis: value of flexion extension views in imaging. Radiology 164:215
- Rieger J, Hosten N (1994) Spinal cord sarcoidosis. Neuroradiology 36:627
- Rizzo JF, Lessel S (1988) Risk of developing multiple sclerosis after uncomplicated optic neuritis a long-term prospective study. Neurology 38:185-190
- Rocca MA, Mastronardo G, Horsfield MA, Pereira C, Iannucci G, Colombo B, Moiola L, Comi G, Filippi M (1999) Comparison of three MR sequences for the detection of cervical cord lesions in patients with multiple sclerosis. AJNR 20:1710-1716
- Ross JS (1989) Inflammatory disease: Modic MT, Masaryk TJ, Ross JS (eds) Magnetic Resonance Imaging of the spine. Year Book Medical Publ, Chicago, pp 167-182
- Ross JS, Masaryk TJ, Modic MT et al (1987) MR imaging of lumbar arachnoiditis. Am J Neuroradiol 8:885
- Ross JS, Masaryk TJ, Schrader M et al (1990) MR imaging of the postoperative lumbar spine: assessment with Gd-DTPA. Am J Neuroradiol 11:771
- Ross PM, Flemming JL (1976) Vertebral body osteomyelitis spectrum and natural history. A retrospective analysis of 37 cases. Clin Orthop 118:190-198
- Sandhu FS, Dillon WP (1991) Spinal epidural abscess: evaluation with contrast-enhanced MR imaging. AJNR 12:1087-1093
- Sandson TA, Friedman JH (1989) Spinal cord infarction. Medecine 68:282-292
- Sapico FL, Montgomerie JZ (1980) Vertebral osteomyelitis intravenous drug abusers: report of 3 cases and review of the literature. Rev Infect Dis 2:196-206
- Schwartz ST, Spiegel M, Ho G jr (1990) Bacterial vertebral osteomyelitis and epidural abcess. Semin Spine Surg 2:95-105
- Shakudo M, Inoue Y, Tsutada T (1999) HTLV-I-associated myelopathy: acute progression and atypical MR findings. Case Report. AJNR 20:1417-1421
- Sharif HS (1992) Role of MR imaging in the management of spinal infections. AJR 158:1333-1345
- Shonstrom NR, Bolender NF, Spengler DM (1985) The pathomorphology of spinal stenosis as seen on CT scans of the lumbar spine. Spine 10:806-811
- Smith AS, Weinstein MA, Mizushima A et al (1989) MR imaging characteristics of tuberculous spondylitis versus vertebral osteomyelitis. AJNR 10:619-625
- Smith CA, Pinalis RS (1982) Optic neuritis in systemic lupus erythematosus. J Rheumatol 9:963-966
- Suzuki K, Meguro K, Wada M, Nakai M, Nose T (1998) Anterior spinal artery syndrome associated with severe stenosis of the vertebral artery. AJNR 19:1353-1355
- Szypryt EP, Twining P, Mulhollan RC, Worthington BS (1989) The prevalence of disc degeneration associated with neural arch defects of the lumbar spine assessed by magnetic resonance imaging. Spine 14:977-981
- Tartaglino LM, Heiman-Patterson T, Friedman DP et al (1995) MR imaging in a case of postvaccination myelitis. Am J Neuroradiol 16:581
- Thompson AJ, Kermode AG, Wicks D et al (1991) Major differences in the dynamics of primary and secondary progressive multiple sclerosis. Ann Neurol 29:53-62
- Thrush A, Enzmann D (1990) MR imaging of infectious spondylitis AJNR 11:1171-1180
- Tippet DS, Fishman PS, Panitch HS (1991) Relapsing transverse myelitis. Neurology 41:703-706
- Toshiro K, Ito K, Maruo Y et al (1987) MR imaging of spinal cord in Devic's disease. J Comput Assist Tomogr 11: 516-517
- Trop I, Bourgouin PM, Lapierre Y, Duquette D, Wolfson CM, Duong HD, Trudel GC (1998) Multiple Sclerosis of the spinal cord: diagnosis and follow-up with contrast-

enhanced MR and correlation with clinical activity. AJNR 19:1025-1033
- Trush A, Enzmann D (1981) MR imaging of infectious spondylitis. AJNR 11:1171-1180
- Uhlenbrock D, Seidel D, Gehlen W et al (1988) MR imaging in multiple sclerosis comparison with clinical, CSF, and visual evoked potential findings. AJNR 9:59-67
- Verbiest H (1954) A radicular syndrome from developmental narrowing of the lumbar vertebral canal. J Bone Joint Surg 36B:230-237
- Waldvogel FA, Medoff G, Swartz MN (1970) Osteomyelitis: a review of clinical features, therapeutic considerations and unusual aspects. N Engl J Med 282: 198-206, 260-266, 316-322
- Wallace CJ, Seland TP, Fong TC (1992) Multiple sclerosis the impact of MR imaging. AJR 158:849-857
- Warren RW, Kredich DW (1984) Transverse myelitis and acute central nervous system manifestations of systemic lupus erythematosus. Athritis Rheum 27:1058-1060
- Wirth EP, Post KD, DiChiro G et al (1970) Foix-Alajouanine disease. Spontaneous thrombosis of a spinal cord arteriovenous malformation: a case report. Neurology 170:1114-1118
- Witham RH, Brey RL (1985) Neuromyelitis optica: Two new cases and review of the literature. J Clin Neuro-Ophthalmo 5:263-269
- Wolfe BK, O'Keeffe D, Mitchell DM et al (1987) Rheumatoid arthritis of the cervical spine: early and progressive radiographic features. Radiology 165:145
- Yamamoto M (1986) Recurrent transverse myelitis associated with collagen disease. J Neurol 233:185-187
- Yetkin FZ, Haughton VM, Papke RA et al (1991) Multiple sclerosis: specifity of MRI for diagnosis. Radiology 178: 447-451
- Yuh WTC, Marsh EE, Wang AK et al (1992) MR imaging of spinal cord and vertebral body infarction. AJNR 13: 145-154

Part III

Spine and cord

4- Traumatic and iatrogenic lesions

a- Vertebral
b- Medullary

Vertebral

Cases 75-79: spinal trauma and burst fractures

Clinical and pathological features

- Evaluation of spinal injuries remains a challenge and the imaging approach differs with the clinical state of the injured patient.
- The vast majority of spinal traumas are direct vertebral traumas.
- It is not common that trauma causes an isolated spinal cord injury without any fracture in the spine, particularly in children.
- Spinal injuries may be divided into 3 major categories: fracture-dislocations, being the most frequent type, isolated fractures, and isolated dislocations.
- The effects of spinal injury vary depending upon the intensity, direction and site of impact of the applied forces.
- The most frequent sites where spinal injuries occur are at C1-C2 (craniocervical junction), C4 to C7 (lower cervical spine), and T11 to L2 (thoracolumbar junction) levels. Such spinal levels correspond to most mobile spinal regions. They involve the spinal segments containing the cervical and lumbar cord enlargements, in which the space left between the canal and its content is reduced.
- Whiplash injuries, as commonly encountered secondary to an automobile accident, are typically due to sudden hyperextension and hyperflexion movements of the neck. The vulnerability of the cord to the effects of anteflexion and retroflexion is increased by a pre-existing spondylosis or constitutional stenosis of the spinal canal.
- A particular type occurs with high-velocity missiles causing direct injury to the spine and cord (stab wounds, bullets and shrapnels). Such trauma may cause sideration of spinal cord functions due to concussion, usually reversible in one or two days.
- Patient transfer should be performed with great care by trained personnel, from the emergency room to the imaging department and explore there using cost-efficient and appropriate imaging algorithms.

MR characteristic findings

Morphology and signal intensity

- Diagnostic imaging algorithms should include an initial evaluation and screening with conventional radiography of the spine beginning with a lateral view. This is particularly true when evaluating the cervical spine. But one must be careful not to cause further injury to the cord with neck manipulation during the exploration. Analysis of the lateral view must include the evaluation of the integrity of 4 craniocaudad lines which correspond, front to back, to: the prevertebral retropharyngeal (less than 7 mm thick) and retrotracheal (less than 22 mm thick) soft tissues; the anterior margins of the vertebral bodies and anterior longitudinal ligament; the posterior margin of the posterior vertebral bodies lined by the posterior longitudinal ligament; the spinolaminar line corresponding to the base of the spinous processes bounding the posterior aspect of the spinal canal. An anteroposterior view and an odontoid view with open mouth, are also of importance to evaluate the craniocervical junction. These 3 incidences improve the detection sensitivity of conventional X-rays averaging 90%. The major limitations are the assessment of the spinal canal and cord, as well as the paraspinal soft tissues.
- Spiral CT replaced conventional tomography in the evaluation of most of complex spinal fractures. It presently stands as the modality of choice due to its superior contrast resolution associated with a marked decrease in examination time. It allows also to minimize the manipulation of patients with unstable spine presenting risks for further neurological deterioration. Conventional tomography may still be of help in the evaluation of Chance fractures and in those involving the base of the odontoid process, if MR appears noncontributive.
- With the advent and development of fast imaging using MRI, noninvasive assessment of the spinal canal and its content, directly evaluated using direct multiplanar and particularly sagittal cuts, myelography and CT myelography become useless.

- MRI directly discloses and evaluates spinal cord injuries which may even occur in the absence of vertebral fractures and/or dislocations, adding data of prognostic importance.
- Osseous, joint, ligamentous, discal and paraspinal soft tissue changes associated with spine injury have similar MR appearance regardless of the site of the trauma to the spine.
- Sagittal approach to the spine using T1w and T2w SE sequences are the most informative, complemented at times by the axial cuts. Gradient echo technique may add important data with respect to diagnosis of non-displaced vertebral fractures and depiction of hemorrhagic component of cord contusion.
- MR remains relatively insensitive to detect fractures of the vertebral posterior arches and even more when the fracture involves C1 or C2 vertebrae.
- Ligamentous rupture or injuries due to overstretching is indirectly at least evidenced on T2w sequences as areas of increased SI due to intensification in free water and hemorrhage at the site of tearing. Evaluation of the anterior and posterior longitudinal ligaments (misalignment and focal discontinuity of the related vertebral lines sagittally) or the ligamenta flava (bulging aspect within the spinal canal), or the interspinous ligaments (widening between spinous processes) is therefore indirectly or directly facilitated.
- Injuries involving the facet joints are demonstrated on lateral sagittal cuts on both sides, showing a loss of alignment or a subluxation. Increase in fluid within the joint space may be evidenced as well (hyperintense SI on T2w).
- Disc herniations may be found secondary to a trauma affecting more frequently the cervical and thoracic spine.

- In the cervical spine, traumas involve mainly the craniocervical junction and the lower cervical spine (C6 and C7 more commonly), the craniocervical region being the most common site of injury in children. A diastasis in the atlanto-dental joint superior to 3 mm in adults and 5 mm in children as assessed on lateral X-ray view, raises the diagnosis of disruption of the transverse ligament of the atlas. Cervical spine traumatic injuries vary according to the causal mechanism and involve frequently multiple cervical levels. The most frequent traumatic conditions are those related to hyperflexion, hyperextension or axial compression.
  - Traumas in hyperflexion, usually unstable injuries, involve preferentially C5 or even C6 or C4, giving rise to compression fracture with loss of height of the anterior part of the vertebral body, widening of the interspinous region due to interspinous ligament disruption, and if severe, avulsion of the antero-inferior endplate of the vertebra (commonly C5) with disruption of the posterior longitudinal ligament and fracture of the posterior arch, leading to marked instability. Bilateral facet dislocation with disruption of both the anterior and posterior longitudinal ligaments occurring at C5-C6 or C6-C7 levels lead to a highly unstable vertebral antelisthesis associated with spinal cord compression.
  - Traumas in hyperextension may lead to tearing of the anterior longitudinal ligament with avulsion of the anteroinferior end plate of C2 or C3 vertebrae, with soft tissue swelling and commonly associated tetraplegia. Anterior subluxation of C2 with respect to C3 is secondary to fractures of the pars interarticularis bilaterally. Such injuries must be suspected in presence of a facial trauma.
  - Traumas secondary to axial compression lead to bilateral fractures of the posterior C1 arch compressed between the superior articular facets of C2 and the occipital condyles above (Jefferson's burst fracture). Fracture of the odontoid process occuring in hyperflexion and hyperextension injuries, usually non associated with neurological damage, are divided into 3 subtypes (Anderson): I- fracture of the tip of the dens (rare and should not be confused with an os odontoideum); II- horizontal fracture through the base of the dens (most common), and III- fracture extending to the upper part of the body of C2. Vascular injuries may be associated with cervical spine trauma leading to a dissection of the vertebral (contained within the foramina transversaria) more frequently than the carotid arteries, in the neck (MR-angiography to perform).
  - Post-traumatic disc herniations occur more frequently in the cervicothoracic spine than in the lumbar region. They seem to be more common in flexion-compression injuries. Their morphological aspect and behavior is similar to non traumatic herniated discs.
- In the thoracolumbar spine the most common fractures are those related to axial compression leading, when severe, to a burst fracture and

hyperflexion fractures also recognized in relation with seat belt injuries.

- Burst fractures are usually found between T9 and L5 (T12 and L1 being the most frequently involved), showing an anterior loss of height of the vertebral body (sagittal cut), a widening of the interpediculate distance (AP view/ coronal MR cut) and a widening of the interspinous distance (sagittal cut) due to disruption of the interspinous ligament. Burst compression fractures are very common in elderly women with osteoporosis including a postmenopausal osteopenic insufficiency. They must be differentiated from burst fractures secondary to a metastasis. In pathologic fractures the vertebral body shows expanded anterior and posterior margins with areas of cortical destruction, and lack to demonstrate residual marrow fat. Other localizations may be found helping to confirm the diagnosis, otherwise a follow-up exam about 2 months later may be needed.
- Almost half of the fractures involving the thoracolumbar junction are associated with neurological deficits. Narrowing of the spinal canal in burst fractures occurs in more than half the cases. MRI is highly efficient in the evaluation, noninvasively, of such fractures visualizing clearly (sagittal cut) the relationship between the posterior vertebral body projection and the cord-conus-cauda complex.
- Fractures secondary to hyperflexion are complex and unstable, horizontally oriented, and divided into 2 types: the Chance's type and the Smith's type. In both, the fracture involves the posterior vertebral body and the posterior elements, whereas in the Smith's type the spinous process is respected.
- Note that thoracolumbar fractures are commonly associated with an epidural hematoma secondary to tearing of the venous plexuses.
- Sacral fractures, usually associated with fractures of the pelvis, may occur as isolated horizontal fractures involving the inferior sacrum, secondary to a direct trauma or as stress fractures commonly found in elderly women without any traumatic event reported. Plain films are frequently unremarkable.

Contrast enhancement

- Contrast medium is useless in the evaluation of spinal trauma.

Case 75: Atlanto-axial subluxation with cord compression

Fig. a: sagittal cut;	SE - T1w + Gd	**Fig. c:** axial cut;	SE - T1w
Fig. b: sagittal cut;	SE - PDw	**Fig. d:** axial cut;	GE (3D) - T2*w

Patient presenting with a recent whiplash injury with acute paraparesis. Sagittal MR slices performed using T1w (**a**) and PDw (**b**) sequences show an anterior subluxation of C1 on C2 (**a**, ↑↑) associated to a significant posterior impingement of the cervicomedullary junction (**a**, *) associated to a hyperintense intramedullary lesion corresponding to mild edema and/or associated myelomalacia well shown on the PD images (**b**, ↑). The axial cuts using T1w (**c**) and T2w (**d**) sequences show a right postero-lateral significant impingement of the cervicomedullary junction with compression of the right hemicord (**c**, ↑↑; **d**, ↑↑). Note the hyperintense signal in the prevertebral soft tissues (**a**) remaining hyperintense on PDw (**b**) and on T2*w (**d**), corresponding presumably to hemorragic perivertebral suffusion.

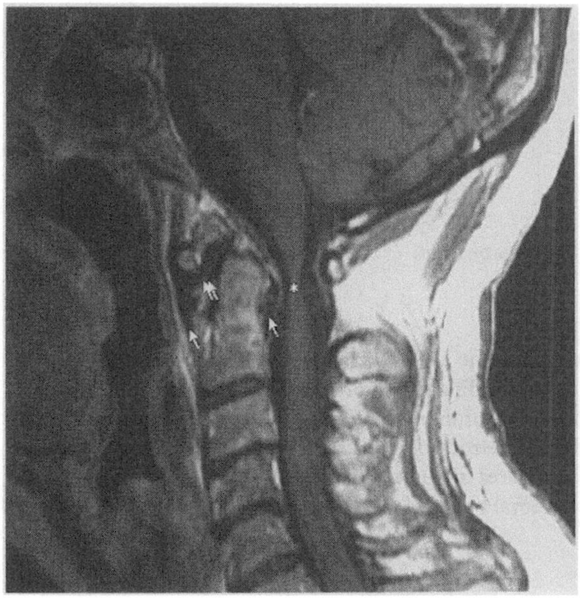

a

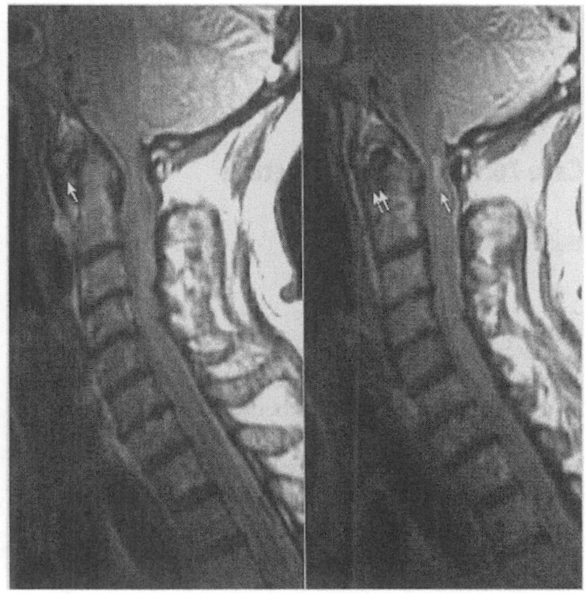

b

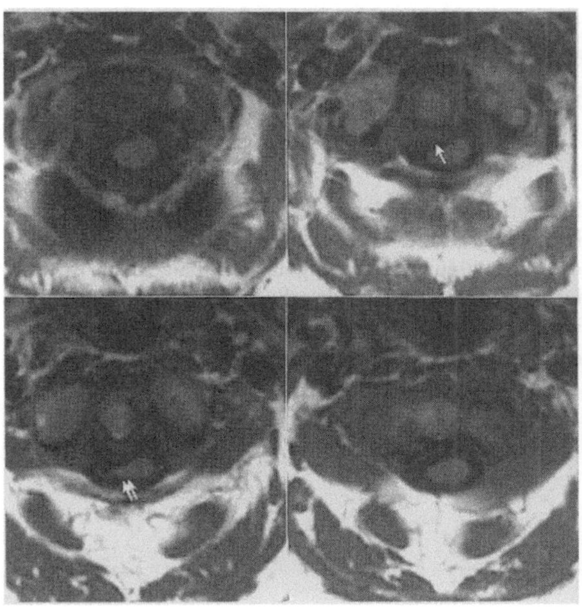

c

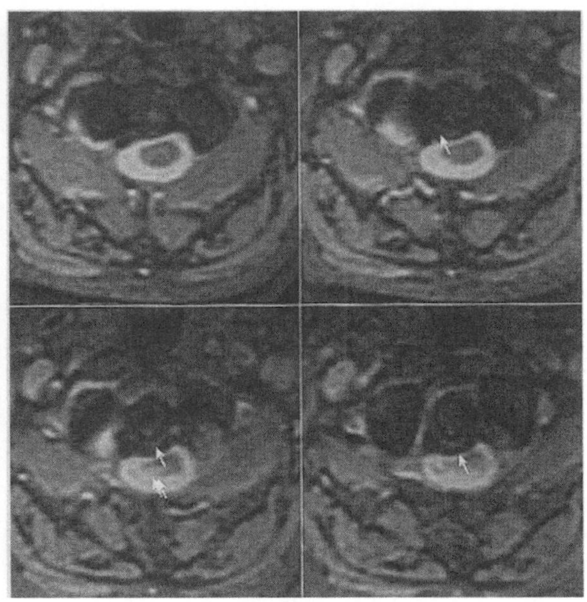

d

Case 76: Acute burst - fracture

Fig. a: sagittal cut (512); SE - T1w

Fig. b: coronal cut; SE - T2w with fat suppression

Patient presenting with back pain consecutive to a trauma. MR exam shows a hypointense signal on T1w involving a major portion of the vertebral body (**a**, right, ↑) with a limited area remaining slightly hyperintense due to the presence of normal fatty marrow (**a**, left, ↑). On T2w with fat suppression, the coronal slices show an increase signal involving the vertebral body in question (b, right) associated to a laterovertebral hyperintense effusion corresponding to hemorrhage (**b**, ↑↑). Note at the upper end plate of the vertebral body, the presence of a very bright band-like zone of increased T2 signal (**b**, ↑), corresponding to the area of burst-fracture. Note the slight decrease of the vertebral body height.

Case 77: Vertebral burst - fissuration

Fig. c: sagittal cut (512); SE - T1w

Fig. d: sagittal cut; SE - T2w

Elderly woman presenting with a subacute osteoporotic fracture of L4. Sagittal MR sequences show slightly irregular band-like area of moderately hyperintense signal on T1w (**c**, ↑) as well as on T2w (**d**, ↑↑) involving the mid-portion of the vertebral body and extending from the anterior to the posterior aspect of the vertebral body, associated on T1w to hypointense adjacent bands superiorly and inferiorly, corresponding to peripheral edema. The hyperintense band is consistent with a subacute fracture defect containing presumably subacute blood. Note the hyperintense signal involving the lower half of the L4 vertebral body on T2w (**d**) corresponding probably to edema.

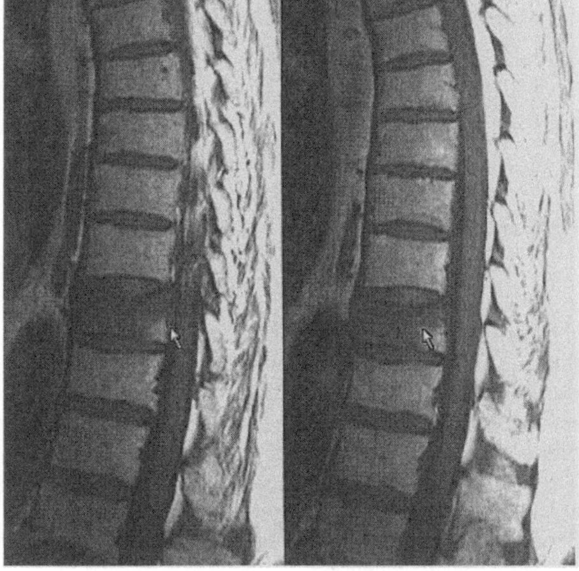

a

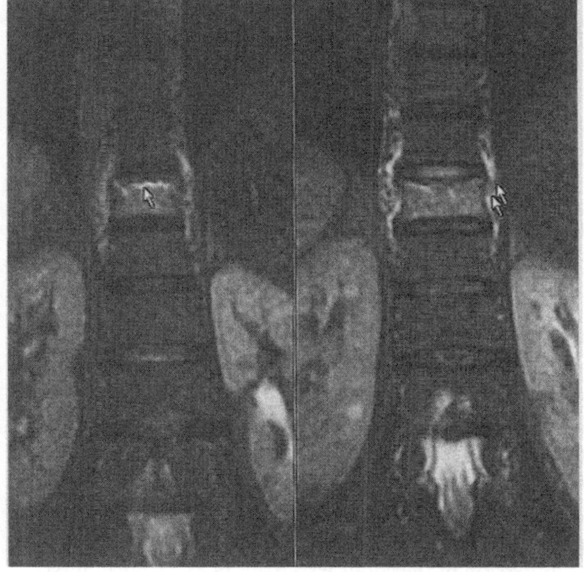

b

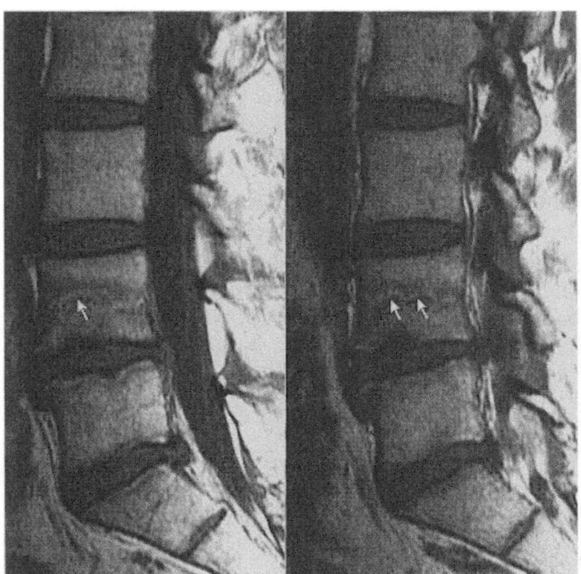

c

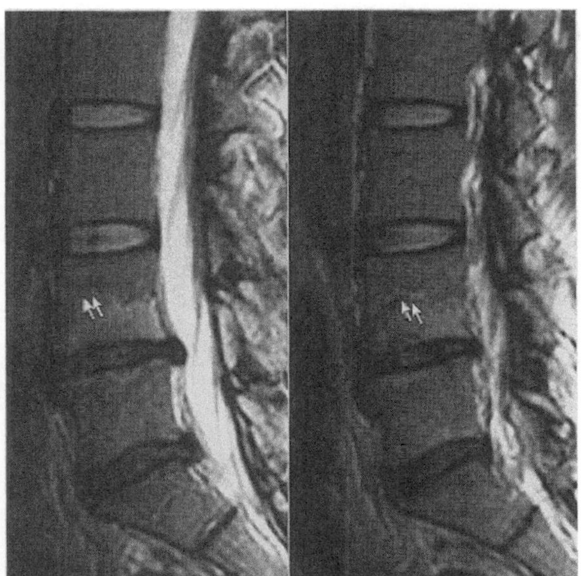

d

Case 78: Subacute osteoporotic burst - fracture

Fig. a: sagittal cut (512); SE - T1w

Fig. b: sagittal cut; SE - T2w

Caracteristic subacute burst-fracture involving the Th12 vertebral body associated with an important decrease in the vertebral body height, presenting a heterogeneous hypointense signal on T1w (**a**) with apparent vertebral fracture (**a**, ↑; **b**, ↑). MR exam shows the persistence of normal marrow areas presenting a high signal on T1w (**a**) and a normal hypointense signal on T2w (**b**) mainly at the posterior portion of the vertebral body. Note the absence of compression of the conus terminalis (**a**, **b**, *). Such burst-fractures are common involving the thoraco-lumbar junction and observed in women with postmenopausal osteoporosis. They must be differentiated from pathological fractures associated to an underlying vertebral metastatic disease. Note in this case that the compression fracture of Th12 is not associated with a retropulsion of the posterior aspect of the vertebra into the spinal canal, and the persistence of normal areas of bone marrow with normal MR signal on all sequences.

Case 79: Cuneiform vertebral body

Fig. c: sagittal cut (512); SE - T1w

Fig. d: sagittal cut (512); SE - PDw

Patient presenting a marked dorsal cyphosis. MR exam disclosed a cuneiform vertebral body with normal signal on T1w (**c**, ↑↑) and PDw (**d**, ↑↑↑), associated to a thinning of the superior vertebral interspace (**c**, ↑; **d**, ↑↑). T2w images showed an abnormal signal involving the correspondent cord segment, hyperintense on PDw (**d**, ↑) and on T2w (not shown) corresponding probably to a sequelar myelomalacia. A hypointense signal was demonstrated on the axial T1w sequence (not shown). Note the antepulsion of the anterior wall of the anteriorly collapsed vertebral body (**c**, right) associated to a slight hypointense signal on T2w.

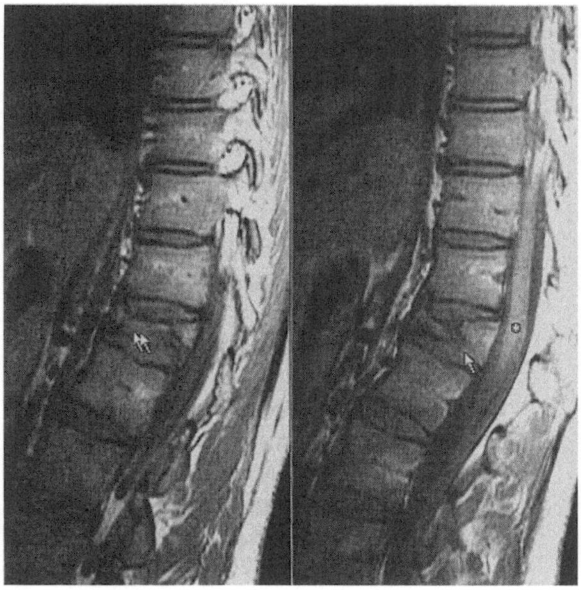

a

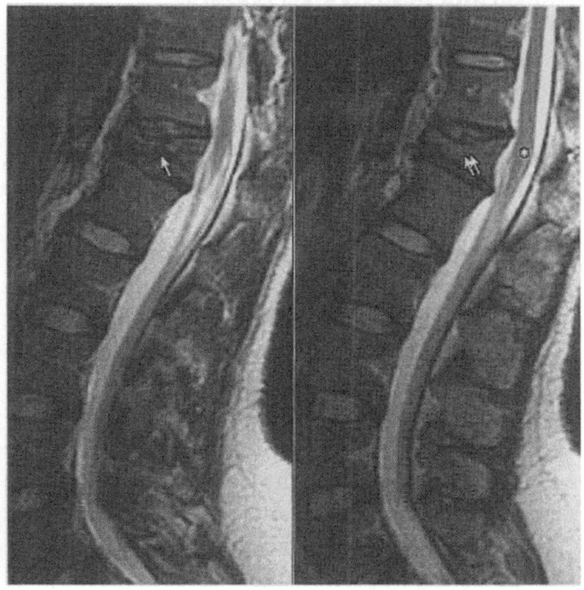

b

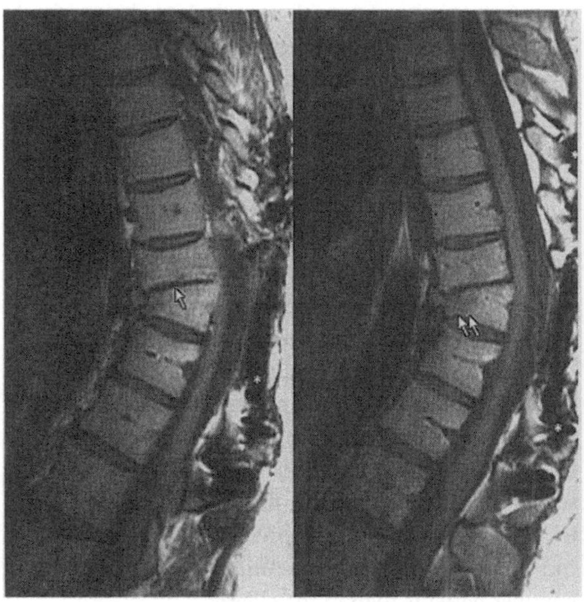

c

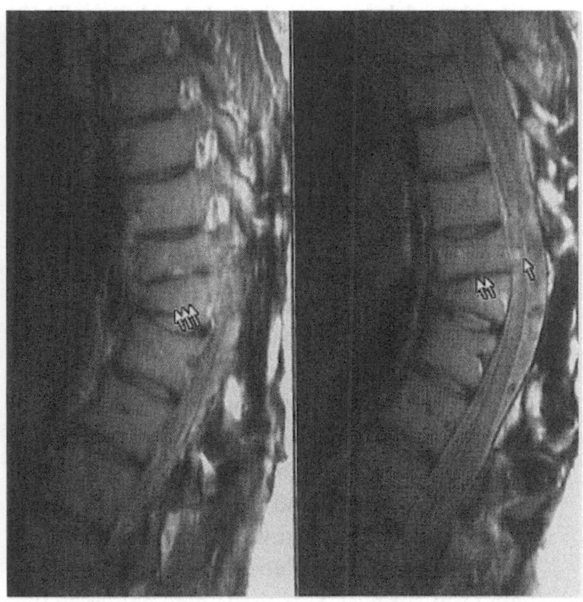

d

Medullary

Cases 80-83: spinal cord injuries

Clinical and pathological features

- Magnetic resonance has much contributed for the evaluation of spinal trauma, giving a direct vision and assessment of spinal cord injury.
- Nearly all post-traumatic patients presenting neurological signs show spinal cord abnormalities on MRI.
- There is poor correlation, if any, between the importance of cord damage or neurological deficit, and the severity of the trauma to bone or soft tissues. But the spinal cord injuries as demonstrated with MR seem to correlate well with the severity of the neurological deficits.
- Spinal cord injuries may occur (about 10% of the cases) without any demonstrated spinal fracture or dislocation. Normal appearance of the cord may not exclude cord concussion.
- The ability of MRI to differentiate between spinal cord edema and hemorrhage is of important clinical and prognostic significance, the patient outcome and neurologic recovery being compromised with spinal cord hemorrhage.
- Correlations between MRI and histopathology in animal models have confirmed the ability of MR to detect and differentiate intramedullary hemorrhage from edema. A sensitivity to detect post-traumatic wallerian degeneration of spinal tracts has been reported as well.
- *Histopathology:* the acute contusional changes secondary to cord shearing, include areas of destruction of gray and white matter with edema and hemorrhagic suffusion, and are maximal at the level of injury extending up to 2 segments up and down. True hematomyelia is uncommon (1%). Rarely a trans-section of the cord may occur. At chronic stages, which may last up to 3 years, there is resorption of the necrotic tissue within the cord leading to areas of cavitation and cicatricial gliosis and fibrosis are found. Hemorrhage is resorbed with hemosiderin deposition. Segmental cord atrophy may be left after long term survival. Scar tissue may obliterate the adjacent leptomeningeal spaces.
- *Signs:* at the acute stage, severe injury to the cord gives rise to a "spinal shock" which may commonly last for weeks to 1 or 2 months, including loss of voluntary movements and reflex functions and all sensations, below the level of the cord lesion, followed by a stage of heightened reflex activity which may last for years. The neurological exam will determine the spinal level involved and may orient the MR exploration. Classically, partial cord injuries may give rise to 5 individualized neurological syndromes:
  - 1-**anterior cord syndrome** (anterior spinal artery territory), frequently due to flexion injuries of the lower spine as secondary to diving. Usually deep pressure and position sense are preserved;
  - 2-**central cord syndrome** (central cervical gray involvement), commonly results from hyperextension injuries. Occurs in elderly patients presenting prior spondylosis of the cervical spine with or without associated hypertrophy of the ligamenta flava. The paretic signs involve mainly the upper limbs with some spasticity in the lower limbs;
  - 3-**brown-Sequard syndrome** (transversal hemicord injury), occurring usually secondary to a direct trauma. It results in an ipsilateral motor paresis with a contralateral loss of pain and temperature sensations;
  - 4-**conus medullaris syndrome** is commonly the result of a compression fracture of T12 or L1 with retropulsion of the vertebral posterior wall into the spinal canal. A loss of bladder and bowel functions and of perineal sensation are found on neurological exam;
  - 5-**cauda equina syndrome** (lumbosacral nerve roots injury) is secondary to spinal and roots compression between L1 to L5.

The clinical outcome and the residual deficits are variable, depending on the severity and localization of the spinal injury, and ranging from complete to incomplete motor paralysis, flaccid or spastic, with sensory impairement in the limbs. These findings tend to be permanent when present for more than 6 months. Neurological outcome is more favorable in younger patients than in the elderly.

Sensorimotor deficits installed years after the traumatic event may occur secondary to cavitation (syrinx) of the proximal cord segment well evidenced on MR examination (sagittal).

MR characteristic findings

Morphology and signal intensity

- MR appearance of spinal cord injuries varies according to the stage considered: acute/subacute post-traumatic period and chronic stage.
- In the acute/subacute stage (up to one month), spinal cord injury include hemorrhagic and edematous changes associated with cord swelling.
  - Hemorrhagic necrosis when present is evidenced most frequently in the central gray matter at the level of the traumatic impact. Its amount and extent is of prognostic significance, as it is reported, to be associated with poor recovery if craniocaudally superior to 10 mm in size. Acute hemorrhage (less than 24 hours) show a hypointense SI on SE-T2w and particularly on GE-T2*w sequences and an iso-intense to hypointense SI on SE-T1w. In the subacute stage, oxidation of deoxyhemoglobin transformed into paramagnetic methemoglobin, becoming hyperintense on T1 and T2w sequences, is delayed occurring in the cord after the first week, apparently due to hypoxia. Hemorrhage predicts a complete cord injury and is usually associated with the most severe neurological status.
  - Edematous intramedullary changes resulting from cord contusion, hyperintense on T2w, develop within the cord with variable extension superiorly and inferiorly. A more severe initial neurological status appears to be correlated with edema extending for more than a vertebral level. Limited edematous area correlated with a better motor improvement in almost 2/3 of the patients.
  - Segmental swelling with increase in cord diameter, is maximal at the level of cord damage tapering cranially and caudally. Swelling does not correlate with the extent of cord damage.
  - Subarachnoid hemorrhage may be associated with acute cord injury, and may give an "absence cord" appearance due to isointensity to the adjacent cord on T1w sequence. Delayed adhesive arachnoiditis may develop (eliciting sometimes a Lhermitte's syndrome).
- In the chronic stage, the damaged cord segment may show areas of myelomalacia including cicatricial gliosis and associated with variable segmental atrophy and hemosiderin deposits (hypointense SI on SE-T2 and mainly GE-T2*w).
- Some patients may experience, several months to years following the traumatic event, a deterioration of their clinical status, presenting a progressive myelopathy. MR may demonstrate (sagittal T1w) a recurrent or a residual cord compression, a syringomyelia (2-3% of the cases but may be much higher in some reports) or a cystic myelomalacia, or a cord tethering secondary to subarachnoid adhesions. A syrinx cavity is well delineated, usually central within the cord associated or not with internal septations, enlarging the cord, and showing iso-intense SI on all pulse sequences. Microcystic myelomalacia (hyperintense on SE-PDw) may contribute to the development of spinal cord cystic cavitation (less than 5 mm in size). The incidence of syringomyelia increases with the severity of the previous cord injury, does not usually correspond with the site of damage and may extend over several vertebral segments in length.

Contrast enhancement

- Contrast medium is useless in the evaluation of spinal cord injury.

Case 80: Cervical post-compressive myelomalacia

Fig. a: sagittal cut; SE - PDw

Fig. b: sagittal cut; SE - T2w

Fig. c: sagittal cut; SE - T1w

Patient referred six months after a major cervical trauma with tetraplegia. MR showed a C4-C5 luxation with a retropulsion of the upper end plate of C5 compressing the spinal cord with major post-traumatic changes. On T1w (c) the cord shows a segmental hypointense signal (**c**, ↑) becoming slightly hyperintense to CSF on PDw (**a**, ↑) and hyperintense close to CSF on T2w (**b**, ↑) associated to a less hyperintense area involving the upper cord segment (**b**, ↑↑) observed also on PDw (**a**, ↑↑). The abnormal signal intensity of the compressed cord segment corresponds to a microcystic post-contusional myelomalacia associated to gliosis and diffuse demyelination involving the remaining portion of the spinal cord and mainly the upper segment.

Case 81: Cervical radicular pseudo-meningocele

Fig. d: sagittal cut; SE - T2w

Young patient presenting a post-traumatic paresis of the arm. Lateral sagittal T2w MR images of the cervical spine and the cervical-thoracic junction disclose a rounded hyperintense signal mass occupying the lowest cervical intervertebral foramen, very well demarcated and fluid-filled (**d**, ↑). The cystic fluid collection shows signal characteristics identical to those of CSF on all sequences performed. The mass corresponds to a root sleeve pseudomeningocele.

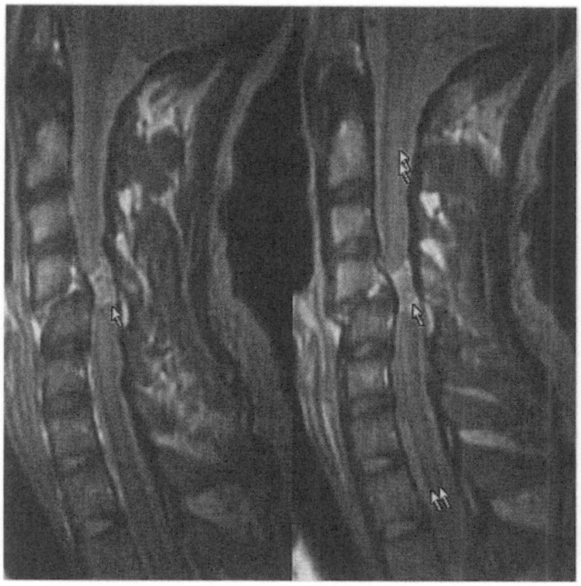

a

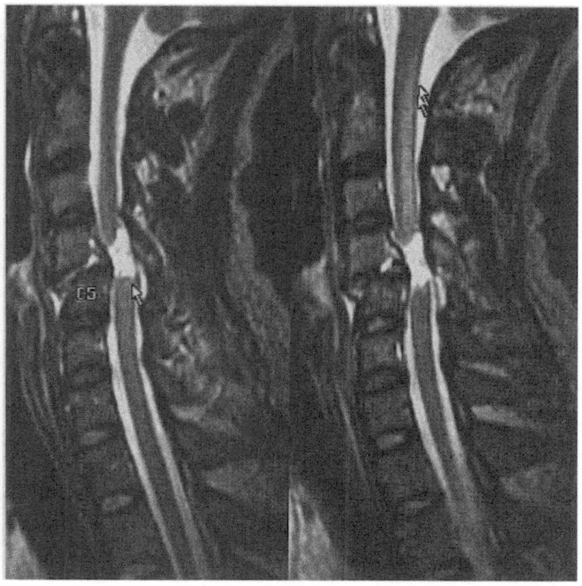

b

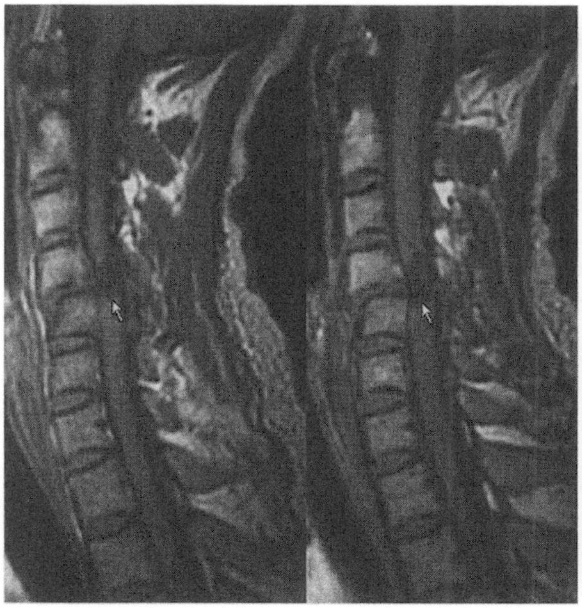

c

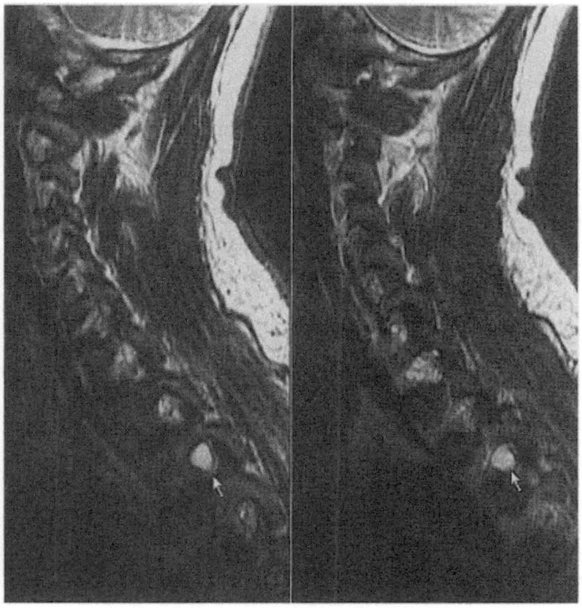

d

Case 82: Post-traumatic syrinx

Fig. a: sagittal cut; SE - T1w and sagittal cut; SE - T2w

Fig. b: axial cut; SE - T1w

Patient presenting with a syringomyelic syndrome. He met many years ago with a motor vehicle accident. MR disclosed a slightly fusiform cord swelling at the thoracic level due to presence of an intramedullary cystic cavity showing a hypointense signal on T1w (**a**, left, ↑) and a hyperintense signal on T2w (**a**, right, ↑), corresponding to a segmental syringomyelia, well demonstrated on the axial T1w cuts (**b**, down, ↑). Note the hypointense signal of the syrinx similar to the hypointense signal of CSF (**b**, ↑↑). Note the thinning of the peripheral laminated normal cord (**b**, *). There was no contrast enhancement of the cord at the level of the syrinx, neither on the remaining apparently normal thoracolumbar cord, after contrast iv.

Case 83: Chance type fracture

Fig. c: sagittal cut; SE-T1w (512)

Fig. d: sagittal cut; SE-T2w

This case evidences the MR aspect of the classically described Chance fracture showing on T1w (**c**) and T2w (**d**) sagittal cuts, an intraspinous fracture at the L1 level posteriorly (**c**, ↑↑) associated to an increased T2 signal within the marrow and the adjacent soft tissues due to edema formation (**d**, ↑↑). Note the extension of this fracture anteriorly to involve the vertebral body showing indirect signs of vertebral marrow edema, hypointense on T1w (**c**) and slightly hyperintense on T2w (**d**). Note the moderate stenosis of the spinal canal at the level of the posterior fracture (**d**, ↑) with slight compression of the cauda equina.

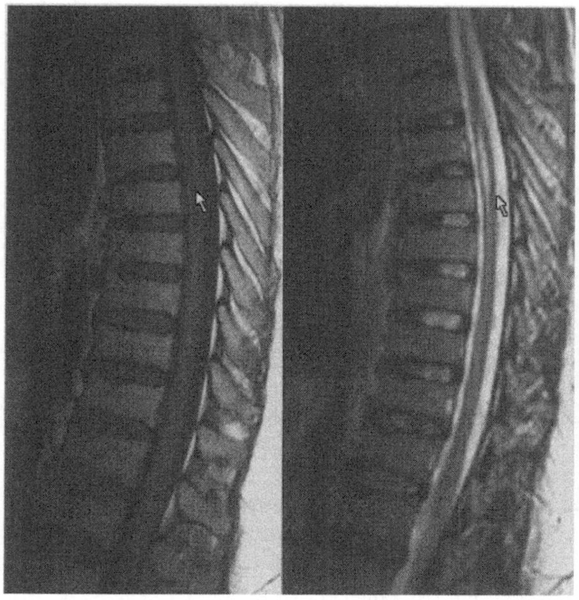

a

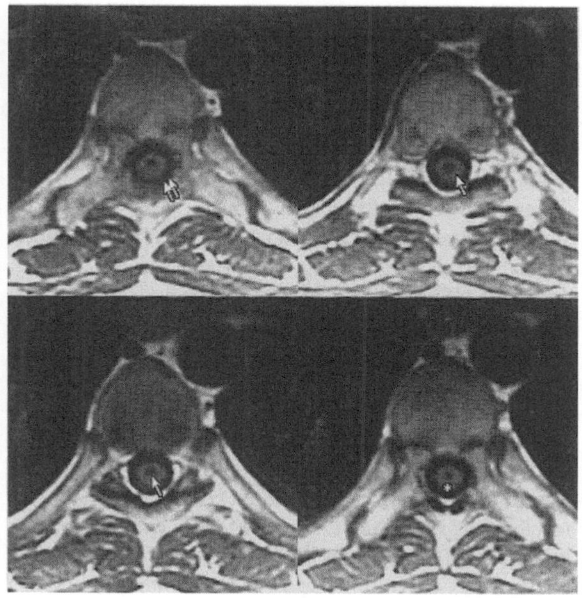

b

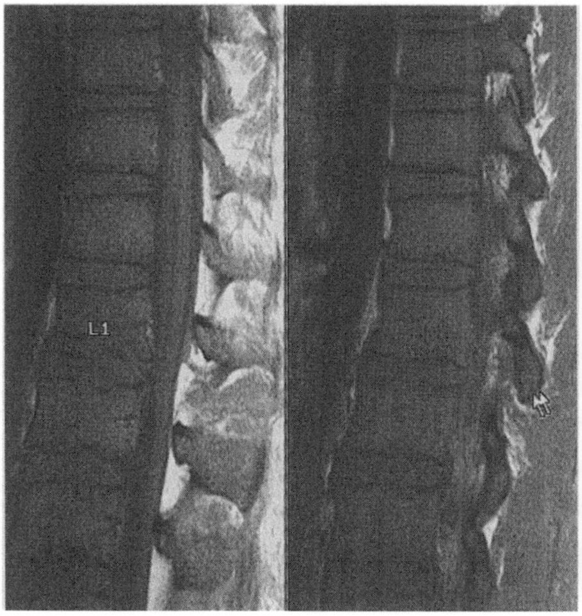

c

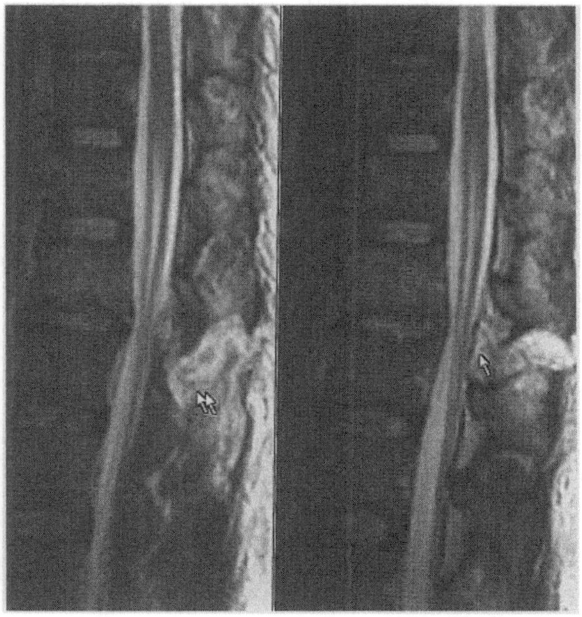

d

Selected References

- Anderson JM, Schutt AH (1980) Spinal injury in children: a review of 156 cases from 1950-1978. Mayo Clin Proc 55: 499-506
- Atlas SW, Regenbogen V, Rogers L et al (1986) The radiographic characterization of burst fractures of the spine. AJR 147:575-582
- Baker LL, Goodman SB, Perkash I et al (1990) Benign versus pathologic compression fractures of vertebral bodies: assessment with conventional spin-echo, chemical-shift, and STIR MR imaging. Radiology 174:495
- Boudurant FJ, Cotler MB, Kulkarni MV et al (1990) Acute spinal cord injury: a study using physical exam and MRI. Spine 15:161-168
- Boukobza M, Guichard JP, Boissonet M et al (1994) Spinal epidural hematoma: report of 11 cases and review of the literature. Neuroradiology 36:456
- Campi A, Filippi M, Comi G et al (1995) Acute transverse myelopathy: spinal and cranial MR study with clinical follow-up. Am J Neuroradiol 16:115
- Cintron E, Gilula La, Murphy WA, Gehweiler JA (1981) The widened disc space: a sign of cervical hyperextension injury. Radiology 141:639-642
- Clark CR, White AA (1985) Fractures of the dens: a multicenter study. J Bone Joint Surg 67A:1340-1348
- Dall BE, Stauffer ES (1988) Neurologic injury and recovery patterns in burst fracture at the T12 or L1 motion segment. Clin Orthop 233:171-176
- Davis SJ, Teresi LM, Bradely WG Jr (1991) Cervical spine hyperextension injury: MR findings. Radiology 181(1): 245-251
- Deeb ZL, Rothfus WE, Goldberg AL, Daffiner RH (1990) Absent cord sign in spinal trauma. Clin Imaging 14:138-142
- Del Bigio MR, Johnson GE (1989) Clinical presentation of spinal cord concussion. Spine 14(1):37-40
- Denis F (1983) The three column spine and its significance in the classification of acute thoracolumbar spinal injuries. Spine 8:817-831
- Emery SE, Pathria MN, Wilber RG et al (1989) Magnetic resonance imaging spinal ligament injury. Spinal Disord 2:229-233
- Finerman GAM, Akai D, Weingarten S (1976) Atlanto-axial dislocation of the spinal cord compression in a mongoloid child. J Bone Joint Surg 58 A:408-409
- Flanders AE, Schaefer DM, Doan HT (1991) Acute cervical spine trauma: correlation of MR imaging findings with degree of neurologic deficit. Radiology 177:25-33; Letter to the editor and reply. Radiology 179:287
- Frankel ML, Hancock DO, Hyslop G et al (1969) The value of postural reduction in the initial management of closed injuries of the spine with paraplegia and tetraplegia. Paraplegia 7:179-192
- Gupta RK, Mehta VS, Banerji AK et al (1989) MR evaluation of brachial plexus injury. Neuroradiology 31:377-381
- Hadley MN, Zabramski JM, Browner CM et al (1988) Pediatric spinal trauma: a review of 122 cases of spinal cord and vertebral column injuries. J Neurosurg 68:18-24
- Janssen L, Hansebout RR (1989) Pathogenesis of spinal cord injury and newer treatments. A review. Spine 14(1): 23-32
- Jefferson G (1920) Fracture of the atlas vertebra. Report of four cases, and a review of those previously recorded. Br J Surg 7:407-422
- Kolkarni MV, McArdle CB, Kopanicky D et al (1987) Acute spinal cord injury: MR imaging at 1.5T. Radiology 164:837-843
- Levinthal HR (1960) Birth injuries of the spinal cord. J Pediatr 56:447-450
- Martin D, Delacollete M, Collignon J et al (1994) Radiation-induced myelopathy and vertebral necrosis. Neuroradiology 36:405
- Matin P (1979) The appearance of bone scan following fractures, including immediate and long-term studies. J Nucl Med 20:1227
- Mawad ME, Rivera V et al (1990) Spinal cord Ischemia after resection of thoracoabdominal aortic aneurysms: MR findings in 24 patients. Am J Neuroradiol 11:987
- Mendelsohn DB, Zollars L, Weatherall PT et al (1990) MR of cord transection. J Comput Assist Tomogr 14:909-911
- Miller MD, Gehweiler JA, Martinez S, Daffiner RH et al (1978) Significant new observations on cervical spine trauma. AJR 136:659
- Nabors MW, Pait TG, Byrd EB et al (1988) Updated assessment and current classification of spinal meningeal cysts. J Neurosurg 68:366
- Pathria MN, Petersilge CA (1991) Spinal trauma. Radiol Clin North Am 29:847-865
- Penning L (1981) Prevertebral hematoma in cervical spine injury: incidence and etiologic significance. AJR 136:553
- Popovich MJ, Taylor FC, Helmer E (1989) MR imaging of birth-related brachial plexus avulsion. AJNR 10:S98
- Quencer RM, Sheldon JJ, Donovan Post MJ et al (1986) MRI of the chronically injected cervical spinal cord. AJNR 7:457-464
- Rizzolo SJ, Piazza MR, Cotler JM et al (1991) Intervertebral disc injury complicating cervical spine trauma. Spine 16 [Suppl 6]:187-189
- Ross JS, Masaryk TJ, Schrader M et al (1990) MR imaging of the postoperative lumbar spine: assessment with Gd-DTPA. Am J Neuroradiol 11:771
- Schapiro R, Youngberg AS, Rothman SLG (1973) The differential diagnosis of traumatic lesions of the occipital atlanto-axial segment. Radiol Clin North Am 11:505-526
- Smoker WRK, Dolan KD (1987) The fat C2: a sign of fracture. AJNR 8:33-38
- Starr JK, Hamley EN (1992) Junctional burst fractures. Spine 17(5):551-558
- Stover SL, Fine PR (1987) The epidemiology and economic of spinal cord injury. Paraplegia 25:225-228

- Traynelis VC, Hitchow PW, Yuh WTC et al (1990) Magnetic resonance imaging and post-traumatic Lhermitte's sign. J Spinal Dis 3:376-379
- Tronnier VM, Steinmetz A, Albert FK et al (1991) Hernia of the spinal cord: case report and review of the literature. Neurosurgery 29:916
- Wittenberg RH, Boetel V, Beyer HK (1990) Magnetic resonance imaging and computed tomography of acute spinal cord trauma. Clin Orthop 200:176-185
- Yamashita V, Takahashi M, Matsuna Y et al (1990) Chronic injuries of the spinal cord: assessment with MR imaging. Radiology 177:25-33
- Yamashita Y, Takahashi M, Matsuna Y et al (1991) Acute spinal cord injury: magnetic resonance imaging correlated with myelopathy. Br J Radiol 64:201-209
- Zwimpter TJ, Bernstein M (1990) Spinal cord concussion. J Neurosurg 72:894-900

Part III

Spine and cord

5- Spinal and vascular malformations

Case 84: congenital syringohydromyelia

Clinical and pathological features

- Pathologically, syringomyelia is a tubular cavitation of the spinal cord, that could represent a dysrhaphic state at minima. It differs from hydromyelia which consists of a congenital dilatation of the central canal, normally obliterated after birth.
- Syringomyelia commonly presents in adulthood and uncommonly in children.
- *Pathophysiology:* several pathophysiologic theories concerning the formation of syringohydromyelia have been proposed. There is strong evidence that syrinx formation is primarily caused by a disturbance in CSF flow.
- *Signs:* symptoms occur usually in late adolescence and show a slow progressive course. The clinical signs include weakness and amyotrophy of the hands and a segmental dissociative loss of sensation over the shoulders and arms (thermoalgesic deficit and preservation of touch). Symptomatology fluctuates over time with long periods of stability. Post-traumatic syrinx may develop symptoms after a very long interval of time that may reach thirty years.

MR characteristic findings

Morphology

- Syringohydromyelia is present in association with more than 2/3 of patients with Chiari II and I malformations and/or with myelomeningocele. It may be associated also with diastematomyelia and/or with tight filum terminale syndrome. Asymptomatic segmental hydromyelia may be found incidentally on MRI (sagittal and axial cuts for confirmation and differentiation from Gibbs artifact).
- MR is the technique of choice in the evaluation of a cord syrinx (sagittal and axial cuts), demonstrating the CSF filled tubular or fusiform cavity that may enlarge the cord. An associated syringobulbia may be found in case of upward extension of the cavity (most often cervical) to the medulla oblongata. The slit is usually located in the lower half or two-thirds of the medulla and is more frequently bilateral. It may be found along the median raphe extending from the 4th ventricle, or may develop anterolaterally between the inferior olive and the pyramid.
- Incomplete septations may be evidenced (sagittal cuts) within the syringomyelic cavity. The syrinx cavity may be central or eccentric within the spinal cord, incorporating the central canal into the syringomyelic cavity.
- Small syrinx cavities may be missed on sagittal cuts and demonstrated only on axial cuts. When very large, the remaining peripheral cord tissue may be almost impossible to depict.
- More than 2/3 of the cases of syringohydromyelia are associated with a Chiari I malformation (case 84), in which the caudal level of the cerebellar tonsils show a downward displacement of more than 5 mm below the basion-opisthion line. The ectopic tonsils may be elongated and/or increased in volume. Dynamic studies using MR (sagittal acquisitions in flexion and extension) and CSF flow studies performed with the head in flexion will demonstrate the obstruction of the foramen magnum with associated bulbomedullary compression.
- Syringohydromyelia is found in association with myelomeningocele in almost 2/3 of the cases, the latter being present in almost all cases of Chiari II malformation.
- Congenital syrinx must be differentiated from intratumoral or post-traumatic syringes. The diagnosis is easier when associated developmental anomalies are found. In doubtful cases or in mixed cases as those associated with a neoplastic condition (primary cord tumors of glial origin or hemangioblastomas), administration of paramagnetic contrast media will be helpful.

Signal intensity

- Fluid contained within the syrinx follows the signal intensity of CSF on all imaging sequences.
- SE-T1w: hypointense SI to cord, iso-intense to CSF.
- SE-T2w: hyperintense DI to cord, iso-intense to CSF.
- Hypointense SI within the cystic cavity may be found on T2w in case of signal loss due to the presence of flow motion, CSF pulsation and turbulences. Such low SI within the syrinx cavity is evidenced when flow compensation techniques are not implemented.

Contrast enhancement

- May be needed in case of suspicion of a syrinx related to a cord tumor.

Case 84: Syringomyelia with Chiari I malformation

Fig. a: sagittal cut; SE - T1w and sagittal cut; SE - T2w

Sagittal T1w (**a**, left) and sagittal T2w (**a**, right) images of the cervical spine and the craniocervical junction showing a mild downward herniation of the cerebellar tonsils reaching the posterior arch of C1 and associated with a large syrinx expanding the cervical spinal cord, and extending downward to reach the upper dorsal cord. The syrinx containing CSF, presents a hypointense signal on T1w (**a**, left, ↑) and a hyperintense signal CSF-like on T2w (**a**, right, ↑). Note the absence of cisterna magna, and the moderate downward displacement of the medulla through the foramen magnum, reaching C1 level and associated to a moderate cervicomedullary kinking. The latter mild hindbrain malformations are more characteristic of Chiari II malformation which shows also malformations of the telencephalon as well as the diencephalon and a beaked tectum, all lacking in our patient.

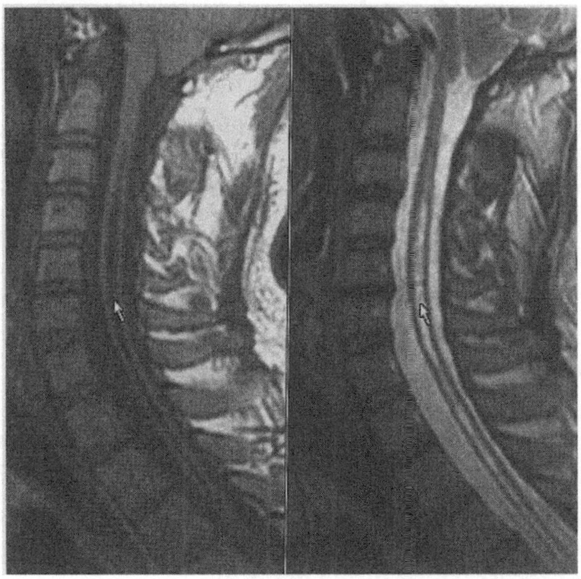

a

Cases 85, 86: Dermal sinus

Clinical and pathological features

- Dermal sinus tracts are congenital squamous epithelium lined tracts extending from the skin into the spinal canal. They connect the CNS from the cranium to the coccyx, with the body surface.
- Embryologically, such anomaly arises between the 3rd and 5th weeks of the fetal life, presumably secondary to an incomplete cleavage between the epithelial and the neural ectoderm.
- A slight male and familial preponderance is observed.
- *Signs:* dermal sinus tract present as a small dimple or opening on the skin usually associated with midline abnormalities such as: tuft of hair at the sinus opening, an area of pigmentation or a capillary hemangioma. The presenting signs may be related to infections of the nervous system (meningitis or abscess formation in the epidural or subdural spaces).
- Dermal sinus tract must be differentiated from a pilonidal sinus occurring in the sacrococcygeal region and found in the intergluteal cleft, extending cephalad before ending blindly.

MR characteristic findings

Morphology and signal intensity

- Congenital dermal sinus tracts are found anywhere from the cranial to the sacrococcygeal region.
- Almost half of the sinuses are considered to penetrate the spinal or cranial cavities and about half of all sinuses are associated with congenital tumors as epidermoids, dermoids, lipomas or teratomas.
- Half to two thirds of the sinuses extend intraspinally and almost half of the sinuses found in the lumbosacral region terminate in intraspinal tumors.
- Associated neural arch defects, spina bifida, hypoplastic or bifid spinous processes (case 85), and even diastematomyelia have been reported.
- The sinus tract extends inward to the related neuroectodermal segmental level. From the lumbosacral region it ascends to the conus, and from the cervical and upper thoracic region it descends one or two segmental levels below to enter the spinal canal.
- On MRI the dermal sinus tract is best seen on SE-T1w sequences (sagittal and axial oblique cuts), appearing as hypointense linear band traversing the bright subcutaneous fat. Its diagnosis is more difficult in the cervicothoracic region due to relative lack of subcutaneous fat. An underlying associated spinal dysrhaphism should be ruled out.

Contrast enhancement

- Contrast administration may be necessary in case of infectious complication.

Case 85: Bifid lumbar spinal process (spina bifida occulta)

Fig. b: axial cut (512); SE - T1w and SE - T2w

Axial T1w (**b**, left, ↑) and T2w (**b**, right, ↑) cuts showing a bifid posterior spinous process at the lower lumbar level without associated intradural malformation.

Case 86: Sacral dermal sinus

Fig. c: sagittal cut; SE - T1w and sagittal cut; SE - T2w

Fig. d: axial cut; SE - T1w

Patient presenting a sacral dermal sinus with concurrent lipoma shown on T1w sagittal cuts (**c**, left, ↑) as area of bright signal due to fat, and on T2w sequence (**c**, right, ↑). Note on the axial cuts on T1w (**d**, down) the low signal tract that crosses the lipoma before merging with the zone of spina bifida. Note the fibrolipoma of the filum terminale demonstrating a hyperintense signal on T1w sequence indicative of fat (**c**, left) and a hypointense signal on T2w (**c**, right, *). Note on T1w in the axial plane the posterior bright signal of the fatty filum at the posterior aspect of the dural sac (**d**, up). Note that the filum is extending all the way down into the terminal lipoma.

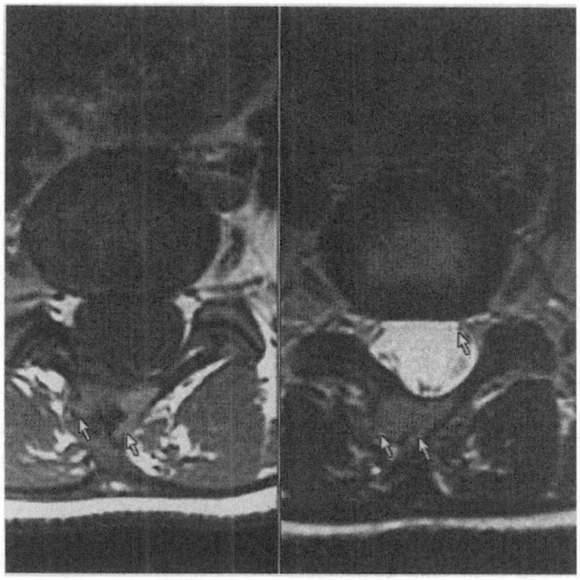

b

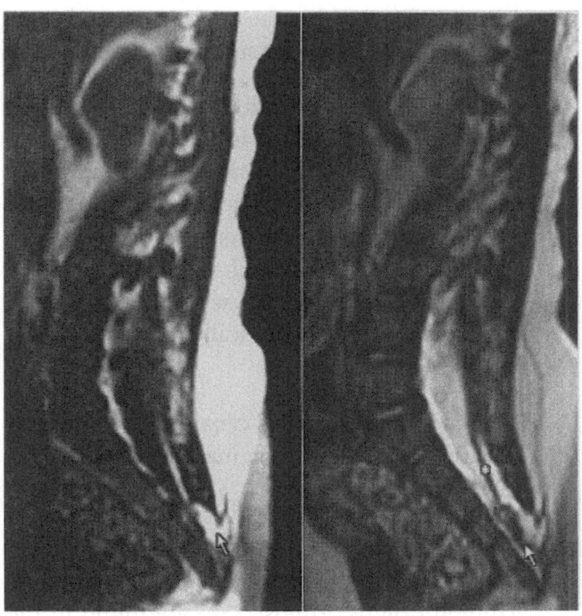

c

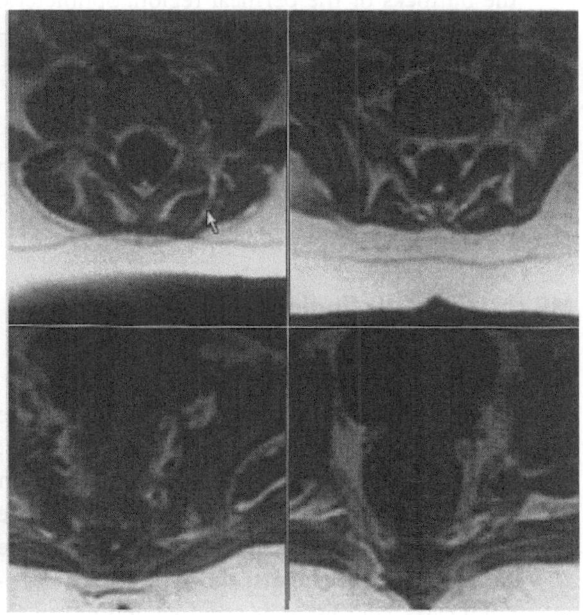

d

Cases 87, 93: spinal meningeal cysts

Clinical and pathological features

- Also called arachnoid cysts, perineurial Tarlov's cysts, leptomeningeal cysts.
- Their incidence is around 1% of all intraspinal masses.
- These cysts are in fact meningeal diverticula communicating in the majority of the cases with the subarachnoid spaces.
- Most of these masses are extradural arachnoid cysts. Some are intradural and very rarely may show both components. The majority are congenital or idiopathic, the remaining being acquired secondary to trauma, surgery, pantopaque myelography, infections, subarachnoid hemorrhage, or epidural anesthesia.
- *Signs:* the clinical presentation depends on the size and location of the cystic formation. Although commonly asymptomatic, signs of nerve root or cord compression may be present. Cervicothoracic arachnoid cysts may cause segmental sensory signs such hyperesthesia (1/3 of the cases) and associated spasticity in the lower limbs (2/3 of the cases). Lumbosacral cysts may give rise to radiculopathy with back and radicular pain, or paresthesias in the inferior limbs, or in the buttocks or the perineal region. Sphincteric disturbances may be present. Typically, the symptomatology may fluctuate with changes in posture and may become worse in the upright position due to cyst filling.
- Cervicothoracic cysts are more common in adolescents and lumbosacral cystse are more common in adults.

MR characteristic findings

Morphology

- Smoothly marginated spinal meningeal cysts have been classified into 3 subtypes by Nabors et al: type I are extradural without nerve root inclusion such as the intrasacral meningoceles; type II are extradural including nerve root such as the Tarlov's perineurial cysts, and type III are intradural cysts, congenital or acquired and unusually associated with spinal dysrhaphisms.
- *Type I cysts* are probably of congenital origin, develop from the dural sac to which they are connected through a pedicle usually found at the entrance of a dorsal nerve root. These cysts are hard to differentiate from type II on imaging. Occult intrasacral extradural meningoceles are included in this category. These cysts may be largely associated with widening of the sacral canal and vertebral scalloping without defect in the anterior or posterior sacral walls. The cyst is attached to the caudal extremity of the dural sac by a pedicle. Such arachnoid cysts have to be differentiated from anterior sacral meningoceles extending through a sacral defect toward the pelvis, and commonly associated with other congenital anomalies (teratoma, dermoid, lipoma, urogenital or anorectal developmental anomalies).

- *Type II cysts* are nerve root cysts, with a large ostium, often multiple and more frequently found in the sacral region. They may further be subdivided into perineurial or Tarlov's cysts and meningeal diverticula. Tarlov's cysts arise distally to the posterior root ganglion, contain nerve fibers and nervous tissue in their wall and usually lack to show patent communication with the perineurial arachnoid space. Perineurial cysts are usually small in the upper spine but may be large in the sacral region (up to 3 cm). Meningeal diverticula arise proximal to the ganglion, contain the nerve fibers within the cystic cavity and communicate largely with the subarachnoid space.

- Almost 75% of extradural cysts, excluding Tarlov's cysts, are located in the thoracic region, the remaining are lumbosacral and even rarely cervical (less than 5%). The majority are found posterior or lateral to the dural sac. Half of them may extend to the neural foramen. They are usually large, communicate with the subarachoid space and may extend vertically over up to 6 vertebral levels.

- Neural foramina or spinal canal bony erosion and widening are found in the majority of the extradural cysts.

- *Type III cysts* are much less frequent than the formers, found intradurally and of congenital or acquired origins. Almost 75% are found in the thoracic spine. Most of the congenital cysts are located posterior to the cord contrary to the acquired cysts which tend to be anterior in location.

Signal intensity

- Arachnoid meningeal cysts show SI similar to the intrathecal CSF.
- SE-T1w: iso-intense to CSF (hypointense SI).
- SE-T2w: iso-intense to CSF (bright SI) or more hyperintense to CSF in case of loculated cyst.
- Decreased SI on T2w sequence may be encountered in case of large communicating cysts, due to flow motion and turbulences causing a spin dephasing with loss of SI on T2w sequences.

Contrast enhancement

- Congenital cysts do not show contrast enhancement.
- Acquired cysts may show mild peripheral and/or nerve roots enhancement.
- In case of marked and thick peripheral enhancement other etiologies must be ruled out, such as: synovial cysts or neurenteric cysts (hyperintense to CSF on T1w), cystic dermoid or epidermoid, teratomas.

Case 88: caudal regression syndrome

Clinical and pathological features

- This rare congenital malformation also named caudal dysplasia, seems to result from caudal mesodermal disturbance occurring in the fetal life, before the 4th week of gestation.

- The syndrome is sporadic, occurring with an incidence of 1/7500 births. In its complete form it is found in more than 1/20 000-100 000 births. Chromosomal aberrations have been found in some cases.

- Maternal diabetes is found in association with the syndrome. The cause is probably multifactorial. Less than 20% of infants with caudal regression syndrome have diabetic mothers.

- Embryologically, some elements point to a maldevelopmental process involving the level of closure of the caudal neuropore.

- *Histopathology:* the spinal cord shows a marked dysplasia of the distal cord, more severe ventrally. The cord ends at L2 without normal tapering in a suggestive "wedge" shape. The anterior horn cells are diminished in number in the distal extremity of the cord.

- *Signs:* the syndrome includes several anomalies and defects involving the spine and visceral organs of the gastrointestinal and genitourinary tracts. Feet deformities and sensorimotor deficits of the lower limbs may be present, with motor weakness usually more severe than the loss of sensation. Bladder and bowel dysfunctions and incontinence are almost invariably present. The clinical aspect depends upon the degree of spinal involvement. In mildest asymptomatic forms, absence of coccyx is found. In most severe forms, there is distal thoracolumbar and sacral agenesis. Spinal agenesis may vary from partial to total sacral agenesis. In hemisacral agenesis, there is frequently an elevation of the ilium and a hip defect. In total agenesis of the sacrum, the pelvis may articulate with the lumbar vertebrae. Atrophy of the lumbosacral spinal roots and nerves may be associated, with secondary atrophy of the perineal, legs and buttocks muscles. In extreme cases, a sirenomelia with fusion of the lower limbs may be present.

MR characteristic findings

Morphology

- Caudal vertebral bodies and sacral agenesis are easily demonstrated radiographically and on sagittal MR cuts.
- MR will depict the cord abnormality, its shape and evaluates its spinal level. The terminal end of the cord is characteristically blunted or wedge-shaped or club-shaped, its dorsal aspect extending further distally to the ventral on the mid sagittal cut. In most cases, a thickened filum terminale is found.
- Two groups, almost equally represented, have been defined according to the end position of the conus terminale. In group I, the conus ends abruptly at T12 or T11 levels with a club-shaped or wedge-shaped aspects. Central canal is slightly enlarged and internal gray and white matter are malformed. In group II, the conus is found below L1 level, tethered by a thickened filum terminale in about 65% of the cases, or associated with a myelocystocele or a lipoma, or is elongated with associated hydromyelia. The sacrum is in this group relatively developed (down to S2 segments or more).
- A spinal stenosis is usually evidenced on axial cuts. Abnormalities of the thecal sac may be found with widening of the dural sac or high termination.
- Other neural malformations may be found, more commonly spina bifida and myelomeningocele. Among those reported diastematomyelia or tethered cord are present.
- Associated bony malformations may be found: hypoplastic or hemivertebrae, scoliosis or kyphosis, spina bifida. The last caudal vertebral bodies (2 or 3) are frequently fused, secondary to absence of segmentation. Associated tumors such as lipomas or dermoids, are easily depicted and their relationship with the thecal sac and cord evaluated on T1w MR sequences.
- Atrophy of gluteal and peritoneal muscles are well shown with evidence of fatty replacement (bright SI on T1w). Anorectal and genitourinary anomalies should be evaluated.

Contrast enhancement

- No need for contrast administration, morphological informations provided by MRI being diagnostic.

Cases 89, 92: tethered cord syndrome

Clinical and pathological features

- Also named tethered conus or tethered filum or tight filum terminale syndrome.
- It corresponds to a conus medullaris in a low position tethered by a short thick filum terminale.
- Its incidence is increased in VATER syndrome and in lumbosacral agenesis.
- *Signs:* they are usually present during the first two decades secondary to rapid growth and increase during adolescence. They include: feet deformities as a pes cavus; sensorimotor disturbances and weakness; gait difficulties; bladder dysfunctions with enuresis; low back pain, stiffness and later calf wasting and atrophy. Cutaneous anomalies found in half the children and which may help to depict asymptomatic child are: lumbosacral hypertrichosis, hemangioma, subcutaneous lipoma, skin dimple and sinus tract. Underlying bony abnormalities are frequent.

MR characteristic findings

Morphology and signal intensity

- MRI clearly demonstrates on T1w sagittal and particularly on axial cuts the abnormally low situation of the conus medullaris (below L2-L3 in a child over 5 years old), and the tubular enlargement of the filum terminale with a diameter greater than 2 mm (at L5-S1).
- The tip of the conus may be difficult to delineate on the sagittal cuts, and is more easily evidenced on axial cuts (SE-T1w).
- A dilated central canal or a signal abnormality may be evidenced within the lower cord in 25% of the cases (hydromyelia or myelomalacia), best shown on axial cuts (T1w or T2w).
- The filum terminale may show in almost 1/4 of the cases a fatty content: fibrolipoma. It may be sometimes difficult to separate the filum from the posterior aspect of the thecal sac to which it is tightly applied. The dural sac is often wide.
- In nearly 10% of the cases the filum is not really present, the cord being elongated reaching the fundus of the thecal sac to end in a small lipoma.
- Bony abnormalities are frequently associated. Nearly all the patients show a spina bifida occulta (plain films). A scoliosis is present in about one fifth of the patients.

Contrast enhancement

- No need to perform any contrast administration.

Case 87: Tarlov's cysts (meningeal perineurial type II cysts)

Fig. a: sagittal cut (512); SE - T1w and sagittal cut; SE - T2w

Fig. b: coronal cut (512); SE - T2w

Patient presenting an incidental finding of multiple sacral perineurial (Tarlov) cystic formations corresponding more exactly to diverticules due to outpouchings which communicate with the subarachnoid space. The cystic formations observed are iso-intense to CSF on T1w (**a**, left) and on T2w (**a**, right, ↑) sequences and show on the coronal high resolution T2w slices the presence of a nerve root within the cystic cavity suggesting a Nabott's type II extradural cyst arising from a nerve root sleeve in these cases. Note that these small cysts are usually asymptomatic and do not require further evaluation neither surgery.

Case 88: Caudal regression of the spinal cord

Fig. c: sagittal cut; SE - T1w and sagittal cut; SE - T2w

Child presenting a mild caudal regression syndrome. Sagittal MR images of the lumbo-sacral region reveal a rather abrupt, characteristically blunted appearance of the distal spinal cord as shown on T1w (**c**, left,↑) and T2w (**c**, right, ↑) sequences. Note also the progressively narrowing of the distal spinal canal (**c**, **d**, ↑↑). Note that the spinal cord ends at the lower thoracic level. Note also the dysgenesis of the lower lumbar and sacral posterior arches. The sacral agenesis is not shown.

Case 89: Tethered cord (tight filum terminale syndrome)

Fig. d: sagittal cut (512); SE - T1w

Sagittal T1w sequence showing a low lying conus medullaris tethered by a tight filum terminale extending all the way down into a terminal lipoma shown at the S2 and S3 levels posteriorly, as a band of hyperintense signal on T1w. Note that it may be difficult to determine where the conus medullaris ends and the filum terminale begins. In this case, the spinal cord appears markedly elongated extending downward at the lower end of the dural sac, ending in this lipoma at the posterior caudal end of the thecal sac as confirmed on the axial slices.

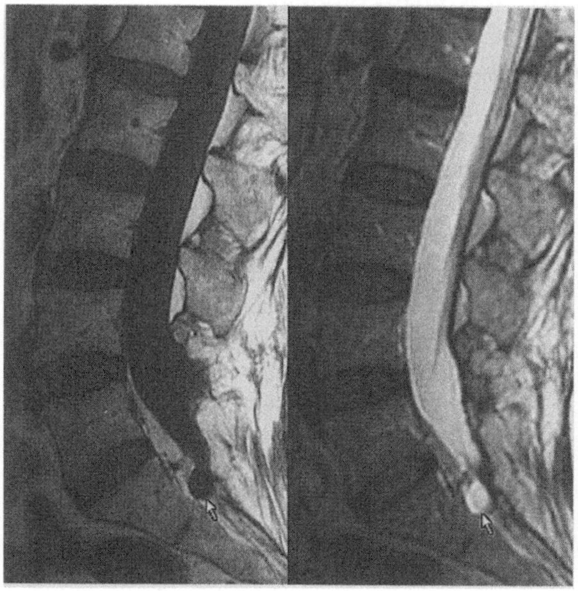

a

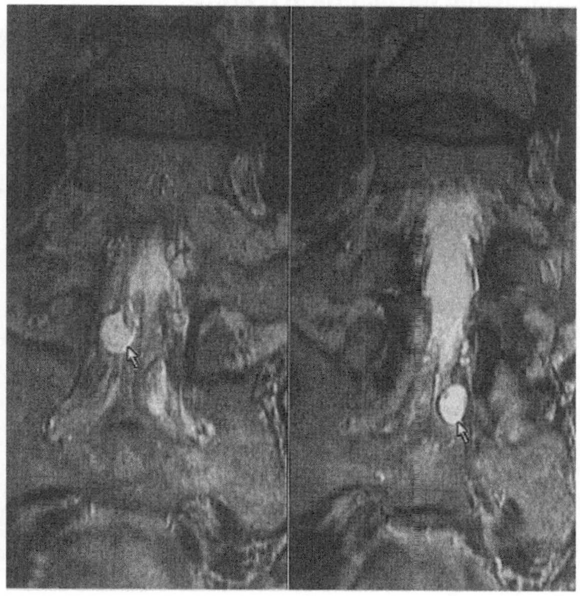

b

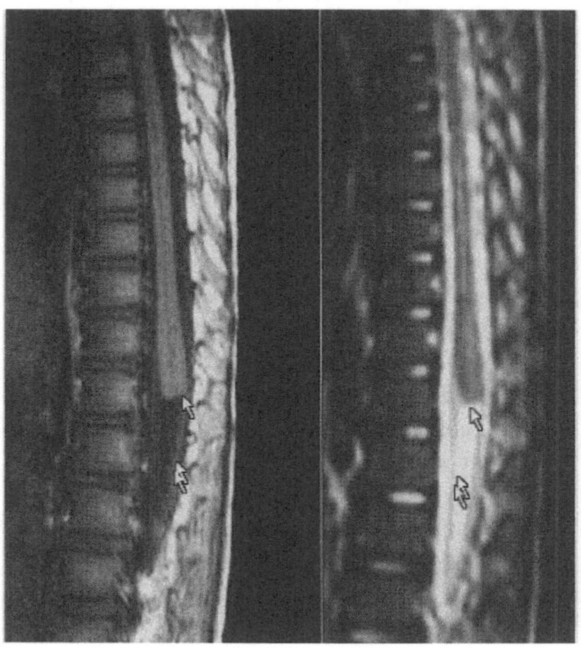

c

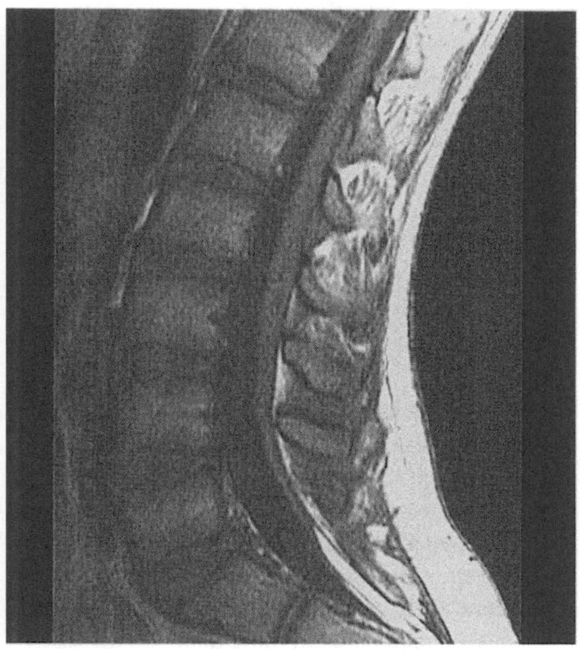

d

Cases 90, 91: diastematomyelia

Clinical and pathological features

- Congenital spinal malformation characterized by a sagittal cleft splitting the cord into two hemicords.

- It is almost always associated with spinal osseous abnormalities or malformations and variable cord tethering.

- Diastematomyelia must be differentiated from diplomyelia consisting of a segmental duplication of the spinal cord.

- It shows a marked female prevalence in more than 2/3 of the cases.

- *Histopathology:* each hemicord is protected by its own pia, the dura and arachnoid being constituted of a single or a double sac. The two hemicords are frequently asymmetric in size and the gray matter is roughly organized into ventral and dorsal columns. The hemicords are rotated about 90 degrees with a ventral median sulcus and ventral gray matter facing each other. Aberrant neural of fibrous tissue, often accompanied by blood vessels, may be found crossing from one hemicord to the other. The cord above the segmental cleft is normal but may show a hydromyelia (50% of the cases). The cord below fuses distally. The conus terminale is often low in situation associated with a thickened filum terminale in almost half the cases.

- *Signs:* patients present with variable deformities in the lower limbs and feet, muscular atrophy, sphincteric incontinence, and gait disturbances. Cutaneous signs are present in almost half the cases: hypertrichosis in the midline in the region corresponding to the diastematomyelia and others such as sacral dimple, lipoma, dermal sinus, hemangioma. In more than half the cases, a scoliosis is associated, usually related to vertebral deformities including hemivertebrae, vertebral block, butterfly-shaped vertebrae, and becoming more severe with age. The diagnosis is commonly made during childhood.

MR characteristic findings

Morphology and signal intensity

- The role of MR, replacing CT-myelography, is to demonstrate the spinal cleft and to evaluate the position of the conus terminale and the associated cord and vertebral abnormalities.
- The spinal cord cleft usually found between levels T9 and S1 in the majority of the cases, involving very rarely the cervical or upper thoracic spine. Two subtypes are distinguished: in the first one (almost 60%), the two hemicords are included in a single common arachnoidal and dural sheaths. Usually no osseous or fibrous spur is present; in the second subtype (more than 40%), the two hemicords are surrounded separately and coaxially by the dura-arachnoid membranes, separated by a sagittally oriented osseous or a fibrocartilaginous spur, found most often between T12 and L5. The spur may be midsagittal or oblique, attaching mainly to the lamina posteriorly and extending to the vertebral body anteriorly, to which it will later fuse. In 5% of the cases a double spur may be found. Coronal and axial MR cuts performed using high resolution morphological sequences (SE-T1w and T2w), will evaluate optimally the extent of the two hemicords and demonstrate the eventual spur (GE-T2* sequence may help to depict the osseous spur).
- In half the cases a hydromyelia, probably due to a dilation of the central canal, is easily evidenced using axial cuts (SE-T2w or T1w), within the cord above the clefting, that may extend to the hemicords.
- The level of the conus terminale (usually below L2) and the tethering of the cord are well shown on the sagittal (SE-T1w or T2w) and axial cuts (SE-T2w).
- Each hemicord may give rise ipsilaterally to the ventral and dorsal nerve roots and when asymmetrical, 3 of the 4 roots may originate from one of the hemicords.
- Associated vertebral malformations and osseous abnormalities must be evaluated, including scoliosis (50-90%), failure of segmentation (50%), hemivertebrae (30%), butterfly vertebrae, decrease in sagittal dimension

of the bodies and spinal canal focal AP narrowing at the level of the spur.

- Abnormalities of the posterior vertebral elements are observed in the majority of the cases: spina bifida (50-90%), widening of the interpediculate distance, and thickened laminae. When found in association with a scoliosis (screening with plain films), an underlying diastematomyelia must be ruled out (MR).
- Other malformations may be associated including a Chiari II or I malformations or a Klippel-Feil syndrome, or Sprengel's malformation or lipomyelomeningocele.
- Tumors may be found such as teratomas or Wilm's tumor.

Contrast enhancement

- There is no need to perform a contrast injection.

Case 90: Diastematomyelia with hydromyelia (diplomyelia)

Fig. a: coronal cut; SE - T2w

Fig. b: axial cut; SE - T2w

T2w SE images in the coronal (**a**) and the axial (**b**) planes, showing a diastematomyelia with splitting of the cord into two hemicords as shown on the coronal T2w sequence (**a**, left) and on the axial slices (**b**), showing moreover hydromyelia as cystic cavities in each hemicord at the upper level with hyperintense signal similar to the subarachnoid CSF (**b**, up). Note that at the upper level, the two hemicords are separated by a fibrous cleft dividing the dural sac into coaxial arachnoid dural tubes surrounding each hemicord (**b**, up, left), the meningeal cleft disappearing at the lowest level (**b**, down, right). Note the associated malformation of the spinal canal whose shape is markedly abnormal as well as the related vertebrae showing segmental anomaly, observed in more than 80% of the cases. Note the absence of posterior neural arch.

Case 91: Diastematomyelia with scoliosis

Fig. c: coronal cut; SE - T2w

Fig. d: axial cut; SE - T2w

Patient presenting an important dorso-lumbar scoliosis as shown on the coronal T2 slices (**c**). On the corresponding spinal cord segment, oblique axial slices using T2w sequence (**d**) showed a splitted spinal cord segment into two hemicords without associated hydromyelia (**d**, ↑). Note the mild deformation of the spinal canal.

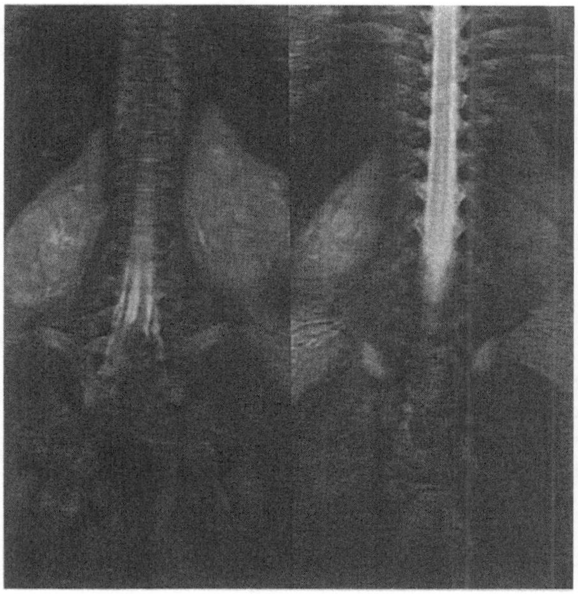

a

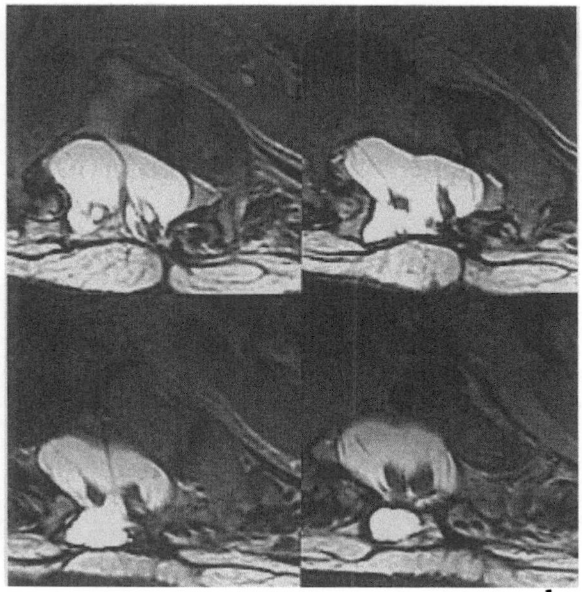

b

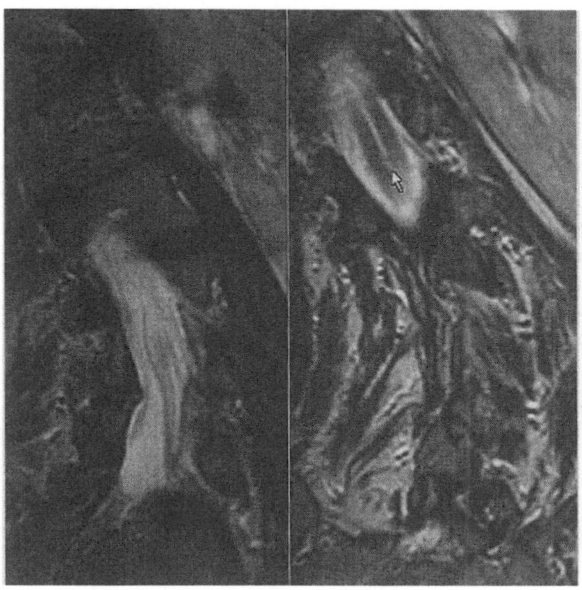

c

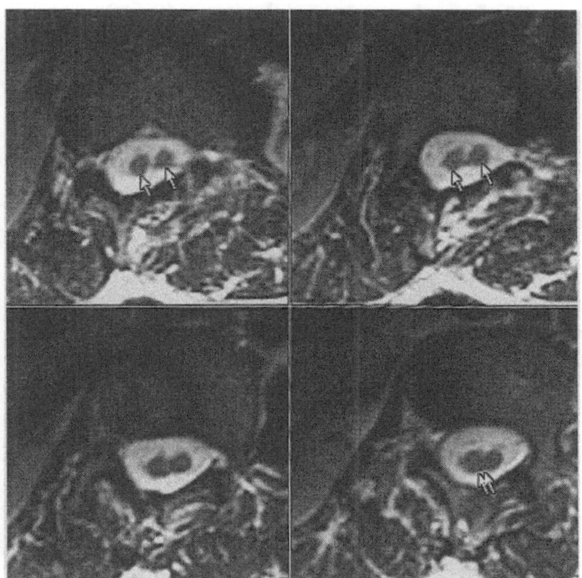

d

Case 92: Tethered cord with hydromyelia

Fig. a: sagittal cut; SE - T1w

Fig. b: coronal cut; SE - T2w

Fig. c: axial cut; SE - T1w with fat suppression

Associated spine and cord malformations consistent with a caudal regression with a tethered cord. Sagittal T1w slices evidence a low ending spinal cord stretched inferiorly, showing at the lumbar level a syrinx with hypointense signal on T1w (**a**, ↑) and hyperintense signal on T2w (**b**, right, ↑) consistent with hydromyelia, also shown on the T1w with fat suppression axial cuts (**c**, ↑). Note the malformed lower lumbar vertebral bodies (**b**, ↑↑) and the dysplastic sacrum and coccyx. Note that a hydromyelia is seen in approximately 25% of the patients presenting a tethered cord requiring no treatement when minimal.

Case 93: Sacral meningocele (type I cyst)

Fig. d: sagittal cut (512); SE - T1w and sagittal cut; SE - T2w

Sagittal cuts with T1w (**d**, left) and T2w (**d**, right) sequences showing a congenital fusion of L2 and L3 vertebral bodies (**d**, ↑↑) associated to a caudal cystic formation in the lower sacral canal and reaching the level of S2-S3 at its upper extremity. This sacral cystic formation discovered on MRI performed for back pain produces enlargement of the lower sacral canal with a mild scalloping of the anterior wall. The cyst lying distally to the thecal sac shows an iso-signal to CSF on T2w sequence (**d**, right) but a slight hyperintense signal to CSF on T1w (**d**, left), involving mainly its lower portion. Note the absence of epidural fat at the level of this extradural cyst and the enlargement of the lower sacral canal due to the gradual enlargement with increased pressure within the cyst.

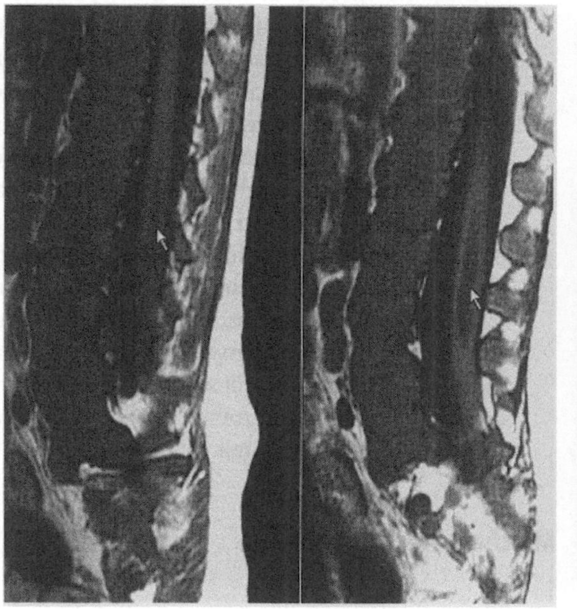

a

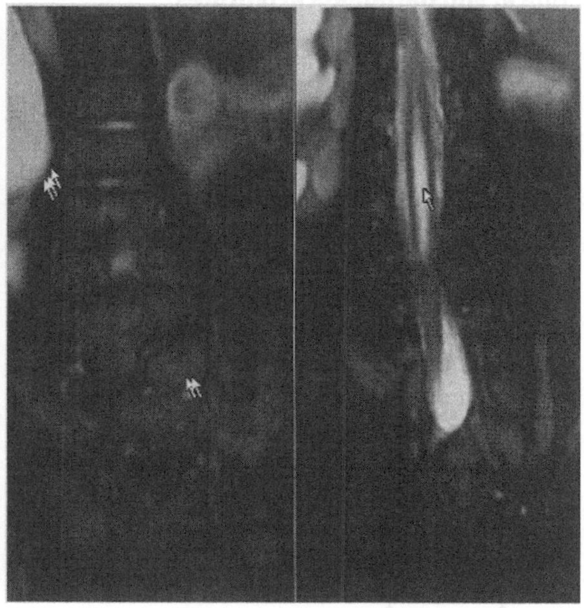

b

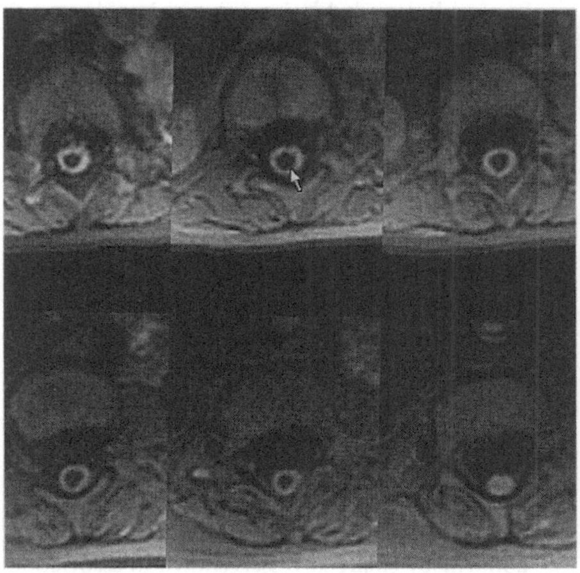

c

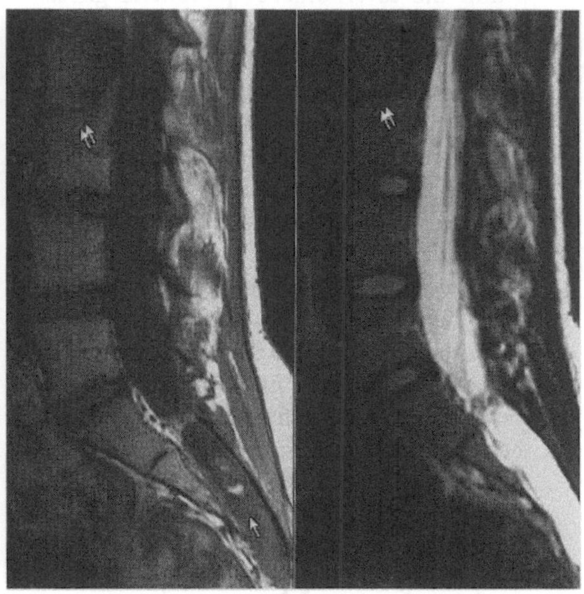

d

Cases 94, 95: spinal vascular malformations

Clinical and pathological features

- Angiomas of the spinal cord are relatively rare lesions occurring with a frequency of less than 5% of that of primary cord tumors.
- Spinal vascular malformations are mainly represented by cavernous angiomas (case 94) or cavernomas and arteriovenous malformations. Cavernous hemangiomas are found more frequently in the vertebral bodies (case 95).
- They may occur at any age, but most of them are found between ages of 15 to 60. Patients with intramedullary lesions tend to develop clinical signs at a younger age (third to fourth decades) as compared with patients with dural arteriovenous malformations (fifth to seventh decades). A predilection for men is observed. Familial occurrence and association to intracranial lesions have been reported in spinal cavernomas.
- *Histopathology:* cavernous angiomas are benign non encapsulated lesions formed of multiple cryptic blood filled spaces lined with endothelial cells. No intervening nervous tissue is evidenced within the cavernoma. A peripheral gliosis is found at the periphery as well as hemosiderin-stained macrophages due to propensity of the lesion to bleed. Arteriovenous malformations are characterized by the presence of abnormal enlarged vessels. The veins are commonly as large as the arteries and the pial arteries are large and abnormal. The vessels show thickened walls with lack of elastic substance. The spinal angiomas have been divided into 5 types on anatomic and angiographic basis.
- *Signs:* the clinical syndrome presented, usually non specific, mimics a myelopathy, but may also show some suggestive features. More than 2/3 of the patients may present a progressive medullary syndrome with fluctuating and intermittent neurological signs in half of them. Signs and symptoms may worsen after exertion or on recumbency. Long standing pain and dysesthesias are very frequent being even of most common initial complaints. A claudicating gait is also an early and frequent complaint. Bladder dysfunction is present. Few years later a progressive spastic paraparesis develops that may lead to a paraplegia with severe micturitional disturbances. Diagnosis may not be done until 2 years or more after symptoms onset in at least half the patients. A sudden onset of a medullary syndrome with paraplegia or tetraplegia and an excruciating pain in the back, may characteristically occur in case of bleeding with hematomyelia a particularly in case of an intraspinal cavernous angioma, or in case of a subarachnoid hemorrhage (10-15%) that may be associated with an arteriovenous malformation. The spontaneous thrombosis of an angioma may also give rise to a similar catastrophic syndrome due to an acute necrotizing myelitis (Foix and Alajouanine syndrome). Rarely, the association with a metameric cutaneous and a vertebral angioma (Cobb's syndrome) orients to the diagnosis of the cord angioma.

MR characteristic findings

Morphology and signal intensity

- MR appearance of intraspinal cavernous malformations is almost similar to intracranial cavernomas (see brain cavernomas). Spinal cavernomas are small rounded lesions measuring about 1-3 cm in diameter. The lesion is well delineated with a heterogeneous center due to its content rich in blood breakdown products at different stages (areas of increased SI on T1w and T2w SE sequences). It is commonly surrounded by a rim of low SI due to hemosiderin deposition from prior hemorrhages. The lesion is usually solitary in the spinal cord contrary to brain cavernomas. The cord adjacent to the cavernoma may show linear areas of low SI on T2w or even more conspicuously on T2*, extending cephalad and caudally, corresponding to hemosiderin deposition from prior hematomyelia (case 94). The cord at the level of the cavernoma may be slightly enlarged. Edema may be present in case of recent bleed (hyperintense SI on T2w). There are no abnormally enlarged vessels shown in association with the cavernous angioma, particularly on the spinal cord surface. Spinal and MR angiography are normal.
- Spinal arteriovenous malformations are very rare, shown as serpentine low-to-absent (flow void) SI images within and/or adjacent to the spinal cord. Five different types of spinal angio-

mas may be distinguished: type I or dural arteriovenous fistulas, probably the most common type, shown as one or two vessels found along the midthoracic or thoracolumbar dorsal spinal cord, which may cause cord ischemia secondary to venous congestion, enlargement of the conus medullaris showing a hyperintense SI on T2w (axial cuts using SE-T2w sequences appear more accurate to depict serpentine vascular flowvoids within the bright CSF); type II or spinal cord arteriovenous malformation (may cause hematomyelia) including a nidus found inside the cord parenchyma over a short segment, feeding one or more feeding arteries arising from the anterior or posterior spinal arteries, and draining into the enlarged coronal venous plexus; type III, a very unusual form, occurring mainly in childhood, named the juvenile type of angioma, characterized by the presence of intramedullary and extramedullary vascular components; type IV or direct arteriovenous fistulas, usually found anterior to the cord, consist of a direct shunt between the anterior spinal artery and the draining vein without interposition of a nidus; type V, which corresponds to the metameric type of arteriovenous malformation, involving the cord, the vertebra and the skin.

Contrast enhancement

- Cavernous angiomas do not usually show contrast enhancement (enhancement is difficult to depict within the lesion due to hyperintense SI from blood). A mild peripheral rim of enhancement may be shown after a recent bleeding. In doubtful cases contrast administration may help to rule out primary or secondary cord tumors which usually enhance and enlarge the cord.
- Contrast administration and high resolution MR images using surface coils may help to depict and better delineate the abnormally enlarged vessels in the intradural space as within the cord. Follow-up MR exams are of interest after therapy to evaluate nidus and vessels thrombosis.

Case 94: Cervical intramedullary vascular malformation

Fig. a: sagittal cut (512); SE - T2w

Fig. b: sagittal cut (512); SE - PDw and GE - T2*w

Fig. c: axial cut; SE - T2w

Small mass involving the left paramedian cervical cord located behind C6 level, showing a heterogeneous signal on T2w represented by an area of low signal rim due to hemosiderin, surrounding a focal area of high signal intensity representing methemoglobin (**a**, ↑). Note on the axial cuts the presence of a small lesion with a target appearance on SE-T2w (**c**, ↑). Note on T2w (**a**) and on Pw (**b**, left) sequences the presence of an area of high signal intensity beneath the lesion associated with a slight focal expansion of the cord that may correspond probably to recent ischemic edema explaining presumably recent clinical exacerbation (**a**, **b**, ↑↑). T2*w GE images showed an extensive hypointense band extending from C6 up to C2 levels best depicted on such sequences and corresponding to hemosiderin deposition due to old hematomyelia. Note that these extensive areas of low signal are better shown on GE sequences than on SE-T2w sequence (**a**) because of marked hypointense signal of secondary to the magnetic susceptibility effect of the blood breakdown products.

Case 95: Vertebral hemangioma

Fig. d: sagittal cut; SE - T2w

High resolution sagittal T2w sequence showing a slightly heterogeneous increased signal intensity from a thoracic vertebral body (**d**, ↑). Note the thickened trabeculae imaged as small foci of decreased signal resulting in vertical striations associated to areas of honeycomb pattern. Note that there is no cortical expansion of the vertebral body walls. The hemangioma involves here the entire vertebral body extending partly into the neural arch (**d**, right). The lesion showed a high signal intensity on T1w sequence (not shown). This high signal on T1w due to the fat fraction indicates an indicative form of benign vertebral hemangioma. Note that plane films are often diagnostic of vertebral hemangiomas. Smaller lesions are frequently observed on MRI and do no require further evaluation, nor treatment.

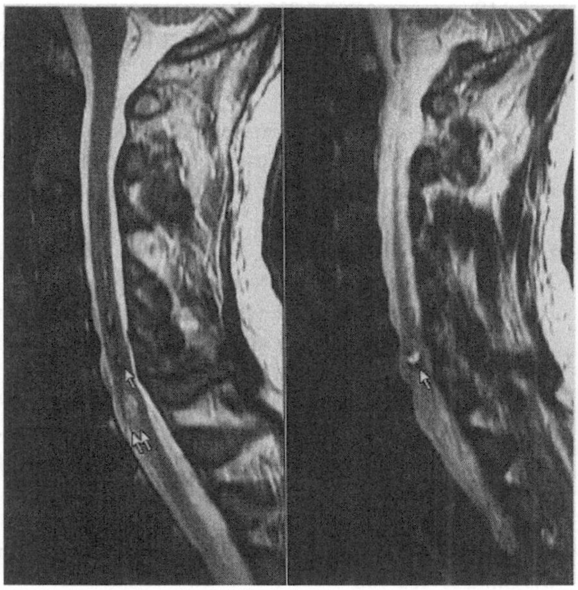

a

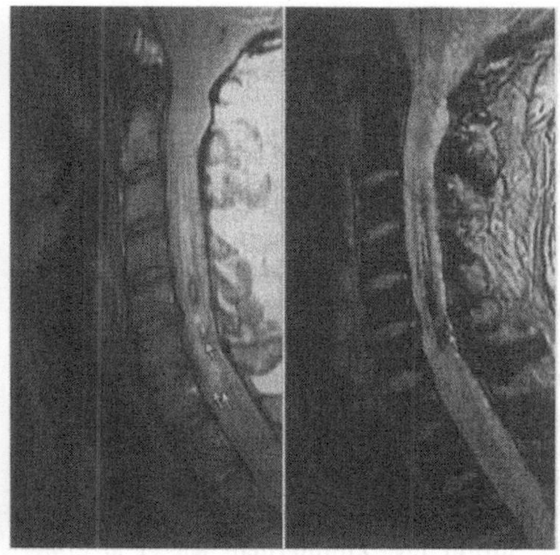

b

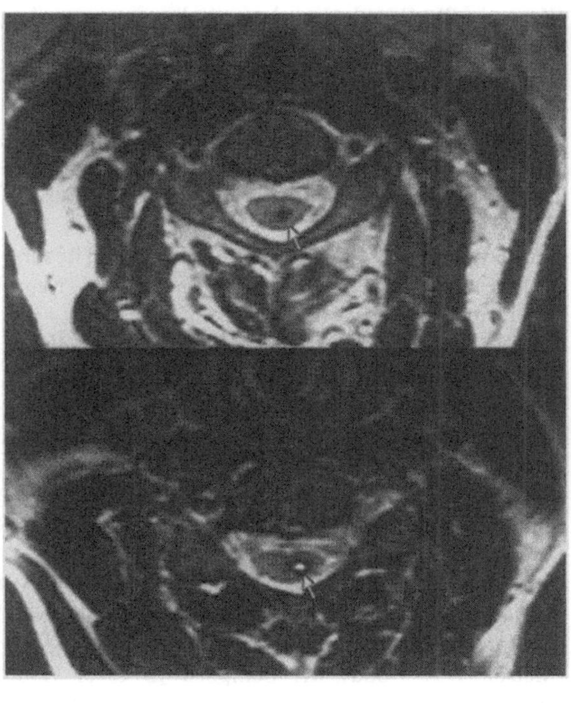

c

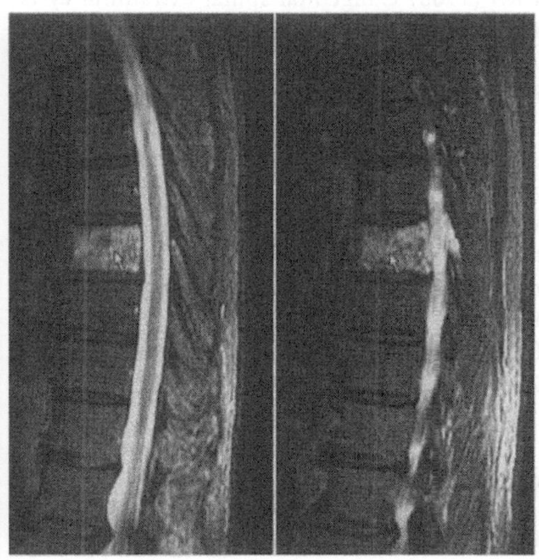

d

Selected References

- Alvisi C, Cerisoli M, Giulioni M, Guerra L (1987) Long-term results of surgically treated congenital intradural spinal arachnoid cysts. J Neurosurg 67:333-335
- Baily IC (1970) Dermoid tumors of the spinal cord. J Neurosurg 33:676-681
- Barkovich AJ, Edwards MSB, Cogen PH (1991) MR evaluation of spinal dermal sinus tracts in children. AJNR 12: 123-129
- Borden JA, Wu JK, Shucart WA (1995) A proposed classification for spinal and cranial dural arteriovenous fistulous malformations and implications for treatment. J Neurosurg 82:166
- Boukobza M, Guichard JP, Boissonet M et al (1994) Spinal epidural hematoma: report of 11 cases and review of the literature. Neuroradiology 36:456
- Brophy JD, Sutton LN, Zimmerman RA et al (1989) Magnetic resonance imaging of lipomyelomeningocele and tethered cord. Neurosurgery 25 (3):336-340
- Browne TR, Adams RD, Roberson GH (1976) Hemangioblastoma of the spinal cord: review and report of five cases. Arch Neurol 33:435
- Byrne TN, Waxman SG (1990) Spinal cord compression diagnosis and principles of management. FA Davis, Philadelphia, pp 116-118
- Cecchini A, Locatelli D, Bonfanti N (1988) Lipomyelomeningoceles: a neuroradiological approach. J Neuroradiol 15:49-61
- Cloward RB (1968) Congenital spinal extradural cysts: case report with review of the literature. Ann Surg 168: 851-864
- Cole GP, Flannery AM, Gulati AK (1989) Intrasacral meningocele. Case report and review of the literature. Spine 14:1418-1420
- Enomoto H, Goto H (1991) Spinal epidural cavernous angioma. Neuroradiology 33:462
- Erkulvrawatr S, Gammal TE, Hawkins J et al (1979) Intrathoracic meningoceles and neurofibromatosis. Arch Neurol 36:557-559.
- Galzio RJ, Zenobii A, Lucantoni D, Cristuib-Grizzi L (1982) Spinal intradural arachnoid cyst. Surg Neurol 17: 388-391
- Garceau GJ (1953) The filum terminale syndrome. J Bone Joint Surg 35A:711-716
- Guidetti B, Gagliardi FM (1977) Epidermoid and dermoid cysts: clinical evaluation and late surgical results. J Neurosurg 47:12-18
- Gundry CR, Heithoff KB (1993) Epidural hematoma of the lumbar spine: 18 surgically confirmed cases. Radiology 187:427-431
- Guthrie BL, Ebersold MJ, Scheithauer BW et al (1989) Meningeal hemangiopericytoma: histopathological features, treatment, and long term follow-up of 44 cases. Neurosurgery 25:514
- Hall P, Albrigh AL, Brunbler JA (1988) Diagnosis of tethered cord by magnetic resonance imaging. Surg Neurol 30: 60-64
- Han JS, Benson JE, Kaufman B et al (1985) Demonstration of diastematomyelia and associated abnormalities with MR imaging. AJNR 6:215-219
- Harrison MJ, Mitnick RJ, Rosenblum BR, Rothman AS (1990) Leptomyelolipoma: analysis of 20 cases. J Neurosurg 73:360-367
- Hasuo K, Mizushima A, Mihara F et al (1996) Contrast-enhanced MRI in spinal arteriovenous malformations and fistulae before and after embolisation therapy. Neuroradiology 38(7):609
- Hendrick EB, Hoffman HJ, Humphreys RP (1983) Tethered spinal cord. Clin neurosurg 30:457-463
- Hilal SK, Marton D, Pollack E (1974) Diastematomyelia in children: radiographic study of 34 cases. Radiology 112: 609-621
- Hoffman HJ, Taecholarn C, Hendrick EB, Humphreys RP (1985) Management of lipomyelomeningoceles: experience at the hospital for sick children (Toronto) J Neurosurg 62:1-8
- James CCM, Lassman LP (1964) Diastematomyelia: a critical survey of 24 cases submitted to laminectomy. Arch Dis Child 39:125-130
- Just M, schwartz M, Ermet JA, Higer HP, Voth D, Pfannenstiel P (1988) Magnetic resonance imaging of dysrhaphic myelodysplasia. Child's Nerv Syst 4:149-153
- Keim HA, Greene AF (1973) Diastematomyelia and scoliosis. J Bone Joint Surg 55A:1425-1435
- Kendall BE, Valentine AR, Keis B (1982) Spinal arachnoid cysts: clinical and radiological correlation with prognosis. Neuroradiology 22:225-234
- Kennedy PR (1979) New data on diastematomyelia. J Neurosurg 51:355-361
- Laredo JD, Reizine D, Bard M et al (1986) Vertebral hemangiomas: radiologic evaluation. Radiology 161:183
- Larsson E-M, Desai P, Hardin CW et al (1991) Venous infarction of the spinal cord resulting from dural arteriovenous fistula: MR imaging findings. Am J Neuroradiol 12:739
- Logue V, Edwards MR (1981) Syringomyelia and its surgical treatment an analysis of 75 patients. J Neurol Neurosurg Psychiatry 44:273-284
- Love JG, Daly DD, Haris LE (1961) Tight filum terminal. Report of condition in three siblings. JAMA176:31-33.
- Lunardi P, Missorti P, Ferrante L, Fortuna A (1990) Long-term results of surgical treatment of spinal lipomas: report of 18 cases. Acta neurochir 104:64-68
- Mascalchi M, Cosottini M, Ferrito G, Salvi F, Nencini P, Quilici N (1998) Posterior spinal artery infarct. AJNR 19:361-363
- Mascalchi M, Bianchi MC, Quilici N et al (1995) MR angiography of spinal vascular malformations. Am J Neuroradiol 16:289
- Matson DD, Woods RP, Campbell JB et al (1950) Diastematomyelia (congenital clefts of the spinal cord). Pediatrics 6:98-112

- Mawad ME, Rivera V et al (1990) Spinal cord Ischemia after resection of thoracoabdominal aortic aneurysms: MR findings in 24 patients. Am J Neuroradiol 11:987
- Mohan V, Gupta SK, Tuli SM et al (1980) Symptomatic vertebral hemangiomas. Clin Radiol 31:575
- Monfarrij NA, Palmer JM, Hahn JF et al (1989) Correlation between magnetic resonance imaging and surgical findings in the tethered spinal cord. Neurosurgery 25(3):341-346m
- Nabors MW, Pait TG, Byrd EB et al (1988) Updated assessment and current classification of spinal meningeal cysts. J Neurosurg 68:366-377
- Naidich TP, Harwood-Nash DC (1983) Diastematomyelia: hemicord and meningeal sheats signs and double arachnoid and dural tubes. AJNR 4:633-636
- Naidich TP, McLone DG, Fulling KH (1983) The Chiari II malformation. Part IV. The hindbrain deformity. Neuroradiology 25:179-197
- Naidich TP, McLone DG, Mutluer S (1983) A new understanding of dorsal dysraphism with lipoma (lipomyeloschisis): radiologic evaluation and surgical correction. AJNR 4:103-116
- Nelson MD, Bracchi M, Naidich TP, McLone DG (1988) The natural history of repaired myelomeningocele. Radiographics 8(4):695-706
- North RB, Kidd DH, Wang H (1990) Occult bilateral anterior sacral and intrasacral meningeal and perineurial cysts: case report and review of the literature. Neurosurgery 27: 981-986
- Palmer JJ (1974) Spinal arachnoid cysts. Report of six cases. J Neurosurg 41:728-735
- Pamg D, Wilberger JE Jr (1982) Tethered cord syndrome in adults. J Neurosurg 57:32-47
- Raghavan N, Barkovich AJ, Edwards M, Norman D (1989) MR imaging in the tethered spinal cord syndrome. AJNR 10:27-36
- Ricolfi F, Gobin PY, Aymard A, Brunelle F, Gaston A, Merland JJ (1997) Giant perimedullary arteriovenous fistulas of the spine: clinical and radiologic features and endovascular treatment. AJNR 18:677-687
- Rindahl MA, Colletti PM, Zee CS, Taber P (1989) Magnetic resonance imaging of pediatric spinal dysraphism. Magn Reson Imaging 7:217-224
- Robinson L, Dominguez R, Cabrera J et al (1989) Multiple meningeal cysts in Marfan's syndrome. AJNR 10: 1275-1276
- Rogers HM, Long DN, Chou SN et al (1971) Lipomas of the spinal cord and cauda equina. J Neurosurg 34:349-354
- Samuelsson L, bergstrom K, Thomas KA, Hemmingsson A, Wallensten R (1987) MR imaging of syringohydromyelia and Chiari malformations in myelomeningocele patients with scoliosis. AJNR 8:539-546
- Satliff JH, Kendall BE, Kingsley DPE et al (1989) Closed spinal dysraphism: analysis clinical, radiological and surgical findings in 104 consecutive patients. AJNR 10:269-277
- Simpson RK Jr, Rose JE (1987) Cervical diastematomyelia: report of a case and review of a rare congenital anamoly. Arch Neurol 44:331-335
- Tarlov IM (1970) Spinal perineurial and meningeal cysts. J Neurol Neurosurg Psychiatry 33:833-843
- Tortori-Donati P, Fondelli MP, Rossi A, Raybaud CA, Cama A, Capra V (1999) Segmental spinal dysgenesis: neuroradiologic findings with clinical and embryologic correlation. AJNR 20:445-456
- Tronnier VM, Steinmetz A , Albert FK et al (1991) Hernia of the spinal cord: case report and review of the literature. Neurosurgery 29:916
- Turjman F, Joly D, Monnet O et al (1995) MRI of intramedullary cavernous haemangiomas. Neuroradiology 37: 297
- Wight RL (1971) Congenital dermal sinus. Prog Neurol Surg 4:175-191
- Wisoff JH, Epstein F (1989) Management of hydromyelia. Neurosurgery 25(4):562-571
- Wolpert SM, Anderson M, Scott RM, Kwan ESK, Runge VM (1987) Chiari II malformation: MR imaging evaluation AJR 149:1033-1042
- Yamada S, Zine DE, Sanders D (1981) Pathophysiology of "tethered cord syndrome". J Neurosurg 54:494-503

Iconographical index

Part III. Spine and cord

Imaging technique and normal anatomy

ALPHABETICAL INDEX

E

F

G

H

I

K

L

M

N

O

P

R

S

T

V

W

X

Composition, photogravure et impression
JOUVE, 11, boulevard de Sébastopol, 75001 PARIS
N° 314535V — Dépot légal : Octobre 2002

MIX
Papier aus verantwortungsvollen Quellen
Paper from responsible sources
FSC® C105338

If you have any concerns about our products,
you can contact us on
ProductSafety@springernature.com

In case Publisher is established outside the EU,
the EU authorized representative is:
Springer Nature Customer Service Center GmbH
Europaplatz 3, 69115 Heidelberg, Germany

Printed by Libri Plureos GmbH
in Hamburg, Germany